Artur Braun
Electrochemical Energy Systems

Also of Interest

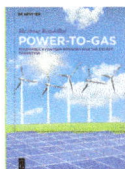

Power-to-Gas.
Renewable Hydrogen Economy for the Energy Transition
Boudellal, 2018
ISBN 978-3-11-055881-4, e-ISBN 978-3-11-055981-1

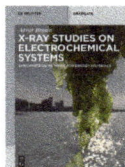

X-ray Studies on Electrochemical Systems.
Synchrotron Methods for Energy Materials
Braun, 2017
ISBN 978-3-11-043750-8, e-ISBN 978-3-11-042788-2

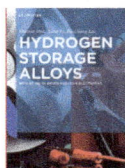

Hydrogen Storage Alloys.
With RE-Mg-Ni Based Negative Electrodes
Han, Li, Liu, 2017
ISBN 978-3-11-050116-2, e-ISBN 978-3-11-050148-3

Biorefineries.
An Introduction
Aresta, Dibenedetto, Dumeignil (Eds.), 2015
ISBN 978-3-11-033153-0, e-ISBN 978-3-11-033158-5

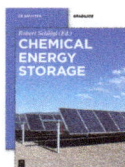

Chemical Energy Storage.
Schlögl (Ed.), 2012
ISBN 978-3-11-026407-4, e-ISBN 978-3-11-026632-0

Artur Braun

Electrochemical Energy Systems

Foundations, Energy Storage and Conversion

DE GRUYTER

Author
Dr. Artur Braun
Empa. Eidgenössische Materialprüfungs- und Forschungsanstalt
Überlandstr. 129
CH - 8600 Dübendorf
Schweiz

ISBN 978-3-11-056182-1
e-ISBN (PDF) 978-3-11-056183-8
e-ISBN (EPUB) 978-3-11-056195-1

Library of Congress Control Number: 2018952439

Bibliographic information published by the Deutsche Nationalbibliothek
The Deutsche Nationalbibliothek lists this publication in the Deutsche Nationalbibliografie; detailed
bibliographic data are available on the Internet at http://dnb.dnb.de.

© 2019 Walter de Gruyter GmbH, Berlin/Boston
Typesetting: Integra Software Services Pvt. Ltd.
Printing and binding: CPI books GmbH, Leck
Cover image: courtesy of Dr. Artur Braun

www.degruyter.com

To Gye Weon, Agneta and Lars

Preface

Electrochemical energy systems are prospective systems for a sustainable, safe and renewable source of future energy. This textbook provides an introduction to the wide field of electrochemistry and energy storage and conversion. This book is an outcome of my lecture notes that I prepared for a course conducted by me in 2017 for graduate and undergraduate students at the Department of Mechanical Engineering, Yonsei University, Seoul, Korea.

Electrochemistry is experiencing a renaissance since more than ten years. A self-standing discipline in science, research and technology, electrochemistry is scattering into various other areas. It is not a domain of only electrochemists anymore. With Professor Ryu who hosted me several times at Yonsei University, I now have three personal collaborators from departments of Mechanical Engineering worldwide, including one computational mechanical engineer who are actively working in electrochemistry.

Why is electrochemistry taking a center stage? Well, batteries are electrochemical energy storage devices and we need batteries for our mobile gadgets. There has been a lot of research and development going on in the past 30 years for supply to the growing market of mobile applications such as video cameras and mobile phones. About 20 years ago, batteries were seriously considered for powering automobiles.

In the late 1990s, hybrid cars became available where a traction battery could power an electric motor in addition to the conventional fossil fuel combustion engine. Batteries have become larger and their use has become more diverse. Electricity from hydropower, solar PV panels and wind farms can be stored in large factory-type batteries.

Fuel cells were also a hot topic for automotive traction. However, in a review on fuel cells published in 1996, fuel cell experts Srinivasan and Mosdale [Srinivasan 1996] wrote:

> To date, the only application of fuel cells has been and continues to be as an auxiliary power source in space vehicles.

It was only in 2015 that the first serial automobile propelled by a hydrogen fuel cell was presented. You can see the car in front of a hydrogen filling station printed on the front cover of this book.

Meanwhile, biological systems and living organisms experience a revival as electrochemical study subject and object. While these are not yet being used in a real technology for energy storage purpose, living organism (cells, plants, animals and humans) by itself is a complex and independent electrochemical (electrophysiological) factory. Bioelectrochemistry, which is originally the earliest field of electrochemistry where electricitiy was discovered, has a bright future.

https://doi.org/10.1515/9783110561838-201

Students can directly relate with batteries, which are the universal and common energy storage product. Through supercapacitors, the electrochemical double layer concept for energy storage will be explained. The field of energy conversion will be showcased by various fuel cell and electrolyzer concepts. Through photoelectrochemical cells, the field of semiconductor photoelectrochemistry and its relation to photosynthesis in nature will be rationalized. This latter example paves the way to the universality of electrochemical processes in all living organisms and thus the link to electrophysiology, life sciences and medicine.

The course will also show how devices and components are made from materials chemistry and engineering.

Various electroanalytical methods will be demonstrated and their link with performance assessment concepts will be explained.

You will observe some portions of this book that have no immediate or obvious relation to electrochemistry. I have done this on purpose because I want to link and compare electrochemical methods with conventional energy conversion and storage technologies. This book contains a number of anecdotes and autobiographical elements. These will help the reader to better understand how science can emerge by serendipity and technology that evolve by economic or societal circumstances.

Artur Braun
Bodega Bay, California, Summer 2018

Acknowledgement

I came to electrochemistry by ways of life. I grew up in rural Germany which provided me with plenty of opportunities and pleasure of finding things out [Feynman 2005]. I owe biggest thanks to my parents who purchased a chemistry set for me when I was 11. Within a year or two, I had built a well-equipped general laboratory [Beuys 1980][1] in our attic, including electronics and radio stuff (with technical expertise and training from two ham radio amateurs in the neighborhood), a wired logic computer, a microscope, an aquarium with *Artemia sp.* Nauplii and eventually a refractory telescope, everything at the lowest cost possible.

A real scientist should have an extensive formal education in mathematics and physics. This is why I worked myself toward university and studied physics. While I first envisioned going into relativistic cosmology and then into colloidal physics, I enjoyed my physics diploma thesis on magnetic films and ultrahigh vacuum physics with Matthias Wuttig at Kernforschungsanlage (KFA) Jülich; however, Harald Ibach, one of the Institute's directors said that the future directions would be the second harmonic generation and the solid–liquid interface. This I considered as a serious career advice.

When I found an opportunity for a PhD project on solid–liquid interface at ETH Zürich and Paul Scherrer Institut (PSI), specifically on electrochemical double layer capacitors, I thought that might be a good choice and thus I moved to Switzerland. My supervisor Rüdiger Kötz, a physicist, told me that work on the solid–liquid interface could be no less clean and pure than work on the solid–vacuum interface. Brief remarks like this one can help you a lot to feel more comfortable about a new line of research.

The research experience that I gained in the Electrochemistry Laboratory at PSI from 1996 to 1999 was very broad, and this experience increased in value in the past 20 years as I further matured. My colleagues at PSI and also the leaders Otto Haas and Alexander Wokaun had a great influence on widening my view on science and technology. A theory is an important, sometimes an invisible backbone but mostly dressed in visible real applications. This book, in part, is a consequence and manifestation of this insight.

My coincidental contacts with Heinz-Günter Haubold in Jülich and Hamburg, and with Rainer Saliger and Hartmut Pröbstle at Universität Würzburg opened me to the world of synchrotron methods on electrochemical systems [Braun 2017]. The sabbatical of Elton Cairns in our lab at PSI and the kindness of my office mate Hans-Peter Brack, who invited me to deliver a talk at the MRS Spring Meeting 1999 at San

---

1 Joseph Beuys: "Aber ich habe aufgrund dieser Unsicherheit, die ich dann für eine spezielle Berufsentscheidung damals hatte, etwa Kinderarzt oder Chemiker oder Physiker … habe ich ein allgemeines naturwissenschaftliches Studium begonnen. Ich habe mich meinen naturwissenschaftlichen Studien eigentlich seit meinem 14. Lebensjahr gearbeitet und hatte ein sehr umfangreiches Laboratorium aufgebaut."

https://doi.org/10.1515/9783110561838-202

Francisco paved my path for a great postdoc fellowship in Berkeley in Elton Cairns' group, where I worked on batteries, and along with Stephen Cramer's group on metalloproteins – with synchrotron radiation methods. This is how I learnt x-ray spectroscopy [Braun 2017].

After an intermezzo of four years in the world of fossil fuels and combustion technology in Kentucky, which is one of the American coal states, I returned to Switzerland and resumed my research in electrochemistry and then in the field of high temperature solid–gas interface at Empa where I still work. I owe this to Peter Holtappels, who introduced me to this quite different field of electrochemistry that finds application in high temperature fuel cells and electrolyzers.

In more than 12 years that followed, my research group grew at times to 12 staff and students, and we explored and sometimes pioneered various fields of electrochemistry, specifically the aforementioned fuel cells, also in addition to batteries and photoelectrochemical cells. Most of the time, back then, funding agencies were favourable to our research and you will find evidence for their support in the many references provided in this book from my workful and successful group members, to whom I am grateful. My gratitude goes also to my many collaborators in Switzerland and worldwide, many of whom are coauthors of the papers listed in the bibliography of this book.

I am grateful to Editorial Director Dr. Konrad Kieling who paved me once again the path for a book with De Gruyter, and Project Editor Lena Stoll, who further smoothened the path. Much of the time that was necessary for my original research provided in this book was taken from my family time budget. I therefore owe much of the success of my work to my family.

Contents

1 Introduction

This book shall give you a very general overview of electrochemistry for energy storage and conversion. It will give also you an introduction in the general field of electrochemistry, including some more exotic and seemingly esoteric aspects of electrochemistry.

What is electrochemistry? I don't want to provide you with a clear-cut definition of the term. There are many books on electrochemistry which you can buy, check out from the library or find online for free. I have no particular recommendation for you here. Since there are so many books you can check out for free, I hope you find the one which suits your needs, taste and interests. However, this book here follows the philosophy of a *studium generale* and you will find that it differs from all other books you find on electrochemistry.

At this very early stage of the book, I want to also issue a safety notice. This book will give you numerous inspirations on experiments that you can do in the laboratory, but experimental work bears the risk of accidents. You must therefore follow good laboratory practices when you deal with chemicals, electricity, machining tools and so on. Look for safety instructions, always obey them and never ignore them.

So, what is electrochemistry?

What is chemistry? That is an easier question. Chemistry deals with the chemical reactions between chemical elements, with passing on of electrons and maybe protons from one chemical element, atom and ion to another one; with reduction and oxidation processes of materials: chemical conversion of materials, either solid, liquid or gas. This is a quick and practical answer for what chemistry is. When you control these chemical reactions with electrical equipment, with cables, electrodes, electrolytes, then, I think, this is electrochemistry.

Well-known applications of electrochemistry are the batteries. Batteries store chemical energy and release it with an electric voltage as an electrical current which we use in daily gadgets like cell phones, cameras and many other portable electric devices. Watches, sensors and so on may be run by batteries. Electric vehicles (EVs) like the new Tesla car run on batteries.

For educational purposes, I have included in this book historical material [Leddy 2004], some of which dates back several hundreds of years. This material is well connected with names that have made it into physical and chemical terminology, such as Volta, Galvani, Ampere, Joule, Watt and so on. I believe it is important that the student readers who are interested in an academic degree understand and see examples how science and how knowledge evolves over time, times and ages.

As the views of the societies change over the decades and centuries, so does the perception of the science and technology. Scientific evolution is not only a struggle about right or wrong, but also a struggle about which theory fits the reality better.

https://doi.org/10.1515/9783110561838-001

Those of you who will make it into an academic career will learn that not only the scientific argument but also the way how and by whom it is being brought forward is important for whether a scientific idea wins or not. This is even more important where science turns into technology and business and social change [Feynman 1955]. Sometimes, a new technology will not make it to the market unless society, represented for example by the government, subsidizes the new technology with incentives [Norberg-Bohm 2000].

Finally, I want to point out that it was good academic practice in all disciplines of arts, science and humanity to seriously allow for views and opinions, particularly with respect to controversial views and opinions. Scientific dispute is part of the essence of science. For example, there are, fortunately, still publications around with the title "Current Opinion," not only in science but also in the field of law.

1.1 Mobility and electromobility

Mobility is nowadays a keyword with positive and negative connotation. Here in this book on "electricity," mobility stands certainly for the mobility of electric charge carriers, be it electrons, holes, ions and so on. Without mobility, the charge carriers cannot transport their charge and hence no electric current, no electric power and no mechanical action would be possible. The electromotive force (EMF, E.M.F., e.m.f., emf) would not come to action. Mobility thus stands for dynamics.

Mobility is also another expression for freedom. The freedom that you can go and that you can move everywhere you want is important in and symbolized by modern western society. This feeling of freedom is resembled for example by the electronic experimental music piece "Autobahn" by German pioneering electronic music group Kraftwerk [Buckely 2012, Nationalpost 2014, Schiller 2014]. The cover image (designed by Emil Schult, a graduate student of Joseph Beuys) of the single piece Autobahn shows an endless highway, the "Autobahn" in German nature setting along with electric power transmission lines.

Certainly, the freedom to endlessly moving around, the constitutional right and freedom to travel [Crusto 2008] needs also endless energy, either from gasoline in the tank of the car for the mobility or for the access to electric energy far away from the power plant – in German: Kraftwerk. Or you can sail with the wind in a modern air plane or on a boat.

Mobility is also an expression for the deployment of machinery and human workforce (stimulated mobility [Papatsiba 2006, Walsh 2010, Wilken 2017] or forced mobility [Beladi 2011, Kerbalek 2010, Lovelock 2016, Nica 2009, Zimmermann 2005]) and in so far can have a rather negative connotation. In this very same context, migration, a term well known in electric phenomena such as electrophoresis, has become also a term of time with respect to human migration [Dessalegne 2001] including forced migration [Lamar 2010].

Another aspect of freedom and mobility comes when I remember the late 1970s when the Walkman was invented [Braidwood 1981, Waysand 1984]. The Walkman was a small portable magnet tape cassette player; it was not much larger than the cassette itself and would play music to your earphones and headphones for several hours. For example, the recording capacity of one cassette would be 240 minutes. You could clip the Walkman to your belt or keep it in a pocket and listen to music while you are walking around in the city or in nature. Two small batteries were sufficient to power the Walkman.

Twenty years later, the iPod was invented. It had one-third of the weight of the Walkman and it could store 100–200 times more music pieces [Xu 2016]. The iPod was powered with even smaller batteries. The miniaturization of electronics to microelectronics [Braun 2017] for portable consumer products provided an incentive to make efforts in miniaturizing the power sources. To be able to listen to your favorite music everywhere you want without the need of being in a concert hall or at your own home, movie theater and entertainment center has consequences on the entire society and cultural development.

When American artist John Rand [Rand 1841] invented the paint tube over 150 years ago, he maybe did not anticipate that the impressionist painters would have the opportunity to leave their ateliers with their painting equipment and start painting anywhere outside where they wanted. Painters became mobile and we experience this today because an entirely new era of paintings became possible, as we can see in the art museums today. Great impressionist painter Pierre Auguste Renoir [Redaction 2012], born in 1841, remarked the significance of the invention of paint tubes for the development of art and impressionism:

> Sans les peintures en tube, il n'y aurait eu ni Cézanne, ni Monet, ni Sisley ou Pissarro, rien de ce que les journalistes appellent "impressionnisme."

Would you guess that Rand's patent [Rand 1841] served as technical reference for a deferred action battery [Oestermeyer 1962]? Such kind of batteries is employed in hearing devices and in torpedoes, for example.

Eberle [Eberle 2012] and Von Helmolt et al. [Von Helmolt 2007] wrote status papers and opinion papers on the hydrogen mobility infrastructure from the perspective of engineers in the automotive industry. Now, in 2017, as I started writing this book, hydrogen mobility has become reality for the consumer.

1.1.1 The fuel cell electric vehicle Hyundai ix35

In Figure 1.1, you see a fuel cell (FC) car that runs on hydrogen. It is a Hyundai ix35 Fuel Cell from model year 2015 [Ashley 2012]. As you see from the top left panel in Figure 1.1, the car looks like a small sports utility vehicle (SUV). It can take four or five passengers and has a comfortably large trunk. The car was purchased from a local

Figure 1.1: (Top left) Hyundai ix35 car in the fuel cell electric vehicle (FCEV) version with 130 hp. (Top right) The car at a hydrogen gas fueling pump station, operated by a local retail store. The hydrogen is produced by electrolysis of water with power from a hydro power plant. (Left bottom) The handle pushed into the hydrogen tank; filling takes few minutes. (Right bottom) The panel shows I have pumped 1.12 kg hydrogen for 12.21 Swiss Francs (around US $12). A full tank costs around US $50 and runs practically 400 km and nominally 596 km.
Photos by Artur Braun in 2016.

Hyundai dealer in Switzerland and is part of a Swiss research program "move" at Empa.

The upper right panel shows the FC car at the hydrogen gas station in Hunzenschwil, Switzerland. At this time when the book is written, it is the only public hydrogen gas station in Switzerland where you can fill hydrogen cars or trucks with hydrogen and pay with credit card like at any other conventional gas or diesel station. The hydrogen filling station is part of a conventional gas station, operated by Swiss retail store chain COOP (Coop pronto).

The left bottom panel in Figure 1.1 shows the FC car while hydrogen is being pumped into the hydrogen tank in the car. The filling looks not much different than any other gas filling into a car. You have a filling terminal at the hydrogen station, a hose and a handle which you connect with the fuel inlet of the car. The hydrogen is pressed to 700 bar into the tank which is in the trunk. The right bottom panel shows the customer display at the filling terminal. It reads that I have filled the car with 1.12 kg of hydrogen and I will have to pay 12.21 Swiss Francs for it.

Table 1.1 lists a number of technical details of the car.

Table 1.1: Technical details of the Hyundai ix35 Fuel Cell for the driver and consumer.

Detail	Value
Maximum power	100 kW (136 PS)
Maximum torque	300 Mn
Battery (lithium polymer)	24 kW
Acceleration (0–100 km/h)	12.5s
Maximum speed	160 km/h
H_2 tank capacity	5.64kg at 700bar
Refueling time	~3 min
H_2 consumption	0.96 kg/100 km
Driving range (NEDC)	594 km
Curb weight	1830 kg

Source: Hyundai.

How does a fuel cell vehicle work? First, it's abbreviated with FCEV, which stands for fuel cell electric vehicle. Therefore, it is an EV. Its motor is an electric motor, and not a combustion engine. The electric motor needs electric energy which can come from a battery or, like here, from a FC.

Why is the fuel cell called fuel cell? What is the fuel? The FC for the FCEV is a PEM-FC, a polymer electrolyte membrane (also called proton exchange membrane) fuel cell. Those run with highly pure hydrogen H_2 gas, which is the fuel.

Isn't hydrogen "pure" anyway? No, not necessarily. Most of the hydrogen that is produced worldwide is produced from water and fossil fuels, specifically by guiding water vapor over hot coal. This triggers the water–gas shift reaction which reads

$$CO + H_2O \Leftrightarrow CO_2 + H_2$$

When this chemical reaction runs incomplete, the produced hydrogen may have carbon monoxide (CO) impurities. When such impure hydrogen is used as fuel for the PEM-FC, the CO can act as a catalyst poison.

In the FC, the hydrogen gas (H_2) reacts with the oxygen (O_2) from the air and forms water (H_2O) according to the following reaction:

$$2H_2 + O_2 \Rightarrow 2H_2O$$

The exhaust from the FC car is therefore water, and nothing else.

The Hyundai that I have been using received the hydrogen from an electrolysis plant that was powered from the water in a river (hydro power). The electric motor can receive electric energy from a buffer battery and from a fuel cell. The FC is supplied with hydrogen which is stored at 700 bar pressure in the H_2 tank in the back of the car.

Figure 1.2 shows two photos from the operation display in the ix35. The car is moving a slight downhill road and needs no power from the FC or from the battery because it is moving with the momentum it had already plus the gravity that acts on the car.

Figure 1.2: Display in the Hyundai ix35 Fuel Cell car showing the real-time energy flow between engine, battery and fuel cell.

The display in Figure 1.2 (left panel) shows a graphic representation of the hydrogen tank on the right, the electric motor and front wheels on the left, and the FC and the battery in the middle. As this car is rolling downhill, the kinetic energy $E_{kin} = p^2/2m$ from the momentum $p = mv$ of the car can be stored as electric energy in the battery, which is called power recuperation. The reading "Regenerativer Bremsmodus" in the display indicates this recuperating situation. The breaking power from the car running downhill is preserved and converted into electric energy by the electric motor and then stored in the buffer battery, as indicated by the pink lines in Figure 1.2.

The right panel in Figure 1.2 shows when the car is in the "Energiehilfemodus," which I would translate with "energy assistance mode." This is the situation when the "gas pedal" is fully pushed for maximum acceleration. The FC receives hydrogen fuel from the hydrogen tank in the trunk (pink line), and the resulting electric power is given to the electric motor (blue line). But the requested acceleration by pushing the pedal in maximum mode is so high that the FC cannot deliver immediately (in very short time) the necessary energy. Therefore, the buffer battery delivers fast power (green line) in addition to the electric power from the FC.

During very dynamic driving, you can follow the concert of these components. The color arrows show which component is delivering or accepting power from other components. You can imagine that numerous sensors and actuators are necessary in order to coordinate the components for the desired driving, and the coordination is done by an onboard computer. As a safe driver, you certainly must pay attention and focus on the traffic around you, and on that display only when the situation permits to do so safely.

Don't all electric cars run on batteries? No. Not necessarily. Most cars run on combustion engines (I am reminded that I was presenting this lecture to mechanical engineering students), and the starter battery supports the ignition of these engines. But the power for driving the car comes from the combustion, from the explosion of the gasoline. The lead acid battery contains not enough energy to run the car, but it does a great job in running the starter.

The Tesla cars have no combustion engine. Rather, they have a large battery which has enough energy to run the cars some hundred miles before the battery needs a recharge (note that the gasoline car needs a refill). Would you believe that 100 years ago in Berlin, there were around 100 electric automobile companies? Tens of thousands of EVs were running in the metropolitan cities all over the world [Maxwill 2012].

Why, then, are nowadays most cars running on combustion engines? That's a long story. Since 20–30 years, there has been increasing interest in getting fuel-cell-powered cars and utility vehicles such as busses in cities. Particularly when in large cities there is air pollution from diesel engine trucks and busses, people may be interested in paying some extra money for new types of cars and busses when the cities become cleaner [Heo 2013].

Other than the cleanliness, an electric engine has a fundamentally principal advantage over the combustion engine (more correctly: heat power machine), and this is the thermodynamic efficiency:

$$\eta_{\text{Carnot}} \leq \frac{T_{\text{high}} - T_{\text{low}}}{T_{\text{low}}}$$

This relation is a result of the Carnot cycle, which we know from the physics class on statistical physics. The difference of the temperature bath, $T_{\text{high}} - T_{\text{low}}$, determines the efficiency.

Electrochemical processes by default run isothermal – at the constant temperature. They are thus not subject to the limitation of the Carnot cycle and therefore in principle allow for higher efficiency than with a heat power machine. This is not withstanding that an electrochemical machine will heat up during operation and thus emit radiation in the form of Joule heat.

The efficiency for an electrochemical process is defined as

$$\eta = \frac{W}{Q_{\text{in}}}$$

where W is the work which we can derive from the electrochemical reaction. W is, for example, the energy which we can derive from the EMF of a battery, fuel cell or dynamo (in V). Q is the enthalpy of formation in the chemical reaction.

Helmholtz transferred the principles of thermodynamics on the electrochemistry. He introduced the term "free energy," which is necessary to predict whether a chemical reaction is thermodynamically possible – by the Gibbs–Helmholtz equation:

$$\left(\frac{\partial(G/T)}{\partial T}\right)_p = -\frac{H}{T^2}$$

By July 2018, I have driven this car over 16,000 km across Europa. On the German Autobahn, I drove it as fast as I could on a few occasions where possible. The number on the speedometer was then 170 km/h. Soon I found that the hydrogen tank would empty very fast, much faster than I am used from gasoline cars. One of my colleagues, a FC expert at Paul Scherrer Institut, told me that in FC systems, the efficiency decreases with increasing load, whereas it is the opposite with the combustion engines. As I am pulling the maximum power of the FC at 150 km/h and above, the FC produces electricity with lesser efficiency than when at much lower speed [Büchi 2018]. This is nicely illustrated in the paper by Büchi et al. [Büchi 2014].

Finally, I want to show you how the FCEV looks inside under the hood. Every little detail of the car is subject to experimental and mathematical studies, for example, the effect of the blower and ejector pressure on the performance [Kim 2016]. Hyundai and Kia have been developing FCs for vehicles for many years [Ahn 2006, Ahn 2005, Kim 2010].

Figure 1.3 shows the Hyundai ix35 Fuel Cell with open hood. The black square block in the center is the housing of the FC stack. For the layman, the FCEV looks not fundamentally different from a car with combustion engine. Behind the block that contains the FC stack are numerous thick cables which lead the electric power to the electric motors that drive the car. It cannot be avoided that the FC and other components become hot during operation, particularly when there is a high power demand on it. Therefore, the car has a cooling system. The FCEV has also a large lithium ion battery but this cannot be seen so easily. It is mounted under the car. A detailed illustration of all components is for example shown in the Emergency Response Guide [Hyundai 2013].

Figure 1.4 shows the rear space of the FCEV. It is quite spacious and allows for taking a number of suitcases or other cargo during travel. On my long ride from Zürich to Berlin in early 2017, where I had experiment beam time at the synchrotron BESSY-II in Adlershof, there was enough space for two passengers, suitcases and scientific instruments which I used for the experiment.

When you lift up the black panel under the black blanket cover (Figure 1.5), you see the white cylinder hydrogen tank at the place where older cars used to have the spare wheel. So, it seems the Hyundai has no spare wheel but this holds for many other cars as well nowadays. The white gas cylinder is quite large as you can see. It can take

Figure 1.3: Car components under the hood of the Hyundai ix35 Fuel Cell vehicle. The large square-shaped block in the center is the housing of the fuel cell stack.
Photo taken in Strassbourg, France, 24 May 2017, by Artur Braun.

almost 6 kg hydrogen gas when compressed to 700 bar. On the left side near the tank is a small lead acid battery that is used to keep some of the onboard memory working. That battery is not used to drive the car. It is not a "traction battery." The traction battery is the aforementioned lithium battery which is mounted under the car, not visible to us.

During the many discussion that I had with various people that I met during and after my cruises with the FCEV, I was asked how safe this hydrogen is. I cannot comment other than "I feel safe." The car comes with some safety features which are specific to hydrogen storage at 700 bar in traffic. On my recent trip to Bruxelles in 2018, I spoke with an engineer from Turkey who told me he had watched the crash tests of hydrogen containers during a visit to the USA, and he said he was convinced about the safety of such tanks.

There is certainly a basic risk of vehicle traffic because of the high speed of vehicles and the energy content which they have in addition to the liquid gasoline or diesel fuel, or the chemical energy in the traction batteries or in the hydrogen tanks.

Figure 1.4: Rear space in the Hyundai ix35 Fuel Cell with the blanket put aside. Under the black panel is the hydrogen tank.
Photo by Artur Braun.

Figure 1.5: Rear space in the Hyundai ix35 Fuel Cell with the back cover opened. The white roundish container is the hydrogen storage tank that can sustain the 700 bar pressure. It can contain around 5.64 kg H_2 at this pressure.
Photo by Artur Braun.

1.1.2 Comparison of efficiency of the combustion engine and fuel cell

Hydrogen has an energy density of 33.3 kW h/kg [Fung 2005]. The Hyundai can be filled with 5.64 kg of hydrogen when a pressure of 700 bar is applied. This makes 187.8 kW h and is sufficient for almost 600 km at low speed. Let us make the assumption that we can make it only to 400 km with a full hydrogen tank, but then at a speed of 100 km/h which is realistic. So, we use around 190 kW h for 400 km. Doing this calculus is an exercise for the student.

Trucks and busses have hydrogen storage tanks which are designed for 350 bar pressure only. The law of Boyle and Mariotte states that the pressure p and volume V of an ideal gas at temperature T correspond to each other as p/V = constans:

$$\frac{p_1}{p_2} = \frac{V_2}{V_1} = \text{constans}$$

When we increase the pressure, the volume decreases proportionally. The temperature dependence of the volume change at constant pressure is given by the law of Joseph Louis Gay-Lussac:

$$V(T) = V_0\left(1 + \gamma_0(T - T_0)\right)$$

The constant $1/\gamma_0$ equals 273.15 K and denotes the absolute zero point of temperature.

According to the equation of state for ideal gases

$$p \cdot V = N \cdot k_B \cdot T = m \cdot R_s \cdot T$$

we have at half the pressure only half the number N of gas molecules or half the gas mass m, when the temperature and the volume are kept constant. k_B is Boltzmann's constant and R_s is the specific gas constant.

We remember also that ideal gas equation of state is quite naïve when compared against real gases that have a molecular interaction. The real gas equation of state gives a better yet not entirely accurate description of the relationship between pressure, volume and temperature:

$$\left(p + \frac{n^2 a}{V^2}\right) \cdot (V - n \cdot b) = n \cdot R \cdot T$$

The constants a and b are the cohesion pressure and the covolume, respectively. They are specific for the particular gas.

As the volume V is the volume of the hydrogen gas tank, the pressure and temperature are the only parameters which should play a role in the hydrogen fuel cell car so far. As we aim for a particular pump pressure at the gas station which is specified by the protocols of the Society of Automotive Engineers [SAE 2016, James 2014, Kinn 2009], the temperature remains a parameter of some uncertainty.

The amount of gas in the tank depends ultimately also on the temperature T. This is why the hydrogen vehicle and the pump station communicate via an infrared information exchange interface during the pumping process.

To end with the long story short, the tank (gas container) can stand only 700 bar or 350 bar pressure p and has only a constant volume of V. The ambient temperature T will have some influence on the maximum mass m of gas in the tank.

As I mentioned already, when I engage in discussion with people over the new hydrogen FC vehicles, some of them want to know whether the pressure tank is safe, in particular, as it contains hydrogen. When an accident happens, the tank being under pressure may explode, and the compressed hydrogen may escape, mix with oxygen from the ambient air and explode as Knallgas. It is difficult for me to properly respond to this other than: the tank is officially approved by the regulation authorities. And driving in a conventional car that carries 50 L of gasoline that hypothetically can also combine to an explosive mixture – this is a situation that has become part of our daily live. Personally, I see no difference between the hydrogen car and the gasoline car when it comes to this particular aspect of safety.

For a spacious vehicle as large as a truck or a bus, it makes no big problem when a larger tank volume compensates for the consequences of a lower gas pressure. For a passenger car, space is more precious and thus tanks that can sustain a high pressure as 700 bar are crucial. For your comparison, when I began writing this section of the book, I was living in an extended-stay hotel during my stay at Yonsei University in Seoul, Korea. That hotel had a coffee and espresso machine from the Swiss manufacturer Schaerer which is equipped with a pressure gauge (Figure 1.6). My coffee was always made with a pressure of around 8 bar. Hence, the pressure in a 350 bar tank (for the hydrogen FC trucks) is over 40 times larger than the pressure in that coffee machine. And the pressure in the 700-bar tank (for the hydrogen FC cars) is almost a 100 times larger than in that coffee machine.

For comparison, the pressure in the tires of your bicycle can range from 2 to 10 bar. Table 1.2 is taken from the recommendations of tire producer Schwalbe. You notice from the table that the tires with the wider diameter have a lower pressure. When you are riding a racing bike, you may find that the most suitable pressure for your sports bike is around 10 bar. Racing bikes typically have very narrow tires. You notice also that Schwalbe recommends a higher pressure when the bike rider has a heavier weight. Why is that? Where do these two relations (more weight → more pressure; less diameter →more pressure) come from? What is the physical rational behind it? First, it may be important to say that 1 bar is approximately the pressure exerted by 1 kg "weight" on 1 cm^2 area. About 1 kgf/cm^2 equals 98.0665×10^3 Pa, where kgf stands for kilogram force. This is an old unit not covered by the standard SI system.

Pressure p is defined as the force F that acts on an area A: $p = F/A$. The force F in this case here is the weight of the rider (plus the bike), which is caused by the gravity acceleration constant g ($g = 9.81$ m/s^2) acting on the mass of the rider (and the mass of the bike) m: $F = mg$. Hence,

$$p = \frac{F}{A} = \frac{m \cdot g}{A} = g \cdot \frac{m}{A}$$

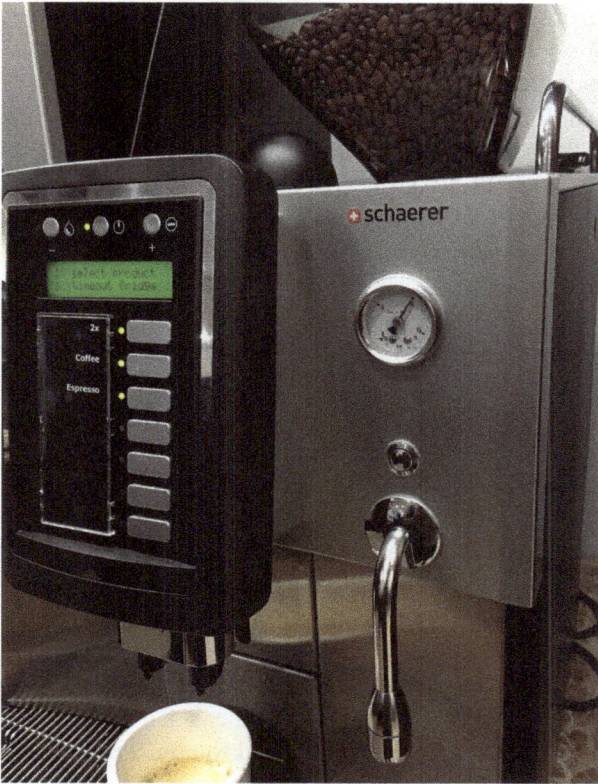

Figure 1.6: This coffee machine (Schaerer, Switzerland) used 10 bar pressure for the processing of my "espresso" coffee, as shown in the round pressure gauge on the right.
Photo by Artur Braun.

Table 1.2: Recommended tire pressures for bicycle wheels depending on the tire width (mm) and the weight of the bike rider (lbs).

Tire width	Body weight		
	Approx. 130 lbs	**Approx. 185 lbs**	**Approx. 240 lbs**
25 mm	85 psi/5.9 bar	100 psi/6.9 bar	115 psi/7.9 bar
28 mm	80 psi/5.5 bar	95 psi/6.6 bar	110 psi/7.6 bar
32 mm	65 psi/4.5 bar	80 psi/5.5 bar	95 psi/6.6 bar
37 mm	55 psi/3.8bar	70 psi/4.8 bar	80 psi/5.5 bar
40 mm	50 psi/3.4 bar	65 psi/4.5 bar	80 psi/5.5 bar
47 mm	45 psi/3.1 bar	55 psi/3.8 bar	70 psi/4.8 bar
50 mm	35 psi/2.4 bar	55 psi/3.8 bar	70 psi/4.8 bar
55 mm	30 psi/2.1 bar	45 psi/3.1 bar	55 psi/3.8 bar
60 mm	30 psi/2.1 bar	45 psi/3.1 bar	55 psi/3.8 bar

Source: Schwalbe North America [Schwalbe 2017].

When you increase the mass m of rider and bike, then the pressure p must decrease. When you decrease the area A of the tire in contact with the road, then p also must decrease. This is a simple exercise for the reader.

The inflation pressure in car tires is typically around 2 bar. The gas pressure in a rifle cartridge of the NATO caliber 7.62 × 51 mm (almost identical to caliber 0.308 Winchester) is around 4000 bar [Mizrachi 2017]. These two extreme numbers between tire pressure of two bar and rifle shot of 4000 bar give us a good idea of what 700 bar hydrogen tank pressure in a car means for us.

Gasoline has an energy density of 12.7 kW h/kg [Golnik 2003]. The Hyundai ix35 1.6 GDI that runs on super gasoline has a tank that contains 58 L gasoline and has a range of 852 km. The density (specific weight) of super gasoline is 0.76 g/cm^3 or 0.76kg/L. The weight of 48 L superbenzine is therefore 44.08 kg. Multiplied with its energy density, we then obtain 560 kW h that is sufficient for 852 km. I do not know at which speed you can go with a full gasoline tank and make it to 852 km. Let us assume you can do so at 100 km/h. For half the distance, roughly 400 km, you need thus half the gasoline which makes 280 kW h. Compare this now with the above Hyundai FCEV, which needs only 150 kW h of energy for that distance. Hence, the FCEV needs half of the energy of the gasoline version. This is an amazing demonstration of how efficient FCEV can outperform cars with standard gasoline combustion engines.

While writing this section, I was reflecting over my own driving style with the FCEV; I had to pay some attention to the new FCEV driving. I had to because hydrogen pump stations were not always in reach and I had to adjust my driving style so as to not consume too much hydrogen fuel by speeding or not driving with foresight. Driving with foresight is what you are supposed to learn in driving school. By driving with foresight, you can anticipate the behavior of the other traffic on the road and see whether you will approach the light when it is red and you must be prepared to stop and waste energy by using the brakes, or green and you can move ahead with constant pace. But still, even when you "own the road," you may notice that after a period of speeding, you unintentionally will slow down somewhat and drive for some extended period with a slower pace. In addition to my own experience over 35 years of driving cars, I found a reference that gives a scientific justification or rational for this behavior [Paoletti 2014]:

> The biological evidence showing that unsteady locomotion might in fact be the norm and steady-state locomotion is the exception begs for a move away from characterizing locomotion based on a mathematically convenient steady-state assumption.

1.1.3 Some notes on the development of battery vehicles

The first time I drove in an electric car was when my mentor at Lawrence Berkeley National Laboratory, an eminent UC Berkeley battery and FC professor, gave me a ride uphill. He drove a Toyota hybrid car back in the late 1990s. At that time, we were working in a battery research program for EVs, funded by the US Department of Energy. Back then, it was still something special when you had a hybrid car.

It was not only the progress in making better batteries that allowed for the development of EVs. It was also necessary to improve the cars, such as making ultralight materials, vehicle configuration and even aerodynamic skin friction. Wyczalek [Wyczalek 1995] projects in his report in 1993 that battery EVs have a brighter future because the automotive manufacturers had been very responsive, creative and synergistic with improvement of acceleration performance, and vehicle styling and configuration. Of course, in the same time, public funds were invested in finding better battery systems for automotive applications [Kinoshita 1996].

Before private electric cars like the ones from Toyota and Nissan and Tesla, for example, were available, tax payers money was invested for sustainability projects in public transportation. Urban planning increasingly takes into account the ways of how people get into the city and through the city and out again [Mitropoulos 2013, 2016]. Densely populated cities with congestion areas and policies do not permit anymore vehicles with excessive engine exhaust. EVs such as battery cars and FC cars do not have any such exhaust. Hybrid cars have a large battery which has enough power for traction of the vehicle, and they still have a conventional gasoline (diesel) engine.

Recuperation of energy from slowing down the car is being stored in the battery and can be released during acceleration. The overall fuel consumption is less, and the well-known diesel exhaust plumes of public busses during acceleration disappear to a large extent.[1] This is one reason why hybrid technology was tested first at the larger scale in public transportation with busses. There have been many programs for development and deployment of hybrid busses worldwide. Some of the readers of this book may have spotted here or there such busses, particularly if they paid attention to their environment.

---

1 I have worked for 4 years in combustion aerosol research at University of Kentucky and the Consortium for Fossil Fuel Liquefaction Science (CFFLS, later CFFS). Major part of my work was the analysis of diesel soot [Braun 2005] Braun A: Carbon speciation in airborne particulate matter with C (1s) NEXAFS spectroscopy. *J Environ Monit* 2005, 7:1059-1065.doi: 10.1039/b508910g. Inhaling diesel soot is not good for your health [Bolling 2012] Bolling AK, Totlandsdal AI, Sallsten G, Braun A, Westerholm R, Bergvall C, Boman J, Dahlman HJ, Sehlstedt M, Cassee F, Sandstrom T, Schwarze PE, Herseth JI: Wood smoke particles from different combustion phases induce similar pro-inflammatory effects in a co-culture of monocyte and pneumocyte cell lines. *Part Fibre Toxicol* 2012, 9:45.doi: 10.1186/1743-8977-9-45, [Totlandsdal 2012] Totlandsdal AI, Herseth JI, Bolling AK, Kubatova A, Braun A, Cochran RE, Refsnes M, Ovrevik J, Lag M: Differential effects of the particle core and organic extract of diesel exhaust particles. *Toxicol Lett* 2012, 208:262-268.doi: 10.1016/j.toxlet.2011.10.025, and I can confirm that the diesel car manufacturers have taken this issue very seriously. Combustion control for reduction of aerosols and volatile organic matter and filters for particulates have progressed very well in the recent decades and modern diesel engine technology thus made the environment much cleaner.

Figure 1.7 shows a snapshot of a hybrid bus that I made during my sabbatical in Hawaii from 2010 to 2011. On my recent trips to Seoul in Korea – a city that suffers from air pollution from many sources, not only vehicle exhaust – I have seen hybrid busses, and lately in Phoenix in Arizona as well. In 2015, 41% of the public transit busses in the United States used alternative fuels and hybrid technology [Metromagazine 2015].

Figure 1.7: Snapshot of the roof top of a hybrid bus in downtown Honolulu. Hybrid busses need extra spacious infrastructure that is noticeable particular on the roof top of the bus.
Photo by Artur Braun, 26 January 2011.

Today, when I frequently go by bus in Switzerland, I may be sitting in a hybrid bus. The VBG Verkehrsbetriebe Glattal AG and Verkehrsbetriebe Zürichsee Oberland AG (VZO), two local transportation utility companies, have been exploring hybrid busses since 2011 [VBG 2012]. Both companies had one hybrid bus each, which turned out to have 30% less energy consumption than their diesel busses. Since 2015, VBG operates five Volvo 7900 HYBRID busses in their fleet, which counts some 120 busses in total.

This Volvo 7900 HYBRID bus has a water-cooled 600 V lithium ion battery unit with 1.2 kW h capacity, in addition to the 240-hp diesel engine. The electric motor has 120 kW. As a passenger, I can say these busses make basically no noise. It is a silent

Figure 1.8: One of the hybrid electric busses (Volvo 7900) of the VBG Verkehrsbetriebe Glattal AG in the Kanton of Zürich, Switzerland.
Photo taken by Artur Braun, 6 July 2017, in Dübendorf.

form of transportation in the city and in the villages. In Figure 1.8, you see one of the Volvo 7900 HYBRID busses in routine operation in Dübendorf, Switzerland.

The electric and hybrid busses and cars mentioned in this section are all operated with lithium batteries, but there may be EVs that have different batteries. In Section 4.8 in this book, I will describe the ZEBRA (ZEolite Battery Research Africa) batteries which were used in cars and busses. I remember that we purchased for our children a battery-powered scooter in the early 2000s. It was a made-in-China product and contained a small lead acid battery (probably a 12 V system), which you could charge at home.

1.2 Hydrogen mobility

Hydrogen mobility is a special case of E-mobility. E-mobility means transportation with EVs. Let us count which kind of vehicles would apply. Certainly, the electric cars like the Tesla, which run on battery power, are included. Electric trucks can run on batteries. There are electric bikes, electric scooters and electric motorbikes. There are the formula E electric racing cars. Table 1.3 is a simple overview of various kinds of vehicles which are propelled by various fuels.

Table 1.3: Table of which vehicles can run on which fuel.

Fuel–vehicle matrix

Vehicles that run on various fuels

	Car	Bus	Truck	Motorbike	Scooter	Bike	Airplane	Boat	Submarine
Gasoline	x	x	x	x	x		x	x	
Diesel	x	x	x					x	x
Natural gas	x	x	x						
Biomass gas	x	x	x						
Wood gas	x	x	x	x					
Battery	x	x	x	x	x	x	x	x	
Hydrogen						x	x		x
Nuclear									x

Figure 1.9: General Motors FCEV loaded on a truck in Honolulu, Hawaii, on December 5, 2010. Photo by Artur Braun.

This table is not supposed to be taken too seriously, too literally. Let us not argue over details whether one particular vehicle can or cannot be propelled with a particular fuel. The overwhelming majority of cars run on gasoline, and almost every plane runs on kerosene, unless it is a glider. Most bikes are powered by human power. But researchers and technologists have tried to widen the scope – with more or less success.

The first time that I spotted a FCEV was in December 2010 on a highway in Honolulu – still loaded on a truck [Motavalli 2010] as shown in Figure 1.9. I did recently some research and found that the Hawaii Gas Company (TGC) could provide sufficient hydrogen for 1400 FC cars per day without any extra work. And with minimum changes, TGC could double that amount. The FCEV was from General Motors and belonged to a project for hydrogen mobility on Oahu.

1.2.1 Hydrogen fuel cell busses and trucks

There have been numerous projects where busses were equipped with fuel cells and hydrogen tanks. We learnt already that the hydrogen tanks in busses are designed for 350 bar pressure. The lower pressure has to be compensated by a larger tank volume of course [Andaloro 2011, De Lorenzo 2014, Dispenza 2012, Heo 2013, Sergi 2014]. The Nikola Motor Company (https://nikolamotor.com) promises electric trucks that are powered by hydrogen fuel cells.

In June 2017, Swiss automotive technology developer ESORO received the general approval for operating fuel cell-powered trucks on public roads in Switzerland. Prior to this, the 18 t weighing cooling truck had been used for experimental purposes by Swiss retail store company COOP for several thousand kilometers. One full hydrogen tank permits for a range of 400 km. The filling of the tank takes less than 10 min. The hydrogen is produced by electrolysis with electricity from a hydro power plant in Switzerland [Bönninghausen 2017].

Let us see how such a truck is built. For this, we have to look into the table that shows the technical specifications of the Truck as provided by ESORO on their webpage [Esoro 2016]. ESORO has built the truck as a prototype from parts and components which they received from other suppliers. They used a basic chassis from well-known German truck manufacturer MAN (Maschinenfabrik Augsburg-Nürnberg). The Dutch and Swiss EV technology firms EMOSS and ceekon AG, respectively, provided the electrical components. Emoss says on its webpage "EMOSS produces electric trucks, busses and vans under its own brand," but they also produce the E-Drivelines, this is, "modular drivelines and battery packages supplied by original equipment manufacturer (OEM) to transform any vehicle into an EV."

Ceekon AS develops EVs for a CO_2 neutral mobility and it is the exclusive representative of EMOSS in Switzerland, as it says on their website (http://www. ceekon.ch/de/Ueber-uns/Hintergrund.17.html). The maximum permissible weight of the truck is 34 t. Together with a trailer, it can load 60 delivery racks which you typically see at the backdoor in supermarkets and which are directly brought to the shelves after unloading from the truck.

As the truck is an EV, it has an electric motor which is built from Canadian producer tm4 Electrodynamic Systems which is connected to the power train with an automatic transmission. The electric engine (motor) has 250 kW power which corresponds to 340 horse powers (hp).

The energy which is necessary to generate this power is stored as chemical energy in the hydrogen which is stored in the gas tanks on the truck. The hydrogen is converted with oxygen (from the air) to H_2O in the fuel cell system in the truck, which delivers an overall EMF of $U = 250$–500 V. The FC system is built from 455 cells (FCs) and can deliver $P = 100$ kW. The electric power P is the product of the voltage U and the current I:

$$P = U \cdot I$$

The overall current in the FC system at 500 V is therefore

$$I = \frac{100\,\text{kW}}{500\,\text{V}} = \frac{100,000\,\text{V A}}{500\,\text{V}} = 200 \ \text{A}$$

When the system has 250 V, the current is 400 A. As the system is built from 455 fuel cells, every single FC delivers therefore on average current of 0.88 A. This is an exercise for the reader.

We notice that the power which the motor can pull from the system (250 kW) is by a factor 2.5 larger than the power which is delivered by the FC system. This is because the FC system is not supposed to do the work alone. If you have ever driven a FCEV, you might feel that it is less speedy than a battery vehicle – when you drive it in the economy mode. We will learn later when we discuss the Ragone diagram why this is so. What I can say here is that fuel cells suffer more from diffusion limitation processes than batteries. In return, batteries cannot store as much energy as the FCs can. This is why the FC system in the truck is supplemented by a battery. The PhD thesis of Kyung Won Suh [Suh 2006] explains in detail how the requirements of EVs can be met by hybrid engineering.

We read in the table of the ESORO Factsheet [ESORO 2016] that the truck has two battery systems with a capacity of 60 kW h each, in total 120 kW h. We do not know whether the two units are connected in parallel or in series. The battery is of the $LiFePO_4$ type (positive electrode) and was produced by Chinese manufacturer CALB Ltd. The battery system can deliver a voltage of 500–750 V. The $LiFePO_4$ battery has an EMF of 3.2 V.

During my time as a battery scientist in Berkeley, the $LiFePO_4$ cathode material was praised as one with a high capacity, but it had a relatively poor electronic conductivity, which amounts in a poor power density. Obviously, the power density of this battery is large enough that it can be used to pull trucks.

When we want to achieve the specified 750 V with cells that deliver 3.2 V each, we need at least 235 such cells (235 × 3.2 V = 752 V). The battery system must therefore have at least 235 such LiFePO$_4$ cells. Those add up to a total capacity of 120 kW h. Let us for simplicity assume that the battery system has not 235 cells, but 240 cells. Then, with 240 cells having a capacity of 120 kW h, two cells have 1 kW h and 1 cell accordingly has 0.5 kW h, that is, 500 W h for one cell.

CALB sells a whole variety of LiFePO$_4$ cells with a nominal voltage of 3.2 V and varies capacities. Let us divide the aforementioned 500 W h per cell (which is 500 V A h) by the 3.2 V and we obtain a capacity of 156 A h. The CALB SE200 Battery has 200 A h and weighs 5.7 kg. This makes an overall battery weight of 1,368 kg, which is huge.

The specific energy of LiFePO$_4$ batteries is known to have around 100 W h/kg. A 500 W h battery would thus weigh around 5 kg. The specific power is around 200 W/kg. A cell of 5 kg would therefore deliver 1000 W (1 kW). With 240 cells, we would thus obtain 240 kW power. This is sufficient to power the electric motor that has 250 kW.

For every 100 km distance driven, the truck consumes around 7.5–8 kg of hydrogen in the fuel cell. The hydrogen is stored in pressurized bottles from Australian manufacturer Luxfer. Each bottle can contain up to 4.93 kg hydrogen. Seven such bottles make up the hydrogen storage system that add up to 34.5 kg H$_2$ (an early comparison of different storage technologies and costs was made by Eklund and Von Krusenstierna [Eklund 1983]; a recent report from NREL is found in Parks [Parks 2014]).

1.2.2 Vehicle range comparison 1994 versus 2017

The ZEBRA battery, a battery type that I come later to in this book, can be used for mobile applications (in trucks and busses [Capasso 2014, 2015a, b, Valero-Bover 2014, Veneri 2017], extended by FCs in busses for which there exist numerous practical examples [Andaloro 2011, Antonucci 2015, De Lorenzo 2014, Dispenza 2012, Ferraro 2011, Sergi 2014] and in stationary applications [Vogel 2015]).

When you think of a bus fleet in a city where the busses are almost constantly on the way, it is easy to maintain the ZEBRA batteries all the time at the high operation temperature. When you think of a passenger car which is only used once or few times a day for commuting to work or for doing the grocery shopping, you see that most of the time the privately owned passenger car (it would be different with a taxi) is at rest, that is, "cold." This is one reason why ZEBRA batteries are not particularly suitable for privately owned passenger cars (I mean consumer cars; it may be different for a Taxi fleet). This is notwithstanding that Daimler Benz tested a whole range of Mercedes 190 at the island of Rügen in Germany in the 1990s [Sudworth 1994]. Daimler's movie archive lets you watch 30 min of the project [Daimler 1994].

Figure 1.10 shows how many kilometers you could drive with the Mercedes 190 at a particular speed with one charge of a particular type of batteries, specifically sealed

Range (Km)

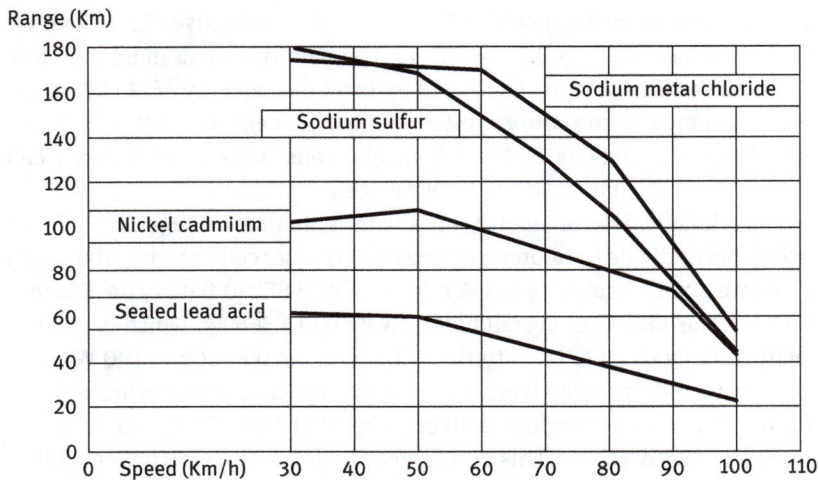

Figure 1.10: Range of Mercedes Benz 190 cars depending on the speed and on the battery type used (sealed lead acid; nickel cadmium, sodium sulfur, sodium metal chloride).
Reprinted from *Journal of Power Sources*, 51, Sudworth JL, ZEBRA Batteries, 105–114, Copyright (1994), with permission from Elsevier [Sudworth 1994].

lead acid; nickel cadmium, sodium sulfur and sodium metal chloride. The lead acid battery lets you drive 60 km as long as your speed does not exceed 50 km/h. With the nickel cadmium battery, the range is 100 km/h at the same driving conditions. You can almost double the range when you use sodium sulfur technology or the ZEBRA battery [Sudworth 1994]. As soon as you begin speeding, the range decreases considerably. At 100 km/h, you do not get further than 25–60 km.

I have made similar experience with the Hyundai ix35 Fuel Cell car, which drives with hydrogen. On my various routes on highways that mandate high speed and country roads that mandate medium speed and in urban and rural areas where only low speed is permitted, I collected sufficient data to obtain a global impression on the relation between range and speed. One important lesson I learnt is that the high thermodynamic efficiency of electrochemical energy conversion can outperform the combustion engine only at low and moderate speed. Following the recommended speed on highways, which is in Germany 130 km/h, the range would drop from a theoretical 600 km to only 200 km for the full tank (5.4 kg H_2 at 700 bar pressure). It will be worse when you go 140 or 150 km/h.[2]

---

2 In June 2018, I attended a summer school in Brixen/Bressanone in Süd-Tirol, Italy. I went there with the Hyundai ix35 Fuel Cell. I went over two Alpine pass roads, the highest (Flüelapass) with 2383 m above sea level, back and forth and I tried to avoid highways. During that trip, I took one day off and went to Venice to see whether it is possible to go as a consumer from the most southern hydrogen pumping station, which is in Bozen/Bolzano to Venice to the Mediterranean Sea and also back to

The dashed line in the data in Figure 1.11 shows some sort of master curve. It indicates the upper boundary of speed and range that I made on my tours. The curvature of this line looks actually quite similar to the curvature of the consumption data of the Mercedes 190, which ran on batteries.

You can quickly check this, as an exercise, when you scan both figures and rotate Figure 1.10 counter clockwise and then flip it on the vertical axis, in a graphics program. This makes that axis for speed and range is exchanged. Now you paste Figure 1.11 onto Figure 1.10 and contract or expand both axes so that the numerals on the axes are identical. Note that you are now comparing the performance of the Mercedes 190 with one-eighth of the performance of the Hyundai ix35 Fuel Cell, because my data in Figure 1.11 refer to one filling segment only. It would be an exercise to plot the data from both figures together into one plot. It would be a "senior exercise" to formulate a mathematical model which produces the curvature of the "master curve" for any type of car: gasoline, battery and FC.

For example, the green data point (31) at the upper left in Figure 1.11 shows 31 km range at 128 km/h speed – for the green segment. You have to multiply this with factor 8 because the car has eight segments, that is, 8 × 31 km = 248 km, and this total range at 128 km/h speed. At 100 km/h speed, we can even make around 76 km range per segment, when we look at blue data points 33 and 3 near the dashed line. This must have been a fortunate segment because with a factor 6, this would make a total range of 608 km per tank. This is 14 km more than the theoretical range disclosed by Hyundai (594 km).

The data points in Figure 1.11 that are very far away from the dashed line were recorded when there was traffic jam, long uphill tours, rain and heavy winds – conditions which impede your ride. It appears that you can make 35 km for sure at 100 km/h. This makes 280 km total range for full tank. Looking at the performance of the Mercedes 190 run on batteries, you can make 20–60 km at best. The range of the FC car (in 2017, model from 2015) is five times longer than the range of the battery car in 1994.

I made a regression curve by least square fitting to some of the data points on or near the master curve and assumed a parabolic relationship between speed and range. The result looks as follows:

$$v(x) = 136.92 \frac{km}{h} - 0.1815 \left[\frac{1}{h}\right] \cdot x \ [km] - 0.00397 \frac{1}{km \cdot h} \cdot x^2 \ [km^2]$$

Bozen. For the entire trip of 1217 km, I had consumed 11.17 kg of H_2. This is a range of 108.95 km/kg. For a full tank of 5.64 kg, this is a range of 615 km. This is 20 km more than the range mentioned by the manufacturer. My day trip from Bozen to Venice Piazzale Roma and back to Bozen was 407 km in total, mostly through the Val Sugana. Upon my return at the H_2 pump station in Bozen, Via Enrico Mattei, the engineer filled up again with 2.83 kg H_2. This yields a range of 143.82 km/kg. For a full tank, this yields a range of 811 km [Paladini 2018]. Wow!

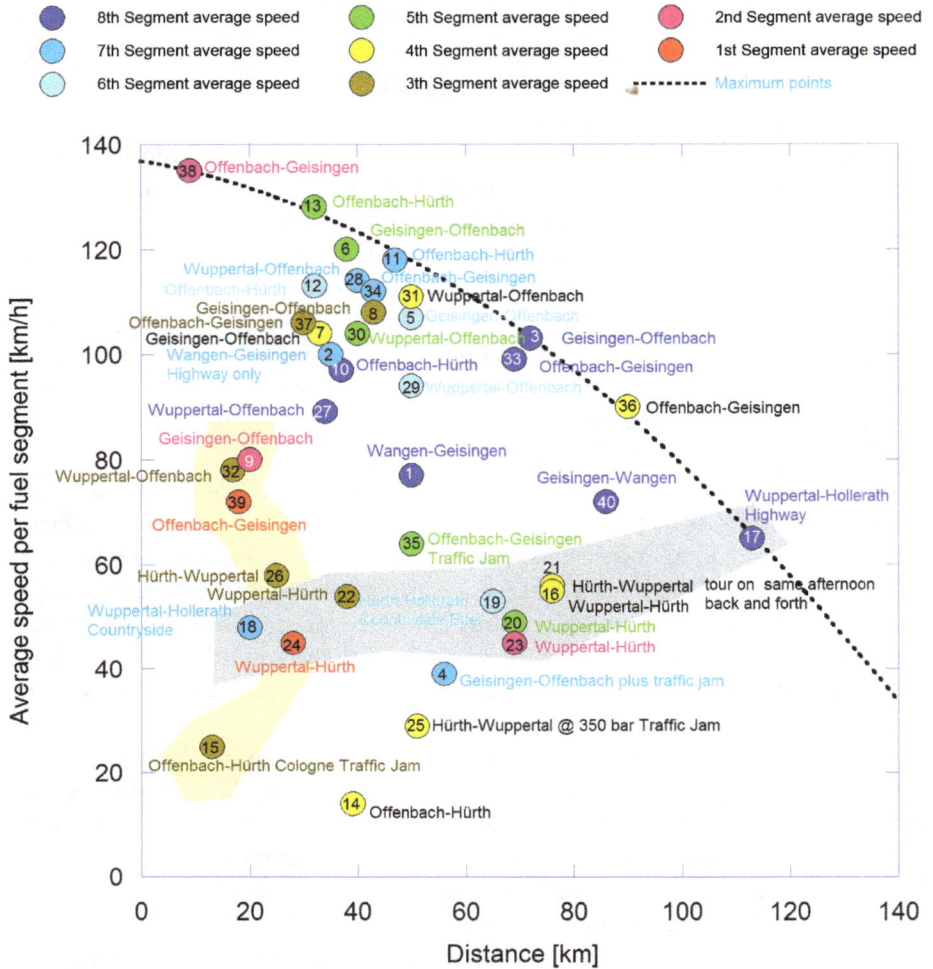

Figure 1.11: Overview of empirically obtained H_2 consumption data with the Hyundai ix35 Fuel Cell on various tours in Germany in 2017. The car shows the fuel filling level in eight segments (indicated with different color bullets). Shown is the averaged speed (km/h) during consumption of one segment and the obtained range (km) for that segment. Numbers in bullets indicate the sequence of segments emptied (1, blue, full tank to 39, red, almost empty tank). The dashed line is a master curve that shows the maximum available range for one segment depending on the speed. My record trip with 143 km/kg H_2 is not included in this dataset.

This regression curve may help us to determine the maximum possible range depending on the speed we chose during cruise when we reformulate and solve for distance x. We probably have to carry out a mathematical integration when we change the speed during cruise. This is a practical exercise for the reader. But it would be way more interesting if we had a functional relationship between $v(x)$

and x based on technical parameters (invariants) of the vehicle and external parameters such as wind, friction and so on. A simple regression like shown above does not substitute for a real mathematical model mentioned before [Webster 1997].

1.2.3 Hydrogen fuel cell bicycle

Now that we have seen trucks and busses as the large scale representatives of hydrogen-based electromobility, let us turn to bikes, to two wheelers, to motor bikes and scooters. You can equip a bike with an electric motor and power it with a battery or with a fuel cell. My colleague Kondo-Francois Aguey-Zinsou at University of New South Wales in Sydney has made a H_2 FC bike [Crozier 2014] which you can see also on YouTube. A small hydrogen container bottle is attached to the bike which can catch 100–200 l of compressed hydrogen. A schematic is shown in Figure 1.12.

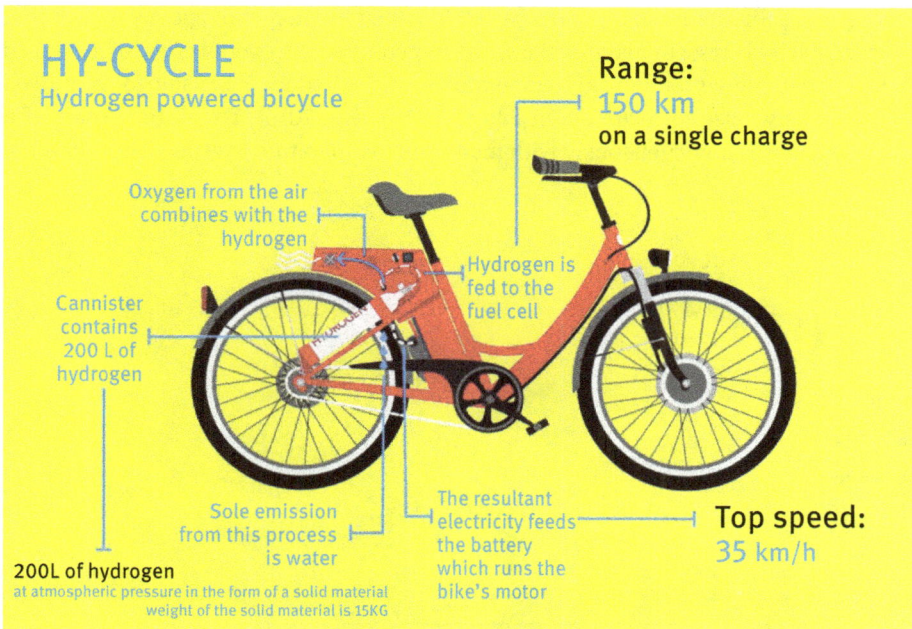

HY-CYCLE
Hydrogen powered bicycle

Range:
150 km
on a single charge

Oxygen from the air combines with the hydrogen

Hydrogen is fed to the fuel cell

Cannister contains 200 L of hydrogen

Sole emission from this process is water

The resultant electricity feeds the battery which runs the bike's motor

Top speed:
35 km/h

200L of hydrogen
at atmospheric pressure in the form of a solid material weight of the solid material is 15KG

Figure 1.12: Schematic of the HY-CYCLE hydrogen powered bike with fuel cell technology. Photo by courtesy of University of New South Wales, Prof. Kondo-Francois Aguey-Zinsou.

We notice from the specifications in Table 1.4 that the range of 125 km for the bike is given for a speed of 20 km/h. Acceleration for an increase in speed requires an increase of power that needs to be drawn from battery and fuel cell, which means

Technical specifications	
Range	125 km at 20 km/h
Maximum speed	35 km/h
Battery	518 W h lithium-ion battery
Battery	6 h on mains power
Fuel cell power	100 W
Canister	738 W h capacity, 100 l H_2
Canister exchange time/refill	30 s

that more energy is used in a shorter time. At higher speed, we feel higher aero-dynamic resistance. Possibly the temperature of battery and FC will increase which might have influence on resistances in the battery and FC system.

For example, the $LiMn_2O_4$-positive electrode is an electronic polaron conductor [Horne 2000, Sheftel 1966, Shimakawa 1997], which phenomenologically means that its conductivity increases exponential with temperature. The overall effect of increased power capability at enhanced temperature and a reduced resistance is indeed observed, as we see for example in Figure 1.13 [Zheng 2016]. There, the power of a battery increases homogeneously with increasing temperature irrespective of whether

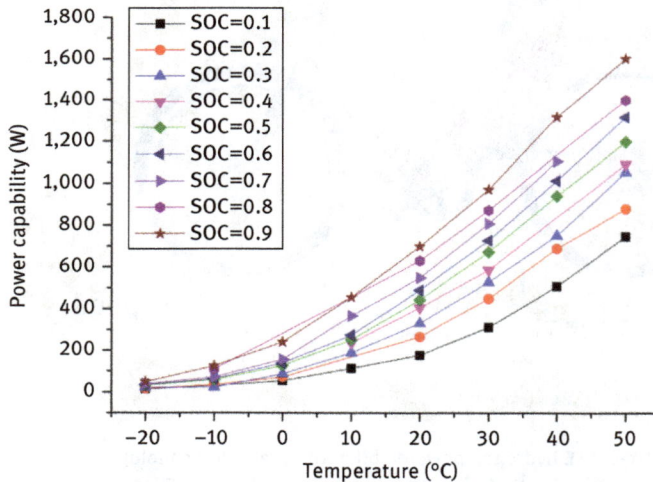

Figure 1.13: Power capability–temperature curves at different state of charge levels (SOC) of a battery cell.

the battery is fully charged of partially charged. Also, increased FC temperature may be beneficial [Song 2006]. Therefore, temperature increase has not necessary adverse effect on energy consumption other than that it allows for higher power and thus secondary increased consumption. We will later see how the Nernst voltage is a linear function of the temperature T.

The chemical kinetics of battery reactions is also enhanced at increased temperature. If we speed up to the maximum speed of 35 km/h, we can expect the temperature in the battery to increase and provide maximum power but it will become discharged very fast.

1.2.4 Hydrogen-powered submarines

When you drive a hydrogen fuel cell car, you will notice that it makes no sound. The chemical conversion in the fuel cell is a silent process, and the produced electricity propels silent electric motors. Only the sound from the wheels on the asphalt can be heard, once the car has an appreciable speed. Such silent system is interesting for military submarines which do not want to be noticed by the enemy. When submarines run on diesel engines, as many of them do, the noise from the combustion engine can be easily detected. Nuclear powered submarines have nuclear reactors that produce silent energy, but at a very high cost. In addition to the near-silence, FC-powered and battery-powered submarines have also no exhaust and no strong heat dissipation profile [Wing 2013].

Submarine technology is also interesting from the general perspective of artificial life support systems in space and outer space [Gitelson 2008, Lehto 2006, Lunan 2002, Nelson 1991, 1992a, b, 2013]. As submarines need air for breathing for the navy men and sailors during undersea operation, electrolysis of water for oxygen generation in submarines has been an early application of electrolyzers. This is a reason why nautical engineers expended their views from electrolyzers to fuel cells, which are inherently parallel technologies.

Some submarines are used as unmanned underwater vehicles [Davies 2007], for example, for deep sea exploration or oceanographic studies. They too can be propelled with FCs.

Some submarines are propelled with a hybrid technology, such as diesel engine for surface operation and lead acid batteries [Amphlett 1997, Suh 2006] or ZEBRA batteries for subsurface operation [Kluiters 1999]. Rolls Royce had a project where rescue submarines can be propelled with ZEBRA batteries [Vogel 2015].

The U 212 class vessel is an attack submarine[3] built in the early 2000s by Howaldtswerke-Deutsche Werft GmbH and Thyssen Nordseewerke GmbH for the

---

3 According to Naval Technology [Technology 2018] U212/U214 Submarines [https://www.naval-technology.com/projects/type_212/], and [Wing 2013] Wing J. Fuel Cells and Submarines – Analyst

German Bundesmarine. PEM-FC for submarines are also subject to studies in other countries such as United States, Russia [Wing 2013] and Korea [Song 2006], but information about this military sector is sparse for obvious reasons.

The propulsion of the entire U 212 has a power of 3120 kW (=4243 hp). It has an air-independent propulsion system of 300 kW which is built from several FCs. The U 212 features a PEM fuel cell (BZM SINAVYCIS) built by Siemens AG, which delivers a 34 kW power per unit. The BZM SINAVYCIS stands for quick switch on/off behavior, low-voltage degradation and long service life [Wing 2013]. About 34 kW corresponds to 46 hp (Pferdestärken PS) only; this is less power than an ordinary car has nowadays. My Volkswagen Käfer (Beetle) from 1984 had 34 hp (PS). Remember the Hyundai ix35 Fuel Cell model in 2015 has 136 PS.

Let us *try* – because we lack the privileged information of the insiders – to get behind the military items by accessing some sources in the internet, specifically the SINAVY PEM FC, which you find in the hyperlink to Siemens [Siemens 2013] or alternatively in Bakst [Bakst 2014]. Figure 12 in Siemens [Siemens 2013] shows a set of nine PEM FCs (nine stacks) in a testing rack. The black front panels are the radiators for cooling the stacks.

The figure on the right in Hauschildt [Hauschildt 2009], and on page 23 in Hoffmann [Hoffmann 2015], shows one such FC module (FCM 120). Visual inspection shows that around half of the length of the module (top portion) is occupied by the PEM FC stack and the other half with support infrastructure connected with the bottom support plate.

The table that is after Figure 12 in the original reference [Siemens 2013] lists the specifications of four different fuel cell modules (FCM), that is, FCM 34, FCM 120, FCM NG 80 and FCM NG 135.[4] The modules have a prismatic shape as we can see and as they are listed with a height, width and length. Let us begin with the FCM 34 but glance at the same time to FCM 120, which has 120 kW, this is 3.5 times higher power than FCM 34. The size of both modules is however almost identical. Height and width of FCM 34 is 48 cm each and the length is 145 cm. FCM 120 has height and width of 50 and 53 cm, respectively, and a length of 176 cm. This is a volume of 334 L for FCM 34 and 466 L for FCM 120. Thus, FCM 120 has a 40% larger volume only but 3.5 larger power.

View. Fuel Cell Today. 3 July 2013. http://www.fuelcelltoday.com/analysis/analyst-views/2013/13-07-03-fuel-cells-and-submarines U 212 is an attack submarine. [Morcinek 2013] Morcinek M. Kriegsspiele im Nordatlantik – U-Boot soll US-Träger angreifen. N-TV. 7 Februar 2013. https://www.n-tv.de/panorama/U-Boot-soll-US-Traeger-angreifen-article10085286.html "Für die simulierte Angriffsfahrt auf den amerikanischen Träger verfügt das deutsche U-Boot jedoch über einen besonderen technischen Vorteil, einen waffentechnischen Trumpf im Uniformärmel."

4 The modules FCM NG 80 and FCM NG 135 are outdated, but the purpose of this book is not showing the newest products but demonstrating how technology and innovation evolves. Note also that not all of the four modules came to operation. Two of them were under development.

The manufacturer specifies the weight (mass) of both modules to 650 and 900 kg. This is an increase of only 40%. The specific weights of both modules are therefore the same (1.95 and 1.93 kg/L). This is only possible when there is a significant improvement in the design of the module and its cells (compare the Solar Impulse (HB-SIA) and Solar Impulse 2 (HB-SIB)).

The length L of the module is probably the stacking height (145 and 176 cm) of the fuel cells. Page 23 in the original presentation of Hoffmann [Hoffmann 2015] shows that half of the length of the FCM 120 is used for the stack, whereas the other half is used for periphery infrastructure.

Consider now Figure 5 in Siemens [Siemens 2013] where the engineers have divided the nominal electrode area of 40 × 40 cm into two separated subunits.[5] Each of these two units would produce 0.7 V and thus add up to 1.4 V. With this design trick, we can increase the voltage on the cost of the current. The power will certainly remain the same: $P = U \times I$. The module with name FCM NG stands for the third generation (NG: New Generation?) of Siemens' FCs and can deliver 80–160 kW. FCM 34 stands for 34 kW and FCM 120 for 120 kW.

The voltage range is specified as 50–55 V for the FCM 34 and 208–243 V for FCM 120; 243 V/50 V = 4.86 and 208 V/55 V = 3.78. The ratio of the voltage ranges between FCM 34 and FCM 120 therefore from 3.8 to 4.9. This is somewhat larger than the ratio of power, which was 3.5. The power scales therefore roughly with the voltage. We can assume that the voltage per cell is 1 V and that therefore 50 cells are used in the FCM 34 and 120 cells in the FCM 120. Inspection of the brochure from Siemens [Siemens 2013] shows they actually calculate with 1 V per cell. FCM 34 could therefore have 50–60 cells in series; one cell therefore should have an averaged thickness of 2.4–2.9 cm. The same calculation for FCM 120 yields an averaged thickness per cell of 0.75–0.85 cm. The thickness ratio of the two cell types ranges therefore around 3.5, which is the same ratio we determined for the two different power classes. Page 25 in the presentation of Hoffmann [Hoffmann 2015] shows that half of the length of the FCM 120 is used for the stack, whereas the other half is used for periphery infrastructure.

Hoffmann [Hoffmann 2015] explicitly says "Brennstoffzellen-Batterie" on page 26 of his presentation, and he does not imply that there is a battery involved in the FC module. Rather, he uses the correct language as I will introduce later in this book where the literal meaning of battery is used for arrangement of units. Here, the "Brennstoffzellen-Batterie" is an array of FC modules.

The cells have a geometric area of 400 × 400 mm [Siemens 2013], which fits the size of the modules specified in Table 1.5. Looking at Figure 5 in Siemens [Siemens 2013], we notice that the square-shaped cells from FCM 34 have

---

5 This figure is meanwhile outdated, but let us continue for the sake of learning how different generations of cell design have different impact. This is notwithstanding that we do not have all necessary information available.

become divided in half size rectangles. This would allow for doubling of the voltage, for example. This would however limit the area per cell and thus limit the current.

The graph in Figure 6 in Siemens [Siemens 2013] shows Siemens testing data of one of their PEM FC cells. The cell voltage U_{cell} in volt is plotted versus the cell current I_{cell} in ampere. The right abscise shows in addition the power output of the cell as P_{cell} in Watt. Naf 115 and Naf 117 refer to two different types of Nafion® proton conducting membranes. Naf 115 has a thickness of 127 µm, and Naf 117 has a thickness of 183 µm.

When no load is employed on the FC, the cell voltage is slightly over 1 V, which is basically the open circuit potential. As soon as they apply a load and a current is flowing through the cell, the cell voltage drops drastically to below 0.9 V. This is in the range of 0–100 A. At 250 A, the cell voltage is decreasing approximately linear with increasing current. As the power is the product of voltage and current, $P = U \times I$, the power increases homogeneous. As we compare the evolution of the power with current, we see that the cell with the thick Naf 117 membrane approaches 800 W as the 1500 A current are approached, whereas the cell with the thin Naf 115 membrane will slightly exceed the 900 W power at the same current. The influence of having a 30% thinner membrane (127 vs 183 µm) is thus obvious. The thinner membrane can for a higher current maintain the necessary cell voltage, which is 0.55 and 0.63 V, respectively. We therefore learn how working on the components of a FC system, such as using an electrolyte membrane with a different thickness, can have a noticeable influence on the FC performance. When you increase the electrolyte path in an electrochemical cell, the resistance will increase correspondingly (compare the last section in Chapter 3, Electrochemical cells). This is the rational for thinning down an electrolyte as much as possible.

When the FCs are assembled to stacks, the voltage correspondingly adds up to the module voltage. The graph in Figure 7 in Siemens [Siemens 2013] shows the voltage transient of a PEM FC Module that was tested in operation for over 5,300 h; this is 220 days or more than 7 months. This is long enough for several submarine patrols.

The module voltage range in Figure 7 in Siemens [Siemens 2013] goes up to 248 V. We are therefore likely dealing with data from the FCM 120. We will see this from Figure 5 in Siemens [Siemens 2013] that compares the design of FCM 34 and FCM 120. The FCM 120 base area is divided into two rectangular bases for the electrodes, whereas the FCM 34 has a quadratic single electrode area. The consequence of such design is that the single quadratic electrode provides around 1.1 V per cell, whereas the FCM 120 with two electrodes would produce 2×1.1 V per area. About 120 such "double cells" would therefore add up to a nominal voltage of 240 V. Because the geometrical area in the FCM 120 is around half of the area of FCM 34, the current will also be half. This is the trade-off we have to pay for such trick.

The FCM 120 module is launched with a constant load at 560 A and the module voltage shows around 216 V. After 500 h of such operation, the load connected to the module is started and stopped. The voltage is not exceeding 220 V during this phase. From 1,200 to 1,700 h, a particular load profile is driven with the module and the voltage remains constant at 215 V. A new start–stop phase is run from 1,700 h to 2,200 h and the voltage fluctuates between 215 and 217 V. At 2,200 h, the load is changed so that only 390 A current is drawn. Immediately, the voltage rises to 234 V. Over the following 500 h, the voltage homogeneously drops to 230 V.

Now, we finished the use of the PEM FC in the submarine, and we treated already the hydrogen car, truck and bus in this chapter. PEM FCs have already been used in spaceflight long back. Large freight ships are considered for using PEM FC for emergency power supply. Jingang et al. [Jingang 2012] state

> Fuel cells can also be an interesting solution for ships power. However, the developments of fuel cell systems for ship are in infancy. The only exception is the PEMFC in the submarines.

The reverse use of PEM FC as electrolyzer is considered for storing solar energy from PV panels. The use of FC on ships is not necessarily to power the entire ship, but to provide on board power. Large ships use heavy bunker oil for their diesel engines which create a lot of pollution. As the number of such ships on the oceans is increasing as much as the international trade worldwide is increasing, there is an increased interest in curbing this pollution without reducing the maritime traffic frequency.

A recent presentation [Hoffmann 2015] shows how Siemens, for example, looks at the suitability of fuel cells in a wide range of mobility technologies. They include decentralized electric power stations, ocean freightliners, busses, trucks, cars, railroads, gas tankers, space shuttle and renewable energy storage systems. Siemens' projects range from stationary applications of >50 kW to mobile and finally portable applications of as low as 2 W.

1.2.5 Fuel cell-powered airplanes

Is it possible to power airplanes with fuel cells? How would this work?

For airplanes that fly with jet engines, fuel cells would be of no use for propulsion. In jet engines, the kerosene (a hydrocarbon liquid fuel) is combusted in a turbine. We need an electric motor when we want to use FCs. Electric motors on planes work as was shown by the impressive around the globe tour of the Solar Impulse by Andre Boschberg and Bertrand Piccard in 2015–2016. The energy for the motors was delivered by onboard PV panels, which was stored in onboard batteries [Batterybro 2015].

We learnt in the section about electric cars that improvement in nonbattery-related materials and design helped to actually bring electric cars on the market [Wyczalek 1995]. We have a similar situation here where the lightness of airplane structure was considerably improved. The lightest sheets of carbon fiber material were used which had only ~30% of the weight printer paper. The wingspan of SolarImpulse is 72 m, which is comparable with a Boeing 747.

The surface of the Solar Impulse 2 aircraft is decorated with 17,248 monocrystalline silicon solar cells with 135 μm thickness. They produce electricity that is stored in four Kokam ultra-high energy NMC (nickel manganese cobalt oxide) lithium polymer battery packs with 150 A h cells of 38.5 kW h each. The total capacity is 154 kW h. The energy density is density of 260 W h/kg [Kokam 2016]. This is enough power to fully charge the batteries in about 6 h and allow the four electric motors (17.5 CV each) to stay airborne all night.

When the Solar Impulse endured fast changes from hot to cold temperatures, the batteries became damaged and forced the team to interrupt their world tour for several months [Rogers 2016]. But the tour was a success in the end and it was demonstrated that it is generally possible to travel around the globe in an electric air plane powered with solar PV – even at night, when the sun is not shining and the electricity was stored in batteries.

The first flight with a fuel cell-powered airplane was probably made in 2009 with a HK 36 Super Dimona [Klesius 2009]. The plane carried a 90 kg hydrogen FC which produced the electricity for an electric motor to turn its propeller. Note that the FC could not deliver the energy (actually, the power, 45 kW) which was required for the takeoff.

For this extra power, a lithium ion battery assisted. You may remember why there is a traction battery in the FCEV that I started this chapter with. Thus, the plane could lift off the runway in Ocaña in Spain. At 3,300 ft altitude, Pilot Barberán disconnected the battery, and for the following 20 min, the Super Dimona flew straight and at requested altitude at around 60 mph on the FC only. This was probably the first time a manned airplane had flown by a FC alone.

Later in this book, we will learn about solid oxide FCs (SOFC), which do not require hydrogen as fuel. Rather, they can operate with the less expansive natural gas and biomass gas and syngas. This is a great advantage of SOFC. They require, however, very high operation temperatures from 600 °C to 1,000 °C. Aguiar et al. [Aguiar 2008] discuss in their paper an unmanned aerial vehicle that is propelled by an SOFC and a gas turbine.

Before ending this chapter, I would like to play again with words. Section 1.3 ended with the mentioning of the space shuttle. FCs have been considered as electric power sources in the early stages of spaceflight. Was the space shuttle, where the space ships *powered* with FCs? Or were they *propelled* with FCs? Was the airplane mentioned in Section 1.4 *powered* or *propelled* with FCs?

The space shuttle was propelled by three rockets. The large tubular rocket contains liquid hydrogen that is cooled to 20 K, a very low temperature for such a large scale installation, and liquid oxygen at 90 K. The hydrogen–oxygen mixture is a strong rocket propellant [Sloop 1978], which burns at over 3,300 K and thus gives the spaceship with total mass m a very large momentum $p = mv$, while the velocity is increasing and the mass, due to loss from fuel combustion, is decreasing. Yet, the spaceship needs two solid booster rockets [Feynman 1988, 2005, Fletcher 1986, Rogers 1986], which are the two thinner tubes at the space shuttle, which contain ammonium perchlorate as oxidizer and aluminum powder as fuel, plus iron oxide as reaction catalyst, plus some polymer binder materials. These three rockets give the spaceship, the space shuttle the thrust and momentum which it needs to come into orbit and stay in orbit. This is not an electrochemical process. Rather it is brutal chemical combustion.

The space shuttle is put into orbit around the Earth with the thrust of its own three rocket engines which are fed by the mixture of hydrogen and oxygen, and with the assistance of the solid booster rockets. This provides a momentum $p = mv$ and a corresponding kinetic energy $E = p^2/2m = 1/2mv^2$ in order to lift the space shuttle from the Earth to some altitude such as 500 km (work against the gravity force by Earth), for example, and they give it also the rotational energy for the circular motion around the Earth, which yields a balance of centrifugal force of the space shuttle in orbital motion and gravitational force from the Earth as the central body. This is how the space shuttle is brought into space and kept into space, without any further energy from the rockets because they are empty once the three large tubes are removed from the space shuttle.

You may have derived in school which minimum speed is necessary to escape from the Earth surface and make it and remain in orbit. This escape velocity is frequently referred to as first cosmic velocity and yields for the Earth with radius r_e and mass m_e and gravitation constant G:

$$v_1 = \sqrt{G\,\frac{m_e}{r_e}} = 7.9\,\text{km/s}$$

There is a small amount of rocket fuel inside the space shuttle that helps it manoeuver in small increments to adjust its position in orbit where necessary, such as tilting and rotating. But once in space, the space shuttle needs to be operated in order to keep the astronauts inside warm and comfortable, to permit them to communicate with the Earth with their radios, and to perform various scientific experiments during its mission. This all requires energy. And this energy is not used to *propel* the space shuttle (bring it into orbit), but to *power* the space shuttle (give it the energy necessary for being useful there in orbit).

Some of us remember that the Skylab had four large wings that were covered with solar panels which provided Skylab with electric power. When the rooftop of space shuttle opens, it reveals its innermost solar panels which then become open for solar radiation. But both Skylab and space shuttle have FCs and some limited amount of necessary fuel.

The FC- and battery-powered air plane is propelled by the FC and battery. The lithium batteries have the role as booster power sources in order to lift the plane above the ground, and after that the FCs can maintain speed of the place. Anything else which operates with electricity in the plane such as signaling, radar and communication radio and transponders will also be powered by the FCs. In contrast to conventional aviation where jet engines or stroke engines, both of which are based on combustion chemistry, are used to provide the necessary speed for the plane to lift off from the ground, these FC and battery-powered airplanes take entirely advantage of electrochemical processes.

You will find much in this book which relates not at all to electrochemistry. I do this for didactic purposes. Providing energy for a community or for individual use is very important, essential that most of us do not reflect over where the energy is coming from and what is necessary to produce, deliver and tap this energy.

1.3 Solid oxide fuel cell for use in cars

German automobile maker BMW was interested in using solid oxide fuel cells for powering the onboard electronics in their vehicles [Carter 2001]. The idea is the following. More and more energy coming from the combusted fuel in the engine is being converted in electric energy for powering all kinds of devices such as the necessary electric controls, board computer, air conditioning and so on. It would be worthwhile to use the heat of the engine under the hood providing an environment for the SOFC, which runs anyway at high temperatures. The necessary fuel for the SOFC could for example come from the engine exhaust when it contains unburnt hydrocarbons or carbon monoxide. BMW eventually abandoned the idea but from the fundamental point of view of energy conversion [Salameh 2008] it makes sense.

1.4 SOFC for laptops and cell phones

It is possible to make SOFC large enough as power plants which can serve a community or an industrial complex with electricity. We can imagine also the opposite way and try to miniaturize SOFC [Evans 2009] so that they are handy for portable electronics. I remember one eminent professor from ETH when he delivered a talk at Empa 10 years ago. He explained to the audience how a portable computer, a laptop might be powered in future not with an expensive and rechargeable lithium ion battery but with cigarette lighter butane gas C_4H_{10} fuel cartridge, which would propel an SOFC that in turn would deliver the electricity for the device.

One argument which you can bring against such use is that SOFC require and develop high temperatures. The professor then explained that a microfuel cell could be based on thin film technology, and when you use thin layers, ultrathin layers of electrolyte, the temperature required to activate the charge transport would be lower than in bulky electrolytes.

This is indeed a valid argument. As we will see in one of the future chapters on solid electrolytes, ultrathin films which you may grow with epitaxy technology can indeed bring some advantages over conventional technology, such as lowering of the proton transport activation energy [Chen 2010, 2011].

It is certainly not easy to establish a new product on the market against all economical and societal odds. MIT Start-up company Lilliputian Systems, who promised SOFC batteries, vanished from the scene [Kanellos 2014]. However, such bad news does not necessarily discourage researchers [Huber 2014].

1.5 Energy storage for stationary applications

Now let me come to stationary energy storage. Stationary means not mobile (mobile includes handheld devices and automobiles and other mobility). And stationary means the storage is a fixed installation which can be residential and large scale industrial, for example. There is a relatively wide range of energy storage technologies available. We can generally distinguish between storage of mechanical energy, chemical energy and electric energy. Table I in Lemofouet [Lemofouet 2006] gives a simple and not all inclusive overview of some storage technologies with life-time and efficiency data.

Hydropower is certainly a stationary installation, because you cannot carry the water power plant with you. The electrochemical and electrical storage principles listed in Table I of the original publication [Lemofouet 2006] are mostly available for mobile applications, but the redox flow battery is typically a larger installation which is not being "carried around."

For example, pumped water hydropower plants have a lifetime of 75 years and an efficiency of 70–80%. A compressed air energy storage system could last 40 years. For redox flow batteries, there are no long term data available but they will endure 1500–2500 charging cycles at 75–85% efficiency. Metal air batteries have an efficiency of 50% and their lifetime is limited to 100–200 cycles. A lead acid battery with 75% efficiency can have 200–300 cycles' lifetime. A sodium sulfur battery with 89% efficiency may run 2000–3000 cycles. Other modern batteries with efficiencies up to 90% and 95% may run 500–1,500 cycles, and supercaps with 93–98% efficiency have 100000 cycles [Bärtsch 1999] or more (see [Lemofouet 2006]).

Later in this book, we will see how wood as a fuel is used as energy carrier, for example. Wood is one sort of biomass which stores chemical energy [Krajnc 2015]. Chemical energy is also stored in fossil fuels such as coal, natural gas and mineral oil. Hydrogen is a chemical fuel which is increasingly getting attention again [Bockris 2002, Gregory 1972]. A more sophisticated energy storage principle is employed in electrochemical galvanic elements that we know as batteries. There is a very wide range of batteries, which we see later in this book. Bocklisch has listed many important technical data on storage systems in Bocklisch [Bocklisch 2015], such as energy density, installation costs, self-discharge rate, lifetime and efficiency (see Table 1.5).

1.5.1 Solid oxide fuel cell power plants

Because of the high efficiency of FCs, electric power plants based on FC technology have been explored quite early. With the availability of huge amounts of natural gas, SOFC are particularly interesting for such application [Watanabe 1996].

Developers and manufacturers are, for example, Rolls Royce [Agnew 2005, Gardner 2000, Magistri 2007, Trasino 2009, 2011] with prototypes in the 1 MW range, which, for example, can be considered as power source for water treatment plants [Siefert 2012]. Siemens and Westinghouse [Bujalski 2008, Casanova 1998, Dollard 1992, Hassmann 2001, Walde 1976] to name only two with a long history in SOFC research and development.

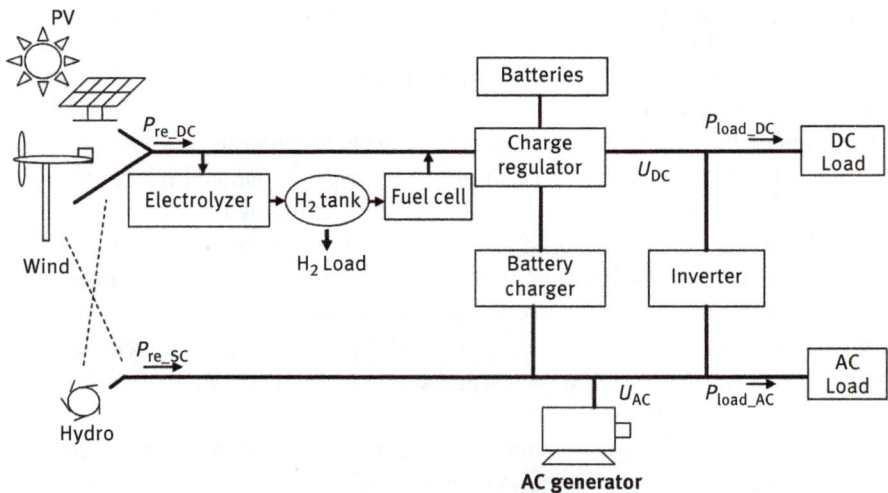

Figure 1.14: This image explains the fundamental concept and control strategy of the system: the use of all that energy that is not consumed.
Reprinted from *Renewable Energy*, 32, Dufo-Lopez R, Bernal-Agustin JL, Contreras J, Optimization of control strategies for stand-alone renewable energy systems with hydrogen storage, 1102–1126, Copyright (2007), with permission from Elsevier [Dufo-Lopez 2007].

Table 1.5: CAES compressed air energy storage. SMES superconducting magnet energy storage. Reprinted from Energy Procedia, 73, Bocklisch T, Hybrid Energy Storage Systems for Renewable Energy Applications, 103 – 111, Copyright (2015), doi: 10.1016/j.egypro.2015.07.582, with permission from Elsevier. [Bocklisch 2015].

Comparison of technical specifications and performance data of energy storage systems

	Supercap	SMES	Flywheel	Lead acid	Lithium ion	NaS	RedoxFlow	Hydrogen	PumpedHydro	CAES
Energy density W h/L	2–10	0.5–10	80–200	50–100	200–350	150–250	20–70	750/250 bar 2400/liquid	0.27–1.5	3–6
Installation costs E/kW	150–200	High	300	150–200	150–200	150–200	1000–1500	1500–2000	500–1000	700–1000
Installation costs E/kW h	10000–20000	High	1000	100–250	300–800	500–700	300–500	0.3–0.6	5–20	40–80
Reaction time	<10 ms	1–10 ms	>10 ms	3–5 ms	3–5 ms	3–5 ms	>1 s	10 min	>3 min	3–10 min
Self-discharge rate	Up to 25% in first 48 h	10–15%/day	5–15%/h	0.1–0.4%/day	5%/month	10%/day	0.1–0.4%/day	0.003–0.03%/day	0.005–0.02%/day	0.5–1%/day
Cycle lifetime	>1 Mill.	>1 Mill.	>1 Mill.	500–2000	2000–7000	5000–10000	>10000	>5000		
Lifetime in years	15	20	15	5–15	5–20	15–20	10–15	20	80	ca. 25
System efficiency in %	77–83	80–90	80–95	70–75	80–85	68–75	70–80	34–40	75–82	60–70
Short-term (<1 min)	XXX	XXX	XXX		X		X			
Mid term (>1 min, <2 days)			X	XXX	XXX	XX	XX	X	XX	XX
Long term (>2 days)				X		X		XXX	XXX	XX

1.5.2 Hydrogen for stationary and residential applications

Whenever there is talking about solar PV and wind power, the question for the storage of this kind of electric energy comes up. You can send (sell) this electric energy to the electric grid utility company and let the "system" decide over the fate of the electricity which you do not use or need for yourself.

You can store the electric energy in your own battery system and then use it later when you need it. In spring 2017, I received an email from an old friend of mine. As he had occasionally learned that I was working as a scientist, he asked me about my opinion on solar electricity storage from a PV panel into a battery system. You can read this up in Chapter 4 on lead acid batteries.

Or you can use the electricity to operate a water electrolyzer that will produce for you hydrogen and oxygen gas. The hydrogen is a fuel and can be stored in containers and later be converted again with a FC into electricity. Michael Strizki, a New Jersey inventor, has pioneered the Hydrogen House [Strizki 2017].

The Hydrogen House is equipped with PV panels which collects solar energy and converts it into electricity for direct use in the house. When there is no electricity used but produced by the PV panels, the electricity is used by an electrolyzer that converts water into hydrogen fuel and oxygen gas. The Hydrogen House has in the front yard a whole battery of gas containers which store the original solar energy now in chemical form as hydrogen fuel. As David Biello [Biello 2008] puts it in his article "No more power bills ever" about the Hydrogen House in Scientific American:

> The only way to get a zero-carbon footprint is to grab the big power plant in the sky.

Italian company Giacomini develops residential utility systems and components for those. Their H_2YDROGEM system uses electric energy for hydrogen production by water electrolysis. The hydrogen is later used in FCs for electricity production and in a catalytic burner for producing heat and for cooking in kitchen, for example [Giacomini 2017]. The general concept of the system is sketched in Figure 1.14.

The Giacomini company has its history in the Italian province of Novara near the Lake Orta. Water therefore plays an important role for the company. In the sketch of their H_2YDROGEM system, we see how water is electrolyzed in an electrolyzer (see the blue line) which takes the necessary electric energy from a low cost source, such as vagabonding electricity from the electric grid (see the pink line) when there is no huge demand for electricity. The thermal energy produced in this process can be used for heating purposes (see the red line). The hydrogen that is being produced is a fuel which can be immediately burnt in a catalytic fuel burner, for example, for heating a room or for being again converted in a PEM FC (see the green lines). When there is no need for immediate use of the produced hydrogen, it can be stored in the hydrogen storage container for later use. The system delivers heat and electricity upon demand.

Figure 1.15: The Giacomini H$_2$YDROGEM system installed in the ground floor of a residential home in Orta San Giulia, Novarra, Italy.
Photo by Artur Braun

Figure 1.16: Researchers viewing through the window of a building which has a Giacomini system installed.
Photo by Artur Braun.

The author of this book came across this system (Figure 1.15) on the occasion of a project meeting on 15 December 2011 for the EU NanoPEC project [Augustynski 2008] in Novara, Italy. The system is located in the first floor of a historic residential building in Orta San Giulia (see Figure 1.16).

We have seen in Figure 1.14 how several components are joined together to make an integrated energy system that has already some complexity. You certainly can become creative and add more components. Think for example about a solar PV that you install on the rooftop of your residential home which you employ for immediate use of electricity during the daytime.

When you have excess energy that you cannot use immediately, you may think of charging a large battery or running a water electrolyzer to produce hydrogen fuel. There may be times when you cannot produce energy with the PV panels on your rooftop and when all your self-generated stored energy is exhausted, but you have still a diesel tank and a diesel engine-based electricity generator and maybe a heating storage system. To harmonize and synchronize all components, you may employ some mathematical principles for the optimization of the system. One way how this could work is shown in the study of Dufo-Lopez [Dufo-Lopez 2007] and in the PhD thesis of Ulleberg [Ulleberg 1998].

The three principal electric energy suppliers are PV, wind turbines and hydro-power turbines and they deliver power P_{re} into the system, which delivers an electric power output on two different kind of loads as alternate current and direct current: P_{load_DC} and P_{load_AC}. The arriving electric DC power P_{re_DC} can be directly used at the load. When there is excess power which cannot be used by the load, a charge regulator will manage whether batteries are charged with P_{re_DC} or – when the batteries are fully charged – whether the water electrolyzer will be operated with the excess energy. The produced hydrogen will then be stored in a H_2 tank.

The hydrogen is stored in a tank and can be supplied to a hydrogen load. This can be for example a hydrogen gas station for a vehicle. The hydrogen from the tank can also be used in a FC which produces DC electricity which will be supplied back to the electric DC part of the system; it may charge the batteries or it may directly be used by the DC load.

The hydropower can propel an AC generator which produces an U_{AC} voltage for the P_{load_AC}. When there is need for P_{re_DC} instead, an AC/DC converter (inverter) is employed. Via the charge regulator, the hydropower can be used to charge the batteries or operate the electrolyzer.

This scheme is realized for example in the "Grid to Mobility Research Center" in Martigny, Switzerland [Ligen 2018]. This research center is actually a power plant which receives electric energy from the electric grid, which we can consider as a black box. The grid can contain electricity from nuclear power plants, hydropower, PV and others. The electric energy produced in Switzerland is 32.8% nuclear, 59% hydro-power, 5% thermal energy and 3.2% diverse renewables [BfE 2016].

The power plant is located at a large waste water treatment facility near the highway and operates a hydrogen electrolyzer based on conventional alkaline technology. The hydrogen is stored in conventional steel bottle batteries (I will explain later in this book what battery means) which are connected with pumping cascades

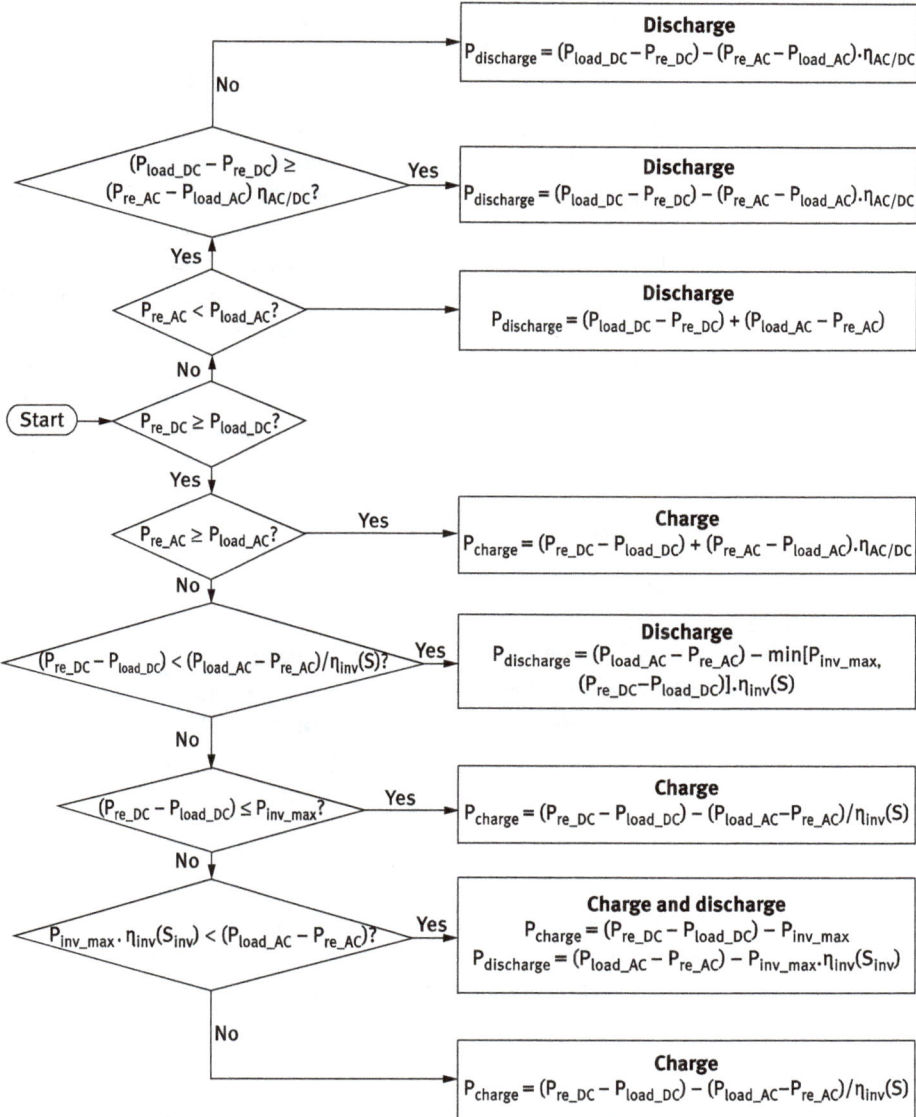

Figure 1.17: Logic flowchart of charge and discharge processes.
Reprinted from *Renewable Energy*, 32, Dufo-Lopez R, Bernal-Agustin JL, Contreras J, Optimization of control strategies for stand-alone renewable energy systems with hydrogen storage, 1102–1126, Copyright (2007), with permission from Elsevier [Dufo-Lopez 2007].

which are connected with the hydrogen gas station. The plant operates also a large vanadium oxide-based redox flow battery which provides power for EVs. This power plant is not made for public use. It is still a research center where scientists including PhD students study systems like the one sketched in Figure 1.14. But upon arrangement, one can access the facility and get hydrogen or electricity for vehicles. I went to this power plant on the 7 March 2018 in order to fill up our Hyundai ix35 Fuel Cell with hydrogen – when I was trying to break another hydrogen mobility record by attending the press and media reception at the Autosalon in Geneva.

The details of the energy supply and energy demand – actually the powers P_{re} and P_{load} – and the technical characteristics of the components of the power plant (electrolyzers, batteries, generators, inverters etc.) are the necessary input data for the management of the system. Figure 1.17 shows a simple flow diagram which begins at "Start" with the question "Is the available renewable DC power at least as large as the required DC power at the load site?" When this question is answered "Yes," the next question is "Is the available renewable AC power at least as large as the required AC power at the load site?" When this question is also answered with "Yes," the charge regulator will charge the system according to following relation, where also the efficiency $\eta_{AC/DC}$ of the AC/DC converter is taken into account:

$$P_{charge} = (P_{re\_DC} - P_{load\_DC}) + (P_{re\_AC} - P_{load\_AC}) \cdot \eta_{AC/DC}$$

Further logic considerations are taken into account for the next steps in the flow diagram. These are the general foundations for the logic management of the power plant. The next step is then to actually process the available data mathematically in order to optimize the overall operation of the plant for example with respect to the costs. Such systems can be programmed with Matlab and Simulink [Lajnef 2013].

1.6 Driving with water?

When I lately met people on several various occasions where the discussions eventually turned toward mobility and renewable energy, I was confronted with the suggestion or conviction that it should be possible to power cars not by conventional fuel, not by batteries and also not by hydrogen, but by water. In short, "don't use gasoline for the car but use just plain water."

I have been in the believe that this topic, this question was settled. You cannot run a car on water as fuel. This is because water is no fuel. You certainly can use water and then produce hydrogen fuel from it – by electrolysis or by steam reformation. But to do so, you need to input energy. At least you need 1.23 V for the electrolysis if we disregard any overpotentials as kinetic barriers.

I met one gentleman who believed it should be possible to have water in a container in the car and then electrolyze it with a battery, for example, the starter battery (lead acid battery) and produce the hydrogen fuel which you can use for a hydrogen combustion engine (BMW made such for a short time) or for a hydrogen FC. This approach would certainly work, but there is no gain for you.

You will not get more energy from this procedure than you would get from the battery directly. It is more reasonable to use the battery and power the electric motor for the car, better than using the battery for water electrolysis and then recombine the produced hydrogen with oxygen in a FC and then again produce electricity which you use for the electric motors.

Recently I came across websites that claimed that some individuals had found a way where water could be used as fuel to run a motorbike or a car. It was not disclosed how this would work. I believe it is sure to say that this is baloney.

More recently, I learnt that some gentleman had purchased a kit where water was added to the fuel system in the car and then this should somehow improve the car or the propulsion or the driving. The gentleman had tested the car after the kit was installed. Over many well-controlled test drives, he could not find anything that would point to less fuel consumption. Experts that he involved found that the car produced less NOx after the kit was used. Well, I do not know what kind of kit this is and what it actually promised for the driver, but it is good to hear that the exhaust becomes cleaner.

Adding water to the fuel (water as an additive) can have some catalytic effect on the combustion. Water (H_2O) is a gentle oxidizer and can help improve combustion. When you add too much water to the fuel, too much energy of the fuel combustion will be used to evaporate the water. As water has a large heat capacity, too much energy would be lost. Therefore, there is no simple scaling law that says the more water you add to the fuel, the better it works for your purpose.

The goal of naïve people would be that you improve the system to an extent that in the end no fuel, no benzine or diesel is necessary anymore but only water is necessary for powering the car. I have not written this book in order to make just this point clear. But the experience which I made in the aforementioned discussions with some people motivates me to close this first chapter in this book with some serious comments about what we can expect from physics and chemistry, and what not.

Water is not an energy carrier like a fuel. Expecting water to deliver you energy is the same like you would expect that you can light a fire from the ash, and not from the wood. Water is formed when you combust hydrogen and oxygen in the proportions 2:1. The energy is stored in the reaction enthalpies of hydrogen and oxygen (or wood and oxygen). When these two components react, you can derive the free energy from the two gases and use it to perform work.

Once the water is formed, the energy is gone. What ash and carbon dioxide are to the wood, water is to the hydrogen. The only remaining energy in water which you can use is when the water is very warm or hot, such as water vapor – the heat energy, for whatever you can use it. Or when you have water stored at a high altitude and use it in a hydro power plant. Other than that, a bucket of water in front of you has not much useful energy. You should go and find energy elsewhere. It's not in the water – with one exception.

I will close the book with nuclear fusion in the last chapter. When the water contains the heavy hydrogen nuclei deuterium and tritium, their nuclear forces contain a huge amount of energy. But we cannot tap this energy yet other than in the hydrogen bomb. Scientists are working toward controlled nuclear fusion which would bring huge energy that we could use. Fortunately, it is not so easy to carry out such nuclear reactions. If it was easy, then all oceans on Earth could easily light up in a giant explosion and become a sun.

2 The electrochemical double layer

Whenever an electrode[1] is inserted into an electrolyte,[2] an electrochemical double layer (EDL) is built up at the interface between the electrode and electrolyte. For metal electrodes, the surface has a negative charge that extends beyond the electrode surface and may attract positive charges, when present in the electrolyte, as illustrated in Figure 2.1.

The formation of a double layer is a general response of a solid when exposed to a fluid. Fluids are liquids and gases. Therefore, even dust particles and aerosols in the atmosphere are surrounded by a double layer from ambient humidity and gases. When two solids are joined together, an interfacial layer is formed instead. The physical origin for both phenomena is the difference in the chemical potential

$$\mu_i = \mu_i^0 + RT \ln a_i$$

of each component i with activity a_i that are present in a system of two different phases I and II, when they are in a direct contact with each other:

$$\mu_i(\mathrm{I}) \neq \mu_i(\mathrm{II})$$

The phenomenon of a charged surface was discovered several hundreds of years ago and was first reported – in German language – on 11 October 1745, by German lawyer and researcher Ewald Georg von Kleist (10 June 1700 in Vietzow, Pommern; 11 December 1748 in Köslin), which soon was referred to as the Leiden Jar [Von Kleist 1745]. The Leiden jar has its name from the city of Leiden in the Netherlands, where Professor Pieter van Musschenbroek conducted in 1746 studies on electrically charged jars. These glass jars had a thin metal coating outside on the glass and inside in the glass. The glass itself was an insulator; the jar resembled therefore what we know today as plate capacitor. When a metal rod was put in a damp glass jar with two such metal coatings outside and inside, it could be electrically charged with an electricity generator. When a person would touch the metal rod later, it was possible that the person received an electric shock. The Leiden jar was therefore the first technology to store electricity in a controlled way.

---

1 By default, electrodes should have a good electronic conductivity. Therefore by "electrode," I mean here first a piece of metal. It is not necessary that this metal piece has the function of an electrode in an electric circuit. If for whatever reason or purpose a metal piece is in contact with an electrolyte, such double layer will form. This can be the body of a ship or submarine in the ocean, or a metal tube through which water or some other liquid is flowing. Next, we can relax the condition of a well-conducting metal and also allow other solid materials in contact with a liquid, for example, a semiconductor. We will come to this chapter later in this book.
2 Normally, we think that an electrolyte is some liquid. But there are also solid electrolytes. We will learn about these in the chapters on fuel cells and proton conductors.

https://doi.org/10.1515/9783110561838-002

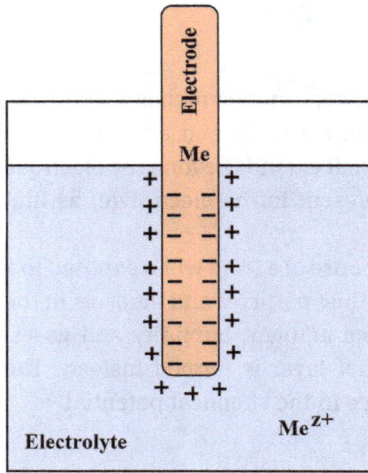

Figure 2.1: Schematic for a metal electrode, for example, copper, inserted into an aqueous electrolyte such as copper sulfate. The free electrons in the copper cause negative charge at the copper surface, which is compensated by positive-charged metal ions Me^{z+}. There is difference of the potential between electrode and electrolyte and electrolytic double layer at the phase boundary metal/electrolyte. The chemical potential in the electrode is lower than in the electrolyte:
$\mu_{Me^{z+}}$ (metal) $< \mu_{Me^{z+}}$ (solution)

An example of a simple system is illustrated in Figure 2.1, where a copper metal rod as the solid electrode (phase I) is dipped in a copper sulfate solution being the liquid electrolyte (phase II).

In general, the thermodynamic equilibrium condition

$$\mu_{Cu^{2+}}(\text{solution}) = \mu_{Cu^{2+}}(\text{metal})$$

will not be satisfied immediately after insertion of the electrode in the electrolyte. However, the equilibrium will be gained by a chemical reaction on the electrode surface, either by an anodic dissolution of the metal Me to metal ion Me^{z+} or by cathodic coating from the solution with the exchange of z electrons,

$$Me^{z+} + ze^{-} \Leftrightarrow Me$$

which is facing an electrical potential difference φ. The condition for the equilibrium yields (with Faraday's constant F)

$$\mu_i(I) + z_i F\varphi(I) = \mu_i(II) + z_i F\varphi(II)$$

In the case of dissolution of the metal,[3] the metal ions will charge the boundary layer (double layer) on the electrolyte side positively, and the electrode remains negatively

---

3 Dissolution may take place even between solids. The Kirkendall effect is one manifestation of this. See, for example, [Kodentsov 1998] Kodentsov AA, Rijnders MR, Van Loo FJJ: Periodic pattern formation in solid state reactions related to the Kirkendall effect. *Acta Materialia* 1998, 46:6521-6528.doi: 10.1016/s1359-6454(98)00309-7.

charged. At the isoelectric point,[4] the dissolution will be stopped by electrostatic forces. The equilibrium can be perturbed by applying an external DC potential with the result that the charge of the double layer can be increased or decreased, or the sign of the charge can change. When the external applied potential does not exceed the electrolysis potential of the electrolyte and the dissolution potential of the electrode material, no faradaic current will occur. In this case, only the EDL will be recharged. This case will be focused on in this chapter. For example, to a milder extent, similar processes occur at solid–solid interfaces of semiconductors. Whether the processes are mild or harsh depends also on the temperature T.

Hermann von Helmholtz has systematically investigated the physics and chemistry and thermodynamics of the double layer 150 years ago [Helmholtz 1879]. The relationship between space charge density ρ and potential φ gilt is followed by the Poisson equation:

$$\frac{\partial^2 \varphi}{\partial x^2} = -\frac{4\pi\rho}{\varepsilon}$$

Between the electrode surface and inner Helmholtz plane (IHP), there is no charge density: $\rho = 0$. Therefore, we yield the homogeneous Poisson equation (=Laplace equation):

$$\frac{\partial \varphi}{\partial x} = \text{constans}$$

Hence, the potential between electrode surface and IHP is a linear function of the space coordinate x. In the metal electrode, there is no charge ($\rho = 0$). In the electrolyte, the average charge density is also zero ($\rho = 0$); in this case, the change of potential is not just constant – it is actually zero. With these boundary conditions, we arrive at the curvature of the potential:

$$\varphi(x) = \begin{cases} \varphi_M & x \leq 0 \\ \varphi_L - \varphi_M \cdot \frac{x}{d} + \varphi_M & 0 \leq x \leq 0 \\ \varphi_L & d \leq x \end{cases}$$

A mathematical complete derivation of the double layer potential is treated by Arnold [Arnold 2004] in his textbook on ordinary differential equations.

It is an important condition that the EDL has little or no electric conductance (it is an insulating layer), so that a drop in the potential, which accounts for an electric field, occurs over the EDL when a voltage is applied.

Looking at the sketched condition in Figure 2.2, we realize that this EDL extends only over the molecular distances at the boundary. I have written 2 Å as an arbitrary

---

4 The isoelectric point (pI, pH(I), IEP) is the pH at which a particular molecule carries no net electrical charge in the statistical mean.

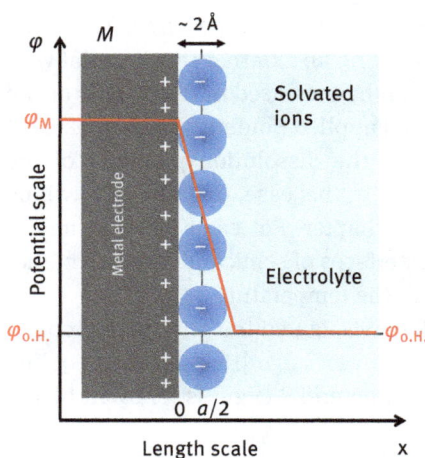

Figure 2.2: Sketch of potential at electrode, double layer and electrolyte across their length scale. The potential in the bulk of the electrode and in the bulk of the electrolyte are considered constant (horizontal black line). Hence, there is no electric field in these regions. At the solid–electrolyte interface, the potential difference amounts in a strong electric field (black line with negative slope).

number. In extreme cases, it may be as "thick" as 10 Å, but the double layer is in general very thin.

Here is an exercise for the reader: When we apply a voltage over this system of say 1 V, then how large is the electric field E across the EDL?

$$E = \frac{\Delta V}{\Delta x} = \frac{1\,\text{V}}{2 \times 10^{-10}\,\text{m}} = 0.5 \times 10^{10}\,\text{V/m}$$

The strength (electric field strength) of this electric field is huge.

The variation of the electrical potential in Figure 2.2 is oversimplified. The actual physical situation is far more complex. But it serves well for the understanding of the situation at the EDL and, as we see later, for the working principle of EDL capacitors (EDLC).[5]

---

5 When you are an expert in your bubble, in your own narrow community, you have little or no confusion with abbreviations. When I researched on supercapacitors during my PhD thesis time, DLC was the abbreviation for double layer capacitor. In 1997, at my first large conference in Paris ([ECS 1997] ECS: ECS and ISE meet in Paris: 1997 joint international meeting program, le Palais de Congrès, August 31-September 5, 1997. *The Electrochemical Society Interface* 1997, 6.), I joined a session about DLC. Every speaker used the term DLC but I was confused what they meant by DLC because there was nothing about supercapacitors. Finally, I asked one speaker what she meant by DLC and she responded – "diamond-like carbon." So I learnt something by asking a question in the audience of a symposium. When you have ever been to a talk with a large audience, think of the many attendants who are silently nodding or remaining calm; they have no questions and it looks like they understand everything. My colleague and previous MRS Meeting Co-chair Prof. Ken Haenen from Hasselt University and IMEC in Belgium ([Braun 2014]; Braun A, Fan HY, Haenen K, Stanciu L, Theil JA: Braun, Fan, Haenen, Stanciu, and Theil to chair 2015 MRS Spring Meeting. *MRS Bulletin* 2014, 39:740-741.doi: 10.1557/mrs.2014.183.) is an expert on DLC – the diamond-like carbon.

The electrode should have a pretty constant potential over its thickness. Why is this so? Well, if it wasn't, we would not use the material as an electrode. By default, an electrode is a good electronic conductor, such as copper, silver, aluminum and gold. These metals have a very high electronic conductivity (for numerical data on conductivities and temperature coefficients, see the many references in [Giancoli 2008, Goldman 1991, Griffiths 1998, JFE Steel Co. 2017, Kayelaby 2018, Kotowski 2015, Matula 1979, Matweb 2018, O'malley 1992, Ohring 1995, Pan 1995, Pashley 2005, Pawar 2009, Pierson 1994, Porter 1991, Serway 1998, Stratton 1914, Transmissionline 2011, Ugur 2006]). Any local differences in the electric potential of the electrode will be immediately equilibrated by the conduction electrodes of the metal. Therefore, the profile of the potential is a straight horizontal line.

The electrolyte has a potential (Redox potential) that is different from the electrode potential (Fermi energy). At the interface (the double layer), a potential gradient exists that yields an electric field. This double layer has a thickness of around one monomolecular or mono atomic layer of around 1–2 Å. The strength of the electric field is

$$\vec{E} = -\varepsilon_r \varepsilon_0 \operatorname{grad} \varphi = -\varepsilon_r \varepsilon_0 \frac{\mathrm{d}\varphi}{\mathrm{d}x} \hat{e}_x$$

In an electric field, we can store electric energy (as physicists, we know that we can store energy in fields). The energy density in the field is

$$W = \int \vec{E} \mathrm{d}\vec{x} = \frac{1}{2} \varepsilon_r \varepsilon_0 \vec{E}^2$$

with ε_r being the relative dielectric constant (water: ≈ 10) and ε_0 the electromagnetic field constant (8.854×10^{-12} A s/V m). The potential difference in an electrochemical cell may not exceed a voltage of around 4–5 V, if the electrolyte is an organic electrolyte (such as acetonitrile). At higher voltages, the electrolyte will be decomposed. If an aqueous electrolyte is used, the decomposition voltage is around 1 V. With above data, an upper limit for the energy density of around 1 kJ/cm^3 is obtained for the double layer. Note that battery engineers are constantly looking for ways to suppress electrolyte decomposition so that a larger potential can be applied to batteries and capacitors. In this context, the term "dielectric breakdown" is important. The electrolyte is an important component in electric systems where no electric leakage may take place. Only the ionic conductivity is the allowed transport process, normally.

2.1 Estimation of the capacitance by geometrical considerations

I show here a simplistic estimation [Braun 1999a, b] of the maximum amount of electric charge which can be deposited on a surface. The surface of choice is that from glassy carbon (GC), which is sometimes used in electroanalytical chemistry as a porous model electrode. Electric screening effects by anions and hydrated ions and interactions between surface and ions are ignored. I consider an ideally smooth crystal surface and charge carriers (ions) with a size that is similar to the size of a carbon atom.

The density of atoms on a surface plane of SIGRADUR® K (ρ = 1.53 g/cm^3, mass number 12) is σ = 1.78 × 10^{15} atoms/cm^2 – the mean atomic distance of which yields around 2.37 Å. Any crystallographic interstitial site can be filled with one charge carrier (ion). With the assumption that each ion carries at least one elementary charge of e = 1.602 × 10^{-19} C at a voltage of 1 V, a minimum capacitance of 286 µF/cm^2 can be assigned to this surface area. This value is by a factor of 14 higher than the usually reported experimental value of 20 µF/cm^2 [Kinoshita 1988], which was experimentally verified also by the author [Braun 1999a] (23 µF/cm^2) on untreated flat SIGRADUR® K and SIGRADUR® G samples.

In the above calculation, the repulsive interaction between charges of even sign was neglected. The potential between two elementary charges at a distance of 2.5 Å is 6 V (Coulomb's law). At a potential difference of 1 V only, the distance increases to 14.4 Å. Taking the repulsion into account, the number density of ions per surface area decreases by a factor of (14.4 Å/2.37 Å)^2 = 36. The capacitance corrected for the repulsion forces then yields around 286 µF/cm^2/36 = 8 µF/cm^2.

The capacitance of the samples was only related to their apparent surface area (geometrical surface). However, the internal surface area is quite larger than this. There follows another estimation of the capacitance per internal surface area. Relating the capacitance per apparent surface area to the internal surface area per film volume, the capacitance should be in the range of 20 µF/cm^2.

Consider a sample of SIGRADUR® K, 2 h activated at 450 °C. The internal surface area of the film material of the sample (Brunauer–Emmett–Teller [BET] measurements) and the film thickness (from scanning electron microscopy, SEM) are known. The active film thickness is around 45 µm (=45 × 10^{-4} cm). A sample with an apparent area of 1 cm^2 then has an active[6] pore volume of around 45 × 10^{-4} cm^3. The internal surface area in this volume has yet to be determined.

The BET internal surface area of the active film material was determined to around 1000 m^2/g. It is necessary to determine the mass density of the active film material (m = ρ × V) so that we know the internal surface area per film volume. Due to the oxidation process during activation, the GC film density is smaller than the GC bulk density. Starting with the bulk density ρ_{bulk} = 1.53 g/cm^3, a mass of m = 1.53 × 45 × 10^{-4} g = 6.885 mg is obtained. The weight loss during activation at

---

6 The term "active" is often used in lieu for a better term that is not known. Here, it was used for active carbon, and active means that the carbon can adsorb and absorb fluids in its pores. For this to happen, the pores in the carbon must be accessible, they must be open. Before activation, the glassy carbon has pores but they are isolated and closed.

450 °C is around 1.573 mg/cm^2 h. In 2 h, the weight loss is therefore 3.15 mg. The remaining mass of the film is therefore 6.885 − 3.15 mg = 3.735 mg.

Film material with this mass has a BET internal surface area of around 1,000 m^2/g. So, the internal surface area is O = 3.735 × 10^{-3} × 1,000 m^2 = 3.735 m^2. The sample with 1 cm^2 geometric area has an internal surface area of around 37,350 cm^2. The capacitance of this sample is around 1 F. Hence, the capacitance per surface area yields 1 F/3.735 × 10^4 cm^2 or 27 μF/cm^2, respectively. This calculated value is in good agreement with the experimentally verified value of around 20 μF/cm^2.

2.2 Electrochemical double layer capacitors (supercapacitors)

Supercapacities depend fundamentally on the supersize surface areas of the capacitor electrodes. I want to focus in this section on the capacities as they arise from the EDL. I have estimated the double layer capacitance for a plain geometrical surface area to around 20 μF/cm^2. Every multiple of surface area by a suitable enhancement of porosity and roughness on the electrode amounts therefore to a corresponding double layer capacitance.

2.2.1 Simple design of an electrochemical double layer capacitor

A macroscopic schematic of a supercap is illustrated in Figure 2.3. The top panel shows two vertical aligned metal current collectors sketched in red color, on the left with a plus pole setting and on the right with a minus pole setting. An electric contact wire is soldered on each current collector. The current collectors can be copper, aluminum, steel, for example. Keep in mind that in electrochemical devices, we always have electrolytes that may pose aggressive environment for involved components. Metals, for example, may corrode in some electrolytes. The Pourbaix diagrams give information on the stability of materials in electrolytes as a function of pH and external electrochemical potential.

In the middle between the two current collectors, we see a blue color component called separator. The separator can be a sheet of paper or any other porous membrane from glass fiber or polymers or cellulose that shall have no electronic conductivity. The separator separates the two electrodes and prevents them from having direct physical contact, preventing thus also any electronic contact. The separator shall have ionic conductivity. The separator material as such normally has no ionic conductivity. But it can be soaked with liquid electrolyte, such as dilute sulfuric acid or KOH or NaOH, for example. The ionic conductivity is thus provided by the electrolyte which penetrates the porous network of the separator and thus has ionic paths from one electrode left from the separator to the other electrode right from the separator.

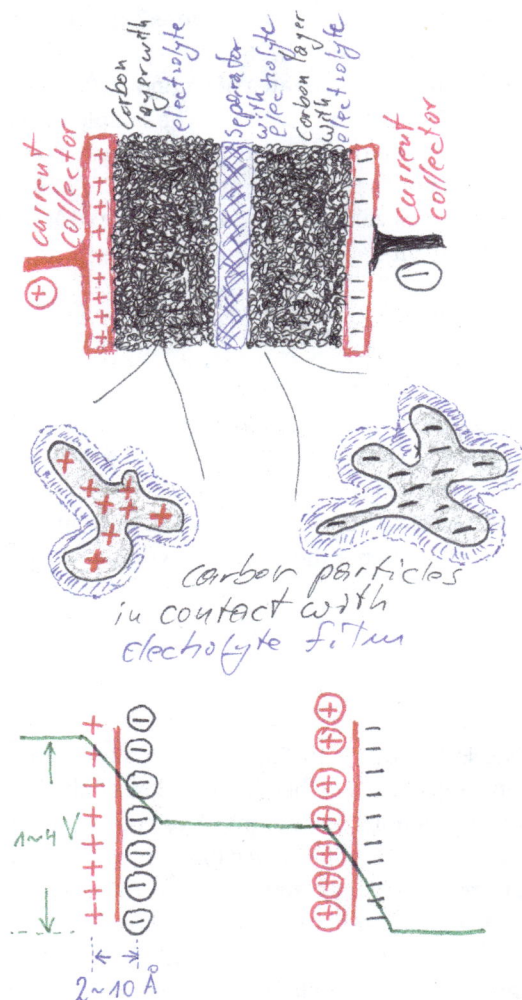

Figure 2.3: (Top) Design of a carbon particle-based double layer capacitor with two current collectors, one separator and two carbon powder layers, all soaked with electrolyte. (Middle) Magnification of polarized (charged) carbon particles surrounded by thin electrolyte film. (Bottom) Spatial variation of the electric potential in the capacitor between the two current collectors.

The black scribbled material between the two red color current collectors and the blue color separator is the two electrodes. Note therefore that the two metal current collectors were on purpose not called electrodes. This black material is carbon, for example, active carbon or carbon black or some other carbon material with some porosity and electronic conductivity. It can also be a mixture of various carbon materials, such as carbon black and graphite. These carbon electrodes are not necessarily monolithic. They only have to have some minimum compactness to provide mechanical stability and the electric connectivity basic function of charge storage and transport. Practically, it is two carbon layers or two carbon films with a thickness of around 1 mm or below.

It is necessary, however, that the two carbon electrodes are also soaked with electrolyte and that the electrolyte is compact distributed throughout the entire capacitor space. This becomes clear when we look into the suggested microstructure of the carbon electrodes, which is shown in the middle range of Figure 2.3. The left side shows an aggregate of carbon particles that is surrounded by the electrolyte film. The electrolyte film is only a few nanometers thin. We assume that the carbon aggregates next to the current collector are in direct electronic contact with this current collector and that the other carbon aggregates in that electrode maintain direct electronic contact with their next neighbors. As the potential of the current collector is set to positive, the particle aggregates show preferentially positive charge. Correspondingly, the carbon electrode attached to the current collector with negative charge will show a negative electric charge.

The electrolyte film surrounding the particles will have the opposite charge from the carbon particles. Note that this electrolyte film extends throughout the entire capacitor volume as it penetrates the separator. This condition holds not for the two carbon electrode films; these are separated by the separator. We can therefore sketch the variation of the electric potential across the capacitor as shown in the bottom part in Figure 2.3, whereas the spatial distance between the two current collectors (this can be 1 mm or less) which enclose the capacitor volume cannot be taken literally for the determination of the electric field.

The strongest change in the potential is actually at the carbon–electrolyte interface, which is in the order of 10 Å. For the left part of the capacitor, this is then half of the maximum potential (1/2 V for aqueous electrolytes and 2 V for organic electrolytes[7]) and for the right part of the capacitor the other half of the potential, as shown in Figure 2.2.

2.2.2 Operation principle of electrochemical double layer capacitors

We remember from physics class that we can build a capacitor from two metal plates that are kept apart by some distance so that they have no direct electric contact. The positive charges on one electrode plate and the negative charges on the other electrode plate cause an electric field E. When we apply a voltage U on the electrode

---

7 The actual limits may be higher depending on the circumstances. If you work on the improvement of electrolytes and have accomplished an increase of the maximum voltage window, you certainly will not agree with these numbers. Therefore, let us not argue too much about actual numbers. For this, I refer the reader to the special literature.

plates, more and more electric charges q are being built up to the total charge Q. We are performing electric work W against the repulsive charges Q:

$$W = \int_0^{Q_0} U(q)\,dq = \frac{1}{C}\int_0^{Q_0} q\,dq = \frac{1}{C}\frac{1}{2}q^2\Big]_0^{Q_0} = \frac{Q_0^2}{2C} = \frac{1}{2}\,C\,U_0^2$$

W is the electric energy stored in the plate capacitor arrangement.

In the previous section, we learnt about the electrostatic field in the EDL. We can use this field in order to store electric energy pretty much the same like in a capacitor. Such capacitors are called EDLC, or supercapacitor (SuperCap) or ultracapacitor (UltraCap).

Figure 2.4 shows schematically the situation of one "electrode." At the bottom, we have a vertical oriented current collector. A current collector is typically a metal with a high electronic conductivity. Often the current collector is a metallic plate so that it provides sufficient electric conductivity to carry, to collect as much current from either the cable on the outer side or from the electrode on the inner side, but provides also sufficient mechanic stability for the capacitor arrangement, or any electrochemical cell.

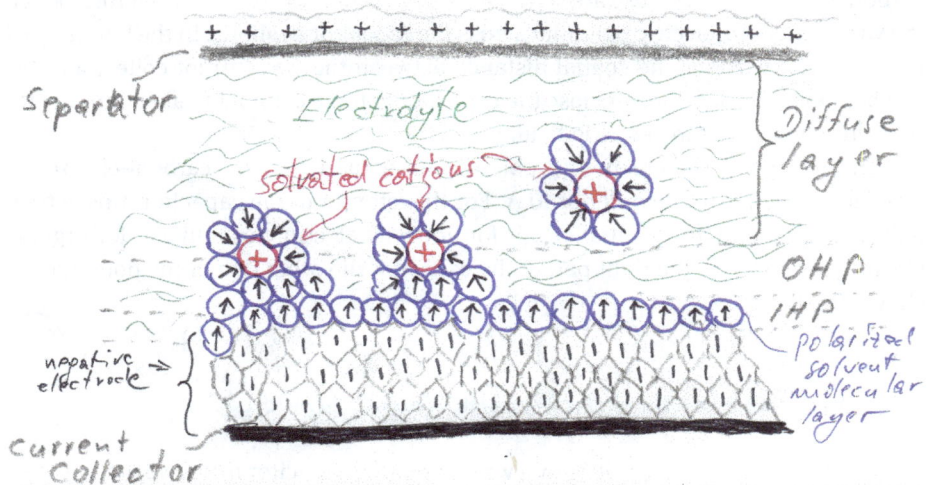

Figure 2.4: Schematic of a supercapacitor electrode assembly inserted in electrolyte with arrangement of electric charges in Helmholtz layers and diffusion layer

Attached to the current collector is the electrode material. The designer of the sketch has intentionally used hexagonal elements to suggest that the electrode material is graphite-like carbon. Unlike the metallic current collector, the carbon is the actual electrode material which has an electrochemical function in the double layer capacitor.

As the current collector is set to a negative potential, for example, by a power supply, by a potentiostat, by a battery or by a PV panel, the electrode underneath becomes negatively charged. Let us say, it becomes negatively polarized. The surface where the solid electrode material meets the liquid electrolyte is then negatively charged by electrons. This is indicated by the bluish disks with the black arrow pointing away from the negative electrode. This negative charge forms a region with a finite thickness which extends to around 3 Å. This is based on an estimate made by W. Thomson in 1870 [Helmholtz 1879]. It is a layer of molecular dimension. This region is called the IHP.

It is a good exercise for the reader to take pencil and paper and redraw the sketches for the double layer and the capacitor arrangement in order to fully understand where which component has what function, and which condition could be necessary to realize this function fully.

Solvated positive charges, ions and cations in the electrolyte will be attracted to these negative charges because of fundamental electrostatic principles. These solvated ions may have a positively charged center and a negatively charged boundary from negative charges in the electrolyte. Overall, however, the cations try to settle at the IHP and tend to assume a rather uniform arrangement. The thickness of the solvated ion is now adding to the thickness of the IHP and forms the outer Helmholtz plane (OHP). As the polarization from the electrode cannot reach through the OHP strongly, the positive charges in the electrolyte will not be attracted as much to the electrode as the ions in the OHP. The concentration is lower in this region than in the OHP. The cations here are not so well arranged and we therefore call it diffuse layer.

Hermann von Helmholtz [Helmholtz 1879] concluded that there must be a double layer because he thought about the microscopic consequences of charged interfaces. Before that, researchers had only looked into electric systems where charges were far apart (theory of the distribution of electricity in conducting bodies).

We will later see that the double layer is way more complex than described here. But the "simple" double layer concept shown here is already a very good model which helps to develop the discussion. In later sections in this book, we will learn about more refined models of the double layer. A review on the theory of the double layer is available for example in the Encyclopedia article from Schmickler [Schmickler 2014]. Its historical development is outlined by the review by Damaskin and Petrii [Damaskin 2011].

2.2.3 Arrangement of charges and ions between supercapacitor electrodes

So far, we have looked at an electrode that was at a negative potential. When we consider an arrangement where one electrode is facing a second electrode in an electrolyte, then we can build the capacitor. This is sketched in Figure 2.5, based on the previous sketch in Figure 2.3. Two nominally identical carbon electrodes, for example GC powder, are coated on titanium foil [Schüler 2000].

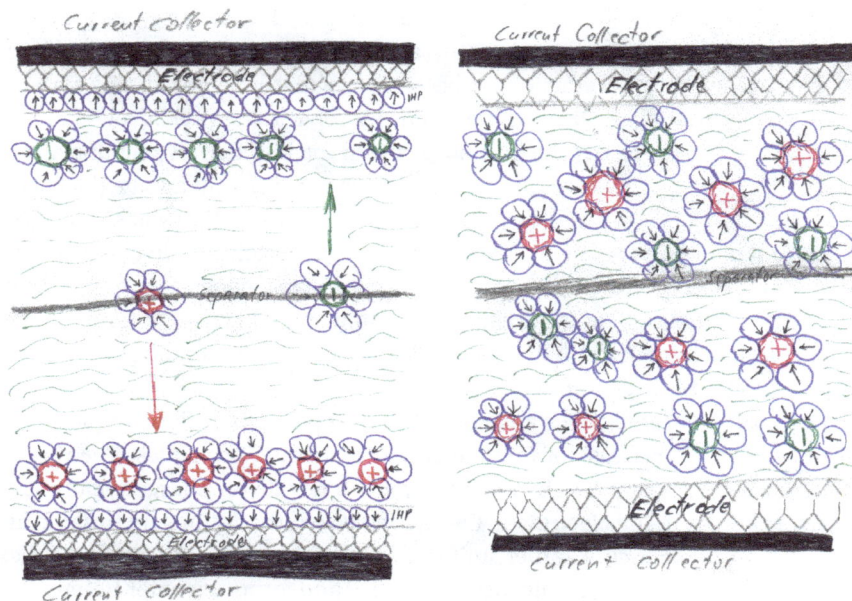

Figure 2.5: Distribution of charges, that is, electrons and solvated ions in the electrolyte and near the electrodes in a supercapacitor in the charged (left sketch) and discharged (right sketch) state.

In the top image in Figure 2.5, we have two carbon electrodes assembled to a capacitor with the separator membrane in between. The solvent and electrolyte occupy in the image a rather wide space that is not in the right proportion to the actual distances between carbon, electrolyte layer and separator in the capacitor. The electrolyte and solvent contain solvated anions and cations. The cations are the red circles with "+" and they are hydrated with water molecules (blue circles) which have a dipole moment as indicated by the gray arrows inside. The anions are green circles with "−," and those have also a hydration shell with the dipole moments from the water molecule pointing away from the negative charge.

As this capacitor is not charged, the hydrated ions and the water molecules are randomly distributed throughout the electrolyte volume. There is no macroscopic electric field present. The situation will change when we charge the capacitor. In the bottom panel in Figure 2.5, the left side current collector is put on a positive potential that will immediately orient the water molecules attached at the electrode surface so that the dipole moments point to the positive charge electrode. The attractive Coulomb force will arrange the water molecules close enough on the electrode surface. The attractive Coulomb force will also attract the negative anions in the electrolyte volume so that an enrichment of negative charges occurs near the positively charged electrode on the left.

At the same time, the electrode on the right is negatively charged. The water molecules near this electrode will orient in a way that their dipole moment points way from the electrode. This means the two protons H^+ of each water molecule will seek close proximity to the negative electrode, whereas the negatively charged O^{2-} of the water molecule will point away from the electrode. The positive charge cations in the electrolyte will migrate to the negative electrode because of the attractive Coulomb force. In the same way as the anions have migrated to positive electrode, a charge separation will cause an enrichment of positive charges at the negative electrode.

2.3 Ragone diagram

Practical question: Why do we need another sort of capacitor, the EDLC, when we have already capacitors, and batteries? Let us therefore look at Figure 2.6, the Ragone diagram. The axes are the specific power in W/kg and the specific energy in W h/kg. We can also call them power density and energy density. Note that the energy W (work) is power times the time: $W = P \times t$.

We see in the Ragone diagram in Figure 2.6 on the upper left a field called capacitors. This means capacitors have a power density of around 2000×10^7 W/kg. At the same time, they have an energy density of 0.01–0.05 W h/kg. These data are not for one specific capacitor. Rather, most available capacitors fall somehow in this range.

Electrical circuits can be mathematically modeled with user-friendly software such as PSpice®. The software contains many electric components such as resistors, capacitors, coils, diodes, transistors, processors and so on in its digital toolbox. They all have specific $I-V$ profiles and characteristics. In the years 1996–1999 when I made my PhD thesis, I had a colleague who modeled circuits. He worked on lithium batteries and was interested in comparing them with supercapacitors. Back then, the software package had no element for supercapacitors included. One reported effort for implementing supercapacitors is shown in Kim [Kim 2011].

On the right, we see fields occupied by the supercaps, batteries and fuel cells. Obviously, the capacitors have a high power density when compared to these three latter devices. But their energy density is rather low. On the other hand, do batteries have a high energy density, but their power density is very low.

The supercapacitors have a power density in the range from 10 to 10^6 W/kg. Their energy density can range from around 0.05 to around 15 W h/kg. Note that their upper border line is not as horizontal as those from capacitors, batteries and fuel cells. Instead, it has a linear decaying profile which is typical for supercapacitors. Supercapacitors need not be based on carbon electrodes. There are also metal oxide-based supercaps [Qi 2016].

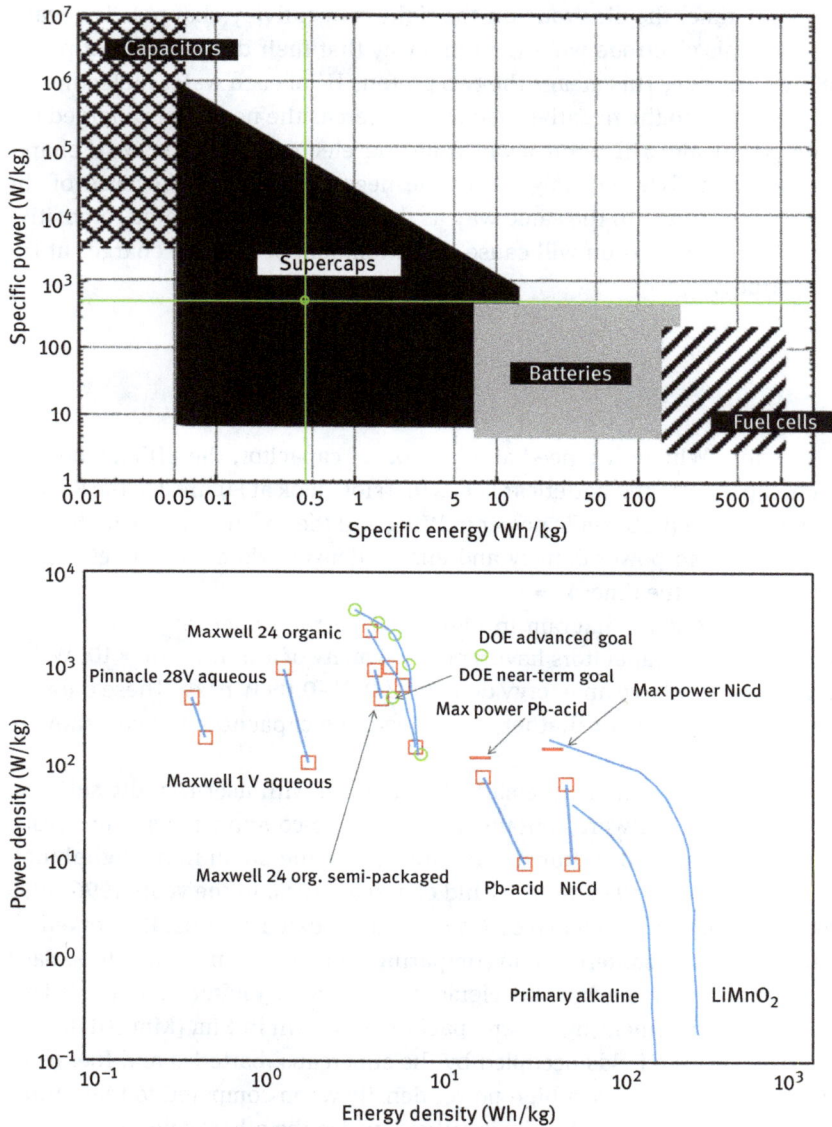

Figure 2.6: Ragone diagram showing the regions of traditional capacitors, supercaps, batteries and fuel cells with respect to specific power and specific energy [Kötz 2000].
Reprinted from Kötz R, Carlen M: Principles and applications of electrochemical capacitors. Electrochimica Acta 2000, 45:2483–2498. Copyright (2017) with permission from Elsevier. Ragone diagram after Pell and Conway [Pell 1996]. Adapted from Pell WG, Conway BE: Quantitative modeling of factors determining Ragone plots for batteries and electrochemical capacitors. *Journal of Power Sources* 1996, 63:255–266. Copyright (2017), with permission from Elsevier.

Batteries range with their specific power between 10 and 500 W/kg and with their specific energy between 8 and 200 W h/kg. The fuel cells have a considerably higher energy density and yet an appreciable power density when compared with batteries. However, we recall from the trips with the Hyundai ix35 FCEV that it has a wider range than any battery vehicle. But when accelerating, it obviously needed a lithium ion traction battery for assistance. This is a practical indication that the batteries likely have a higher power density than the fuel cells.

For a long time, the region between capacitors and batteries was empty in the Ragone diagram. Figure 2.6 (top panel) is more schematic in order to highlight the general concept, whereas Figure 2.6 (bottom panel) shows actual technical data which were relevant 20 years ago [Pell 1996]. You learn from this plot the authors referred to performance targets that had been set by the US Department of Energy.

A generalized theory of the Ragone plots can be found in a paper authored by my colleagues from ABB Corporate Research [Christen 2000]. What we learn is that EDLC fill a gap for which previously there was no storage device available. When I finished my PhD thesis in 1999 [Braun 1999a], supercapacitors were a hard-to-find component in electronics. This is because they range between capacitors and batteries [Simon 2014].

To date, almost every catalogue of electric and electronic components has super-caps in their capacitor portfolio. They have become so common that they are even not (not necessarily) distinguished from conventional capacitors anymore. There is one point however which should be mentioned, and this is the different discharge and charge characteristic of conventional capacitors and supercapacitors.

Figure 2.7 shows a collection of supercaps that were commercially available in and before the year 2000 [Kötz 2000]. The very large capacitor on the left side looks similar to those developed by Canadian company TAVRIMA Ltd. I may be wrong but I believe the design originates from Russia. I was told such huge capacitors were used in Siberia in order to crank start diesel engines of locomotive trains and tanks in the military when there is extreme cold winter.

The reader finds more background in the ultracapacitor book by Deshpande [Deshpande 2014]. Tavrima Ltd. used such large capacitors also for experiments with racing cars [Braun 2005]. I remember one of their supercaps had a capacitance of 60 F at the maximum voltage of 28 V. This is an energy content of 60 kJ (1 J = 0.000277778 W h; 1.58 kJ = 0.4386 W h/kg). The maximum power is 20 kW and the equivalent series resistance (ESR) is $R_{ESR} = 0.01\,\Omega$. It had a cylinder shape with over 23 cm diameter and over 56 cm length. Its weight is 38 kg.

With these data, it is possible to put it in its place in the Ragone diagram: 526.3 and 0.4386 W h/kg, as I have done with the green line in the left image of Figure 2.6. With the known capacitance $C = 60$ F and ESR $R_{ESR} = 0.01\,\Omega$, the time constant can be determined to $\tau = R \times C$.

Note that the Ragone diagram shows the specific power and specific energy with respect to the weight (mass). It is not always clear what those who put a device in the Ragone diagram mean by weight. Is it the weight of the entire device? Is it the weight

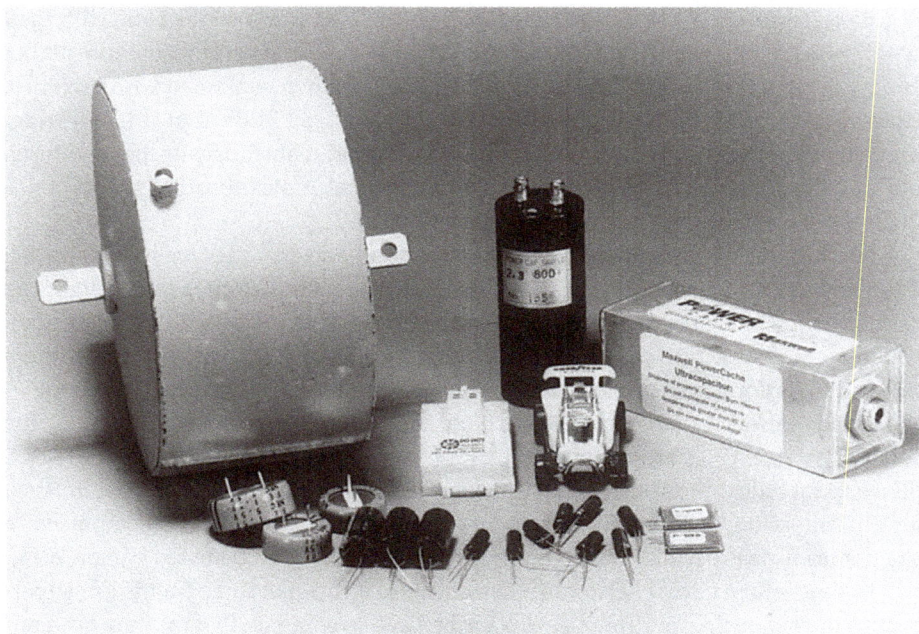

Figure 2.7: Commercial electrochemical capacitors and toy equipped with an electrochemical capacitor. Reprinted from *Electrochimica Acta*, 45, Kötz R, Carlen M, Principles and applications of electrochemical capacitors, 2483–2498, Copyright (2000), with permission from Elsevier [Kötz 2000].

of the electrode? Or is it the weight of the active material only? You have to ask this question when you want to compare capacitors, batteries and fuel cells in the Ragone diagram.

Sometimes when materials are very light but the systems are very bulky, it may be more relevant when you provide the specific power and specific energy with respect to the volume and not with respect to the weight (mass). Vendors tend to make their products as attractive as possible and you might run as a buyer in unpleasant surprises if you do not investigate the product properly before you decide to buy.

You notice that the Ragone diagram has the axes given in the double logarithmic scale. This is because the power and energy of the various energy storage devices extend over an extremely wide range covering several orders of magnitude. This is the only rational for plotting the data on the logarithmic scale. A useful side effect is that the time constant τ can be shown as a straight diagonal in the Ragone plot.

Compared to a battery, you cannot put so much energy into a capacitor. Why do we have capacitors, then? The great advantage of the capacitor is that it can store and release electric energy very fast. So, time is an issue. Particularly, capacitors are therefore ranked with respect to their time constant τ. Within the time t, the current I, the voltage V or the electric charge Q decreases to $1/e$ of its original value. We will see in Section 2.4 how to determine the time constant τ.

As the power P is defined as $P = dW/dt$, time plays a decisive role as P and W are plotted in the Ragone diagram. The ratio of P and W therefore accounts roughly as the time constant t, which in the Ragone diagram then shows up as linear curve, as a straight line intersecting the entire diagram. Figure 2.8 shows a modified Ragone diagram [Wee 2012] with the discharge time plotted over the rated power. This diagram compares not only electrochemical energy storage devices but also mechanical systems such as flywheels, compressed air and hydropower storage, along with the corresponding lines (dotted lines) for the time constants.

Figure 2.9 shows a conventional Ragone plot but with converted axes. Here, the energy density is plotted versus the power density. Hence, the fuel cells are placed in the upper left corner and the electrolyte capacitors in the lower right. The vertical lines which denote the time constants dissect the Ragone plot nicely in the ranges with slow storage with large time constants such as the fuel cells, the range with the many batteries and time constants between 1 and 100 s and the capacitors with the time constants in the millisecond range but very low energy densities.

2.3.1 A supercapacitor with carbon powder electrodes

Supercapacitors with carbon as the active material are attractive because many carbon materials have the necessary high surface area (active carbon) and electronic conductivity. Typically, carbon powder, such as carbon black and graphite, and some binder are mixed and the slurry is poured on some metal foil, which is the current collector. Also, carbon is a low-cost material, which makes it attractive for the consumer market.

The electric contact between the active carbon particles and the current collector is quite important. When the electric resistance between both phases is too large, the electric losses are too large. Adding graphite to the active carbon serves to minimize the resistivity between the active carbon particles.

One way of improving the electric conductivity between carbon and metal current collector is the chemical formation of a carbide layer provided that it has a high electric conductivity. Titanium carbide (TiC) is such compound. During my PhD thesis, I made experiments with titanium metal as current collector, and carbon powder and organic solvents which I pyrolyzed to the extent that the carbon chemically bonded to titanium and thus formed electrically conductive TiC [Schüler 2000].

We used GC powder as electrode material. The reason for this was a secondary one. Later in this chapter, I will show how we used monolithic GC electrodes. GC [Jenkins 1976] is a monolithic material by default. Monoliths broken in the factory by accident can be shredded and milled and then used as GC powder for various applications.

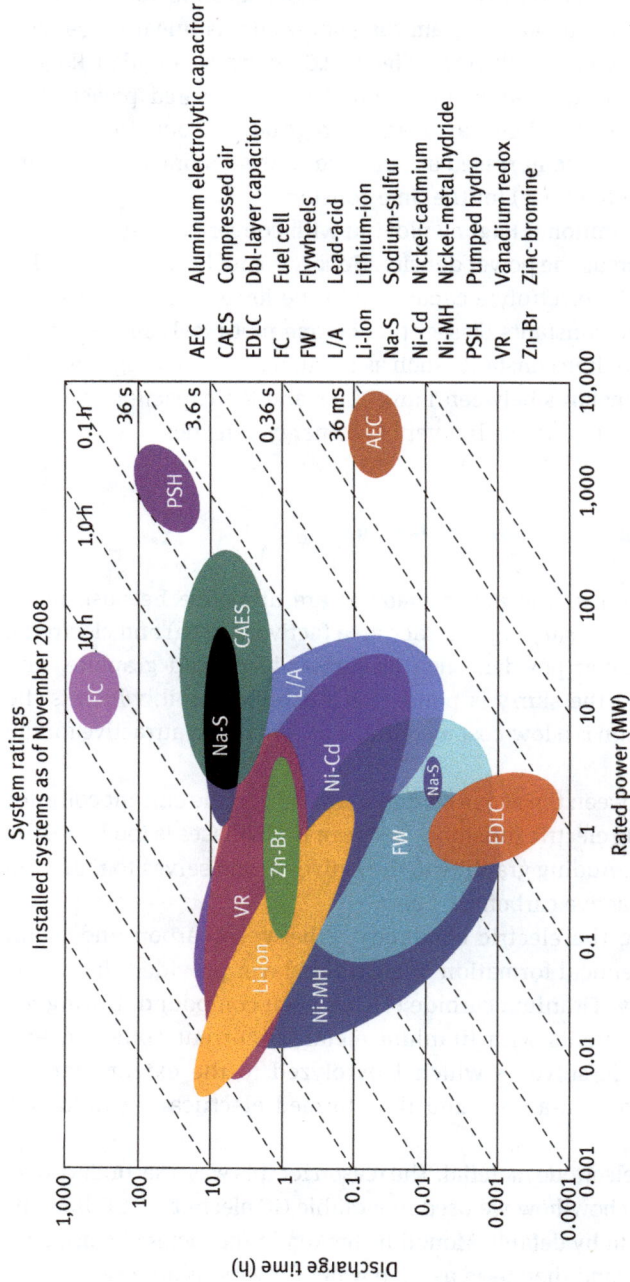

Figure 2.8: Ragone diagram for several energy storage systems, including mechanical systems [Wee 2012]. The dotted lines with positive slope denote the time constants ranging from 36 ms to 1,000h.

Reprinted from *Renewable & Sustainable Energy Reviews*, 16, Wee HM, Yang WH, Chou CW, Padilan MV, Renewable energy supply chains, performance, application barriers and strategies for further development., 5451.5465, Copyright (2012), with permission from Elsevier.

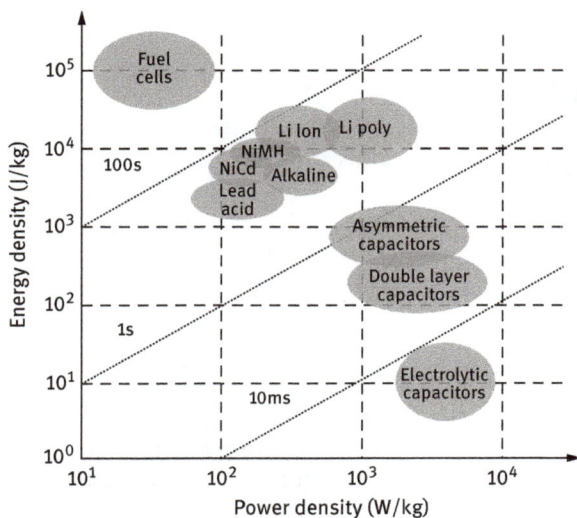

Figure 2.9: Ragone chart for capacitors, supercapacitors, batteries and fuel cells [Knight 2008]. © 2008 by the authors; licensee Molecular Diversity Preservation International, Basel, Switzerland. The figure is taken from an article that is an open-access article distributed under the terms and conditions of the Creative Commons Attribution license (http://creativecommons.org/licenses/by/ 3.0/): Knight C, Davidson J, Behrens S: Energy Options for Wireless Sensor Nodes. Sensors (Basel) 2008, 8:8037–8066. doi: 10.3390/s8128037.

The GC powder per se has only a low internal surface area and thus can hardly be used as electrode material. But the powder can be oxidized in a furnace and then a film with open porosity and high surface area forms on the particle [Braun 2002]. This process of active film formation is meanwhile well understood and has been modeled with mathematical accuracy [Braun 1999a, Braun 1999b, 2000, 2003a , 2006, 2011].

Figure 2.10 shows the BET surface area of GC powder from plates pyrolyzed from phenolic resin at 2,200 °C (SIGRADUR® G) and 1,000 °C (SIGRADUR® K). The G-type produces a surface area of around 200 m^2/g and the K-type produces a surface area of almost 1,000 m^2/g. It was therefore decided to use the activated K-type powder. The curvature of the increase of surface area can be mathematically modeled based on the geometrical parameters of particles and film with open porosity [Braun 2002].

Figure 2.11 shows the cyclic voltammograms (CVs) of a supercap electrode where GC powder was bonded on a titanium current collector. The CV shows that a sufficiently large surface area is available that produces a correspondingly large area in the CV, indicative of a large electrochemical capacitance.

After the electrode was subject to cyclic voltammetry, it was electrochemically reduced by applying an electric bias of –300 mV for some short time. This causes the electrochemical reduction of surface functional groups in the pores of the GC and

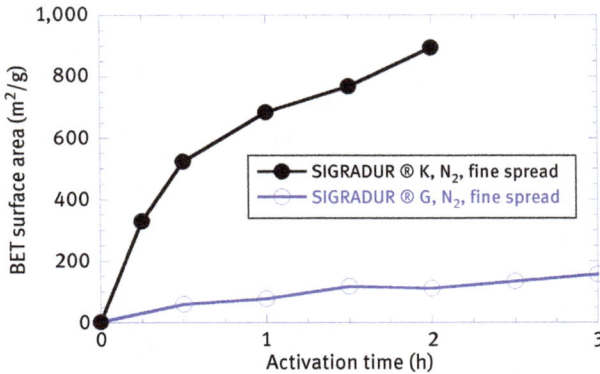

Figure 2.10: Growth of BET internal surface area of glassy carbon powder of SIGRADUR® type K (filled symbols) and G (open symbols) as a function of oxidation in an air furnace at 450 °C. For activation, the powder was finely spread in a monolayer on a ceramic dish with smooth surface. The BET experiment was done with nitrogen gas.

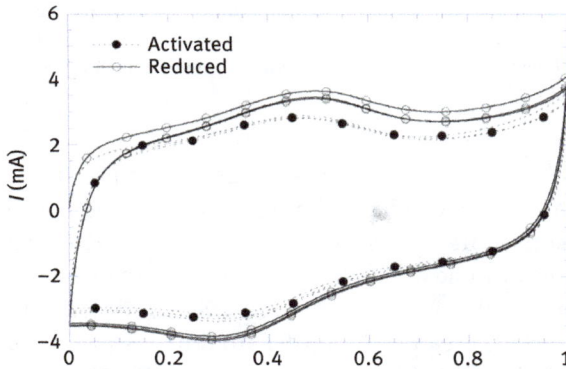

Figure 2.11: Sequence of cyclic voltammograms of a supercapacitor electrode made from glassy carbon powder bonded on a titanium current collector in preparation for patent in Schüler [2000]. The CVs with the filled bullets are from thermally gas phase-activated glassy carbon powder. The open bullets denote the CVs which were recorded after the electrode was electrochemically reduced in the electrolyte

yields a somewhat higher current density, which manifests in a larger area in the CV [Braun 2004]. The closed symbols (bullets) in the CV were recorded from the thermal gas phase-oxidized (activated) GC powder, and the CV with the open symbols (circles) was obtained after electrochemical reduction (Table 2.1).

Column 1 is total BET area per sample mass in m^2/g. Column 2 is the total BET area per electrode in m^2. Column 3 is the total BET area in m^2 per geometric electrode

Table 2.1: Glassy carbon deposited on 200 µm thick titanium sheet current collector.

| | BET surface area of Ti/C electrode assemblies | | | | | | | |
| | Not activated | | | | Activated | | | |
	m^2/g	m^2	m^2/cm^2	m^2/cm^3	m^2/g	m^2	m^2/cm^2	m^2/cm^3
Sand blasted	12.5	0.53	1.17	200	16.3	0.63	0.76	126
Not sand blasted	17.6	0.77	1.51	250	32.5	1.64	3.05	500

area in cm^2. Column 4 is the total BET area per carbon film volume in m^2/cm^3, where on both sides of the Ti sheet, a carbon film thickness of 30 µm was assumed.

The titanium substrate was coated with the carbon film either as received or after it was sand blasted in order to clean and roughen the titanium surface for better bonding with the carbon. Then, the BET surface area of the entire electrode was determined with gas adsorption measurements. The total BET surface area per electrode was 0.77 m^2 for the not sand-blasted electrode. With known size and weight of the electrode, the area can be normalized to the size and mass and we gain 12.5 m^2/g surface for the sand-blasted electrode and 17.6 m^2/g for the not sand-blasted electrode – when the carbon powder is not activated. Sand blasting had therefore no positive effect for the internal surface area, as far it could be determined with BET.

The same measurements were done with electrodes that were coated with activated GC powder. The sand-blasted electrode has 16.3 m^2/g surface area, but the not sand-blasted electrode has double that area, that is, 32.5 m^2/g BET surface area. It is not always predictable whether a treatment of a material yields better or worse results. The sand blasting was considered an idea worthwhile to try in order to promote adhesion between the titanium metal and the coating. Obviously, it did not work out as far as the BET surface area of the electrode was concerned. However, we did not do a systematic study. A systematic study would include also other cleaning steps for the substrate and variation of the sand-blasting parameters. All this was not done because it is time consuming and needs a lot of manpower.

Another question was whether the carbon coating would be so compact, so dense, so sealing that it would cover the entire titanium surface. To check this, the titanium surface was coated with silver paint before the carbon coating was applied. If the carbon film had holes, the electrolyte would leak through and react with the silver layer underneath – more so than with the titanium substrate. Silver is a very good marker in this respect.

Figure 2.12 (left panel) shows the CV of the electrode which was coated with a silver layer before the carbon coating was applied. The CV is overshadowed by a huge peak at around 550 mV versus the reference of the SCE (saturated calomel electrode) in anodizing direction, which is countered by the reduction peak at 300 mV. The CV

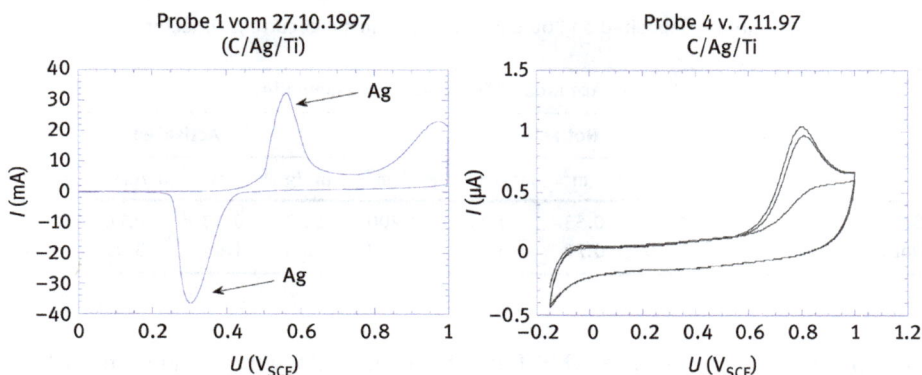

Figure 2.12: Cyclic voltammogram recorded from a supercapacitor electrode assembly according to the patent of Schüler [Schüler 2000] with an interfacial silver metal layer. The left side panel shows a large peak from Ag dissolution, overshadowing the double layer capacitance from the carbon electrode. The right side panel shows three cycles of a CV from the same type electrode with no Ag peak visible, revealing improved coating of the substrate by the carbon film.

in Figure 2.11 shows only a current of 4 mA for the entire electrode. The huge silver peak is indication that the carbon layer is not compact and does not cover entirely the titanium substrate underneath. The right panel in Figure 2.12 shows a CV of the same type of electrode where the Ag peak is absent. This shows that the new electrode has improved carbon-coating properties.

The gas phase oxidation of GC causes a considerable alteration of its microstructure and crystallographic structure. For the development of the GC powder-based electrode assembly [Schüler 2000], I activated powder in the furnace and recorded an X-ray diffractogram (XRD) from it when deposited on a silicon wafer. The sample looked like a carbon cake. Figure 2.13 shows the two diffractograms of the nonactivated and the activated powder. Oxidation causes a depletion of the diffracted intensity at the range of 20°–26°. This amounts to a shift of the Bragg reflection from 24° to 26° (see [Braun 1999a]).

2.3.2 Supercapacitors with monolithic carbon electrodes

We have learnt how carbon powder can be used for supercapacitor electrodes. A different approach was chosen in the 1980s by researchers from Siemens [Miklos 1980], who made experiments with GC electrodes. GC is made by pyrolysis of phenolic resins. Pyrolysis is the heat treatment under vacuum or under inert atmosphere gas such as nitrogen or argon or helium so that no chemical reaction can take place between the pyrolyzed material and the environment. The pyrolysis product is a solid monolithic carbon plate with a shiny black surface; GC is also known as vitreous

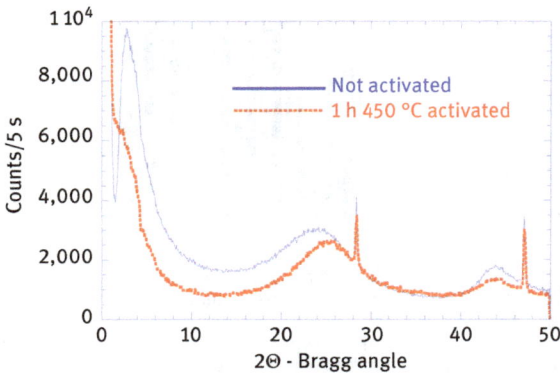

Figure 2.13: Powder X-ray diffractogram (Cu *K*-alpha) of two supercapacitor electrodes of activated (1 h at 450 °C in a muffle furnace in air) glassy carbon powder and nonactivated powder, coated on Ni. The powder cake was removed from the Ni current collector. The sharp Bragg reflections at 28° and 47° are from silicon marking powder.

carbon [Jenkins 1976]. GC contains a huge number of very small pores which show that the solid plate has a lower specific weight than for example graphite, although GC is in some ways reminiscent to graphite. The XRDs in Figure 2.13 show intensity around 26° where graphite has its prominent (0 0 2) Bragg reflection.

The pores in as-pyrolyzed GC are not in contact with each other. Rather, they are isolated to an extent that even helium cannot pass through a GC plate. When GC is oxidized, the pores become however opened and connected. Such oxidized GC has a large internal surface area which is accessible for fluids (liquids, gases), including electrolytes. This is one reason why such oxidized (activated) GC monoliths (rods) have been used as electrodes in electroanalytical chemistry.

Miklos et al. [Miklos 1980] used flat plates as electrodes for supercapacitors. They had in mind to use them for medical applications such as pace makers. A very detailed study on the activation process was carried out at the Paul Scherrer Institut (PSI) in the 1990s for electrochemical oxidation [Kötz 1998a, Sullivan 1997, 2000] and for thermal gas phase oxidation [Braun 1999a, b].

Figure 2.14 shows two supercapacitor prototypes which were built from oxidized GC plates by Dr Martin Bärtsch at PSI in the late 1990s. The GC plates (K-type) with 5 and 12 cm diameter were purchased from HTW (Hochtemperatur-Werkstoffe). The plates were oxidized at a temperature around 450 °C in a muffle furnace supplied with air [Bärtsch 1999].

The GC plates must be oxidized carefully to prevent that the film with open porosity on both sides of the plate becomes so thick that there is no nonoxidized separation layer anymore between the films [Braun 2000, 2003b, 2006]. Every oxidized electrode plate is thus a bipolar plate which can be stacked to a capacitor of

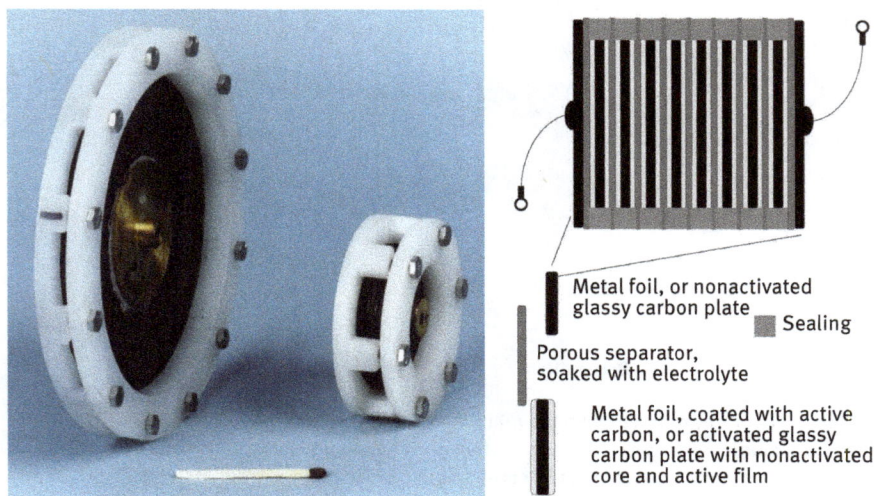

Figure 2.14: Two supercapacitor prototypes with activated glassy carbon electrodes. The larger one on the left is made from two electrodes with 12 cm diameter. The smaller one on the right is made from 10 activated glassy carbon electrodes with 5 cm diameter. Mechanical stability of the caps is maintained via polypropylene frames which are pressed together with stainless steel screws. Current collectors are made from brass disks which are brazed with silver paint on the outer faces of the glassy carbon electrodes. The electrodes had a nominal thickness of around 1 mm and were thus relatively thick. Photo by courtesy of the Paul Scherrer Institut.

desired voltage and capacitance when the electrodes are kept apart with separators which can be soaked with electrolyte.

The electrodes need to be glued to provide good electric contact and sealed to prevent leakage of electrolyte. The white polyethylene frames in Figure 2.14 warrant mechanical stability for carrying the device around and handle it for demonstration purposes. The two endplates from GC have brass plates as current collectors glued with silver paint to warrant the necessary good electric contact for wiring.

To the best of my memory, it was Dr Martin Bärtsch who made the first supercap at PSI. I do remember when he had this device prepared for the first time even without the polyethylene frame. He carried it in his pocket when we travelled to a project meeting [Kötz 1996] to Swiss battery company Leclanche in Yverdon Les Bains. Dr Hans Desilvestro, who was awaiting us in order to bring us to the meeting room, was surprised that we had already built the supercap. Immediately he wanted to know whether it would work. If I remember correctly, my colleague had not even tested it yet as it was made just the day before.

So Dr Desilvestro went into his laboratories, while our group was still standing outside on the campus of Leclanche S.A. When he returned few minutes later, he had a battery and a small light bulk and some cables in his hand. Dr Bärtsch and Dr Desilvestro exchanged hands and cables and battery and supercap to charge the supercap for maybe 10 s. Then, they removed the battery and held the small light bulb on the cables: The lamp was lighting for 10 or 20 s out there in the sunshine, or even longer. Our supercap worked, and it worked very well!

Over time, we made larger and thicker supercapacitors which in the end looked nice like the one shown in Figure 2.15. This supercapacitor can be charged to 24 V, although it looks quite thin. This is because we moved from GC plates with 1 mm nominal thickness to much thinner electrodes with 100 or 150 µm nominal thicknesses. This is another example how improvement and optimization of components can yield a more attractive product.

Figure 2.15: Supercapacitor prototype developed by PSI [Hahn 2001] for the Lok2000 project [Schönborn 1998]. Thirty electrodes with 150 µm nominal thickness and 12 cm diameter. Photo by courtesy of the Paul Scherrer Institut.

We remember we made as similar observation with the fuel cells from Siemens which were produced for submarines. There we looked at two different Nafion membranes, Naf115 and Naf117 with different thicknesses. The thinner electrolyte membrane provided a better power of the device. I made a similar observation recently when I visited the hydrogen filling station of EPFL in Martigny, where Dr Heron Vrubel and PhD student Yorick Ligen showed me long conventional alkaline electrolyzers with thick membranes and a very short electrolyzer based on proton exchange membrane (PEM) technology with thin membranes, both of which had the same power [Ligen 2018].

2.4 Activation of glassy carbon electrodes

The properties of the GC should be homogeneous throughout the entire material, but this is not necessarily the case. Material near the surface can have a different structure from material in the bulk. We tested whether the bottom of the plate would yield the same capacitance like the top of the plate, when activated. For this, you break the GC plate into pieces of roughly same size and turn half of them upside down and activate them in the furnace. Then you remove from the furnace two samples from upside and downside-placed specimen after 15 min. After another 15 min, you do the same, and so on. This is how you obtain a set of samples with activated top and bottom over a particular activation time. Figure 2.16 shows the capacitance of these samples. The data points are close together which shows it makes no difference whether you "misplace" the electrodes with upside versus downside.

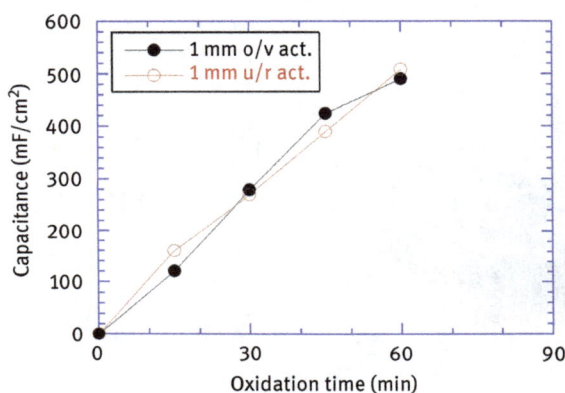

Figure 2.16: Increase of double layer capacitance on 1 mm thick K-type glass carbon electrodes with activation (gas phase oxidation in air in a muffle furnace) time. Capacities determined from the high frequency intercept of the impedance imaginary part. Filled symbols denote capacitance from the top side of the electrode. Open symbols denote the capacitance from the bottom side of the electrode.

GC is sold as K-type, which is obtained by pyrolysis of 1,000 °C, and G-type, which is obtained by pyrolysis of 2,200 °C. Experience shows that the G-type can hardly be activated by thermal gas phase oxidation. Capacities obtained that way with range around 10 mF/cm^2. The K-type GC can be thermally activated very well; it is easy to obtain 1,000 mF/cm^2. The situation is different when we carry out the activation by electrochemical oxidation (anodization). Then, the G-type produces very high capacities [Kötz 1998a, Sullivan 1997, 2000].

The GC plates are also sold as very thin sheets of say 60 and 100 μm, and 1 cm. We learnt that the thickness too has an effect on the electrochemical properties. The actual difference in performance is not just a result of thickness, but a result of different structure from different thickness [Braun 1999a].

2.5 Electrical characteristics of the double layer capacitor

Supercapacitors can be measured and assessed by standard electrochemical testing hardware. A description of how to design a custom built system is shown in the paper by Kopka and Tarczynski [Kopka 2013].

The small supercap prototype shown in Figure 2.14 was charged and discharged over 100000 times (cycles). One charge cycle took around 40 s. Altogether, this was 4 million seconds, which makes around 45 days. Figure 2.17 shows part of the voltage transient that was extracted at about half time of this long campaign (28 days). Cycle number 59954 was extracted. The time axis was therefore corrected from 59954 to 0 s.

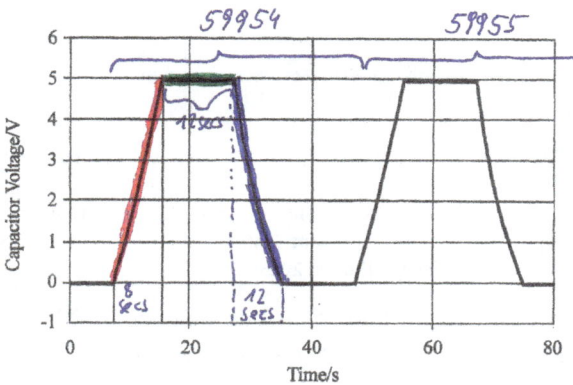

Figure 2.17: Voltage transient during constant current charge and discharge of the 5 V bipolar GC stack. Cycles 59,954 and 59,955 out of 100,000 cycles [Bärtsch 1999].

At around 6.5 s, the charging sets on from 0 V with an approximately linear profile to arrive after 8 s at 5 V. The automatic capacitor cycler keeps the 5 V for 12 s and then discharges the capacitor during 12 s to 0 V. This is a pretty reproducible and symmetric charge and discharge pattern. You can test the symmetry by flipping Figure 2.18 horizontal and overlay them. They would perfectly match only when they are perfectly symmetrical.

In my research, I frequently use the following method in order to compare graphics and other scientific data. The quickest way is to copy the figure with some graphics software and paste the figure in a new graphics panel. Then, you copy this figure and paste it so that you have the same figure twice on your graphics panel. Now you use the flip command in the graphics software which rotates the graphics by 180° on the vertical axis. Now you try to align the vertical and horizontal axes so that the two figures overlap and match all contours perfectly.

If you have this book in the printed form, you can scan this page with a scanner or make a photo with a smartphone and load Figure 2.17 on your computer and

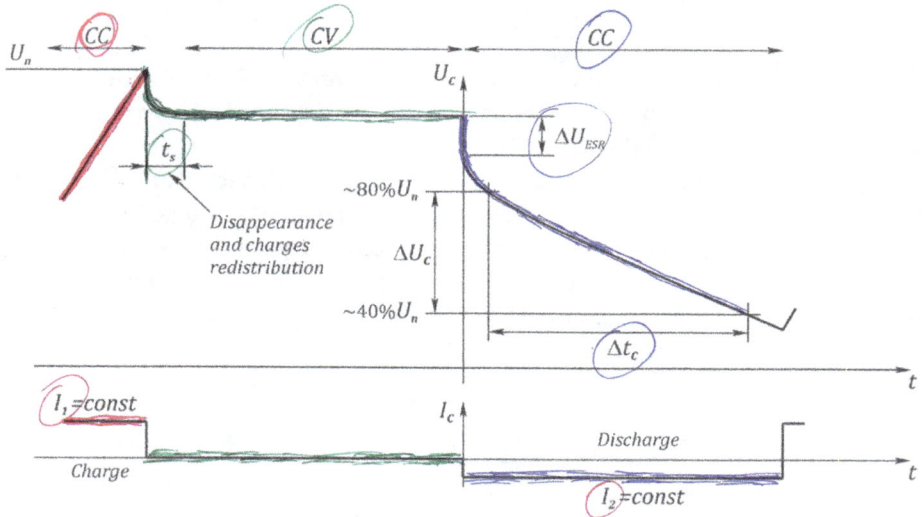

Figure 2.18: The method of determination of the values of supercapacitor parameters with constant current CC and constant voltage CV.
Reproduced and adapted from Kopka R, Tarczyński W: Measurement System for Determination of Supercapacitor Equivalent Parameters. Metrology and Measurement Systems 2013, 20:581-580.doi: 10.2478/mms-2013-0049, with Creative Commons License [Kopka 2013].

process it there. When you have this book as e-book or as pdf file on your PC, you can directly copy the graph and paste it in your graphics software and you can start right away.

If you want to look deeper into the graph in Figure 2.18, you can use tracing software which reads the graphics file into your PC and converts it into a data file with columns and rows so that you can plot the data in a way which you would like to plot it with your data processing software. My favorite tracing program is DataThief [Tummers 2006]. This is very helpful when you want to compare your own spectra (or any graphical data) with spectra from other researchers that you find in literature and visualize this for example in publications.

Sometimes, your own data have a different reference frame than the ones you want to compare with. Then, you can transform your data to the reference frame of the data in publications of other researchers, or the other way around. One example for this is known in X-ray diffractometry, when the Bragg angle is transformed to the scattering vector [Braun 2017a].

Close inspection of the overlapping of the original Figure 2.18 and its mirror image shows that there is a slight gap between the charge and discharge curve. The profile of the charge branch has a straighter flank than that of the discharge branch. The reason for this is not known to me. The difference may be because of the supercapacitor or because of the data acquisition system. It is just a noteworthy observation, the resolution of which would need further investigation.

Figure 2.17 shows only the profile of the voltage. We would be interested also in the evolution of the current during charge and discharge because the product of voltage and current constitutes the power which we can draw from the supercapacitor.

In electrical engineering, it is common to distinguish between constant current (CC) and constant voltage operations. Just think of a power supply which would deliver you (the device) a constant voltage, irrespective of how much power is drawn from the device. Or the device would require a CC from the power supply. This is irrespective which voltage would evolve in the circuit. A battery is laid out for the delivery of a particular constant voltage. For most of its lifetime, it will deliver a constant voltage and current.

Figure 2.18 shows an exercise how to determine the values for the parameters of a supercapacitor [Kopka 2013]. We have here two panels. The upper panel shows the voltage U over the time t. The lower panel shows the current I over time t. Let us begin with a CC I_1 = constant, which I have marked with red color in the lower panel. As the current is constant, it is a horizontal straight line. When we are filling the super-capacitor with electrons during CC charging, a potential is being built up in the supercap which gives rise to the linear increase of the voltage U_n, which I have marked in red color as the steep straight line in the top panel. The time phase during which the supercap is being charged is labeled with CC.

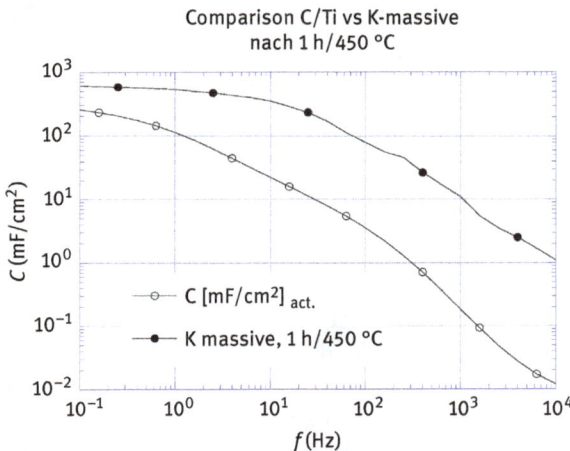

Figure 2.19: Impedance spectra of two electrodes from activated glassy carbon. The upper one from monolithic and the lower one from powder coated on titanium substrate. The activation procedure for the glassy carbon was the same: 1 h at 450 °C in air.

Now, we disconnect the power supply so that the current is 0. No further electrons per time are now flowing into the supercapacitor, but still with the condition constant current, which may be 0. The evolution of the current is shown as the green long horizontal line in the bottom panel. After this, you will experience a change of the current. The voltage is abruptly following the changed situation by redistribution of the electric charges which has the consequence that the voltage drops accordingly. This process of charge redistribution takes the time t_s and has an exponential characteristic (exponential decay). In the ideal case, after the time t_s, the voltage will remain constant at level U_c.

As Kopka and Tarczynski [Kopka 2013] point correctly out, the processes that take place in supercapacitors require that measurements are sufficiently long so that the supercapacitor can obtain a steady state situation. Unlike conventional dielectric capacitors, EDL capacitors display processes which are comparably slow. For the relaxation to proceed and finish, longer time is necessary than with normal capacitors.

Let us now reconnect the power supply, but change the polarity so that we basically have a sink, not a source, for the electrons charged in the supercap. We control the power supply so that the current in reverse direction is I_2 = const., this is the second CC phase. We see that the voltage sharply drops by a considerable amount and then assumes an exponential decay profile. The sharp voltage drop – basically a vertical drop (vertical part of the blue line) – is a drop due to the equivalent series resistance: ΔU_{ESR}.

The ESR is then the ratio of two numbers we read from the experiment:

$$\text{ESR} = \frac{\Delta U_{ESR}}{I_2}$$

The ESR causes the voltage drop ΔU_{ESR} over the ESR resistance when the discharge current I_2 flows through it. While the supercapacitor keeps getting discharged at current I_2, the voltage continues to drop here with a linear profile. You can derive from this behavior an equivalent capacitance that derives from the relative change of the electric charge ΔQ_C over the voltage drop ΔU_C:

$$C = \frac{\Delta Q_C}{\Delta U_C} = \frac{I_2 \cdot \Delta t_C}{\Delta U_C}$$

The authors of the work [Kopka 2013] define the voltage drop over the range where the voltage decreases from 80% of its nominal value to 40%.

Until now, we have looked at the transients of current and voltage. We extend our assessment of the supercap by looking at its impedance which is determined with the AC (alternating current) method, that is, impedance spectroscopy [Barsoukov 2005]. This method is in more detail explained in the chapter on analytical methods. The impedance spectrum shows the real part and the imaginary part of the complex resistance (impedance) as a function of the electric excitation frequency. In electrochemical measurements like here, the frequency ranges typically from 0.1 Hz to 1 MHz, notwithstanding that some researchers may chose higher or lower boundaries.

The capacitance C can be determined from the electrochemical impedance spectrum via the relation:

$$C(f) = -\frac{1}{2\pi f \cdot \text{Im}(Z)}$$

We notice here an additional parameter, and this is "f", the frequency with which the system is electrically excited with an AC signal [Barsoukov 2005]. This implies that the capacitance C is not a constant number but a dispersion quantity which depends on the frequency.

From the real part Re(Z) of the impedance spectrum, we can determine also the ESR, as simple as

$$\text{ESR} = \text{Re}(Z)$$

Two impedance spectra of two GC electrodes are shown in Figure 2.19. The capacitance C is plotted versus the frequency f. The upper curve with filled bullets was recorded from a monolithic GC electrode which had been gas phase oxidized in air for 1 h at 450 °C in a muffle furnace. At a low frequency of 0.1 Hz, the capacitance is 600 mF/cm^2. With increasing frequency, the capacitance decreases considerably. With 100 Hz excitation, the capacitance is below 100 mF/cm^2. At 10 kHz, it is only 1 mF/cm^2.

The second impedance spectrum (open symbols) was obtained from one of the electrodes where activated carbon powder was coated on a titanium substrate. We notice immediately that the capacitance values for all frequencies are lower than from the monolithic electrode. At 0.1 Hz, we read 250 mF/cm^2, and at 10 kHz, we read 0.01 mF/cm^2. The overall trend of decreasing capacitance with frequency is attributed to a so-called diffusion resistance which originates from the decreased mobility of ions in very small pores in the highly porous carbon [Barcia 2002].

The graph in Figure 2.20 shows the CV of a GC electrode. The electrochemical cell which allowed for the measurement of this electrode was a laboratory glass beaker filled with 3 M sulfuric acid as electrolyte. The working

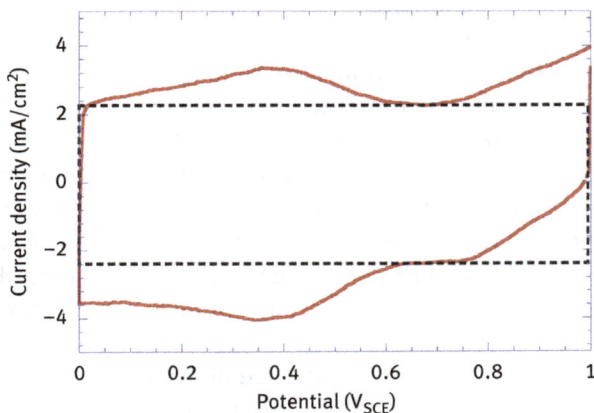

Figure 2.20: Cyclic voltammogram of an activated glassy carbon electrode in 3 M sulfuric electrolyte. This was a single electrode measurement with a Kalomel reference electrode [Braun 1999a].

electrode (WE) was a GC slide of 1 mm thickness and around 3 cm length. The width was around 1 mm. This electrode was gas phase activated in an air-vented muffle furnace at 450 °C for 90 min. It was clamped in a Teflon sample holder which had a brass stick as electric contact and which was connected to the WE contact of the potentiostat.

The counter electrode (CE) was a complete GC disk of the same type K, but with 5 cm diameter. This was the size and shape of the GC plates as they were purchased from the supplier (HTW in Thierhaupten, Germany). The electrodes for studies were cut with a small electric Dremel tool into pieces of rectangular shape.

The CE was also clamped in a Teflon sample holder and contacted to a brass stick which was cabled to the CE contact of the potentiostat. The distance between both electrodes was around 4 cm and they were arranged roughly coplanar, or parallel. There was no reference electrode. Instead, the reference electrode contact from the potentiostat was short connected to the WE.

The potentiostat imposes now a bias potential on the electrochemical cell which I just described from 0 to 1 V. This may be in steps of 10 mV/s. Thus in 100 s, we achieve 1,000 mV which is 1 V. This is the anodic scan. The red curve in Figure 2.20 is the measured current, begins at 0 V, and shoots in the first or second, in the first 10 mV from 0 to around 2 mA. With increasing bias potential, the current increases with a virtually linear slope until 0.35 V, where the current has increased from 2 to 3 mA. From 0.4 to 0.65 V, the current decreases again to 2 mA and then increases again with linear slope to 4 mA at 1 V.

Now, we terminate (actually, the potentiostat does) the voltage scan at 1 V and decrease the voltage by 10 mV/s toward 0 V. This is the cathodic scan. On the time axis, we increased the bias linear and then decreased the bias again to the origin potential. We have thus achieved one CV.

2.6 Determination of the time constant

Let us consider the discharge of a capacitor which we monitor with two multimeter; one for the current and one for the voltage. You connect the capacitor with a power source such as a power supply or a battery. You connect the multimeter parallel with the capacitor and this will give you the voltage reading. You connect the capacitor on one pole with the power supply with one cable. You take two more cables, one of which you connect with other pole the capacitor and the other cable you connect with the other pole of the power supply. The two loose ends you connect with the second multimeter; when you charge or discharge the capacitor, this multimeter will tell you the current which is flowing.

Consider now that the capacitor is charged to 5 V and you disconnect the power supply and replace it with a resistor or some bulb. The capacitor will discharge and release its energy as electric current into the resistor. With a watch or stop watch, you read the current of the multimeter and write it down every couple of seconds. It may be helpful when you have a second person assisting you so that you can share the work.

Nowadays as many people have a smart phone with digital camera, you can record your experiment in the video mode or stroboscope mode and then have photos of the current reading and voltage reading along with the time stamps of the photo. Then, you can acquire more data points than you would be able to in the full manual mode. Certainly, when you work in a laboratory which is equipped with a computer controlled potentiostat and digital data acquisition, experiments become easier.

The manually recorded discharge current data of a capacitor with electrode size of 12.5 cm² are shown in Figure 2.21. We notice immediately that there is a rapid decay of the current in the first few seconds. When you actually do the experiment, you are overwhelmed with the speed and will likely repeat the experiment a couple of times. With the proper experience and training, you may obtain better results.

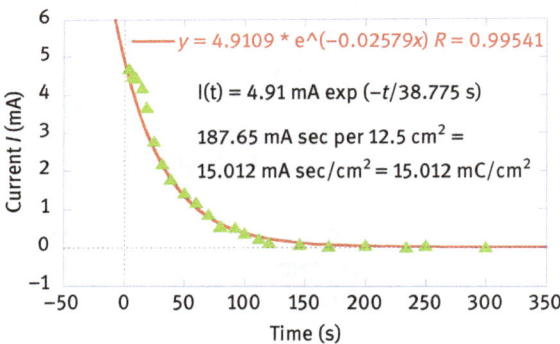

Figure 2.21: Transient of the current of a capacitor during discharge. The solid red line is the exponential least square fit through the manually obtained data points.

The data curve can be fitted with a simple exponential of the form:

$$I(t) = I_0 \, e^{-t/\tau}$$

with time constant τ as the invariant:

$$I(t) = 4.91 \, \text{mA} \cdot \exp\left(\frac{-t}{38.775 \, \text{s}}\right)$$

The time constant t for this discharge process is determined here to $1/0.02579$ [1/s] = 38.775 s. This seems to be a rather "slow" capacitor because it takes a long time before it is discharged (=empty). The large time constant suggests that there could be a moderately fast faradaic process taking place in the capacitor. This could be a hybrid between capacitor and battery. The time constant is also defined as the product of resistance R and capacitance C: $\tau = R \times C$.

The second invariant which we get from the least square fit is the value for $I_0 = 4.91$ mA. We cannot read it right away from the multimeter because in the very instance where we start discharging, the value on the multimeter decreases already rapidly. Using the least square fit, we can extrapolate to $t = 0$ (extrapolation to zero) and this yields I_0. The uncertainty of the values for the current for times smaller than the first measured data point can be very large. Every effort for measuring data accurately in this region will be rewarded by a more confident extrapolation to zero.

We can also extrapolate to $t \to \infty$ (extrapolation to infinity) in Figure 2.21. This should not be so difficult because we see a clear flat plateau for very large discharge times. In the ideal case, the current should be zero after infinite time. We can refine the least square fit function by allowing for a constant value or even for a linear decay in addition to the exponential decay. Then, certainly we have another invariant that we have to deal with. When the current for large times is not zero but finite, even it is very small, this could be an indication of some leakage current in the system.

The time constant (with R, C) gives you a first quantitative indication of the charge carrier dynamics of the system (capacitor, battery, fuel cell and any other system). You can transform the time (time domain, time space) into the frequency domain (frequency space) via a Fourier transformation. Then, you obtain the equivalent of an impedance spectrum. For the Fourier transformation to be accurate, you must extrapolate the discharge curve (or charge curve) to 0 and to ∞ because you will have to carry out an integration within this range.

Impedance spectra are typically recorded with a frequency response analyzer (FRA). Often, these are limited to a specific range of current, say, 1 A. When you want to investigate a large capacitor with currents larger than the limit of the FRA, here larger than 1 A, then it will not work. In this case, you can digress and do a discharge measurement as shown here and use the Fourier transformation.

2.7 Determination of electric charge and capacitance

The electric charge Q is measured and defined as 1 C = 1 A/s and is determined by the integration of the current transient shown in Figure 2.21. We are basically counting the electrons by measuring current over time; this is called chronoamperometry. By integration of the current $I(t)$ over time t, its electric charge Q is determined here to 187.65 mA/s for the entire device, this is: Charge $Q = 187.65$ mC.

Now imagine, the capacitor has a weight (mass) of 10 µm (10×10^{-6} kg); this is very small but it could be an application for microelectronics on a circuit board. Then, the gravimetric charge is 187.65×10^{-3} C/10×10^{-6} kg = 18.765×10^{3} C/kg.

The electric capacitance unit is 1 F = 1 C/V and calculated as ratio of electric charge Q per applied voltage U. At a voltage of 5 V, the specific gravimetric capacitance C is then 18.765×10^{3} C/kg/5 V = 3,753 C/kg/V = 3,753 F/kg or 3.75 F/g. Hence, the specific gravimetric capacitance is $C = 3.75$ F/g.

The corresponding energy density is then $1/2 \times (18.765 \times 10^{3})^{2} \times C^{2}$ kg^{2}/3,753.

Following relations between electric current I, charge Q, capacitance C, energy W and voltage U are relevant [Braun 1999a]:

$$1\,C = 1\,A/s$$

$$1\,\text{Farad} = 1\,F = 1\,\frac{C}{V} = 1\,\frac{A \cdot s}{V} = 1\,\frac{A^2 \cdot s^4}{kg \cdot m^2}$$

Capacitance of the capacitor is

$$C(U) = \frac{Q}{U}$$

Energy in the capacitor

$$E(U) = \frac{1}{2}\,CU^2 = \frac{1}{2}\frac{Q^2}{C}$$

$$1\,J = 1\,V\,A\,s = 1\,W\,s = 1\,C\,V = 1\,kg\,m^2/s^2$$

With this, the basic procedure for obtaining charge and energy of the device is explained.

2.8 Equivalent series resistance of capacitors

Kingatua provides a free available section for the reader on the determination of the ESR [Kingatua 2017]; an alternative is found in Rix [Rix 2003]. The ESR is the same like the double layer capacitance, a nonideal characteristics of a capacitor. It is an essential characteristic of an electric component in devices. We typically want to decrease the ESR as much as possible in order to minimize I^2R losses [Beaty 2013]. A high ESR may cause excessive and dangerous heat in the device. Its correct determination is therefore necessary. The ESR has not necessarily a constant value; its value needs to be determined for a particular frequency or frequency range.

The ESR of capacitors can be very small in the range of milliohms. A significant effect arises usually only in electric circuits which serve radiofrequency applications and microelectronic circuits with low power, and not so very much for large capacitors. When the capacitors are part of energy harvesting systems, where accumulated charge needs to be released quickly, it is critical that the capacitors have low ESR values. System designers may however opt for supercapacitors with a large ESR and potential leakage rather than for conventional ceramic capacitors, because the former offers considerably higher energy density.

2.8.1 Determination of the ESR with an ESR Meter

You can purchase an ESR meter in an electronics store. This is an instrument specifically designed to measure the electrical series resistance with a reasonable accuracy. The meter applies an AC on the capacitor with a voltage divider network. The frequency of the AC is typically chosen at a value where the reactance (the reactance is basically $1/\omega C$) of the capacitor is negligible. An ESR meter is rather a diagnostic tool and not an analytical tool.

The ESR meter imposes a current through the capacitor for a period so short that it does not become fully charged. As the current is flowing, a potential difference, a voltage is generated over the capacitor. The voltage is the direct product of the imposed current and the ESR of the measured capacitor; in addition, the charge in the capacitor constitutes a small negligible voltage. The ESR is determined by the division of the experimentally measured voltage over the current.

2.9 Determination of energy density

The electric energy from the total charge stored in the capacitor can be related to its volume (volumetric energy) and its mass (gravimetric energy).

Example: Let the volume of a thin film capacitor be 10×10^{-4} cm^3. The charge per volume (volumetric charge) is 188×10^{-3} C/10×10^{-4} cm^3 = 188 C/cm^3 = 188×10^3 C/L.

Let the mass of the thin film capacitor be 10 µg. The specific charge (charge per mass; gravimetric charge) yields Q.

The energy which we can store in a capacitor depends parabolic on the charging voltage that we apply: $E = 1/2CU^2$.

When we apply a voltage of 5 V rather than 1 V, the energy in the battery at 5 V would be $5 \times 5 = 25$ times higher than in the case of 1 V. Battery engineers and capacitor engineers therefore would like to apply a as high as possible voltage. A problem is that the electrolyte in the battery (or capacitor) may face a dielectric breakdown when too high a voltage is applied.

2.9.1 A word on electrolytes and electrolyte decomposition

In electrochemical energy converters, this is usually the chemical decomposition of the electrolyte; when this contains water, it will electrolyze at a voltage higher than 1.23 V (thermodynamic limit). Some kinetic barriers and overpotentials may make that you actually can apply 1.5 or almost 2.0 V before this practically occurs. But at $U >$ 1.23, we so-to-speak basically "drive without safety belt" and must be prepared for unpleasant surprises.

When you use an electrolyte that contains no water such as an organic electrolyte, this might be way more stable against decomposition or electrolysis. You may charge it up to 3 V or even 5 V depending on the chemical composition of the electrolyte. Electrolytes are therefore very important for batteries [Li 2016] and supercapacitors [Zhong 2015].

However, if the organic electrolyte contains traces of water, this trace water might still be electrolyzed when the applied voltage in the device is high enough. What happens then is that the H_2O molecules would become H_2 and O_2 gas, at the least. There could be even more complex reactions.

A practical man might now say: "Bingo! This is how I remove the water in my organic electrolyte; I electrolyze it."

This could actually work, in my opinion. Problem is that the evolved gas in a sealed battery could make other problems, mechanical problems over time. The integrity of the entire battery cell could be jeopardized.

So maybe you want to start without any water right from the beginning and therefore use drying methods for battery materials synthesis and processing. For that you would use a drying furnace and glove box and so on.

The other problem when you do not use a glove box is that other molecules in the ambient air can do harm to your experiment. Air contains 80% nitrogen and this will react immediately with lithium metal to black crusty Li_3N if that is part of your battery mixture. Or the lithium metal will react to white fluffy LiOH if the atmosphere is very humid. Lithium perchlorate sounds to me very hygroscopic; it will absorb a lot of water. Therefore, it needs to be dried in an oven. Careful – it is a strong oxidizer. Maybe it is even not very shock resistant.

2.10 Determination of power and power density

We know that power P is defined as work per time; $P = W/t$, or energy per time; $P = E/t$. The electric power of the capacitor is determined by the energy which is stored in it and released during a particular time. Typically, we refer the released energy to the aforementioned time constant τ which is determined from the charging and discharging characteristic. Depending on the complexity of the capacitor or battery or other electrochemical device, we may notice from their charge and discharge curves that their characteristics do not allow for one simple time constant which can be determined from least square fitting of one exponential to the data points. It may be necessary to include a second exponential or more.

Hence, as the current during a discharging and charging varies over time, so does the power density. The technical specifications of capacitors do usually not account for such fine details. The time over which the PSI capacitor in Figure 2.17 [Bärtsch 1999] is charged and discharged is 8 s and 12 s, respectively. It is an exercise for the reader to determine the power and power density of this supercap based on the data given in Bärtsch [1999].

2.11 Fine structure of the double layer

2.11.1 Helmholtz' approach

When researchers in the nineteenth century tried to understand the nature of electricity, they found that

> Electricity disappeared in the inner volume of a body and would only occur in the infinitesimally thin surface of the body.

as Helmholtz writes it in his paper from 1879 [Helmholtz 1879].

This finding follows also from what I remember as "Satz von Gauss" (Gauss Law) which states that the flux from a vector field v inside a space V with boundary ∂V through a closed surface A ($=\partial V$) equals the sum of the sinks and sources which are enclosed or included by that surface:

$$\oint_A \vec{v} \cdot d\vec{A} = \int_V \text{div}\,\vec{v}\,dV$$

For electric charges enclosed by a surface A over a volume V, this "Satz von Gauss" states that the flux of the electric field through the closed surface equals the enclosed total electric charge, divided by the constant $1/\varepsilon_0$, and takes following specific form:

$$\oint_A \vec{E} \cdot d\vec{A} = \frac{q}{\varepsilon_0}$$

Helmholtz calls the two layers adjacent to some fictitious plane area from one and the opposite side in infinitesimal distance from each other "double layer" and assumes that one layer has as much positive electricity as much as the other has negative electricity [Helmholtz 1853]. With "electricity" he means what we understand today as electric charge.

I mentioned in Section 2.1 that the double layer is usually understood as the interface layer between an electronic conductor and ionic conductor. But also between two solids and between two immiscible liquids, such double layer may form. Typically at the phase boundary of two charged phases, there are two charged layers with opposite signs of charge (polarity).

The discharged double layer carries the so-called potential of zero charge (pzc), where the metal surface is not charged and where also the electrolyte has no net charge. This does however not imply that the pzc would be zero, that is, pzc = 0. Water is a very important phase (dry, liquid vapor) for the physics and chemistry of the double layer and at some point readers may be interested to understand the interaction of water with other phases at the molecular level. Jordi Fraxedas provides an extensive review book on water at surfaces [Fraxedas 2014].

Figure 2.2 has already visualized the physical model. The metal conductor M (gray rectangle) is sketched as a lateral extended electrode and has at the surface a positive electric charge when it is lifted to a positive electric potential φ_M by an external electric circuit. Solvated ions with negative charge (blue disks) in the electrolyte are attracted by the positive-charged electrode surface and arrange densely as the inner Helmholtz layer. The solvated ions have a diameter of around 2 Å and the thickness of the inner Helmholtz layer is then defined as $a/2$.

The next step is now translating this physical image into mathematical language. The electric potential of the system is therefore sketched versus the spatial coordinate x (right panel of Figure 2.13). The reference for zero (0) on the space coordinate x is given at the solid–liquid interface or solid–electrolyte interface. The thickness of the charged layers, the Helmholtz layers, can be inside metals in the range of 1 Å and in solutions in the range of 1–100 Å. In the bulk of the electrolyte, the electric potential is φ_{oH}. This large extended layer is called outer Helmholtz layer . We recall from Section 2.1 that the electric field E which forms over a difference of the potential (gradient) is determined as

$$E = \frac{\Delta\varphi}{\Delta x} = \frac{\varphi_M - \varphi_{oH}}{(a/2)}$$

In first approximation, we have a linear decrease of the potential over the inner Helmholtz layer, which is indicated in the figure with a linear slope. Inside the electrode and inside the electrolyte, the potential is constant φ_M and φ_{oH}, respectively. When that inner Helmholtz layer has a thickness of 2 Å and the potential difference is 1 V, the strength of the electric field is 0.5×10^{10} V/m as we recall from Section 2.1.

For the determination of the electric capacitance C and electric energy which is charged in such double layer, we only need now the dielectric constants ε_0 and ε_r:

$$W = \int \vec{E} d\vec{x} = \frac{1}{2} \varepsilon_r \varepsilon_0 \vec{E}^2 = \frac{1}{2} CU^2$$

In its most primitive stage, the model of the Helmholtz double layer describes therefore only a differential capacitance which is dependent from the dielectric constant and the thickness of the double layer and not depending on the charge density. This model is sufficient for a description of the charge separation.

2.11.2 The improved model for the double layer by Gouy and Chapman

Let us now deepen the view into the solid–liquid interface. We have so far considered only the first tight layer of ions arranged at the electrode surface and ignored any further structure of the electrolyte past that layer. Now we account for the diffusion and mixing of ions in the solvent. At later stages, we may also look into the possibility of adsorption of ions at the electrode surface and the interaction of potential dipole moments and multipole moments between solvent and electrode. These contributions would lead to a further refinement of the double layer structure (Figure 2.22).

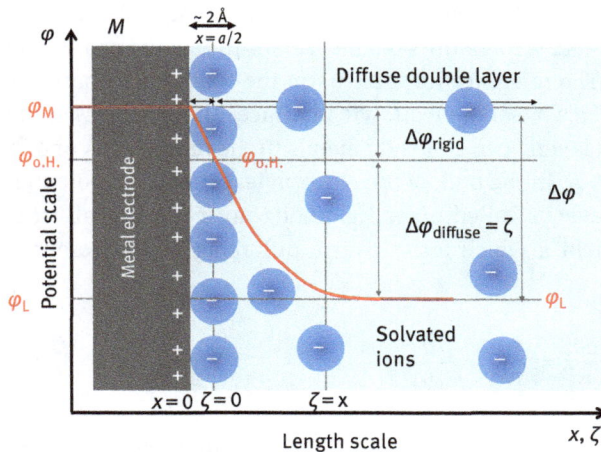

Figure 2.22: Schematic of the Gouy–Chapman model of the electrochemical double layer, where M is the metal electrode, oH is the outer Helmholtz plane, $a/2$ = radius of solvated ions and $\Delta\varphi$ is the Galvani voltage. For the Stern model, x is the thickness of the diffuse double layer and $\zeta = \zeta$ potential.

Louis Georges Gouy in 1910 and David Leonard Chapman in 1913 extended Helmholtz' theory by considering the thermal motion of counter ions in the electrolyte, which would cause an extended and diffuse layer in the electrolyte; extended as in that it would be as large as several molecular layers. We therefore talk about the

Gouy–Chapman layer which depends on a potentially external applied voltage and also on the electrolyte ion concentration. The spatial charge distribution of ions is rationalized with the Maxwell–Boltzmann statistics and thus as a function of the distance from the electrode surface. It is mathematically described and modeled with a Maxwell–Boltzmann distribution. The derivative of this distribution yields an electric potential with an exponential profile, decreasing with increasing distance from the electrode surface.

Figure 2.22 shows a physical model for an arrangement of electrode and electrolyte in the spirit of Gouy and Chapman. Now, we loosen up the tight anion layer right on the surface and allow for a diffusive arrangement of ions. The concentrated negative electric charge at the electrode is still maintained, but somewhat relaxed and hence also the corresponding attractive electrostatic force toward the anions. The graph in Figure 2.22 shows consequently that the variation of the electric field in the inner Helmholtz layer is not linear anymore.

We can define now a characteristic length over which the electric potential will decrease to $1/e$ of its maximum value measured at a particular point. This is the Debye length and is known for plasmas, such as electrons in clouds, or in semiconductors and in electrolyte solutions. The mobility of ions and their concentration in electrolytes is relevant for the Debye length λ_D. For electrolytes with relative dielectric constant ε_r and ionic strength I, we obtain

$$\lambda_D = \sqrt{\frac{\varepsilon_0 \varepsilon_r k_B T}{2 N_A e^2 I}}$$

We see that the Debye length scales parabolic with electrolyte temperature T; with increasing temperature, the Debye length and thus the double layer thickness are widening.

2.11.3 The model by Otto Stern

It has turned out that the improved model by Gouy and Chapman will deliver inaccurate results when the double layer is charged very strong. Otto Stern realized around 1924 that such very strong charged electrode surface would restore the strict first layer of ions as Helmholtz had originally suggested in his simple model.

I have sketched this situation in Figure 2.23. We begin with a negatively charged electrode (charged interface) which is inserted in an electrolyte which contains anions (green circles with a "−" sign) and cations (red circles with a "+" sign). These ions are solvated and thus contain the hydration shell from water molecules (H_2O, blue circles with a black arrow for the dipole moment of the water molecule).

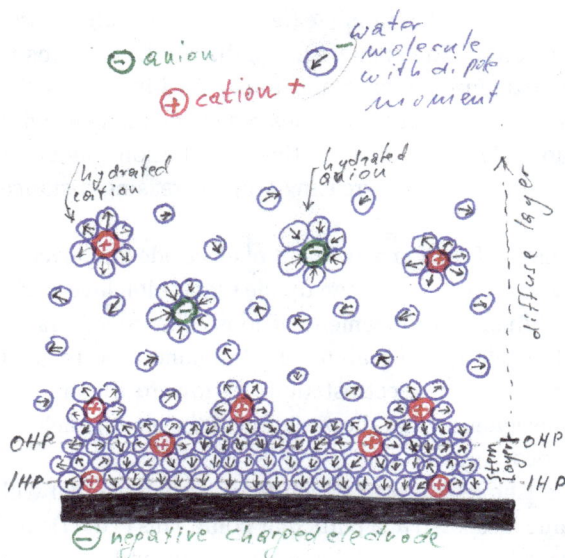

Figure 2.23: Simple visual model for the electrochemical double layer after Stern. Green circles denote anions, red circles denote cations and blue circles denote water molecules with a black arrow dipole moment, the tip of which has the positive charge. IHP is inner Helmholtz plane; OHP is outer Helmholtz plane.

The water molecules with their dipole moments are surrounding the ions. The dipole moment is pointing toward the anion and pointing away from the cation. The dipole moment of the water molecule is also pointing to the negatively charged electrode. And so do the cations, because they are positively charged. When we charge the electrode even more negatively, the cations will strive toward the electrode and the dipole moments of the water will arrange firmly on the electrode and constituting the original IHP.

When the potential is strong enough, more cations will arrange toward the electrode. The last densely packed cation layer will then form the OHP. The IHP and OHP will then constitute the Stern layer. The ions beyond that layer will arrange loosely as the diffuse layer.

Stern thus combined the Helmholtz model with the Gouy–Chapman model and arrived what we call today the Stern double layer. Specifically, the Stern double layer takes into account that the ions have a finite size. Therefore, the closest proximity of the ions to the electrode surface would be half the ion diameter.

Stern based his work on Nernst's simple considerations of the double layer, but Stern remarks in his paper [Stern 1924] that his work is motivated by Debye's works on electrolytes [Stern 1924]. Stern specifically asks what the structure of the double layer is and how deep it would reach into the solution (electrolyte).

Stern remarks that the specific adsorption of ions on the electrode surface can be decisive in this respect. In the further outline of his work, Stern introduces literally the concept of the molecular capacitor (molekularer Kondensator), in which he models the double layer by a plate capacitor, through which he smears[8] the electric charges homogeneously.

Stern determines the capacitance C of this molecular capacitor by assuming constant values for the dielectric constant ε_r of the double layer and the distance δ between the "center of gravity" of the electric charges and the surface of the metal electrode. This distance δ is by good approximation the ionic radius. The capacitance is defined as

$$C = \frac{\varepsilon_r \cdot U}{4 \cdot \pi \cdot \delta}$$

with the capacitance per area A being

$$\frac{1}{A} \cdot \frac{Q}{U} = \frac{\varepsilon_r}{4 \cdot \pi \cdot \delta}$$

Stern finds a double layer capacitance of 24 μF [Stern 1924]. This is a reasonable value for a smooth surface. The ratio of the ion thickness over the dielectric constant yields a value of 0.33×10^{-8} cm, which is 0.33 Å.

2.11.4 Bockris–Müller–Devanathan Model

The various models of the EDL contain inconsistencies which have been addressed in a seminal paper by Bockris et al. [Bockris 1963]. A further refinement of the structure of charged interfaces is accomplished by taking into account the influence of the electrolyte or solvent. They included the action of specifically adsorbed anions on the electrode surface. "Specifically adsorbed" means that the chemical specificity of the electrode ions and electrolyte ions has an influence on the adsorption of these ions on the electrode material. This includes particularly chemical reactions, redox reactions where ions in the solvent or electrolyte can be oxidized or reduced when electron transfer occurs on the electrode surfaces.

---

8 With a given set of charges which you can take as point charges, located at particular positions, you can consider them in a coordinate system with specific coordinates from which you can calculate the electric potential and electric field. The term "smear" is sometimes used when the scientists simplify the arrangement of a number of particular charges at non-particular positions with continuum theory. Then, you add up the charges and divide them by the volume and arrive at some statistical mean value. You have symbolically and mathematically smeared (distributed) the charges homogeneously over the volume.

Such redox reactions are detected by extra current waves at specific electric potentials and are called pseudocapacitance, a term which was first used by Conway in 1962 [Conway 1962].

We recall that at the charged electrode surface, the adsorbed solvent ions form the inner Helmholtz layer, see Figure 2.23. The solvated cations arrange immediately at the inner Helmholtz layer and thus form the outer Helmholtz layer. These are the counter ions with respect to the electrons in the electrode. They cause the double layer capacitance. When one single specifically adsorbed cation has penetrated the inner Helmholtz layer, it might receive one electron from the electrode which constitutes a current which shows up as a pseudocapacitance. The cation has become an anion.

2.11.5 Dependency from the electrode material

By choice of drawing the electrode material with a hexagonal pattern in some previous sections, I implied that carbon would be the electrode material. This is not a real requirement. We can chose between the wide range of metals and semiconductors. But the result is not the same. The double layer capacitance may vary noticeable with the electrode material [Trasatti 1981]. For this effect to be understood, we need the distribution of the charge carriers in the electrode.

A very early treatment on a related matter (electrocapillarity, i.e., the surface tension between a mercury drop electrode and the electrolyte depending on the adsorbed species) was published by Oscar Knefler Rice 90 years ago. Rice assumed that electrons in the mercury electrode would follow the Fermi statistics and thus he could explain experimental electrocapillary curves [Rice 1928]. The mercury electrode and electrodes from second and third row metals can be well explained by a model where the electron density is considered homogeneously distributed in the electrode like a continuum (the so-called Jellium model). The electric potential of the charges within the electrode reaches through the electrode–electrolyte interface and extends to some extent into the electrolyte [Godula-Jopek 2015, Schmickler 1984]. The situation may become more complex when the electrode is a semiconductor.

2.11.6 The pseudocapacitance in RuO_2

In 1970, Trasatti and Buzzanca submitted in a Preliminary Note about their observations made with ruthenium oxide RuO_2 [Trasatti 1971]. A preliminary note is a publication in which researchers and the editor want to rapidly share important findings before the time for a full and complete publication passes by. The two authors investigated a thin polycrystalline film of RuO_2 and a single crystal monolith of the nominally same material because they were appealed by the metal-type conductivity of this metal oxide, and by the fact that RuO_2 films could be deposited at relatively low temperatures with a simple process which would allow for sharp chemical interfaces between the metal current collector and the RuO_2 film on top.

This allowed comparison of a low-temperature-grown thin polycrystalline film with a high temperature-grown monolithic single crystal. According to their Preliminary Note, the XRD showed the same crystallographic phase but differences in the Bragg peak position and peak width. This is not surprising. It is well known that increasing the temperature promotes crystallization of materials which certainly manifests in improved crystallographic ordering of atoms, increase of coherence domains and which is clearly visible in Bragg reflections with a higher intensity and small full width at half maximum.

Figure 2.24 shows the (0 0 2) Bragg reflection of RuO_2 synthesized at temperatures from 250 °C to 800 °C. The intensity for the peak from the sample synthesized at 250 °C is 34,000 counts per second at the maximum position of around 58.7°. Prepared at 300 °C, the maximum shifts to 58.8° with 37,500 counts per second. With 400 °C, the maximum shifts further to 59.2° with 40,000 counts per second. This trend continues homogeneous but there is a jump in the peak position and intensity between 500 °C and 600 °C. You can verify this as an exercise when you

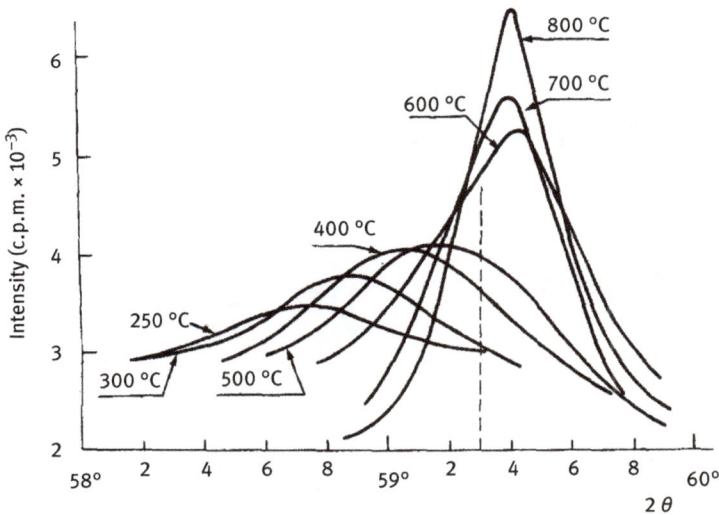

Figure 2.24: Influence of the temperature on the shift and broadening of the 0 0 2 peak. The dotted line indicates the position of the Schafer peak.
Reprinted from *Materials Research Bulletin*, 7, Pizzini S, Buzzanca G, Mari C, Rossi L, Torchio S, Preparation, structure and electrical properties of thick ruthenium dioxide films, 449–462, Copyright (1972), with permission from Elsevier [Pizzini 1972].

determine the positions and intensities of the peaks and plot these versus the heat treatment temperature (HTT[9]).

The changes in the crystallinity are accompanied by considerable changes in the electric conductivity as determined by the 4-point DC conductivity measurement. Pizzini et al. [Pizzini 1972] prepared the samples with a HTT from 400 °C to 800 °C. Figure 2.25 shows the conductivity of the RuO_2 film annealed at 400 °C, 450 °C and 550 °C, with the conductivity measured from ambient temperature to around 1,000 ° C. The conductivity is plotted versus the reciprocal temperature like in an Arrhenius plot. The three samples show the same steep slope in conductivity at high temperatures from around 800 °C to 1,200 °C which indicates the same activation energy for a particular charge transport.

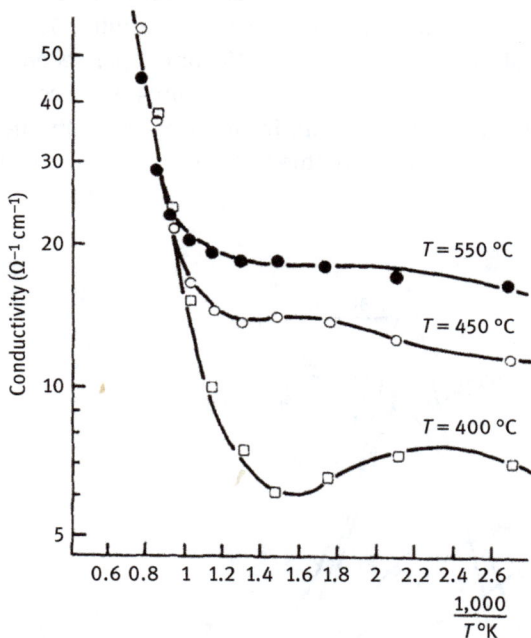

Figure 2.25: Electrical conductivity of type I films grown on silica glass and annealed at different temperatures.
Reprinted from *Materials Research Bulletin*, 7, Pizzini S, Buzzanca G, Mari C, Rossi L, Torchio S, Preparation, structure and electrical properties of thick ruthenium dioxide films, 449–462, Copyright (1972), with permission from Elsevier [Pizzini 1972].

---

9 Some researchers mistake HTT as "high temperature treatment," but to best of my knowledge, "heat treatment temperature" is the correct wording and meaning.

For the lower temperatures, we find an intermediate minimum in conductivity and with increasing temperature some plateau-like behavior. The height of the plateau at low temperatures increases with increasing HTT.

We must distinguish in this experiment from the synthesis temperature which I also called HTT, and the measurement temperature. I want to make you aware of the following. One sample was synthesized at 400 °C, and then it was subject to the four-point DC measurement which was from ambient temperature to over 1,000 °C. Therefore, although the sample was synthesized at a low temperature and thus had the low-temperature structure indicated in Figure 2.24, the temperature had to be raised for the conductivity measurement to way over 800 °C. This is a fundamental dilemma which means that while we measure the sample at temperature T_m, we may also modify the structure of the sample when it was synthesized and processes at $T \gg T_m$.

The films were synthesized by the decomposition of $RuCl_3$. The authors found that residual chlorine was incorporated in the film and likely acts like a dopant which has influence on the transport properties of the film including the electric conductivity. Table 2.2 lists the chlorine content and oxygen content along with the lattice distance for the (0 0 2) reflection planes (Bragg planes) and the distance between the Ru–Ru planes and the crystallite size.

Table 2.2: Thermal treatment, composition, lattice constants and average crystallite size of RuO_2 films.

Annealing temperature (°C)	Chlorine content weight (%)	Oxygen content weight (%)	d_{002} (Å)	d_{Ru-Ru} (Å)	Average crystallite size (Å)
250	26.43	16.18	1.572	3.137	(410)
300	4.77	23.98	1.569	3.131	(325)
400	3.84	23.29	1.565	3.123	203
500	3.39	24.12	1.562	3.117	215
550	2.84	22.85	–	–	–
600	2.08	2.37	1.555	3.103	367
650	2.07	23.16	–	–	–
700	0.89	24.72	1.556	3.105	606
800	0.40	26.12	1.556	3.105	875
RuO_2 single crystals	–	24.05	1.557	3.107	–

Reprinted from *Materials Research Bulletin*, 7, Pizzini S, Buzzanca G, Mari C, Rossi L, Torchio S, Preparation, structure and electrical properties of thick ruthenium dioxide films, 449–462, Copyright (1972), with permission from Elsevier [Pizzini 1972]. The data for the RuO_2 single crystal were taken from Boman [Boman 1970].

With increasing HTT, the Cl content is decreasing from >25% at 250 °C to <0.5% at 800 °C. The distance between the Ru cations decreases from 3.137 to 3.107 Å, and Trasatti and Buzzanca begin their Preliminary Note with the understanding that the proximity of Ru atoms causes an overlap of the Ru d orbitals with corresponding charge transfer. This is in contrast to the assessment of Boman [Boman 1970] who carried out the structure refinement on RuO_2 and concluded that the distance between the Ru atoms in RuO_2 would be rather large. It is not so uncommon that researchers disagree over particular issues for some time.

The further research of Sergio Trasatti and Giovanni Buzzanca on the electrochemical properties of RuO_2 led to the insight that the electrochemical charging behavior of specifically adsorbed ions is comparable to capacitors [Trasatti 1981]. The specifically adsorbed ions provide a charge transfer between the ion and the electrode which is called pseudocapacitance. Frequently, the charge transfer in this context is rationalized by the Marcus theory of electron transfer [Marcus 1997].

Figure 2.26 shows the CVs of a RuO_2 thin film deposited on a titanium metal substrate and a RuO_2 single crystal. The polycrystalline film produces a smaller area in the CV than that of the single crystal. The polycrystalline film has a CV with a rectangular shape which confirms that we are dealing in principal with a double layer capacitor. The CV of the single crystal however shows a typical redox-like behavior. The additional capacitance derived from this redox process is called pseudocapacitance.

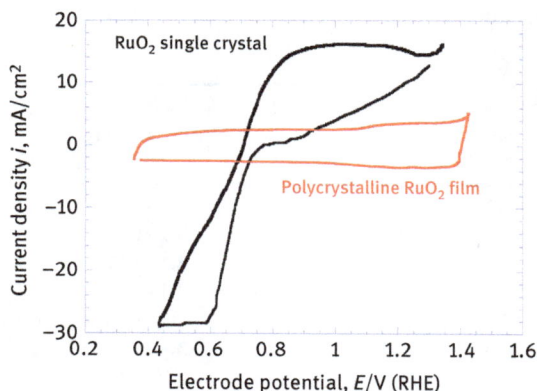

Figure 2.26: i/E curves (cyclic voltammograms) for a film (small red CV) and single crystal (large black CV) of RuO_2 in 1 M $HClO_4$ at 40mV/s.
Reprinted and adapted from *Journal of Electroanalytical Chemistry*, 29, Trasatti S, Buzzanca G, Ruthenium dioxide: a new interesting electrode material, A1–A5, Copyright (1971), with permission from Elsevier [Trasatti 1971].

Trasatti et al. produced films with various thicknesses and found that this pseudocapacitance scaled with the film thickness. Thicker films had a higher pseudocapacitance. This reveals that the pseudocapacitance is a bulk property. They speculated therefore that the redox capacitance was based on solid state chemical reaction, that is, phase transformation where proton H^+ would participate in the reduction of the Ru^{4+} to Ru^{3+} with concomitant formation of a hydroxide or oxy-hydroxide (or hydrous ruthenium oxide):

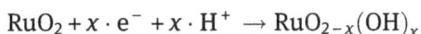

$$RuO_2 + x \cdot e^- + x \cdot H^+ \rightarrow RuO_{2-x}(OH)_x$$

To know whether we are really dealing with Ru^{4+} and Ru^{3+} is not so trivial. Answering this question requires a chemical analysis which can be carried out with chemical titration or, this is a more modern method, with X-ray spectroscopy. Unfortunately, Ru belongs to the elements which make it somewhat difficult to distinguish the two ions Ru^{4+} and Ru^{3+} with X-ray and electron spectroscopy [McKeown 1999]. It is sometimes helpful for the spectroscopic analysis when the oxidation of the ruthenium can be done *in situ* [Stefan 2002a, b].

For the analysis of a CV, it is necessary that the geometrical surface area is accurately known. Then, you can refer the current versus the geometrical area and determine the current density as A/cm^2. When you have a porous electrode, it may be necessary that you determine the internal surface area of the electrode and then relate the current to the internal surface area. This way, it is possible to determine whether a chemical reaction is taking place only at the surface or also in the volume. The intercalation of lithium into battery electrodes is a chemical reaction taking place in the volume, for example.

Nowadays, the CVs are digital information on a computer and can be analyzed with various software. The area enclosed by a CV corresponds to the electric charge passed through the WE. While the CV is plotted as current (or current density, when the electrode area is known) versus the potential, the time is a "hidden" variable in the CV because the scan of the CV is done with a particular speed, with a particular rate given in mV/s. The CV in Figure 2.26 was recorded with 40 mV/s, for example. You therefore have the time axis which you need to determine the electric charge from the integration of the current over time.

I still do remember the times when CV and other data were not stored in a computer memory. Instead, they were plotted with a pen on a sheet of paper. When I was a diploma student at KFA Jülich, the Auger spectra for element determination and composition of the sample surface were plotted with a pen on a large DIN A2 paper.

The photo in Figure 2.27 shows me at my work bench at KFA Jülich next to the UHV (ultra-high vacuum) system where I studied ultrathin magnetic films. To my left side, you see a tall laboratory electronics rack with power supply and electronic controls for a LEED (low energy electron diffraction) system and an Auger spectrometer. On the top of the rack, in the upper right corner in the photo, you see part of an A2 size sheet with pen plotted Auger spectra. These were always good enough for quantitative analysis. We would remove the sheet from the plotter and then hang over it and use rulers and pens for further quantitative analysis [Braun 1997].

During my time as doctoral student at Paul Scherrer Institut, our lab had an old potentiostat which plotted the CV also on paper with a marker pen. With accurate calibration, you can make a quantitative electrochemical analysis of a CV plotted on paper the following way.

Figure 2.27: The author in the laboratory of IGV in KFA Jülich, Germany 1995 during his Physics diploma thesis research. The portrait photo was recorded – because of the opportunity – with a video camera available and necessary for digital data acquisition of LEED I/V curves with the AIDA-PC video data system [Braun 1996, 1997, Heinz 1996]. Auger spectra were still recorded with a pen plotter, as visible on the sheet on the upper right corner in the photo. Photo credit by Artur Braun.

You cut out the CV from the paper with a scissor and then weigh the mass of the cutout CV on a fine balance. Then, you weigh an entire uncut sheet from the paper block and determine its weight and its weight per area. As the sheet is rectangular, you measure the area A with a ruler by the length a and width b: $A = a \times b$. Using a fine balance, you can then determine the size of curved areas which you often have in a CV.

A discharge/charge curve of a RuO_2 film is shown in Figure 2.28. Refer for comparison to Figure 2.17 in this chapter where we charged and discharged a super-cap based on GC electrodes [Bärtsch 1999] and to Figure 2.18 for a method on how to determine supercapacitor parameters [Kopka 2013]. The electrode potential is plotted versus the time during charging and discharging (linear scales for E and t). At first glance, we see five ranges.

In the beginning, the potential is just below 1.5 V, and during around 72 s, the potential drops to 0.25 V. The thin straight line through this range shall demonstrate the linear decay over the time range τ_1. We cannot use the term "time constant" here for τ because it is not the time which would be constant but the rate E/t [(1.45 – 0.25

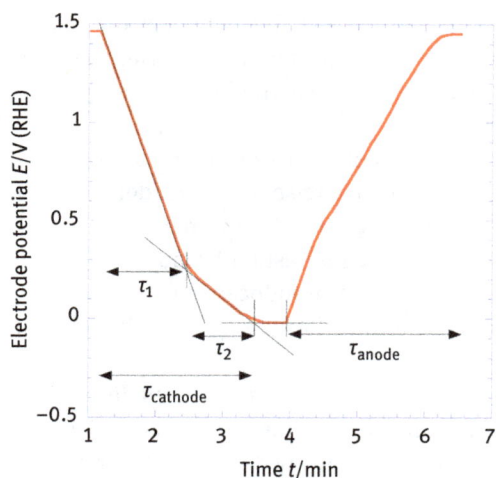

Figure 2.28: Cathodic and anodic charging curves for RuO^2 films in 1 MHClO$^{4\cdot}$ Reprinted and adapted from *Journal of Electroanalytical Chemistry*, 29, Trasatti S, Buzzanca G, Ruthenium dioxide: a new interesting electrode material, A1–A5, Copyright (1971), with permission from Elsevier [Trasatti 1971].

V)/72 s = 16.67 mV/s]. In the 60 s to follow, τ_2, the potential drops further to 0 V, with a much lower slope [(0.25 − 0 V/60 s) = 4.17 mV/s]. This is a factor four between the two slopes. So, we have a fast discharge process over τ_1 and a slow discharge process over τ_2.

The electric charge q passed through the electrode in this cathodic direction was determined under different current densities [Trasatti 1971] (see Figure 2.29: q_{total} = $q_{\tau 1}$ + $q_{\tau 2}$). It is an interesting observation that the charge transferred in the fast

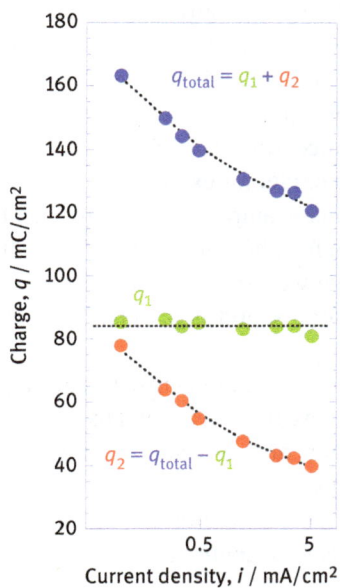

Figure 2.29: Charges measured according to Figure 2.21, as a function of current density.
Reprinted and adapted from *Journal of Electroanalytical Chemistry*, 29, Trasatti S, Buzzanca G, Ruthenium dioxide: a new interesting electrode material, A1–A5, Copyright (1971), with permission from Elsevier [Trasatti 1971].

process during τ_1 does not depend on the current density, whereas the charge transferred in the slow process over τ_2 depends very strong on the current density. The lower the concentration of protons H^+, the lower was the limiting value of the current density for $q_{\tau 1}$ to be constant.

You notice that Trasatti and Buzzanca have plotted the ordinate value 0.5 and 5 mA/cm² , but not the 0. It might be an interesting exercise for the reader to find in which way the ordinate is plotted, in linear or in logarithmic or in parabolic form.

The ruthenium dioxide supercap has been very interesting because of its huge capacitance which it derives predominantly from its pseudocapacitance, which is around 720 F/g. The theoretical capacitance of a RuO_2-based supercap should be in the range of 1400–2000 F/g [Chen 2013].

The high price of ruthenium is a major reason why it has not made it to the consumer market. The energy content in such a ruthenium-oxide-based electrode arrangement is so large that it resembles a battery. Conway described the transition from the double layer capacitance to the rechargeable battery as a supercapacitor and found that the pseudocapacitance does not rest only on specifically adsorbed ions but is rather a result of redox reactions and intercalation [Conway 1991, 1999]. Brousse et al. explain why the capacitance of metal oxide electrodes with a redox peak in the CV shall not necessary be termed having a pseudocapacitance [Brousse 2015].

2.12 The electrolytic capacitor

We learnt in the previous section about the ruthenium oxide supercapacitor. The capacitor is made by depositing a ruthenium oxide layer on a metal current collector. Ruthenium oxide is not the only material with this kind of redox-based capacitance. The previously mentioned carbon supercapacitors have quinone and hydroquinone as redox couple which yields a minor pseudocapacitance. And there are other metal oxides which can form oxy-hydroxides such as nickel oxide, for example.

In the scientific literature, you will find numerous examples where very good battery electrode materials may be tried as materials for photoelectrodes, fuel cell electrodes and supercapacitor electrodes, and the other way around. Photoelectrode materials were to be used as intercalation battery materials, and so on. Metal oxides are quite versatile, but not entirely interchangeable.

We learnt also that a carbon film can be joined with a titanium current collector by TiC formation [Schüler 2000]. The next level is that we can form a metal oxide on a current collector by direct electrochemical anodization of the metal. Consider an aluminum foil which you connect as an anode to a power supply. Then, you oxidize the aluminum to Al^{3+} in the form of Al_2O_3, which forms on the aluminum anode as a thin layer, typically 1.5 nm/V. Aluminum oxide is a dielectric material with a relative permittivity ε_r of around 10.

Practically, you take two aluminum foils between which you place a sheet of paper which you soak with a liquid electrolyte such as boric acid dissolved in ethylene glycol. With anodization of one of the aluminum foils, you form the aluminum oxide layer (a formation step). This is the anode which is formed *in situ*, so-to-speak. The cathode is typically a gel electrolyte containing boric acid or glycol. As this is a capacitor with a polarity of anode and cathode, it cannot be operated with AC current.

The electrolytic capacitor is widely used in electronics and electric engineering [Venet 2013], particularly the aluminum electrolytic capacitor [Epcos 2013]. It can also be made with niobium oxide and tantalum oxide, which are considered costly materials, however. For the reader more interested in this kind of capacitor, I recommend the book from Alexander Georgiev, which you can download for free from www.archive.org [Georgiev 1945].

2.13 The Interface between semiconductors and electrolytes

We assumed in the previous section that the electrode is made from a good conducting metal or metal oxide, and the distribution of the electrical potential is therefore very homogeneous. The situation is somewhat complicated when the conductivity of the electrode is not so good. Then, there are not enough electronic charges which can bring the electric potential in special equilibrium upon a perturbation. This is for example the case when we use a semiconductor as electrode material, such as silicon or various metal oxides.

First, we should recall some basic properties of semiconductors [Peter 2016]. Unlike metals, they have no free charge carriers (the idea of "free" charge carriers is an illusion anyway). Charge carriers are seeded by doping of the semiconductor material with foreign elements. Doping with electron donor atoms makes the material then an n-type semiconductor. Upon thermal excitation, the charge carriers (electrons) are generated by ionization of the donor atoms. These electrons are lifted from the valence band to the conduction band where they occupy vacant levels.

These electrons follow the Fermi–Dirac statistics and assume, with donor concentration N_c (say, less than 10^{18} foreign atoms per cm^3), a density of states (DOS) n which depends on the temperature T of the semiconductor:

$$n_{(E_c, T)} = N_c \left[\frac{1}{1 + \exp((E_c - E_F)/k_B \cdot T)} \right]$$

At temperature $T = 0$ K, the shape of distribution yields the well-known Fermi ice-block, a rectangular shape of the DOS versus energy, which ends at the Fermi energy $E_F = 0$. The Fermi energy is a measure for the free energy of the electron. With increasing temperature, the Fermi ice-block melts and the probability for an electron to assume a particular energetic state E_c in the conduction band is given by $n(E_c, T)$.

The DOS for holes follows analog when we dope the semiconductor with acceptor atoms

$$p_{(E_c, T)} = N_V \left[\frac{1}{1 + \exp((E_F - E_V)/k_B \cdot T)} \right]$$

The energetic distance between the conduction band E_C and the valence band E_V is the band gap energy E_G. The Fermi energy E_F ranges, depending on the system, somewhere in the band gap.

Figure 2.30 illustrates [Gelderman 2007] a solid–liquid interface or, in the context of this section, a solid–liquid junction. In the beginning (on the top left side; A), we have a solid n-type semiconductor in contact with a liquid electrolyte. The semiconductor has randomly distributed positive and negative charges, and there is electric neutrality in the semiconductor. The semiconductor is characterized by the energy band structure which has a valence band with energy E_V, and a conduction band with energy E_c, which enclose the energy band gap. The band positions can be determined with X-ray and photoemission spectroscopy [Braun 2017a].

On the side in Figure 2.30 (top right side; a), we see the energy scale of an electrolyte with an arbitrary redox couple and redox states *Red* (donors) and *Ox* (acceptors). The energy positions of these states can be determined with cyclic voltammetry. Cyclic voltammetry can also be used, alongside with impedance spectroscopy, for the determination of defect states and surfaces states on the semiconductor electrode [Peiris 2014]. Note how the authors [Gelderman 2007] used Gaussian profiles for the broadening of the energy positions of the redox states. The same broadening is used for the states in the electrode.

As the semiconductor comes in contact with the redox couple in the electrolyte, the Fermi energy E_F and the redox potential E_{Redox} are typically not at the same energy. As the Fermi level of the solid (electrode) and the redox potential of the liquid (electrolyte) are usually not on the same energy level, the Fermi level of the semiconductor will shift to meet the redox potential of the electrolyte.

When the semiconductor is of n-type, then the condition $E_F > E_{Redox}$ holds. Energetic equilibrium is then achieved that electrons flow from the electrode to the electrolyte so that the new condition $E_F = E_{Redox}$ holds. Note, however, that the redox potential does not adjust; it is the Fermi energy of the semiconductor which adjusts to the situation. This is shown in Figure 2.30, panel b, where the dashed line displays the energetic equilibrium. The transfer of electric charge produces a region on each side of the junction where the charge distribution differs from the bulk material, and this is known as the space-charge layer.

As the electrode has delivered electrons to the electrolyte, the electrode becomes positively charged. The electrode becomes pronounced positively charged at the surface and lesser underneath the surface; this is a diffusely charged region called the space-charge region. This has the effect that the conduction band and the valence

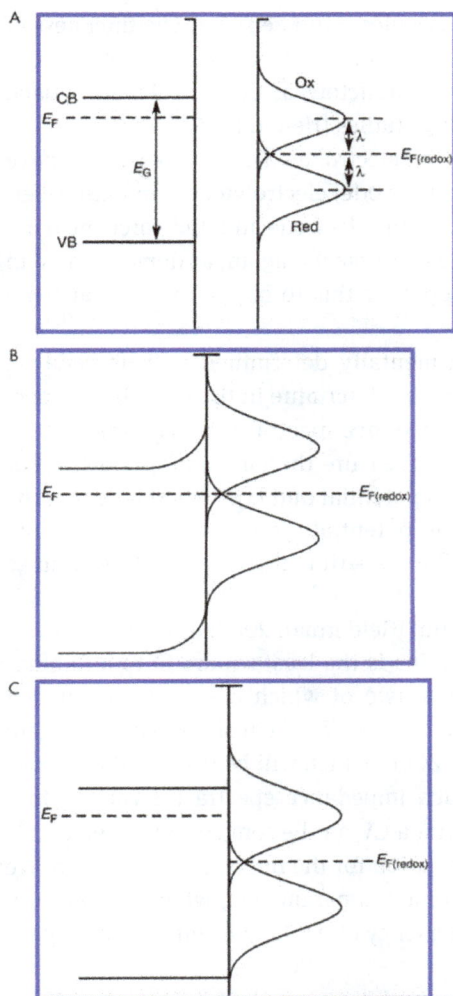

Figure 2.30: (a) Schematic of an n-type semiconductor showing the valence and conduction bands (VB and CB, respectively), Fermi level (E_F), band-gap energy (E_G), and the redox states in solution (Ox and Red), with their corresponding Fermi level ($E_{F(redox)}$) and solvent-reorganization energy (λ). (b) Electronic equilibrium between the n-type semiconductor and redox couple in solution. (c) Situation when the semiconductor is at its flat-band potential V_{fb}.

Reprinted (adapted) with permission from Gelderman K, Lee L, Donne SW: Flat-Band Potential of a Semiconductor: Using the Mott–Schottky Equation. *Journal of Chemical Education* 2007, 84. doi: 10.1021/ed084p685. Copyright (2007) American Chemical Society [Gelderman 2007].

band move up[10] – energetically. This effect is called band bending. The thickness of the range where the charge can accumulate or deplete can be in the range of 40–80 Å for example for ZnO (a very good model semiconductor), as Gerhard Heiland [Ibach 2005] has worked out in one of his pioneering studies [Heiland 1957].

When we apply an external potential to the electrode – we assume that we have an electrochemical cell with semiconductor electrode, electrolyte, CE and also reference electrode – we can push more electrons in the electrode and thus force the band bending back and thus have the energy bands become flat again, as demonstrated in Figure 2.30c. The potential necessary to apply for this to happen is the flat band potential, V_{fb}.

The flat band potential can be experimentally determined with impedance spectroscopy [Barsoukov 2005]. Practically, you determine in the impedance spectrum a surface capacity. Then, you record impedance spectra over an entire range of DC bias potentials that you apply while you measure the impedance spectra. For every impedance spectrum, you select a bias potential and keep it fixed during the measurement. In order to control this bias potential, you need to monitor the potential with a reference electrode. The experiment needs a FRA and a potentiostat.

Figure 2.31 shows an impedance spectrum [Gelderman 2007] of ZnO which was recorded with +0.8 V versus SCE. The solid line is the least square fit to a Randles circuit [Randles 1952] with four components, two of which are a charge transfer resistance R_{CT}, in parallel with a capacity C (actually a constant phase element Z_{CPE}). Deconvolution of the spectrum yields ZCPE, which will be used further below. We need to measure a whole range of such impedance spectra for various bias potentials. It is therefore necessary to first run a CV on the semiconductor electrode to see where the interesting range of investigation for the impedance spectra is. We are probing here the system with two experimental parameters, which we can control, this is, the DC bias potential and the frequency of the AC current that we impose with the FRA.

The spatial distribution of charge carriers in the semiconductor electrode follows the Poisson equation. The energetic distribution follows the Fermi–Dirac statistics. It is possible and allowed to simplify this by a simple Boltzmann distribution. Together

---

10 The bands "move up" energetically, not spatially. There is sometimes the misunderstanding about deep states in condensed matter in this context. These states are not located somewhere deep in the electrode underneath the electrode surface somewhere. Deep means that they have low energy with respect to the Fermi level E_F. This is not withstanding that there may be spatial regions in the electrode material where electrons may have a higher or lower energetic level. The band bending is an energetic band bending but it occurs with respect to the energy levels at the surface where the electrolyte is met, and with respect to the energy levels in the bulk of the electrode. Consequently, there is a region beneath the surface where the band bending takes place, as illustrated in Figure 2.30b.

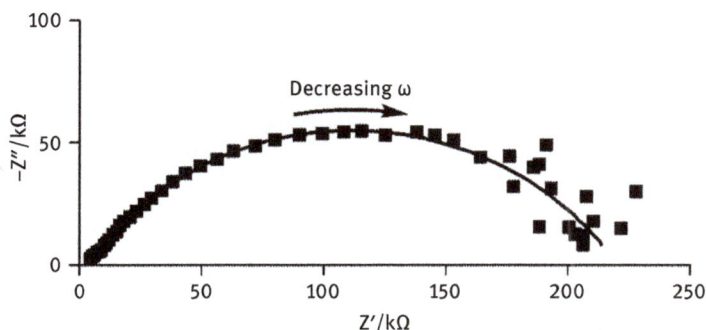

Figure 2.31: Typical EIS response for ZnO immersed in 7×10^{-4} M $K_3[Fe(CN)_6]$ (+0.8 V vs SCE). Reprinted (adapted) with permission from Gelderman K, Lee L, Donne SW: Flat-Band Potential of a Semiconductor: Using the Mott–Schottky Equation. *Journal of Chemical Education* 2007, 84.doi: 10.1021/ed084p685. Copyright (2007) American Chemical Society [Gelderman 2007].

with Gauss' law, one can solve the Poisson equation and arrive at the Mott–Schottky equation [Gelderman 2007], which relates the flat band potential V_{fb} with the bias potential V, surface capacitance C and temperature T:

$$\frac{1}{C^2} = \frac{2}{\varepsilon \varepsilon_0 A^2 e N_D} \left(V - V_{fb} - \frac{k_B T}{e} \right)$$

When you plot the experimentally derived capacity values C as $1/C^2$ versus the bias V, you obtain the Mott–Schottky plot from which you derive as the invariants the charge carrier concentration N_D and the flat band potential V_{fb}, as exercised by Gelderman et al. in Figure 2.32.

Heiland has shown in some of his works how the oxygen adsorbed on ZnO and its exposure to radiation can alter the electric conductivity [Heiland 1952, 1954, 1955], and how electronic states in semiconductors can cause electric compensation which affects and explains the transport properties of the semiconductor [Heiland 1978].

Small changes in the stoichiometry of the semiconductors can cause noticeable, considerable and significant changes in their transport properties and optical properties, for example. Their influence on the electrochemical properties can be investigated to some extent with Mott–Schottky analysis. We will see later in this book how information from this analysis can be linked with the density of defect states obtained with X-ray and electron spectroscopy.

There is a beautiful method which allows for probing the EDL and the solid–liquid interface in general with an X-ray method called X-ray standing waves [Hussain 2016, Materlik 1984, 1987, Zegenhagen 1991, 2013]. The method allows for the determination of the crystallographic structure of the solid surface and adsorbates and also for their chemical analysis at the same time.

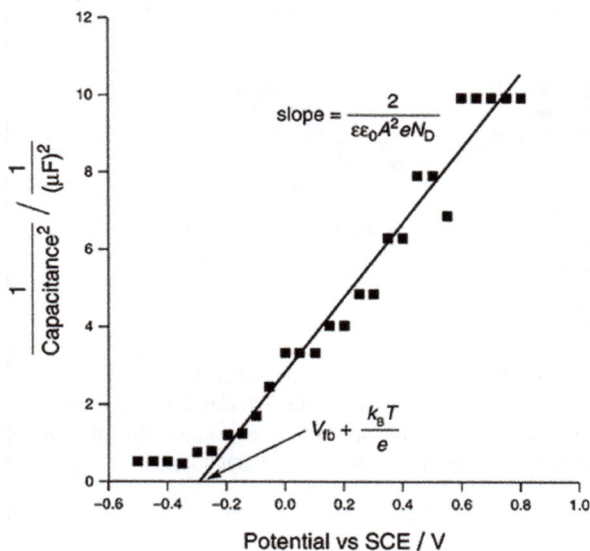

Figure 2.32: Mott–Schottky plot for ZnO in 7×10^{-4} M $K_3[Fe(CN)_6]$ (1 M KCl).
Reprinted (adapted) with permission from Gelderman K, Lee L, Donne SW: Flat-Band Potential of a
Semiconductor: Using the Mott–Schottky Equation. *Journal of Chemical Education* 2007, 84.doi:
10.1021/ed084p685. Copyright (2007) American Chemical Society [Gelderman 2007].

XPS is often used for surface analysis, a method very relevant for this chapter. But it is not so trivial to investigate an electrode surface with XPS while it is in contact with electrolyte [Bökman 1992, Braun 2017a, b, Shchukarev 2006a, b, 2008].

The aforementioned "small changes in stoichiometry" are very important for engineers who develop materials for energy converters. Changing the precursors for electrode and electrolyte synthesis, changing the processing methods for components can introduce such small minute changes which we may find ridiculous but they can affect the electronic structure in the bulk and on the surface.

Too often technician and engineers in industry, in factories make negative experiences with their formulations and processes and then do not know why this is so. Sintering an electrode in oxygen rich or in reducing atmosphere can have drastic consequences. It can take many years before they know what parameter, sometimes hitherto unknown or identified, corrupted their product or made it prone to failure.

3 The electrochemical cell

In the naïve understanding, the electrochemical cell is a container that includes two electrodes and an electrolyte so that at least one electrochemical reaction can take place in that cell. The fuel cell is an electrochemical cell. Batteries are electrochemical cells. When you want to do electrochemical experiments in the laboratory, you need a container in which you can place electrodes, electrolytes and other necessary components. This also would be called as an electrochemical cell. Practically it is therefore a special kind of chemical reactor vessel or container. Also, the electrochemical cell is a theoretical concept of any environment that can be reduced and idealized to a limited space where electrochemical reactions can take place. Much of the theory of electrochemistry becomes clear once we understand how an electrochemical cell looks like.

At some point, engineers want to increase the size of electrochemical cells because then they can be used for the production of large amounts of materials. Then, we would call the cell as "electrochemical reactor." Industrial electrolyzer plants use such reactors. The cells can also be used for the storage of large amounts of energy. A car starter battery would fall in this category, but also large storage units for residential home and industrial scale applications. And electrochemical cells can be scaled up for the production of large amounts of energy, such as fuel cells in a small portable device or utilities, or fuel cells in cars, trucks, busses and large-scale fuel cell plants. These fuel cells produce electricity on demand by electrochemical conversion of fuels with an oxidant.

And I want to mention several extremes. Biological organisms such as plant cells can be considered as electrochemical cells because electrochemical reactions take place in plant cells (or mammal cells). Their architecture warrants that electric charges such as electrons, protons, electron holes and ions and polarons can move the way it was designed for.

To some extent, our whole planet bears similarities with an electrochemical cell. Electric charges flow in the inner regions of the Earth and between Earth and atmosphere, where to some extent electrochemical reactions (ionization) take place when we consider lightning and polar light as effects that arise from electrochemical processes.

3.1 The lemon battery

The simplest electrochemical cell that I can think of is the "lemon battery." But this lemon battery is not a "battery". It is just an electrochemical cell. A "battery" is a set of cells, a collection of cells. We will see this in the next Chapter. How does a lemon battery work? Figure 3.1 shows you a battery of four lemon fruit, in each of which a copper penny and a zinc nail are inserted. The citric acid in the fruit (maybe assisted by some amino acids) will dissolve Cu atoms and Zn atoms so that we have an

https://doi.org/10.1515/9783110561838-003

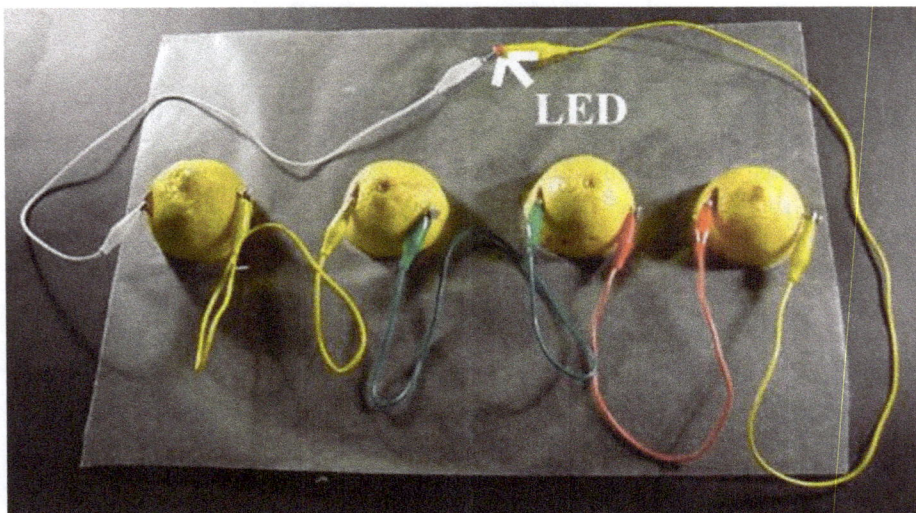

Figure 3.1: Four half lemons in which a copper penny and a zinc nail are pressed into, connected in series with cables and connected to a light emitting diode (LED).
Reproduced with kind permission from Hila Science Centre, Canada [Hila 2018] and http://www.platform21.nl/id/3753.html.

electrolyte containing Cu^{2+} and Zn^{2+} cations. The copper penny is the plus pole, and the zinc nail is the minus pole.

In the beginning, not the entire juice volume is filled with Cu and Zn ions. Only a thin region near the metal electrodes will contain the Cu and Zn ions, respectively. The two electrodes constitute a redox couple which will yield practically an electromotive force (EMF) of around 0.9 V. It is hardly enough voltage to light a small light bulb and it works better for light emitting diodes (LED).

The EMF is determined by and specific to the chemical constituents, which form the redox couple. The references [Atkins 1997, 2010, Aylward 2008, Bard 1985, 2001, Connelly 1996, Cotton 1999, Courtney 2014, Greenwood 1997, Leszczynski 2013, Lide 2006, Pourbaix 1966, Vanýsek 2011, Winter 2018] show numerous half-cell reactions and redox couples with the corresponding standard potentials. We will see later how the voltage of a cell is built up from the standard potentials of the participating electrode materials.

When we need a higher voltage, we can connect several lemons in series, as is shown in Figure 3.1. This should yield 3.5–4 V. This could be too much voltage for some delicate bulbs which can only sustain 1.5 V, for example. Connecting the lemon cells in parallel will yield the same low voltage of one lemon but the current would be quadrupled. And it seems you can upscale the battery without adding more lemons, but by just pushing in more nails and more pennies in series, or in parallel, depending whether you want to increase the voltage or the current.

3.2 The Daniell element

What is going on inside such battery element? The archetype battery cell is the Daniell element [Daniell 1839]. Let us take it for its conceptual simplicity, as shown in Figure 3.2. We have on the left a container which contains an aqueous $ZnSO_4$ solution as electrolyte, in which a Zn slab is dipped. This is a situation similar as shown in Figures 2.1 and 2.2. The Zn ions in solution are in equilibrium with the Zn on the slab. On the right side, we have another container with a copper slab dipped into a $CuSO_4$ solution. Here is also equilibrium between solvated Cu^{2+} ions and Cu in the slab. The chemical potential of the Zn slab, however, is different from the potential of the Cu slab. Their potential difference amounts practically to a galvanic voltage. The Daniell element is thus a galvanic cell.

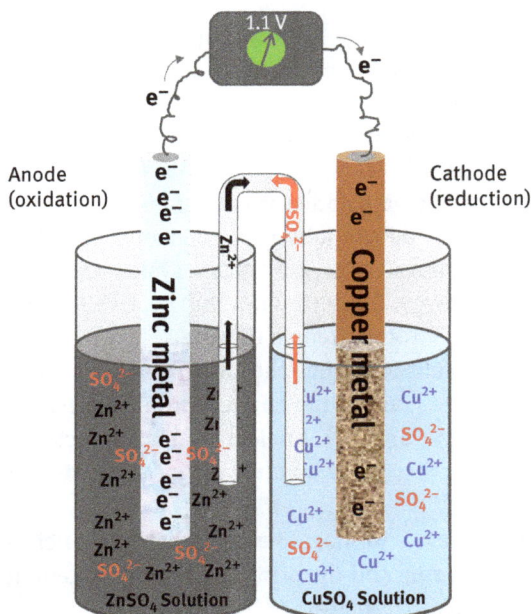

Figure 3.2: Daniell element. A zinc and copper metal rod are inserted in zinc sulfate and copper sulfate electrolyte and yield a potential difference (voltage) of 1.1 V. A salt bridge combines both electrolytes and closes the electric (electronic and ionic) circuit and allows for electric current flow.

As zinc has a lower standard (reference) potential ($E_0(Zn) = -0.76$ V) than copper ($E_0(Cu) = +0.34$ V), a larger concentration of Zn ions are being dissolved into the electrolyte, and a lower concentration of Cu ions are being dissolved in the copper sulfate. Copper has therefore effectively less electrons left in its bulk than zinc. When we now connect the Zn and Cu slab with a cable (an electronic conductor), we will measure a potential difference of $+0.34$ V $- (-0.76)$ V $= +1.1$ V, as listed in Table 3.1.

Table 3.1: Half-cell reactions, net reactions and corresponding electro-chemical standard potentials and net voltage of the redox couple.

Reaction steps	Location	Potential
$Cu^{2+} + 2\,e^- \Leftrightarrow Cu$	Anode; reduction	+ 0.34 V
$Zn \Leftrightarrow Zn^{2+} + 2\,e^-$	Cathode; oxidation	−0.76 V
$Cu^{2+} + Zn \Leftrightarrow Cu + Zn^{2+}$	Cell reactions	+ 1.10 V

When we close the electric circuit by connecting the electrolytes in the containers with an ionic conductor (typically referred to as a salt bridge), the excess electrons in the Zinc slab can move through the wire to the Copper slab. These electrons in the copper help chemically reduce the Cu^{2+} ions in the electrolyte and deposit on the copper slab surface. This is called cathodic deposition. In the Zn container, the Zn atoms are meanwhile going to be dissolved. As the system is striving for electric charge balance, sulfate $SO_4{}^{2-}$ will move from the copper sulfate solution through the salt bridge to the zinc sulfate solution. At the same time, Zn^+ will move to the copper container. Thus, the mass balance and the charge balance are maintained.

The two parts of the redox reaction can be spatially separated in a cell. The electrons will then not move directly from the system Zn/Zn^{2+} to the system Cu/Cu^{2+} through the cell container. Rather, the electrons move through the metal wire from the zinc rod to the copper rod. Hence, an electron current is flowing.

Both separated systems are called half cells. Overtime the zinc rod will dissolve; this is called corrosion. In as much does the mass of the copper rod increases. We have thus two redox couples: (Zn/Zn^{2+} and Cu/Cu^{2+}). In electrochemical technology, the arrangement of the electrodes in the cell is written in the cell diagram notation: $Zn/Zn^{2+}//Cu^{2+}/Cu$.

We can demonstrate on the number beam with negative and positive branches that how the negative standard potential of zinc and the positive standard potential of copper add up to 1.10 V. It is the potential difference −0.76 V − (+0.34 V) = −1.10 V, which "add up" to the cell voltage. Just "adding" the potentials − 0.76 V + 0.34 V = − 0.42 V leads to the wrong cell voltage. See how this is done in Figure 3.3.

The Daniell element is a primary element. It is designed for the one-time use in a battery operation and then discarded when the zinc is used up. It would be worthwhile, however, to be able to recharge the cell and so to be able to start over again with new 100% charge. The electricity necessary for that could come from a charging station at home or from solar cells anywhere outside.

The problem is that the chemical changes in the zinc are accompanied by irreversible structural changes. The integrity of the electrode suffers to an extent that the cell is not usable anymore when used up. Yet, battery engineers are trying to control the operation and structure of the cell in a way that it is possible to recharge

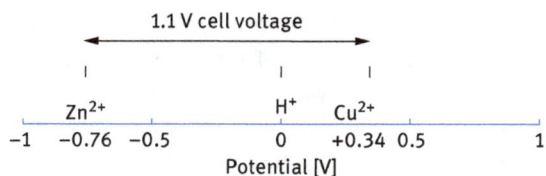

Figure 3.3: The potential scale extends from negative to positive potentials. The negative potential of Zn^{2+} (−0.76 V) and Cu^{2+} (+0.34 V) add up to 1.10 V cell voltage. Hydrogen with 0 V is the reference potential.

the battery after use. This requires analytical studies which are not yet considered as state-of-the-art.

Nakata et al. [Nakata 2015] have developed an X-ray spectroelectrochemical cell, which can host a Daniell element of Cu and Zn and which allows for the recording of x-ray fluorescence spectra of the electrochemically active components in the cell during operation with a high spatial resolution. It is thus possible to record with a charged coupled device (CCD) camera images of the electrodes and follow the structural disintegration and reformation.

3.3 Reference cells

This book is about electrochemical energy systems. Their purpose is to provide electric power for systems and for individuals. Usually, we want as much power as possible, hassle-free, anytime and for the lowest price. The batteries and fuel cells are therefore optimized for efficiency and stability.

Now I will present electrochemical cells that were not designed as power sources. The two cells that I describe below were designed as reference cells [Hoffman 2016] for analytical purposes. I do not want to imply here that the purpose was only for academic reasons. In the further development of electric engineering 150 years ago, it became necessary to accurately determine EMF. The two researchers mentioned below who built such reference cells were working on real-life systems in an engineering and business environment. It is therefore not surprising that the topic of reference cells or standard cells was summarized in a monography [Hamer 1965] by the US National Institute of Standards and Technology (NIST), which is part of the Department of Commerce.

3.3.1 The Clark cell

In the early times of electrical engineering, the need for stable reference power sources arose. Josiah Latimer Clark (10 March 1822–30 October 1898) was a British engineer who worked on the "connectivity of people," specifically bridges, tube mail and telegraphy. In the latter field, he researched on the delay of the electric signals in undersea cables for telegraphy. Being an engineer, he was interested in standardization of physical units and introduced the units Farad, Ohm, Volt and Watt. There had been difficulties in finding a material standard for the determination and measurement of the EMF and the resistance which is involved with it. The Daniell element provided a voltage that typically varied for no apparent reason by 5%. This is why Clark spent time and efforts in finding a more stable electric power source, a "normal element."

Clark [1872] explains in his paper the motivation for his work

> The Daniell's element, which has been most frequently used for this purpose, commonly varies five per cent or more without apparent cause.
>
> From a conviction that if similar conditions could be ensured similar combinations would always give the same electromotive force, the author was led to institute a series of experiments, extending over four years, which led to the discovery of a form of battery that is sensibly constant and uniform in its electromotive force.

and the technical details how he prepared the electrodes. He refers to the electrodes as "negative element" being pure mercury and the "positive element" being pure zinc. He needed some practical tricks with preparing pastes from mercurous sulfate and zinc sulfate to contain the liquid mercury. A glass container contains the entire cell, and electric contact with the liquid mercury inside is made via a platinum wire pushed through the solid sulfate.

When we consider the architecture of the aforementioned lines (the reader is advised to read into Clark's original paper which is publically available now as open access), we can write it up in following scheme of active components:

$$Zn|ZnSO_4|Hg_2SO_4|Hg$$

Now we can further "spread it out" and see the minus pole is at the solid zinc electrode, which is in contact with crystalline zinc sulfate, which is in contact with saturated zinc sulfate solution (the electrolyte), next to another concentrated zinc sulfate solution. There we have the solid mercury sulfate in contact with the liquid mercury plus pole. The mercury sulfate is a depolarizer, which lowers the overpotential to the mercury:

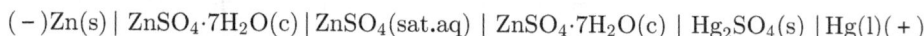

$$(-)Zn(s) \mid ZnSO_4 \cdot 7H_2O(c) \mid ZnSO_4(sat.aq) \mid ZnSO_4 \cdot 7H_2O(c) \mid Hg_2SO_4(s) \mid Hg(l)(+)$$

At 15 °C, the open circuit potential (OCV) of the Clark element is 1.4328 V. The cell is particularly stable when only low currents are drawn. It is the general requirement that a reference cell delivers no or little current. Clark [Clark 1872] writes in his original paper:

> The element is not intended for the production of currents, for it falls immediately in force if allowed to work on short circuit.

Table 3.2 lists the EMF values that Clark has determined with an electrodynamometer over a period of 2 weeks before Christmas in 1871. The averaged EMF determined with

Table 3.2: Electromotive force E measured by Clark [Clark 1872] with his normal element in the days between December 8 and 21, 1871.

By the electrodynamometer		
Date	Value of E (V)	Remarks
8 December 1871	1.4583	3 cells.
9	1.4651	3 cells.
14	1.4616	3 cells.
15	1.4561	3 cells.
15	1.4579	2 cells.
16	1.4586	3 cells.
16	1.4517	3 cells, coil turned 180°.
16	1.4552	3 cells, coil turned back 180°.
16	1.4555	3 cells.
16	1.4535	2 cells.
16	1.4564	3 cells.
18	1.4649	3 cells.
19	1.4562	3 cells, coil turned 180°.
19	1.4558	3 cells, coil turned back 180°.
20	1.4615	3 cells.
20	1.4539	3 cells.
20	1.4551	2 cells.
21	1.4549	3 cells.
Mean	1.45735	Temperature 15.5 °C

this method was 1.45735 V. You can see from the right column in the table that Clark used several cells and checked for potential influence of the orientation of the wire coil of the electrodynamometer.

Clark repeated the experiments two months later with a sine galvanometer and obtained 1.45621 V as mean value. Again, Clark made consistency tests by changing experimental conditions with respect to the galvanometer. I have plotted the data in graphical representation in Figure 3.4 to check how the trend for the EMF evolves overtime. Figure 3.4 is divided into two panels, the left of which shows the data from the electrodynamometer recorded in December and the right side shows the EMF values measured with the sine galvanometer.

We can see globally that the EMF values are slightly decreasing. At second glance, it appears that the EMF is slightly increasing around the weekend and decreasing during weekdays. The trend seems to have a fine structure with the weekends. For that purpose, I have specifically highlighted the Saturdays and Sundays so that you can see for yourselves [Timeanddate 2018]. Weekends are generally highlighted because no work is done on these days.

Obviously, Clark (or one or more of his associates) worked on Saturdays and when we look into the second phase of the experimental campaign in February 2017, he also worked on Sundays. A passionate researcher may sometimes break with the cultural conventions for the sake of science. On weekends, the central heating (if there was any in 1871–72) may be turned down to safe energy costs. On weekends, there may be fewer people in the building and thus the humidity and CO_2 concentration may be different from those days where people are working and thus present in the building.

As we inspect the right panel in Figure 3.4, we see that the suggested trend of increase of EMF over the weekend is indeed recovered, whereas the global trend of decrease of the EMF over the two weeks experiment is maintained. Clark reports that for both parts of the campaign, the temperature was 15.5 °C. I therefore cannot further speculate why the EMF follows a periodic pattern where the weekend plays a role in Clark's studies. In the last chapter of this book, I make some general comments on reproducibility of experiments which should be considered if you in your own studies and research encounter "irregularities." Often there is a practical reason for such, and sometimes you may find out actually where such irregularities originate from. When an experimenter reports "data from week such and such are missing because the potentiostat got damaged" or, "in the first half of August I was on vacation, this is why no data are recorded for these two weeks," then I trust such data.

Later in this chapter about batteries, I will explain how lithium metal would corrode in ambient air on my desk either to white fluffy LiOH or to black brittle Li_3N. The different degree of humidity in my office and lab would be the cause for the different "oxidation" behavior.

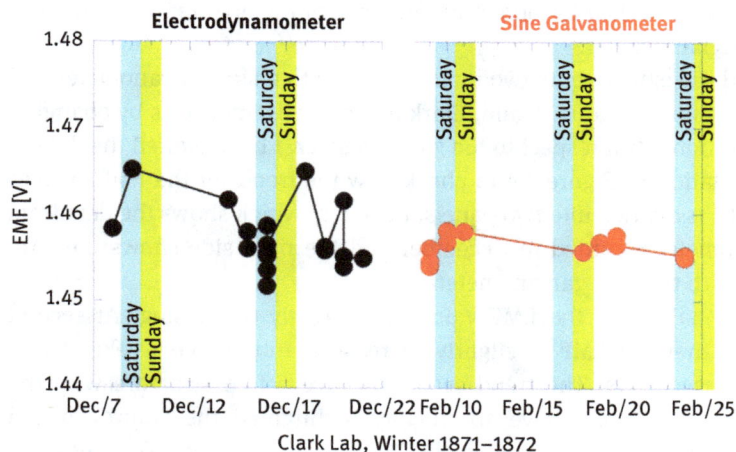

Figure 3.4: Variation of the EMF of Clark's normal cell in the period of 8–21 December 1871, and 9–24 February 1872, in his laboratory. Data taken from Clark's original work [Clark 1872]; see Table 3.2. The data were recorded at 15.5 °C.

3.3.2 The Weston cell

While Clark's work was material for the development of references and a reference cell, it was another researcher who actually established the reference cell. Edward Weston (9 May 1850–20 August 1936) was a physician from Great Britain who emigrated to the USA right after graduation in medicine. A young immigrant in a dynamic new world, he found a job in the new established electroplating industry and became entrepreneur in the electric power and lighting industry almost as illustrious as Thomas Edison.

For the development of the Weston cell, Weston [Weston 1893] built on the success of Clark and used mercury as the negative electrode but used cadmium as positive electrode and cadmium sulfate as electrolyte. Following the same advice given for the Clark cell, Weston amalgamated the positive electrode with mercury. Weston received the patent on his invention, a "voltaic cell," in 1893. He waived the rights on his patent[1] in 1911 when the Weston cell became the International Standard for the EMF.

The anode reaction reads

$$Cd_s \rightarrow Cd^{2+}(aq) + 2e^-$$

The cathode reaction reads

$$(Hg^+)_2 SO_4^{2-}(s) + 2e^- \rightarrow 2Hg(l) + SO_4^{2-}(aq)$$

In analogy to the Clark cell, the negative electrode is amalgamated with mercury. The electrolyte is the sulfate of the cadmium. Mercurous sulfate is used as depolarizer. Weston has improved the design of the cell by moving from one compartment to two separated compartments for the two half reactions in an H-shaped glass vessel.

One advantage of the Weston normal element is its stability over a wide temperature range. This is important because normal elements with a considerable temperature dependence would deliver different values for the "normal" EMF when the measurement is made at locations with different temperature. Imagine the temperature in your own laboratory on a Monday morning and on a Friday afternoon. They may be different. When your laboratory is a modern climatized room, then they

---

1 On rare occasions, patent holders waive their rights. In my book on X-ray methods, I presented two examples [Braun 2017b] Braun A: *X-ray Studies on Electrochemical Systems - Synchrotron Methods for Energy Materials*. Berlin/Boston: Walter De Gruyter GmbH; 2017. Konrad Röntgen did not patent his discovery of x-rays because he felt a patent would block the further development of x-ray technology. A more recent and prominent example is that of Tesla-founder Elon Musk who announced on Twitter *"all patents are belong to you"* [Musk 2014] Tesla Motors 2014 [https://www.teslamotors.com/de_CH/blog/all-patents-are-belong-you], for the same reasons like Röntgen's.

may be the same. When not, then a thermometer may help you to at least determine the actual temperature.

When you find that results from an experiment differ, the different temperature could be the reason for that. This is why often in scientific publications, the temperature during the experiment is disclosed in the "Experimental" section of the paper. This is also why it can be sometimes important to not only *measure* the temperature but also *control* the temperature. Here it is important that the thermometer is at the same position where the experiment is made, preferably at sample position.

The EMF of the original Weston normal element is 1.018638 V. This value is "lower" than the EMF of the Clark normal element, but the amplitude of the EMF of a normal cell is not important here. Important for a normal element is the reproducibility and stability of the EMF.

It has turned out that EMF of the Weston normal element does not vary strong with temperature at about ambient temperature. This is important for practical laboratory applications. A Taylor expansion for the range from 0 °C to 40 °C is given by

$$E_{(T)}[V] = E_{(T=20°C)}[V] - 0.0000406 \ (T-20) - 0.00000095 \ (T-20)^2 \\ + 0.00000001 \ (T-20)^3$$

3.4 Faraday's law

The relation between the converted mass and the electric charge was derived by Michael Faraday [Faraday 1834]. A derivation of Faraday's law as we use it today can be found in Strong [Strong 1961] and Ehl [Ehl 1954].

When Dr Faraday carried out his "Experimental researches in Electricity", he did not provide modern mathematical equations but instead explained the proportions of physical quantities literally (§ 13 On the absolute quantity of Electricity associated with the particle or atoms of Matter in Faraday [Faraday 1834]). Frederick C. Strong checked at the book exhibition at the 136th American Chemical Society (ACS) Meeting in September 1959 in Atlantic City NJ [Archives 1959] several dozens of books for their wording of Faraday's Law of Electrolysis, which states the corresponds of used electrons and electrochemically converted mass, with a necessary constant or proportionality which is known as Faraday's constant of 96489.9 C.

Use of Faraday's constant actually allows phrasing Faraday's law as the [Strong 1961] "statement that 96489.9 Coulombs passed through an electrolytic cell will cause one gram equivalent weight (it should be called equivalent mass) mass of material to react or form at an electrode." Strong sees that the two different forms of Faraday's law contain three variables, that is, the electric charge q as the amount of electricity, the amount of material with mass m and the gram equivalent weight m_e.

Faraday stated that q is proportional to m when m_e is constant, whereas the second law states that when q is constant, m is proportional to m_e. Strong writes this as

$$q \propto \frac{m}{m_e}$$

Strong writes the proportionality then as an equation as he introduces a constant

$$q = \frac{m}{m_e} \times \text{constant}$$

which he calls F for Faraday's constant. He writes the ratio of m/m_e as the number of equivalents n, so that Faraday's law simply reads

$$q = nF$$

with the definitions

$$n = \frac{m}{m_e} \text{ and } q = \int_{t_1}^{t_2} i\, dt$$

"Faraday took great pains to distinguish between *quantity* and *intensity* of the electric current.[2] He used the words carefully and correctly and was seemingly well aware of the distinguishing characteristics and effects of each. This is a matter of no small importance, for it was precisely confusion on this point which helped to lead Berzelius astray [Ehl 1954]:"

> When electrochemical decomposition takes place, there is great reason to believe that the quantity of matter decomposed is not proportionate to the intensity, but to the quantity of electricity passed. [Faraday 1833].

Observe that we have here an interesting analog from modern physics to this observation and proposition by Faraday: The photoelectric effect [Hertz 1887] is the emission of electrons from atoms once excited with photons (ultraviolet radiation or x-rays or even γ-rays) of sufficient energy. It is not the intensity of the radiation, which causes the emission, but the wavelength λ, which is related with their frequency v and energy E_{photon} via the relations $E_{photon} = hv = hc/\lambda$ of the photons [Hallwachs 1888]; see [Braun 2017a].

---

2 Here is an analogy to the photoelectric effect [Einstein 1905] Einstein A: Generation and conversion of light with regard to a heuristic point of view. *Annalen Der Physik* 1905, 17:132–148. [Hertz 1887] Hertz H: Ueber einen Einfluss des ultravioletten Lichtes auf die electrische Entladung. *Annalen der Physik und Chemie* 1887, 267:983–1000.doi: 10.1002/andp.18872670827. For the creation of a photoelectron by excitation of an atom with light, it is not the number of photons, that is, the light amount that counts. Rather, the photon has to have a sufficient high energy $E=hv$ to be able to eject an electron from the atom.

Nowadays, Faraday's law is written (with constants and parameters listed in Table 3.3) as

$$m = \left(\frac{Q}{F}\right)\left(\frac{M}{z}\right)$$

Note that M/z is the same as the equivalent weight of the substance altered.

Table 3.3: The constants and parameters used in Faraday's law.

m	m is the mass of the material (chemical element) liberated at an electrode in grams
Q	Total electric charge passed through the material in Coulomb
F	$F = 96485$ C/mol is the Faraday constant
M	Molar mass of the substance in g/mol
z	Valence number of ions of the substance (electrons transferred per ion).

For Faraday's first law, M, F and z are constants, so that the larger the value of Q, the larger m will be. For Faraday's second law, Q, F and z are constants, so that the larger the value of M/z (equivalent weight), the larger the mass m will be.

In the case of electrolysis at constant current I, the electric charge Q

$$Q = I \cdot t$$

leading to the converted mass m

$$m = \left(\frac{It}{F}\right)\left(\frac{M}{z}\right)$$

and then to

$$n = \left(\frac{It}{F}\right)\left(\frac{1}{z}\right)$$

where n is the amount of substance ("number of moles") which is liberated: $n = m/M$ and t is the total time the constant current was applied.

For a battery, for example, we determine the total electric charge Q by integrating the current over time:

$$Q = \int_{0}^{t} I(\tau)d\tau$$

Faraday's law is known as one of the four Maxwell equations. It appears that the fundamental relationship between electric charge and converted mass was discovered by Carlo Matteucci [Matteucci 1847], as was communicated by no lesser than Michael Faraday himself for the Royal Society:

We admit as clearly demonstrated by experiment that electro-magnetic, as well as electro-chemical action, give the measure of the electric current; in other words, that different quantities of electricity produce chemical and magnetic effects proportional to these quantities.

Carlo Matteucci, who is considered the scientific heir of Galvani [Moruzzi 1996], used to work on animal electricity and is considered as the founder of electrophysiology [Wade 2011]. He discovered the relation between electric charge and converted mass independently from galvanic experiments. Matteucci was the discoverer of the so-called action potential in electrophysiology [Moruzzi 1996, Schiff 1876].

What is Faraday's law useful for? Consider the Daniell element. Here you have an anodic dissolution of zinc and a cathodic deposition of the copper on the copper rod. This means there will be a weight loss of the zinc rod and a weight gain by the copper rod. With a multimeter, you can monitor the current through the Daniell element overtime, which allows for determination of the charge converted Q. With the known equivalent weight, we can quantify the mass changes on the zinc and copper rods, for example. This method has also been applied for *operando* lithium battery studies with X-ray spectroscopy at the synchrotron [Braun 2001, 2003].

3.5 The electrochemical equivalent

In Faraday's law, we learnt about the proportionality between electric charge and converted mass in an electrochemical reaction. It is of technical and economic interest that how much material is deposited when an electric charge of 1 Coulomb (1 C = 1 Ah) is working on it. This quantity is called the electrochemical equivalent:

$$Eq = \frac{M}{z \cdot F}$$

and tells you the maximum possible yield of the process. The origin of the electrochemical equivalent comes from electroplating and galvanization, but it has relevance for the industrial production of aluminum and chlor-alkali syntheses, and the electrolysis of water. For this relation and formula to apply, you need to know the elements, which are involved in the chemical reaction, specifically the molecular mass, and you need to know how the valence of the elements changes. For aluminum production, for example, you have to consider that the valence changes from Al^{3+} to Al^0.

For the electrochemical oxidation or reduction of 1 mol of an element by one valence unit, an electric charge of 96485.336 As/mol (Faraday's Constant) is necessary. This is a theoretical value. The actual necessary charge for deposition, dissolution, or synthesis of an element can be larger than this value because of parasitic side reactions. A summary for electrochemical equivalents for frequently occurring elements is shown below in Table 3.4.

Table 3.4: Selected data and examples for the electrochemical equivalent Eq. after [John 2007].

Table for the electrochemical equivalent of some selected chemical elements

Element	Molecular mass (g/mol)	Change of Valence	Eq (µmol/As)	Eq (g/Ah)	Examples in technology
Hydrogen	1.0079	1 ↔ 0	10.364	0.0376	Water electrolysis
Oxygen	15.999	2 ↔ 0	5.1821	0.298	
Fluorine	18.998	1 ↔ 0	10.364	0.709	Fluor production
Sodium	22.990	1 ↔ 0	10.364	0.858	Natrium production
Aluminum	26.981	3 ↔ 0	3.4548	0.336	Aluminum production
Chlorine	35.451	1 ↔ 0	10.364	1.32	Chlor-alkali electrolysis
Chromium	51.996	6 ↔ 3; 3 ↔ 0	3.4548	0.647	
Chromium	51.996	6 ↔ 0	1.7274	0.323	
Manganese	54.938	4 ↔ 3; 3 ↔ 2	10.364	–	Zinc-manganese cells
Manganese	54.938	4 ↔ 2; 2 ↔ 0	5.1821	1.02	
Manganese	54.938	7 ↔ 4; 3 ↔ 0	3.4548	0.683	
Manganese	54.938	7 ↔ 0	1.4806	0.293	
Iron	55.845	3 ↔ 2	10.364	–	
Iron	55.845	2 ↔ 0	5.1821	1.04	
Iron	55.845	3 ↔ 0	3.4548	0.695	
Nickel	58.693	2 ↔ 0	5.1821	1.09	
Cobalt	58.933	2 ↔ 0	5.1821	1.10	
Copper	63.546	2 ↔ 1; 1 ↔ 0	10.364	2.37	
Copper	63.546	2 ↔ 0	5.1821	1.19	Copper purification
Zinc	65.409	2 ↔ 0	5.1821	1.22	Zinc-manganese cells
Rhodium	102.91	3 ↔ 0	3.4548	1.28	
Palladium	106.42	2 ↔ 0	5.1821	1.99	
Silver	107.87	1 ↔ 0	10.364	4.02	
Cadmium	112.41	2 ↔ 0	5.1821	2.10	
Tin	118.71	4 ↔ 2; 2 ↔ 0	5.1821	2.21	
Tin	118.71	4 ↔ 0	2.5911	1.11	
Platin	195.08	2 ↔ 0	5.1821	3.64	
Gold	196.97	1 ↔ 0	10.364	7.35	
Gold	196.97	3 ↔ 0	3.4548	2.45	
Lead	207.2	4 ↔ 2; 2 ↔ 0	5.1821	3.87	Lead battery
Lead	207.2	4 ↔ 0	2.5911	1.93	

Now, we have noticed the similarity of Faraday's law and the electrochemical equivalent. The former is a fundamental law which provides the important link and quantitative relationship between electric current and chemical reaction. The latter is important for the yield of the product and to some extent the cost and price on the market for that product. The yield is an economic parameter. I will give an example on that in the Chapter on Batteries which deals also with recycling of batteries.

3.6 Electrolyzer cells

3.6.1 Hydrogen fuel from water electrolysis

In January 2018, I had two meetings in Leiden in the Netherlands and in Bruxelles in Belgium, both of which were related to hydrogen. My employer had, in 2017, begun assessing the carbon footprint. In this context, the impact of international air travel (flights) on the CO_2 budget certainly plays a role. I therefore chose to not fly to the Netherlands and Belgium, but use our fuel cell electric vehicle (FCEV).

I must admit that I considered this travel with the FCEV a further opportunity for me to test whether the action radius, the range of the FCEV and the hydrogen gas station network in Europe would work for long range travel. This is not withstanding that I had made my previous two travels to Leiden in 2017 by airplane via Amsterdam; whereas, my trip to Bruxelles in September 2017 was already made with our FCEV. You know also from the previous Chapter in this book that I extended the range also to the Mediterranean Sea, specifically Venice.

On my way from Zürich to Leiden in January, I had my second hydrogen refill at the gas station from French TOTAL near Karlsruhe, which was established in September 2017 [Herdlitschka 2017]. When I was finished with pumping and taking photos of the gas station, I met two gentlemen who had just exited the hydrogen installation, which is present at every hydrogen gas station. I went to them and learnt they were engineers from the European Institute for Energy Research (EIFER) in Karlsruhe, who had made some maintenance visit to the station. They told me the station produced the hydrogen on-site using a high temperature solid electrolyte electrolyzer from sunfire GmbH in Dresden, Germany. The power for the electrolysis would come from the solar panels on the roof of the TOTAL gas station or from the electric grid. Well this was interesting news to me. We will see later in this book how such solid electrolyte cells work, but in a slightly different context.

A snapshot of the current electrolyzer activities in industry in early 2017 [Geitmann 2017a] was provided by Sven Geitmann, a publisher (HZwei) who also runs a blog on hydrogen and fuel cells. As hydrogen storage is taking center stage, the production of hydrogen by electrolyzers is following. Since November 2017, electrolyzers which produce hydrogen on-site for public gas stations are eligible for receiving government subsidies in Germany to the extent of 60% of the investment costs, writes Geitmann [Geitmann 2017b]. Currently, a H_2 gas station costs around 1 Million Euros. The electrolyzer of such station can thus be subsidized to 60%, which is a substantial support. For the Karlsruhe hydrogen station, the government support was 970000 Euros [Herdlitschka 2017].

3.7 Water electrolysis

Electrolysis is when you use an electric current with a voltage large enough that can split molecules. You can try this at home with two 1.2 V batteries (which makes 2.4 V) that you connect in series; connect them with two metal nails and put them into water that contains some table salt. You will notice gas bubble formation. We know electrolysis as the splitting of the water molecule into oxygen gas and hydrogen gas in an electrochemical cell. It follows the chemical reaction

$$2H_2O + 2e^- \rightarrow H_2 + 2OH^-$$

which requires a substantial amount of energy, but yields then a highly valuable fuel H_2 and oxygen O_2:

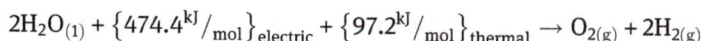

$$2H_2O_{(1)} + \left\{474.4^{kJ}/_{mol}\right\}_{electric} + \left\{97.2^{kJ}/_{mol}\right\}_{thermal} \rightarrow O_{2(g)} + 2H_{2(g)}$$

Hydrogen is a very important gas in chemical industry for many processes [Stiller 2014]. Hydrogen, at large still to date, is produced from fossil fuels [Gaudernack 1998] by steam reforming of water over a hot carbonaceous reservoir. The chemistry of the process is governed by the watergas shift reaction and the Boudouard equilibrium. This process is based on the availability of carbon and hydrocarbon fuels and thus not necessarily a sustainable and environmental-friendly process. We read about this later in this book.

When you carry out the electrolysis with only one 1.2 V battery, likely you will not notice any electrolysis. This is because 1.23 V which are thermodynamically sufficient to split the water molecule are practically not sufficient for the oxygen evolution reaction (OER). There are kinetic barriers, which keep the aforementioned reaction from proceeding. To overcome these barriers, either a higher voltage is necessary, or an electrocatalyst is necessary, which lowers the overpotentials.

The electrocatalysts that are well known are, for example, platinum or iridium oxide. Often it is very costly noble metals. It depends on the type of reactor type, components and materials including the electrolyte which kind of electrocatalyst is necessary or suitable. A comparison of various electrocatalysts for the OER are find in the review of McCrory et al. [McCrory 2013].

3.7.1 Water electrolysis from renewable energy

Using green electricity from an electrochemical cell appears to be a more sustainable route for hydrogen production [Fang 2015] then producing hydrogen from fossil fuels that power the water steam reformation. When all hydrogen is produced with green energy by electrolysis, then the chemical industry (power to X) and not only the transportation sector will become greener as well (power to gas) [Lehner 2014].

In 1890, Danish teacher and inventor Poul La Cour developed a plant which produced electric power from a wind mill. He used the electricity to run an electrolysis plant which produced hydrogen gas from water. The gas was used for lighting in school and also for welding. His vision was the electrification of rural areas and he was therefore a pioneer of decentralized energy supply from renewable sources [La Cour 1905].

Norway, a fully electrified country, has used its rich and powerful water supplies very early to make hydropower plants. Because these produced more electric energy than could be used in the electric grid, water electrolyzers for hydrogen production were found to be a suitable solution for using the excess electric power. An array of electrolyzers in a huge electrolysis plant in Norway is shown in Figure 3.5. The hydrogen produced in the plant was used for fertilizer production.

Figure 3.5: Electrolyzer battery at the ammonia fertilizer plant in Glomfjord, Norway, operated by Norsk Hydro in the years 1953–1991. The 168 electrolyzer units manufactured by Nel produced 30,000 Nm^3 hydrogen per hour (Nm^3 = normal cubic meter; Nm^3 determined at 1.01325 bar pressure at 0 °C temperature with 0.0446158 kmol). The power requirement was 135 MW.
Reproduced from [Nel 2017] with kind permission from Nel.

3.7.2 Deuterium from water electrolysis

Some portion of the naturally occurring water contains water molecules which contain not protons but deuterons, the hydrogen isotope which has double the mass of the proton. Water with deuterons is therefore called heavy water. As chemical bonds with the heavy deuterons have a considerable lower zero point vibration energy than chemical bonds with the light protons, higher activation energy must be overcome to break this bond. The practical consequence is that during electrolysis with a well-controlled electric potential, the light water will be preferentially electrolyzed. This yields to a concentration of heavy water in the electrolyte.

To enhance the deuterium yield, one can combine the water electrolysis with chemical exchange. Rather than burning off the hydrogen, one can extract the heavier component from the hydrogen deuterium mixture of higher separation stage by catalytic exchange with the condensation water from earlier stages. At 80 °C, where a nickel or platinum catalyst warrants a significantly high reaction rate, the equilibrium constant of the exchange reaction

$$HD + H_2O \Leftrightarrow H_2 + HDO$$

yields

$$K = \frac{[H_2]\,[HDO]}{[HD]\,[H_2O]} = 2.74.$$

Thus, the heavy hydrogen isotope has a strong tendency for enrichment in water [Becker 1956]. During the Second World War, the Vemork plant in Norway could produce around 1,500 kg heavy water in one year.

Heavy water (D_2O) played a role as neutron moderator material for nuclear experiments on atomic projects in the Second World War. Heavy water can slow down the speed of neutron to a limit which is necessary to cause nuclear fission. Germany tried to accrue large amounts of heavy water from the Vemork hydropower plant in occupied Norway, which belonged to the Norsk Hydro Company. Due to sabotage in the night of 27 to 28 February 1943 by British special operation soldiers, Germany could not get sufficient supplies of heavy water [Helberg 1947]. It is still a matter of debate whether Germany was working on the atomic bomb in the Second World War, or not.

3.7.3 Three major water electrolysis concepts

There are three major technologies for water electrolysis, the
I) Alkaline electrolysis with a liquid caustic electrolyte at around 80 °C operation temperature;
II) Electrolysis with a proton conducting polymer electrolyte membrane (PEM) under acidic ambient conditions and
III) High temperature electrolysis in a solid electrolyte electrolyzer cell above 700 °C.

The type (I) electrolysis has been used commercially for over 100 years with great success. In the mid of the last century, electrolyzer with over 100 MW output power was installed at large plants to provide hydrogen for fertilizer production. The basal planes of these huge electrolyzers have areas of several square meters.

Electrolyzers from type (II) have become popular in the last 20 years mainly because of progress in PEM technology. The electrolyzers are used in technical niches or in industrial applications where only small amounts of hydrogen (10 m^3/h) are used. One of my colleagues works in atmospheric chemistry and frequently measures air samples at his lab at the Jungfrau Joch at the top of the Swiss Alps. His gas chromatograph needs hydrogen for operation, and the hydrogen is produced on site by a small portable PEM based electrolyzer. For a review on PEM electrolyzers, see Carmo [Carmo 2013].

The high temperature electrolysis in type (III) is industrially developed, as is for example demonstrated by its use at the Karlsruhe H2 filling station. Single electrolyzer stacks operate in the lower kW range at 750–1,000 °C. One advantage of this high temperature conversion principle is that the cells can be operated in the electrolyzer mode and in the fuel cell mode with the same cell (stack) [Chen 2015]. This is a considerable synergy in the development of materials for electrodes and electrolytes. This concept can also be used for the combination of steam electrolysis and Fischer Tropsch synthesis of hydrocarbon fuels [Verdegaal 2015].

An extensive book on the production of hydrogen by electrolysis has recently been authored by Agatha Godula-Jopek [Godula-Jopek 2015]. Despite the growing importance of electrolyzers for the coupling of various energy sectors in energy economy [Smolinka 2012], the world market for electrolyzers is rather small. According to Hydrogeit's assessment in 2017, business models which allow for a profit are currently missing [Geitmann 2017a].

Swiss manufacturer IHT used to sell the S-556 electrolyzer [Iht 2018]. The high pressure electrolyzer was made from four electrolyzer blocks. IHT worked with electro-lyzer units which include 139 electrolysis cells each. This was called S-139 cell block. They offered either one S-139 cell block or they assembled them to electrolyzers of two cell blocks or four cell blocks – which made the S-556.

It is the size and number of cells in such a block, which determine the amount of hydrogen that can be produced. According to IHT, one cell produced 1.367 Nm^3 H_2 per hour. Thus, the S-556 could produce in the past per hour 760 Nm^3 hydrogen. Meanwhile, IHT has improved its products and the electrolyzers' efficiency has grown to nearly 15%. Their new electrolyzers are being used now in the new EU project [Symvouloi 2017].

High pressure electrolysis is based on pressurizing the water (electrolyte) in the electrolyzer during electrolysis, with the result that the hydrogen extracted from the electrolyzer is pressurized as well. This saves not really the pressurizing step of the hydrogen in the end, because beforehand we have already a pump attached to the electrolyzer. But pressing water uses less energy than pressing hydrogen gas. The energy savings range around 3% to 5%, which eventually favor this high pressure electrolysis over conventional electrolysis [Onda 2004].

Figure 3.6 shows two small electrolyzer systems used at the hydrogen gas filling station in the city of Martigny near the Swiss Alps, established by EPFL Lausanne and the City of Martigny. The researchers there are testing and implementing two alternative systems, this is a long alkaline electrolyzer and a short PEM electrolyzer. While you may want to have a small as possible PEM fuel cell in your car because of performance requirements versus the space requirements, you will still feel comfortable with a large and traditional alkaline electrolyzer in stationary applications such as at the hydrogen filling station. We will see in the Chapter on batteries that for stationary applications, size and weight are not necessarily important. A "large and heavy" solution may be a worthwhile pick in comparison with a maybe more expensive and delicate novelty with lesser weight, smaller size but higher cost.

Figure 3.6: Custom-built alkaline electrolyzer system with two stacks (50 kW, 10 bar, "long"), and a custom built PEM electrolyzer (25 kW, 40 bar, "short") at the hydrogen filling station in Martigny, Switzerland [Ligen 2018].

The aforementioned three different concepts of water electrolysis are illustrated in Figure 3.7 [Manage 2011]. Table 3.5 lists some technical and conceptual details for the three different water electrolyzer technologies [Manage 2011].

The alkaline electrolyzer is shown in panel (a). Potential cathode materials can be, for example, Ni metal, and Ni–Co–Fe alloys can be used as anodes. They can be directly coated on a diaphragm (zero gap geometry), which allows for the migration

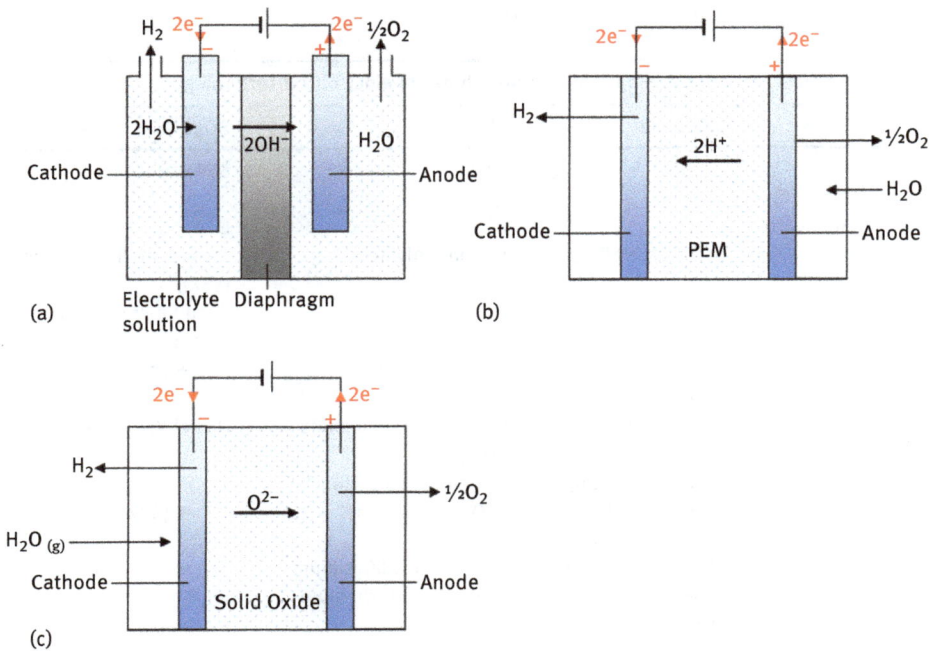

Figure 3.7: Schematic of different electrolyzers and their operation.
Reprinted from International Journal of Hydrogen Energy, 36, Manage MN, Hodgson D, Milligan N, Simons SJR, Brett DJL, A techno-economic appraisal of hydrogen generation and the case for solid oxide electrolyzer cells, Pages 5782–5796, Copyright (2011), with permission from Elsevier. Compare also [Sapountzi 2017].

of ions but warrants that the hydrogen and oxygen, which are formed at the cathode and anode respectively, are not mixing to become Knallgas (the explosive concentrated mixture of two parts H_2 and one part O_2). The alkaline electrolyte can be three molar KOH or NaOH. A well-suited diaphragm material was asbestos $Mg_3Si_2O_5(OH)_4$ because of its porosity and mechanical and chemical stability in strong caustic electrolytes at up to 85 °C working temperature. But asbestos was banned for any application since 1990 because of its malign effects on human health. Typical diaphragm thickness was 2.5 mm.

The filler mineral is crunched and mixed with polysulfone (a binder material) and some solvent and then mixed with a binder and processed to a fabric for example by tape casting. Barium sulfate $BaSO_4$ and $CaSiO_3$ Wollastonite are good mineral alternatives to the banned asbestos. Synthetic filler materials are for example $BaTiO_3$ and ZrO_2 (3YSZ, 3% yttrium stabilized zirconia).

The well-known electrolysis with polymer electrolyte membranes (PEM) can take place around ambient temperature and usually does not reach 100 °C. At higher temperatures, the PEM do not have the required moisture anymore.

Table 3.5: Comparison of three different electrolysis concepts.

Electrochemical reactions and data for different types of electrolyzer technology

	Alkaline	PEM	SOEC
Operation Temperature(°C)	70–90	< 100	500–1000
Ionic charge carrier	OH^-	H^+	O^{2-}
Material	$KOH_{(aq)}$, $NaOH_{(aq)}$	Sulfon. polymers, e.g., Nafion™	Yttria-stabilized zirconia (YSZ), Scandia-stabilized zirconia
Cathode: reaction material	$2H_2O + 2e^- \rightarrow H_2 + 2OH^-$ Nickel with Pt catalytic coating	$2H^+ + 2e^- \rightarrow H_2$ Pt black, iridium oxide (IrO_2), ruthenium oxide (RuO_2)	$H_2O + 2e^- \rightarrow H_2 + O^{2-}$ Nickel-YSZ cermet
Anode: reaction material	$2OH^- \rightarrow \frac{1}{2}O_2 + H_2O + 2e^-$ Nickel or copper coated with metal oxides	$H_2O \rightarrow \frac{1}{2}O_2 + 2H^+ + 2e^-$ Platinum black, iridium oxide (IrO_2) ruthenium oxide (RuO_2)	$O^{2-} \rightarrow \frac{1}{2}O_2 + 2e^-$ Perovskite oxides (e.g., lanthanum manganate)

Reprinted from International Journal of Hydrogen Energy, 36, Manage MN, Hodgson D, Milligan N, Simons SJR, Brett DJL, A techno-economic appraisal of hydrogen generation and the case for solid oxide electrolyzer cells, Pages 5782–5796, Copyright (2011), with permission from Elsevier.

Alkaline electrolysis is typically done at increased temperatures of 70–90 °C. For electrolysis in solid oxide electrolyte cells (SOEC), temperatures between 500 °C and 1000 °C are necessary. At lower temperatures, the oxygen ions will not conduct in the solid electrolyte.

All three concepts of electrolysis have their advantages and disadvantages. Figure 3.8 [Manage 2011] illustrates the role of thermodynamics for the electrolyzer concepts. The required energy per mole and also the electric potential necessary for the water splitting are plotted versus the water temperature. With increasing temperature, the heat energy Q of the water increases linear. The electrical energy necessary for water splitting (in the form of the potential ΔG) decreases linear with increasing water vapor temperature.

The total energy that we gain from the process is $\Delta H = \Delta G + Q$. Note that we have ignored here any influence of pressure, which would also enter the thermodynamic equation. Note also the effect of the phase transition of water from liquid to gas at 100 °C. And note that we are considering here only the energy budget from the thermodynamic perspective. Kinetic barriers do exist and the lowering of those is the task of electrocatalysis.

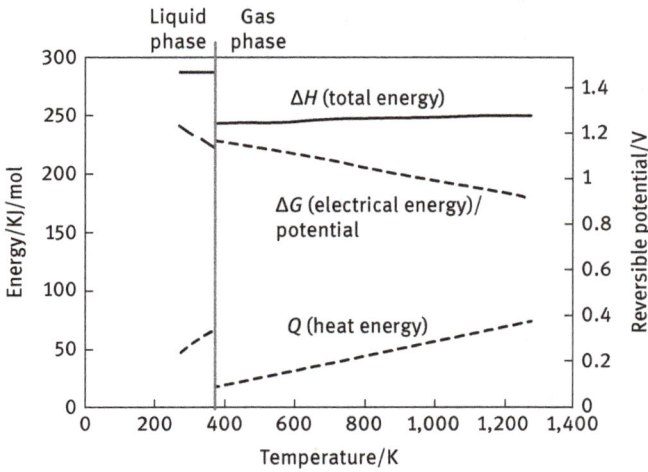

Figure 3.8: Relation of thermodynamic quantities during water electrolysis over a wide temperature range, including the water–vapor phase transition. With increasing temperature, the heat energy Q increases linearly, whereas the electric energy ΔG decreases linearly. The total energy is $\Delta H = \Delta G + Q$.
Reprinted from International Journal of Hydrogen Energy, 36, Manage MN, Hodgson D, Milligan N, Simons SJR, Brett DJL, A techno-economic appraisal of hydrogen generation and the case for solid oxide electrolyzer cells, Pages 5782–5796, Copyright (2011), with permission from Elsevier.

The materials used for the electrolysis reactor have to withstand, like in any electrochemical reactor, the harsh conditions that arise from the pH of the electrolyte, any applied electric bias potential and the temperature. In addition, the materials should be environmentally benign and affordable. This is why there is still research and development going on in electrode and electrolyte development.

3.7.4 Combined hydropower and PEM electrolyzer plant for hydrogen production

The hydrogen that I have pumped for our Hyundai FCEV at the COOP hydrogen gas station in Hunzenschwil, Switzerland, was supplied from the IBAarau hydropower plant. For over 100 years, this hydropower plant delivered a power of 12.5 MW of electricity. Since 2016, around 2% of the power is used for H_2 production in the world's first PEM electrolyzer directly connected to a hydropower plant [IBAarau 2016], with an anticipated annual H_2 production of 20000 kg. Table 3.6 lists the technical specifications of the hydropower plant in Aarau.

Table 3.6: The table is split into four blocks with specifications of the hydropower plant, the PEM electrolyzer, the hydrogen compressor and the hydrogen delivery trailer.

Specifications of the hydrogen production system in Aarau, Switzerland	
Hydropower plant IBAarau	
Established in year	1895/1912, Renewed in 1957
Turbines	11 Kaplan turbines
Average annual production	109 GWh
Production in summer	60 GWh
Production in winter	49 GWh
Maximum power	16 MW
Average power	12.5 MW
Water flux averaged over a year	300 m^3/s
PEM electrolysis	
Supplier	Diamond Lite S.A.
Manufacturer	Proton OnSite (USA)
Type	C Series, C 30, Proton Exchange Membrane (PEM)
Electric power	5.8 kWh/Nm^3
Hydrogen yield	30 Nm^3/h H$_2$, i.e., 2.7 kg H$_2$/h
Base pressure	30 bar
Purity	99.9998%
Max. water usage	30 L/h
Gas Compressor	
Supplier	Sera ComPress GmbH
Type	Metal Membrane Compressor
Suction	27–31 bar
Outlet pressure	211 bar
Power	30 Nm^3/h
Propulsion	Crankshaft with flywheel
Hydrogen transport trailer	
Supplier	Messer Schweiz AG
Pressure vessel	10 steel containers
Operating pressure	200 bar
Geometric volume	23 m^3
Hydrogen transport capacity	338 kg
Trailer size	Length: 12.7 m
(without truck)	Width: 2.5 m
	Height: 3.6 m
Mass of trailer	32 t

Reproduced from [IBAarau 2016].

Here is a simple exercise for the reader: find out how many Hyundai ix35 fuel cell cars can be filled per day with this capacity. For comparison, in around 2010, Hawaii Gas Company was able to deliver at Oahu per day 7,000 kg H$_2$ without any extra effort [Motavalli 2010]. How many cars can you fill per day with this amount?

3.7.5 Electrolyzer powered with solar cells

The fastest way to produce solar hydrogen is currently by the direct combination of a photovoltaic cell with an electrolyzer. Arriaga et al. [Arriaga 2007] have shown a system where a photovoltaics PV plant is combined with a PEM EC system that produces "solar hydrogen." Note that this is not a photoelectrochemical cell (PEC), but "only" a combination of two independent systems. Figure 3.9 shows the electric installation diagram of the PV plant. It shows four rows of nine PV panels. You see that four panels are connected in series and those provide 48 V. This means one panel (module) should provide 12 V. Then, nine of such series are connected in parallel.

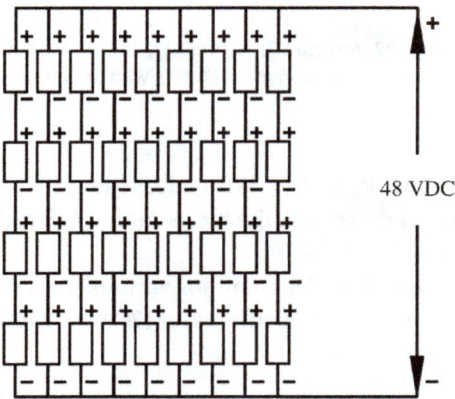

Figure 3.9: Electric installation diagram of the solar panel installation from [Arriaga 2007]. Reprinted from *International Journal of Hydrogen Energy*, 32, Arriaga LG, Martinez W, Cano U, Blud H, Direct coupling of a solar-hydrogen system in Mexico, 2247–2252, Copyright (2007), with permission from Elsevier. [Arriaga 2007].

The peak power of each panel or module is 75 W. The plant of $4 \times 9 = 36$ panels can provide therefore a peak power of 36×75 W = 2700 W. The working point is often determined by the maximum power point (MPP). Figure 3.10 shows the I–V characteristic of a solar cell which was the basis for the determination of the performance of the plant. The open circuit potential

I_{SC} would be the current of the PV panel when it is short circuited. E_{OC} is the voltage of the panel at open circuit when no current is flowing and thus no power is delivered. R_S is the series resistance and V_t is the temperature dependent voltage. The equation

$$I = I_{SC} \left[1 - exp \left(\frac{E - E_{OC} + I \cdot R_S}{V_t} \right) \right],$$

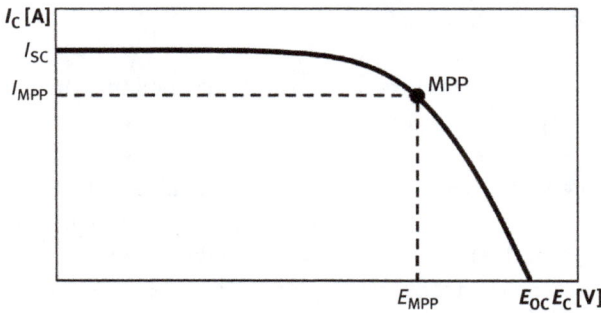

Figure 3.10: Typical curve of a photovoltaic cell: EOC and ISC are voltage at open circuit and short circuit current of the cell, respectively. Point of maximum power (MPP), current (IMPP) and voltage (EMPP) at this point.
Reprinted from *International Journal of Hydrogen Energy*, 32, Arriaga LG, Martinez W, Cano U, Blud H, Direct coupling of a solar-hydrogen system in Mexico, 2247–2252, Copyright (2007), with permission from Elsevier. [Arriaga 2007].

was used for the determination of the unknown R_S and V_t. With a least square fit, they were determined. It was also necessary to determine the temperature T of the panels and the current I_{MPP} and voltage E_{MPP}.

Figure 3.11 shows the simulated I–V curves of the PV plant for various powers (200–1000 W) for temperatures from 25 °C to 60 °C. The filled squares (■) over each

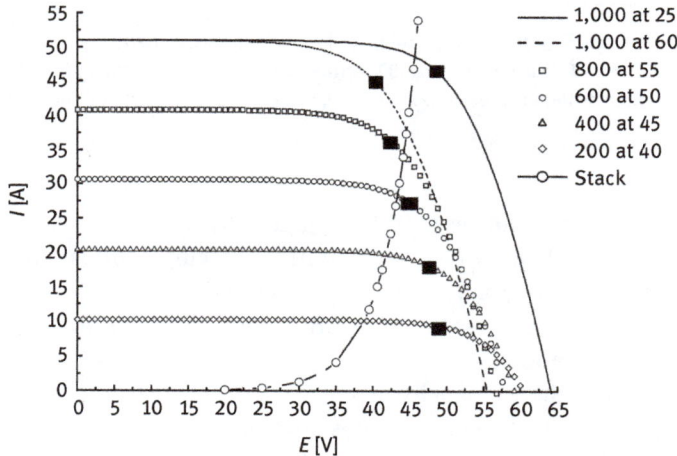

Figure 3.11: (a) Coupling of electrolyzer stack curve and PV system curve at different conditions of irradiance and temperature, values on graph are in units (W/m^{2}at°C), also the MPP is shown as label (■).
Reprinted from *International Journal of Hydrogen Energy*, 32, Arriaga LG, Martinez W, Cano U, Blud H, Direct coupling of a solar-hydrogen system in Mexico, 2247–2252, Copyright (2007), with permission from Elsevier. [Arriaga 2007].

curve denote the maximum power point. At 1000 W, there is a considerable difference in the voltage at 25 °C (~50 V) and 60 °C (~40 V).

The $I-V$ curve for the electrolyzer stack is shown in the same figure with open circles (o). The produced hydrogen flow for this system was 20±10 normal L/min [Arriaga 2007].

3.7.6 SHINE – a project for a solar hydrogen electrolyzer

The Swiss research program NanoTera (Nano-Tera: engineering complex systems for health, security, energy and the environment) had the aim to use nanoscience and nanotechnology for the making of solutions for real-world and real-life problems [Leblebici 2019]. My colleagues at EPFL in Lausanne and CSEM in Neuchatel won a grant in this program where we were supposed to make a solar hydrogen reactor based on reverse fuel cell technology. The acronym of our project was SHINE, which stood for Solar Hydrogen Integrated Nano Electrolyzer [Moser 2014].

We wanted to use a reverse fuel cell concept for the development of an electro-lyzer. The necessary energy would come from concentrated solar light which was focused on a solar cell that had a smaller area than the reflector. By this way, you can catch sunlight over an area of say 100 m² and project it on a PEC reactor of an area of say 0.5 m². The PEC reactor uses a semiconductor to absorb the projected light and produce a Nernst potential large enough to oxidize water into hydrogen and oxygen in an integrated electrolysis cell. One important aspect was the use of low-cost electrocatalysts for the reactor, but this is not withstanding that the cost for expensive noble metal electrocatalysts for the oxygen evolution reaction do not add significantly to the overall cost of a PEC system.

A comparable small-scale pilot production plants in Switzerland were for example at the Michelin research center near Fribourg. An area of 55 m² was covered with PV panels with 15% efficiency which powered an alkaline electrolyzer at 75% efficiency. This system could produce per day 1.56 kg hydrogen compressed to 30 bars.

3.8 Ammonia synthesis for fertilizer production

When the huge Assuam (Aswam) dam in Egypt was planned after Second World War, an ammonia fertilizer production plant [El Nashar 1982] was part of the projection. Heavy water was also considered as a by-product by electrolysis, for example, [Thayer 1959].

The invention of the ammonia synthesis by Haber and Bosch over 100 years ago has given the growth of human population worldwide an unprecedented impact, because fertilizers could be made from the abundant nitrogen in the air based on following simple chemical reaction:

$$N_2 + 3\ H_2 \Leftrightarrow 2\,NH_3$$

Nitrogen can be obtained by fractioned distillation of liquid air. Hydrogen used to be taken from the chlor-alkali electrolysis, which has H_2 as a waste product. Ammonia synthesis is very energy intensive process. Access to low-cost energy means therefore also to provide affordable food for the population, for the people.[3] It is therefore not surprising ammonia production was part of the Aswan dam tasks in the vision of Egypt's economic strategy for independent food production.

The Aswan dam was used first as a huge electric power source for the electrification of the country to the best of its capacity. Next it was a power source for hydrogen production by electrolysis. Hydrogen is a very valuable, high energy chemical raw product which can be used for example for ammonia production. Ammonia can be produced electrochemically [Ali 2016, Bicer 2017, Cui 2017, Grayer 1984, Kyriakou 2017, Shimoda 2017, Shipman 2017, Singh 2017].

Even to date, ammonia production is an important topic not only for developing countries but also for industrialized countries, as is demonstrated by a recent round table meeting on sustainable ammonia synthesis, organized by the U.S. Department of Energy [Norskov 2016]. The importance of food and food policies is therefore not to be underestimated. It plays a vital role in geopolitics (LaRouche 1995, Ghoneim 2012, Pinstrup-Andersen 2014, Kneen 2003) as much as the access to energy and important minerals and metals.

3.8.1 Case study: Fertilizer production at the Assuan dam in Egypt

It is interesting for the reader to learn some more details about how a country like Egypt made detailed steps toward a developed and industrialized country using renewable energy. The Egyptian Chemical Industries KIMA in Aswan was founded

---

3 Says Lyndon LaRouche: "This is artificial. It is not the harvest. The problem has been as with the US free trade policy and with the US domestic and foreign agriculture policy ... that the United States has been deliberately dropping the level of food production of the United States below the equivalent of a 100% of our own requirements. We are a net food importing nation if we don't count our grain which is produced for the international markets. Most countries on this planet are food shortage countries. This is a condition which has been deliberately created for political reasons in order to establish what is called political food control of policies of government. For example, Egypt produces 40% of the food required to feed its population. ... [Larouche 1995] Larouche LH. Tarpley WG. Interview. 1995. LaRouche Warns of Food & Metals Hoarding. August 22, 1995.
http://www.larouchepub.com/tv/tlc_programs_1995.html
https://archive.org/details/LaRoucheWarnsOfTheOligarchyHoardingMetalsAndFood
https://archive.org/details/LaRoucheWarnsOfHoardingMetalsAndFood "

in 1956 as an Egyptian joint stock company which belongs to the chemical industries. The stock capital was permitted to a maximum of two billion Egyptian pounds, and 1.2 billion Egyptian pounds were actually paid through shareholders (55% by the chemical industries holding company, 39% by public bodies, banks and insurance companies and 6% by individual shareholders).

Today one Egyptian pound counts 0.056 US $. So, simplified, in 1956 the company had a 67 Million US $ available; back in 1956 this was a substantial capital. The fertilizer production plant started in May 1960 with a production capacity of almost 1600 t of fertilizer per day. Tables 3.7 and 3.8 list a brief history of the KIMA production capacity. It appears that while the maximum fertilizer production capacity decreased from 1960 to 1988 by 40%, the relative nitrogen content of the factory increased from around 20% to 33%, probably by improved processing and technology.

Table 3.7: Ammonia fertilizer production capacity of the KIMA plant in Assuan from 1960 to 1980

Production capacity of the KIMA fertilizer plant		
Date, year	Maximum production capacity in tons of ammonia fertilizer per day	Nitrogen content
22 May 1960	1593	20.5%
12 November 1964	1256	29%
7 November 1968	1053	31%
20 June 1988	975	33.5%

Source: http://www.kimaegypt.com/.

Table 3.8: Details on the chemical production and products of the KIMA plant.

Maximum capacity	Production launch	Addition of supplementary factory
2.5 t per day	16 March 1964	Hydrochloric acid production
7,500 t per year with proportion 75%	1 October 1967	Ferrosilicon production
270 gas cylinders with capacity 7 m^3	2 January 1973	Oxygen mobilization – first and second compressor
200 t per day, raised to 300 t	17 December 1998	Pure ammonium nitrate production 34.80%
Daily 5 ton of silica dust withproportion 92:94	2006/7/10	Filter unit of ferrosilicon factoryand collection of silica dust

Source: http://www.kimaegypt.com/.

The KIMA plants and its residential premises for their staff took an area of 4.5 km² in the south east of Aswancity. Back then, the water dam power plant delivered 280 MW, around 200 MW of which was used for electrolysis and fertilizer synthesis, with water and air being the raw materials. Later, 12 Francis turbines with 175 MW power each produced a total power of 2100 MW, which was around 10% of the electric energy consumption in Egypt.

KIMA is actually a large industrial complex; this is why it occupied over 4.5 km². It contained the H_2 production unit with 37000 m³/h capacity from water electrolysis. The nitrogen production unit separated the nitrogen from liquefied air, with a capacity of 13000 m³/h. With the hydrogen and nitrogen available, they could feed the ammonia production unit which produced 400 tons of NH_3 (fertilizer precursor) per day. This is the capacity of 16 large trucks per day.

Half of the ammonia produced was further used in a nitric acid HNO_3 production unit, with a capacity of 1400 t per day (this is almost 60 large trucks). Note that the HNO_3 contains mostly water; this is why there is so much gain in weight. The amount of ready fertilizer (ammonium nitrate) per day was 665 t, which means every hour one large full truck would leave the factory on a 24/7 operation basis. The factory complex had also fertilizer bagging facility. These are the core operation facilities, but the industrial complex had also water and electricity utility facilities, general management buildings, laboratories, machine shops, workshops, training and computer facilities.

3.8.2 Electrochemical ammonium production

The ammonium production that I wrote about so far is done via the Haber–Bosch process, which requires high temperature (400–500 °C) and high pressure (130–170 bar) [Kyriakou 2017]. Notwithstanding there has been interest to produce ammonium by electrochemical processes [Furuya 1990]. The review by Kyriakou et al. [Kyriakou 2017] lists an impressive number of studies which were carried out at high temperatures ($T > 500$ °C), intermediate temperatures (100 °C $< T < 500$ °C) and low temperatures ($T < 10$ °C).

The electrochemical reaction for ammonia synthesis in a high temperature cell with oxygen ion conductor reads at the anode

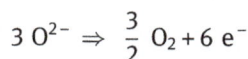

$$3\,O^{2-} \Rightarrow \frac{3}{2}\,O_2 + 6\,e^-$$

and at the cathode

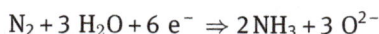

$$N_2 + 3\,H_2O + 6\,e^- \Rightarrow 2\,NH_3 + 3\,O^{2-}$$

At the cathode, there is a water electrolysis taking place alongside with the synthesis of the ammonia. Oxygen is a side product in this process. A simple exercise for the reader is the formulation of the overall reaction from the two above half-cell reactions.

Ammonia synthesis can be done with various cell architectures. Figure 3.12 shows the schematic of an ammonia reactor with a proton conducting electrolyte membrane, where synthesis takes place in a gas phase reaction at high temperature [Kyriakou 2017]. An example for such cell was studied by Shimoda et al. [Shimoda 2017], who used electrodes and electrolytes which are also know for high temperature solid oxide fuel cells (SOFC), specifically the yttrium substituted barium cerate (BCY) as electrode and nickel oxide as anode. Note that this cell accepts because of the high temperature process gases, not liquids. Higher temperatures improve the reaction kinetics.

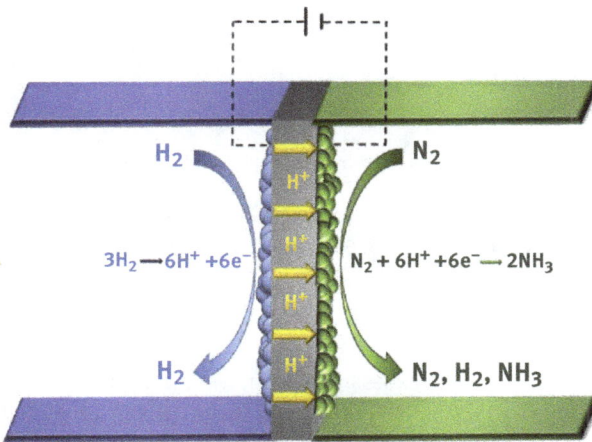

Figure 3.12: Schematic diagram of the solid-state ammonia synthesis (SSAS) process in a double chamber proton conducting reactor cell.
Reprinted from *Catalysis Today*, 286, Kyriakou V, Garagounis I, Vasileiou E, Vourros A, Stoukides M, Progress in the Electrochemical Synthesis of Ammonia, 2–13, Copyright (2017), with permission from Elsevier. [Kyriakou 2017].

An example for an electrolyzer cell that works with liquid reaction partners is shown in Figure 3.13 [Kyriakou 2017]. The first important step in this cell is the production of hydrogen by water electrolysis. A membrane from a silver palladiumalloy warrants the separation of hydrogen produced in the center container in Figure 3.13 from the electrolyte, which is phosphoric acid H_3PO_4.

The membrane is shaped as a tube and constitutes also the working electrode. In the center of the tube is a co-axial platinum wire as the counter electrode. The operation temperature of this reactor is 100 °C. A ruthenium electrocatalyst promotes the reaction of gas phase N_2 which is blown into the electrolyte, with the hydrogen produced in the cell.

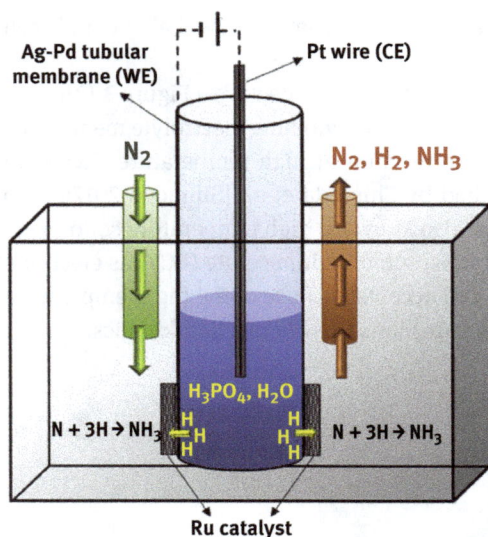

Figure 3.13: Schematic representation of the apparatus used by Itoh et al. [Itoh 2007]. Hydrogen is produced *in situ* from the electrolysis of water.
Reprinted from *Catalysis Today*, 286, Kyriakou V, Garagounis I, Vasileiou E, Vourros A, Stoukides M, Progress in the Electrochemical Synthesis of Ammonia, 2–13, Copyright (2017), with permission from Elsevier. [Kyriakou 2017].

In this simple design, the cell will emit excess nitrogen, excess nonconverted hydrogen and the desired reaction product NH_3. It is the work of the chemical reaction engineers to improve this process and the reactor design for more efficient conversion and separation of gases. Ammonia is an aggressive and toxic gas that requires specific safety regulations. It may not be released in the atmosphere in high concentrations [Sissell 1998].

The bond breaking of the N–N bond in N_2 and the formation of the N–H bonds toward NH_3 is a delicate process, the balance of which depends on the applied electric potential in the cell and on the kind of electrocatalyst [Singh 2017]. Electrochemical ammonia synthesis, in particular the kinetic of the reaction, is governed by the principles summarized in the Butler–Volmer equation, which we will learn about later in this book.

The thermodynamic limit for the energy E^0 (electrochemical potential) where the onset of ammonia synthesis begins is given by the difference of the free energy G_0 of the reaction

$$E^0 = \frac{-\Delta G_R^0}{n \cdot F}$$

At some potential, the hydrogen evolution reaction will dominate the processes, as is shown in the plateau. The Butler–Volmer kinetics is shown in Figure 3.14 as the ammonia formation rate per electrode area versus the electrode potential. Only the trend and curvature is shown; no specific numbers are supplied. A potential temperature study would yield a similar trend like shown in Figure 3.8.

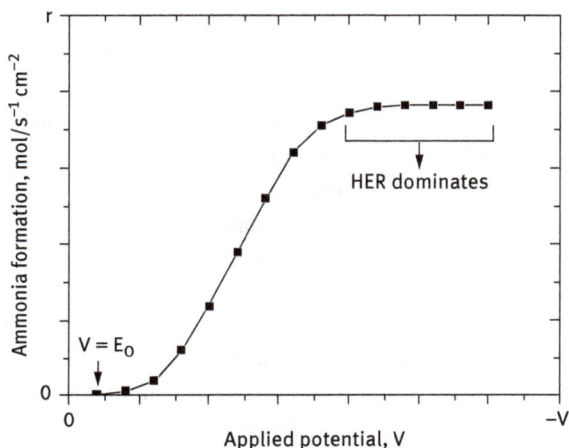

Figure 3.14: Dependence of electrocatalytic rate of ammonia synthesis on the applied potential. E_0 denotes the onset potential calculated from equation $E^0 \cdot n \cdot F = -\Delta G_R^0$.
Reprinted from *Catalysis Today*, 286, Kyriakou V, Garagounis I, Vasileiou E, Vourros A, Stoukides M, Progress in the Electrochemical Synthesis of Ammonia, 2–13, Copyright (2017), with permission from Elsevier. [Kyriakou 2017].

3.9 Chlor-alkali electrolysis

The chlor-alkali electrolysis is used to decompose rock salt NaCl into chlorine gas Cl_2 and sodium hydroxide NaOH. Chemical industry worldwide has a huge demand for these two products. This process is done at a large industrial scale with membrane electrochemical reactors with tremendous energy consumption [O'brien 2005].

Chloride ions from the brine (salt solution) are oxidized at the anode by losing two electrons and forming chlorine gas

$$2\,Cl^- \Rightarrow Cl_2 + 2\,e^-$$

In the cathode half reaction, water H_2O provides protons H^+ which becomes with the electrons reduced to hydrogen gas H_2

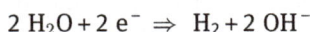

$$2\,H_2O + 2\,e^- \Rightarrow H_2 + 2\,OH^-$$

Hydrogen here is a valuable side product, which was, for example, used for the classical Haber–Bosch ammonia synthesis (a nonelectrochemical process).

It requires an ion exchange membrane that allows passage for the Na^+ ions so that they can react with the hydroxyl groups OH^- and form NaOH:

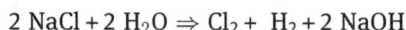

$$2\,NaCl + 2\,H_2O \Rightarrow Cl_2 + H_2 + 2\,NaOH$$

3.10 Aluminum production

Aluminum electrolysis is an application of electrometallurgy. A number of metals are refined by electrochemical methods, but aluminum is a metal (like Na, K, Mg, Li) which cannot be produced with any other process than electrochemical reduction of aluminum oxide or similar aluminum minerals. As their electrochemical potential is below that of water H_2O, water would decompose in electrolysis before any of the metals would be reduced. Hence, the electrolysis is done with molten salt electrolytes.

The mineral bauxite Al_2O_3 is molten and then subject to a simple electrolysis where graphite constitutes the anode and the cathode. We use here a liquid electrolyte, but it is not an aqueous solution containing Al^{3+}. As Al_2O_3 has a very high melting point, it has to be mixed with another compound called cryolite Na_3AlF_6. You may recall from thermodynamics that a mixture of two compounds has a lower melting point that any of the two constituents of the mixture. This is a trick we take advantage of in the melt flow electrolysis. The electrolysis temperature is then only 960 °C. The necessary voltage is around 5.5–7 V.

At the anode, oxygen is formed that reacts with the graphite electrode to carbon dioxide or carbon monoxide. The carbon is therefore consumed.

3.11 Reactions and potential distribution in the electrochemical cell

We have now seen a number of electrochemical cells and reactors and also some of their applications. I want to conclude this chapter with the sketch of a solid-state proton conducting cell which can be used as fuel cell or electrolyzer cell.

Figure 3.15 shows a schematic of a solid oxide fuel cell with a ceramic proton conductor, this is the anode layer, the electrolyte layer and the cathode layer, arranged versus the spatial ordinate (distance). The abscise reads the electric potential. The anode and the cathode are electrodes and therefore – in the ideal case – have no resistivity and therefore the potentials ϕ°_A and ϕ°_C for anode and cathode are constant over the thickness of these electrodes (horizontal lines). This holds for the electrodes when the cell is under load or at open potential. The potential difference is either the open circuit potential $E = \phi^\circ_C - \phi^\circ_A$.

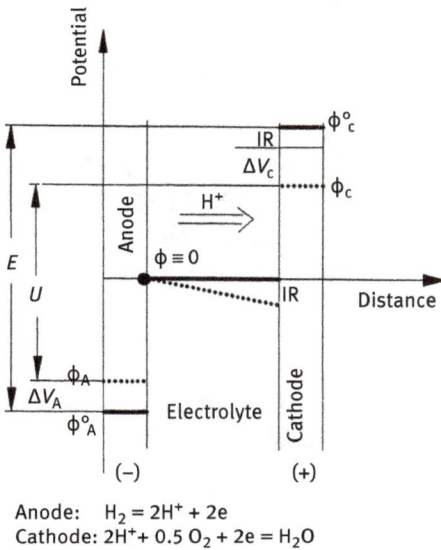

Figure 3.15: Simplified picture of the principles of solid-state proton conductor (SSPC) fuel cell operation and electric potential relations at open circuit voltage (solid thick line; ____) and under load (thick dotted line....). E is the open circuit potential; U is the cell voltage; ΔV is the overpotential; ϕ is the cathodic (C) and anodic (A) potential.
Reproduced from [Jewulski 1990].

Anode: $H_2 = 2H^+ + 2e$
Cathode: $2H^+ + 0.5\,O_2 + 2e = H_2O$

When you connect an electric load to the cell, which may be an electric resistance for analytical reasons or an electric motor or a light bulb or an electric heating element for practical application, then the cell voltage which helps us power the aforementioned utilities is $U = \phi_C - \phi_A$, which you collect from the current collectors. The reason why U<E is that there are overpotentials, that is, kinetic barriers which prevent the chemical reactions at anode and cathode to take place at the potentials given by thermodynamic principals.

The reaction at the anode is the oxidation of the hydrogen $H_2 = 2H^+ + 2e^-$, but it will not take place at the potential ϕ°_A, but at $\phi_A < \phi^\circ_A$. The protons H^+ move from the anode–electrolyte interface through the ceramic proton conducting electrolyte to the cathode.

The same suppression of reaction by the overpotential happens at the cathode, where the protons become reduced and form water with the oxygen that entered the porous cathode: $2H^+ + \frac{1}{2}O_2 + 2e^- = H_2O$, so that $\phi_C < \phi^\circ_C$. Under load, the potential across the electrolyte will vary because of the electric current that flows through the electrolyte meets the electrolyte resistance. This sequence of events governs basically all electrochemical cells.

It is only since 10 years that the overpotentials in solid-state electrochemical reactors (cells) with solid–gas interfaces can be experimentally assessed with element-specific and chemical orbital-specific accuracy [Grass 2010, Whaley 2010, Zhang 2010].

4 Batteries

I grew up with the understanding that a battery is the *"energy thing,"* which you need to power the flashlight (torch) or the portable radio, and certainly the car starter battery. As kids, we frequently got flashlights here and there. Some flashlights were very small and contained one small cylindrical battery of the IEC-R6 size type (IEC: International Electrotechnical Commission). Batteries of this shape and size were called Mignon batteries. It would deliver 1.5 V. Back then, there was no light emitting diode (LED) available.

The light source was a small bulb that had a metal wire (tungsten filament) in the evacuated glass bulb and a metal cylinder with some thread for mechanical fixture and electric contact to the battery, and a soldered tip at the bottom for the second electric contact. With two wires, the battery, the bulb and two skilled hands, we could light the bulb with the battery. As you see from Table 4.1, there are many kinds of Mignon batteries. They can deliver a voltage from 1.2 to 1.7 V. The voltage depends on the underlying chemistry of the battery. Their capacities range from 800 to 3000 mAh. Today, batteries are omnipresent but even over a 100 years ago batteries were well known. As this book is not a book on batteries only, I advise the reader to consult also specialist literature on batteries [Kiehne 2003, Trueb 1998].

4.1 Batteries from the perspective of the consumer

Table 4.1 shows numerous battery types which officially rank as Mignon batteries, but they have different chemistries, which means they have different electrochemical Redox couples with consequently different resulting Nernst voltages. The batteries may go by different names as far as their size is concerned. The capacity may also be different. Some couples are environmentally more benign, others are more economical and cheaper, other couples may have a necessary higher voltages or longer lifetime. The "Dura" in Duracell® suggests, for example, a longer lifetime. Some batteries are rechargeable; they are called secondary batteries. The nonrechargeable batteries cannot be used anymore after use and those are called primary batteries. So the consumer has a choice. Note that the batteries here are called cells. I will explain later the meaning of cell and battery, and why there is a difference.

Larger flashlights contained larger batteries which too provided 1.5 V. But the batteries had a larger volume (same height like Mignon but wider diameter) and had more weight. Three such batteries were stacked together in the handle of the flashlight, so that the total voltage was 4.5 V. With the higher voltage, one could operate a bulb which produced more light because it would release more power (Watt). Because of the higher mass of the battery electrodes, the capacity was higher as well.

https://doi.org/10.1515/9783110561838-004

Table 4.1: Comparison of various Mignon-type battery cells with different chemistries, voltages and energy densities. Given are also their commercial trade names.

Some nonrechargeable Mignon cells (primary cells)						
System	Voltage (V)	IEC label	ANSI label	Other names	Capacity (mAh)	Energy (Wh)
Alkaline	1.5	LR6	15A, 15AC	AM3, AM-3,	2000	2.71
Zn-Coal	1.5	R6	150, 15CO	UM3, 2006, 3006	1200	1.8
Li iron sulfide	1.5	FR6	15LF	L91	3000	4.5
NiOOH	1.7	ZR6		NX1500		

Rechargeable Mignon cells (secondary cells)						
System	Voltage (V)	IEC label	ANSI label	Other names	Capacity (mAh)	Energy (Wh)
Ni MH	1.2	HR6	1.2H2	NH15	2500	3.0
NiCd	1.2	KR6 KR157/51	1.2K2	CH15, 5006	1000	1.2
Lithium ion	3.7		LS14500	14500	750	2.8
Ni Zn	1.6				1500	2.4

There was one battery available which had a rather flat appearance. It was as large or as small as a hand. The electric contacts were brass stripes 5 mm wide and 2 cm and 4 cm long – both at the top side of the battery. It was very easy to contact them for experiments with wires; no particular battery container was necessary. Once I disassembled such flat battery and I learnt that it contained inside three Mignon batteries (actually, it contained three cells), the contacts of that were soldered together to arrive at 4.5 V total voltage.

Another type of battery was actually quite small but it provided 9 V (Figure 4.1, left image). It had no cylindrical shape but a rectangular cross section. We call it prismatic shape. As opposed to cylindrical shape. These are the two basic design shapes for many electrochemical energy storage and converter devices. The two contacts were on the top of the battery; they had a particular shape so that one could not by mistake or by accident change the polarity when connecting them with a device, such as a portable radio. When I opened that battery, I learnt that it contained six prismatic pellets of battery cells. One such pellet looked actually very handy and would basically be a 1.5 V element for whatever purpose.

Several years later, a new type of battery appeared on the market, the flat round cylindrical button cells (or coin cells because of their shape like buttons and coins), which you would use in your electronic watch. Maybe, it was the development of digital wrist watch which paved the path for the development of these small flat batteries, and vice versa. We learn that some part of the battery technology is the proper shaping and packaging of the battery. With continuing miniaturization of electric equipment and utilities, the batteries had to adopt size and shape and also voltage requirement. This is why more and more standards arose. The right image in Figure 4.1 shows a short cylindrical 3 V lithium ion primary battery, which is particularly recommended for digital cameras.

Figure 4.1: (left) DURACELL® 9V alkaline battery of type (IEC Nomenclature) MN1604, 6LP3146 (6 cells to 1.5 V each, alkaline (L) electrolyte KOH, Prismatic (P) shape with dimensions 48.5 mm × 26.2 mm × 17 mm. The cells contain Zn/MnO_2 elements. (right) Energiezer® 3V battery. Notation CR means it is a lithium manganese-oxide-based battery with cylindrical C shape.

As a young boy, I also knew that cars had a battery somewhere, the so-called starter battery. I remember that these were all very heavy, particularly those in the heavy equipment of the farmers (tractors). These starter batteries were all prismatic blocks which had two metal knobs on the top, one with a plus sign and the other with a minus sign on it, see Figure 4.2. The knobs were made from soft lead metal and had the size of a thumb. They had also a closure/opening where acid of distilled water could be refilled. When you lifted the battery and shook it, you would hear some noise and feel the shaking of a liquid inside.

A car starter battery is designed to provide 12 V. Trucks and heavy equipment may require 24 V. A motor bike engine may require 6 V only. The power of the starter batteries is used to crank the combustion engine. Large motors need a larger crank unit, which requires a higher voltage. Figure 4.2 shows on the left a car starter battery ("miocar" built for the Swiss retail stores MIGROS. The dark plastic

Figure 4.2: (left) 12 V starter battery with 60 Ah capacity. (right) Two different batteries from the VARTA POWERSPORTS "Freshpack" series.

casing contains the lead and lead oxide electrodes and the liquid sulfuric acid electrolyte. The casing is sealed so as to prevent electrolyte leaking. The battery requires no maintenance. Its specifications are listed in Table 4.2. There you see that the weight of the car starter battery is more than 16 kg. Lead batteries are heavy because lead is a very heavy element with 11.342 g/cm^3 density. The heavy weight is one reason why battery researchers have been looking for battery materials lighter than lead.

Table 4.2: Technical specifications of the car starter battery miocar shown in Figure 4.2 (left battery, black).

Specification	
Packaging size	Length/depth: 17.3 cm, width: 23.2 cm, height: 22.5 cm
Size of product	Length: 232.0 mm, width: 173.0 mm, height: 225.0 mm
Weight	16.1 kg
Charging voltage	12 V
Capacity	60 Ah
Current during cold start	510 A

Lead acid batteries are also built for other applications than car starting. Some are small and very handy. They can be used for small motorbikes, boats and scooters. The VARTA POWERSPORTS "Freshpack" batteries come with a separate container of battery acid and tubing, which shall be used when demand for the battery has come. Figure 4.2 (right, white batteries) shows such batteries for motor bikes and similar sports utilities. They deliver 12 V at 12 Ah capacity [Varta 2018]. The specifications for this type of battery are listed in Table 4.3.

Table 4.3: Technical specifications of the utility battery "VARTA POWERSPORTS."

Specifications	
Model	512 011 012
Capacity	12 Ah
Cold start current	160 A
Width	82 mm
Length	136 mm
Height	161 mm
Short code	YB12A-A (12N12A-4A-1)

4.2 Where does the name "battery" come from?

When I was in my late teens and rented my apartment, the architect showed me around and in the bathroom he mentioned the word *"Mischbatterie"* (which I translate here to mixing battery) when he showed me something like this, see Figure 4.3:

Figure 4.3: A water mixing "battery" in a washing room. The "battery" is the set of valves which control the mixing of hot (red button) and cold (blue buttons) water.

A *mixing battery* is a set of valves which are in place to mix gases or liquids. Later I remembered that I had come across the word "battery" in the context of military and warfare, like in "battle," the English word for French "Bataille." Battery therefore originally meant an arrangement of canons or guns as shown in the painting in Figure 4.4. This painting shows the battle of Waterloo, Napoleon's defeat. In the front, we are looking at three aligned canons – a battery of canons. So we learn battery in general is an arrangement of single units. "Battery" in electrochemistry is therefore nothing else than an arrangement of single electrochemical cells, galvanic cells,[1] fuel cells, photoelectrochemical cells and so on.

Figure 4.4: Painting from a scene from the battle of Waterloo (18 June 1815), showing a battery of three cannons. (Bild: Rundgemälde Waterloo, Photo reproduced with kind permission from Mag. Uwe Schwinghammer, http://www.wopic.at/

And so the definition reads in the book by Ayrton in Section 118 [Ayrton 1891]:

> "A battery is the name given to a collection of galvanic cells, arranges so as to produce a larger current than could be obtained with a single cell under the particular circumstances."

---

1 We will learn in the fuel cell chapter that fuel cells are not called batteries, but stacks. This is because the single units, the single cells, are stacked upon each other together.

The Recommendations on the Transport of Dangerous Goods by the United Nations [UN383 2013] define, for example, "Battery means two or more cells which are electrically connected together and fitted with devices necessary for use, for example, case, terminals, marking and protective devices." And "Button cell or battery means a round small cell or battery when the overall height is less than the diameter."

It is important that we become clear about the proper terminology. As early as 1891, Ayrton in his book brought some order and consistency in the popular field of electricity generation and storage, as can be seen from Figure 4.5, where I displayed

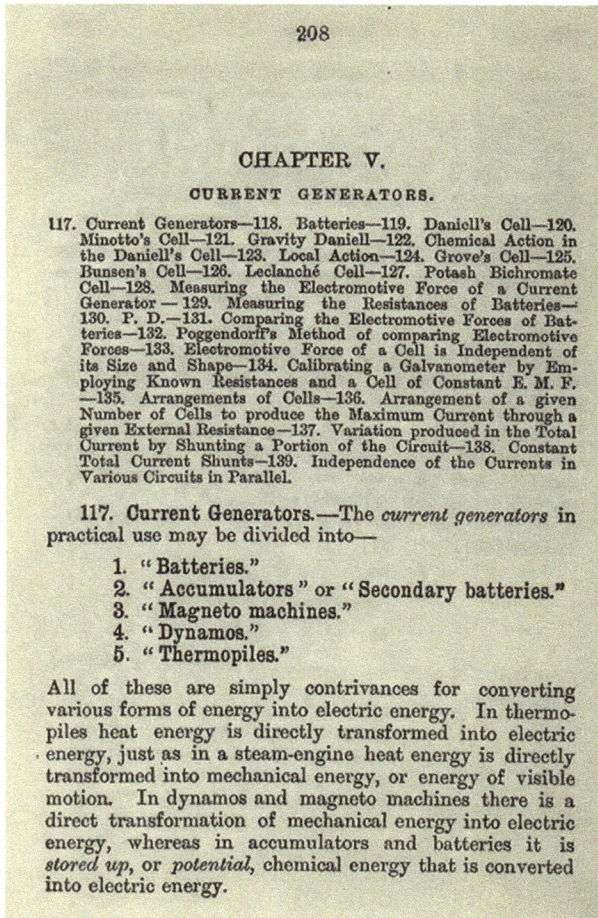

208

CHAPTER V.

CURRENT GENERATORS.

117. Current Generators—118. Batteries—119. Daniell's Cell—120. Minotto's Cell—121. Gravity Daniell—122. Chemical Action in the Daniell's Cell—123. Local Action—124. Grove's Cell—125. Bunsen's Cell—126. Leclanché Cell—127. Potash Bichromate Cell—128. Measuring the Electromotive Force of a Current Generator — 129. Measuring the Resistances of Batteries— 130. P. D.—131. Comparing the Electromotive Forces of Batteries—132. Poggendorff's Method of comparing Electromotive Forces—133. Electromotive Force of a Cell is Independent of its Size and Shape—134. Calibrating a Galvanometer by Employing Known Resistances and a Cell of Constant E. M. F. —135. Arrangements of Cells—136. Arrangement of a given Number of Cells to produce the Maximum Current through a given External Resistance—137. Variation produced in the Total Current by Shunting a Portion of the Circuit—138. Constant Total Current Shunts—139. Independence of the Currents in Various Circuits in Parallel.

117. **Current Generators.**—The *current generators* in practical use may be divided into—

    1. " Batteries."
    2. " Accumulators " or " Secondary batteries."
    3. " Magneto machines."
    4. " Dynamos."
    5. " Thermopiles."

All of these are simply contrivances for converting various forms of energy into electric energy. In thermopiles heat energy is directly transformed into electric energy, just as in a steam-engine heat energy is directly transformed into mechanical energy, or energy of visible motion. In dynamos and magneto machines there is a direct transformation of mechanical energy into electric energy, whereas in accumulators and batteries it is *stored up*, or *potential*, chemical energy that is converted into electric energy.

Figure 4.5: Page 208 from the book "Practical Electricity: A Laboratory and Lecture Course" [Ayrton 1891] "The author W.E. Ayrton differentiates in Section 117 between (1) "Batteries" and (2) Accumulators. In (5) he refers to "Thermopiles". Note that the term "Pile" was used by Matteucci [Matteucci 1847] where he differentiates between Grove's piles, Faraday's piles and Wheatstone's plates, for example.

the page 208 of his book. There he distinguished batteries from accumulators, which he names secondary batteries, for example.

Figure 4.6 is also taken from the book of Ayrton (there Figure 75, page 209) and shows such "battery" of five galvanic Cu-Zn cells (Daniell elements). Since they are connected in series, the total voltage should be $5 \times 1.1 \text{ V} = 5.5 \text{ V}$. It shows five separate containers filled with electrolyte (diluted sulfuric acid), each of which has a Zn electrode and Cu electrode. The Zn electrode of the first cell is connected with a wire, which constitutes the current collector at one pole (defined as the negative pole). The Cu electrode in the first cell is connected via a copper wire with the Zn electrode of the second cell. The electronic current flows from the negative pole through the "battery" to the positive pole. Furthermore, the electronic current flows from the positive pole through the external circuit (which is not shown in Ayrton's Figure 75) to the negative pole.

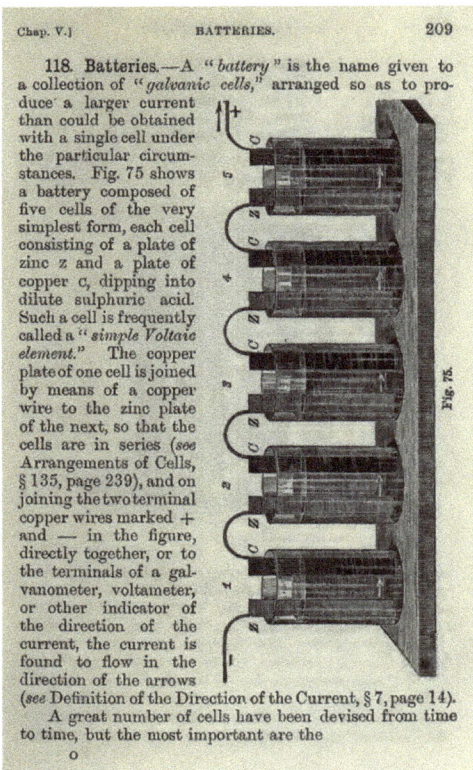

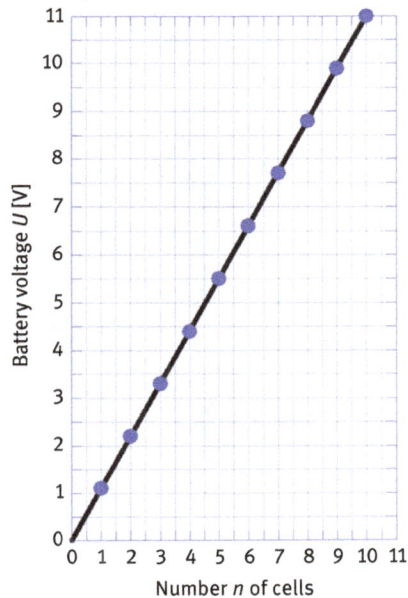

Figure 4.6: Page 209 from the book "Practical Electricity: A Laboratory and Lecture Course" [Ayrton 1891]. It shows how five Daniell elements are assembled to one battery. Ayrton defines "battery" as a collection of galvanic cells.

The purpose of this section about terminology is not to point you to a particular standard term and definition. After all, standard terms are a matter of consent and convention, which may anyway change overtime. Rather I want to make the reader aware that terminology evolves overtimes and may be subject to change. It is therefore important to make sure that we are always "on the same page" with students, teachers, colleagues, project partners, audience, readers and so on. This requires critical thinking and language skills.

As we are adding the cells to a battery in series, as shown on Ayrton's page 209, the electromotive forces (EMF) are adding up linearly. I have made simple linear plot for the readers in the right panel of Figure 4.6, which shows how the EMF of n cells are adding up to the total voltage of the battery with a number n of cells. The EMF of Daniell's element is roughly around 1.1 V, the slope in the plot is therefore 1.1 V/cell.

On the next page 210 (Figure 76) in his book, Ayrton comes to the design details of the cells (see my Figure 4.7). Figure 76 in that book shows the design of the

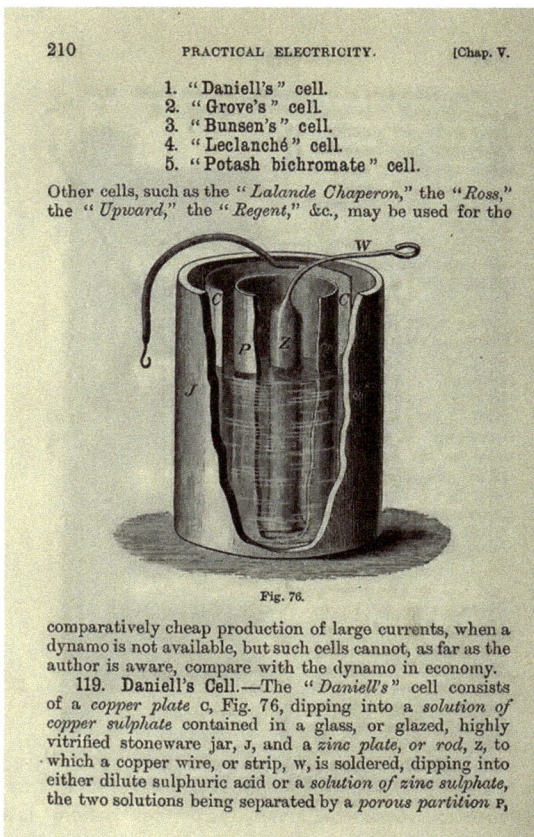

210   PRACTICAL ELECTRICITY.   [Chap. V.

1. "Daniell's" cell.
2. "Grove's" cell.
3. "Bunsen's" cell.
4. "Leclanché" cell.
5. "Potash bichromate" cell.

Other cells, such as the "*Lalande Chaperon*," the "*Ross*," the "*Upward*," the "*Regent*," &c., may be used for the

Fig. 76.

comparatively cheap production of large currents, when a dynamo is not available, but such cells cannot, as far as the author is aware, compare with the dynamo in economy.

119. **Daniell's Cell.**—The "*Daniell's*" cell consists of a *copper plate* c, Fig. 76, dipping into a *solution of copper sulphate* contained in a glass, or glazed, highly vitrified stoneware jar, J, and a *zinc plate, or rod*, z, to which a copper wire, or strip, w, is soldered, dipping into either dilute sulphuric acid or a *solution of zinc sulphate*, the two solutions being separated by a *porous partition* P,

Figure 4.7: Page 210 from the book "Practical Electricity: A Laboratory and Lecture Course" [Ayrton 1891] with Figure 76 showing the design of the Daniell's element.

Daniell's cell with a highly vitrified (no leakage of copper sulfate liquid electrolyte) stoneware jar (J) as principal container, which contains copper sulfate electrolyte solution. In the center of the jar stands a beaker (P), which has porous walls so that electrolyte is soaked through, providing ion transport. The copper electrode (C) is dipped in the electrolyte in the jar (J) and positioned between jar and beaker (P). On the top boundary, a wire (W) is soldered for electric contacting. Inside the beaker (P), a Zinc rod is dipped in zinc sulfate electrolyte solution. The actual EMF of Daniell's cell may change from 1.07 to 1.14 V depending on the electrolyte concentrations. Ayrton points out that the EMF of the various cells is at large independent from the size of the cell.

4.3 The Leclanché element

The Leclanché element (element = cell) was patented on 1866 [Leclanché 1866] by its inventor George Leclanché. In its simplest design, it has a zinc metal anode and a carbon cathode. For an entire century, the Leclanché element was an economical success because of its low cost, good performance and ready availability (Linden 2002). The Leclanché S.A. Company is based near Lausanne in Yverdon Les Bains in the French-speaking part of Switzerland. Paul Rüetschi, a battery expert from Leclanché S.A., has authored a book on batteries [Trueb 1998]. The 9V DURACELL battery shown in Figure 4.2 is a zinc alkaline battery, which has basically the same chemistry like the Leclanché element. The battery is made from six stacked prismatic elements.

Figure 4.8 shows the original design of the Leclanché element. In Chapter II of Bottone's book "Electric Bells and All About Them – A Practical Book for Practical Men" [Bottone 1889], he explains how to build a Leclanché cell at home. The Leclanché element is a primary element, which contains a liquid electrolyte. A primary element or primary cell is designed to be used only once until it is empty and then it is discarded.

Rechargeable elements such as those used in the lead-acid accumulator are rechargeable. To prevent leakage of a liquid electrolyte, it was thickened with gelating agents. It was one of the forerunners of the so-called dry element (Zinc coal element and alkaline manganese battery), which are nowadays standard batteries. When I was working on supercapacitors, we were thinking about immobilizing the electrolyte using Cabosil (Cab-O-Sil®) as gelating agent. This powder from fumed metal oxides can be mixed with liquids (electrolyte) and will swell and gelate it. This prevents the electrolyte from leaking. However, it is not necessary to have excess electrolyte. A highly porous electrode can be soaked very easily with little electrolyte.

The Leclanché element delivers 1.5 V. You can consult a list of Standard Reference Potentials of the involved battery materials (see, e.g., [Atkins 1997, 2010, Aylward 2008, Bard 1985, Bard 2001, Connelly 1996, Cotton 1999, Courtney 2014, Greenwood 1997, Leszczynski 2013, Lide 2006, Pourbaix 1966, Vanýsek 2011, Winter 2018]). Its

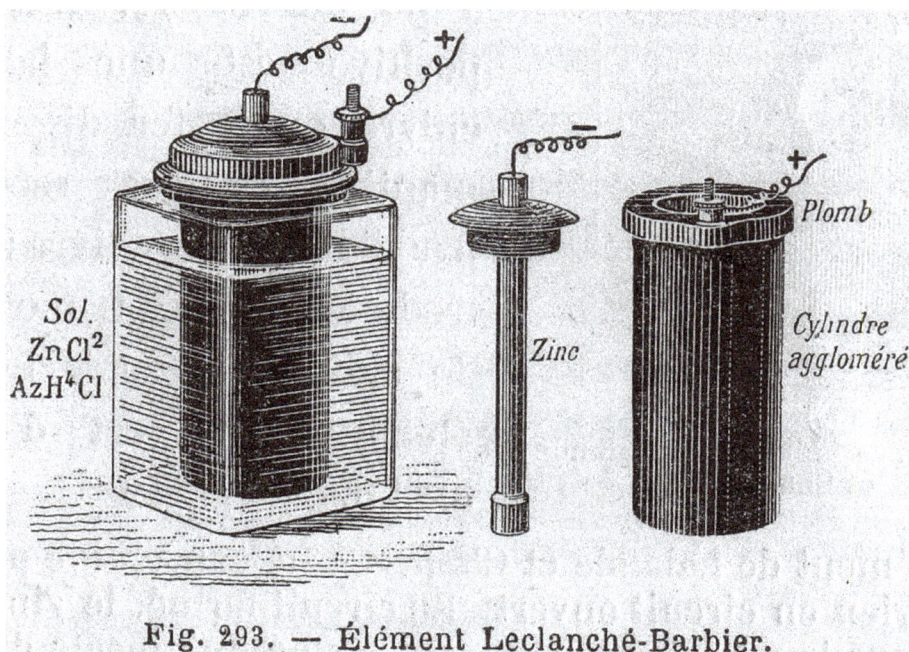

Fig. 293. — Élément Leclanché-Barbier.

Figure 4.8: Schematic of the Leclanché element. (Az = azote, French for nitrogen).
Image with kind permission from Éditions Vuibert.

anode (negative pole) is made from zinc metal ($Zn^{2+}+2e^- \Leftrightarrow Zn$ yields -0.76 V). The cathode is a compressed blend of manganese oxide (MnO_2, Braunstein) and carbon, such as from graphite and acetylene soot. A graphite rod in the mass represents the positive pole of the element. The electrolyte is an aqueous ammonium chloride solution (NH_4Cl x H_2O). The Leclanché element belongs therefore to the alkaline batteries. The electrolyte was in the course of the element development mixed with flour, starch or methylcellulose so as to have a fixated electrolyte. These are the same materials which are used to glue wall paper on a wall. Over the years in the late nineteenth century, the Leclanché element was steadily improved. Separator papers helped to section the electrolyte. Addition of $ZnCl_2$ helped to increase the energy density. The Zn anode was alloyed, which helped to limit the hydrogen formation during discharge (note that this is an application of electrocatalysis). The shelf time and life time of the element were extended by improved container technology against exposure to air.

The cathode is surrounded by manganese oxide MnO_2 (Braunstein). This kind of battery was also called "Braunsteinzelle" (German) and typically used in flashlights. The function of the MnO_2 is a special one. During battery operation, it is possible that the EMF from the standard reference potentials will cause oxidation of the water in the aqueous electrolyte, including the decomposition of the ammonium chloride, both of which can cause hydrogen evolution at the cathode.

The hydrogen evolution is a parasitic process which will lower the EMF and increase the internal resistance. So-to-speak, the hydrogen evolution is an electric shortcut, which will lower the voltage of the Leclanché element. The MnO_2 prevents hydrogen evolution at the cathode. This effect is called depolarization. MnO_2 is therefore used as depolarisator [Depolarisator 1998].

Because of the many components, the electrochemistry of the Leclanché element is not trivial when it comes to adding the standard reference potentials for all reaction that take place in the cell.

The standard potentials for the manganese-based components are not easy to be found. I suggest that the reader looks as an exercise into the reference book by Bard, Parsons and Jordan, and the Electrochemical Series by [Bard 1985, Vanysek 2010].

The anode reaction is the dissolution of the zinc metal by liberation of two electrons:

$$Zn(s) \rightarrow Zn^{2+}(aq) + 2e^-$$

At the cathode, several reactions take place. Manganese in manganese oxide becomes reduced from Mn^{4+} to Mn^{3+}.

$$2\,MnO_2(s) + 2H^+ + 2e^- \rightarrow 2\,MnO(OH)\ (s)$$

An alternative writing of this reaction is the formation of manganite Mn_2O_3. Note that the Mn in this compound has the oxidation number (III), that is, Mn^{3+}. This Mn^{3+} ion is chemically not very stable and has the tendency to disproportionate to Mn^{2+} and Mn^{4+}. When in contact with water, the Mn^{2+} may dissolve. This may be the reason why the form $MnO(OH)$ is preferred. The grains of the MnO_2 are exposed to the aqueous electrolyte only at their surface. Not all matter in the battery is immediately converted.

$$2\,MnO_2(s) + H_2(g) \rightarrow Mn_2O_3(s) + H_2O(l)$$

The zinc ions will form with the ammonia a complexion

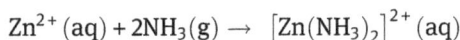

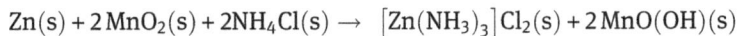

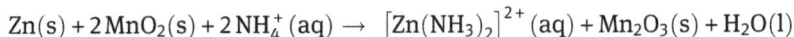

$$Zn^{2+} + 2NH_4^+ + 2Cl^- \rightarrow \left[Zn(NH_3)_2\right]Cl_2 + 2H^+$$

$$Zn^{2+}(aq) + 2NH_3(g) \rightarrow \left[Zn(NH_3)_2\right]^{2+}(aq)$$

$$Zn(s) + 2\,MnO_2(s) + 2NH_4Cl(s) \rightarrow \left[Zn(NH_3)_3\right]Cl_2(s) + 2\,MnO(OH)(s)$$

$$Zn(s) + 2\,MnO_2(s) + 2\,NH_4^+(aq) \rightarrow \left[Zn(NH_3)_2\right]^{2+}(aq) + Mn_2O_3(s) + H_2O(l)$$

In modern times, the zinc coal elements became a different design. The battery container is made from zinc metal, which gives the cell (battery) its cylindrical shape and mechanical and structural stability. This container is also the anode.

The container contains the gelated electrolyte which may look like tooth paste. In the center of the element, along the rotation axis is a graphite rod, which serves as current collector. It is mixed with carbon and manganese oxide. The long-term problem with the Leclanché cell is that the zinc eventually corrodes away. You can see this when the container disintegrates. You may also smell the ammonia from the leaking electrolyte.

Today, battery engineers have succeeded to make zinc coal batteries rechargeable. Since 1993, secondary zinc coal batteries are on the market, which are officially termed as alkaline manganese cells. AccuCell is one brand [Müller 2001]. I mentioned already in Chapter 3 (Daniell element) that it was desirable to extend the lifetime by batteries by turning primary cells to secondary cells. An example of how structural and morphological changes during charging and discharging of a Zn electrode which were monitored with X-ray imaging is shown by Nakata et al. [Nakata 2015].

Manke et al. [Manke 2007] have combined X-ray and neutron methods for the elucidation of structural changes in the anode and cathode parts of the Zn alkaline battery while it was in operation. The neutron method was used to identify hydrogen in the manganese oxide matrix.

Manganese oxide is one important ingredient for batteries as we see now and as we will see later. The production of high-quality battery grade manganese oxide is a technology field of its own. While disproportionation may be seen as a potentially malign process during battery operation, particularly in lithium ion batteries, disproportionation can also be used as a trick in chemical technology [Kanungo 2007].

Manganese oxide is also used as electrocatalyst for the oxygen evolution reaction in alkaline electrolyzers. The chemical processes occurring at manganese oxide surfaces are of interest but not very well understood [Braun 2002]. Even today, researchers carry out fundamental studies with synchrotron radiation, such as *in situ* soft X-ray absorption at the X-ray core levels of O 1s and Mn 2p [Risch 2017].

Battery engineers are constantly on the search for improved battery materials. Doping is sometimes a suitable way of increasing activity of materials or durability; Sharma et al. used Mo-oxide and found improved performance [Sharma 1999].

You can notice from the reaction equations in this Section that the coal does not participate in the battery chemistry. It is possible to electrochemically convert coal directly in solid oxide fuel cells [Deleebeeck 2017].

4.4 The rechargeable lithium ion battery

The Danielle element is a one-way "battery." When all zinc is consumed, the element cannot be recharged. The reason for this is that it was not designed for that. The electrode materials undergo structural and morphological changes during operation and lifetime, and these are not conformal reversible. You cannot bring the battery back in its original shape using an external current the reverse way and then reduce the zinc. However, researchers are working on that. The AccuCell [Müller 2001], for example, is a successful implementation of this for the Leclanché element.

There exist rechargeable batteries as you all know and they are called accumulators. These rechargeable batteries are called secondary batteries. Those batteries which cannot be recharged are called primary batteries. The lithium ion battery is in principle of a rechargeable battery (secondary battery; accumulator). But the first generation of lithium batteries were primary batteries, and many of the lithium batteries are still primary batteries.

Its working principle is as follows. The cathode (which in terminology of rechargeable batteries is always called positive electrode) is made from lithium manganite $LiMn_2O_4$ or some other metal oxide which contains also lithium in its crystal lattice. Prominent are those materials which have a spinel crystal structure, such as $LiMe_2O_4$ with "Me" being a 3d metal like Co, Ni, Mn.

Another interesting cathode material is the olivine crystal structure material $LiFePO_4$[2] [Kulka 2015], and more and more other materials are being explored. $LiFePO_4$ gives a high capacity, but it has a poor conductivity. Poor conductivity will result in poor power density. A material with a high capacity does not necessarily make a good battery. And a very good light absorber does not necessarily make a good solar cell. Therefore, some tricks are necessary to use it as positive electrode, for example, making nanoparticles and coating them with graphite. See in this context also a study on photoelectrochemical cell materials [Chang 2011, Gaillard 2012, 2013].

Let me stay here with the $LiMn_2O_4$. The anode materials (negative electrode) are typically lithium metal, lithium alloys (tin Sn is a favorite alloying partner) or lithium carbon Li_6C. The particular attraction of the lithium battery is that lithium has a very low molecular weight of 6.94 g/mol and a very large potential of −3.04 V versus the standard hydrogen electrode, by virtue of which the lithium battery has a high energy density [Hellwig 2013].

4.4.1 Some basic Li battery cathode chemistry

3d-transition metal oxides are useful for positive lithium battery intercalation electrodes. By topochemical reaction, they can accept or release "foreign" ions in their

---

2 When you do a literature search on $LiFePO_4$ batteries, you may come across a number of studies where Mössbauer spectroscopy is used. This is because the $LiFePO_4$ contains iron Fe, and Mössbauer spectroscopy is a very important method for the study of the hyperfine interactions specifically and almost uniquely for Fe, and for Sn, Sb and Te.

crystal host such as Li^+, Mg^{2+} or H^+. Charge balance is maintained by a redox reaction, and electron flux yields the electric current.

$$2MnO_2 \underset{-Li^+, -e^-}{\overset{+Li^+, +e^-}{\rightleftharpoons}} LiMn_2O_4$$

My standard system during my time in Berkeley was $LiMn_2O_4$ spinel, obtained by a high temperature ceramic reaction, then mixed with carbon black and graphite and polymer binder, cured and dried, then assembled to a cell with Li metal as counter electrode, and polymer separator, and $LiPF_6$ as electrolyte. Such a cell has 3.05 V open circuit potential (OCV) and a gravimetric capacity of 148 mAh/g. The cell can be additionally charged by applying 4 up to 4.3 V, while most of the Li-ions are removed from the spinel host. Redox reactions and structural changes during charging and operation limit the lifetime.

Figure 4.9 displays a disassembled lithium primary button cell battery from Maxell. The nominal voltage is 3 V. You can identify in the photos the two stainless steel endcaps, the cathode and the corroded lithium anode, and the white separator. These components are stacked together, as indicated in the sketch on the right side in Figure 4.9.

Maxell CR2016/3 V lithium primary battery (not rechargeable)

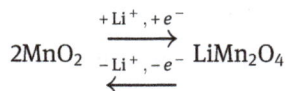

3/4 inch

| End cap with negative Lithium metal electrode (LiOH, already corroded) | Separator membrane (glass fiber fabric) | Positive electrode with (unknown) active material, carbon, binder, and metal matrix for better conductivity | Endcap |

Cathode, $LiMn_2O_4$

Electrolyte, $LiPF_6$

Anode, Li, LiC_6

LiOH, Li_3N

Figure 4.9: Disassembled primary lithium battery from Maxell. Top row shows stainless steel current collector and end plate from inside with corroded lithium metal. The white disk is the separator. The black smaller disc is the active electrode material. The black metal disk shows the other stainless steel end plate from inside, with traces from the black active material.

My colleague Erich Deiss from PSI had shown how average voltage, energy density and specific energy of lithium ion batteries can be determined from first principle calculations [Deiss 1997]. I still remember when he presented in our monthly PSI electrochemistry seminar (1996 or 1997) what he called a "Blumenstrauß" of differential equations, which he typically treated with FORTRAN programming language.

It is possible to determine the voltage for lithium insertion by total energy calculations, which is a tool of condensed matter physics. The method (density functional theory) is well established and the mathematical apparatus is available either as commercial software package or freeware. You begin with assuming a crystal structure where you type in the lattice positions of either atom of the composition that you want to study, and the mathematical formalism determines the total energy. The mean insertion voltage is then [Mishra 1999]

$$\bar{V} = -\frac{\Delta G_r}{F} = \frac{1}{F} \cdot (\Delta E_r + p\Delta V_r + T\Delta S_r) \cong -\frac{\Delta E_r}{F}$$

The insertion and extraction of lithium from the spinel crystal host cause changes in the volume V and entropy S, but these changes are so small that they are negligible. Therefore, it is sufficient to determine the internal energy E_r only.

Deiss et al. [Deiss 1997] showed that the total energy of lithium ion batteries calculated at 0 K (this is typically the temperature for which one can calculate condensed matter systems. This is why we physicists like 0 K temperature.) differ not more than 30 meV. The calculation at 0 K is therefore a very good approximation for the inner energy at 300 K [Mishra 1999].

According to the above chemical reaction equation, we see that the lithium stoichiometry is changing during the charging and discharging of the cell. The actual stoichiometry is thus parameterized with an x such as $Li_{(1-x)}Mn_2O_4$.

You synthesize the material in your laboratory as a powder and then check with X-ray diffractometry whether your material indeed represents the crystallographic phase that you want. Then, you mix this spinel powder with some amount of graphite (which is made from graphite mineral or from synthetic graphite) and carbon black (soot from acetylene combustion; soot is used in car tires which gives them stability and black color).

Graphite increases the electronic conductivity between the spinel grains, and carbon black helps distributing the graphite flakes and spinel particles. Then, a polymer binder such as polyvinylidene fluoride (PVDF), which is dissolved in some organic solvent, is mixed to the powder mixture. This is then poured over aluminum foil and fine distributed with a doctor blade to a homogenous film. Let it dry in air and bring it then in a drying furnace where the solvent evaporates at 70–100 °C. Then take the aluminum foil with electrode film (Figure 4.10) out and use a punch knife to make small electrode disks with identical diameters. The black film is the active cathode material and the aluminum foil is the current collector.

Figure 4.10: Aluminum foil coated with dried LiMn$_2$O$_4$ slurry. The rectangular shape of the dark grey slurry is obtained by putting plastic space holders and scotch tape on the aluminum foil. The space holders assist in defining a height over which you can distribute with a ruler the slurry to equal height. Produced by Dr. D.K. Bora at Empa.

Lithium batteries run on organic electrolytes. The decomposition potential of organic electrolytes can be as high as 4 V. This means that you can apply to the battery a voltage up to 4 V before the electrolyte decomposes. Because the stored energy in a battery scales with the square of the voltage, battery engineers welcome organic electrolytes. Going as high as 4 V means the energy density will rise by a factor of 4 × 4 = 16 when compared to an electrolyte which has a decomposition potential of 1 V only.

Organic electrolytes for lithium batteries must not contain any residual water, otherwise the water will decomppose already at the water electrolysis conditions of theoretically 1.23 V and practically 1.8 V. The evolving gas can damage the battery. Also, such water can form aggressive HF when H$_2$O and LiPF$_6$ react, and thus can corrode the electrodes. There are however efforts to make water-based electrolytes for lithium batteries [Suo 2015].

Note therefore that when you exceed the voltage over the electrolyte, you invest your energy in the decomposition of the electrolyte and not into storing your energy in the battery. You will also lose your electrolyte and thus the functionality of the battery. When the decomposition builds up gas, the battery may bulb out and get damaged and it may even explode. Therefore, the decomposition voltage of the electrolyte is an important technical constant.

4.4.2 Excursion: Production of battery graphite with hydropower in Switzerland

Graphite is a carbon mineral, which we use, for example, in the "lead" pencils. There are many graphite producers worldwide. Graphite, for example, is made by the Swiss company Timcal Switzerland, which a few years ago changed its name to Imerys Carbon and Graphite. The company receives the *natural graphite* from mineral mines in several countries in the world. The company produces *synthetic graphite* from the thermal decomposition of silicon carbide (SiC) at a factory in Bodio the Ticino Valley in Switzerland.

The thermal decomposition of SiC is an electrothermal process. Huge piles of SiC slack from steel factories are arranged in long sausages of maybe 10 m long (or longer) and around 3 m diameter. Two huge electrodes are attached in front and end of the SiC "sausage." SiC is electrically conducting and will be heated up by the strong current (many Amperes), which is applied through the electrodes. The silicon will separate and the carbon remains inside as graphite, which is then packed in bottles and sold for battery production, for example. The graphite product that I know was called TIMREX®.

When I visited Timcal I was surprised about how simple the electrothermal process looks like. I wondered how they can make a business, can make a profit in Switzerland, which is a country with fairly high salaries. This is because of the cheap electric power, which one can get from the local hydropower plant (http://www.aet.ch). Figure 4.11 shows an aerial photo of the small city of Bodio in the Ticino Valley. Ticino is the name of the river and also the name of the Swiss state, where Italian is the main language. On the left you see the Highway which brings you from Switzerland to Italy, Milano. On the right you see the railway. By the way, the new Gotthard Basistunnel (57 km through the Alps) has its southern entry in Bodio. And in the middle front you see the factory buildings of Timcal. Being next to such major arterial road and railway connection is certainly important for a factory, which wants to deliver its product to customers worldwide.

How does it work with the hydropower? Let us look at Figure 4.12. There is lake "Lago Tremorgio" up in the mountains some 10 km away from Bodio, as illustrated on the map, and its water can rush through large pipes steeply down where it hits on turbine blades which run an electric power generator in the shown building at Centrale Tremorgio, which delivers electricity in the valley. When there is no natural lake, engineers can build a dam and have the power plant also built right in the dam or near the base of the dam. For example, Germany has 270 dams, many of which feed electric power plants [Köngeter 2013].

Figure 4.11: Timcal changed recently its name to Imerys Graphite & Carbon. http://www.imerys-graphite-and-carbon.com. The Timcal company has its graphite factory in the city of Bodio in the Ticino Valley, Switzerland. Photo by kind permission of Imerys Carbon and Graphite.

Lago Tremorgio has an altitude of 1,850 m. The power plant Centrale Tremorgio is located at 1,000 m above sea level. The height difference of around 800 m is the potential energy which is converted into kinetic energy which drives the turbines. This is sketched in Figure 4.13. At peak times (lunch time and evening time) when there is high demand for electricity in Europe, many such mountain lake power plants in the Alps produce and sell the electricity at a high market price. When there is low demand in Europe for electricity, its price is low. The mountain lake power companies use this cheap electricity from the grid and pump the river water up in the mountain. This has been a very lucrative business model for Switzerland for a

Figure 4.12: (top left) Lago Tremorgio which delivers water to Azienda Elettrica Ticinese, the mountain lake power plant building which houses the electric generators (lower left). (top right) Map showing Alp mountains, position of Lake Tremorgio and City of Bodio. (lower right) Electricity generator from Brown Boveri. Photos from Azienda Elettrica Ticinese, CH-6513 Monte Carasso.

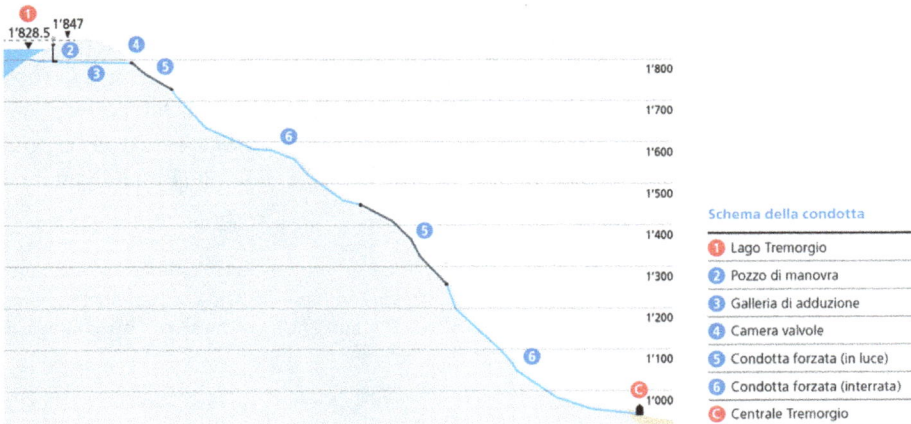

Schema della condotta

1 Lago Tremorgio
2 Pozzo di manovra
3 Galleria di adduzione
4 Camera valvole
5 Condotta forzata (in luce)
6 Condotta forzata (interrata)
C Centrale Tremorgio

Figure 4.13: Height profile of the Lago Tremorgio and power station in Centrale Tremorgio at the ground in Ticino Valley, Switzerland. A height difference of 800 m provides the potential energy which is converted into kinetic energy through the fall on the turbine blades on the ground. Figure from Azienda Elettrica Ticinese, CH-6513 Monte Carasso.

long time. Only the cheap solar power in Germany is now making competition in Switzerland. Anyway, the low electricity cost in this Valley makes it possible that energy intensive factories in the valley can produce at low cost. The Oleftalsperre, in Germany at the border to Belgium where I grew up, has 3 MW power and produces 2.6 GWh/a electric energy with two Francis turbines from water which falls 52 m. The water area of the dam is 1.05 km^2 [Köngeter 2013].

4.4.3 Carbon materials in battery electrodes

The search for "better" carbon materials in battery applications is an ongoing quest. This is not withstanding that there is a very wide range of low cost and high performance carbon materials available for that purpose. Soot from the burning of acetylene in fat conditions (low oxygen concentration) is produced. Carbon black is produced by pyrolysis of liquid hydrocarbons to about 8 million tons per year worldwide, 90% of which for use in rubber (tire) industry. Less than 5% are used for batteries [Collection 2012]. The economic volume of carbon black is way over 10 billion dollar per year [Notch 2015]. Let us assume then we have 10 million tons which cost 20 billion \$. Then, for 1 cent you get 5 g of carbon black. Compare this to the relatively high costs of a battery. This means there is a huge amount of carbon black available at relatively low cost.

However, the carbon must have the right structure and condition for batteries and supercaps and fuel cells. It should have a high electrochemically available surface area with the right pore size distribution and geometry.[3] It should have the necessary electronic conductivity.

Charcoal is a carbon material from pyrolysis of trees, leaves and other plants. The tiny channels which are necessary in plants for transport of water are hierarchically arranged and quite useful for applications in battery electrodes and supercaps. Carbon derived from biomass has been the subject of numerous studies for battery and capacitor applications. Some researchers consider how organic waste such as coconut shells are rice shells can be pyrolyzed to battery carbon, for example, find Kalyani and Anitha [Kalyani 2013]:

> Particularly, carbon (unactivated) derived from pyrolyzed peanut shells exhibited a maximum specific capacity of 4765 mAh/g in the case of lithium-ion batteries and coconut shell derived carbon in KOH electrolyte gave capacitance of 368 F/g and $ZnCl_2$ activated carbon from waste coffee grounds exhibited 368 F/g in H_2SO_4.

---

3 To some extent, it is possible to "guess" the pore geometry from impedance measurements [Keiser 1976] Keiser H, Beccu KD, Gutjahr MA: Abschätzung der porenstruktur poröser elektroden aus impedanzmessungen. *Electrochimica Acta* 1976, 21:539-543. doi: 10.1016/0013-4686(76)85147-x.

Other researchers investigate how fancy synthetic carbon materials such as nano-tubes [Kumar 2018], fullerenes and graphene can be used for the same purpose. The latest approach that I learnt of was using soot from heavy diesel engines on marine ships [Lee 2017].

4.4.4 Assembly of the lithium ion battery

Now we put this electrode assembly in a Swagelock® cell, which is prepared for battery experiments, with the active film on the top and the aluminum at the bottom in direct electric contact with the stainless steel cell. Figure 4.14 shows such Swagelock® cells. They contain two massive steel cylinders between the two of which you place the lithium negative electrode, the separator and the spinel positive electrode. Two poly-propylene jackets warrant that the cell is properly sealed. A middle piece from steel with threads forms the container.

With two screws and threads you squeeze the cylinders and the jackets together. A spring loaded sample table warrants that there is still good enough electric contact

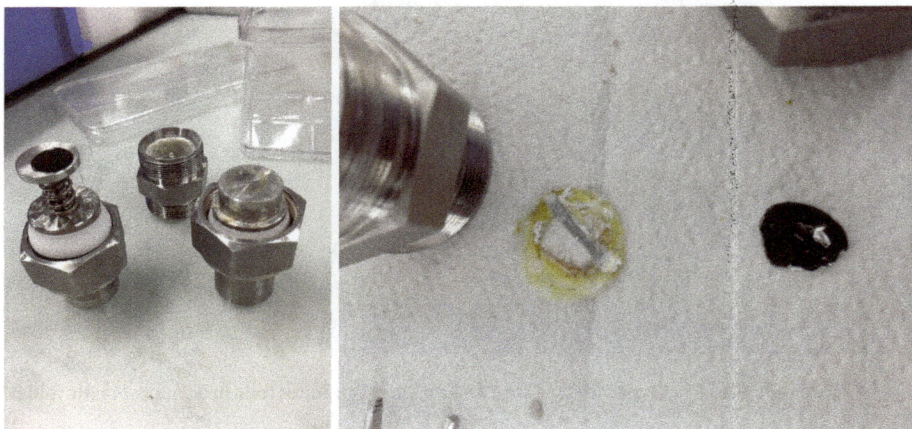

Figure 4.14: Swagelok Cell for battery measurements. Stainless steel parts can be assembled to a cell and the interior is gas tight sealed. (Left) Center piece from stainless steel is with threads and an inner tube from polypropylene showing residues from electrolyte. This inner piece is electrically insulated from the two top and bottom stainless steel cylinders and allows for connection with a lithium metal stripe as reference electrode. (Right) Spring load mechanism with a current collector and spinel electrode disk on top are attached to a stainless steel cylinder. The cylinder shown on the very right would carry the lithium metal electrode. The center piece in the middle connects the steel cylinders with a screw thread mechanism. The polypropylene warrants electric insulation. You may want to wrap an extra sticky scotch tape around the cylinders to improve electric insulation. Small threads drilled in the steel cylinders allow for metal screws inside which can be connected with crocodile clamps to the potentiostat.

between the electrodes and the current collectors (this is the two aforementioned cylinders). With properly insulated electric contacts you have a cell which can be operated in a typical 3-electrode arrangement. You have to practice assembling and handling this cell and eventually you can run nice CVs and charge–discharge curves, impedance and so on.

Now take a Celgard polymer separator sheet and punch disks with a diameter 2 mm larger than the electrodes (e.g., Celgard 3401 or, as shown in Figure 4.15, Celgard 2400). Insert it over the electrode in the Swagelock cell. No later than now should you continue with the work in an argon-filled glove box or glove bag. As we will be working with metal lithium, the nitrogen in the air will react with the lithium and make Li_3N. With the humidity in the air, it will react to LiOH. Therefore, we need to work in inert atmosphere, and nitrogen in this work is no inert atmosphere. Now with the first Celgard separator on top, take a strip of lithium and place it on the separator. It should be long enough to reach over the separator boundary so that you can contact it with the reference electrode cable from the potentiostat.

Figure 4.15: A thin pile of folded white Celgard 2400 polymer separator foils in a yellow plastic folder.

Now take another separator disk and place it over the lithium stripe, concentric with the electrode, and the first separator. When you have experienced difficulties with placing the separator properly, you may want to add one drop of battery electrolyte (Merck) on them. Polymer separators are sometimes electrostatically charged; a drop of electrolyte will make them stick on other surfaces and then it is easier to handle them. Finally, you punch out a disk of lithium metal and place it over the top separator. Now add several drops of battery electrode in the battery cell and observe how the white separator becomes opaque because it is wetted. The battery electrolyte is a mixture of di methyl carbonate in which $LiPF_6$ crystals are dissolved.

The top piece of the Swagelock Cell is now pressed on the lithium piece that it has direct electric contact. Now close the cell firmly and lock it. You can check now the voltage of the battery cell with a multimeter. It should show around 3.05 V. This open circuit voltage (OCV) is what we would expect from the Redox couple $LiMn_2O_4$ versus Li metal.

Now we are charging the battery. For this, we connect it with a power supply or with a battery cycler or potentiostat (galvanostat) and impose a current on the battery. We connect two multimeters (one in parallel for the voltage, and one in series for the current) so that we can monitor the current and the voltage over the lithium battery cell. Certainly you can use a full automatic battery cycler for this purpose, but you understand and learn things better when you assemble everything step-by-step by yourself with simple tools and devices.

Tonight my wife showed me "rice paper" which is used for making Vietnamese food (Figure 4.16). I had never seen it before but its shape and structure and consistency reminded me of 12 inch wafers used in semiconductor technology. I thought this rice paper, when pyrolyzed, might give a good carbon electrode. I did some Internet search and immediately found this publication by Zhang et al. [Zhang 2012] who used the rice paper as a separator in their lithium battery and compared it with a Celgard 2400 separator. Soon I found another paper in which the authors carbonized the rice paper and used it as an electrode component for lithium batteries [Zhang 2013].

Figure 4.16: (left) A rice paper wafer for food applications, comparable to the one used in [Zhang 2012, 2013]. (right) Shrimp and vegetable wrap with cooked rice paper.

4.4.5 Electrochemical characteristic of lithium battery spinel cathodes

Figure 4.17 shows overlaid two cyclic voltammograms of one $LiMn_2O_4$ positive electrode in the potential range from 3.1 to 4.4 V from a fresh and from a 100 times cycled battery [Tucker 2002]. The CV was run at an extremely low scan rate of 1 week per half scan. The two peaks reveal two current waves which originate from the oxidation of the cathode [Rougier 1998, Striebel 1999]. My colleague Mike Tucker had recorded the first CV (fresh electrode) immediately after he immersed it into the electrolyte to record its condition as "fresh" as possible.

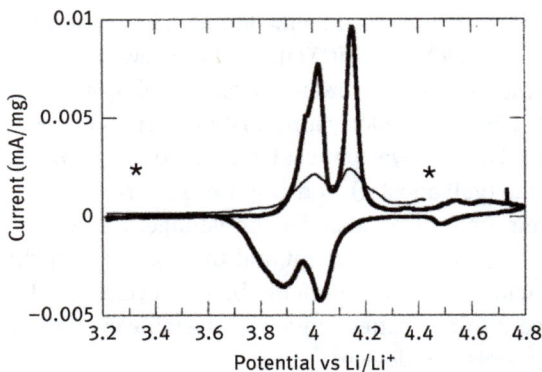

Figure 4.17: Slow scan cyclic voltammogram of fresh $LiMn_2O_4$ (thick line) and $LiMn_2O_4$ cycled 100 times (thin line) on the 4 V plateau. The potential limits imposed on long-term cycling are indicated with asterisks.
Republished with permission of The Electrochemical Society Inc., from Journal of The Electrochemical Society, A⁷Li NMR Study of Capacity Fade in Metal-Substituted Lithium Manganese Oxide Spinels, Tucker MC, Reimer JA, Cairns EJ, 149, 5, A574 – A585, 2002; permission conveyed through Copyright Clearance Center, Inc. [Tucker 2002].

The electrode was weighed on a microbalance so that the current can be related to the mass of the positive electrode. The current density is smaller than 0.01 mA/mg. Tucker et al. noticed immediately two changes, that is, the capacity as integrated over the scanned range is considerably decreased (to around 60%), and the two oxidation peaks become broader. When the CV scans were made faster (e.g., 100 mV/s would be faster than 10 mV/s), the capacity was lower. This suggests that there were some diffusion limited losses.

Tucker et al. give a rational for the broadening of the oxidation peaks during de-lithiation of the cathode, that is, the substitution of lithium for manganese in the crystallographic 16d site and the resulting problems for a necessary order–disorder phase transition according to Gao et al. [Gao 1996].

When you start out with a fresh electrode, the initial CV sweep may first show one large single peak and then evolve over several cycles to the well-known shape with double peaks, as shown in Figure 4.18 [Striebel 1999]. I would call these first few cycles as the "formation cycles." These two peaks correspond to the reversible deintercalation of lithium from the spinel with two corresponding energy levels of 4.03 and 4.17 V. The spinel film was made by pulsed laser deposition (PLD). This relatively novel deposition method became established in the mid-1990s and certainly was explored from battery researchers.

The conjugated peaks for the intercalation of lithium show in cathodic direction at 4.11 and 3.99 V. The four peaks are labeled 4.17 V (A1) and 4.03 V (A2) for the anodic sweep (charging) and 4.11 V (C1) and 3.99 V (C2) for the cathodic sweep (discharging) in Figure 4.19. Striebel et al. [Striebel 1999] report that the total charge for three

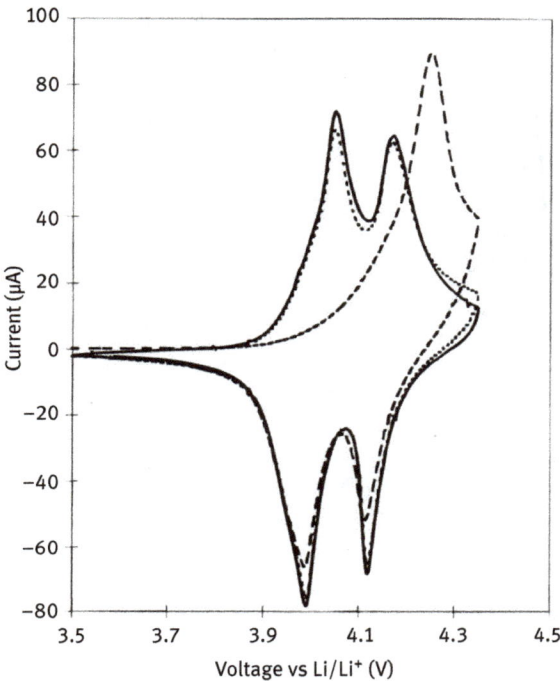

Figure 4.18: The approach to a steady-state cyclic voltammogram for a 300 nm thin $LiMn_2O_4$ film immediately after immersion in 1 molar $LiPF_6$/EC/DMC (1:2), 1 mV/s: 1st cycle (– – –), 2nd cycle (.......), 3rd cycle (——).Republished with permission of the Electrochemical Society Inc., from Electrochemical studies of substituted spinel thin films, Striebel, K. A. Rougier, A. Horne, C. R. Reade, R. P. Cairns,E. J., 146, 12, 4339–4347, 1999; permission conveyed through Copyright Clearance Center, Inc. [Striebel 1999].

Note: I want to make here a footnote remark on the difference between thick films and thin films. This depends on the position of the observer. A surface scientist who monitors how single atoms are deposited on a surface and arrange as a film when one flat layer of atoms with one atom thickness is completed, may call this layer as thin film or ultrathin film. When the surface is not completely covered with atoms and when there are "hole patches" in the incomplete film, or when the coverage is so small that we speak rather of islands formed by atoms, then we talk about fractions of a monolayer (ML), such as 0.7 ML, for example. From this perspective, a film of 100 atomic layers (around 100 nm thick [Braun 2003a] Braun A: Conversion of Thickness Data of Thin Films with Variable Lattice Parameter from Monolayers to Angstroms: An Application of the Epitaxial Bain Path. *Surface Review and Letters* 2003a, 10:889-894. doi: 10.1142/s0218625x03005761.) is already thick or even extremely thick for the surface scientist. Engineers who work with films made from particles with a primary size of 10 μm may arrive at a 25 μm film thickness for a battery electrode [Braun 2015] Braun A, Nordlund D, Song S-W, Huang T-W, Sokaras D, Liu X, Yang W, Weng T-C, Liu Z: Hard X-rays in–soft X-rays out: An operando piggyback view deep into a charging lithium ion battery with X-ray Raman spectroscopy. *Journal of Electron Spectroscopy and Related Phenomena* 2015, 200:257-263. doi: 10.1016/j.elspec.2015.03.005. With this experience from a standard battery electrode of say 10 μm thickness, an electrode deposited with PLD and with only 0.3 μm thickness is then a thin film. Therefore, it is often a matter of perspective, and a film of 100 nm thickness may be thin for one researcher and thick for another researcher.

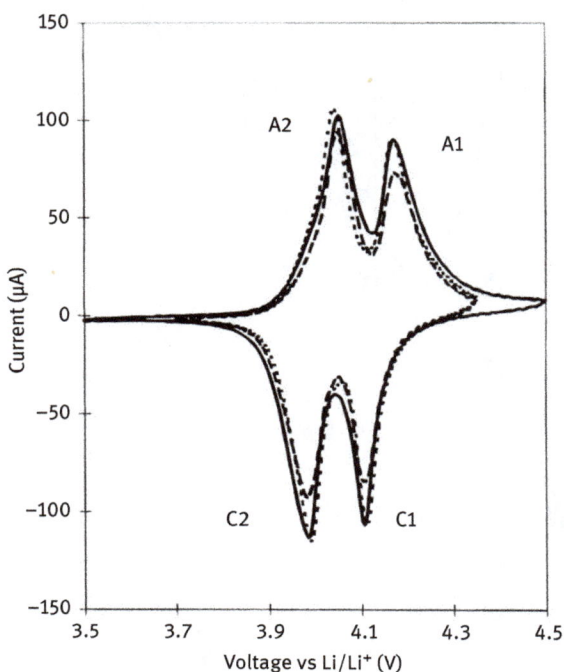

Figure 4.19: Cyclic voltammograms of three 300 nm thin $LiMn_2O_4$ films in 1 M $LiPF_6$/EC/DMC (1:2), 1 mV/s showing reproducibility of the steady-state shape.
Republished with permission of The Electrochemical Society Inc., from Electrochemical studies of substituted spinel thin films, Striebel, K. A. Rougier, A. Horne, C. R. Reade, R. P. Cairns, E. J., 146, 12, 4339–4347, 1999; permission conveyed through Copyright Clearance Center, Inc. [Striebel 1999].

consecutive CVs varies by less than 3%, as is indicated by the change of the height of the peaks in the CV with increasing sweep number. It is possible, as already mentioned to model the lithium intercalation at the quantitative level, that this includes also accounting for the microstructure of the electrode and not only the electronic structure [Deiss 2001].

Note therefore that the CVs shown in Figures 4.17–4.19 are specific to the measured sample only. Slight changes in composition (stoichiometry) can cause changes in the CV. When you exchange precursors for electrode synthesis, you may get such which have more or less impurities depending which quality you purchase from which vendors. Having impurities in a product that you purchase as precursor does not necessarily mean you get a worse electrode. The term "impurity" has a negative connotation but its effect may be throughout positive on electrode performance.

Note also that we are here dealing with the cathode (positive electrode). The anode (negative electrode) may have different characteristic. Lithium carbon or lithiated carbon such as Li_6C can be used as anode. The lithium atoms are intercalated in between the graphene sheets of the carbon. Here in the spinel cathode, the lithium is extracted (de-intercalated) from the crystal lattice during charging and inserted, that is, intercalated.

You may intentionally add foreign elements in trace concentrations to the mixture which is known as doping. When you add a significant concentration of such elements you may call it as substitution. We can discuss and debate what is the difference between doping and substitution (see, e.g., [Merz 2016]). The term doping is typically used in semiconductor technology and refers to the use of foreign elements with a concentration of not more than 1%. The term substitution is typically used in inorganic chemistry and materials science and bears some similarities to alloying from metallurgy. The difference in concentration is here in the order of 10% and certainly can be also 50% (exceeding 50% means you have to re-name your compound from the beginning) or 2%. At 1%, you may call it as doping again.

It may be difficult to understand how 1 in a 100 atoms can make a difference. Look it from the social perspective. One person in a crowd of 100 or more can be destructive and stop cooperation, or it can launch new cooperation of the crowd just by behavior and speech, such as in a football stadium, in a parliament or on a marketplace. In condensed matter, there are numerous ways of cooperation such as phonons, spin waves and polarons. Transport properties can be substantially affected by doping, and this is why some material engineers need very high purity materials and others intentionally dope it with foreign elements.

A case with substitution of Mn in the parent compound $LiMn_2O_4$ by foreign elements Ni and Co is shown in Figure 4.20. We see here substitution levels of 5% and 20% on the B-site in the AB_2O_4-type compound (spinel) with A = lithium and B = manganese, cobalt and nickel. $LiMn_2O_4$, (– – –) $LiMn_{1.90}Ni_{0.10}O_4$, (........) $LiMn_{1.75}Co_{0.25}O_4$, (.–.–.–.) $LiMn_{1.75}Ni_{0.25}O_4$. The position and relative "spectral weights" of the oxidation peaks change when the composition of the positive electrode (cathode) changes. This electrochemical response reflects the different oxidation and reduction levels of the B-site elements, particularly when two or more different elements are on the B-site positions.

When you impose a current on the positive electrode in anodizing direction along the potential axis so that you oxidize it, the lithium ions are extracted from the electrode (de-lithiation). You can determine the electric charge involved in this process by integrating the current over the time. When you return to the original potential then you reverse this process as a reduction on the cathodic path. Here, too, you can determine the charge involved by integration. The area under a CV, where you do both processes one by one, corresponds therefore to the electric charge which is put into the electrode and which is taken out from the electrode. The relative concentration x of lithium in the electrode with composition $Li_xMn_2O_4$ is therefore a parameter over which the current and potential can be related to.

The height and width of the peak therefore determine which is the largest contribution to the capacity, if we can ignore the double layer capacity which often

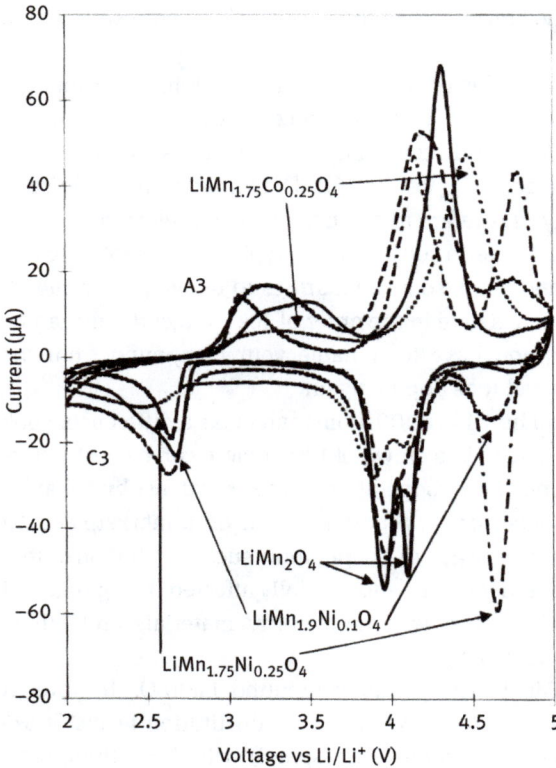

Figure 4.20: Extended range cyclic voltammograms of 0.3-μm-thick pure and substituted films in 1 M LiPF$_6$/EC/PC, 1 mV/s: (——) LiMn$_2$O$_4$, (– – –) LiMn$_{1.90}$Ni$_{0.10}$O$_4$, (........) LiMn$_{1.75}$Co$_{0.25}$O$_4$, (.–.–.–.) LiMn$_{1.75}$Ni$_{0.25}$O$_4$.
Republished with permission of the Electrochemical Society Inc., from Electrochemical studies of substituted spinel thin films, Striebel, K. A. Rougier, A. Horne, C. R. Reade, R. P. Cairns, E. J., 146, 12, 4339–4347, 1999; permission conveyed through Copyright Clearance Center, Inc. [Striebel 1999].

extends homogeneously and flat over the entire potential range. Striebel et al. [Striebel 1999] can therefore assign a capacity not only to the entire electrode charging range, but also limit the capacity to particular states of charge. You see this in Table 4.4 for the aforementioned compositions. LiMn$_2$O$_4$ has a theoretical capacity of 148 mAh/g when the battery is fully charged and thus all Li extracted. The stoichiometry reads then "Mn$_2$O$_4$" and the Mn has the oxidation state Mn^{4+}. As there is no current wave for the potential larger than 5 V, there is also no specific capacity in this range (NA, Table 4.4).

When the battery is only have been charged, the Li content is 50% so that we can speak about a range where the Li stoichiometry parameter x is larger or smaller than 50% ($x > 0.5/x < 0.5$). The substituted electrodes have the CV peaks at different

Table 4.4: Theoretical capacity calculations for $Li_xMn_{2-y}Me_yO_4$.

Starting Composition ($x = 1$)	Full capacity to Mn^{4+} (mAh/g)	Capacity ratio $x > 0.5/x < 0.5$	Capacity at $E < 3.5$ V (mAh/g)	Capacity for $5 > E > 4.5$ V (mAh/g)
$LiMn_2O_4$	148	50/50	148	NA
$LiMn_{1.90}Ni_{0.10}O_4$	118	60/40	163	30
$LiMn_{1.75}Ni_{0.25}O_4$	74	100/0	184	74
$LiMn_{1.75}Co_{0.25}O_4$	111	67/33	147	0

Republished with permission of the Electrochemical Society Inc., from Electrochemical studies of substituted spinel thin films, Striebel, K. A. Rougier, A. Horne, C. R., Reade, R. P. Cairns, E. J., 146, 12, 4339–4347, 1999; permission conveyed through Copyright Clearance Center, Inc. [Striebel 1999].

positions. Their stoichiometry includes Ni and Co and this has effect on the valence state of the Mn. It is left as an exercise for the reader to calculate the oxidation state of Mn when all Li is extracted or present in the electrode. Keep Ni constant as Ni^{2+} and Co as Co^{2+}. The full capacity for the low-level Ni substituted electrode is 118 mAh/g. The aforementioned capacity ratio around half-charged electrode is 60%/40%. For the lower potentials <3.5 V, its capacity is 163 mAh/g, and for the potentials between 4.5 and 5 V, it is 30 mAh/g.

The theoretical possible capacities of many battery electrodes have been determined overtime. Table 4.5 lists the ratio of actual measured capacity over the theoretical capacities. The doping or substation of Mn with Ni and to a very minor extent for Co brings obviously a success because these exceed the capacity of $LiMn_2O_4$ parent compound.

Table 4.5: Integrated capacities as percentage of theoretical values.

Composition	"4-V" capacity (%)	"4-V" peak ratio ($x > 0.5/x < 0.5$)	"3-V" capacity* $E < 3.5$ V (%)	Capacity for 4.5 $> E > 5$ V (%)
$LiMn_2O_4$	103	55/45	70	NA
$LiMn_{1.90}Ni_{0.10}O_4$	123	61/39	60	107
$LiMn_{1.75}Ni_{0.25}O_4$	132	100/0	53	86
$LiMn_{1.75}Co_{0.25}O_4$	106	64/36	24	0

*Calculated from the cathodic sweep.
Republished with permission of the Electrochemical Society Inc., from Electrochemical studies of substituted spinel thin films, Striebel, K. A. Rougier, A. Horne, C. R. Reade, R. P. Cairns, E. J., 146, 12, 4339–4347, 1999; permission conveyed through Copyright Clearance Center, Inc. [Striebel 1999].

Often, the diagnostics of a battery is the charge and discharge curve where the voltage is monitored over the charge, the latter ranging from 0% to 100%. Often matters also the speed at which the battery is charged and discharged in cycles.

This relevant quantity is then termed as rate and indicated as the C-rate: The rate 1C means that the battery is charged in 1 h. The rate xC means that the battery is charged in $1/x$ h. The rate 2C therefore means the battery is fully charged in ½ h, that is, 30 min. When we charge the battery slower than in 1 h, such as in 20 h, then we will call the rate as $C/20$ or $0.05C$.

Figure 4.21 shows the charging and discharging of a rechargeable lithium battery where the current and voltage are plotted versus the charging and discharging time [Braun 2015]. We begin charging with a constant current of 0.1 mA, over which the voltage is increasing from 4 to 4.3 V. You can transform the time coordinate to charge coordinate by integration of the current overtime; then we can plot the voltage over the charge, as mentioned before in the explanation of the C-rate.

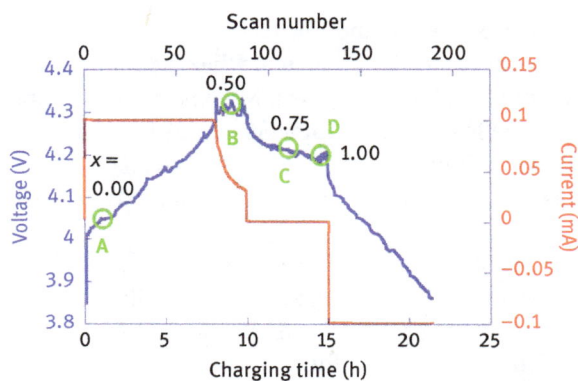

Figure 4.21: Transient of cell voltage (blue line) and cell current (red line) of an LiMn$_2$O$_4$-based model battery for synchrotron experiments [Braun 2015].

The curves in Figure 4.21 do not look as nice and smooth as those from battery cycle instruments. The reason for this is because the cell was an *in situ* cell which was specifically designed for X-ray Raman synchrotron experiments. Emphasis therefore had to be put not only on the proper electrochemical operability but also on the suitability for a quite novel kind of synchrotron measurements [Braun 2015].

In the following Section, I will explain how such cells (X-ray spectroelectrochemical cells, *in situ* cells, *operando* cells) can be used to study processes that take place in batteries can be monitored at the molecular scale with X-rays.

4.4.6 *Operando* and *in situ* X-ray spectroscopy on a lithium ion cell

We have speculated in the previous Section over the oxidation states of the 3d metal ions on the cathode material. I say "speculated" because it is not so obvious from electrochemical or electroanalytical investigations whether an ion has a lower or higher valence.

Oxidation states are typically determined with chemical titration, which is a classical method of analytical chemistry. The method is well established but also somewhat laborious. An alternative method which has become more and more established for that purpose since around 30 years is X-ray spectroscopy. The atoms are absorbing X-rays and excite electrons when the X-ray energy is identical with a core-level energy of the atom. When you scan the X-ray energy by which you irradiate the sample, you can record an X-ray absorption spectrum, which shows particular spectral features which allow you to tell the type of element present in the irradiated sample, its oxidation state and also its spin state. Generally you can determine much of the molecular and electronic structure of materials this way [Braun 2017].

The easiest way to perform such analysis is by directly exposing the sample (an electrode such as a cathode) to the X-ray beam, after you have disassembled the battery cell. In the Section about supercapacitors, I have shown one experiment where glassy carbon electrodes were studied with small angle neutron scattering [Braun 2004]. The electrodes were prepared electrochemically under different conditions, then removed from the electrochemical cell and transferred "in dry condition" to the new Swiss Spallation Neutron Source SINQ[4] at the Paul Scherrer Institute [Atchison 1986, 1989, Bauer 1998a, b, 1999, Fischer 1986, 1988, 1989, 1997, Hegedus 1986]. So, the samples which I brought there were measured in dry condition and not in their genuine electrochemical environment. This simple method of analysis is called *ex situ*. Most analyses of materials are done *ex situ*, although the term *ex situ* is hardly ever mentioned.

Now let us get back to the battery electrodes. During exposure of the sample to the ambient environment, the oxidation state of the electrode components could be altered – by exposure to oxygen nitrogen, humidity, for example. Some researchers worried about this and wanted to determine the electrode properties while the material is contained in the battery, so that no unwanted changes can happen to the electrode structure. Such kinds of experiments are called *in situ* studies. You study the material without taking it from its supposed environment.

For example, measuring a piece of lithium with X-ray spectroscopy typically requires an *in situ* arrangement because lithium will react with oxygen, nitrogen and humidity from the atmosphere and then undergo phase transformations from metal Li to LiOH or Li_3N, for example. It is simply not possible to get a correct X-ray absorption spectrum (or XPS spectrum) from pure Li because it will corrode in air

---

4 SINQ is not a neutron reactor; it does not need fissile nuclear material for neutron production. Rather, it uses a proton accelerator ring which shoots protons on a liquid metal (mercury Hg) bath, which triggers a nuclear reaction which produces neutrons.

(recall the sketch in Figure 4.9 where the lithium metal is surrounded with corrosion layers of LiOH or Li_3N.). Therefore, it must be brought into a vacuum or inert gas (not nitrogen though, because Li will form Li_3N with nitrogen) before it can react with the ambient. Such measurement is then an *in situ* experiment.

The next step of "complification" is when you operate the battery while you measure it with X-rays. After all, you measured it already with electrochemical methods while it was operating, or charging and discharging. Such methods are called *operando* methods. Below I will explain how such *operando* X-ray spectroscopy experiment is done on a working battery.

For such battery *operando* experiment with X-rays, you will have to make your own battery cell which allows the X-rays to come into the cell and hit the materials of interest (cathode, anode or both, or just electrolyte) and also returns the X-rays after interaction with the matter. The design of such special cell is not a trivial task. I have outlined the problems in detail in my book [Braun 2017] and suggest everybody to design and build spectroelectrochemical cells for *in situ* and *operando* measurements. You can have way more fun with your studies when you have and use an *in situ* cell.

Figure 4.22 shows on the left a polypropylene plate with a black O-ring (BUNA®) and a shiny lithium metal negative electrode, which is covered by an electrolyte moist and translucent separator foil (Celgard®). On the right you see the stainless steel end plate with two O-rings and a black spinel positive electrode in the center. A 6-inch-long ruler indicates the size of the components. What you cannot see in the photo is that the two plates have holes (apertures) which allow the X-rays to come through and hit the spinel positive electrode. There is however a 380-μm-thick beryllium window attached over each hole to prevent electrolyte from evaporation and to prevent air to come in [Braun 2003b].

Figure 4.22: Disassembled *in situ* X-ray spectroelectrochemical cell with lithium metal piece on the left polypropylene plate and cathode active layer on the right stainless steel end plate Design from 1999/2000.

The electrodes were prepared at my lab in Berkeley, and so was the spectroelectrochemical cell [Braun 2003b]. I brought my materials and cell parts to the Stanford

Synchrotron Radiation Laboratory SSRL in Menlo Park, California, X-ray beamline 2–1. I assembled my cell on site in the chemistry lab of SSRL. As there was no glove box available for my purpose (theirs was filled with nitrogen – not good for lithium), I brought a plastic glove bag. I purchased Argon gas at SSRL and filled my glove bag with Ar and assembled the battery cell in protective Ar atmosphere

Figure 4.23 shows exemplary a disassembled cell which I hold in the Argon filled glove bag. The end plate is made from PEEK® plastic. The shiny disk is the aluminum foil with the spinel electrode on its back side, facing to the PEEK® plate. Note two lithium metal stripes one of which is the counter electrode and the other of which is the reference electrode. It requires two separators to make sure that the reference electrode has a short electric contact neither with the positive electrode ($LiMn_2O_4$) nor with the negative electrode (Li). The right image in Figure 4.23 shows how I have removed and turned upside down the aluminum foil disk; now the back side with the spinel is visible, held with a tweezer.

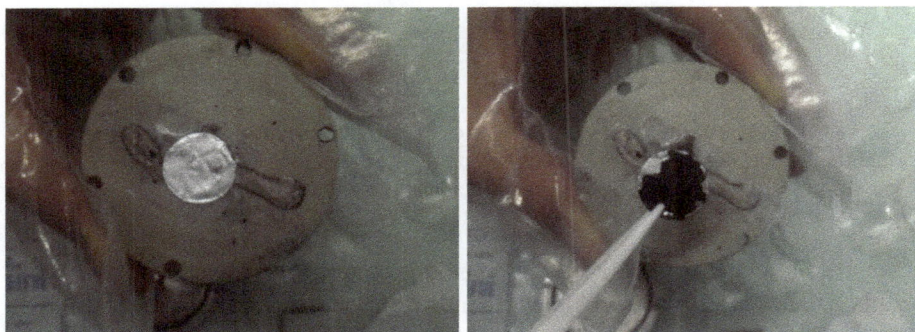

Figure 4.23: Disassembled *in situ* X-ray spectroelectrochemical cell (design from 2012) inside a glove bag. The cell was used for the study reported in [Braun 2015].

Figure 4.24 shows the evolution of the electric charge imposed on the cell from Figure 4.23 by the potentiostat. The charge was obtained by integration of the measured electric current over the time of the experiment, that is, mA min. You can carry out such integration with many software programs; I typically use KaleidaGraph®. You have two data columns from the potentiostat; that is, current in mA and time in seconds. The current showed a very zigzag variation which can originate from instabilities in the cell. When you integrate with your software current overtime, you will obtain the electric charge Q:

$$Q = \int_0^t I_{(t)} \cdot dt$$

then the data curve will look much smoother, when you plot it versus the time t, which is clearly visible in Figure 4.24. We begin with 30 mA min and notice a sigmoidal increase to 1800 mA min after 1600 min. I am saying "sigmoidal" because it is reminiscent to a square root curve, or parabolic, depending whether you refer to the abscissa or to the ordinate axes. Such square root law overtime is an indication of a diffusion limited process.

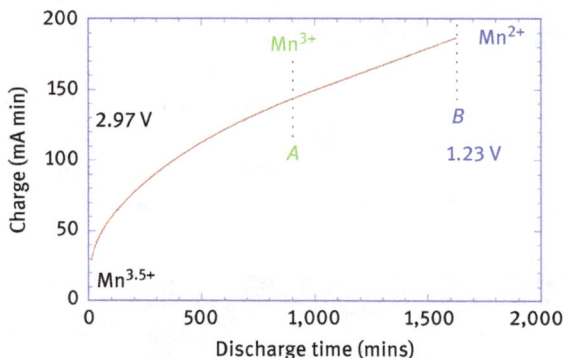

Figure 4.24: Evolution of charge of a $LiMn_2O_4$-based battery cell, as obtained from integration of current through the discharge process. X-ray absorption spectra at the K-shell core level were recorded at OCV (2.97 V) before discharging, and at states A and B.

When the battery *in situ* cell was assembled, the open circuit voltage (OCV) was 2.97 V; this is 0.08 V less than anticipated from theory. By imposing a negative current in the cell, we extract lithium ions from the spinel host (de-lithiation). This causes the battery voltage to considerably decrease with the consequence that the Mn becomes substantially reduced. We are basically forcing a deep discharge of the cell. The current was monitored with a portable potentiostat from Pine Instruments [Braun 2001] which we brought to the synchrotron beamline. The cell was discharged for 27 h with a current not exceeding 0.2 mA.

We have recorded X-ray absorption spectra at the Mn K-shell absorption edge at three different states of charge. The first spectrum was recorded when the *in situ* cell was in OCV condition. The cell was not connected to the potentiostat. The OCV was checked with a simple multimeter. Then the K-edge XANES (X-ray absorption near edge spectroscopy) spectrum was recorded. From the nominal stoichiometry $LiMn_2O_4$, we can assume that the average oxidation state is formed by one Mn^{3+} and one Mn^{4+} per formula unit which yields $Mn^{3.5+}$. The corresponding XANES spectrum is shown in Figure 4.25.

The black XANES spectrum in Figure 4.25 was recorded when the *in situ* cell was at OCV condition. The abscissa shows the X-ray absorption μD according to the relation

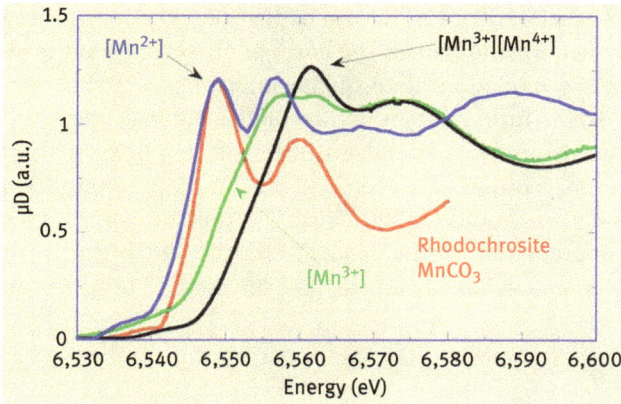

Figure 4.25: XANES spectra of $Li_{(1-x)}Mn_2O_4$ for three different oxidation stages/states of Mn [Braun 2003b]. Included as reference is the spectrum of the mineral rhodochrosite (red spectrum) after [Shiraishi 1997].

$$I_{(E,D)} = I_{0(E)} \cdot e^{-\mu_{(E)}D}$$

I_0 is the X-ray intensity (basically the number of X-ray photons of one particular energy E) delivered by the synchrotron beamline before it hits on the sample which has thickness D and linear absorption coefficient $\mu(E)$. The linear absorption coefficient is the characteristic of the material. The negative logarithm of the ratio of X-ray intensity before (I_0) and after sample ($I_{(E,D)}$) is plotted as the spectrum in Figure 4.25 versus the X-ray energy E. You see this by rearranging the equation. The X-ray absorption is very low from 6530 to 6547 eV and then has a steep increase.

The energy at which the increase is steepest is called "absorption edge." You can determine the energy position of the absorption edge by forming the first derivative of the spectrum with respect to the energy E and then look for the maximum of the derivative, or you can form the second derivative of the spectrum and search where it becomes 0. This is the typical way to determine the X-ray absorption edge.

When you compare the energy position of this steepest onset with the onsets of the other three spectra, then you see they are shifted toward lower energies. The shift of X-ray spectra for a particular chemical element reveals that the element may have different oxidation states. The spectrum of a metal has a lowest absorption edge position. The more you oxidize an element, the more you shift its absorption edge to higher X-ray energies. This phenomenon is called chemical shift and can be used for the determination of the oxidation states of chemical elements [Braun 2017].

This black spectrum is representative to the mixed Mn^{3+} and Mn^{4+} ions in the spinel electrode as it persist in OCV condition. The green spectrum was recorded 900 min after discharging (state A), so we expect that the manganese has become somewhat reduced. This is indeed the case, as the absorption edge is slightly below 6545 eV. The oxidation state of the manganese in the electrode is on average Mn^{3+}. Upon further de-lithiation of the battery, the manganese becomes more reduced, and this is indeed observed when we look at the blue spectrum which was recorded at the end of the lithiation (state B). I suspected that the oxidation state would be Mn^{2+}, and the spectral coincidence with the features of the reference spectrum from [Shiraishi 1997] $MnCO_3$, a mineral with Mn^{2+}, reveals that we have reduced the Mn during de-lithiation from $Mn^{3.5+}$ to Mn^{2+}.

4.5 Lead acid battery

To most of us, the lead acid battery (in German: Blei-Akku, Blei-Akkumulator) is known as the auto *starter* battery. It is also used in motor bikes and boats for starting the crank. It is also used for *powering* small scooters and toys. The lead acid battery is built from metallic lead plates as electrodes. The electrolyte is typically sulfuric acid with 37% weight concentration. Lead is a toxic heavy metal with atomic mass 207.2 u. Because of the high density of 11.342 g/cm^3 of lead, lead acid batteries are hardly considered for mobile applications. Those of you who have already lifted a car starter battery in a store or garage have experienced the heavy weight of such batteries.

The lead acid battery was supposedly invented by Wilhelm Josef Sinsteden in 1854 [Euler 1980] when he experimented with lead plates and sulfuric acid. By repeated electric cycling of the lead plate coupling with a power source, Sinsteden found that on one of the plates a white film was grown whereas the other remained as metallic lead. He noticed that after this forming (formation cycle), the cell had a capacity for charge accumulation.

The lead acid accumulator is particularly attractive because it has a well-established technology and because they are low cost. Where weight and space play no superior role, the lead acid battery can yet be of great service, for example for solar energy storage. The electric current delivered by solar panels can easily be stored in a large and heavy lead acid battery in the basement of a residential home or at a large-scale electric storage complex. There used to be a large lead acid accumulator power plant in Aachen, Germany, long time ago. The paper by Hunt describes a 5 MW lead acid battery power plant for use in a lead smelting factors as uninterrupted power supply [Hunt 1999].

The operation principle of the lead acid battery can be demonstrated by looking at the chemical processes, which take place during charging and discharging. Note that the lead acid battery is an accumulator. In its uncharged and pristine state, the sulfuric acid forms a lead sulfate $PbSO_4$ layer on both lead plates. Before it delivers power, it needs to be charged. When the battery is connected with a power source, the positive electrode will form a lead oxide PbO_2 layer, whereas the negative electrode will be a metal lead.

The negative pole is the metallic lead electrode in contact with the sulfuric acid, which will then form a solid lead sulfate layer at the surface of the negative electrode while two electrons per formula unit are delivered into the electric circuit:

$$Pb + SO_4^{2-} \rightarrow PbSO_4 + 2\ e^-$$

The positive pole is also a lead electrode but with a considerable lead oxide layer on the surface. In contact with the sulfuric acid electrolyte including the hydronium ions, it will accept the two electrons from the electric circuit and transform to lead sulfate, while six moles of water are released:

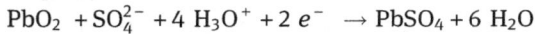

$$PbO_2 + SO_4^{2-} + 4\ H_3O^+ + 2\ e^- \rightarrow PbSO_4 + 6\ H_2O$$

When the battery is charged, the two reactions above go the opposite way. The overall reaction during charging and discharging therefore reads

$$Pb + PbO_2 + 2\ H_2SO_4 + \Leftrightarrow 2\ PbSO_4 + 2\ H_2O$$

ased on the electrochemical equivalent, we can determine for the electric energy a total charge of $2 \cdot 96485$ As. This yields 53.6 Ah for a lead acid accumulator.

The voltage of the lead acid battery can be determined from the difference of the two relevant electrochemical standard potentials (a long list is available in the references [Atkins 1997, 2010, Aylward 2008, Bard 1985, Bard 2001, Connelly 1996, Cotton 1999, Courtney 2014, Greenwood 1997, Leszczynski 2013, Lide 2006, Pourbaix 1966, Vanýsek 2011, Winter 2018]). The formation of the lead sulfate from lead in sulfuric acid yields −0.36 V.

$$Pb + SO_4^{2-} \rightarrow PbSO_4^{2-} + 2\ e^-\ |-0.36V$$

The reaction of the lead oxide layer with sulfate and protons in the electrolyte yields + 1.68 V.

$$PbO_2 + SO_4^{2-} + 4H^+ + 2\ e^- \rightarrow PbSO_4 + 2H_2O|+1.68V$$

This makes in total an EMF of + 2.04 V:

$$E^0_{total} = +1.68V - (-0.36V) = +2.04V$$

as we can also read from the number beam in Figure 4.26:

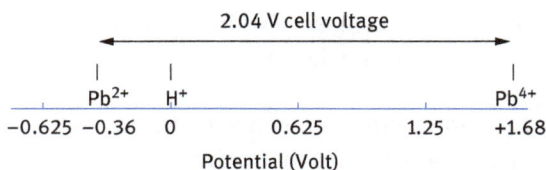

Figure 4.26: Determination of cell voltage of the lead acid accumulator from the potential difference −0.36 V − (+1.68 V) = −2.04 V.

The lead acid battery provides therefore around 2 V per cell. Lead acid batteries are available in multiples of 2 V, such as 6 V (for motorbikes), 12 and 24 V for cars, sports utility vehicles, tractors, trucks and so on.

Readers with an interest in chemistry may now wonder whether lead oxide PbO_2 is stable in sulfuric acid. The Pb(IV) ions tend to be dissolved. This is the thermodynamic fate of lead, but the overpotential of the hydrogen prevents this dissolution [Nernst 1900].

120 years ago, it was not so obvious that hydrogen evolution would be favored over a platinum electrode but not over a lead electrode [Reed 1901]. In response to Nernst's and Dolezalek's paper on the gas polarization of the lead acid battery (which in 1900 existed already, but was still a matter of debate with respect to the underlying electrochemical theories, as we witness here in this dispute), Reed published a polemic paper where he challenged the validity of the experimental observations of Nernst and Dolezalek [Dolezalek 1901, 1904, Nernst 1900] and also the interpretation. Nowadays, it is rare that such polemic scientific disputes are carried out in public.

In his paper, Reed [Reed 1901] explains the rational for his response:

> "The results which would follow from such a theory, if it could be established, are so momentous, and the sponsors who stand for the theory are of such eminent authority, that I shall consider it unnecessary to apologize for going rather minutely into the details of the experimental facts and arguments, which seem to entirely refute it."

Reed is actually acting up against authority in his paper, but rather than going against the authority of a contemporary understanding of a scientific theory, he challenges "the sponsors who stand for the theory." This was not so much in the spirit of Richard P. Feynman's philosophy of the freedom to doubt [Feynman 1955]. Feynman was also good at challenging authorities, but from the perspective of scientific advancement one should rather doubt a theory and not so much doubt a supporter of a theory. The author of this book knows only one case where one reader dug particularly deep and detailed into the work of some other researcher and published it as a correspondence article [Kreuer 2012].

When dealing with the electrochemical stability of a material or ion, it is worthwhile to look into the Pourbaix diagram [Delahay 1951, Hyde 2004, Pourbaix 1974, Vasquez 2006]. This is shown for lead (Pb) in Figure 4.27. We are dealing here again with an example where a kinetic barrier prevents – or delays – a reaction which otherwise is thermodynamically favored. The fact that the hydrogen has an overpotential is a

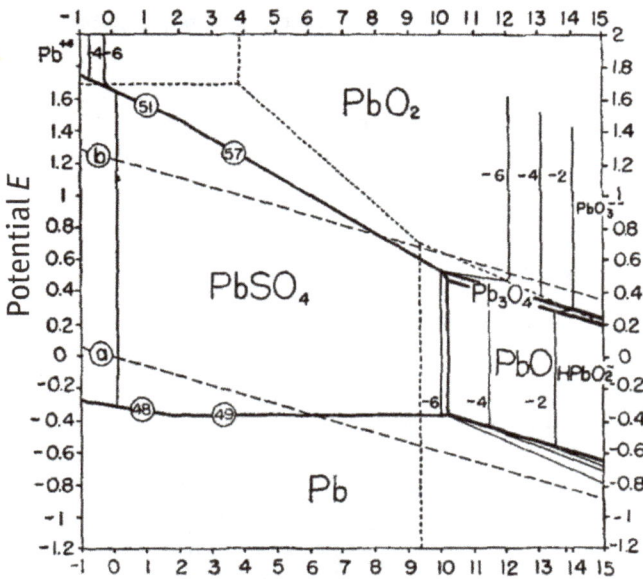

Figure 4.27: Potential-pH diagram (Pourbaix diagram) of lead Pb in aqueous solutions in presence of sulfate ion. Note that line 52 is erroneously indicated as line 57. The area between lines a and b corresponds to thermodynamic stability of water. Thin lines represent equilibrium conditions between a solid phase and an ion at activities $1, 10^{-2}, 10^{-4}, 10^{-6}$. Heavy lines represent equilibrium conditions between two solid phases. Dotted lines represent equilibrium conditions between two ions for a ratio of activities of these ions equal to unity. Circled figures and circled letters refer to equations in the original paper in [Delahay 1951].

Republished with permission of *Journal of the Electrochemical Society*, from Potential-pH Diagram of Lead and its Applications to the Study of Lead Corrosion and to the Lead Storage Battery, Delahay P, Pourbaix M, Van Rysselberghe P, 98, 2, 57–64, 1951; permission conveyed through Copyright Clearance Center, Inc.

technical precondition that the charging of the lead acid battery is possible without dissolution of the lead oxide layer. Otherwise, the water in the electrolyte would be split/ oxidized and then the energy for charging would be lost on the water electrolysis.

The Pourbaix diagram [Pourbaix 1974] is a phase diagram which shows which phases form from a particular material, metal under particular pH and applied external bias potential. Pourbaix has created an entire collection of such diagrams for numerous systems [Pourbaix 1974]. You see in Figure 4.27, the bias potential plotted from −1.2 to 2.0 V over the entire pH range from −1 to 15 (!). At strong negative potentials, the lead Pb is metal over the entire pH range. In acidic condition, the lead will convert to lead sulfate, and also until pH 9 from 0 to 1.2 V. With increasing potential, the lead sulfate will further be anodized, oxidized to lead oxide PbO_2; over the entire pH range. The diagram shows also in which oxidation state the Pb persists under which thermodynamic conditions.

Whenever a metal is exposed to other ions and electric fields, not only batteries, fuels cells, but also pipelines, oil tankers, submarines, your amalgamized teeth with the saliva in your mouth, the Pourbaix diagram gives a very good estimate about which compounds and phases may eventually form. It is therefore a good practice for everybody who engages in electrochemistry to consult the corresponding Pourbaix diagrams.

4.5.1 Self-discharge

Overtime, a charged lead acid accumulator can discharge according to following chemical reaction

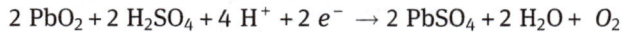

$$2\ PbO_2 + 2\ H_2SO_4 + 4\ H^+ + 2\ e^- \rightarrow 2\ PbSO_4 + 2\ H_2O + O_2$$

We see that hydrogen and oxygen gas can be formed during this discharge. This is basically the explosive Knallgas. It is therefore necessary to keep batteries in a well-vented environment. As the formation of the gas can be accompanied by a considerable build-up of gas pressure, lead acid accumulators were not designed in closed containers. Loss of hydrogen and oxygen during self-discharge causes loss of water (H_2O) in battery electrolyte. They always had holes on the top of the containers through which the gas could escape and through which the liquid level could be adjusted by refilling with distilled water or battery acid. The hole also permits to insert a hydrometer for the measurement of the electrolyte concentration. However, you may not tilt such battery or turn it upside down, so as to avoid the electrolyte drips out and causes harm to man and material.

In the last 30–40 years, battery producers found ways to absorb the pressure and make accumulators which do not need the aforementioned holes and further maintenance. Leakage of liquid electrolyte can be prevented using gel type electrolytes [Ferreira 2002, 2006] (fixed electrolyte). For instance, you can mix your liquid with hydrophilic silica powder (German: Kieselsäure) such as CAB-O-SIL®. 1 used this material for thickening of sulfuric acid for our supercapacitor projects at PSI. Nowadays AERO SIL® is advertised for battery gel applications.

As the pH of the electrolyte (sulfuric acid) is related with the electrochemical potential, the concentration (actually, the density) of the sulfuric acid in the lead acid accumulator can give account of the state of charge. When the accumulator is fully charged (100%), the density of the sulfuric acid is 1.28 g/cm^3. When the accumulator is fully discharged, the density is 1.10 g/cm^3. The relationship between density of the electrolyte and charge is linear. With a hydrometer, which can be inserted in a hole which is left in the accumulator container for that purpose, the density can be directly measured and thus the state of charge is determined.

As the pH of the electrolyte (sulfuric acid) is related with the electrochemical potential, the concentration (actually, the density) of the sulfuric acid in the lead acid accumulator can give account of the state of charge. When the accumulator is fully charged (100%), the density of the sulfuric acid is 1.28 g/cm^3. When it is fully discharged, the density is 1.10 g/cm^3. The relationship between density and charge is linear. With a hydrometer, which can be inserted in a hole in the accumulator for that purpose, the density can be directly measured and the state of charge determined.

4.5.2 Jump starting a car

How do you aid another driver whose starter battery died? In 2004, I visited Maxwell Technologies in San Diego CA. I flew there from Lexington KY. Maxwell was gentle enough to pay me the flight ticket, the hotel and a rental car. When I landed in San Diego, I picked up the rental car at the airport and drove to the hotel to check in. The next morning after breakfast I went out to get my car which I had parked outside at the parking place. I could not start it! The battery was empty. Then I found I had left the lights on overnight. The light switch was still in operation position, and the light had consumed all electric energy over night until none was left in the starter battery.

I went to the reception desk and asked whether they could help me. My conversation was overheard by an older gentleman, an American, and he told me he could help me. He had a starter cable in his hotel room. I replied somewhat surprised "Oh. Thank you that would be great!" It may sound unbelievable, but soon later the old Gentleman was back and he held a starter cable, in his hand. He told me he was travelling a lot and he had always a starter cable with him when he flies in. Unbelievable, but there he stood with the starter cable for me. He gave me the cables and went to his car and drove it next to my car. Then we both opened the hoods of our cars. The front ends of the cars were facing each other.

Then I took the red cable and connected it with the plus pole of my dead battery and then the other red end of the cable with the plus pole of the Gentleman's battery. Meanwhile, he had connected the black cable with his car's negative pole and was about to connect the other black end with the negative pole of my dead battery, but he asked me to get into my car and turn the starter key while he would be closing the battery circuit and power my battery.

I was familiar with this procedure because I had learnt it from a young lady when I had a dead battery at home in Germany. I could not get to work to PSI in Switzerland (I lived with my family across the border in Germany) in the morning. A young lady from the neighborhood – I had never seen her before – offered me her help. I think I was the one who had the starter cables but I never used them before. She drove her car close to mine, we opened the hoods and when I stood there clueless with the cables, she said "Ich glaube es gehört Plus nach Plus und Minus nach Minus." This is how it worked.

Few days later, I spoke with one of my neighbors, an electric engineer from Poland, over my experience with the dead battery. And he told me he had just purchased and installed a larger battery for his car: a battery with more Amp hours. What does that mean? What did my neighbor mean by "more Amp hours"?

Wouldn't a larger battery imply a higher voltage? Why would we need a higher voltage anyway?

We do remember the experiment with the lemon battery in the previous Chapter where we had inserted metal strips, copper and zinc into the lemon. We could immediately read a voltage with the multimeter. When we increase the size, the area of the electrodes, we do not get an increase in the voltage [V], but we get an increase in the current [A]. The electric current scales linear with the area [cm^2] of the electrodes when they are in contact with the acid of the lemon (electrolyte). When we divide the current by the geometrical area of the electrode, we get a current density in A/cm^2. The lead acid battery provides therefore around 2 V per cell. Batteries are available in multiples of 2 V, such as 6 V (for motorbikes), 12 and 24 V for cars, sports utility vehicles, tractors, trucks and so on.

We see therefore that a stacking of the electrodes in parallel will increase the current and thus also the stored charge in the battery. The charge is the current integrated overtime

$$Q = \int Idt$$

and yields Ah (Ampere hours). So my neighbor purchased a battery where simply more single positive and negative electrodes were stacked together in parallel so as to increase the total electrode area in the electrolyte and thus to increase the electric charge stored in the battery. The voltage would remain the same. The necessary voltage of the battery depends on the specifications of the starter crank or any other device which needs a battery.

4.5.3 Lead acid battery for solar energy storage

Last year a friend of mine sent me an email and told me his daughter and her husband were planning on getting a PV system on the roof of their home, and they also wanted to store the electricity. He wanted to know from me whether I could recommend him any technology for that. The solutions they had been looking into were a high performance lithium ion-based system and a lead acid-based system.

I felt that the "old" lead acid battery would be a good solution because it is based on a well-established science and technology. Certainly lead acid batteries are not used for modern transportation, mainly because the lead electrodes are so heavy and this brings down their specific energy and specific power in the Ragone diagram. But the technology is quite low cost when compared with the modern lithium ion technology.

Since the family of my friend was going to install this system in a residential home, they are not going to carry and move it around anymore after installation. It is not a solution for mobility, but for stationary application. You can buy all these systems already from many providers. Figure 4.28 shows a photo from Energy

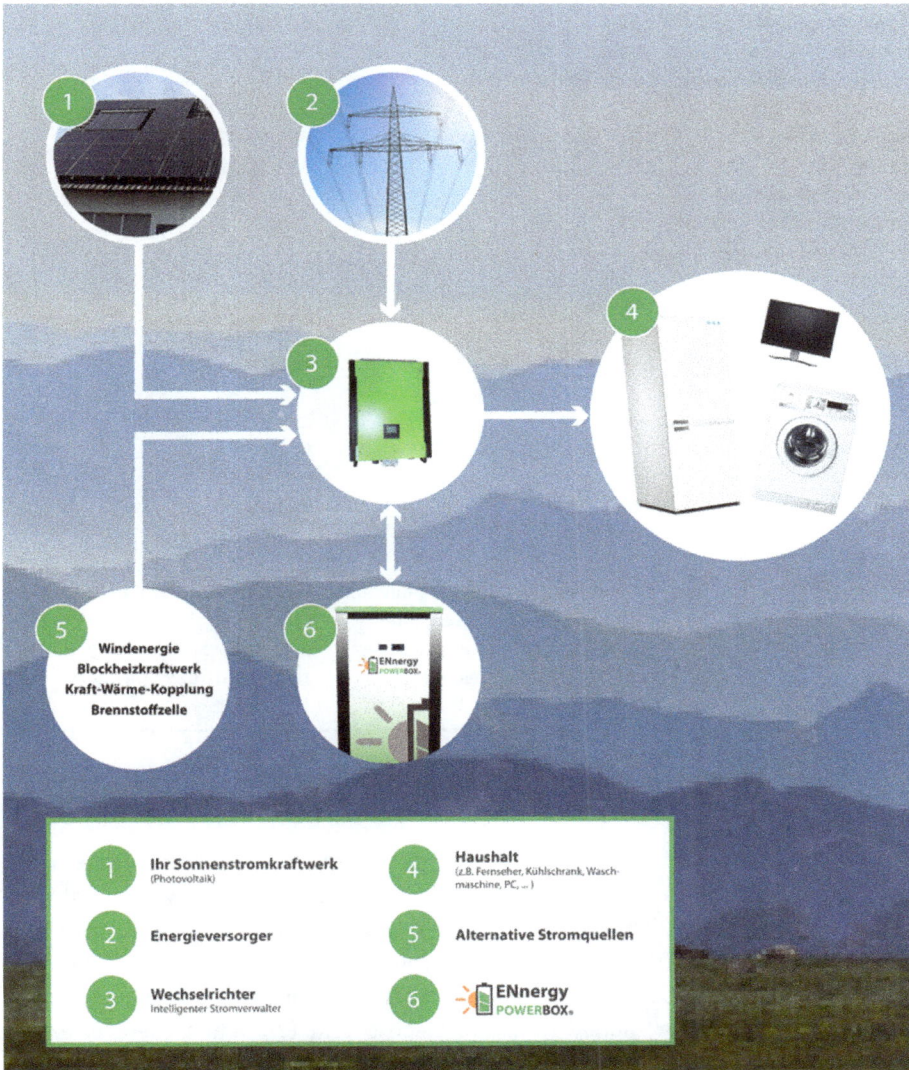

Figure 4.28: Advertising page from system provider ENnergy GmbH in D – 58256 Ennepetal, Germany, showing how electric power from the grid or from solar PV panels along with the POWERBOX® storage system. Reproduced with kind permission from ENnergy.

GmbH in Germany. They sell the Energy POWERBOX® system, the specifications of which are listed in Table 4.6.

Table 4.6: Technical data from an electric storage system used for use with photovoltaic panels.

Speicher type	EPB menoria 9	EPB menoria 18
Storage type	TPPL/Blei-Gel	TPPL/Blei-Gel
Usable capacity Kapazität	8 kWh	16 kWh
Maximum short power	12 kW	24 kW
Maximum constant power	10000 W	20000 W
Therm. efficiency	90%	90%
Geometric size $L \times B \times H$	600 × 600 × 635 mm	600 × 600 × 1,082 mm
Anzahl Zyklen bei 30% (bei 25 °C) Entladungstiefe	5500 Zyklen	5500 Zyklen
Anzahl Zyklen bei 50% (bei 25 °C) Entladungstiefe	3000 Zyklen	3000 Zyklen
Nominalspannung (DC)	48 bis 56 V	48 bis 56 V
Maximaler Ladestrom bei 12 V	100 A	200 A
Maximaler Entladestrom bei 12 V	275 A	1500 A
Rapid charging system	ja in 2 Std. *	ja in 4 Std. *
Sicherungslastschalter	IP 20	IP 20
Ambient temperature	5 °C bis 40 °C	5°C bis 40°C
Kühlung elektronisch gesteuert	Vorhanden	vorhanden
Notstrom	Ja	ja
Umschaltzeit von Netz in Notstrombetrieb	<7 – 10 ms in Verbindung mit von ENnergy zertif. Wechselrichter	<7 – 10 ms in Verbindung mit von ENnergy zertif. Wechselrichter
Speicherschutz	automatische Erhaltungsladung zum Schutz des Speichers	automatische Erhaltungsladung zum Schutz des Speichers
Herstellergarantie	10 Jahre**	10 Jahre**

*Depending on specific operation conditions.
**Check current warranty regulations.

Item (1) in Figure 4.28 shows the PV panel on the rooftop of a residential home which can deliver some electric power of some quantity. This energy will go into the transformer (3) which delivers the electricity to your utilities at home, such as a washing machine or refrigerator (4), for example. When the performance of your PV panels is not strong enough, for example at night, then the transformer captures the electricity delivered for money from the electric grid (2). When there is excess electricity arriving at the transformer (3) which you cannot use in your utilities (4), the energy will go into the (6) POWERBOX® and be stored in there until you need it for the utilities (4). You can feed the transformer (3) and thus the utilities (4) or (6) POWERBOX® with electricity from other renewable or alternative energy (5), such as

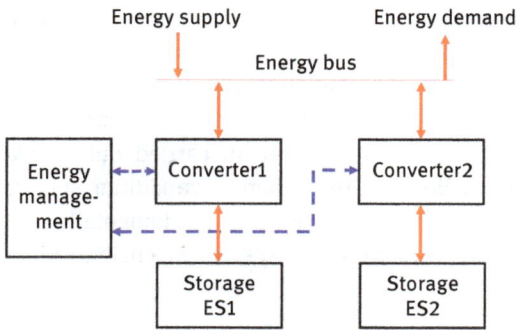

Figure 4.29: Basic schematic of a HESS Hybrid Electrochemical Storage System. Reprinted from Energy Procedia, 73, Bocklisch T, Hybrid Energy Storage Systems for Renewable Energy Applications, 103 – 111, Copyright (2015), doi: 10.1016/j.egypro.2015.07.582, with permission from Elsevier. [Bocklisch 2015].

wind power, fuel cells and so on. This scheme is basically generalized in Figure 4.29 as published in a work by Bocklisch [Bocklisch 2015].

Technical specification for a small and large electric storage unit for residential home applications. Data from ENnergy GmbH, D-58256 Ennepetal, Germany.

4.6 Redox flow battery

The redox flow battery (RFB) is some sort of hybrid between rechargeable battery and fuel cell where energy is stored electrochemically in redox couples in two liquids. We are dealing therefore with a misnomer, when we call it battery; but this is the official term. The two aforementioned liquids can be stored in virtually unlimited large containers which is an old concept currently experiencing a revival. Hence, as the electric energy can be charged continuously in the containers, it can be called accumulator. This is why the RFB belongs in this Chapter. In the last couple of years, the RFB has gained more interest as energy storage solution [Perry 2016].

The earliest record [Kangro 1949] on the RFB which I found dates back to the 28 June 1949, which was a patent application by Dr Walther Kangro for "Verfahren zur Speicherung von elektrischer Energie," which was granted 5 years later on 26 June 1954. A second patent with reference to liquids in the title was filed in 1954 by him [Kangro 1954]. Kangro states in his paper [Kangro 1962] that he began to think about storage of energy in the electrolyte, rather than in the electrodes already as early as 1941, independently from M. Volmer. There is also a PhD thesis on RFB [Pieper 1958] from this early time. In the 1970s, NASA began experimenting with and development of RFBs [Bartolozzi 1989]. In the 1980s, there was also rising interest in RFB for commercial applications. An early paper was reported at the ECS Meeting in 1987 [Shimizu 1987]. In 2016, a Focus Issue with title "Redox Flow Batteries – Reversible Fuel Cells" was published by the Electrochemical Society, Inc. [Weber 2015], many of which as open access articles.

Since around 20 years, the RFB is ready for applications such as power quality control, emergency power, back-up power, stabilization of renewable energy, for example [Miyake 2001]. An early synthesis procedure for a low-cost industrial scale RFB is presented in Nakajima et al. [Nakajima 1998].

Figure 4.30 is taken from a publication by Jervis et al. [Jervis 2016] because it is an open access paper and because it shows how a complex flow cell was designed for synchrotron-based X-ray tomography studies. Timofeeva et al. carried out *in situ* XANES studies in 2013 [Timofeeva 2013]. Because the different vanadium species have different and specific optical absorption characteristics, it is fundamentally possible to determine the state of charge of a RFB from the uv-vis spectra [Petchsingh 2016].

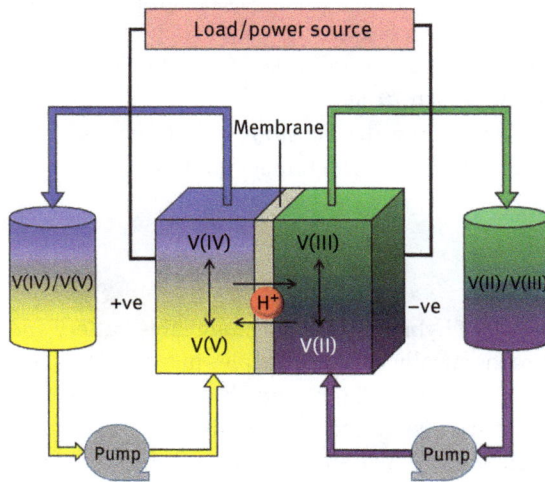

Figure 4.30: A schematic of a VRFB showing two tanks containing electrolyte with electroactive vanadium species, the positive and negative electrodes (usually carbon felts) and the proton conducting membrane. The power of the battery is dictated by the size of the electrodes, while the energy storage is decoupled and dependent on the size of the electrolyte tanks. In charging mode V (IV) is oxidized to V(V) at the positive electrode and V(III) is reduced to V(II) at the negative electrode. The reactions are reversed for discharge. Reproduced from Jervis R, Brown LD, Neville TP, Millichamp J, Finegan DP, Heenan TMM, Brett DJL, Shearing PR: Design of a miniature flow cell for *in situ* x-ray imaging of redox flow batteries. *Journal of Physics D: Applied Physics* 2016, 49:434002. doi: 10.1088/0022-3727/49/43/434002.

In the center of the RFB is a proton-conducting membrane (The proton conductor membrane is a scientific matter of its own. Su et al. [Darling 2016] studied NAFION with respect to nonaqueous electrolytes with small angle X-ray scattering.), which separates the electrolyte compartment in two reservoirs which contain the liquid electrolyte at the anode side (anolyte) and at the cathode site (catholyte). The best-known case is where vanadium ions are in the electrolyte. The left compartment contains V^{4+} and V^{5+}, the highly oxidized vanadium ions. The right compartment contains V^{2+} and V^{3+}, the less oxidized vanadium ions.

You notice that this system considers the entire wide range of vanadium oxidation states from V^{2+} to V^{5+}. The electrochemical activity of vanadium depends on this property. Alternative systems still need cations with a wide range of oxidation states, such as Ti/Mn [Kaku 2016]. In charging mode, an electric power supply is connected to the electrodes so that V^{4+} is oxidized to V^{5+} at the positive electrode, and V^{3+} is reduced to V^{2+} at the negative electrode. The redox reactions read [Maruyama 2017] for the positive electrode

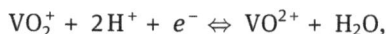

$$VO_2^+ + 2H^+ + e^- \Leftrightarrow VO^{2+} + H_2O,$$

and for the negative electrode the reaction reads

$$V^{2+} \Leftrightarrow V^{3+} + e^-.$$

It is left as an exercise for the reader to calculate the oxidation states for the vanadium ions in VO_2^+ and VO^{2+}. Be reminded that the oxygen ion typically goes by O^{2-}.

Overall, during charging we have an enrichment of the V^{5+} with a corresponding depletion of the V^{4+} in the anolyte container. A pump carries the anolyte away into a reservoir, where the concentration of V^{5+} is steadily increasing. The analog occurs with the catholyte at the negative electrode where the concentration of V^{2+} is increasing. We notice in Figure 4.30 that two pumps are necessary for the RFB. You can think about alternative flow directions but Trainham and Newman calculated that the established geometry in the RFB is the efficient and practical one [Trainham 1981].

A research group from Washington State [Liyu Li 2014] patented recently a RFB which has a supporting solution that contains chloride ions, in contrast to the common sulfate ions. The anolyte contains thus Cl^- as anion and V^{2+} and V^{3+} as the cations. The other container contains Cl^- and Fe^{2+} and Fe^{3+} ions. It is possible according to the patent that the electrolyte contains also SO_4^{2-} anions instead of Cl^-, and it is possible that all cations and anions are mixed.

The RFB has a comparably low energy density and it is worthwhile to look for ways to increase the energy density. The total energy content certainly scales with the volume of electrolyte, but the energy density scales with the concentration of the electroactive, that is, redox active ions. The CV shown in Figure 4.31 speaks about 1.5 and 2 molar concentrations of vanadium. Roe et al. [Roe 2016] showed how the right preparation of electrolyte chemistry would allow for an increased concentration of vanadium ions using precipitation inhibitors.

The cyclic voltammogram of their system with vanadium as cations only is shown in Figure 4.31. One CV was recorded with sulfate containing electrolyte. The second CV was contained with chloride only as electrolyte. In the sulfate electrolyte, the vanadium undergoes three oxidation processes which are nicely demonstrated by three redox peaks. The first peak originates from the oxidation of V^{2+} to V^{3+} at -0.3 V (Ag/AgCl electrode). The second peak for V^{3+} to V^{4+} at around 0.45 V, and the third

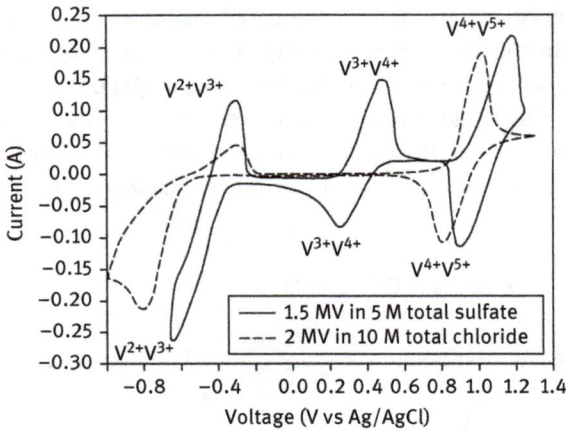

Figure 4.31: Cyclic voltammogram of vanadium redox flow electrode-electrolyte system with sulfate and chloride anolytes, respectively. 1.5 and 2 MV means 1.5 molar V concentration and 2 molar V concentration, respectively. Reproduced from US Patent US 8,771,856 B2 Jul. 8, 2014 [Liyu Li 2014].

peak for the oxidation from V^{4+} to V^{5+} at 1.15 V. These are the positions of the current maxima. The onset of the oxidation currents is at −0.45, 0.25 and 0.9 V. The corresponding cathodic peaks are also well developed.

It is noteworthy that when the electrolyte is changed to chloride then the middle peak in the CV is missing. It appears that the oxidation from V^{3+} to V^{4+} is not present in this chloride containing electrolyte.

The handling of the liquids does require mechanical pumping; RFB are therefore typically not "handy." Figure 4.32 shows a 200 kW RFB which is located in Martigny, Switzerland [Ligen 2018]. It is based on vanadium chemistry. It requires two containers for the actually electrochemical units and the liquid storage. The primary energy for RFB comes from the electric grid. The battery is large enough to serve as a fast charging stage for electric vehicles.

An alternative chemistry for RFB is using titanium and manganese [Dong 2015, Kaku 2016]. These materials are less costly than vanadium.

$$Ti^{3+} + H_2O \leftrightarrow TiO^{2+} + 2H^+ + e^-$$

$$Mn^{3+} + e^- \leftrightarrow Mn^{2+}$$

$$2Mn^{3+} + 2H_2O \leftrightarrow Mn^{2+} + MnO_2 + 4H^+$$

In Chapter 3, I mentioned the importance of carbon for electrochemical applications. Dong et al. [Dong 2017] present an improved carbon paper-based electrode for this type of RFB. When you inspect the microstructure of the electrode you notice similarities with the porous fibrous membranes used in alkaline electrolyzer. Qiu et al. carried out

Figure 4.32: Vanadium-based redox flow battery plant at the EPFL research lab in Martigny, Switzerland. The two large white containers feed that fast electric charging terminal for electric vehicles. 200 kW/400 kWh Vanadium Redox Flow Battery connected to a 80 kW charger in Martigny. Photo by Artur Braun, 7 March 2018.

a combined experimental and computational study where they investigated the fluid transport properties of the RFB electrodes [Qiu 2012].

Most of my scientific presentation which I deliver at conferences, symposia, seminars and workshops contain one particular slide where I point out the importance of the correlation of structure, transport properties and function, in particular for electrochemical energy storage and conversion materials, components, devices and systems. The comprehension and the control of the transport properties are essential for the functionality and integrity of electrochemical energy storage and conversion devices including the batteries, fuel cells, capacitors, electrolyzers and photoelectrochemical cells (Figure 4.33).

These transport properties include the charge transport, the mass transport, the heat and radioactive transport and the light transport. The electric charges in this respect include the electrons, electron holes, ions including protons and polarons [Alexandrov 2010, Emin 1982, Firsov 2007]. The mass transport includes the fluids, which can be liquids, gases and plasma. When we think of electrolytes, the fluids can certainly include also ions, which would make that we have a coupled problem of mass coupled with charge. The heat and radioactive transfer is a topic of its own and typically dealt with by mechanical engineers. As soon as we are working on devices,

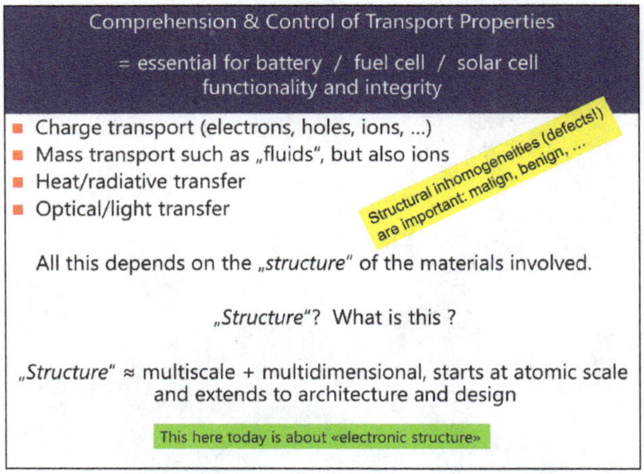

Figure 4.33: PowerPoint slide from most of my presentations, highlighting the importance of transport properties and structure in electrochemical energy and storage devices.

in particular devices with an appreciable size, the formation of Joule heat needs to be considered and dealt with.[5] Thermal management then becomes a part of the design of the device and system. It is elegant when you can solve the heat or cold problem with passive solutions (e.g., using natural convection). Otherwise you need to attach an active solution, such as pump or a fan, for example.

It is very advantageous when one can mathematically model a system. One advantage is that one can make simulations and try out how to improve the design for better

---

5 Here is a practical example. When you charge and discharge supercapacitors, they may become warm or even hot depending how long you use them. Most of the supercapacitors come in a cylinder shape. When you assemble them in a large "battery" of caps, you may have to think about how to guide the heat away so that the heat does not destroy the cap. Inside it has liquid electrolyte which may boil and then explode the capacitor housing. As they have cylindrical shape, there is enough open space between the cylinders even if you pack them close. This open space helps to guide the heat away. When you look at the ZEBRA battery, you see that all its elements have a prismatic geometry. You can pack them together with no open space inbetween them. There is nothing wrong with that because the ZEBRA battery operates at elevated temperatures. The closer you pack the elements, the less heat loss takes place. It is easier to operate the ZEBRA battery that way. Hence the design of the storage media should, actually must, take into account the requirements for thermal management. If you cannot optimize thermal management by simple geometrical arguments, then you may have to include radiators or cooling elements in the storage system which occupies space, creates extra costs, and needs extra maintenance. Modern cars with combustion engines have cooling liquid which cools the engine before it runs too hot. The cooling liquid can be water, a mixture of water with an organic such as glycol, which extends the temperature range of being liquids, or even some oil. The old Volkswagen beetle—a low maintenance vehicle—had a radiator fan attached to the engine so that no cooling liquid and no cooling pump was necessary.

performance [Dennison 2015]. A simple mathematical model for the RFB accounting for the mass transport and charge transport was developed by You et al. [You 2009]. I will explain further below why I am following here the paper by Qiu et al, who made a combined experimental and computational study [Qiu 2012]. The mass transport is, for example, the flow of liquid electrolyte through the connected pore space. This transport is governed by the continuity equation ("the divergence of the flow u is 0")

$$\nabla \cdot u = 0$$

and by the Navier–Stokes equations

$$\frac{\partial u}{\partial t} + (u \cdot \nabla)u = -\frac{1}{\rho}\nabla p + v\nabla^2 u$$

where u is the velocity vector of the liquid, ρ the density of the liquid (its specific weight, specific mass), p the external pressure exerted on it and v its kinematic viscosity. ∇ is the differential operator (nabla). To address this problem of fluid dynamics, knowledge in differential calculus is required.

As for the practical approach to solve the Navier–Stokes equations, Qiu et al. chose to employ the rather new field of Lattice–Boltzmann method (LBM) and not the conventional and established computational fluid dynamics methods. There is a practical reason for that: Qiu et al. have experimental microstructure data from tomography which comes in voxels. The LBM is particularly suited to solve problems which can be expressed in voxel structure. We used this approach in a project on small electrochemical energy converters [Karlin 2006]. Note therefore that the mass transport is not only a matter of liquids but also includes gases, which is important for the air electrodes in lithium air batteries, for solid oxide fuel cells and PEM fuel cells and so on.

The transport of the charge carriers of the type j (vanadium with various oxidation states, H^+, SO_4^{2-},) with charge z_j, concentrations C_j and diffusivities D_j follows the convection–diffusion equation plus an additional term which accounts for the gradient of the electric potential φ (though negligible small for RFB when compared to other devices [You 2009]):

$$\frac{\partial C_j}{\partial t} + u \cdot \nabla C_j = D_j \nabla^2 C_j + \nabla \cdot \left[\frac{z_j C_j D_j}{RT} \nabla \varphi \right]$$

The concentration of the sulfate ions SO_4^{2-} is experimentally not known, but can be determined via the neutrality condition [Qiu 2012]

$$\sum z_j C_j = 0$$

The aforementioned lines described the processes which take place in the pore space of the electrodes and membrane. Darling et al. have prepared a study on transport requirements for separators for RFB [Darling 2016]. The electrode in the RFB is typically made from carbon fibers which carries the electrons to or from the current collectors. The electric current density J produced by the gradient of the electric potential φ reads

$$J = -\kappa_s \nabla \varphi.$$

The electronic conductivity of the electrode material is κ_s. In analogy to mass conservation, we have also charge conservation (the current density has no sources, no sinks, no divergence: div $J = 0$):

$$\nabla \cdot (\kappa_s \nabla \varphi) = 0$$

The effective electrolyte conductivity (ionic conductivity) κ_{eff} is derived from the sum of all species with diffusivities D_j; note that species with positive or negative sign (cations, anions) require such treatment:

$$\kappa_{\text{eff}} = \frac{F^2}{RT} \cdot \sum_j z_j^2 D_j C_j.$$

As the ionic current has also no divergence, we get a closed expression

$$\nabla \cdot \left[\kappa_{\text{eff}} \nabla \varphi + F \sum z_j D_j \nabla C_j \right] = 0$$

based on which we are able to determine the potential field. But we are not done yet. The next step is to include the chemical reactions between electrolyte and electrode surface, which are governed by the Butler–Volmer equation. I refer the reader at this point to the original works [Qiu 2012, You 2009].

A further refinement of the analysis can be done by making spatially resolved measurements of the electric properties. Note that Qiu et al. did already a great structural analysis work based on tomography. It would be worthwhile to read the currents and voltages not only from the current collectors as global quantities but from sectioned electrodes which are placed at particular places in the electrode and membrane and maybe electrolyte to get the "electric" data with a high spatial resolution. For that you would have to place probe electrodes at particular locations in the device. You can see this, for example, in the work of Gandomi et al. [Gandomi 2016].

The so far largest RFB is going to be installed in El Cajon near San Diego in California as a joint project by Utility San Diego Gas and Electric (SDG&E) and Sumitomo Electric (SEI) [Kenning 2017]. The system can deliver 2 MW power and 8 MWh energy which is considered sufficient to power 1,000 residential homes for 4 h.

4.7 Zinc air battery

The zinc air battery is well known for powering hearing aids and also other portable devices. They typically come therefore as small coin cells. The zinc air battery is originally a primary battery with 1,050 Wh/L volumetric capacity and 340 Wh/kg gravimetric capacity and cannot be recharged, but rechargeable systems have been developed and could deliver 215 Wh/L and 170 Wh/kg and thus outperform other high-performance battery systems [Tinker 2000]. Zinc air batteries have a high energy density and a long shelf life, but their relative inferior cycling stability are a fundamental problem for the development of rechargeable zinc air batteries [Huot 1997]. Zinc air batteries can be used for other portable devices and then are called disposable displacement batteries. One example was the development of a 3,300 mAh battery by ElectricFuel®, which could deliver 2 A peak pulses for GSM cell phones and 500 mA continuous current; note that four of such cells were necessary to emulate one lithium battery of 3.6 V [Koretz 2001].

For the basic concept of a zinc air battery, see the sketch in Figure 4.34 [Arlt 2014]. They typically come as small coin cells. You can see X-ray radiography images of such coin cells in Arlt et al. [Arlt 2014]. The anode is metal zinc and the cathode is a porous electrode support which is coated with an oxygen catalyst which promotes the dissociation of oxygen O_2 from the air; this is why this cathode is called as air electrode. Both electrodes are separated by a membrane which is soaked with an alkaline electrolyte. It is possible to determine the distribution of species in such cells

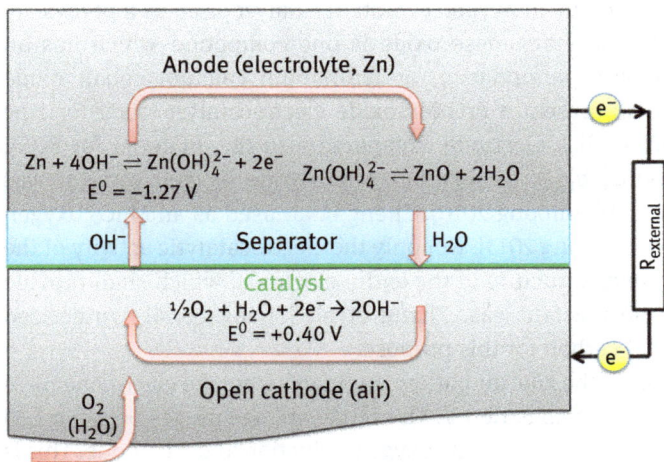

Figure 4.34: Sketch of reactions in a zinc air battery during discharging. Reproduced from [Arlt 2014] Arlt T, Schroder D, Krewer U, Manke I: In operando monitoring of the state of charge and species distribution in zinc air batteries using X-ray tomography and model-based simulations. Physical Chemistry Chemical Physics 2014, 16:22273–22280. doi: 10.1039/c4cp02878c. Published by the PCCP Owner Societies.

with some spatial resolution with X-ray tomography [Arlt 2014]. A 20-page review on recent advances in zinc air battery technology was authored by Li and Dai [Li 2014b].

The zinc air battery was invented 100 years ago and further developed in the Second World War in response to the lack of resources. Bunsen refers in one of his letters to the "Zink-Kohle-Batterie" as early as 1841 [Bunsen 1841]. The energy density of the zinc air battery is quite high and its discharge curve is virtually horizontal. This makes it a very good power source for analog and digital hearing aids, particularly because it can be produced as small button cells. The zinc electrode is provided in powder form so that a large electrode–electrolyte interface is made which provides a good power density. For a master thesis on zinc electrode development see for example [Schutting 2011]. The battery is designed in a way that air can enter the cell so that the oxygen can react with the electrocatalyst of the cathode.

Here is a principal disadvantage of the zinc air battery: you cannot switch it off. When the vent of the battery is open end, the cell reactions will progress until the battery is empty. Its use for hearing aids is therefore quite justified because you do not really switch a hearing aid, in contrast to flashlights, walkie talkies or other mobile devices. The air electrode is typically made from a porous mixture of carbon and graphite and an electrocatalyst powder. The latter is typically a metal oxide, sometimes with perovskite structure. During my time as doctoral student at PSI, we had a research group that specialized in the development of rechargeable zinc air batteries [Müller 1993, 1996]. I remember that much of their work was the development of high performance cathode layers.

These cathode layers contain an oxygen evolution catalyst often as a perovskite layer. Praseodymium calcium manganese oxide is one compound which has an appreciable oxygen catalytic performance, or lanthanum calcium cobalt oxide [Lippert 2007]. Hyodo et al. present a PrCaMn-oxide electrocatalyst study for zinc air batteries with a rotating disk electrode, compared also this material for SOFC applications [Richter 2008a, b]. A lanthanum cobalt oxide oxygen catalyst was studied by Boonpong et al. [Boonpong 2010]. Cheng et al. used as an MnO_2 oxygen catalyst in zinc air batteries [Cheng 2013]. Not only the electrocatalytic activity of the metal oxide but also the microstructure of the entire electrode, which shall provide optimized diffusion of the reactant gas. Zhang et al. [Zhang 2004] synthesized composites from MnO_2 and carbon for this purpose.

The theoretical voltage of the zinc air battery originates from the electrochemical reaction of zinc with oxygen which is 1.60 V. The actual voltage ranges between 1.35 and 1.4 V because of the overpotential for the oxygen reduction at the cathode. This is the same voltage range like for the zinc mercury batteries, which are not produced nowadays anymore.

The anode reaction forms zinc in alkaline environment to zinc hydroxide; the standard potential for this reaction is −1.199 V:

$$2\,Zn + 8\,OH^- \rightarrow 2\,Zn(OH)_4^{2-} + 4\ e^-$$

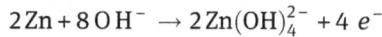

The zinc hydroxide complex (it is actually a so-called zincate complex, unlike many other colorful complexes, it has no color) reacts further in the electrolyte toward zinc oxide with release of water:

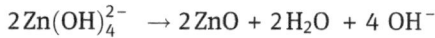

$$2\,Zn(OH)_4^{2-} \rightarrow 2\,ZnO + 2\,H_2O + 4\ OH^-$$

The reaction at the cathode involves not the cathode material except for its oxygen catalytic properties, which does not enter the chemical reaction equation; the standard potential for this reaction yields +0.401 V:

$$O_2 + 2\,H_2O + 4\ e^- \rightarrow 4\ OH^-$$

In summary, we have the following relevant electrochemical reaction with a net voltage of 1.6 V. Have a look in the necessary standard reference potentials:

$$2\,Zn + O_2 + 2\,H_2O \rightarrow 2\ Zn(OH)_2$$

The zinc corrodes in the electrolyte alongside with the formation of zinc oxide and zinc hydroxide, which is subject to fundamental kinetic theory [Szczesniak 1998]. A simple geometrical model for the progression of the ZnO layer on the anode is shown in Figure 4 of [Bhadra 2015] and in Arlt et al. [Arlt 2014].

The proper choice of zinc powder from various suppliers with different characteristics may have influence on the battery performance because of different structure and purity [Perez 2007]. Even adding additional compounds to the electrode can have an influence on battery structure and performance, as Moser et al. show in an *operando* X-ray study [Moser 2013] and [Gallaway 2014]. As zinc is a limited resource, its recycling form batteries needs to be considered, but in competition with other battery, chemistries readily established on the market. An "old" paper from the 1990s [Wiaux 1995] addresses the economical challenge in recycling and waste management for zinc batteries. Not only the zinc but also the manganese is the component of the battery and is subject to recovery and recycling [Freitas 2007].

Rather than doping or alloying the zinc with some foreign element, you think of completely replacing the zinc by a different metal. Of interest for metal air batteries are cadmium, iron, zinc, aluminum, magnesium, sodium [Ha 2014] and lithium [Girishkumar 2010, Kuboki 2005, Semkow 1987]. There are also efforts in replacing zinc by aluminum and thus making an aluminum air battery [Sun 2015], which has a theoretical capacity of 1032 Ah/kg. The lithium air system has a theoretical capacity of 1120 Ah/kg, which is almost as double as high as the zinc air system with 659 Ah/kg [Haas 1996].

El-Sayed et al. [El-Sayed 2010] showed with a detailed analytical study how the alloying of zinc with traces of nickel can improve the battery performance, such as the lowering of the overpotentials. Cai et al. have allied the zinc with indium and observed a similar effect on the suppression of corrosion and hydrogen production [Cai 2009]. Figure 4.35 shows a Tafel plot of the I/V curves of the pure Zn anode and an anode doped with 0.5% Ni. You will see later in this book in the Chapter on electroanalytical methods how the Tafel plot is used as approximation to the Butler–Volmer equation which models the electrode kinetics in electrochemical reactions. For this, one has to form the logarithm of the current density versus the potential, which is taken from the so-called polarization curve (I(V curve). El-Sayed et al. argue that the addition of trace amounts of Ni to Zn forms an alloy which promotes in the passivating region the electrochemical reaction, which they refer to as "self-catalysis" [El-Sayed 2010]. It suppressed the hydrogen evolution and thus corrosion of the zinc.

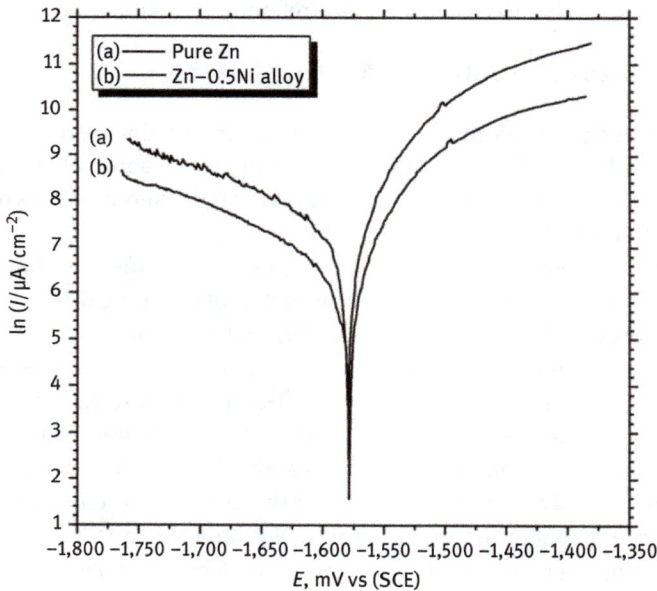

Figure 4.35: Comparison between Tafel polarization curves for (a) pure Zn and (b) Zn–0.5Ni alloy in 7 M solution of KOH at 25 °C.
Reprinted from *Journal of Power Sources*, 195, El-Sayed AR, Mohran HS, El-Lateef HMA, Effect of minor nickel alloying with zinc on the electrochemical and corrosion behavior of zinc in alkaline solution, 6924–6936, Copyright (2010), with permission from Elsevier [El-Sayed 2010].

El-Sayed et al. investigated the pristine and Ni-doped zinc electrode also with impedance spectroscopy. Figure 4.36 shows the Nyquist plot of the impedance spectra of the pristine zinc anode at three different bias potentials, this is at the

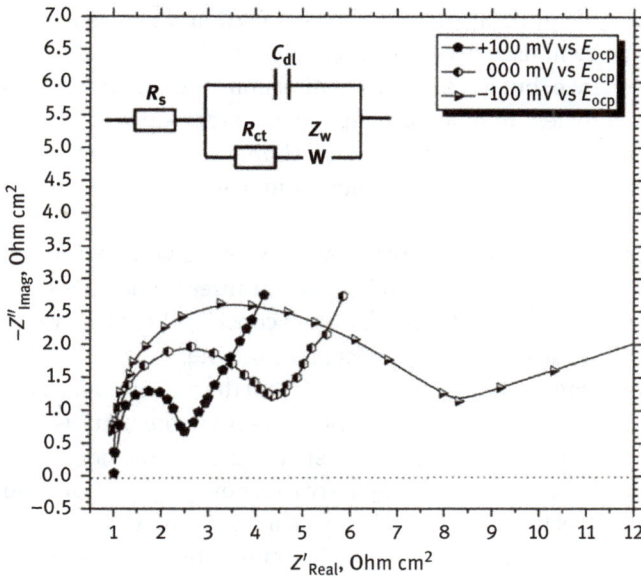

Figure 4.36: Nyquist plot for pure Zn in 7 M KOH at applied different potentials, AC amplitude 5 mV, the frequencies from 100 kHz to 5 Hz, and at 298 K. Inset (top) is the Randle's equivalent circuit. Reprinted from *Journal of Power Sources*, 195, El-Sayed AR, Mohran HS, El-Lateef HMA, Effect of minor nickel alloying with zinc on the electrochemical and corrosion behavior of zinc in alkaline solution, 6924–6936, Copyright (2010), with permission from Elsevier [El-Sayed 2010].

OCV and +100 and −100 mV above and below the OCV, respectively. Note that the axes are not isometric; 1 unit on the real axis (ordinate) is slightly larger than 1 unit on the imaginary axis (abscissa). This is somewhat unfortunate and prevents the reader from seeing whether potential semicircles are indeed present in the plot. It is therefore advised that Nyquist plots are always made isometric.

The spectrum recorded at OCV is the one with the circle symbols. It shows a nice semicircle (despite the nonisometric plot) with a diameter extending from 1 Ohm cm^2 to around 3.5 Ohm cm^2, which is a radius of $(3.5-1)/2 \ \Omega cm^2$, which yields 1.25 Ωcm^2 charge transfer resistance. The 45° slope straight line from 4.5 to 6 Ωcm^2 is the Warburg impedance. El-Sayed et al. have modeled the spectra with a Randles circuit from four components, this is the double layer capacity C_{dl}, the charge transfer resistance R_{ct}, the Warburg impedance Z_W and a serial resistance for electrolyte and connections R_s.

Let us now move to the next spectrum which is recorded at +100 mV DC bias, as indicated with the filled symbol. Its semicircle has a smaller radius which we can estimate by visual inspection to $(2.5-1)/2 \ \Omega cm^2 = 0.75 \ \Omega cm^2$. Note that the Warburg impedance branch is nicely developed and parallel to the one recorded at OCV.

The third spectrum was recorded at −100 mV versus OVC and shows a wide semicircle, but with a shape reminiscent of a mouse. It looks like a shape from a Gerischer [Lewis 1997] impedance, but in the opposite direction. The shape suggests that one semicircle is not a sufficient number of components to reproduce the actual processes taking place. But in a first step, we can be simple and pragmatic and continue with the simple calculation like done before and estimate the charge transfer resistance to $(6-1)/2$ $\Omega cm^2 = 2.5$ Ωcm^2.

We see therefore that the charge transfer resistance decreases with increasing potential in the range from −100 to +100 mV in 7 M KOH. As the three semicircles meet at the highest frequency points (100 kHz) at 1 Ωcm^2, we can consider this as the value for the serial resistance, such as electrolyte resistance, for example.

The impedance spectra in Figure 4.37 were recorded from the nickel-doped zinc electrode, again at −100, 0 and +100 mV versus the OCV. We first note that the real axis extended from 0 to 24 V, whereas the impedance spectrum from the nondoped zinc electrode extended only to 12 V. And the imaginary axis now extends from 0 to 12 V, whereas the nondoped electrode needs 7 V only. This is the first and simple indication that the alloyed electrode yields an overall larger resistance. Table 4.3 in

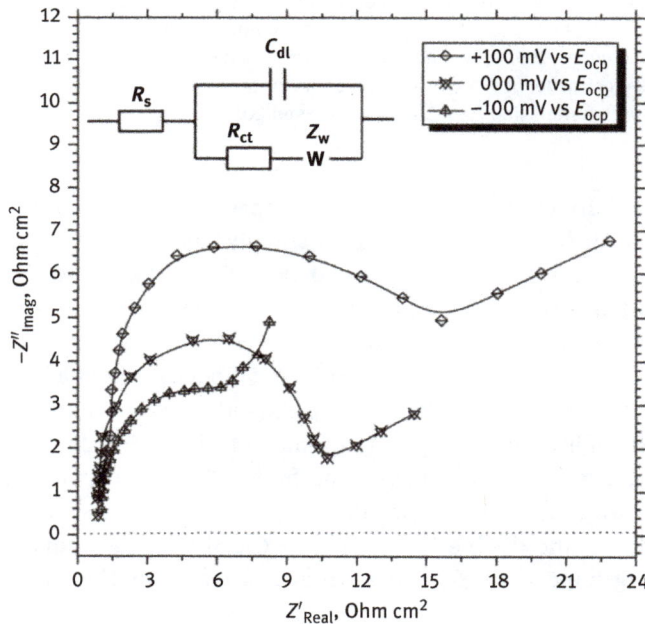

Figure 4.37: Nyquist plot for Zn−0.5Ni alloy in 7 M KOH at applied different potentials, Ac amplitude 5 mV, the frequencies from 100 kHz to 5 Hz, and at 298 K. Inset (top) is the Randle's equivalent circuit. Reprinted from *Journal of Power Sources*, 195, El-Sayed AR, Mohran HS, El-Lateef HMA, Effect of minor nickel alloying with zinc on the electrochemical and corrosion behavior of zinc in alkaline solution, 6924–6936, Copyright (2010), with permission from Elsevier [El-Sayed 2010].

reference [El-Sayed 2010] shows the fit parameters which the authors have obtained by deconvolution of the impedance spectra with the Randles circuit shown in Figures 4.36 and 4.37.

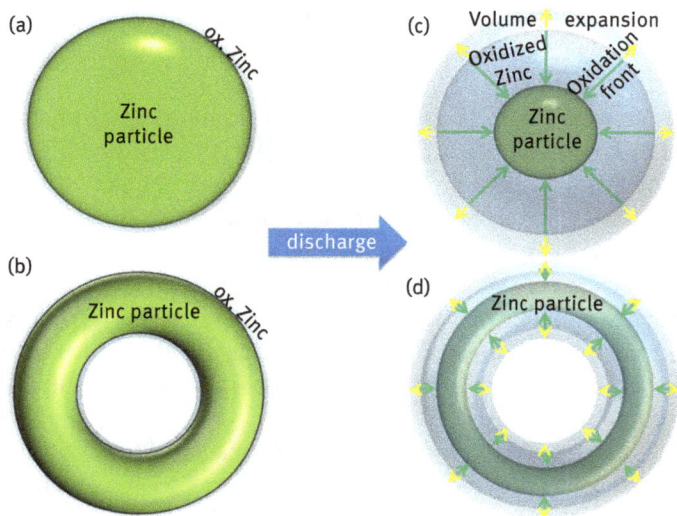

Figure 4.38: Shrinking-core-concept for zinc oxidation during battery discharge for (a) and (c) sphere-like zinc particles and (b) and (d) torus-like particles. (a) and (b) represent particles in a fully charged battery while (c) and (d) represent particles in a partly discharged battery (at a SOC of around 60% in our measurements, see Figure 4.5). Oxidation front (green arrows) starts at the particle surface, while volume changes due to the higher density of oxidized zinc are indicated by yellow arrows. Reproduced from [Arlt 2014] Arlt T, Schroder D, Krewer U, Manke I: In operando monitoring of the state of charge and species distribution in zinc air batteries using X-ray tomography and model-based simulations. Physical Chemistry Chemical Physics 2014, 16:22273–22280. doi: 10.1039/c4cp02878c. Published by the PCCP Owner Societies.

The charge transfer resistance Rct for the Zn electrode is 9.5, 6.0 and 2.5 Ωcm^2, but for the Zn-0.5Ni alloy 21.5, 12.5 and 10.6 Ωcm^2 for the potential range from −100 to +100 mV, respectively. This corresponds basically to the observed increase of the real axis range by a factor, and the general increase of charge transfer resistance by alloying. Also, the Warburg impedance Z_W doubled when the electrode was alloyed.

4.7.1 Efforts for zinc air secondary batteries

Because zinc air batteries have a high energy density and a long shelf life [Huot 1997], there was an interest in having rechargeable zinc air batteries [Haas 1996]. Problems with their electrochemical cycling stability were therefore a reason for more research. The cycling stability is directly connected with the corrosion properties of the zinc anode in the electrolyte. These properties are characterized in the Pourbaix diagram. Minakshi et al. [Minakshi 2010] have remodeled a primary zinc air battery into a secondary, rechargeable one for degradation studies.

It is therefore possible to build rechargeable zinc air batteries and the selection of the proper electrolyte for best results depends on the particular application, but as of yet, such batteries are not ready for the consumer market or for any serious commercial application [Mainar 2018].

The research group around Stefan Müller at PSI worked many years on the improvement of electrodes for secondary zinc air batteries [Müller 1993, 1996]. To a large extent, this concerned the electrode processing such as tuning the pore size of pasted zinc electrodes (around 1 mm thickness); for example, adding 10% cellulose increased the cycle life substantially and also the peak power drain [Müller 1998b]. They worked also on the upscaling of electrodes and arrived at a 200 cm^2 large battery and obtained a battery with a capacity of 30 Ah. The air electrode was based on the $La_{0.6}Ca_{0.4}CoO_3$ activated bifunctional oxygen electrode [Holzer 1998]. They built a 100 W rechargeable zinc air battery where the zinc was deposited on high surface area porous copper foam [Müller 1998a].

We learnt already in the previous Chapter on supercapacitors how carbon particles were oxidized. The zinc particles in the zinc air battery become oxidized during battery operation. As the density of ZnO and $Zn(OH)_2$ is smaller than Zn, there will be a volume expansion for the reacting particles while the oxidation front propagates into the interior of the particle, step-by-step consuming the zinc. This system follows the principles described in the well-known textbook by Levenspiel [Levenspiel 1962]. Arlt et al. carried out X-ray radiography analyses on actual zinc air batteries and found spherical Zn particles and toroidal Zn particles in the electrode [Arlt 2014].

The zinc anode becomes oxidized with oxygen from air in an alkaline electrolyte toward zinc oxide or zinc hydroxide. When you remove the converted zinc by new fresh zinc, you can continue the zinc air battery to run. Basically, this means you give the battery a "refill," you fill the battery with a new pellet of fuel. This is comparable to the concept of extreme case that can yield formation of dendrites that can cause electric shortages in the electrode. Such dendrite formation is not specific to zinc air batteries only.

Dendrite formation can be a problem in lithium batteries and it has been experimentally verified for platinum during fuel cell operation conditions with X-ray microscopy (STXM) [Berejnov 2012]. An X-ray *in situ* study on the zinc air battery is reported by Nakata et al. [Nakata 2015]. Chen et al. used calcium as an additive to the zinc anode to suppress dendrite formation [Chen 2004]. They found that the formation of calcium zincate provides improved electrochemical performance of the entire battery.

The rechargeable zinc air battery then requires a specifically designed bifunctional gas diffusion electrode which can oxidize the oxygen from the air and which can oxidize the hydroxide layer at the triple phase boundary of air, solid electrode and liquid electrolyte.

Chen et al. used calcium as an additive to the zinc anode to suppress dendrite formation [Chen 2004]. They found that the formation of calcium zincate provides improved electrochemical performance of the entire battery.

4.7.1.1 The rechargeable high temperature lithium air battery

Finally, I want to mention in this section the rechargeable lithium air battery, which was first submitted as an Accelerated Brief Communication to the *Journal of the Electrochemical Society* by May 5, 1987, and subsequently published in August 1987 [Semkow 1987].

The design of the cell is sketched in Figure 4.39. This is a high temperature electrochemical cell which operated at 600–850 °C because a solid electrolyte membrane (stabilized ZrO_2) was used. The configuration of the cell was a lithium

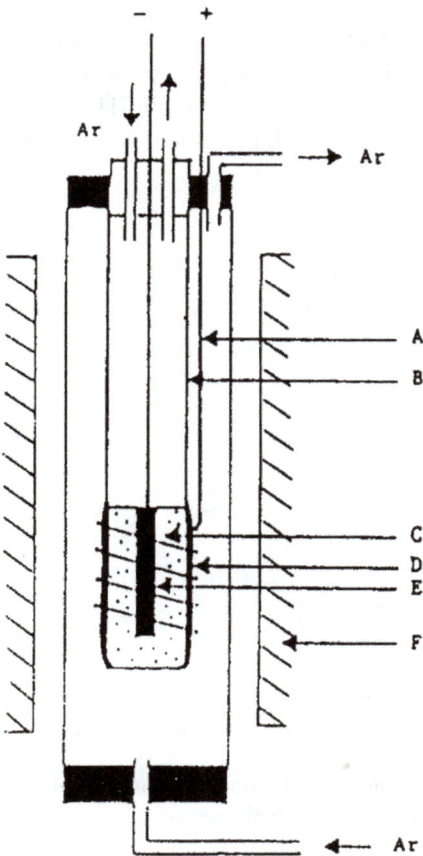

Figure 4.39: Schematic drawing of lithium-oxygen secondary cell. (A) Current collection from oxygen electrode. (B) Stabilized zirconia solid electrolyte. (C) LiCl-LiF-Li_2O molten salt. (D) $La_{0.89}Sr_{0.10}MnO_3$ oxygen electrode. (E) Lithium alloy negative electrode. (F) Furnace.
Republished with permission of *Journal of the Electrochemical Society*, from A lithium oxygen secondary battery, Semkow KW, Sammells AF, 134, 8A, 2084–2085, 1987; permission conveyed through Copyright Clearance Center, Inc.

alloy (Li_xFeSi_2) as the negative electrode, then a molten salt from LiF-$LiCl$-Li_2O, then the zirconia electrolyte membrane and as the positive electrode an oxygen catalyst with composition $La_{0.89}Sr_{0.10}MnO_3$, which was coated on a platinum current collector.

The electrocatalyst layer was formed *in situ* by heating the precursors (lanthanum acetate, manganese carbonate and strontium carbonate and ethylene glycol and citric acid) in the cell up to 1250 °C. It may sound strange that the term "in situ" is here used for the synthesis and processing of a material or device, and not for the analysis of a system. But frequently this term is used in this context. For example, the water oxidation catalyst from Ru, W and Co which my group used in the study is [Toth 2016] needed to be prepared in the condition where it was supposed to be used (*in situ*). Same holds for the copper dye sensitizer molecule which we are working currently on [Braun 2016, Constable 2009, Hernández Redondo 2009]. For the latter project, some obstacles needed to be overcome with the preparation of the photoelectrode when we carried out *operando* and *in situ* NEXAFS spectroscopy in the liquid cell first described in [Jiang 2010]. The dye x had to be synthesized *in situ* in that liquid cell before it could be measured.

The high temperature lithium air battery produces an OCV of 2.4 V as shown in Figure 4.40, at a current density of 20 mA/cm^2. The figure shows the evolution of the cell potential during the operation of 25 h. The stoichiometry of the anode (Li_xFeSi_4) is changing with the increasing lithiation and arrives at lithium when the potential (cell voltage) of 2.4 V is reached.

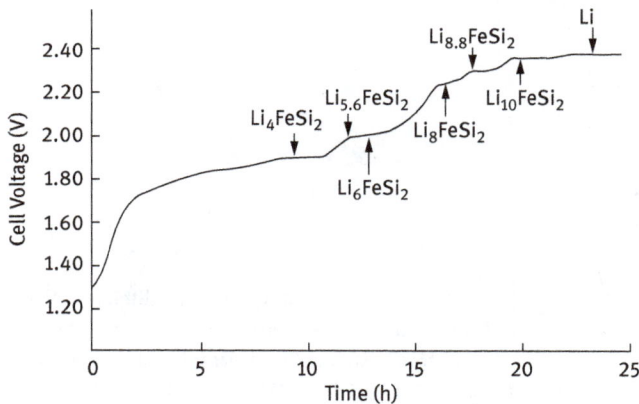

Figure 4.40: Galvanostatic IR free changing curve for the cell $FeSi_2$/$LiCl$-LiF-Li_2O/ZrO_2(5w/o CaO)/ $La_{0.89}Sr_{0.10}MnO_3$ at 650 °C. The compositions of some Li_xFeSi_2 alloys formed are shown. Current density 20 mA/cm^2 (versus the Li_xFeS_2 electrode).

The overall chemical reaction

$$FeSi_2Li_x + O_2(air) \Leftrightarrow Li_2O(molten\ salt) + FeSi_2$$

has the lithium content x as a parameter which manifests in the formation of different phases during charging and discharging as indicated in Figure 4.41. I do not know how Semkow and Sammells have determined the actual compositions or phases other than by integrating the current during charging and discharging and then determine via Faraday's law how much lithium was inserted into or extracted from the electrode. With the known weight of the negative electrode and the current density, this should be possible.

The curvature of the charging curve in Figure 4.40 is not homogeneous. There are humps which indicate that there are several energetic plateaus[6] where the insertion or extraction of lithium would not cause a change of the cell voltage. From the materials perspective, this is very interesting because we want to know by which processes energy can be stored and electric potential can be built in a battery. It certainly would be interesting to run the same experiment several times and remove the electrode after 10 h where the anticipated composition is Li_4FeSi_2 and run an EDAX or XRD on this electrode to confirm the composition and the crystallographic phase, respectively.

As the cell has to be disassembled (destructed) for that purpose, it cannot be used again to continue the charging and bring to the composition Li_8FeSi_2 after 16 h. For every new EDAX and XRD measurement, an extra new cell must be made and measured. It is thus more convenient when we are able to make *in situ* and *operando* X-ray measurements where we can change the condition of the electrode and measure it without destruction of the cell. This is one reason why nondestructive analytical methods are so valuable.

The intercalation and de-intercalation of lithium in the negative electrode is verified according to the charge and discharge curves shown in Figure 4.41. During 20 h, the cell was charged and discharged in the voltage range from around 1.95 to 2.4 V.

Semkow and Sammells at Eltron in Illinois had their work partially funded by NASA. It is therefore not surprising to learn that they also researched on how electrochemical processes could be used to produce chemicals on the Moon – also funded by NASA [Sammells 1988].

---

6 Graphite can be intercalated by lithium, for example. The lithium then seeks place between the graphene layers. The occupation of the space between graphene layers follows specific mathematical laws, or rules may be a better term, and the electronic structure of the Li_xC compound is reflected by the geometrical arrangement subject to these rules [Safran 1980] Safran SA: Phase Diagrams for Staged Intercalation Compounds. *Physical Review Letters* 1980, 44:937-940.doi: 10.1103/ PhysRevLett.44.937. Accordingly, the electrochemical potential follows the electronic structure.

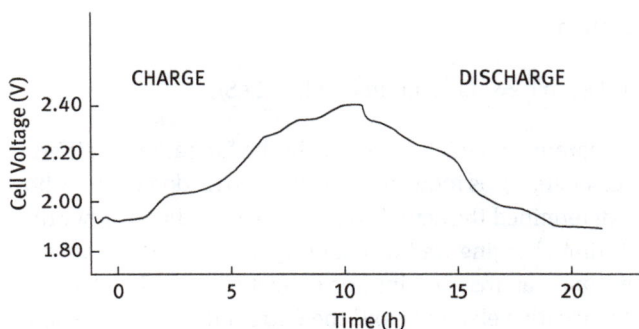

Figure 4.41: IR-free charge-discharge curve for the cell $Li_xFeSi_2/LiCl-LiF-Li_2O/ZrO_2$(5w/o Cao)/ $La_{0.89}Sr_{0.10}MnO_3/Pt$ at 20 mA/cm^2 (at Li_xFeSi_2 electrode). Total cell resistance 24 Ω. Temperature 650 °C. Republished with permission of *Journal of the Electrochemical Society*, from A lithium oxygen secondary battery, Semkow KW, Sammells AF, 134, 8A, 2084–2085, 1987; permission conveyed through Copyright Clearance Center, Inc.

4.8 The ZEBRA battery

ZEBRA stands for ZEolite Battery Research Africa (or for the favorite animal of the inventor of the ZEBRA battery, Johan Coetzer [Batteriesinternational 2016, Coetzer 2000]). The ZEBRA battery operates with a molten salt electrolyte at medium temperatures (200–350°C), for example, sodium aluminum chloride $NaAlCl_4$. The energy density which can be achieved with this kind of is around 120 Wh/kg and the power density is around 180 W/kg [Dustmann 2004]. This internal operation temperature requires particular attention for thermal insulation and cooling [Frutschy 2013]. As the ZEBRA battery is a rechargeable battery, its electrodes are not named as cathode and anode but positive and negative electrodes.

The negative electrode is sodium metal. The architecture of the battery is as follows. Sodium metal is contained in a metal beaker. The positive electrode is a blend of $Ni/NiCl_2/NaCl/NaAlCl_4$, which is contained in a beaker from beta-alumina β-Al_2O_3. Beta-alumina is a solid electrolyte which provides high ionic conductivity with cations [Kummer 1967] such as Na^+, K^+, Li^+, Ag^+, H^+, Pb^{2+}, Sr^{2+} or Ba^{2+} at temperatures from 250 to 300 °C. As the electrolyte is solid, it is therefore also a separator for the two liquid or molten phases of the sodium and the molten $NaAlCl_4$ electrode.

The relatively high temperature promotes some degradation processes in the cell. It would therefore be worthwhile to have the battery running at more ambient temperatures [Gerovasili 2014].[7] Unfortunately, lower temperature reduces the necessary wettability of beta-alumina by molten sodium and thus the electrochemically active

---

[7] We have a similar issue with the SOFC high temperature fuel cells which operate with oxygen ion conductors as solid electrolytes from 600 to 1,000 °C. A ceramic proton conductor would already work at half this temperature.

geometric area. But adding a Pt grid, for example, by screen printing to the negative electrode, can neutralize this effect [Li 2014a].

Because sodium is a major component of the ZEBRA battery, it belongs to the class of sodium batteries [Hueso 2013]. As there are no chemical side reactions in the ZEBRA battery, its efficiency is virtually 100% [Sakaebe 2014]. The different operation principles and architectures of the various sodium batteries are sketched in Figure 4.42 [Ha 2014]. The complete reaction during battery discharge can be written as described in Table 4.7. The phase diagram of $NiCl_2$ and NaCl shows existence of two phases in the range 150–400 °C [Bones 1989]. Spatially resolved phase analysis can be made with X-ray and neutron methods [Braun 2017, Hofmann 2012, Zinth 2015, 2016].

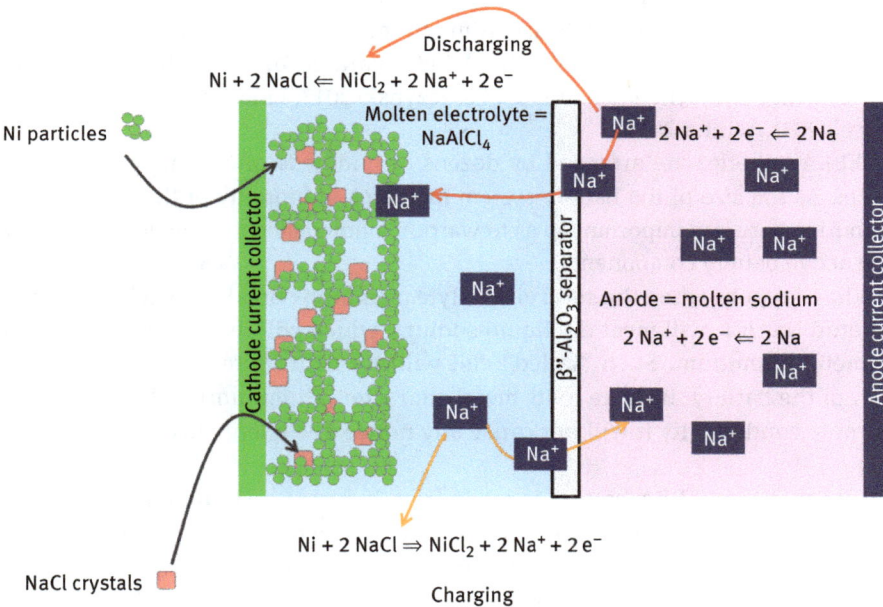

Figure 4.42: Schematic diagrams of design, charge flow and mass flow of the Na-NiCl$_2$ cell; inspired by a graphics in [Ha 2014].

Table 4.7: Electrochemical reactions and potentials of the $NiCl_2$/NaCl-based ZEBRA battery at 300 °C.

Electrode	Reaction	Potential
Positive electrode	$NiCl_2 + 2\,Na^+ + 2\,e^- \leftrightarrow Ni + 2\,NaCl$	
Negative electrode	$2\,Na \leftrightarrow 2\,Na^+ + 2\,e^-$	−2.71 V
Net cell reaction	$NiCl_2 + 2\,Na \leftrightarrow Ni + 2\,NaCl$	2.58 V

The positive electrode is sometimes doped with $Fe/FeCl_2$ which can be resulting in a secondary electrode reaction to gain extra power [Bones 1987]. Adding sodium sulfide Na_2S makes this a hybrid battery with an additional charge plateau [Lu 2013a]. It is an exercise for the reader to estimate the thermodynamic implications which arise from the coexistence of such two phases in the battery when compared to ZEBRA batteries with more complex chemistries such as described in Lu et al. [Lu 2013b].

Upon battery charging, the electrochemical reactions run in the opposite direction. The OCV is typically 2.58 V. Depending on the design of the ZEBRA cell, the electrode materials and the charging rate, the OCV may vary. When we take sintered iron as electrode and not nickel, the OCV will be only 2.35 V. It is an exercise for the reader to consider alternative chemistries for ZEBRA batteries where other 3d metals are used instead of Ni, and then determine the OCV based on the standard redox potentials according to the long list of references with data [Atkins 1997, 2010, Aylward 2008, Bard 1985, Bard 2001, Connelly 1996, Cotton 1999, Courtney 2014, Greenwood 1997, Leszczynski 2013, Lide 2006, Pourbaix 1966, Vanýsek 2011, Winter 2018].

ZEBRA batteries are arranged by dozens or hundreds of cells in series and in parallel. As the size of the battery system increases, uniformity of the temperature distribution becomes important so as to warrant a homogeneous chemical reactivity of the active battery components.

What happens when the solid electrolyte membrane breaks in such cell? Then the liquid $NaAlCl_4$ will meet the liquid sodium, which will react to NaCl rock salt and metal aluminum. Such "failed" cell will not contribute to the energy and power of the battery. But due to its metal character (aluminum) and thus its high electronic conductivity it will not cause any noticeable energy loss in the system either.

Note that the ZEBRA battery is made from low cost and abundant materials such as Ni, Fe, Al_2O_3, NaCl, which can be recycled easily as well. The battery has an inherent safety because it will turn off once the temperature decreases. Frutschy et al. have provoked overcharging of particular cells in a ZEBRA battery and investigated how safe the battery would remain in case of such failure [Frutschy 2015].

4.8.1 ZEBRA battery for submarines

The ZEBRA battery in a submarine would certainly be a mobile application. This was studied by Kluiters et al. 20 years ago for the Royal Dutch Navy [Kluiters 1999]. Many submarines are propelled with diesel engines at the surface and with electric motors when submerged in the ocean. The electric motors are powered by lead acid batteries, which are recharged by alternators by the diesel engines during surface operation. Because lead acid batteries can produce hydrogen, there is a risk-by-design that such hydrogen can mix with oxygen from air and eventually explode as Knallgas. A ZEBRA battery would not have this kind of danger.

There have been speculations [Evans 2017] that the loss of the Argentinian submarine ARA San Jose (a German built TR1700 class submarine) on or after the 15 November 2001, was caused by an explosion of such explosive gas. Before the explosion, the submarine notified the naval base that water had swapped over during surface operation and leaked into the battery room and caused a short circuit. Three hours later, sonar stations detected an explosion in the area where the submarine went missing. Vessel and crew have not been found since.

Let us look at the ZEBRA battery for the Dutch Navy – with specifications listed in Table 4.8. They chose a battery pack from AEG Anglo Batteries which operates on sodium/nickel chloride chemistry. The cells are considered "monolithic" and they are assembled as 110 cells in series, and two of these units assembled in series to constitute the "battery." The 110 cells add up to theoretically 110×2.58 V = 258 V, but the authors disclose an actual OCV of 183.5 V.

Table 4.8: Specification for the AABG (AEG Anglo Batteries) ZEBRA sodium/nickel chloride battery type Z5/171.

Description	Data
Battery type	Sodium/nickel chloride
Manufacturer's code	Z5/171
Cell type	ML1C (monolith)
Cell configuration	2 × 110 cells in series/parallel
Size (incl. controller) $L \times B \times H$	810 mm × 541 mm × 315 mm
Size (excl. controller) $L \times B \times H$	730 mm × 541 mm × 315 mm
Weight	200 kg
Open-circuit voltage (OCV)	110×2.58 V = 183.5 V
Minimum voltage (2/3 OCV)	189.2 V
Maximum discharge current −60 s	2 × 80 As = 160 A
Maximum continuous discharge current	80 A
Maximum voltage during recuperation	110×2.85 = 313.5 V
Charging voltage	110×2.67 = 293.7 V
Maximum charge current	No practical limit
Capacity	64 A h
Energy	18 kW h
Specific energy	90 W h/kg
Energy density	132 W h/dm^3
Working temperature	270–350 °C
Cooling	Air
Cell resistance (0% DOD to 80% DOD)	17 mΩ
Battery resistance (0% DOD to 80% DOD)	9 mΩ

Note: Data for the ZEBRA battery with specifications for a ZEBRA battery powered submarine. The battery was not actually tested in a submarine (DOD = depth of discharge)) [Kluiters 1999].

The reader can compare as an exercise the design and performance of this ZEBRA battery from AEG Anglo Batteries with the Siemens fuel cell which is used in submarines.

4.8.2 ZEBRA batteries for stationary applications

Stationary applications include electric power back-up plants in remote areas where no or little connectivity to an electric grid is given. One system developed by General Electric is the Durathon™ battery which can be used, for example, in telecom applications [Rijssenbeek 2011]. What is meant here is that the relais towers for mobile communication are operated with diesel engine generators which charge storage batteries. Often it is lead acid batteries which are used for this purpose (similar example for submarines, see [Kluiters 1999]), but the ZEBRA battery may to some extent be a better storage solution according to Rijssenbeek et al. [Rijssenbeek 2011] than the lead acid battery. On Prince Edward Island in northern Canada, wind power is being stored in ZEBRA batteries [Bunker 2015].

4.8.3 Design of the ZEBRA battery

The cells of the ZEBRA battery have a prolate geometry; they typically look long and slim. Cord-Henrich Dustmann of Battery Consult in Switzerland, along with his colleagues, have designed a ZEBRA battery built from cells with a flat oblate geometry [Vogel 2015]. One cell has a capacity of 1.5 Ah at 2.58 V. About 126 such cells add up to 325 V and can store 7 kWh electricity. This could be a suitable system for storing solar power.

The website http://www.energystorageexchange.org/projects shows an impressive number of electricity storage projects across the world. A quick glimpse on this large database shows that not all projects which I am aware of are included in this list. It is also possible that the list shows projects which are not current anymore. The list is therefore in my opinion not complete, but it gives a very good overview on the location of projects, storage technology and size and capacity of the plants and details on the kind of ownership.

We read from Figure 4.42 that $NiCl_2$ becomes reduced and metal Ni clusters precipitate and grow when the positive electrode becomes discharged. The crystallization and dissolution kinetics have a decisive influence on the performance of the battery, which Rock et al. [Rock 2016] have studied at the quantitative level with electron microscopy and cyclic voltammetry and chronoamperometry.

As there are structural and morphological changes in the materials of electrodes, electrolytes and current collectors, it is often worthwhile to conduct analytical studies which go beyond the mere electrochemical and electroanalytical assessment.

Otherwise, the battery will always be a "black box" which nobody really understands. The same holds for fuel cells, capacitors and so on. X-rays and neutrons are helpful probes which allow you to study the devices without destructing them [Braun 2017]. It is very rewarding when you then have a "picture" in your mind what is going on in the material during operation and be able to make quantitative mathematical models which are not necessarily the exclusive domain of electrochemistry; I recommend the text book of Octave Levenspiel [Levenspiel 1962].

Let us look at the charging of the accumulator. We have to remain distinct about the terminology. The cathode is the positive electrode on the left. It is comprised of a three-dimensional network of nickel metal particles which includes rock salt NaCl particles between the pores of the metal network. The network is organized in a way that there is a maximum interface area between the NaCl and the Ni metal which react as

$$Ni + 2\,NaCl \Rightarrow NiCl_2 + 2\,Na^+ + 2\,e^-$$

The metal network is necessary to provide electronic contact into the electrolyte region. The electrolyte is the molten $NaAlCl_4$ salt which extends from the β-Al_2O_3 separator in the middle and the current collector on the left. Note that this separator has ionic conductivity and thus is an ionic conductor, but the actual "electrolyte" is the molten $NaAlCl_4$ salt electrolyte which faces the cathode (positive electrode) current collector. This molten electrolyte penetrates the porous positive electrode network so that there is a maximum interface area between the electrode and the electrolyte. This is important to provide a maximum area where the Na^+ ions from the positive electrode through the molten salt electrolyte to the β-Al_2O_3 separator. This solid ceramic separator prevents mixing of the molten salt electrolyte on the left with the molten Na anode (negative electrode) material on the right. But the separator is an Na^+ ion conductor [Kummer 1967] which provides the ionic path between positive and negative electrode. Upon charging, the Na+ from the left migrates to the right where it solidifies (to molten Na) according to

$$2\,Na^+ + 2\,e^- \Rightarrow 2\,Na$$

We have therefore a liquid/molten anode and liquid/molten electrolyte which are separated by the β-Al_2O_3 separator, a ceramic membrane. The accumulator is now charged and still at a temperature of around 300 °C (Figure 4.43).

Nickel is not the only 3d element which can be used for the ZEBRA technology. Various other pure systems and mixed systems have been explored overtime. Figure 4.44 compares the change of the OCV (e.m.f.) versus temperature for ZEBRA cells with Cu, Ni, Co, Fe and Cr as the positive electrode active ions [Sudworth 1994]. This was the order of falling OCV. With a tracing program [Tummers 2006] you can read in and digitize the figure and then determine the

Cell E.M.F. (V)

$2Na + 2AlCl_3 + NiCl_2 \leftrightarrows Ni + 2NaAlCl_4$

30.5 V

Overcharge

$2Na + NiCl_2 \leftrightarrows Ni + 2NaCl$

2.58 V

Cell reaction

$3Na + NaAlCl_4 \leftrightarrows 4NaCl + Al$

1.58 V

Overdischarge

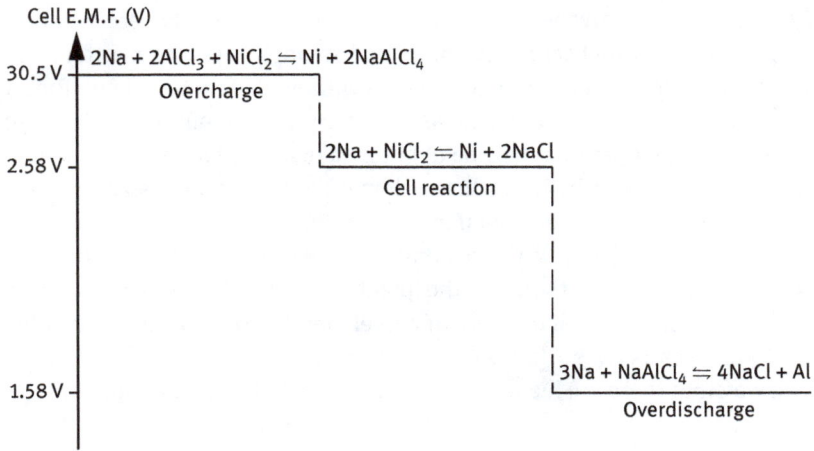

Figure 4.43: Electrochemical reaction mechanism including overcharge and overdischarge for Na–NiCl$_2$ at 300 °C (green, solid line) and Na–FeCl$_2$ at 250 °C (red, dashed line) systems with molten NaAlCl4 as a supporting electrolyte.
Reprinted from *Journal of Power Sources*, 51, Sudworth JL, ZEBRA Batteries, 105–114, Copyright (1994), with permission from Elsevier [Sudworth 1994].

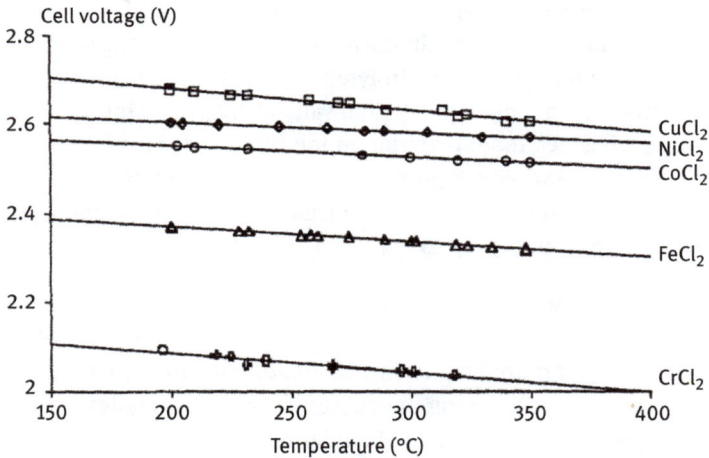

Cell voltage (V)

CuCl$_2$
NiCl$_2$
CoCl$_2$

FeCl$_2$

CrCl$_2$

Temperature (°C)

Figure 4.44: Variation of the OCV of ZEBRA cells for CuCl$_2$, NiCl$_2$, CoCl$_2$, FeCl$_2$ and CrCl$_2$ in the temperature range from 150 to 400 °C. Solid lines are guides to the eye.
Reprinted from *Journal of Power Sources*, 51, Sudworth JL, ZEBRA Batteries, 105–114, Copyright (1994), with permission from Elsevier [Sudworth 1994].

numerical values for the OCV in the range from 150 to 400 °C. $CuCl_2$ has the highest OCV with 2.71 V which drops linearly to 2.59 V at 400 °C. $NiCl_2$ follows as the next best system with 2.62–2.56 V.

We notice that the slope for this system is quite flat and thus shows less temperature variation than the other systems except for $CoCl_2$, which has the same low slope like $NiCl_2$. As nickel, copper and cobalt are quite expensive metals, iron in $FeCl_2$ can be considered a valuable alternative for ZEBRA batteries even if the OCV ranges from 2.39 to 2.30 V. $CrCl_2$ does obviously work as salt for positive electrode in ZEBRA cells, but its low OCV of 2.1–2.0 V in the range from 150 to 400 °C yields no advantage over the other aforementioned systems.

A cell for electroanalytical studies on components for ZEBRA batteries (this includes also the Na-Cu-I battery) is shown in Figure 4.45. It is a container made from two glass tubes which are connected (or separated, if you like) with a NASICON electrolyte membrane in-between. The tubes are filled with a liquid catholyte on one side and with molten sodium on the other side. A metal with current collector is inserted in the molten sodium, and another metal wire with a copper current collector is inserted in the aforementioned catholyte. This is the level of complexity that you need to elaborate practically when you want to work with a molten salt battery such as the ZEBRA battery.

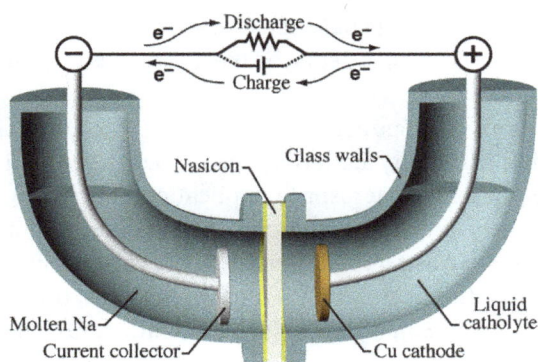

Figure 4.45: Schematic representation of a laboratory-scale cell using liquid Na as the anode, NASICON (Ceramatec Inc.) as the separator and copper as the cathode.
Reprinted from Electrochimica Acta, 112, Zhu H, Bhavaraju S, Kee RJ, Computational model of a sodium–copper–iodide rechargeable battery, 629–639, Copyright (2013), with permission from Elsevier [Zhu 2013].

This cell served also for the development and application of a mathematical model and simulation for the description of various batteries [Zhu 2013]. I would like to recommend this study as an exercise for the reader to follow the principles of charge transport and mass transport in a complex battery type like this one.

4.9 Zinc/silver oxide batteries

The zinc silver oxide cell is an expensive accumulator with short life. It has a nominal voltage of 1.5 V OCV. Its high energy density and power density make the zinc silver oxide cell useful for defense and space applications. One of the zinc silver battery pioneers, Frenchman Henri Andre, built an electric vehicle which he powered with this kind of battery in the 1950s. This type of battery has an illustrious history of over 75 years [Fleischer 1971, Karpinsky 2002]. For some time, this kind of battery was considered a realistic power source for electric vehicles [Editors 1960]. Even today engineers are trying to improve the performance of zinc silver oxide batteries [Ozgit 2014].

The zinc silver oxide battery, for example, powers the German DM2A4 Seehecht torpedo built by Atlas Elektronik [Forecast 2002]. According to this market intelligence from the year 2002, one such torpedo may cost between 500,000 $ and 2,000,000 $. The torpedo has a length of over 6 m, a diameter of around 50 cm and a total mass of 1,370 kg, 260 kg of which is from the explosive warhead. The battery powers the electric motor of 374 PS and the electronic controls in the torpedo. The maximum range of the torpedo which it can run with the battery depends on the speed. At 65 km/h, the range is 20 km; the torpedo is kept "on a leash" from glass fiber, so that it can be controlled from the operator at the warship or submarine which has launched the torpedo. When the speed[8] is reduced to 42 km/h, the range increases to 28 km. The new torpedos have a range of 50 km and a maximum speed of 90 km/h. The Bundesmarine (Germany Navy) has deployed these torpedos in their U212 submarines.

When you do an Internet search you will come across a report from 1962 [Lander 1962] where sealed zinc silver oxide secondary batteries were investigated and developed in a 2-year study (three phases) for aerospace applications, including zero gravity conditions. It was necessary to make hermetically sealed batteries for applications in satellites. You read in the Abstract of this study that they began as the first phase of the project with a literature review in five important topics, this is (1) the silver migration in the cell [Dirkse 1964], (2) the voltage regulation, (3) zinc

---

8 It appears when you reduce speed you can gain more range. I have made a strong experience with the hydrogen fuel cell car Hyundai ix35 fuel cell. Its nominal range is reported to 594 km with 5.4 kg H_2, the amount of fuel pressed at 700 bar into the tank. Practically, I could drive around 400–450 km with a full tank when I did not speed. At German highway speed of 130 km/h you can make it only to 200–250 km range. Later, I learnt I can make it by 30% further than specified by the car manufacturer when I drive at 50 km/h only. When I was a kid in the late 1970s, my neighbors had a personal computer with games that were stored on a magnetic tape recorder. One game was about spaceflight, where you launched a rocket to the moon and you had to be careful with the fuel budget; when you were too fast the fuel would not last to the moon; when you were too slow you could not exit the Earth's orbit. The game was to find out the optimum speed to make it to the moon and back from the moon to the earth. Or maybe the program was called Jupiter Lander. I don't remember.

particle size and displacement in the cell during cycling, (4) gas evolution and (5) terminal sealing.

In the second phase, they designed, built and tested around 100 cells electrically and environmentally until they failed. Note that they cycled the batteries until failure. The failure is essential part of understanding and improving batteries or any devices. In the extreme case you have to make "crash tests" where you force the failure.

Based on the experience from phase two, the researchers prepared one dozen further cells with improved design and better performance for further environment testing. Phase three concluded with a speciation for battery construction. The researchers – a group from he Delco-Remy Company which builds electric equipment – built ten optimized batteries and sent them to the NASA Flight Accessories Laboratory.

So this was not only an academic study on some particular battery type, but a targeted study to provide a hi`gh power battery which would last long enough in a satellite in space where you have vacuum conditions and extreme temperature changes. The principal requirement for the battery was that it can sustain 5,000 cycles at around 28 V in the temperature range from −18 to + 38 °C in vacuum under zero gravity. A cycle was 35 min discharge at 20 A with a subsequent charging of 85 min. Practically, the project was permitted to make a battery for 500 cycles between 0 and 38 °C, with an anticipated energy density of 66 Wh/kg. And there is still research on this matter nowadays, such as improving performance and cyclability of zinc-silver oxide batteries by using graphene as a two-dimensional conductive additive [Ozgit 2014].

Figure 4.46. shows the variation of the voltage of cells that had been cycled at three different temperature: that is −1, +27 and 38 °C with depths of discharges from the maximum by 21%, 30% and 40% of that value. We are looking at a wide range of 600 cycles. All curves begin with the nominal and actual 1.5 V cell voltage [Lander 1962].

Let us begin with looking at the cells cycled at room temperature (+27 °C) with discharge by 21%. The cell voltage decreases almost linear with consecutive cycle number until the 400th cycle is finished, when the voltage has decreased to around 1.48 V. With the subsequent 100 cycle, the voltage decreases rapidly to 1.3 V. When the cells are discharged to 30%, the slope with which the cell voltage decreases from cycle to cycle is much steeper than when the cell is discharged by 21% only. After 180 cycles, the cell voltage drops rapidly to 1.3 V with the next subsequent 100 cycles. The cycling stability suffers even more when we carry out the discharge by 40%. After 100 cycles, the voltage drops to 1.45 V and falls to 1.3 V at cycle # 200. The deeper the discharge, the worse is the cycling stability.

When we raise the temperature from +27 to +38 °C – this is only in the order of 10 K – the cycling stability increases considerably as we notice that the shape of the curves for the voltage is extended by around 100 cycles. The cell was discharged by 21%. The voltage drops to 1.47 V during the first 500 cycles and hits the 1.3 V line

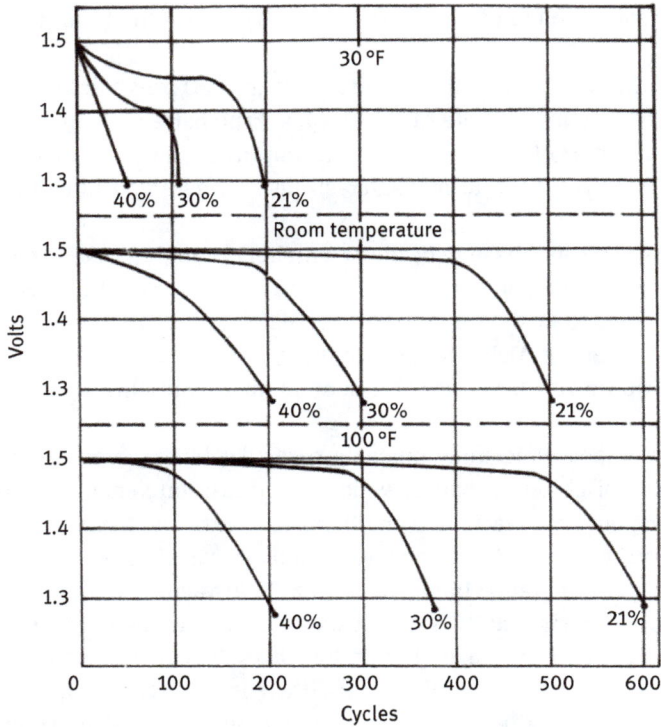

Figure 4.46: End of discharge voltages at each 100 cycle interval for the indicated temperatures and depths of discharges. Reproduced from Figure 13 from Lander JJ, Keralla JA. Development of Sealed Silver Oxide-Zinc Secondary Batteries. 1962. Flight Accessories Laboratory, Aeronautical Systems Division, Air Force Systems Command, Wright-Patterson Air Force Base. Ohio. ASD-TDR-62–668. October 1962 [Lander 1962].

after the 600th cycle. Also upon discharging by 30%, the cycle stability is given for the 280th cycle. And also at 40% discharge, the voltage is relatively stable for the first 100 cycles, whereas at room temperature, this stage was already reached after the first 50 cycles. Hence, raising the temperature by 10 K has a benefit for cycling stability.

Cooling from room temperature (27 °C) down to freezing temperature of +1 °C has a drastic malign effect. The curves that we know from the two higher temperatures are now compressed toward lower cycle numbers. Gentle discharge by 21% causes a decrease of the voltage to 1.45 V after 100 cycles. After the 150th cycle, a sharp decrease of the cell voltage sets on. The 1.3 V value, which we know by now, is reached at the 200th cycle. Upon deeper discharge by 30%, the cell voltage hits 1.3 V after 105 cycles. And only 50 cycles are necessary to discount the cell to 1.3 V when the discharge is 40%. Hence, the cold battery performs worse than the warm battery.

Today, the nominal capacity is defined for +20 °C. At +30 °C, the capacity drops to around 85%, and at −20 °C, it drops to around 50%.

The authors of this study [Lander 1962] argue that the "cause of failure at low temperatures is the inability of the cells to accept sufficient recharge within the 85 minute charge period." While we may be disappointed about the failure at low temperature, the authors of the study find that when the cells which "failed" at +1 °C F were later recharged and cycled at +27 °C, they worked good for 400–600 cycles. The irreversible failure of the cells cycled at room temperature and above was attributed to the disintegration of the nega-tive electrode material [Lander 1962].

The charge and discharge profile of a representative zinc silver oxide battery is shown in Figure 4.47. Here, the battery voltage is plotted versus the time for

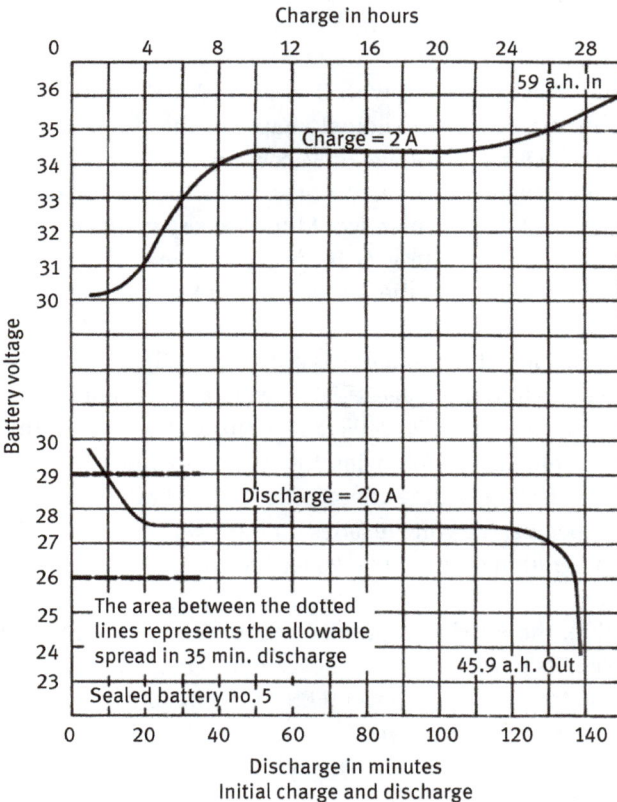

Figure 4.47: Initial charge and discharge. Reproduced from Figure 33 from Lander JJ, Keralla JA. Development of Sealed Silver Oxide-Zinc Secondary Batteries. 1962. Flight Accessories Laboratory, Aeronautical Systems Division, Air Force Systems Command, Wright-Patterson Air Force Base. Ohio. ASD-TDR-62–668. October 1962 [Lander 1962].

charging and discharging. Note the two different ordinate axes for the plot. The bottom ordinate and bottom graph shows the discharge time from 0 to 150 min. The top ordinate and top graph shows the charging time from 0 to 30 h. Note also that the voltage is in the range of 30 V. This means we are not dealing with one single zinc silver oxide cell, but with a complete battery of cells. Because one cell has around 1.5 V, we are looking at a battery of more than 20 cells.

In the graph, it says that "Sealed Battery No. 5" is shown. The battery for the satellite, its later use, needs to be sealed because the satellite is operating in vacuum in outer space. When the batteries are manufactured under ambient pressure, this ambient pressure is basically worked into the cell. When the battery is launched with the satellite into space, with increasing altitude, the external pressure will decrease and the pressure in the cell will become relatively higher with respect to outer pressure, which eventually amounts to vacuum. The pressure difference will cause evaporation of the electrolyte and possibly other mechanical issues affecting integrity of the cell. To prevent this, the battery needs to be thoroughly sealed.

The charging of the battery begins at $t = 0$ with OCV of 30 V with an imposed current of 2 A. The voltage increases in a sigmoidal shape during the first 10 h and then arrives at a plateau which has a voltage of 34.4 V. During the next 10 h, the voltage remains at this plateau value but then lifts of and increases gradually to 36 V, which the battery achieves at 30 h. Where would the battery get its current from when in space? In space, the satellite gets its primary energy from solar panels. The electric charge accumulated in the battery after 30 h of charging at 2 A makes 60 Ah.

While the charging has taken 30 h, the discharge is done in less than 2 ½ h. The discharge current is 20 A from the initial voltage of 30 V, which drops rapidly and linear to 27.5 V in 20 min. Averaged over the 20 cells that comprise the battery, this is a voltage drop of 0.125 V per cell. For the following 100 min, the voltage remains constant at 27.5 V. This is a brilliant discharge characteristic for a battery in 1962. Two hours after the discharge begins, the voltage fades rapidly away to below 24 V, which is 1.2 V per cell. The integration of the current overtime yields a total charge of 49.5 Ah.

The first manmade satellite, the Russian "Sputnik" was brought into space on October 4, 1957. The satellite got its power from three silver zinc batteries, developed by a research group around electrochemistry pioneer [ECS 2013] Vladimir Sergeevich Bagotsky. You notice from the Report by Lander and Keralla in 1962 [Lander 1962] that the military was interested in satellite technology. The military is also always interested in energy, to the extent that it wants its dismounted soldiers to have enough energy without carrying too much weight [Rowe 1997]. The first commercial satellite was launched into space in 1962. It was built by the Bell Telephone Company and served for transatlantic communication. You know now that a seemingly ridiculous component like a battery or a solar cell is instrumental for allowing modern technologies to progress which shape the world and society. Being able to broadcast, for example, a fashion show in Paris to all continents can have impact on the entire society [McClendon 2015]:

> On the following day, July 24, 1962, CBS used the same Telstar satellite to broadcast a live preview of Dior and Balmain haute couture models across America. In so doing, Telstar changed the dynamic of the fashion industry by disrupting the traditional flow of information between its two most powerful centers: Paris and New York.

It is always helpful in science when you can compute and calculate the experiment that you perform in the laboratory. The first important check is the consistency check. There you ask yourself "does this result make sense?" A further step is that you compare your experimental results with a physical model which you can explain at the qualitative level. A further step is that you actually mathematically model the experiment. Scientists usually have a different approach than engineers. The latter often uses commercial simulation packages, one of which, for example, is Comsol®.

A COMSOL application example for the zinc silver battery is based on the one-dimensional model of Torabi et al. [Torabi 2012], which can be found at Comsol [Comsol 2018]. With such so-called "Multiphysics" packages you can simulate the entire device and system.

I will follow here the elaboration by von Sturm [Von Sturm 1981], which is based on first principles and helps understanding the thermodynamics of batteries. The zinc silver oxide cell is designed as follows [Von Sturm 1981]:

$$(-)Zn|Zn(OH)_2||electrolyte||Ag|Ag_2O(+)$$

Note that we need here a metal zinc electrode. Metals typically have an oxide layer at their surface. When the metal is oxidized at the surface, the chemistry in the above equation will certainly be a different one and the battery might not perform so well anymore. It is therefore important to have a clean electrode. This is important for battery fabrication and certainly subject to intellectual property in industry – or in the military [Denison 1957].

It is possible to determine the theoretical open circuit voltage of a battery cell from thermodynamic data obtained from calorimetry. The enthalpy ΔH_0 and entropy ΔS_0 of the reaction under standard thermodynamic conditions are related with the free enthalpy of reaction ΔG_0 [Von Sturm 1981]:

$$\Delta G_0 = \Delta H_0 - T\Delta S_0$$

Calorimetry will yield $\Delta G_0 = -307.6$ kJ/mol. Now it is a simple application of Faraday's law where we form the ratio of the free enthalpy of reaction and Faraday's number and the number of involved electrons, which yields for the cell voltage

$$E_0 = -\frac{\Delta G_0}{n \cdot F} = +1.594\text{V}$$

We can determine the cell voltage also from the standard potentials of the half-cell reactions, which reads at the anode

$$\text{Zn} + 2\text{OH}^- \Leftrightarrow \varepsilon - \text{Zn(OH)}_2 + 2\,e^-\,;\; E_0 = -1.249\text{V}$$

and at the cathode (Mansour et al. used a suite of analytical methods to study the ageing of silver oxide cathodes [Mansour 1990])

$$2\,\text{Ag} + 2\,\text{OH}^- \Leftrightarrow \text{Ag}_2\text{O} + \text{H}_2\text{O} + 2\,e^-\,;\; E_0 = +0.345\;\text{V}$$

The complete cell reaction reads thus

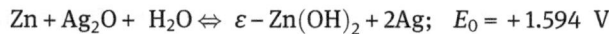

$$\text{Zn} + \text{Ag}_2\text{O} + \text{H}_2\text{O} \Leftrightarrow \varepsilon - \text{Zn(OH)}_2 + 2\text{Ag};\;\; E_0 = +1.594\;\text{V}$$

The reaction-free enthalpy ΔG will increase depending on the concentration (mole number) v_i and activity a_i of the components i of its constituents according to the well-known thermodynamic relation

$$\Delta G = \Delta G_0 + R \cdot T \cdot \sum_i v_i \ln a_i$$

We determine therefore the concentration-dependent cell potential via the Nernst equation

$$E = E_0 - \frac{R \cdot T}{n \cdot F} \cdot \sum_i v_i \ln a_i$$

Using the computational methods of classical statistical physics, we can determine the temperature dependency of the cell voltage ($\Delta S = -66$ J/K mol)

$$\left(\frac{dE}{dT}\right)_P = -\frac{1}{n \cdot F}\left(\frac{d\Delta G}{dT}\right)_P = \frac{\Delta S}{n \cdot F} = -0.34\text{mV/K}$$

We see that, with increasing temperature, the cell voltage will decrease for every Kelvin by 0.34 mV.

The military often needs batteries which are stored for many years and then suddenly be used. Most batteries have the electrolyte already included by manufacturing process but the corrosion of electrodes and other components silently sets on immediately after manufacture. This is why batteries have a particular shelf life. When they become too old, they cannot be used properly anymore.

One way to avoid this and to extend shelf time is to not provide the electrolyte at battery manufacture. Some battery designs have the electrolyte separated in a specific container, and only when the battery is needed for operation, the electrolyte becomes injected into the battery and then you have a fresh battery without the shortcoming from silent shelf operation. Such a battery is called reserve battery. The study by Smith and Gucinski deals with the improvement of such reserve batteries by tuning the electrode chemistry, for example, by alloying [Smith 1999] .

Some of the torpedo batteries operate that way: Right before the torpedo is fired from the submarine, the electrolyte is injected into the battery compartment in the torpedo. The liquid electrolyte is already in the torpedo and also in the battery casing, but not between the electrodes yet. This warrants that the battery is virtually new even after long residence time in the torpedo, as Saft advertises [Saft 2013a, b]:

> "For reasons of safety and performance (shelf life more than eight years), the batteries are only activated by electrolyte injection at the last minute."

4.10 Toxicity of batteries and their materials

Many chemical elements important for battery chemistry are toxic. The lead acid battery contains lead, which is a toxic heavy metal [Body 1991]. The general risk for environment and danger for health when recycling lead batteries has been recognized [Collivignarelli 1986]. Also the battery electrolyte sulfuric acid is an aggressive liquid which can do harm when your skin or eyes get in contact with it.

There are publications which report hazardous lead exposure of families in rural North Carolina who recycled lead acid batteries at home [Dolcourt 1981], lead contamination around a Kindergarten near a battery recycling plant in Taiwan, the exposure of children and adults to lead [Wang 1992] and a severe lead-poisoning of children in Trinidad from similar operations [Changyen 1992].

The protection of our environment has become one of the most important policies in the last 50 years. Product policy has been developed for batteries [Scholl 1995]. Asks Paul Ruetschi in his paper in 1993 [Ruetschi 1993]:

> Worldwide, about 15 billion primary batteries, and well over 200 million starter batteries are produced per year. What is the impact of this widespread use of batteries on the environment?

When I was a summer intern with Philips Research Laboratories in Aachen in Germany, I was supposed to work with metalorganic lead compounds. I had to synthesize lead zirconate titanate (PZT) coatings for DRAM electrodes and ink jet electrodes. I had to mix the precursors in the chemistry lab and dilute them with various organic solvents. I built an ultrasonic evaporation reactor which would generate moist damp vapor from the solution and guide it on the surface of glass or silicon wafers.

Philips as a responsible employer sent me to see a doctor before I would start work at Philips. They had a doctor under contract in Aachen, and I believe they took me a blood sample. I do not remember more of that. But when my 3 months internship with Philips was over, I had to see the doctor again and she made another blood test. But most interesting was that she inspected my gum in my mouth and wrote down in her report "no lead line." Apparently long-term exposure of humans to lead may cause visible traces in the mouth on the gum.

When you deal as a consumer with a car starter battery, then there should be no harm during normal customer operation. You do not get in touch with the lead unless you damage the battery and open it and seek contact with the lead. It plays a role whether the lead is bioavailable. Lead in molecular form can therefore be more harmful than lead in solid metal form. From the consumer perspective, lead batteries are no problem. But from the manufacturer perspective it is. Some of the workforce in the mines and factories may face long-term exposure with bad consequences for their health if there are no precautions taken from employers and workers [Chen 2010, Ekinci 2014].

Usually there is government oversight that the employer follows the corresponding worker protection laws and environmental health and safety (EH&S) regulations. Similar holds for recycling factories where batteries may be shredded to bring to surface valuable metals in the batteries. When working in the laboratory, you work with toxic materials that follow the laboratory safety rules.

Lead has been used for very long times as tubing for the residential water supply. In modern time, society learnt that these lead pipes bear a health problem. Pb_3O_4 is a red color lead oxide which was used in the past as a dye for red and yellow paints. As such it also posed a serious health risk.

Lead is necessary in PZT actuators, and there has not yet been an environmental more benign alternative identified. Lead is used as shielding against radiation, for example, as lead line gloves in gloveboxes where radioactive materials are handled [Landsberger 2005]. Until 30 years ago, lead was used as an anti-shock additive in fuels for combustion engines; engine exhaust contained therefore traces of lead. Therefore, lead acid batteries are not the only items of our technological civilization which contain toxic lead. Lead is used in bullets and shooting ranges contain a lot of lead.

Mercury, Hg, and cadmium, Cd, are also toxic heavy metals which are used in batteries. Ni sounds not as harmful but to many countries it is known as a cancer-causing agent. Efforts to replace Ni by other 3d metals such as Mn or Fe are founded not only in the high price of Ni but also in government EH&S regulations which can make it more difficult to get a battery factory approved. It can just cost more time to satisfy all legal requirements to build a new battery factory. Fewer toxic components in the production line may speed up the approval process.

Zinc as it comes in zinc air batteries is no problematic component from the perspective of toxicity. At the biological cell level, it has critical properties causing cell death, but depletion of zinc in the body can also have negative consequences [Plum 2010]. Zinc oxide is often used as ointment for medical applications.

4.11 Battery transportation

As the demand for more energy and more power in containers with smaller volume and lesser weight is more and more increasing, we have to become aware that such containers with high power density and energy density become a considerable risk when they become larger. High performing lithium batteries, for example, contain flammable components such as carbon and polymer separator plus organic electrolytes [Li 2016]. It is possible that they catch fire. When they catch fire while unattended, large fires may break out. And even small fires may end into a catastrophe when the fire starts in the cargo area of a passenger airplane, or in the passenger cabin of an airplane. In so far, batteries and battery components are sometimes considered as dangerous goods for which government and public bodies have passed regulations, such as the Recommendations on the TRANSPORT OF DANGEROUS GOODS by the United Nations, which has in Section 38 an Amendment on the testing of lithium batteries before their transport can be considered safe [UN383 2013].

4.12 Recycling of batteries

I once visited a battery recycling factory in Switzerland in the late 1990s. The factory received most of the batteries from a recycling program [Jordi 1995]:

> "Swiss legislature stipulates that manufacturers and dealers of batteries are required to take back all types of used batteries, free of charge, and to dispose them of according to the regulations on dangerous waste"

As Jordi expresses in his paper [Jordi 1995], around half of the 3,500 tons of batteries were believed to be disposed by export to Eastern Germany. This way of disposal allowed to keep battery disposal costs in Switzerland to 770$ per ton. From 1991 on, no battery waste export licenses were granted anymore, and Switzerland was supposed to establish its own battery environmental benign waste disposal and recycling industry, which amounted to costs of 4750$ per ton in 1995.

The capacity of the factory exceeded the volume of available batteries. There was actually competition among few factories for the access to the old batteries. The factory was a small company which had hardly over 12 employees, when I remember correctly. The batteries were mechanically shredded and then cooked to some temperature so that many of the components could be separated. The factory had a professional Environmental Health & Safety (EH&S) management. I learnt from the factory owner that the recovered mercury was the only component of the entire processing which could be sold for a good profit at the metal stock market.

Note that the mercury is a heavy metal which is of particular concern with respect to EH&S. The other components, if sold, were sold with a loss. But the fee which was paid by the battery buyers and brought into the battery recycling program from the vendors was distributed among the battery recycling companies. This together with the net profit from mercury sale made the recycling business somewhat lucrative. It would have been more lucrative if the factory had received more batteries. An example of battery collection from households in Minnesota with detailed calculations is provided in an assessment report prepared for the U.S. Department of Energy [Seeberger 1992].

The financing system for battery recycling is therefore an important part for the entire battery business. This is, for example, discussed for zinc batteries by Wiaux and Waefler [Wiaux 1995]. Recycling technology and concepts exist also for nickel cadmium batteries [David 1992, 1995].

Zinc has a moderate toxicity and this needs to be taken into account in battery production and waste management [Plum 2010]. In the zinc air battery, zinc will become zinc oxide. It is possible to reduce ZnO to metallic zinc by thermal reduction at 1,200 °C. This can be done in a solar thermal process [Steinfeld 1995]. The refinery and processing of zinc is certainly not an original matter of batter recycling. Rather, the mining and production of zinc is an old technology.

The processing of the mineral ores in the smelter factories has created environmental and health problems for long times and caused air pollution; for example, the minerals that need to be roasted, such as the sulfides, form sulfur gases. Alfaro and Castro presented in 1998 a zinc refinery in Mexico built in 1982 where a sequence of roasting, leaching and electrolysis was used for zinc production of more than 100,000 tons per year in that factory [Alfaro 1998]. It is noteworthy that zinc electrolysis is a very attractive method for technical and economical reasons [Mattich 1998].

Electrochemical methods in metallurgy, ore processing and also recycling are continuously developed and improved. Bartolozzi et al. present a process where they can recover not only zinc but also manganese from used batteries. The zinc can be separated by electroplating. Manganese (II) can be oxidized to solid MnO_2 and manganese (IV) can be solved in acidic H_2O_2 and then brought to precipitation in alkaline liquid [Bartolozzi 1994].

The maturity and development of recycling of lead from lead acid batteries can be considered excellent. I remember the very early 1970s when metal scrap dealers in villages in Germany went with a loud bell from door to door asking for scrap metal and specifically asking for "Autobatterien." These are car starter batteries, which are full of lead. This is why they are so heavy. Just recently when I attended the MRS Spring Meeting 2018 in Phoenix Arizona in April 2018, during my drive through Phoenix I saw a sign on a major arterial road where a small enterprise advertised that they were looking for car batteries. About 25 years ago Biedcharreton showed already that the collection success of lead acid batteries was over 80% and way more ahead than the collection of paper, glass, steel cans and aluminum cans – at least for Western Europe [Biedcharreton 1993].

Biedcharreton outlines in his paper [Biedcharreton 1993] a closed-loop system for lead acid battery materials recovery and recycling (Figure 4.48).

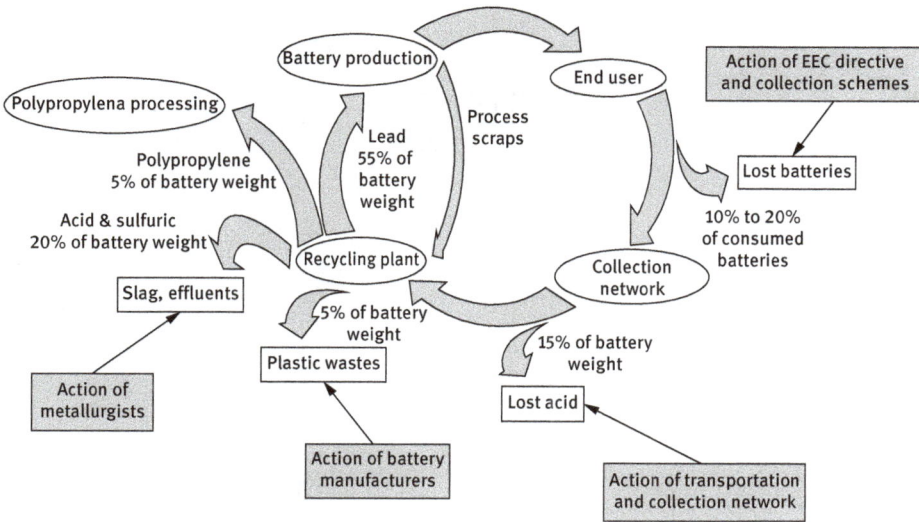

Figure 4.48: The recycling loop of lead acid batteries.
Reprinted from *Journal of Power Sources*, 42, Biedcharreton B., Closed-Loop Recycling of Lead-Acid-Batteries, 331–334, Copyright (1993), with permission from Elsevier. [Biedcharreton 1993]

The principal loop is formed by four elements; this is the battery production plant and the end user. And then the loop is completed by the recycling plant, which needs a battery collection network. The relevant materials which can be recovered are the lead and lead compounds including slag, the battery acid, plastic parts such as polypropylene and effluents. To improve the collection efficiency, losses have to be minimized. For this it needs government action directives with the necessary oversight, action from the collection network participants, action from the battery manufacturers, and also action from the engineers such as metallurgists [Biedcharreton 1993].

One such collection network materialized in the town of Pecos, Texas, with the Recovery & Reclamation Inc, which is authorized to manage hazardous waste from various battery technologies. The project is presented in a publication by Meador [Meador 1995]. The overall process flow in the Pecos project is illustrated in three flow diagrams in Figure 4.49. After the batteries are ceiled on-site, they are sorted, then shredded and then pyrolyzed. The lithium batteries, after the sorting process in the first step, are deactivated (discharging for safety purposes; compare [Sloop 2017]), then dried and shredded.

The town of Pecos was founded on 1881 near the Pecos river when the Texas & Pacific Railroad was built. Major share of the Pecos battery recycling project dealt with batteries from the railway. For the recycling of these zinc-carbon batteries (railroad carbonaire, Edison carbonaire), the electrolyte (KOH) was drained and then the battery bodies are sawn into two parts, top and bottom. Then, further processing of all the aforementioned parts and materials proceeds.

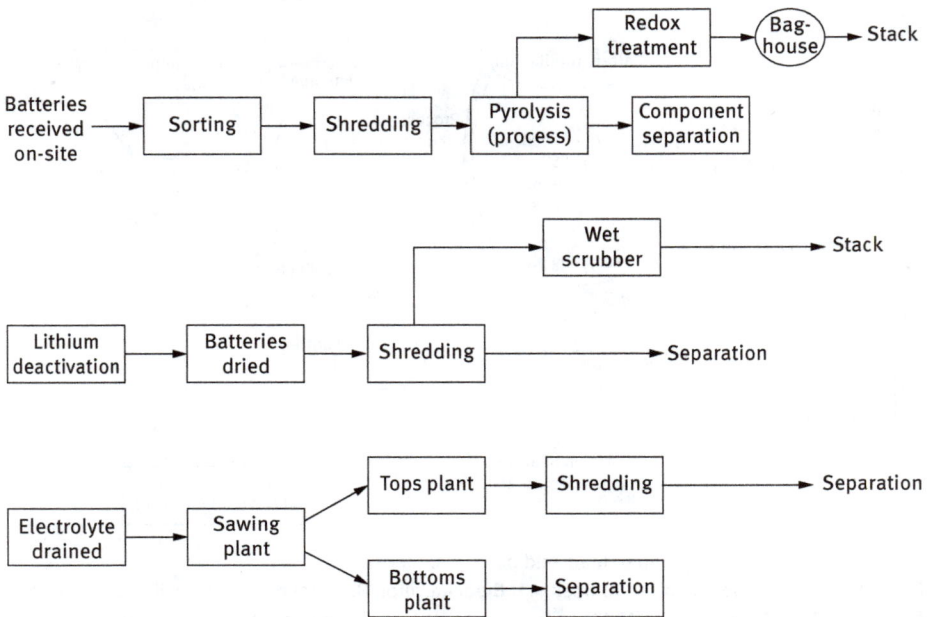

Figure 4.49: Overall process flow for battery recycling in Pecos, Texas. The flow is shown in three parts along with the treatment of lithium batteries and zinc-coal batteries.
Reprinted from *Journal of Power Sources*, 57, Meador WR, The Pecos project, 57:37, Copyright (1995), with permission from Elsevier, [Meador 1995].

Finally, I want to refer the reader to a paper from Cai et al., with and from ClearWaterBay Technology Inc. in California and ClearWaterBay Technology Ltd. in Hong Kong, who have developed a process for the chemical precipitation of lithium from used lithium batteries [Cai 2014]. As the chemistry of the material mixture from shredded and cooked lithium batteries is quite complex, balanced methods are employed to separate the components. The solid-liquid equilibrium governs the phase behavior of the components which is exploited for separation. The authors show a very nice application of Jädecke projections and diagrams in battery waste technology.

4.12.1 Economical aspect of battery recycling

I have partially addressed already economical aspects of battery recycling. Recycling does not have to be profitable. There are many businesses and enterprises in this world that operate nonprofit or not-for-profit. In this case, the revenues created from the business can cover the cost of the operation, like paying salaries for the workers, rent for buildings, operation costs, insurances and taxes. Often such enterprises are exempt from taxation.

We even can go further and think of subsidized businesses where the government supports recycling (or any other purpose) with the money from the tax payers, or where the government forces consumers to pay extra taxes for recycling. Sometimes the question of fairness arises (e.g., "energy justice", see [Olson 2016a, b]): how much burden can you put on tax payers or consumers to arrive at a particular situation which is wished for by a majority or any other portion in a society. It is a political decision of the society whether it goes that way.

The internal rate of return (IRR) is an important metrics for those who want to invest in a business for profit. There is nothing fundamentally wrong with making a profit. Insurance companies, retirement funds and pension funds, for example, need to invest their premiums and expect the necessary return. Many utility companies operate this way.

In his detailed assessment of economical issues, Zabaniotou shows – for the example for Greece – when the recycling of lead acid batteries becomes profitable or at least economically feasible [Zabaniotou 1999]. He looks at various factors that influence the venture worth and the IRR. Figure 4.50 shows how the venture worth and IRR scale with the percentage of lead recover from used lead acid batteries. The graph looks in the range of 54–61% lead recovery and here IRR and venture worth scale are roughly linear. When the lead recovery is 58% or higher, then the IRR is 13% (the interests) or higher. Then, the enterprise is considered as acceptable.

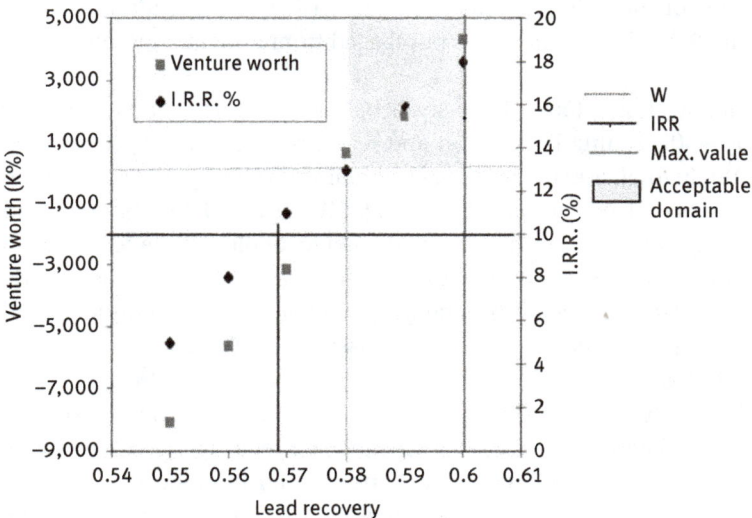

Figure 4.50: Influence of maximum lead recovery on W and IRR (Q°17 000 t:year, P°776$:t). Reprinted from Resources Con-servation and Recycling, 25, Zabaniotou A, Kouskoumvekaki E, Sanopoulos D, Recycling of spent lead/acid batteries: the case of Greece, 301–317, Copyright (1999), with permission from Elsevier, [Zabaniotou 1999].

Whether such enterprise is worthwhile to pursue from the IRR depends also on the size (production capacity) of the plant. With a capacity of around 11,000 tons per year, the business becomes interesting. The business becomes more interesting when the price of lead when sold on the market is increasing. When the ton lead cost 600$, the business would not work for investors because the IRR was 0. When the price for the lead rises to around 680$ per ton, the IRR is 13%. When the price further rises to 770$ per ton, the IRR is a pleasant 28% [Zabaniotou 1999]. This is an example how a lead acid battery recycling project can become a success or a failure just because of the market situation, irrespective of the performance of engineers or managements.

4.13 Large-scale stationary batteries

The small town of Feldheim in Germany hosts the largest battery in Germany. It is the electric energy storage component in a self-reliant village of 40 residential homes near German capital Berlin [Colthorpe 2015, Nandakumar 2017, Steel 2015]. Community participation and involvement was found to be a key driver for the success of such residential community renewable energy projects [Islar 2016]. This may be another driving force why street artists Christopher Wehr and Philipp Vogel, and children had been invited to style and decorate the battery complex in their town. The battery has a 10 MW power and can exchange electricity with the grid. The actual batteries of the Feldheim storage facility (Regional Regulating Power Station, RRKW) are 3,360 battery modules purchased from Korean technology firm LG Chem.

Aachen, Germany, hosts too a large-scale battery which uses five different storage technologies (including lithium ion and lead acid); this is why it is called M5BAT [Stöhr 2016]. In total, 600 batteries, each containing 40 L liquid and weighing 150 kg, occupy a space of 900 m^2 and deliver 5.5 MW power. Low-cost lead-acid technology and costly lithium ion technology and other technologies are combined to a system which can rapidly respond to fast and unforeseen changes in the electric grid which is partially fed from renewable energy carriers such as PV plants and wind power plants. The system is hosted on the premises of RWTH Aachen and costs around 11 million Euros.

The redox flow battery is particularly suited for large-scale battery storage because it is relatively cheap and because it is designed for requiring large electrolyte volumes. Sumitomo has built the largest flow battery so far in California [Kenning 2017]. German utility company EWE has plans to use a giant underground salt mine as a storage system for salt electrolyte and use it as a giant redox flow battery. The size of the plant is large enough to power a large city such as Berlin for 1 h with electricity [Pentland 2017].

Table 4.9: Selected large scale battery storage projects in Germany.

Location	Operator	Site specs	Capacity (MW)
Jemgum (planned for 2023)	EWE	Salt cavern	120
various sites (planned)	LUNA/Steag	Pool system	100
Steag coal plants	Steag	Coal plants	90
Leag's BigBatt (planned)	Leag/Siemens	Lignite plant	50
Jardelund (planned)	Eneco/Mitsubishi	Single system	48
Schwerin 2	Wemag	Dual system	15
Luenen	Daimler	Car batteries	13
Pfreimd	Engie	Pump storage	13
Bordesholm	VBB/RES	Island solution	10
Feldheim	Enercon/Vattenfall	Wind	10
Herdecke	RWE	Pump storage pump	6
Aachen	E.On, RWTH Uni	Hybrid	5
Doerverden	Statkraft	Run-of-river	3

Data from [Franke 2018]. Reproduced with kind permission from S&P Global Platts, https://www.spglobal.com/platts/en.

Large-scale electric storage systems are being developed and installed worldwide. A list of such systems readily installed in Germany or planned for Germany is given by Franke and Loades-Carter [Franke 2018], see Table 4.9. The largest plant is the one that should go underground in a salt mine, with 120 MW power. Various plants are planned with 100 MW power each.

5 Electroanalytical methods

5.1 Simplistic analogy of hydrodynamics and electricity

As this book is to a large extent about electric phenomena, I have to remark on some observations on the analogy between hydrodynamics and electricity. The theory of electricity is quite abstract and not so easy to comprehend for many people. The main reason for this is, in my opinion, that we cannot "touch" electricity – except for the electric shock we may get when we touch an electrically charged body or conductor. And this experience counts more as a sensation and not so much for understanding.

Last year when working in the garden, I used once again the new and cheap hose, a gardening water hose. It is made from a very elastic polymer, like rubber, which is inside another flexible tube from woven, strong plastic fabric. The outer tube serves for mechanical stability, whereas the inner serves as the waterproof, leakage-free pipe; this is the water conductor. You connect this hose to the water faucet at the wall of the house in your garden. You open the faucet and the water from your home will soon pump up the hose, which is getting stiff.

The other end of the hose accepts a handle which is shaped like a gun, like a "pistol." The other end of that pistol comes with different multipurpose holes and jets such as shower, mist, spray and so on. When you pull the trigger, the water will come out of one of the multipurpose holes that you can select. When you release the pistol trigger, the water flow will stop. When you stop the water flow that way with the pistol in your left hand, you will notice that the hose is pumping up in your right hand, provided you hold the elastic hose in your hand. The hose is getting stiff.

When you lay down hose and pistol in that condition, with the water faucet on the wall open and the pistol released and walk away without turning off the water at the faucet, you run in danger that the high water pressure built up in the hose will eventually cause leakage somewhere inside the rubber hose.

There are always some structural inhomogeneities[1] in a material where such leakage may occur first. When you pick up the hose and the pistol and pull the trigger and start spraying the water, the "tension" in the loaded, "charged" hose will ease and go down. I notice this when I keep the hose in my right, tightly closed hand.

When I switch the spraying mode on the water pistol from "mist" to "jet," the flow through the pistol is much larger and there will be more water coming out of the pistol and hose per second and the hose immediately feels softer in my hand. Hence, the pistol imposes a "resistance." This resistance is in the beginning

---

[1] Electric breakdown (dielectric breakdown) in materials is an often observed phenomenon; there is always one first location where it sets on. Think of the lightning in the atmosphere. The entire atmosphere it charged against the Earth. But only at specific locations, the discharge will occur.

https://doi.org/10.1515/9783110561838-005

infinitely high because no water is coming out from the pistol. Only when I pull the trigger, I open the gate for the water, and the resistance decreases from infinity to some finite value.

The closed pistol is like an insulator that blocks the water "current." When I pull the trigger, then there is a water-current, but its strength depends on the mode I have set on the pistol. The pistol works like an adjustable or a switchable "resistor" or "conductor" for the water flow. Note that it is a matter of perspective whether we call something resistor or conductor. In electric engineering, the terms resistivity and conductivity are chosen unambiguously.

But the hose by itself is a resistor because it limits and controls the water flow. Or you can call the hose the conductor, either way, because it allows the water to flow through it when it is connected to the open faucet and when the water pistol trigger is pulled so that water can be released at the end. When you remove the pistol and have an open end hose where the water comes out, a resistance is still there. When you increase the water flow too much by wide opening the faucet, the hose becomes again very stiff. Even an open end in many physics problems can constitute a considerable resistance.

You can "feel" with your hand the "tension" (the German word for tension is "Spannung" and means in the context of electricity simply "voltage") in the water flow when the hose is swelling when you switch to a pistol mode with higher resistance, such as from jet to mist, when less water is coming out through the pistol.

Now you can think of Ohms law that is a linear relation between the voltage U or V (analogue the tension in the hose) and the current I (analogue the water flow through the hose). The proportionality constant is the resistance $R: U = R \times I$. Maybe the analogy shown in this chapter helps somewhat in understanding the relation among current, potential and resistance, in electric theory.

For the analysis, assessment and diagnostics of electrochemical systems, we use typically "electroanalytical" methods. In principle, they are very simple: We are measuring voltages and electric currents. The best known method is probably the "$I–V$" curve. Before coming to the actual methods, it is necessary that we recall some fundamental and simple relations in electric circuits known as the Kirchhoff laws.

As we can measure electric signals nowadays with a very high precision and accuracy, we are put in the position of making the best science out of such data. This includes also the mathematical modeling of the systems that we measure. For this, it is necessary to be able to master the underlying mathematics. For the aforementioned difficulty that some people have with electrical engineering, it is often helpful to solve mathematical problems and practice as many as possible of them. For this purpose, I can recommend, for example, the well-known Schaum's Outline Series [O'malley 1992].

5.2 The Kirchhoff laws[2]

As we can reduce many problems in electroanalytical measurements to the determination of voltages and currents, it is important to recall two fundamental relationships summarized as Kirchhoff's laws.

5.2.1 Kirchhoff's current law

Figure 5.1 shows a connection of four wires that meet at a central knot point. Electric currents i_1, i_2, i_3 and i_4 are running through the knot and through the wires. The currents have a particular direction: either they flow into the knot (currents i_2 and i_3) or they flow away from the knot (currents i_1, i_4). The currents through the knot follow a particular physical law.

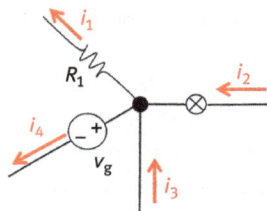

Figure 5.1: According to Kirchhoff's current law, the current entering any junction is equal to the current leaving that junction. $i_2 + i_3 = i_1 + i_4$.

This law is also called Kirchhoff's first law, Kirchhoff's point rule or Kirchhoff's junction rule (or nodal rule). The principle of conservation of electric charge implies that at any node (junction) in an electrical circuit, the sum of currents flowing into that node is equal to the sum of currents flowing out of that node or equivalently: The algebraic sum of currents in a network of conductors meeting at a point is zero. Recalling that current is a signed (positive or negative) quantity reflecting direction toward or away from a node; this principle can be stated as

$$\sum_{k=1}^{n} I_k = 0$$

---

2 The terms "law" and "rule" are often used unambiguously in science. A law is typically considered stricter than a rule. A rule most of the time will be followed, or at least very often, such as Vegard's rule, which states that a physical parameter of a compound will often change and scale according to the stoichiometric change of that compound. This is not always true, and therefore, it is called a rule. Kirchhoff's law is strict because the electric current must follow the conservation law of electric current.

n is the total number of branches with currents flowing toward or away from the node.

This formula is valid for complex currents:

$$\sum_{k=1}^{n} \tilde{I}_k = 0$$

The law is based on the conservation of charge whereby the charge (measured in coulombs) is the product of the current (in amperes) and the time (in seconds).

5.2.2 Kirchhoff's voltage law

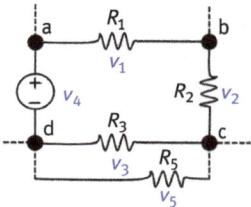

Figure 5.2: According to Kirchhoff's voltage law, the sum of all the voltages around one loop is equal to zero. $v_1 + v_2 + v_3 - v_4 = 0$.

In an electric wire loop, such as shown in Figure 5.2, the sum of all voltages add up to 0. This law is also called Kirchhoff's second law, Kirchhoff's loop (or mesh) rule and Kirchhoff's second rule. The principle of conservation of energy implies that the directed sum of the electrical potential differences (voltage) around any closed network is zero, or more simply, the sum of the electromotive forces (e.m.f., or emf) in any closed loop is equivalent to the sum of the potential drops in that loop, or the algebraic sum of the products of the resistances of the conductors and the currents in them in a closed loop is equal to the total emf available in that loop. Similarly to Kirchhoff's current law, it can be stated as

$$\sum_{k=1}^{n} R_k I_k = 0$$

Here, n is the total number of voltages measured. The voltages may also be complex:

$$\sum_{k=1}^{n} R_k \tilde{I}_k = 0$$

This law is based on the conservation of energy whereby voltage is defined as the energy per unit charge. The total amount of energy gained per unit charge must be equal to the amount of energy lost per unit charge, as energy and charge are both conserved.

5.2.3 Resistivities, capacities and inductivities in electric circuits

There are some important relations when we consider electric components in electric circuits such as resistors, capacitors and inductors (coils) which are arranged in series or in parallel. Figure 5.3 shows a sketch of three resistors, which you can purchase in an electronics store, soldered together in series.

Figure 5.3: Illustration of three resistors R_1, R_2 and R_3 soldered together in series between two wires. The colors on the resistors specify their resistivity including the tolerance (error). The color code is shown in the Appendix and can be used to find out the resistance.

Resistors R_i, which add up in a series circuit as shown in Figure 5.3, can be taken like the sum of the individual resistors:

$$R_{total}^{series} = R_1 + R_2 + \cdots \sum_i R_i$$

Figure 5.4 shows how three resistors are soldered together in a parallel circuit. It's a fundamental difference whether the resistors are in series as in Figure 5.3, or in parallel as in Figure 5.4.

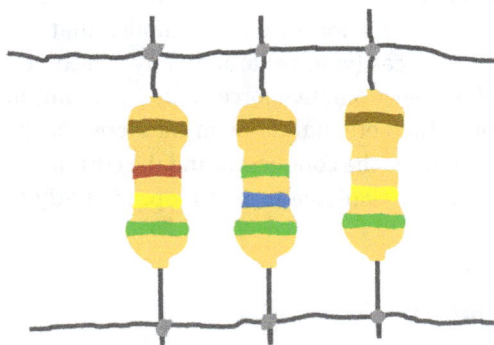

Figure 5.4: Illustration of three resistors soldered together on two parallel wires. The colors on the resistors specify their resistivity including the tolerance (error). The color code is shown in the Appendix and can be used to find the resistance.

However, resistors connected in a parallel circuit add up as follows:

$$\frac{1}{R_{total}^{parallel}} = \frac{1}{R_1} + \frac{1}{R_2} + \frac{1}{R_3} + \cdots = \sum_i \frac{1}{R_i}$$

Note that if we used the conductivity $S = 1/R$, rather than R, then the aforementioned equation would read much simpler. We would not have to deal with the reciprocals, and this could simplify the calculus.

When we want to understand electrochemistry at the molecular level, then we must abstract from the macroscopic electric networks and adopt them to the molecular level. We will see later in this book how at the solid–liquid interface electric charges can pass through the electrochemical double layer or not. Chemical bonds at this interface can be interpreted as resistors or as capacitors.

It is important to note that the total capacity of a set of capacitors that are lined up in a parallel circuit will add up linearly:

$$C_{total}^{parallel} = C_1 + C_2 + \cdots \sum_i C_i$$

where the total capacity of a set of capacitors lined up in a series circuit will add up in the more complicated formula with the reciprocals of each capacity:

$$\frac{1}{C_{total}^{series}} = \frac{1}{C_1} + \frac{1}{C_2} + \frac{1}{C_3} + \cdots = \sum_i \frac{1}{C_i}$$

The next important component in electric engineering is the inductivity L, which is typically exhibited by wire coils. In electrochemistry, inductivities hardly show up other than in long connection cables. When the current in a conductor makes an abrupt change dI over time dt, the conductor will respond with a voltage U. The relation between the change of the current and the induced voltage determines the inductivity L of the conductor:

$$U = L \frac{dI}{dT}$$

The inductivity L depends on the geometry of the conductor. When you carry out impedance spectroscopy and have long cables between the frequency response analyzer (FRA) and the sample, the inductivity L of the cables can become significant and overshadow signatures which originate from the sample. Therefore, one should try to make the connection leads short. When this is not possible, one should make the impedance spectra with the long cables with the sample connected and make an extra measurement with long cables but without the sample. By mathematical impedance modeling, one can then determine the L and subtract it later when the impedance of the measurement with the sample is modeled.

Another reason why leads should be short for is that electric stray fields from the ambient may induce a current in the electric sample setup. This can be fields from the electric grid in the laboratory, fields from furnaces, lamps and so on. The 50 Hz frequency can induce a signature on your experiment either by direct connectivity to the grid via the FRA, because it is plugged in, or by the aforementioned stray fields. This is why electrochemical cells are sometimes kept in a Faraday cage (this stops the electric field component) or in containers from μ-metal (this stops the magnetic field component).

5.3 Redox processes

Redox processes are at the heart of all chemistry. The term *redox* is composed form reduction and oxidation. *Oxidation* is the chemical reaction, where an atom, an ion or a molecule gives away one or more electrons. Even when a fraction of the charge of an electron is given away, this is an oxidation. The opposite is called *reduction* and means that an atom, ion or molecule is receiving electrons.

The redox reaction shows that one reaction partner accepts an electron which is donated by the other reaction partner. We can easily formulate this as we know already from middle school:

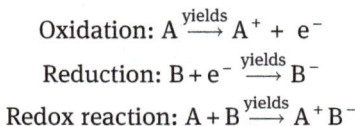

$$\text{Oxidation: } A \xrightarrow{\text{yields}} A^+ + e^-$$
$$\text{Reduction: } B + e^- \xrightarrow{\text{yields}} B^-$$
$$\text{Redox reaction: } A + B \xrightarrow{\text{yields}} A^+ B^-$$

The role of oxygen (which lent its name to the term "oxidation," and oxygen is taken from ancient Greek and means "acid maker") in the development of chemistry as a field of science is not to be underestimated. The experience and discoveries which French researcher Antoine-Laurent Lavoisier [ACS 1999] made with combustion experiments were ground breaking because he considered oxygen a chemical reaction partner in processes where energy was released.

Before Lavoisier, the energy content in matter was attributed to *"phlogiston,"* a mysterious substance which would reside in energetic materials. Everything that would burn would be ranked in terms of phlogiston content. Phlogiston was a substance which was not discovered yet but it was assumed to be there. The phlogiston theory dates back to German philosopher Georg Ernst Stahl and was an established fact in the contemporary science in the eighteenth century.

Let us not look down on those researchers who established a theory which was overturned sooner or later by other researchers. When I studied Physics in Aachen, a colleague of mine, a theoretical physicist, took everything that was established scientific knowledge for granted, which surprised me a lot. From hindsight, everything may look clear unless groundbreaking novel concepts are posed (compare, e.g., [Nernst 1922, Schottky 1922]), but some humility in view of the intellectual achievements made by researchers who spent their entire life on science would be fine. For example, Hermann Minkowski, who died at age 44 from an inflammation of his appendix, worked to finishing a couple of manuscripts – hours short before he passed away [Hall 2010, Hilbert 1910].

It is easy to throw a critique on other people's work. It is not so easy to discover something new in science, to uncover an old secret and mystery. Feynman [Feynman 1955] made some general remarks in his speech on the value of science, specifically the degrees of certainty in scientific theory and the freedom to doubt, which I referred to already in two other chapters in this book.

The fact that we can measure electric currents and determine electric potentials with high accuracy allows us, by means of Nernst equation (the relation between electrode potential and redox couple, Walther Nernst; see [Bonhoeffer 1943, Ertl 2015, Mendelssohn 1964, Van Der Kloot 2004]) and Faraday's law (the relation between converted mass and electric charge, Michael Faraday [Ehl 1954, Faraday 1834]),

> Now it is wonderful to observe how small a quantity of a compound body is decomposed by a certain portion of electricity.

that we can use the tools of electroanalytical chemistry for the monitoring of redox reactions with very high accuracy and subject these to the underlying theory with mathematical equations.

This adds great confidence that we are doing it right. The concept of the redox reaction itself and the insight that charges are shuffled back and forth – in electrochemistry even with a very precise control – is however a relative primitive one, notwithstanding that all chemical technology and chemical engineering is based on Lavoisier's works, which opened a new era in civilization of mankind [ACS 1999].

The idea that chemistry happens when integer quanta of electron charges or say better, electric charges are passed on to some atom, ion or molecule is a simplification we become aware of when we undertake spectroscopic experiments which indicate that one change of charge can incur a change in a system which is larger than one atom, ion or molecule only. A simple example is the observation that the oxidation state of Fe in some seemingly simple compounds like Fe_3O_4 cannot be expressed by an integer number. On average, it is a mixture from Fe^{2+} and Fe^{3+}.

How does one single lithium atom react with the two large extended graphene layers during intercalation into graphite? Is it sharing its electron with the entire graphite lattice? Chemists have been aware of this dilemma for a very long time and this is why chemists rather use the term "formal charge." What really happens at the molecular and atomic scale is then subject to further investigations.

5.4 *I–V* curves

I–V curves (current–voltage curves) are important diagnostic information on electrical systems. We apply a voltage on our electrical system and measure the electric current which is flowing through the system. We need for this a power supply which can regulate the voltage and the current. And we need two multimeter, one for the voltage and one for the current. This is a standard arrangement for measuring any electric systems. We are increasing the voltage from V_0 to V_{max} in either positive ($V > 0$) or negative ($V < 0$) direction and measure the current. What we gain is the I–V characteristic of the system.

5.4.1 Metallic conductivity – Ohm's law

When our system is a metal such as copper plate or steel nail, we will find a linear relationship between voltage and current. This is the experimental and empirical manifestation of Ohm's law. The slope of the I–V curve equals the resistance $R = U/I$. This is the value which we read for example from a multimeter. The resistance of the conductor (we can call it resistor) depends on the geometry of the conductor and its specific resistance ρ [Ω/m]. When you have a conductor with a known geometry, you can determine the specific resistance ρ from the relation (consider a geometry like shown in Figure 5.5)

$$R = \rho \frac{l}{A}$$

A well-known experimental way for the determination of R is via the four-point method. You take a long cylinder or slab with prismatic cross-section so that you can easily determine the length and diameter. Figure 5.5 shows a schematic with a slab with rectangle cross-section A and length l. At the front and end, you can impose a current with known magnitude I in ampere. The middle panel in Figure 5.5 shows a rectangular slab pressed and sintered from LaSrFeNi-oxide powder. This metal oxide has a metal-type electronic conductivity. It has at the front and at the end some platinum paint which provides a good electric contact for imposing an electric current. In the middle of the slab, you see two more platinum-painted electric terminals. These two middle terminals allow you to read a voltage drop.

Note: For the determination of the resistance, not the entire length of the slab is relevant, but the distance between the two inner terminals.

For a quick snapshot, it may be sufficient when we only measure the absolute current and the voltage of the system. At some point, however, it may be necessary that we characterize the size of our system, such as electrode size, weight, electrolyte volume, electrode surface area and so on. This is necessary so that we can assign a capacitor a specific capacity or an electrode material a specific conductivity.

5.4.2 Semiconductor conductivity – the Shockley equation

Experience shows that not all I–V curves measured on materials are linear and thus do not follow Ohm's law. For a semiconductor, we often find a characteristic relationship as shown in Figure 5.6.

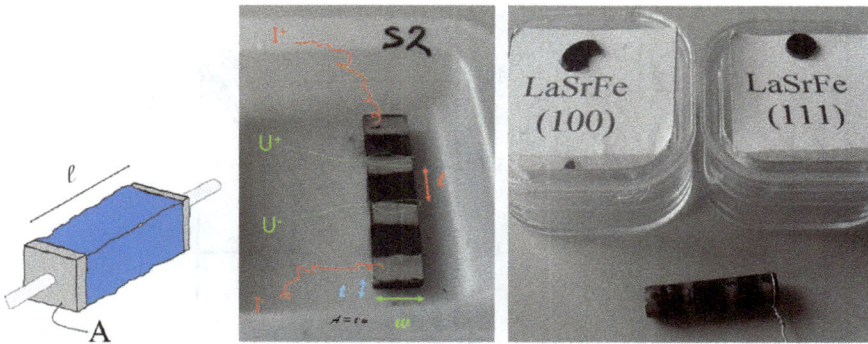

Figure 5.5: (Left) Schematic of a resistor (dark gray) with length *l* and cross-section area *A* between two metallic contacts (light gray current collectors) through which a voltage or current can be applied. (Middle) Sintered ceramic slab of LaSrFeNi-oxide with four painted metallic platinum contacts. The two middle contacts with distance *l* are for voltage reading U^+ and U^-; the two outer contacts are for imposing the current I^+ and I^-. The cross-section area *A* is the product of thickness *t* and width *w* of the slab: $A = t \times w$. (Right) A single crystal of LaSrFe-oxide with cylinder geometry, painted with four metallic platinum contacts.

Note: The platinum for this purpose is provided by various vendors as a liquid paste, where colloidal platinum particles are dispersed in an organic solvent. This ink is brushed over the relevant positions on the ceramic material and dried for some minutes in the ambient. Then, it must be sintered in a furnace at a temperature as high as 950 °C. This warrants that the organic solvent is completely combusted and the colloidal platinum particles become sintered with the porous ceramic and provide a firm mechanical and very good electrical contact. This method is known to those researchers who work in high temperature electrochemistry (SOFC, SOEC) and high temperature solid state chemistry. This is for example different from the use of silver paint, which often, not always, needs no thermal treatment.

This particular platinum paste may not be mistaken with the silver paint, gold paint or platinum paint which is used, for example, for electron microscopy work. There, you typically do not sinter the sample plus ink at 950 °C. This is because many samples would not survive this treatment. Another important aspect is that various colloidal metal inks contain different solvents. Some solvents may be so robust that they act as insulators between the metal particles and do not provide electric conductivity at all. They would only do so once you burn this sample with ink at the aforementioned high temperature so that the solvent is combusted in the hot oxygen in the air in the furnace. You should always check the conductivity of your samples and the contacts with a multimeter. This often helps avoiding waste of time.

Figure 5.6 shows the *I–V* curve of "Kupferoxydul," a copper oxide with Cu(I): Cu_2O. A hundred years ago, it was used as a semiconductor for electric rectifiers. The rectifier is based on a metallic copper plate which is thermally treated and oxidized so that a thin Cu_2O layer is grown on the surface. Then, a lead (Pb) layer is put on the Cu/Cu_2O plate. This is the electric rectifier which lets the current pass only in one direction like a diode. In the early 1920 when the material was produced in large amount for the electronics industry, the term "diode" was not yet used.

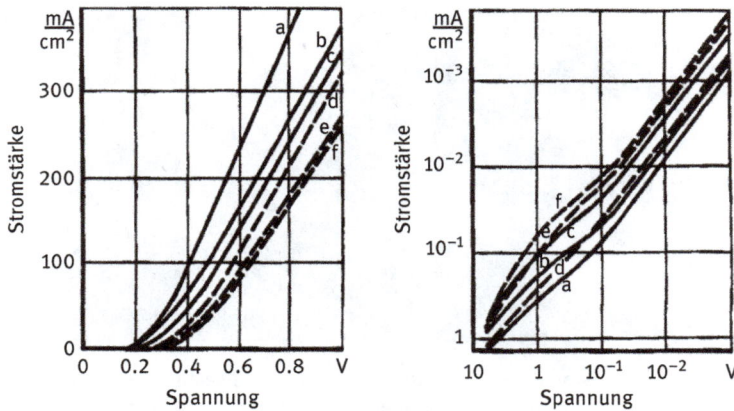

Figure 5.6: The *I–V* curves of Cu_2O in forward (a) and reverse (b) direction. The letters a–f denote the thermal treatment of the copper material before it is converted to Cu_2O. Curves *a–c* are from samples preheated in vacuum at 1,000 °C before oxidation at 500 °C. Curves d–f are from samples directly oxidized at 500 °C.

Reproduced with kind permission from John Wiley and Sons from Nieke H: Über die Halbleitereigenschaften des Kupferoxyduls. XVI. Kennlinie und Kapazität von Kupferoxydul-Gleichrichtern. Annalen Der Physik 1969, 478:251–270. doi: 10.1002/andp.19694780506. [Nieke 1969]

When a voltage is applied in forward direction, the current shows a curved, slightly exponential profile before it grows virtually steep linear. The onset potential and the slope vary with the thermal treatment procedure. Pre-annealed copper (*d–f*) has a less steep slow than the normal copper (*a–c*). At 1 V, the current densities reach values of 250–300 mA/cm^2 where the onset potentials vary from 0.2 V to 0.35 V.

When the polarity of the voltage is switched or, say, reversed, so that it continues to reverse direction left side from the aforementioned onset potential, the current density ranges from 10^{-4} to 1 mA/cm^2. The voltage range for this current range is 10^{-3}–10 V. Note that the conductivity data (current vs voltage) are plotted in double logarithmic scale.

When the semiconducting properties of germanium and silicon were discovered, Kupferoxydul was not used anymore. The *I–V* characteristic of a Schottky diode can be modeled after Wagner [Goucher 1951, Wagner 1931] like a modified exponential, which is called Shockley equation

$$I_D = I_S(T)\left(e^{U_F/nU_T} - 1\right)$$

Interesting here is how the current depends via $U_T = kT/q$ on the temperature of the electrode.

A very important detail in semiconductors is that their electric transport properties are discussed in terms of the electronic band model. Semiconductors have an energy gap between the conduction band and the valence band. This makes that charge carriers first need to be generated by some excitation, to make them actual charge carriers [Ashcroft 1976, Ibach 1988]. This charge carrier generation works that an electron e^- is excited to a particular energy level in the conduction band, and the resulting electron hole h^+ is the conjugated charge carrier. Both contribute to the total current.

The electron holes are for example important in iron perovskites for potential solid oxide fuel cell (SOFC) cathode materials, where holes are doped into the parent material $LaFeO_3$ by substitution of the La^{3+} with Sr^{2+}. The Fe^{3+} will then be oxidized to Fe^{4+}, which looks spectroscopically like $3^d5\underline{L}$ with the ligand hole L [Braun 2009, 2012a]. In semiconductor photoelectrodes such as α-Fe_2O_3, an electron–hole pair is created by optical excitation but will annihilate after some time. It is not a permanent hole but a remnant hole [Braun 2012b].

5.4.3 Ionic conductivity

In this chapter, I implied that the electric charge carriers are the electrons. When we deal with semiconductors, there will be also (electron) holes that are charge carriers. And the ions are charge carriers in the electrolytes. In the chapter on fuel cells, I will make some general remarks on electric charge carriers. The concept of the electro-chemical cell rests on the assumption that the electronic conduction and ionic con-duction are separated by geometric arrangement of electrodes and electrolyte.

We know that the electrolytes can be liquid or solid. In the liquid electro-lytes, the dissociation of molecules plays a significant role. When the mole-cules that form the charge carriers are completely dissociated, we call this a strong electrolyte.

To give the reader an impression about the huge dynamic range of resistivities and conductivities between ionic and electronic conductors and insulators, look at a comparison made by Zyryanov [Zyryanov 2011]:

> A column of pure water 1 mm in length at 18°C has the same electric resistance as a copper wire with the same cross-section and a length of 16×10^6 km (400 times the Earth's equator) or a column of seawater with a salinity of 35‰ and a length of 1–3 km.

Table 5.1 gives you an impression on the specific conductivities of a number of important cations and ions in water at ambient temperature. The proton H^+ is a well-known charge carrier and has obviously a very high conductivity. Li^+ has the lowest specific conductivity in this long list. Note that the conductivity of Li^+ can be certainly different in a different milieu. In lithium ion batteries, we have to consider the conductivity of lithium in an organic liquid electrolyte, such as $LiPF_6$ dissolved in dimethyl carbonate, and the conductivity of lithium in the positive electrode, such as $LiMn_2O_4$ or $LiFePO_4$, and its conductivity in the negative electrode, when it is for example Li_xC.

Table 5.1: Limiting (maximum) ionic conductivity of a number of important ions in H_2O at 298 K.

Λ mS/m²/mol	Ion	Λ mS/m²/mol	Ion	Λ mS/m²/mol	Ion
40.2	$HC_2O_4^{1-}$	7.68	I	6.8	ClO^{4-}
34.982	H^+	7.68	Cs^+	6.192	Ag^+
20.88	La^{3+}	7.64	Rb^+	5.6	$HCOO^-$
19.8	OH^-	7.63	Cl^-	5.5	F^-
15.96	SO_4^{2-}	7.40	$C_2O_4^{2-}$	5.011	Na^+
12.728	Ba^{2+}	7.352	K^+	5.0	HSO_3^{2-}
11.9	Ca^{2+}	7.34	NH_4^+	4.5	Be^{2+}
10.612	Mg^{2+}	7.2	SO_3^{2-}	4.09	CH_3COO^-
10.2	$Co(NH_3)_6^{3+}$	7.2	CO_3^{2-}	3.869	Li^+
7.84	Br^-	7.144	NO_3^-		

The data are listed in descending order.
Data from Adamson [Adamson 1973].

The term dissociation is to the best of my information not used in solid electrolytes. The charge carriers for example in ceramic membrane electrolytes are typically oxygen ions or the oxygen vacancies. The ceramic proton conductors have protons as charge carriers. These charge carriers need typically an elevated temperature in order to have the necessary mobility as charge carrier. From this perspective, the term dissociation would be qualified because chemical bonds are broken and newly formed through the course of conduction of current in the solid [Braun 2017b, Chen 2013].

5.4.4 Separation of electronic and ionic conductivity

One problem in solid state electrochemistry (important for SOFCs and electrolyzers) is that electric currents can be composed of ionic currents and electronic currents. The total current is then a composition, and from a conventional I–V curve, we cannot tell to which extent a material, be it an electrode or an electrolyte, is an electronic conductor or ionic conductor.

It is anyway a good practice in science to not blindly assume that a system may either be this one or that one and not a "mixture" of both.[3] Some materials have to be very good electronic conductors, such as current collectors and electrodes. Electrolytes should have a very high ionic conductivity. And, other materials should have no conductivity at all; these are the electric insulators.

---

3 Be reminded of the particle wave dualism for electrons, for example. It is not that you have to decide whether an electron is a particle or a wave. Wave and particle – these are two different concepts and models which we need in order to comprehend our empirical observations.

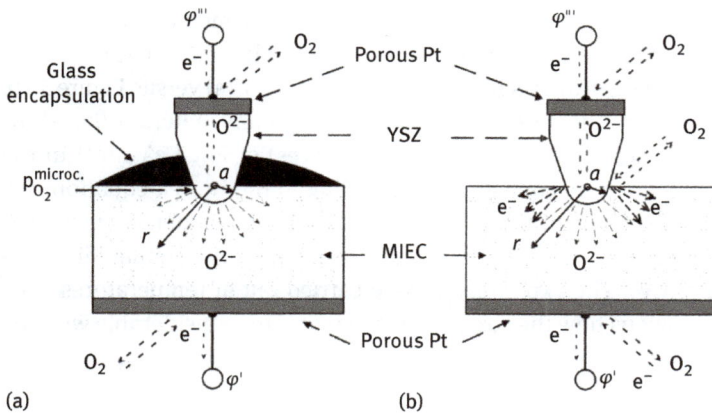

Figure 5.7: Schematic representation of the electrochemical cell with electron-blocking microcontact (YSZ) on a dense sintered pellet of a mixed electronic and ionic conductor (MIEC): (a) encapsulated surface around the microcontact preventing the exchange of gaseous oxygen with the MIEC and (b) nonencapsulated free MIEC surface giving rise to oxygen exchange as well as to a change of the near-surface stoichiometric composition of the oxide.

Reprinted from *Solid State Ionics*, 150, Wiemhöfer H-D, Bremes H-G, Nigge U, Zipprich W, Studies of ionic transport and oxygen exchange on oxide materials for electrochemical gas sensors, 63–77, Copyright (2002), with permission from Elsevier. [Wiemhöfer 2002]

Figure 5.7 shows an arrangement of electrodes which allow ultimately for the separation of ionic and electronic conductivity contributions. The setup is based on very early work on the conductivity of silver sulfide by Wagner [Wagner 1933] and Hebb [Hebb 1952]. See also some general and specific ideas on blocking electrodes in Riess [Riess 1991a, b].

The sample that you suspect to have a mixed ionic electronic conductivity (MIEC) is contacted on one side with an electronic current collector such as gold or platinum, and on the other side with true ionic current collector such as yttrium-stabilized zirconia (YSZ).

Because the ionic conductivity of YSZ is considerable only at high temperatures, it is necessary to anneal the sample setup accordingly. Then, the ambient environment which has an oxygen concentration of around 20% from the air will interact with the YSZ and thus inflect on the conductivity measurement. It is therefore necessary to seal the setup against the ambient so that no oxygen from outside can alter the sample during measurement. This can be a quite laborious preparation.[4]

---

4 In my own Hebb–Wagner experiments, I never made it to the sealed cells. I learnt from an eminent ETH professor that the making of such sealed cells was accompanied by disappointment. Maybe one out of six cells would finally be working because the sealing procedure required a lot of care and also skills. Therefore, it is my opinion that in a time – unfortunately – where graduate students are expected to finish their degree soon, there is not always enough time to learn and practice a delicate

Wiemhöfer et al. [Wiemhöfer 2002] have perfected this method by applying microcontacts through the glass encapsulation as shown in Figure 5.7. The difference of this approach versus a non-sealed cell is exercised in Figure 5.7a) versus Figure 5.7b.

Praseodymium calcium manganese oxide is one of the many materials that show MIEC. Figure 5.8 shows the polarization curves(I–Vcurves) of $Pr_{0.7}Ca_{0.3}MnO_3$in the aforementioned cells – when it is not sealed. According to the Nernst equation, the voltage changes with the concentration of the oxygen. The current density increases with increasing voltage (respectively the oxygen concentration) similar like in a cyclic voltammogram (CV). The experiments were carried out at temperatures from 550 °C to 750 °C. It is obvious how the current density increases considerable with the oxygen concentration and temperature.

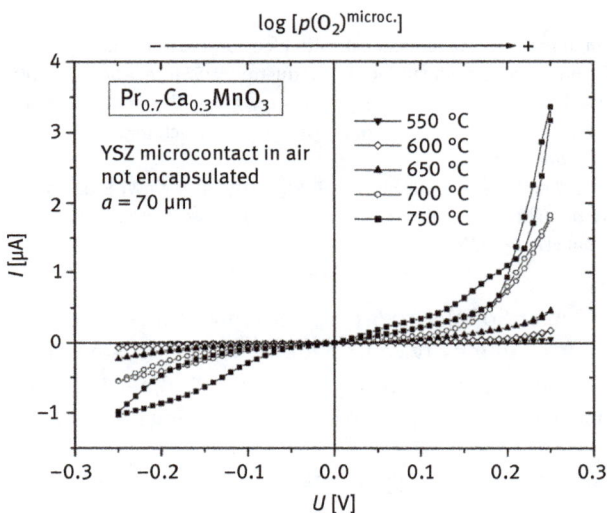

Figure 5.8: Current–voltage curves of a nonencapsulated YSZ microcontact on $Pr_{0.7}Ca_{0.3}MnO_3$ at various temperatures.
Reprinted from *Solid State Ionics*, 150, Wiemhöfer H-D, Bremes H-G, Nigge U, Zipprich W, Studies of ionic transport and oxygen exchange on oxide materials for electrochemical gas sensors, 63–77, Copyright (2002), with permission from Elsevier. [Wiemhöfer 2002].

However, this experiment does not help us for the assessment of the ionic conductivity, because the oxygen from the ambient interacts with the sample. We assume a stoichiometry for the sample of $Pr_{0.7}Ca_{0.3}MnO_3$ but the "O_3" is not necessarily true during exposure of the sample at high temperature in the ambient. The manganese

---

method such as the Hebb–Wagner method to the extent that one can run large series of experiments with ease.

(Mn) in the material can be oxidized or reduced and vary between Mn^{3+} and Mn^{4+} or even to more extreme values.

Figure 5.9 shows the *I–V* curve of the $Pr_{0.7}Ca_{0.3}MnO_3$ when it is encapsulated by a glass seal. The current density from the sample in the glass seal is much lower than the current density of the sample not protected against exchange with oxygen from the ambient. The area within the *I–V* curves (voltammograms) gives account for the electrochemical exchange which occurs during the cycling at high temperature. It gives account of the chemistry but not of the conductivity. Only when we suppress the redox reaction by the sealing, we can access the true ionic conductivity.

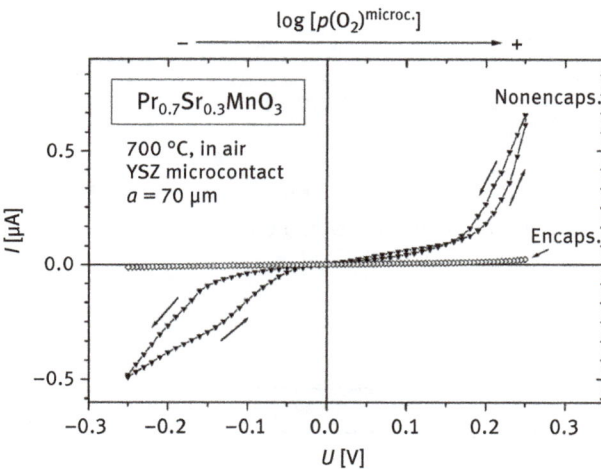

Figure 5.9: Comparison of current–voltage curves of an encapsulated and a non-encapsulated YSZ microcontact on $Pr_{0.7}Sr_{0.3}MnO_3$ at 700 °C.
Reprinted from *Solid State Ionics*, 150, Wiemhöfer H-D, Bremes H-G, Nigge U, Zipprich W, Studies of ionic transport and oxygen exchange on oxide materials for electrochemical gas sensors, 63–77, Copyright (2002), with permission from Elsevier [Wiemhöfer 2002].

Under some conditions, the need for blocking electrodes is not given and separation of ion conductivity becomes somewhat easier. It is based on a "zero driving force" and obtained by marking a short circuit on the MIEC during the measurement [Riess 1991a].

5.4.5 Electrochemical kinetics – the Butler–Volmer equation

When you run an *I–V* curve of an electrochemical system (electrochemical cell), you will likely notice no linear relationship between current and potential. This is because various independent processes in the cell will superimpose so that you cannot make out a linear function. Only in some instances, for example, when electrodes in your electrochemical cell are not well connected, a significant ohmic contribution may show up in the I–V curve, in addition to contributions with electrochemical kinetics. An early study on the question whether Ohm's law is valid in electrolytes was done by Cohn [Cohn 1884].

The Butler–Volmer equation describes the fundamental relationships in electrochemical kinetics [Bockris 2000, Mayneord 1979]. It describes how the electric current through an electrode depends on the electrode potential, considering that both a cathodic and an anodic *chemical reaction* occur on the same electrode:

$$j = j_0 \cdot \left\{ \exp\left[\frac{\alpha_a zF}{RT}(E - E_{eq})\right] - \exp\left[-\frac{\alpha_c zF}{RT}(E - E_{eq})\right] \right\},$$

or in a more compact form [overpotential $\eta = (E - E_{eq})$]:

$$j = j_0 \cdot \left\{ \exp\left[\frac{\alpha_a zF\eta}{RT}\right] - \exp\left[-\frac{\alpha_c zF\eta}{RT}\right] \right\}.$$

This is an equation which we cannot solve analytically for the overpotential η, for example. The constants and variables of the Butler–Volmer equation are explained in Table 5.2.

Table 5.2: Explanation of the constants and parameters in the Butler–Volmer equation for electrochemical kinetics.

Symbol	Description	Dimension
J	Electrode current density	A/m^2
j_0	Exchange current density	A/m^2
E	Electrode potential	V
E_{eq}	Equilibrium potential	V
T	Absolute temperature	K
z	Number of electrons involved in the electrode reaction	
F	Faraday constant	96,485.3329 s A/mol
R	Universal gas constant	8.3144598 (48) J/K/mol
α_c	So-called cathodic charge transfer coefficient	
α_a	So-called anodic charge transfer coefficient	
η	Activation overpotential	$\eta = (E - E_{eq})$

Figure 5.10 shows I–V plots for $\alpha_a = 1 - \alpha_c$. The red curve (bottom line) shows the positive part of the reaction (anodic), and the blue curve (top line) shows the cathodic reaction. Both curves added together make the green curve (middle line). Note that this reflects the situation of electrodes in an electrolyte where chemical reactions are taking place.

What can happen during the course of a chemical reaction is that reaction products build up in front of the electrodes and provide a barrier for the other ions

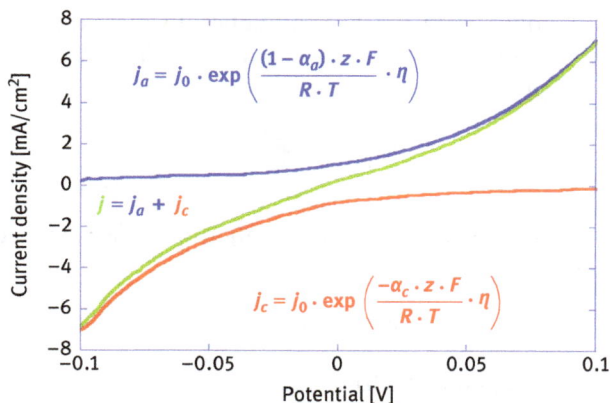

Figure 5.10: Graphical representation of the Butler–Volmer equation. The current density is plotted as a function of the overpotential η. The anodic and cathodic current densities are shown as j_a and j_c, respectively, for $\alpha = \alpha_a = \alpha_c = 0.5$ and $j_0 = 1\ \mathrm{mA/cm^2}$ (close to values for platinum and palladium, two noble metals often used as reference and electrocatalysts in electrochemistry).

to reach the electrode surface. Or reaction partners such as ions can become depleted. This region of accumulation of products or depletion of reaction partners can constitute a diffusion barrier which causes a time delay t for the proceeding of the reaction because the species have to diffuse over the length L of the barrier, subject to the Fick's law of diffusion with diffusion constant D:

$$L = \sqrt{D \cdot t}$$

Under this situation with the parameters explained in Table 5.3, the limiting current is simply

Table 5.3: Explanation of limiting current parameters.

Symbol	Meaning
D	Is the diffusion coefficient
L	Is the diffusion layer thickness
C^*	Concentration of electroactive (limiting) species in the bulk of the electrolyte

$C(0,t)$ would be the time-dependent concentration at the distance zero from the electrode surface

$$i_{\mathrm{limiting}} = \frac{zFD}{L} \cdot C^*$$

Upon close inspection of the Butler–Volmer equation, we realize that we can make two approximations.

In the region of low overpotential when $E \approx E_{eq}$, the Butler–Volmer equation simplifies to [as an exercise, you can try to make a Taylor expansion in the Butler–Volmer equation for small $(E - E_{eq})$]

$$i = i_0 \cdot \frac{zF}{RT} \cdot (E - E_{eq})$$

In the region for high overpotentials, the Butler–Volmer equation simplifies to what is known as the Tafel equation:

$$E - E_{eq} = a - b \log(i)$$

$$E - E_{eq} = a + b \log(i)$$

Here, a and b are the constants for a given electrochemical reaction and cell temperature. The equation is called Tafel equation and the two constants (a, b) are known as Tafel constants. The theoretical values of a and b are different for the cathodic and anodic processes. A plot of the logarithm of the current versus the potential is called Tafel plot.

Figure 5.11 shows a Tafel plot, where the logarithm of the current density is plotted versus the potential in linear coordinates. The Tafel plot is given for three different charge transfer coefficients.

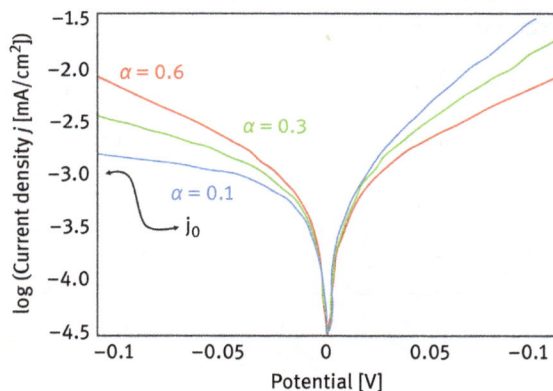

Figure 5.11: Logarithmic plot for different values of α; Tafel plot.

5.5 Cyclic voltammetry

In cyclic voltammetry, we apply a voltage on a working electrode (WE) in a conventional three-electrode cell (with counter electrode CE and reference electrode RE) and then increase the potential from a minimum ϕ'_u of the WE with a particular scan speed (scan rate v, such as $v = 10$ mV/s) to a maximum ϕ''_u. The potential of the WE than follows the function: $\phi(t) = \phi'_u + vt$. When the maximum potential is reached, the same scan rate is applied in reverse direction. This is one cycle for the potential. The potential thus goes up and down as shown in Figure 5.12.

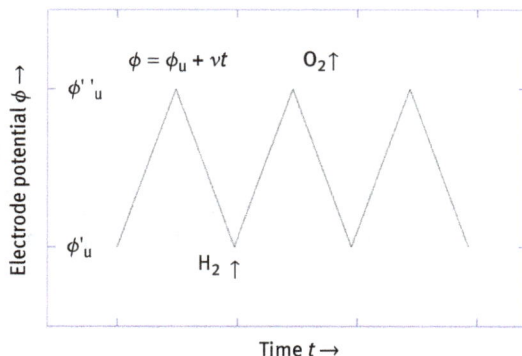

Figure 5.12: Potential/Time relationship at a working electrode during cyclic voltammetry. Negative ($\phi''u$) and positive ($\phi'u$) reversal potential are usually chosen between the evolution of hydrogen and oxygen.

The potentiostat warrants that the potential is correct and the current between the WE and CE is accurate. The software for the potentiostat typically has programmable macros where the operator can set the desired parameters like ϕ'_u, ϕ''_u, v and also the number of cycles. Figure 5.12 shows three cycles for the potential ϕ.

Meanwhile the potentiostat has recorded the current while the potential was scanned. You thus obtain a dataset which has the time as parameter in one column, the potential in a second column and the current in a third column. When you plot the current versus the potential, you obtain what is called a CV. The scan to maximum positive potential is the anodic direction, and the scan to maximum negative potential is the cathodic direction.

Figure 5.13 shows one CV of an oxidized glassy carbon electrode (WE) which was recorded from 0 to 1 V versus the saturated calomel reference electrode . The current density ranges between 2 and 4 mA/cm^2.

The CV contains information on how easy it is to oxidize or reduce a surface of a material. Surfaces of single crystal electrodes with different Miller indices may have differently shaped CVs. A very dense packed surface may easily release an electron than one with a set of lower Miller indices.

In order to study this, we need to run the CVs over single crystal surfaces with well-known hkl surfaces. This is nicely illustrated for gold surfaces in Hamelin and Martins [Hamelin 1996a, b]. It is possible to model and simulate CV with mathematical code, for example in the textbook by Gosser [Gosser 1993]. An online resource can be found on the homepage of Dr Gareth Kennedy (http://www.garethkennedy. net/MECSim.html#Usage).

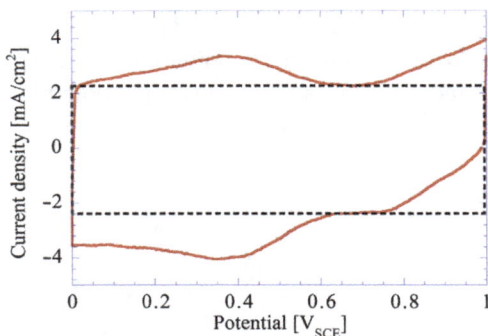

Figure 5.13: Cyclic voltammogram of an activated SIGRADUR_K sample. The scan rate was 10 mV/s. The dashed line depicts a rectangular CV of an ideal capacitor with only double layer charging and discharging.

5.5.1 Randles–Sevcik equation

When you play around with the potentiostat parameters when you scan a CV, you will notice that the magnitude of the current depends also on the scan speed (1 mV/s, 10 mV/s, 100 mV/s and so on). This becomes particularly obvious when you have distinct current peaks i_p from redox couples.

A quantitative relation between the peak current density and scan rate v is given by the Randles–Sevcik equation:

$$i_p = 0.4463 \, n \, F \, A \, C \, \sqrt{\frac{n \, F \, v \, D}{R \, T}}$$

Here is n the number of electrons which are transferred in the redox reaction. This is typically $n = 1$, for example in the reaction $Fe^{3+} + e^- = Fe^{2+}$. F is Faraday's constant, A is the geometric area of the electrode (cm^2, m^2 etc.), C is the concentration of the species which accept or donate electrons, R is the gas constant in J/K/mol, T is the absolute temperature in °K, D is the diffusion constant in cm^2/s and v is the scan rate – the experimental parameter which you hold in the potentiostat.

We see from this equation that the current which we measure goes linear with the electrode area A, and also linear with the concentration C of the species which we reduce or oxidize in the electrochemical cell. In the argument of the radical, we have the temperature in the denominator; this means with increasing T, the peak current is getting smaller. In the numerator of the radical, there is the diffusion constant D, which we cannot change, unless we chose a different solvent. The scan rate v is also in the numerator of the radical; hence with increasing scan rate v, we see the peak current i_p increasing, but not linear. The peak current i_p and v have a parabolic relation.

It is an exercise for the reader to determine this equation for $T = 300$ K (ambient temperature) and to formulate the equation in a Cottrell plot (compare also [Cottrell

1903]) which allows the direct determination of the diffusion constant D from a set of peak currents $i_p(v)$ determined for various scan rates v.

The Randles–Sevcik equation is reminiscent to the Cottrell equation. The latter is used in experiments where a potential is applied in a step function. This means the potential can be applied, or it can be taken away. In both cases, the change is an abrupt step.

Here is an analogue for you to understand what I mean: When there is a marathon running competition, you have a handful of runners at the start line and a bulk of runners waiting behind them. When there is the acoustic start signal, the first line of runners will start and also the second and third line, but there is already some inertia in the fourth and fifth line. The bulk of runners heard the signal at the same time (the speed of the acoustic signal is around 300 m/s) and wants to start but the runners before them are hindering them somewhat. This inertia of the whole crowd with some internal friction is what we call diffusion limitation.

It is not much different when you have the wild crowd running in line and then immediately stop the first leading row. The ones who follow will not notice immediately what makes the front line stop and this causes collisions and friction.

The aforementioned abrupt change of potential in a step function can be expended in a Fourier series. The step can then be approximated by a spectrum of Fourier series and be subject to Fourier transformation. This means that the time domain, in which the potential is switched on or off, is subject to a coordinate transformation in the frequency domain. You probe in the frequency domain with impedance spectroscopy. In so far, the switch on/off experiment and the impedance spectroscopy can be considered canonically conjugated methods because they give you the same kind of information, this is, the charge carrier dynamics. We can close the circle when we consider the scan speed, the scan rate in a CV. When we change the potential at an extremely high speed, say 1000 V/s, then the rate for a change over a 20-mV increment in potential would correspond to 50 kHz. Therefore, we can consider changing the scan rate in a CV comparable to employing impedance spectroscopy.

5.5.2 Determination of HOMO and LUMO from the cyclic voltammogram

When we manipulate the potential of an electrode material in an electrolyte by applying an external potential, we may arrive at a potential where a charge transfer takes place either by accepting or by donating an electron. Cyclic voltammetry is therefore an electroanalytical method for probing the electronic structure of electrode materials and the molecular structure of molecules adsorbed or chemisorbed on electrodes.

Koopmans Theorem [Koopmans 1934] states that "the first ionization energy of a molecular system is equal to the negative of the orbital energy of the highest occupied molecular orbital (HOMO)." From the position of the redox peaks in a CV, we can therefore in general conclude on the HOMO and lowest unoccupied molecular orbital (LUMO) of a molecular species involved in the charge transfer.

Figure 5.14 shows the UV–visible spectrum of the organic semiconductor poly(3-hexylthiophene) (P3HT) in solution and deposited as a film on indium tin oxide (ITO). The spectrum of ITO is taken as a spectral reference and also for background subtraction. P3HT has a maximum luminescence at 600 nm, which is red color. The optical band gap of the film is $E_{g,o} = 1.9$ eV, as determined from the absorption onset at 655 nm.

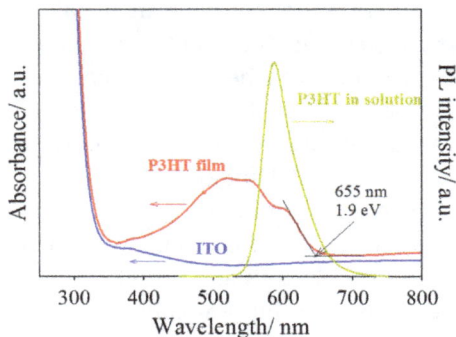

Figure 5.14: UV–vis spectra obtained for P3HT in aqueous solution, ITO and P3HT/ITO film. Reprinted by permission from Springer, *Journal of Solid State Electrochemistry*, 21, 2407–2414, Measurements of HOMO-LUMO levels of poly(3-hexylthiophene) thin films by a simple electrochemical method, Acevedo-Pena P, Baray-Calderon A, Hu HL, Gonzalez I, Ugalde-Saldivar VM, 2017. [Acevedo-Pena 2017].

It is an exercise for the advanced practitioner to (a) subtract the blue ITO spectrum from the red P3HT film spectrum and (b) subtract the yellow P3HT-in-solution spectrum from the red P3HT film spectrum. For this, you have to digitize the spectra with a tracing program such as DataThief [Tummers 2006] or engauge-digitizer [Mitchell 2018], for example.

The CVs in Figure 5.15 are those from P3HT grown in ITO and the empty ITO substrate for comparison. The authors of that study [Acevedo-Pena 2017] point out that they made for every CV new film electrodes because every scan in oxidative or reductive direction can alter the condition of the film.[5] The P3HT has a sharp current wave onset at about +0.12 V in the anodic scan direction. The onset position is determined by extrapolating the flank of the current wave to the current density of 0 mA/cm^2; the intercept is at 0.12 V.

---

5 I would say this is a very important message for every experimentalist who is not very familiar with electrochemistry. The power that 1 V – or a fraction of it – can have on the chemical condition of a sample or sample surface is often underestimated. In this respect, cyclic voltammetry is not a nondestructive analytical method. As a general rule about the nondestructive testing, the reader is advised to give a very deep thought about the impact a method can have on a sample.

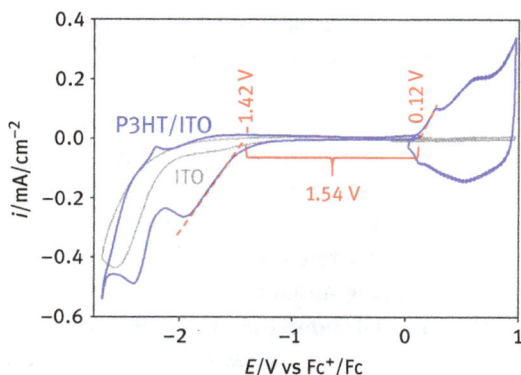

Figure 5.15: Cyclic voltammetry characterization ($v = 0.1$ V/s) of P3HT/ITO and ITO samples in acetonitrile 0.1 mol/L Bu$_4$NPF$_6$ previously bubbled with N$_2$. Anodic and cathodic scans were performed using different films.
Reprinted by permission from Springer, *Journal of Solid State Electrochemistry*, 21, 2407–2414, Measurements of HOMO-LUMO levels of poly(3-hexylthiophene) thin films by a simple electrochemical method, Acevedo-Pena P, Baray-Calderon A, Hu HL, Gonzalez I, Ugalde-Saldivar VM, 2017. [Acevedo-Pena 2017].

The potential is taken with reference to ferrocene Fc$^+$/Fc. Ferrocence is a well-studied electrolyte with well-known reference [Koepp 1960, Tsierkezos 2007] to calculate its oxidation and reduction potentials E_{ox} and E_{red}. The conversion of the Fc$^+$/Fc potential versus any other reference is, for example, demonstrated in Pavlishchuk and Addison [Pavlishchuk 2000]. The HOMO of ferrocene is at –5.22 eV versus the vacuum level. The LUMO of ferrocene is at –3.68 eV versus the vacuum level [Jiang 1997].

Looking again at the CV in Figure 5.15, we notice the current wave onset on the cathodic scan at +1.42 V. This position was again determined by extrapolation of the onset flank to 0 current density. The energy gap across this CV is therefore $E_{g,t} = 1.42$ V + 0.12 V = 1.54 V. As this gap is derived from an electroanalytical method, it is called transport gap. It is interesting to see that this transport gap is by 0.36 V smaller than the optical band gap which was $E_{g,o} = 1.9$ V.

The molecular structure of a dye molecule like P3HT or ferrocene can be determined in addition with impedance spectroscopy, electron spectroscopy and density functional theory (DFT). An example is given in the paper by Barlow et al. [Barlow 1999].

5.5.3 Catalyst turnover frequency

The efficiency of a catalyst (electrocatalyst) is given by its turnover frequency (TOF). The TOF is the number of chemical conversions that a catalyst can make in 1 s. A detailed description how to determine the turnover number in gas phase heterogeneous catalysis is provided in Bonzel [Bonzel 1977].

For the determination of the TOF in electrochemistry, I want to present a study carried out by Yagi et al. [Yagi 2005]. They have compared three different samples: bare ITO, on which they deposited iridium metal Ir and iridium oxide IrO_2. ITO is a well-known transparent conducting oxide used as substrate and current collector in semiconductor photoelectrochemistry experiments. Its electronic conductivity is high enough to serve as current collector and yet lets some visible light shine through.

Yagi et al. first carry out conventional CV in order to learn about the electrochemical properties of the substrate, basically as a reference measurement. Then, they coated the ITO with a thin layer of IrO_2 using an electrochemical deposition method. Figure 5.3 in their original work [Yagi 2005] shows the CVs of the bare ITO, Ir and IrO_2-coated substrates, where the current density ranges to 5 $\mu A/cm^2$ in the potential range from −0.35 to 0.6 V versus the Ag/AgCl reference.

At above 1 V (to 1.3 V), the current density for the Ir and IrO_2-coated ITO increased considerably to 0.15 and 2.3 mA/cm^2, whereas the bare FTO shows virtually no current density even at 1.3 V. The current corresponds to the water splitting reaction with concomitant oxygen evolution which Yagi et al. measured with a gas chromatograph (GC)[6] [Yagi 2005].

Yagi et al. prepared films of different thicknesses; this means the electrocatalyst coverage was different. Figure 5.16 shows the amount of evolved oxygen versus the coverage of metal iridium (top, panel a) and iridium oxide (bottom, panel b). It is very instructive to note that the amount of evolved oxygen scales linear with the coverage or Ir and IrO_2.

The GC data are shown in Figure 5.17 as a function of applied potential. On the bare ITO, no oxygen is evolved, but the IrO_2 film is clearly an electrocatalyst [Koetz 1985, 1984, Silva 1998] because it allows for the oxygen evolution reaction at a moderately low potential of 1.3 V versus Ag/AgCl reference, that is, 3 μmol. The current waves at −0.1 and +0.3 V in Figure 5.3 in Yagi et al. [Yagi 2005] originate from the redox couple Ir^{4+}/Ir^{5+}; the integration of the peak at +0.3 V yields 2.4×10^{-10} mol/cm^2 of electroactive iridium species.

For a quantitative assessment of the performance of the coated electrodes, it was necessary to determine the catalyst coverage. The authors did this by recording a visible light absorption spectrum of the catalyst precursor solution prior to coating

---

6 The experiment by Yagi et al. ([Yagi 2005] Yagi M, Tomita E, Kuwabara T: Remarkably high activity of electrodeposited IrO_2 film for electrocatalytic water oxidation. *Journal of Electroanalytical Chemistry* 2005, 579:83–88.doi: 10.1016/j.jelechem.2005.01.030.) is already quite complex from the instrumentation point of view. You need a power source which can control the potential and also two meters that allow for measuring voltage and current. This can be done with ease when you have a potentiostat and a simple electrochemical cell. In addition, Yagi et al. employ a gas chromatograph. Here, too, we need a power source such as a potentiostat in order to bring the working electrode to the desired potentials. In addition, do we need an electrochemical cell which is sealed and gas tight so that we can control the gas volume which is measured by the GC.

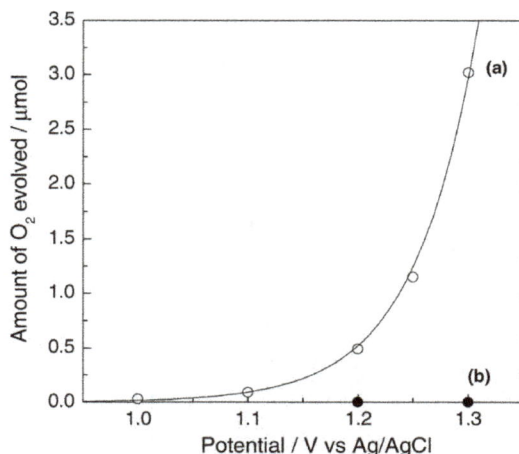

Figure 5.16: Potential dependence of the amount of O_2 evolved during electrochemical water oxidation in 0.1-M KNO_3 aqueous solution (pH 6.3) under the potentiostatic conditions for 1 h using (a) IrO_2-coated ITO electrode [coverage of $IrO_2 = (2.6 \pm 0.2) \times 10^{-10}$ mol/cm²] (open symbols) and (b) bare ITO electrode (filled symbols).
Reprinted from *Journal of Electroanalytical Chemistry*, 579, Yagi M, Tomita E, Kuwabara T, Remarkably high activity of electrodeposited IrO_2 film for electrocatalytic water oxidation, 83–88, Copyright (2005), with permission from Elsevier. [Yagi 2005].

and after coating (electrodeposition). The precursor solution was H_2IrCl_6 dissolved in water. The concentration and concentration change can be measured from the absorbance spectra using the Lambert–Beer law. The authors assume that the difference in the absorbance accounts for the iridium (amounting in iridium oxide IrO_2) coated on the ITO substrate. With this information, you know how many iridium atoms or mol IrO_2 is deposited on the ITO substrate.

The current at 1.3 V versus Ag/AgCl reference for the IrO_2 film is by a factor of 660 higher than for the bare ITO. The current density for the Ir film was by a factor 14 lower than that for the IrO_2 film, although the iridium coverage was in the metal film by a factor 34 higher than in the IrO_2 film.

The amount of oxygen evolved during 1 h of electrolysis was determined for various coverages of Ir and IrO_2, as shown in the two graphs in Figure 5.17. The top panel in Figure 5.17 shows the oxygen evolved per hour over the metal iridium layers with coverage ranging from 0 to 1.4×10^{-7} mol. The amount of oxygen can be well fitted with a linear function, as is shown by the solid straight line. The turnover frequency TOF is the amount of oxygen produced per hour divided by the coverage of the iridium catalyst (the slope of the linear least square fit curve).

$$\text{TOF(Ir)} = \frac{5.5 \times 10^{-6} \, \text{mol/h}}{1.4 \times 10^{-7} \, \text{mol}} \approx \frac{40}{\text{h}}$$

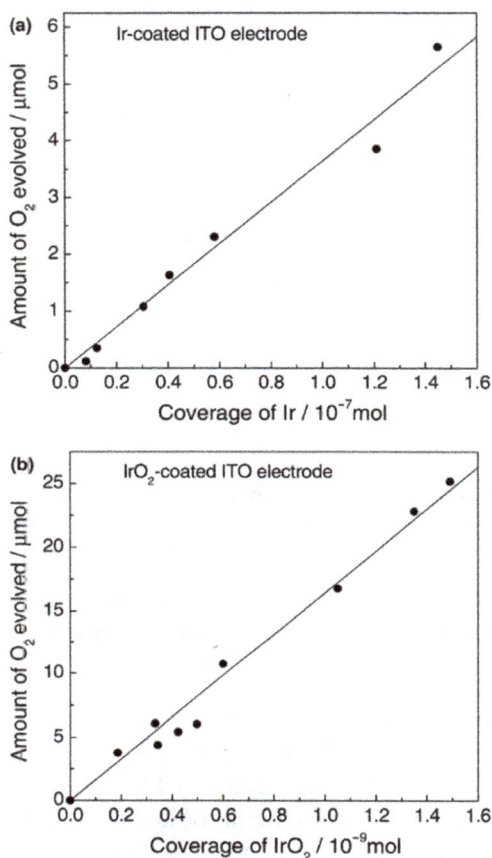

Figure 5.17: Plots of the amount of O_2 evolved during electrocatalysis versus coverage of the catalysts: (a) Ir-coated ITO electrode (electrodeposited at −35 µA/cm^2) and (b) IrO$_2$-coated ITO electrode (electrodeposited at +35 µA/cm^2). The electrocatalysis was conducted in a 0.1-M KNO$_3$ aqueous solution (pH 6.3) at 1.3 V versus Ag/AgCl for 1 h.

Reprinted from *Journal of Electroanalytical Chemistry*, 579, Yagi M, Tomita E, Kuwabara T, Remarkably high activity of electrodeposited IrO$_2$ film for electrocatalytic water oxidation, 83–88, Copyright (2005), with permission from Elsevier. [Yagi 2005].

For the iridium oxide film (lower panel in Figure 5.17), we find also a linear increase of the evolved oxygen over the catalyst coverage. Note however that here the coverage was two orders of magnitude smaller than with the iridium metal film. Therefore, the TOF is considerably higher:

$$\text{TOF}(\text{IrO}_2) = \frac{25 \times 10^{-6}\,\text{mol/h}}{1.5 \times 10^{-9}\,\text{mol}} \approx \frac{16700}{\text{h}}$$

Yagi et al. prepared films of different thicknesses; this means the electrocatalyst coverage was different. Figure 5.16 shows the amount of evolved oxygen versus the coverage of metal iridium (top, panel a) and iridium oxide (bottom, panel b). It is very instructive to note that the amount of evolved oxygen scales linear with the coverage or Ir and IrO_2.

Yagi et al. prepared the electrocatalyst films by electrodeposition. They varied also the current densities for this deposition. Table 5.4 shows the TOF for both type of electrocatalysts (Ir and IrO_2) but grown with current densities of 35, 100 and 350 µA/cm^2. For both types of electrocatalysts, the TOF decreases when the current density for electrocatalyst deposition increases.

Table 5.4: Summary of TOF of IrO_2 and Ir catalysts with various galvanostatic conditions for electrodeposition.

Galvanostatic conditions (µA/cm^2)	Turnover frequency (TOF/h)	
	IrO_2-coated electrode	Ir-coated electrode
35	16400 (±450)	36.4 (±1.4)
100	13800 (±770)	29.6 (+1.1)
350	12700 (±360)	15.7 (+0.5)

TOF was calculated based on the amount of O_2 evolved in electrocatalysis at 1.3 V versus Ag/AgCl for 1 h.
Reprinted from *Journal of Electroanalytical Chemistry*, 579, Yagi M, Tomita E, Kuwabara T, Remarkably high activity of electrodeposited IrO_2 film for electrocatalytic water oxidation, 83–88, Copyright (2005), with permission from Elsevier. [Yagi 2005].

The current density has likely some influence on the morphology of the electrocatalyst particles grown during electrodeposition. It would require further structural analysis and investigation what exactly makes the different TOF.

5.5.4 The need for a reference electrode

We can assign a device such as a battery or a capacitor, a charge and a capacity by measuring current and voltage. This is typically a diagnostic procedure. But as scientists, we are often interested in the specific electrode properties when they are under polarization.

In that case, it is necessary that we have a way to inspect only the electrode and not the entire reactor or system. For example, we want to know at which exact potential, a chemical reaction is taking place. We discussed this already in the previous section on the Butler–Volmer equation. A potential, unlike a voltage, needs always some reference potential. For this, we apply a RE (REF) in very close position to the electrode under consideration, the WE.

The REF is a high ohm electrode so there is no current flowing through it. This is because it is not possible to assign an electrode a potential through which a current is flowing. Important for the RE is that it has a stable potential and a well-known potential.

The RE is often kept in a separate container in an electrolyte which can be brought into contact with the electrolyte of the electrochemical cell via a ceramic diaphragm. The RE is therefore kept like an electrochemical half-cell.

One example for this is the well-known silver/silver chloride RE, $Ag^+/AgCl$. This contains a metal silver wire which is kept in a glass tube which is filled with a saturated aqueous solution of AgCl. The bottom of the glass tube is closed with a ceramic frit (diaphragm) through which the Cl^- ions can flow into the electrochemical cell which contains the WE and CE. The equilibrium condition for this RE is given by

$$AgCl\,(s) + Ag\,(s) + e^- \Leftrightarrow Ag\,(s) + e^- + Cl^- + Ag^+$$

The concentration of the chloride anions determines the Nernst potential of this electrode as follows:

$$E = E^\circ - \frac{R \cdot T}{F} \cdot \ln a_{Cl^-}$$

The silver chloride RE has a standard potential $E^\circ = 0.230$ V versus the standard hydrogen electrode (SHE). The potential of the SHE is by default 0.00 V. When the concentration of the chloride ions changes, the potential will change according to the Nernst equation. Note that also the temperature of the electrolyte has an influence on the potential.

The RE is typically considered an analytical add-on, a tool. But for the accuracy of data, it can play an important role. The RE is a science of its own and reviews of – for example – the silver chloride have been published already in the early 1950 [Janz 1953, Taniguchi 1957]. Sometimes, the RE is connected via a Luggin capillary [Shchukin 1995, Tokuda 1985] for making as close contact as possible with the WE. I recommend the textbook on REs by Inzelt et al. [Inzelt 2013] for the interested readers.

It is not always possible to place a RE properly. This holds particularly when you work with actual devices such as batteries and fuel cells. In the best case, you may be able to place an extra piece of lithium metal near the WE and consider this the reference. The potentiostat has an extra connection for the REs.

When you use REs, then you list the potential of the WE versus this RE. It is common to convert that potential to other potentials, such as the reversible hydrogen electrode . An online source which helps you easily convert one reference scale to another one is http://www.consultrsr.net/by Robert Stanleigh "Bob" Rodgers (2010).

5.6 Impedance spectroscopy

We have so far dealt with direct current (DC) and voltages. When we apply a voltage or impose a current, the change of voltage and current is "immediately." Such rapid change of magnitude of I or V imposes a dynamics which can affect the entire electrical or electrochemical system. What electric engineers then notice in the circuit, in the grid, is a force which acts against the original direction of the current or voltage, literally an "*impedance*."

The impedance is a "dynamic" electric resistance as a response of an external electric or magnetic excitation. From the mathematical perspective, the impedance is a complex quantity with a real and imaginary branch. From the physical perspective, the impedance is a frequency-dependent resistance which has a dissipative and dispersive component in analogy to its mathematical nature. The impedance is abbreviated with Z.

The measurement of complex resistances (Impedance Z) with AC voltage techniques (electrochemical impedance spectroscopy, EIS) is an important tool for the investigation of interface and volume properties of materials [Barsoukov 2005]. Impedance spectroscopy can be used as an analytical method and also as a diagnostic tool [Boukamp 2004].

An early work on electrochemical systems with a small AC current imposed on an electrode was presented in the first paper [Randles 1947] of Randles' series on electrode kinetics [Randles 1947, 1952a, b, c]. When interfaces are studied, such as electrode–electrolyte interfaces, adsorption rates and reaction rates can be studied as well as double layer capacitance. When a system is excited with an AC voltage, say

$$U(\omega) = U_0 \cdot \exp(i\omega t)$$

U_0 being the amplitude, f being the frequency and

$$\omega = 2 \cdot \pi \cdot f$$

Then, the current $I(\omega)$ flowing through the system in general will be shifted against the voltage by a phase shift φ:

$$I(\omega) = I_0 \cdot \exp(i\omega t + \varphi)$$

I_0 being the amplitude of the ac current. The impedance $Z(\omega)$ follows from Ohm's law:

$$Z(\omega) = \frac{U(\omega)}{I(\omega)} = Z_0 \cdot \exp(i\varphi) = Z_0 \cos\varphi - iZ \sin\varphi$$

A schematic of the impedance, showing the phase shift between current and potential, is illustrated in Figure 5.18. The real part of the complex impedance, $Z_0\cos\varphi$, will be abbreviated with Z'. The imaginary part of the complex impedance, $Z_0 \sin \varphi$, will be abbreviated with Z''.

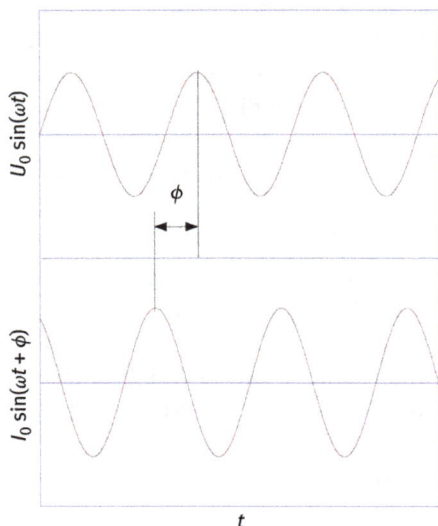

Figure 5.18: Phase relationship between voltage U and current I in the complex impedance. The phases of current and voltage are shifted by the amount of $\Delta\varphi$ against each other: $\Delta\varphi = \omega\Delta t$.

In many cases, systems can be modeled by an array of resistances R, capacitances C and inductances L. Table 5.5 displays the frequency response of these simple elements: The impedance of a combination of R, L and C can be calculated using Kirchhoff's rules. For a series circuit of two impedances Z_1 and Z_2, the impedance is

$$Z_{1,2} = Z_1 + Z_2$$

For a parallel circuit of two impedances Z_1 and Z_2, the impedance is calculated as follows:

$$\frac{1}{Z_{1,2}} = \frac{1}{Z_1} + \frac{1}{Z_2}$$

In impedance spectroscopy, the current is typically recorded versus the excitation frequency. It is possible to record it via the time domain from charge and discharge experiments, when a coordinate transformation is done from the time space to the frequency space using a Fourier transformation.

It was shown that the Onsager relations can be used as additional fundamental information in impedance spectra analysis [Grafov 1971]. It is possible to further enhance the analytical investigation by making dynamic EIS spectroscopy with Fourier transformations [Darowicki 2004].

Table 5.5: Resistivity of various electrotechnical elements as a function of frequency.

Complex resistances	
Ohmic resistance of a resistor	$Z_R = R$
Capacitance of a capacitor	$Z_C = \frac{1}{i\omega C}$
Inductance of a coil	$Z_L = i\omega L$

Coil and capacitor depend on the frequency ω of the applied ac voltage. Ohm resistance is not dependent on frequency in the ideal case.

5.6.1 A word about the misnomer between resistivity and conductivity

Unfortunately, resistors are confused with conductivity. I do remember one of my colleagues saying – when we were sitting over impedance spectra which showed a typical semicircle plus a straight vertical line: "and there is a resistance." He attributed the resistance to the semicircle, which is correct. But what he actually meant was that there was conductivity, a charge transfer.

This is important because in the same spectrum, but at a different frequency range, we saw a vertical straight line. Capacitors cause such as straight line. Capacitors are virtually insulators. They can store electric energy, but they do not pass it on to the other side. Therefore, capacitors have no conductivity (no DC conductivity). Resistors however have conductivity. We are dealing therefore with a misnomer and should not use the word resistivity, but conductivity.

5.6.2 Representation of electrochemical systems by electric circuits

Figure 5.19 displays a schematic of an electrochemical interface, which can be represented by an electric circuit [Hamann 2005]. In the schematic, R_T denotes the charge transfer resistance, which is attributed to chemical reactions occurring on the electrode surface. This resistance is infinite, when no redox processes occur on the electrode. The electrochemical double layer has a capacitance C_D parallel to R_W. If chemical reactions occur, the concentration of ions may decrease in front of the electrode with the result that an additional resistance R_W in series with a capacitance C_W also occurs in the system, in series to the charge transfer resistance and parallel to the double layer capacitance.[7] This element is known as the *Warburg impedance* and is characterized by a straight line with an angle of 45° against the real axis in the complex plane:

---

7 Although termed as *occurrence of a resistance*, the correct speech means the occurrence of conductivity.

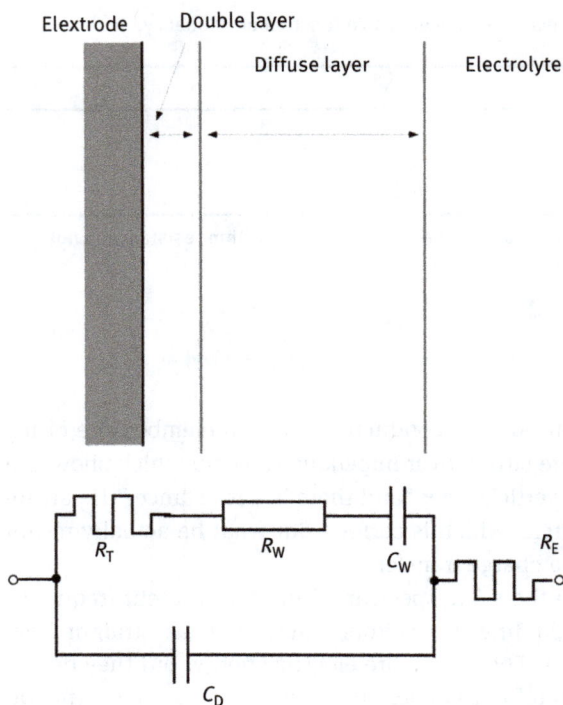

Figure 5.19: Schematic of the electrochemical interface and its corresponding electric circuit.

$$R_W \rightarrow R_W + \frac{1}{i\omega C_W}$$

Finally, the resistance of the electrolyte, R_E, contributes to the overall resistance of the system in series.

Using the complex resistances in Table 5.5 and Kirchhoff's rules, one finds for the impedance of the circuit in Figure 5.19:

$$Z = R_E + \frac{1}{i\omega C_D + (1/(R_T + R_W + (1/i\omega C_W)))}$$

Note that for high frequencies ω, the limit of the impedance Z is the electrolyte resistance R_E. In the case of an electrochemical double layer capacitor electrode, if redox reactions do not take place and if charge transfer resistances are very high, there will also be no remarkable concentration gradients of electrolyte ions. Then, the situation can be represented by a simple series of a resistance and a capacitance, as displayed in Figure 5.20.

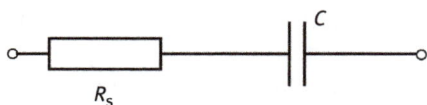

Figure 5.20: Series circuit of a resistance R_S and a capacitance C, representing the most simple capacitor electrode (single electrode).

The impedance of the circuit in Figure 5.20 equals

$$Z = R_S - \frac{i}{\omega C}$$

The complex impedance Z can be split in the real part Z' and imaginary part Z'':

$$Z = Z' + iZ''$$

For the imaginary part, we find

$$Z'' = -\frac{1}{\omega C}$$

Thus, the double layer capacitance of a system that is represented by the circuit in Figure 5.20 can be determined from the imaginary part of the measured complex impedance:

$$C = -\frac{1}{\omega Z''}$$

The capacitance values can often be determined from this equation with a reasonable accuracy. In real capacitors with two electrodes, often a leakage current at open circuit may occur with the result of a self-discharge of the capacitor. The leakage current is taken into account by an additional parallel resistance R_p, as displayed in Figure 5.21.

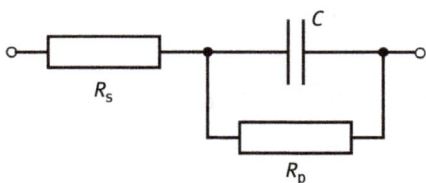

Figure 5.21: Representation of the electric circuit of an EDLC.

The real part and imaginary part of the complex impedance of this circuit are

$$Z' = R_s + \frac{1/R_p}{\omega^2 C^2 + 1/R_p^2}$$

$$Z'' = \frac{-\omega C}{\omega^2 C^2 + 1/R_p^2}$$

From experimental impedance data, all information on R_S, R_p and C can be extracted.

From the high frequency intercept of the impedance plot, the series resistance R_s can be extracted:

$$\lim_{\omega \to \infty} Z' = R_S$$

The information on R_s is necessary to solve the aforementioned equations for the values for R_p and C:

$$C = \frac{Z''}{\omega \left(-2Z'R_s + R_s^2 + Z''^2 + Z'^2 \right)}$$

$$R_p = -\frac{-2Z'R_s + R_s^2 + Z''^2 + Z'^2}{-Z' + R_s}$$

Often experimental data deviate from the theoretical relations in so far as a pure capacitive contribution of C_D is not sufficient to describe the true behavior of the circuit in Figure 5.19. Then, phenomenologically a so-called constant phase element p is introduced:

$$Z_p = \frac{1}{(i\omega)^p C}$$

where p is a real number $0 \le p \le 1$. Often, p is close to 1.

The expressions for R_p and C can be calculated in the same way as in the case without p. However, these expressions are more complicated:

$$R_p = \frac{(Z'^2 - 2Z'R_s + R_s^2 + Z''^2) \sin\chi}{R_s \sin\chi - Z' \sin\chi + Z'' \cos\chi}$$

$$C = -\frac{Z''}{(Z'^2 - 2Z'R_S + R_S^2 + Z'^2) \exp((1/2)\,p \ln \omega^2) \sin\chi}$$

with

$$\chi = \frac{1}{2}p \; \text{signum}(\omega)\,\pi.$$

There exist a number of software packages which you can use to fit and deconvolve impedance spectra into their components. Sometimes, these programs are delivered with hardware, that is, they come with the potentiostat and the FRA.

At this point, I would like to make the reader interested for a software development by Kobayashi, Sakka, Suzuki and associates at NIMS in Japan [Kobayashi 2016]. They have based their program on the general computational platform of Igor Pro, a software platform developed by Wavemetrics, Inc. The advantage of using macros and other software based on Igor Pro is that it is typically an open source [Schmid 2018] which you can develop further for your needs. Even without specific macros, Igor Pro is an excellent data analysis and plot program which you can use for all and any of your scientific data.

5.6.3 Diffusive resistance of porous electrodes

Let me for practical reasons stay with the glassy carbon for a while. Due to the porous active film, the GC electrodes exhibit a typical porous electrode behavior. While an ideal capacitor electrode shows a vertical line in the Nyquist representation of the impedance spectra (Z'' vs Z'), a porous electrode bends over toward smaller impedance values at very high frequencies [De Levie 1963]. The regime of bending over is indicated ideally by a straight line with an angle of 45° with respect to the real axis of impedance. This behavior of the impedance can be attributed to the limited diffusion of ions in the pore filled with electrolyte in a high frequency field [De Levie 1963] and is in so far a transport resistance (similar to a so-called Warburg impedance). This resistance is a series resistance that is schematically assigned as an ohmic series resistance to the capacitive resistance (Figure 5.22). The diffusive resistance of a single porous electrode can be obtained by evaluating the high frequency intercept of the impedance spectra in the Nyquist representation.

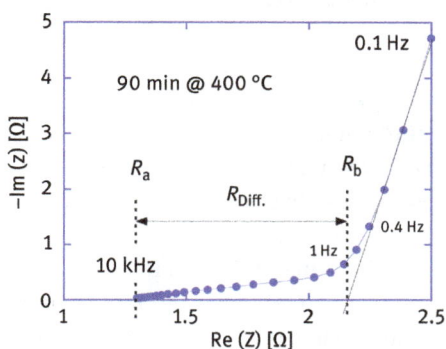

Figure 5.22: Determination of diffusive resistance of porous electrodes. The intercept of the extrapolated tangent to the impedance curve with the real axis toward high frequencies and the high frequency intercept of the actual curve determine the diffusive resistance R_{Diff}.

Figure 5.22 displays an impedance curve of a thermally activated GC sample in Nyquist representation. Measured data points are connected by a solid line. The tangent on this line extrapolating to high frequencies has an intercept with the real axis at position R_b. The actual impedance curve has an intercept at position R_a. The difference $R_b - R_a$ is the diffusive resistance of the electrode, a measure for the electric resistance due to ion transport limitation [Delnick 1993]. This resistance must be related to the electrode surface or even to the active film volume, so that a specific or volumetric diffusive resistance is obtained.

The Warburg impedance is an indication for a diffusion-based resistance of ionic conductivity, see for example [Kaiser 2000]. An example for SOFC is shown in the next section.

5.6.4 Flat band potential and Mott–Schottky plot

We have learnt in the previous chapter that the energy bands in electrodes (any solid) are subject to bending when they are brought in contact with a second phase. These second phases are typically the liquid electrolytes, but can be any fluid, which includes gases, the molecules of which are adsorbed on the surface of the electrode.

The electrochemical potential necessary to reverse this band bending is the flat band potential V_{fb}. The flat band potential can be determined with the Mott–Schottky plot, in which the surface capacitance values C_{ss} are plotted as $1/C_{ss}^2$ versus the potential V like shown in Figure 2.32. From the slope of the Mott–Schottky plot, we can determine the charge carrier concentration, the signum of the charge carriers and the flat band potential. The surface capacitance values are determined typically by deconvolution of impedance spectra in electric circuits.

It is quite interesting to see that the Mott–Schottky plot is known usually to electrochemists who work on semiconductor systems, such as photoelectrochemical cells [Chandra 1984, Chandra 1987, Salvador 1980]. Fuel cell engineers do typically not consider Mott–Schottky plots for the analysis of their materials. And, it is a tool not only used by electrochemists but also by semiconductor researchers [Nagaraj 2002].

5.6.5 Density of states

The impedance spectra may contain various capacities, such as capacities from surface states and from bulk states. You may find defect states or hole states on the surface of electrodes which you can "visualize" when you plot them like a Mott–Schottky plot versus the potential scale. It can be possible that these hole states are also visible by other methods such as with X-ray spectroscopy. A very illustrative example is shown in Bora (2013), Braun (2012b) and Braun (2017a) where two types of hole states found in the oxygen NEXAFS spectra pre-edge region are on the same energy position like capacitive states found and predicted by optical and electrochemical methods [Kennedy 1977]. It can therefore be possible to sketch with impedance data a density of states (DOS) which is comparable to the one known from condensed matter physics.

The DOS is one of the most important quantities of solid matter in condensed matter physics. Particularly, the DOS near the Fermi energy has influence on the electric, magnetic, optical and catalytic properties of the material. The DOS is therefore sometimes sketched as cartoons in some battery papers [Tarascon 2004]. The DOS can be conceptually sketched and also calculated with DFT, but you hardly find any actual empirically determined DOS. One of the very few examples can be found in the thesis and paper of Ensling et al. from TU Darmstadt [Ensling 2006, Ensling 2010].

5.7 Chronoamperometry

You may set on the potentiostat a particular voltage (V = constans) and then monitor the current which is passing through your electrode system versus time. It can help to monitor the electrochemical stability of a system. Figure 5.23 shows the current transient of an iron oxide photoelectrode in 1-M KOH electrolyte. At $t = 0$, the current density is 38 µA/cm^2. There is a rapid decrease in the first few seconds to below 5 µA/cm^2.

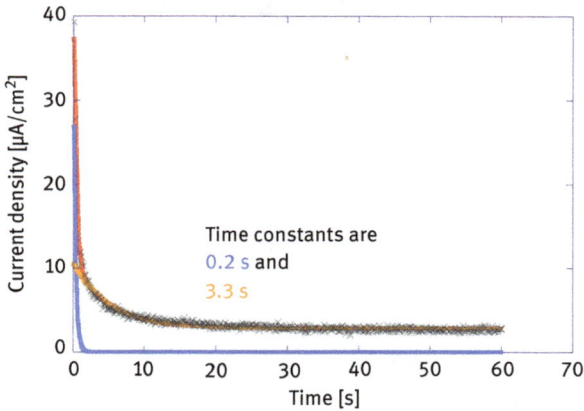

Figure 5.23: Dark current transient of a *photoelectrode* with deconvolution in two exponential discharge curves with two different time constants of 0.2 s and 3.3 s.

The current density is modeled with two exponentials which have different time constants τ_A and τ_B:

$$I_t = I_A \cdot \exp\frac{t}{\tau_A} + I_B \cdot \exp\frac{t}{\tau_B}$$

It turns out that the transient of the charge and discharge curve current can have also other contributions, such as a diffusion term with square root behavior over t, see for example the MSc thesis of G.-J. Moore and the paper by Klahr et al. [Klahr 2012, Moore 2017]:

$$I(t) = \frac{E}{R_{DL}} \cdot \exp\left(\frac{-t}{R_{DL}C_{DL}}\right) + \frac{E}{R_{SS}} \cdot \exp\left(\frac{-t}{R_{SS}C_{SS}}\right) + \frac{n \cdot Fc^0}{\sqrt{\pi}} \sqrt{\frac{D_0}{t}}$$

The values for the double layer capacitance C_{DL} and resistivity R_{DL} and the surface capacitance C_{SS} and resistivity R_{SS} can be obtained by impedance spectroscopy. E is the potential and D_0 is the diffusion constant.

You can integrate the current over t (line integral) and then get the evolution of the electric charge in your system.

$$Q_t = \int_{t_a}^{t_b} I_t \, dt$$

Figure 5.24 shows the discharging of the battery which I have mentioned already in the previous section. Note that here a constant current was imposed on the system, and the voltage of the system is changing during discharging. This plot is further explained in the chapter on batteries in this book.

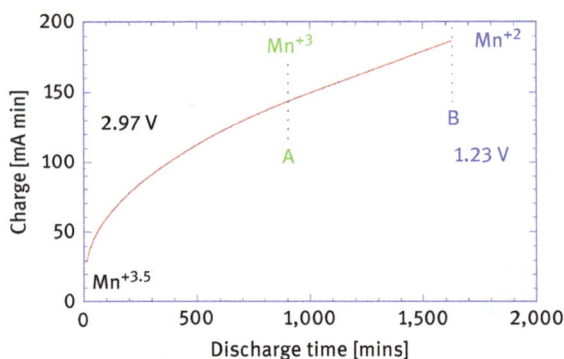

Figure 5.24: Evolution of electric charge during battery discharging. The voltage is changing during this process.

5.7.1 The open circuit potential

I said the first and most basic analytical method in electrochemistry was CV. This is not entirely true. An even prior method is the determination of the open circuit potential (OCV). When you check a battery for its voltage with a multimeter, you are basically checking its OCV. Open circuit means that you are not connecting the battery to a lamp or other sink of energy. Hence, no electric current is flowing.

Consider the Daniell element cell where we had a Zn metal slab and Cu metal slab as electrode. In the beginning, it may show some potential, but as you monitor it, you find that the potential is changing over time, either increasing or decreasing. This means that some electric charge is flowing in the system. Often, it results from chemical processes taken place on the electrode surface.

If you want to understand the behavior of your system, you may want to monitor under OCV conditions. For this, you maybe want to make sure that the temperature of the system is constant and also the darkness or illumination is constant or otherwise under control.

5.8 Boundary layers: A trivial explanation

When you sit calm and silent in a cool pool and pay attention to the water around, you may realize that when you slightly move your arm or a leg away from its rest position, you will notice a colder temperature in that new region of water where you moved your body part – a few millimeters only. Here is an explanation of what happens.

Your body has a temperature of around 36.8 °C. When you sit in the pool with a water temperature of say 28 °C, your body will heat up the water in the pool – at least to some extent ("My. Listen to that splashing. Must be doing the breaststroke. I hope the pool is heated." Dean Martin (referring to swimming Marilyn Monroe): "It's being heated right now" [Clements 2001, Specht 2001]). At the boundary of your skin where it meets the water, body and water pool tend to get into temperature equilibrium.

Your body dissipates heat energy into the water and a thin layer on your skin will get a slightly higher temperature as it is being heated up. This is why you will feel not so cool or cold anymore after same time in the pool. Now, when you slightly move your arm, you destroy this thin warm water layer around your arm and that warm water will mix with the colder water in the outer regions. Meanwhile, your arm will sense the lower temperature of the colder water say 1 cm away from where your arm rested before. The two temperature regions – body and pool – are not separated by a sharp border. Rather, the border extends over a finite region with a diffuse, continuous temperature gradient. In remote regions away from your body, the temperature may be still 28 °C. One centimeter from your skin the temperature might be 35 °C. Five centimeters away, it might be 29 °C. Layers with a gradient profile form also at electrodes. Here, the gradient may be formed over the ion concentration, for example. We then call it concentration gradient. And there is a film over the electrode which extends to the interior of the electrolyte.

5.9 Rotating disk electrode

When you stir the electrolyte in an electrochemical cell, you will immediately notice that the current is increasing abruptly. This is because the concentration gradient across the electrolyte–electrode interface will be destroyed by the agitation of the stirrer. Such concentration gradients form usually at electrodes, and they typically hinder charge transfer. Chemical engineers who design reactors therefore typically provide stirring mechanisms in the vessels in order to enhance the efficacy of the process. Stirring may also prevent reactor vessels from overheating and subsequent explosion [Sambeth 1983a, 1983b, Sambeth 2004].

One can use the stirring effect also for analytical purposes. This is realized with the rotating disk electrode. The electrode has the shape of a cylinder, the axis of which rotates with frequencies f of 100–10000 rounds per minute in the electrolyte, which causes a convective diffusion planar to the electrode surface. The electrode surface dips horizontal into the electrolyte. The rotation axis of the electrode is vertical. The diffusion layer extends along this vertical axis with a thickness $\delta_{(\omega)}$ and depends on the frequency $\omega = 2\pi f$, on the kinematic viscosity v of the electrolyte and diffusion constant D according to following relation:

$$\delta_\omega = 1.61 \cdot D^{1/3} \cdot v^{1/6} \cdot \omega^{-1/2}$$

The electric current between electrolyte and electrode is affected by this concentration gradient layer [Bard 2001, Newman 1966, 1967]. As you can change the rotation speed or frequency as an experimental parameter, it is possible to determine the limiting diffusion current i_d according to the Levich equation:

$$i_d = 0.62 \cdot \frac{v_e}{v_i} \cdot F \cdot D^{2/3} \cdot v^{-1/6} \cdot c_0 \cdot \sqrt{\omega}$$

v_e and v_i is the number of electrons transferred in and out in the half reaction, F is Faraday's constant and c_0 is the electrolyte concentration. A small portion of the charge running through the diffusion layer may be charged in the electrochemical double layer, except when the steady state condition is reached.

5.10 Probe beam deflection

As the concentration of ions near an electrode may change during electrochemical operation, the optical properties and thermal properties of the electrolyte may change accordingly. With a dual laser method, it is possible to optically excite an electrode–electrolyte interface during electrochemical cycling and read the deflection of a second laser beam which runs parallel to the electrode surface. This method, developed at the EETD of Berkeley National Laboratory, can determine the concentration gradient near an electrode [Russo 1987]. For a recent review, see Láng (2012).

5.11 The Nernst equation

The Nernst equation is a relation between the electrode potential and a redox couple:

$$E = E^0 + \frac{RT}{z_e F} \ln \frac{a_{Ox}}{a_{Red}}$$

Table 5.6: Explanation of the symbols used in the Nernst equation.

Symbol	Meaning	Value
E	Electrode potential	
E^0	Standard electrode potential	
R	Universal molar gas constant	$R = 8.31447$ J/mol/K $= 8.31447$ C/V/mol/K
T	Absolute temperature (Kelvin)	
z_e	Number of transferred electrons (also equivalent number)	
F	Faraday constant	$F = 96485.34$ C/mol $= 96485.34$ J/V/mol
a	Activity of the Redox partner	

The Nernst equation can be used for the determination of the electromotive force of a battery and fuel cell and gas sensor, for example. The meaning of the symbols is listed in Table 5.6.

We recall that every combination of two electrodes is an electrochemical (galvanic) cell. This includes batteries, accumulators, supercapacitors, but also biological cells.

The OCV or electromotive force U_0 is the potential difference ΔE of the electrodes, which can be calculated from the Nernst equation. In analogy, it allows the calculation of the equilibrium activities, when an external bias voltage is applied.

5.11.1 The concentration cell

We can demonstrate the function of the Nernst equation in a concentration cell. Consider the sketch of the Daniell element where we had the anode with zinc sulfate and the cathode with copper sulfate. Replace now the zinc sulfate in the anode container by copper sulfate but with a concentration different from the concentration in the cathode container. Here, we have two containers with two same electrodes and same electrolyte but with different concentrations.

When the electric current flows, the concentrations in the electrolytes become equal. The chemical reduction in one half cell with the higher copper concentration in the electrolyte works as follows:

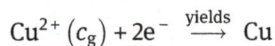

$$Cu^{2+}\left(c_g\right) + 2e^- \xrightarrow{\text{yields}} Cu$$

The chemical oxidation of the metal slab in the half cell with the lower concentration goes as follows:

$$Cu \xrightarrow{\text{yields}} Cu^{2+}\left(c_k\right) + 2e^-$$

The Nernst equation for this arrangement then yields with $z = 2$ for Cu:

$$\Delta E = \frac{RT}{z_e F} \ln \frac{c_g(\text{Cu}^{2+})}{c_k(\text{Cu}^{2+})}$$

By this method, we can utilize the electromotive force ΔE.

5.11.2 Nernst equation in biology

The cell membranes in biological cells separate ranges with different ion concentrations. When the membrane is permeable for a particular kind of ions, the ions will diffuse along the concentration gradient. As the ion is a charged particle, there will be an electric potential difference. This equilibrium can be described with the Nernst potential.

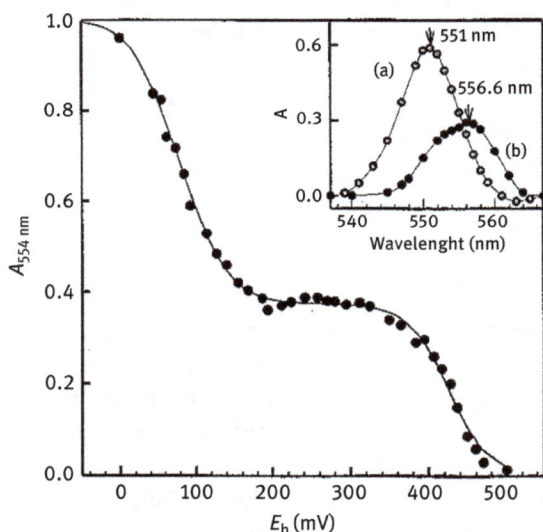

Figure 5.25: E_h dependence of the absorption measured in the R-band for LHI–RC complexes suspended in 25 mM Tris–HCl, pH 8, 1% Deriphat-160. Absorption measurements were performed at 554 nm, a wavelength located between the λ_{max} of the LP hemes (551 nm) and those of the HP (557 nm) hemes. The solid line was obtained with the theoretical redox titration curve using the following values for the redox parameters: E_m(HP1) = E_m(HP2)= 420 mV, E_m(LP1) = 110 mV, E_m(LP2) = 60 mV. The dotted line was obtained when assuming a sum of four n = 1 Nernst curves with the same redox parameters. (Inset) Difference absorption spectra recorded in the R-band at different redox potentials: (a) +45 to +240 mV (LP hemes) and (b) +385 to +470 mV (HP hemes).
Reprinted from *Biochemistry*, 36/40, Menin L, Schoepp B, Garcia D, Parot P, Vermeglio A, Characterization of the reaction center bound tetraheme cytochrome of Rhodocyclus tenuis, 12175–12182, Copyright (1997), with permission from the American Chemical Society [Menin 1997].

The potential difference U_m across a biological membrane with potential outside U_o and potential inside U_i is given by the Goldman–Hodgkin–Katz equation [Bowman 1984]:

$$U_m = U_i - U_o = \frac{R \cdot T}{z \cdot F} \cdot \ln \frac{P_{Na}[Na^+]_o + P_K[K^+]_o + P_{Cl}[Cl^-]_i}{P_{Na}[Na^+]_i + P_K[K^+]_i + P_{Cl}[Cl^-]_o}$$

which is reminiscent of the Nernst equation. The activities of the constituent ions P_{Na} etc. are summed up and taken in the argument of the logarithm like in the Nernst equation [Ohki 1984]. A simple sketch of an experimental setup is shown in Thiel [Thiel 1995]. A generalization of this equation with an analytical expression was derived by Pickard [Pickard 1976].

The conventional Nernst equation is used for the titration of systems known in photosynthesis, such as reaction centers bound to cytochrome with light harvesting complex LHI–RC, as shown in Figure 5.25. A similar Nernst curve was fitted on titration data on a thylakoid membrane [Nedbal 1992].

5.11.3 Lambda sensor

Now that we know we can determine the EMF that results from a concentration gradient, we may reverse the application and use an electric measurement in order to determine a gas concentration. This principle is for example used in high temperature oxygen sensors. This works by a zirconium oxide ceramic electrolyte, two platinum electrodes painted on it and the exposure to an oxygen atmosphere. This method is used for the determination of the oxygen concentration in exhaust in gasoline engines. The oxygen concentration in the exhaust gas is compared with the oxygen concentration in the air. It yields a Nernst voltage, which can be compared with a calibration table.

5.12 Other supporting analysis methods

Electrochemical energy converters and storage devices are quite complex systems. I have pointed out in the previous chapter the link between transport properties of materials and components and their structure. The electroanalytical methods allow for directly measuring the electrical transport properties of materials and components. To some extent, we can make a conjecture with transport properties when we measure valence band and conduction band spectra with various spectroscopy methods.

The structure of materials and components is measured with other methods as listed below. And there are a number of other methods which measure other materials properties. For a full assessment of performance, function and operation and degradation of conversion and storage devices, a wide range of methods may be necessary. My list of methods here in this book is only the result of personal preference and professional experience. The reader is advised to keep eyes open for other methods and approaches as well.

5.12.1 Gravimetric analyses

We learnt that use of the electrochemical equivalent in calculations does not necessary return the correct amount (mass) of materials which was subject of electrochemical conversion. It is therefore necessary to conduct experiments and tests to verify the actual converted amount.

Faraday's law provides a relation between the converted mass of some species and the involved electric charge. While you can measure the electric charge using an ampere meter and a stopwatch, you will have to measure the mass with a balance.

One example is shown in this book where proton conductor ceramics were measured in dry condition and in humid condition. The mass difference was below 1% of the sample weight but the results were plausible.

Let us consider again the Cu and Zn metal stripes and sticks in the Daniell element. We learnt that during operation of the Daniell element, Zn from the slab will be dissolved into the electrolyte, and Cu ions in the electrolyte will deposit on the Cu slab. We can therefore use a fine balance and weigh the mass of the slabs before and after operation and thus obtain the converted mass gravimetrically. Certainly, we have to record the current which flows through the cell using an ampere meter which is connected in series with the cell circuit. At the same time, we should monitor the cell voltage with a volt meter connected in parallel to the circuit. And we need a stopwatch in order to clock the time necessary for the integration of the flown current for the determination of the total charge Q.

Using the fine balance would be a traditional and conservative method for the gravimetric determination of the mass change. Nowadays, you would deposit a very thin copper film or zinc film on a small quartz crystal and use this as a mechanical resonator. The quartz crystal has an Eigen frequency f, which depends on its mass. Specifically,

$$f = \frac{1}{2 \cdot l} \sqrt{\frac{E}{\rho}}$$

with elasticity modulus E, mass density ρ and length l in the direction of the Quartz oscillation. When the thickness l of the Quartz increases, the frequency f decreases. This relation roughly holds also when the thickness of the Quartz increases when some film is coated on it. As frequencies of Quartz crystals can be determined with high accuracy and precision, it is possible to measure the mass gain during film deposition [Sauerbrey 1959].

5.12.2 Particle size and pore size analysis

Battery electrodes are often made from particulates. The particle size is therefore an interesting and important information [Braun 2015]. Similar holds for SOFC electrodes which are made by sintering of pellets from compacted particles (for analytical purposes) or from sintered slurries which were screen printed. Here, too, particles are the primary components. When the particle size is in the submillimeter to submicrometer range, light scattering is often used as diagnostic method.

Related with the particle size analysis is the pore size analysis. To some extent, pores can be considered objects like particles. In scattering theory, pores cannot be distinguished beforehand from particles. Comparable to light scattering, small angle X-ray scattering and neutron scattering can be used for the determination of particle sizes, particularly nanoparticles, pore volumes and internal surfaces and fractal dimensions of hyperspaces in materials.

We learn in this book that electrodes for energy conversion and storage are typically materials with a large porosity and large internal surface areas. The CV can to some extent be used for determination of the internal surface area of the porous electrode (via the double layer capacity). Gas adsorption measurements can be used for particle size and pore size determination [Brunauer 1938].

5.12.3 Structure analyses

The visual inspection of a specimen is probably the most important very first test you do after synthesis or processing or testing. The bare eye is already a very good scientific tool. A magnifying glass might be the next step if you want to look into details of your specimen. When you have access to an optical microscope or an electron microscopy, you can acquire visual images of the surface of the sample with better and better spatial resolution.

Structure is a very broad term and various researchers have various ideas about what structure means. This holds particular if the researcher is a specialist in one particular field who was never exposed to any other field. This is why an electron microscopist may limit the broad term structure to that what you see with electron microscopy. In contrast, a neutron scatterer may consider only those features "structure" which you observe with one or few particular methods in neutron scattering, such as the electronic structure or density of phonon states. The same might hold for someone who is an expert in photoemission spectroscopy and prefers to look at materials along the lines of density of electronic states.

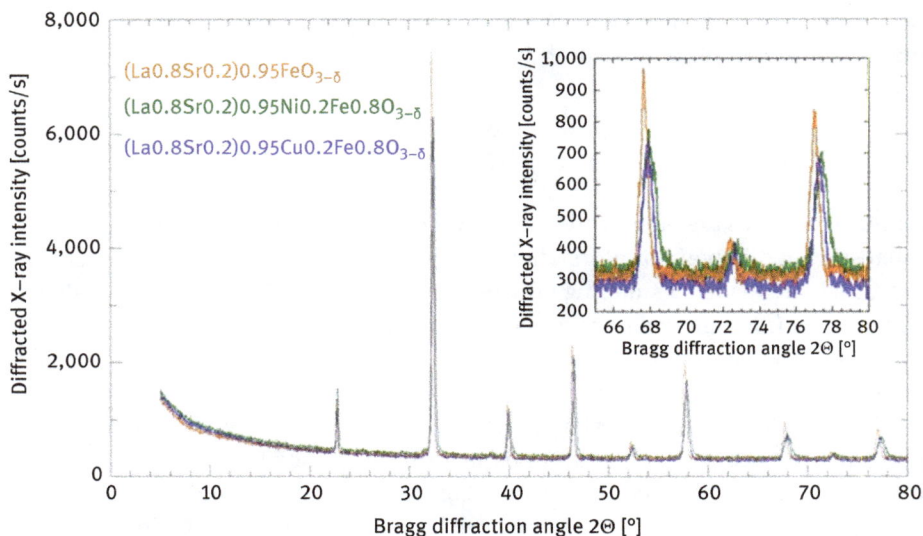

Figure 5.26: X-ray diffractograms of three powder samples for SOFC cathode studies with stoichiometry ($La_{0.8}Sr_{0.2})_{0.95}FeO_{3-\delta}$, $(La_{0.8}Sr_{0.2})_{0.95}Ni_{0.2}Fe_{0.8}O_{3-\delta}$ and $(La_{0.8}Sr_{0.2})_{0.95}Cu_{0.2}Fe_{0.8}O_{3-\delta}$. The sharp peaks are the Bragg reflections characteristic for lattice plane distances. The inset shows the profile and their differences in the Bragg angle range 65°–80°. The shift of the peak positions indicates a difference in the lattice spacing of the material. The X-ray wavelength is given by a Cu-K_α X-ray source.

A well-known structure determination method is X-ray diffraction (XRD). It takes advantage of the diffraction theories by von Laue and Bragg [Bragg 1915a, b, 1922, Ewald 1962] and allows for the determination of the crystallographic phase of crystalline materials and assessment of the amorphous structure at the atomic level. Many laboratories have an X-ray diffractometer and are thus able to carry out a crystallographic phase analysis on materials, see for example materials for SOFC cathodes in Figure 5.26. You can do the same experiment with neutrons. This is called neutron diffraction (ND).

With a microscope, you can investigate the surface topology and morphology of the samples. The resolution of an optical microscopy is good enough to see and count particles with 1 mm size. An electron microscope has a higher resolution and may show you finer features of a sample surface. You may find particles with a size of 50 nm. When you have a monolithic sample such as a ceramic pellet or solar cell electrode, you have to break it or cut it when you want to look into its interior.

This is a way to study the size and thickness of multilayers, for example. When you have multilayers with a very homogeneous arrangement of the layers, you may use X-ray reflectometry or neutron reflectometry. This method bears some similarities with XRD and ND. Combined with mathematical modeling, reflectometry allows us to measure how sharp or how diffuse the interfaces between layers are.

5.12.4 Spectroscopic analyses

We have assumed that the metal electrodes would dissolve in the electrolyte or ions in the electrolyte would re-solidify in the Daniell element. This is accompanied by the change of the valence, the oxidation state of the elements. When you apply an external potential to such cell, you can reverse the process and control with the sign and magnitude of the applied potential the deposition and dissolution of the electrodes.

When you design the electrochemical in a way that you can guide an X-ray beam through it so that it attenuates the electrode which you are interested in to study, you may collect the X-ray fluorescence signal of the electrode. The fluorescence spectrum contains chemical information of the electrode. When the molecular structure of the electrode is changed, for example by oxidizing or reducing it, the fine structure of the spectrum may change accordingly [Yee 1993]. Analysis of the spectra allows therefore to obtain chemical information of the electrode before and after electrochemical treatment [Braun 2017a].

Similar analyses can be done with electron spectroscopy for chemical analysis (ESCA) [Braun 2017a, Hüfner 1995, 2003, Kowalczyk 1976], optical spectroscopy, vibration spectroscopy, nuclear magnetic resonance (NMR) spectroscopy and so on.

5.12.5 Dilatometry

Devices which operate at high temperature or components which need to be processed at high temperature before they can be used in devices will undergo sometimes phase transition or simple thermal expansion. When the components are multilayers (heterostructures) from materials with different thermal expansion constants, it is possible that thermal stress is built up in the component which can cause delamination of the heterostructures. You may notice that by just looking at the structures with a microscope or even with the bare eye. But for the quantification of thermal expansion, typically a dilatometer is used.

A dilatometer holds a sample in a compartment which can be heated or cooled. The sample is supposed to be long in particularly one direction. So typically, the shape of the sample is like a slab. When the slab is heated, it will expand (unless in the unlikely case of contraction, which is also possible for some materials). The slab is held between two plates, one of which is part of a plate capacitor.

The expansion of the plate will change the distance of the plates of the plate capacitor and thus change its electric capacity C. The capacity of a plate capacitor with two plates of area A which are at a distance d is $C = \varepsilon_0 A/d$. Here, $\varepsilon_0 = 8.854 \times 10^{-12}$ C/V m. This capacitor is part of an electric circuit which has a frequency of $f = 1/RC$. As the capacity C changes, the frequency f changes. Frequencies can be measured with an extremely high accuracy. Therefore, the change of the frequency measured with a meter and caused by the change of the distance in the capacitor due to thermal expansion of the slab during temperature change is a measure for the change of the length of the slab.

Figure 5.27 shows the dilatometer curves of nine ceramic slabs with different stoichiometry. The samples are iron perovskites for SOFC studies, where the iron is partially substituted with copper and nickel. The length of the specimen was around 20 mm. The relative length change over the temperature range from 350 K to around 1200–1250 K is around 0.07 mm for four of the samples (P7, P8, Z6 and Z10) and 0.22–0.28 mm for the other samples.

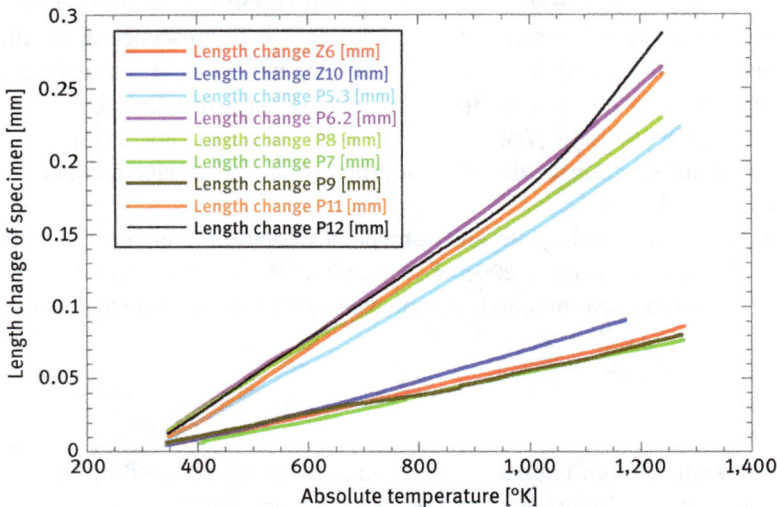

Figure 5.27: Length changes of a set of ceramic slabs (substituted iron perovskites) with different stoichiometry from 350 K to 1200 K.

Figure 5.28 shows the dilatometer curve (red line) of sample Z6 $(La_{0.8}Sr_{0.2})_{0.95}Ni_{0.2}Fe_{0.8}O_{3-\delta}$. The thermal expansion of the sample is over a wide range linear and can be fitted accordingly, as shown by the blue line. The length of the slab versus temperature is

$$Y(T) = L_0 + \alpha \cdot T$$

L_0 is the length at ambient temperature, and α is the thermal expansion coefficient. Williford et al. presented a method for the mathematical modeling of Dilatometry when excessive experimental would be too costly [Williford 2001]. Input data would be the stoichiometry of the materials.

As indicated in the beginning of this section, the method is not restricted to high temperature applications. There may be applications where low and very low temperatures are of interest [Liu 1997], though not necessary for energy applications.

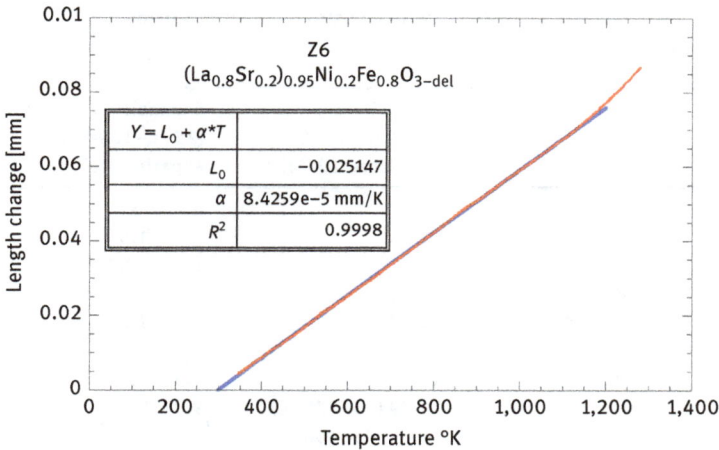

Figure 5.28: Linear thermal expansion (dilatometer curve) of 20% Ni-substituted $(La_{0.8}Sr_{0.2})_{0.95}FeO_{3-\delta}$ (red curve) with linear least square fit (blue curve). The thermal expansion coefficient for the linear range is 8.4259×10^{-5} mm/K.

Table 5.7 shows a list with dilatometer data from 12 different iron perovskites. The thermal expansion is given in one column as "calculated;" this is the thermal expansion coefficient α as obtained by linear regression. The "experimental" value α is a range of values obtained by visual inspection of the dilatometer curves.

Figure 5.29 shows the linear thermal expansion coefficient of a different but comparable set of iron perovskite samples. The visual presentation in a bar plot shows immediately how large the span is for the 12 samples. These samples were considered for SOFC cathode applications. It is necessary to find electrolyte samples which have a comparable or, say, compatible thermal expansion.

We can now try to bring some more systemic in the thermal behavior of the SOFC cathode samples by plotting the thermal expansion coefficients versus a meaningful parameter. In Figure 5.30, this is done with respect to the stoichiometry parameter y, specifically the relative lanthanum content in the material. The expansion coefficients α of three samples are compared.

Sample P15 has the lowest and sample P5 the highest value for α in this series. The expansion scales linear with the lanthanum content. This is an interesting finding. The strontium content is kept constant at 0.2 formula units. We do not know how this linear behavior changes when the lanthanum content becomes extreme, that is, when α becomes less than 0.5 or when it becomes close to unity.

The trend for α looks different when we compare the lanthanum content (stoichiometry parameter y) when it is kept in relation to the strontium content with stoichiometry parameter $(1 - y)$. It appears that the thermal expansion coefficient has maximum when the lanthanum content is 0.65, see the ordinate in Figure 5.31. It should be possible to show a mechanism which explains how the thermal expansion

Table 5.7: Original dilatometry data of substituted Lanthanum strontium ferrous oxide (LSF), a potential SOFC cathode material.

Composition	Name	Expansion α (25 °C–1000 °C) calculated 10^{-6} K^{-1}	Expansion α (25 °C–1000 °C) experimental 10^{-6}K^{-1}
$(La_{0.8}Sr_{0.2})_{0.8}FeO_{3-\delta}$	P 13 T	9.2	8.7–9.5
$(La_{0.8}Sr_{0.2})_{0.9}FeO_{3-\delta}$	P 7 T	11.55	7.5–11.5
$(La_{0.8}Sr_{0.2})_{0.95}FeO_{3-\delta}$	P 6.2 T	12.03	10.0–12.3
$(La_{0.8}Sr_{0.2})FeO_{3-\delta}$	P 5.3 T	11.05	7–11
$(La_{0.8}Sr_{0.2})_{1.05}FeO_{3-\delta}$	P 8 T	11.32	10.5–13.0
$(La_{0.8}Sr_{0.2})_{1.1}FeO_{3-\delta}$	P 9 T	9.91	9–14
$(La_{0.7}Sr_{0.2})FeO_{3-\delta}$	P 14 T	10.0	8.5–10.5
$(La_{0.7}Sr_{0.3})FeO_{3-\delta}$	P 11 T	11.99 (25–775 °C) 14.5 (775–1000 °C)	9.0–13.0
$(La_{0.6}Sr_{0.2})FeO_{3-\delta}$	P 15 T	12.1	8.0–13.7
$(La_{0.6}Sr_{0.4})FeO_{3-\delta}$	P 12 T	12.65 (25–750 °C) 18.63 (750–1000 °C)	11.5–5.0
$(La_{0.8}Sr_{0.2})_{0.95}Ni_{0.2}Fe_{0.8}O_{3-\delta}$	Z 6 T	10.47 (25–800°C) 13.35 (800–1000°C)	10.5–11
$(La_{0.8}Sr_{0.2})_{0.95}Cu_{0.2}Fe_{0.8}O_{3-\delta}$	Z 10 T	10.65 (25–900 °C)	9.8–11.8 (25–900 °C)
Sapphire reference	–		5.8–8.4 (6.8–8.6) Literature
YSZ 8 mol% Y_2O_3 electrolyte	Bannister [Bannister 1992]	10.5×10^{-6} cm/K	

Data were recorded by Dipl.-Ing. Maik Thünemann, Empa for the EU FP6 project Real-SOFC [Steinberger-Wilckens 2007]. The P materials are manipulated on the A-site (ABO3 perovskite structure); the Z materials are substituted with Ni and Cu on the B-site.

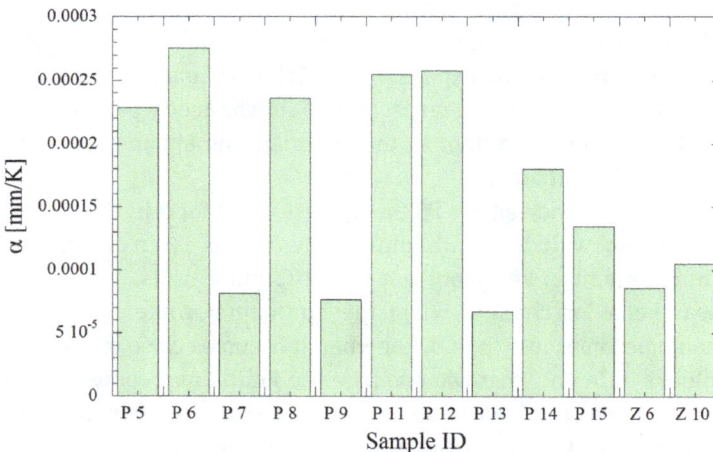

Figure 5.29: Linear thermal expansion coefficients for various SOFC cathode materials with base stoichiometry LSF, substituted with according to Table 5.7.

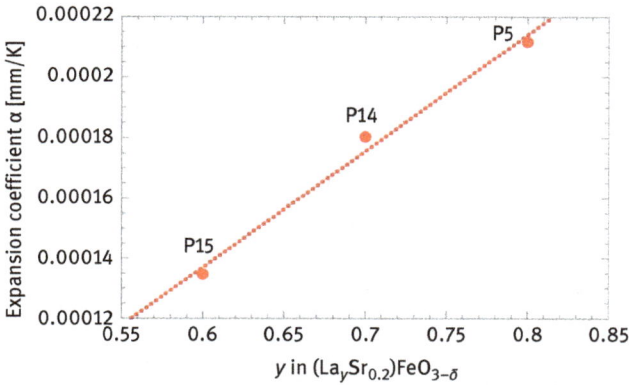

Figure 5.30: Thermal expansion coefficient as a function of the relative La content y in $(La_ySr_{0.2})$ $FeO_{3-\delta}$, with linear least square fit through three data points.

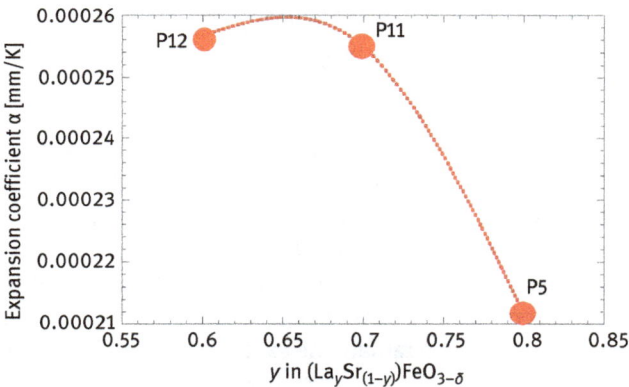

Figure 5.31: Thermal expansion coefficient of La content y in $(La_ySr_{(1-y)})FeO_{3-\delta}$ for three different samples. The dotted line is a cubic spline function, pointing to a maximum for the thermal expansion coefficient at around $y = 0.65$.

is linked with the stoichiometry. For this, however, structural analyses should be necessary.

The thermal expansion coefficient for Cu (LSFC) and Ni (LSFN)-substituted iron perovskites are shown in Figure 5.32, along with those from $(La_{0.8}Sr_{0.2})_zFeO_{3-\delta}$ (LSF). It is possibly a mere coincidence that the smoothing function accounting for the position of the data points from the data points for LSF coincides with the two data points for the B-site doped samples.

When a cathode is operated in a SOFC, it is typically in an oxygen-rich environment at high temperature. The oxygen milieu has an impact on the oxidation state of the ions in the cathode. The iron can have the oxidation states Fe^{3+} or Fe^{4+}, and this

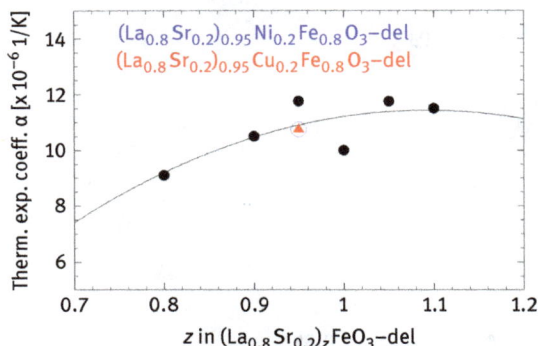

Figure 5.32: Thermal expansion coefficient of $(La_{0.8}Sr_{0.2})_zFeO_{3-\delta}$ (LSF) and 20% Ni substituted and 20% Cu substituted samples for SOFC cathode studies. The solid line is a smoothing function for the data points [Braun 2006].

can alter the spin states and via Pauli's principle the distances between ions. This can have influence on the thermal expansion, as was for example shown in a recent study [Braun 2012a].

5.12.6 Thermal conductivity

Materials for SOFC will be subject to very high temperatures – up to 1000 °C. Interest in the thermal properties of such materials is therefore justified.[8] The high temperature is necessary for the oxygen ions in the electrolyte to migrate and diffuse – to conduct. The heat energy comes from the auto combustion of the fuel in the SOFC with the oxygen from the air. It is necessary to radiate the excessive heat off from the SOFC components.

The thermal conductivity in Watt per Kelvin is obtained by laser flash analysis [González 2010]. Practically, you paint a thin layer of black color on the ceramic sample and direct a laser flash on the sample. A thermal radiation detector positioned behind the sample will record the signal and the time necessary for the heat pulse from the laser spot on the front side of the sample to pass through the sample to the detector.

Shown in Figure 5.33 is the variation of the thermal conductivity κ from ambient temperature to 900°C of pellets with six different compositions of 3D metal-doped lanthanum strontium iron oxide. I had made pellets of around 2 mm thickness and 1 cm diameter. The parent compound is sample P5 with composition $(La_{0.8}Sr_{0.2})_{1.00}FeO_3$. Its thermal conductivity decreases from 0.6 W/m K to 0.35 W/m K.

---

8 For example, the thermal expansion of components such as electrolyte and cathode and anode must be compatible. When the thermal expansion of the electrolyte is larger than that of the anode or cathode, the two latter components may break and peel off.

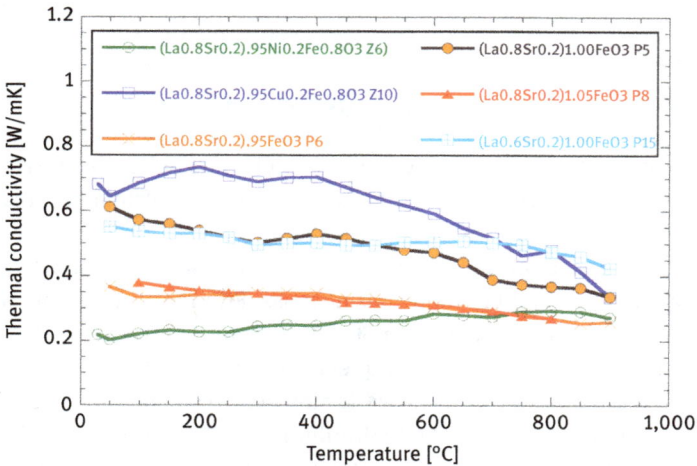

Figure 5.33: Thermal conductivity of $(La_{0.8}Sr_{0.2})_{1.00}FeO_3$ and five substituted derivates in the temperature range 25–900 °C.

Samples P6 and P8 are manipulated on the A-site to 95% and 105%. Their thermal conductivity is virtually identical and begins with 0.4 W/m K and then *decreases* virtually linear to 0.25 W/m K.

Substitution of the iron with 20% copper (Z10) makes a considerable enhancement for κ – close to 0.7 W/m K. The thermal conductivity for the Ni-substituted pellet (Z6) at room temperature is 0.2 W/m K and *increases* almost linear to 0.3 W/m K at 800 °C.

The thermal conductivity is related with the electric conductivity via the Wiedemann–Franz law [Wiedemann 1853]. In there, the electrons are the conducting species. It is interesting to learn that also the ions become current-conducting and thermal-conducting species at very high temperature (Wiedemann–Franz law for ions) [Scott 2000].

6 Fuel cells

6.1 A general remark about electric charge carriers

Ceramic proton conductors are materials which are of interest as solid electrolyte membranes in fuel cells. The electricity-conducting species is here the proton. At this point, it is necessary to talk about electric charge carriers in general. In electronics and electric engineering, the electron is the principal electric charge carriers. The electron has the charge unit -1 (e^-) and virtually has no mass. From the first chapters in this book, we know that the salts such as $CuSO_4$ dissociate in water into Cu^{2+} cations and SO_4^{2-} anions and thus form a liquid aqueous electrolyte.

These ions are positive (cation) and negative (anion) electric charge carriers, respectively. In acidic solutions, there may be hydronium ions H_3O^+ or protons H^+. Therefore, chemistry knows more kinds of charge carriers than electronics. Condensed matter theory knows not only electrons e^- but also the positive charged electron holes h^+. These holes exist particularly in semiconductors such as PV materials, but also in many metal oxides which are used as active electrodes in batteries, solid oxide fuel cells (SOFCs) and photoelectrochemical cells.

I should mention one particularly exotic charge carrier, that is, the polaron [Alexandrov 2010]. A polaron is considered a quasi-particle and constitutes for many materials the dominant charge transport process. Imagine when an ion or an electron, a proton or a hole is trapped in the crystal lattice so that it cannot move, except for a very short distance like on a rubber tape.

When you thermally excite the material, the crystal lattice will start vibrating because its thermodynamic degrees of freedom are excited. These lattice vibrations eventually will constitute phonons. The aforementioned ions, electrons etc. are coupled to these phonons and participate in the phonon dispersion. At some particular activation energy, the ion may be kicked out from its local lattice position to a nearby or even remote other lattice position. This constitutes also a charge transport.

6.2 Variety of fuel cells

It appears that the very first fuel cell was developed by William R. Grove in the mid-1850s. Grove published his work on a gas voltaic battery in the *Philosophical Transactions of the Royal Society*, which was recently republished in Grove [Grove 2009, 2012]; some of the historic papers are available for free download in https://archive.org/details/jstor-108377.

We learnt in the chapter on electrochemical cells that the dissociation of the electrolyte into cations and ions forms electric charge carriers. In contact with the electrodes, electrons will either be given to the electrolyte or taken from the electrolyte, which is known as electrode reactions.

https://doi.org/10.1515/9783110561838-006

Such electrode reactions are fundamental for the operation of the fuel cell, where a fuel such as hydrogen, natural gas, syngas (a mixture of hydrogen and carbon monoxide), methane, basically any combustible gas reacts with oxygen at an anode–electrolyte interface with production of an electromotive force according to the Nernst equation.

There are many ways to carry out such controlled electrochemical reaction. It has turned out that the variety of systems can be categorized in terms of temperature at which the fuel cell system can operate. Figure 6.1 provides a list of various fuel cell systems along with their operation temperature and advantages and disadvantages. Fuel cells are generally classified by operation temperature and type of electrolyte [Rowe 1997].

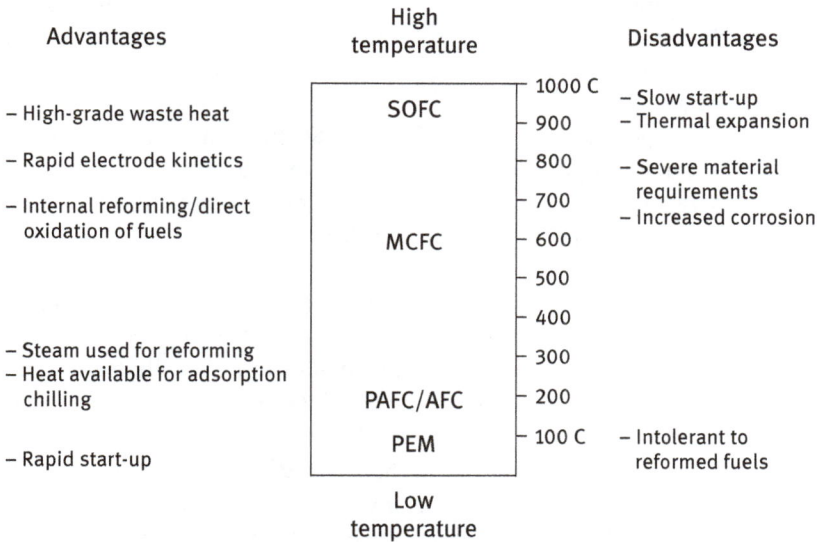

Figure 6.1: Types of fuel cells and advantages and disadvantages of higher temperatures [Perry 2002]. Reproduced with permission from the Electrochemical Society from Perry ML, Fuller TF: A Historical Perspective of Fuel Cell Technology in the 20th Century. Journal of the Electrochemical Society 2002, 149. doi:10.1149/1.1488651 [Perry 2002].

The proton exchange membrane fuel cell (PEM-FC) works at ambient temperature up to 100 °C and it can withstand cold temperatures as well. Preferred are operation conditions where the water is present in the vapor phase which facilitates the transport of water. The PEM-FC can start up immediately, which is very useful for fuel cell cars, for example.[1] But then, PEM-FC can

---

[1] When you push the start key in a fuel cell car it, will start immediately and you can drive. I do remember the times when the cars and heavy equipment that ran with diesel engines needed a couple of seconds heating time. The gasoline engines have spark plugs for ignition. The diesel engines had

only run on pure hydrogen. It is intolerant against hydrocarbon fuels and hydrogen with impurities. Particularly, carbon monoxide is a catalyst poison for this fuel cell.

The phosphoric acid fuel cell (PAFC) has concentrated phosphoric acid H_3PO_4 as electrolyte and operates at medium temperatures from 130 to 200 ° C. The fuel is hydrogen, but it needs not be highly pure hydrogen. But the acidity of the electrolyte and higher temperature are harsh for the fuel cell components.

The molten carbonate fuel cell (MCFC) uses sodium carbonate and potassium carbonate molten in a solid lithium aluminum oxide matrix. The operation temperature of 600 °C and above is necessary for the mobility of the oxygen ions as ionic charge carriers in the electrolyte. The fuel is obtained by reforming methane into hydrogen and carbon monoxide. The carbon monoxide and oxygen from air become CO_2, which is used in the electrolyte for the formation of carbonate.

At even higher temperatures, 650–1200 °C, the SOFCs can combust natural gas and produce a considerable fraction of fuel energy into electric power and heat. The high operation temperatures require ceramic components, the thermal expansions of which must be compatible. The reaction kinetics is quite high due to the high temperature.

As we have learnt about the PAFC and MCFC, which use a liquid electrolyte, we understand why the SOFC is called "solid oxide" – a term which would make beforehand not much sense to us because we feel oxides are typically solid. With liquid water (H_2O) being liquid and carbon dioxide (CO_2) being a gas, we understand that the term oxide is not only used for solids.

Sir Francis T. Bacon was the first researcher to develop a working fuel cell system which turned out as an energy utility system. Bacon, born 1904, a trained engineer and employed by a steam turbine manufacturer, worked as amateur from 1932 to 1941 on the development on a fuel cell which would consume and convert hydrogen and oxygen to electricity in an caustic electrolyte, catalyzed by low-cost nickel, rather than by costly platinum.

With an inherited fortune, Bacon could since 1941 form a critical mass of manpower and teamwork, which amounted to a benchmark fuel cell with 6 kW power in 1959. His fuel cell was eventually licensed by Pratt & Whitney, an aircraft engine company, and then used to power the Apollo space module (Figure 6.2), [Perry 2002] and later with improved technology also the Space Shuttle.

---

heating bulbs which heated the combustion space in the engine. When the space was hot enough, the diesel–air mixture would allow for self-ignition under compression.

Figure 6.2: Apollo command and service modules.
Reproduced with permission from the Electrochemical Society from Perry ML, Fuller TF: A Historical Perspective of Fuel Cell Technology in the 20th Century. *Journal of the Electrochemical Society* 2002, 149. doi:10.1149/1.1488651 [Perry 2002].

6.3 The proton exchange membrane fuel cell

There exist a large number of different fuel cell concepts, most of which people are not familiar with at all. The FCEV from Hyundai with which I started this book is running on a polymer electrolyte membrane fuel cell, and this is a well-known fuel cell type.

The PEM-FC runs on pure hydrogen gas as fuel. It enters the FC through the anode layer where electrons are driven to the electric circuit. The anode is coated with an electrocatalysts, such as platinum Pt. The catalyst lowers the kinetic barrier for the dissociation of the H_2 into protons. This results in positively charged protons, which will diffuse into the PEM. The PEM is a proton conductor. These protons will diffuse further to the cathode side, where they will meet with O^{2-} ions from the oxygen in the air.

The production of O^{2-} ions from oxygen gas requires energy. Therefore, the cathode is also coated with an electrocatalysts. There is currently much research going on for metal oxides as oxygen evolution reaction catalysts. The polymer electrolyte membrane is also called proton exchange membrane.

The schematic in Figure 6.3 shows how hydrogen gas (H_2) enters the anode and becomes oxidized to protons, which travel through the proton-conducting electrolyte membrane. The electrons from the oxidation of the hydrogen travel through the anode to

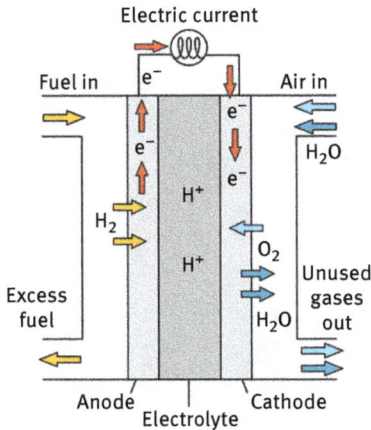

the current collectors. Meanwhile, oxygen from the air enters the cathode and diffuses toward the other side of the electrolyte membrane, where it reacts as O^{2-} with two protons H^+ to H_2O.

Now, we can apply the Nernst potential calculation. The electrochemical reaction at the anode reads a zero energy EMF because it is identical with the hydrogen standard reference potential:

$$H_2 \xrightarrow{\text{yields}} 2H^+ + 2e^-$$

The reaction at the cathode is the formation of water and yields the well-known ~1.23 V:

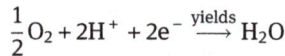

$$\frac{1}{2}O_2 + 2H^+ + 2e^- \xrightarrow{\text{yields}} H_2O$$

The reaction product from oxygen and hydrogen is water vapor. The water is therefore the exhaust, the "ash" of the combustion process. This is why this kind of fuel cell is considered a "green technology." The electromotive force of this reaction yields

$$E^0 = 1.2291 \text{ V}$$

The temperature dependency of this reaction is given by the invariant with a negative sign

$$\frac{dE^0}{dT} = -0.8456 \text{ mV}/\text{K}$$

which means that the electromotive force decreases with decreasing temperature. This is why the favored operation temperature for the PEM-FC is around 80 °C. One important engineering aspect is however that the PEM must contain some minimum humidity but there shall be no drop formation – because drops can block the transport of the water vapor. Blocked pores can build up pressure and this would enter the thermodynamics equations (for the underlying mathematics, see, e.g.,

[Winkler 2008]) which govern ultimately the electromotive force of the fuel cell. Humidity management is therefore critical.

The chemical processes which take place in the fuel cell are not always benign to the components of the cell. Today, it is possible to investigate the structure and the structural changes in fuel cell materials from synthesis through processing to the final fuel cell architecture, along with their molecular speciation. The availability of X-ray methods with a high spatial and spectroscopic resolution is shown to be of great service for this task [Braun 2017, Lee 2014].

6.4 Solid oxide fuel cell (SOFC)

6.4.1 An SOFC for residential home applications

Solid oxide fuel cells operate at very high temperatures from 500 to 1000 °C.[2] At such high temperature, it is difficult to operate with metals. They would easily oxidize at the surface and change their structure. Therefore, the anodes and cathodes are made from metal oxides. Even the electrolyte is made from metal oxides. We know already that the anodes and cathodes must be electronically conducting, and the electrolyte must be ion conducting.

Advantages of the SOFC principle: It can run on hydrocarbon fuels such as natural gas or biomass gas. It therefore needs no pure, expensive hydrogen.

Disadvantages: The high operation temperature makes that it is a heat-producing fuel cell. Only around half of the fuel energy can be used as electricity. The other half of the energy is emitted as heat.

Figure 6.4 shows the Galileo SOFC system from Swiss company HEXIS AG in Winterthur, Switzerland. The system is as tall as an adult person (62 × 58 × 164 cm) and can be plugged in a residential home in the natural gas (city gas) pipe. The electric power is sufficient for one home. The system is silent and does not smell.

Specifically, Galileo has a thermal efficiency of 95%, which means 95% of the energy in the natural gas can be exploited. The electric output is 1 kW_{el} and the thermal output is 1.8 kW_{therm}. The electric efficiency is therefore $1 \text{ kW}_{el}/(1 \text{ kW}_{el} + 1.8 \text{ kW}_{therm}) \cong 36\%$.

6.4.2 Design and architecture of the SOFC stack

The "heart" of a SOFC system, the SOFC "stack" from a much older version of a fuel cell, is shown in Figure 6.5. The stack is built from 60 SOFC electrode assemblies,

---

[2] Some expert readers may argue that the upper and lower temperatures are different, such as 650–1200°C. I believe 500–1000°C gives the right order of magnitude.

Figure 6.4: "Galileo" SOFC system from the HEXIS AG in Winterthur, Switzerland. The system has the size of a refrigerator. It can be plugged to the natural gas supply in a residential home and produces electricity sufficient to power a home. Around two-thirds of the consumed energy is heat which is used to heat the home and supply warm water. Around one-third of the energy is converted to electric power of 1 kW.

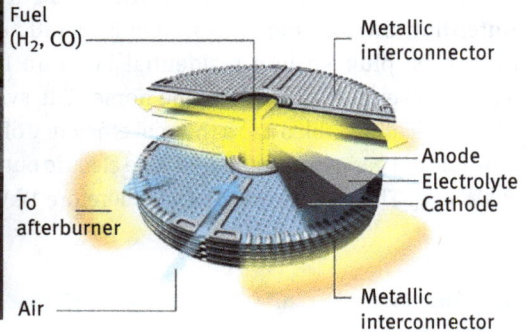

Figure 6.5: (Left) A fuel cell stack from a retired series of the Galileo system, ready for neutron tomography experiments at the Swiss Neutron Spallation Source. (Right) Schematic of an SOFC stack. The electrode assembly is a cathode from a perovskite-type metal oxide, a CGO or YSZ electrolyte layer and an anode cermet. The assemblies are connected with chromium steel plates (interconnects). Stack and Sketch provided by HEXIS AG.
Photo by Artur Braun.

which are stacked together to a thickness of approximately 30 cm. The diameter of the electrodes and stack is around 12 cm.

The SOFC electrode assembly is made as follows: The cathode is typically a LaSrMnO$_3$ metal oxide (LSM) which serves as an oxygen catalyst. The electrolyte is typically a "dense"[3] layer from yttrium-stabilized zircon oxide (YSZ) or from cerium gadolinium oxide (CGO). This electrolyte works on oxygen ion conduction, which sets on at around 600 °C. This is why an SOFC has to work at very high temperatures. At lower temperatures, the electrolyte would not be conducting ions. The selection of high performance materials is still today an ongoing quest in research; for a review on materials for SOFC, see [Mahato 2015].

The anode is a layer made from very porous CGO which is decorated with nickel metal particles. The nickel works as an electrocatalyst which lowers the reaction temperature of the fuel gas with the oxygen. Let us now look at the design, the architecture of the SOFC. The right panel in Figure 6.5 shows a sketch of a disassembled SOFC stack.

The three layers that form anode, electrolyte and cathode are actually a thin monolithic assembly manufactured from screen printing of ceramic slurries and then heating and sintering. This assembly is situated between two chromium steel interconnect plates. These plates have a channel structure so that the gases can diffuse. The chromium is necessary in order to have an interconnect which will not lose its integrity over high temperature over long operation times.

Close inspection shows that the electrode assembly has a large hole in its center. This hole extends throughout the entire stack and allows for feeding the fuel gas from inside the stack. The fuel gas reacts through the anode at the electrolyte with the oxygen ions and produces water vapor and CO_2 gas, which are emitted through channels. The air comes from outside to specific air channels. The management of flow of fluids, radiation and charge carriers can be very well demonstrated in fuel cells.

The three panels in Figure 6.6 show a neutron tomography image from the stack shown in Figure 6.5. You can easily count the 60 cell disks in the left panel in Figure 6.6. The neutron tomography[4] was carried out for no other reason than being able to

---

3 I frequently lecture in seminars on the controversial meaning of the term dense in the context of ceramic electrolytes. The electrolyte membrane needs not have a high mass density. In so far it does not have to be dense. What is required is that the electrolyte membrane is gas tight so that gases coming from the cathode and anode do not mix. And the electrolyte membrane shall not conduct electrons. When a material satisfies these requirements without being "heavy" (dense), then there is no problem with being not dense.

4 In summer 2017, I attended the Eberhard Lehmann Farewell Lectures at the Paul Scherrer Institute. Dr. Lehmann has been instrumental in building a new neutron tomography beamline at SINQ which has been very successful. Particularly noteworthy for this book is that Lehmann and colleagues pioneered the use of this neutron tomography and radiography method for electrochemical systems including fuel cells and batteries, see [Boillat 2017] Boillat P, Lehmann EH, Trtik P, Cochet M: Neutron imaging of fuel cells – Recent trends and future prospects. *Current Opinion in Electrochemistry* 2017, 5:3-10.doi: 10.1016/j.coelec.2017.07.012.

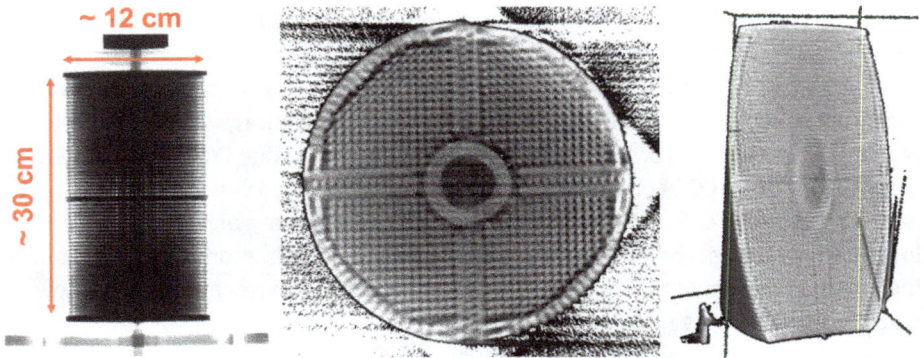

Figure 6.6: Neutron radiography images from a SOFC stack from HEXIS shown in transversal (left) and coronal (second left) and coronal (second right) view. Neutron tomography was carried out at NEUTRA beamline at SINQ, Swiss Spallation Neutron Source.
Data were collected by George Necola, PSI.

see whether we can look into the stack without opening and disassembling it. When the stack is being operated, it heats up from ambient to high temperature. This creates a thermal stress in the materials and components, and also some small irreversible changes in the components [Braun 2016].

When the stack is switched off and when it cools down, some parts like cells may brake because of strain and stress relief. But you do not see this unless you open the stack. But by opening the stack, you may also inadvertently cause some damages. In the end, you can never be sure whether the damages occurred during and after operation, or because you opened it for diagnostics. It was therefore interesting to make a diagnostics of the stack without opening it.

Neutrons have a very large penetration depth to many materials and therefore can be used for three-dimensional imaging like shown here. It would be very difficult to achieve the same results with X-rays. Figure 6.6 shows neutron tomography images from the stack from transversal and coronal direction and from a tilt angle. The actual digital dataset is very large. The computer software allows you to move through the entire stack and inspect it carefully.

The design of this stack is based on stacking single electrode assemblies together. Some designs use a relatively thick electrolyte disk which supports the two electrodes. Some designs instead use a relatively thick anode disk which supports the electrolyte and cathode layers.[5] This is the design for small SOFC.

---

5 If you want to sound like an expert, you may be using the words electrolyte-supported SOFC and anode-supported SOFC.

Large SOFC installations such as power plants use an entirely different design. They use ceramic electrolyte tubes which are coated inside and outside with the anode and cathode layers.

6.4.3 Operation principle of the SOFC

SOFC is an electrochemical cell or even an electrochemical reactor which can convert fuel (H_2, NH_3, CO, CH_4) and oxygen into electric power via the Nernst voltage and Joule heat. We have seen the physical SOFC and its components. In Figure 6.7, you see a schematic which shows how cathode, electrolyte and anode are connected and how the oxygen and the fuel react electrochemically at the anode, which provides the Nernst voltage.

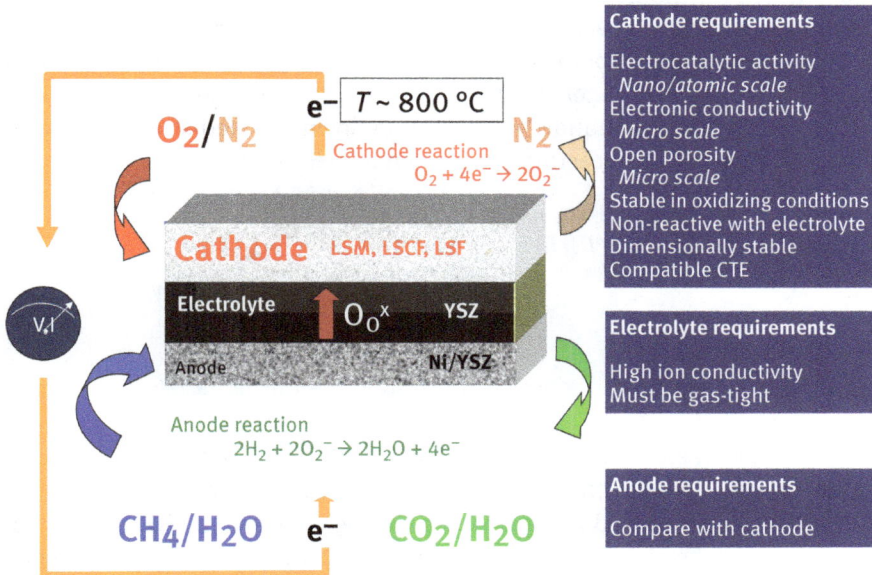

Cathode requirements

Electrocatalytic activity
Nano/atomic scale
Electronic conductivity
Micro scale
Open porosity
Micro scale
Stable in oxidizing conditions
Non-reactive with electrolyte
Dimensionally stable
Compatible CTE

Electrolyte requirements

High ion conductivity
Must be gas-tight

Anode requirements

Compare with cathode

Cathode reaction $O_2 + 4e^- \rightarrow 2O_2^-$

Anode reaction $2H_2 + 2O_2^- \rightarrow 2H_2O + 4e^-$

O_2/N_2 $\quad$ e^- $\quad$ $T \sim 800\ °C$ $\quad$ N_2

Cathode $\quad$ LSM, LSCF, LSF

Electrolyte $\quad$ O_O^x $\quad$ YSZ

Anode $\quad$ Ni/YSZ

CH_4/H_2O $\quad$ e^- $\quad$ CO_2/H_2O

Figure 6.7: Chemical and electrical operation principle of the solid oxide fuel cell.

Let's keep it simple and use hydrogen H_2 as fuel and oxygen from the air as oxidant. We begin with the cathode reaction. The cathode is exposed to the ambient air, which is composed of around 20% oxygen O_2 and 80% nitrogen N_2. The cathode is a metal oxide with perovskite structure, such as LaSrMn-oxide. This material has catalytic properties. It helps decompose molecular oxygen O_2 into ionic oxygen O^{2-}:

$$O_2 + 4e^- \Rightarrow 2\,O^{2-}$$

These oxygen ions O^{2-} can then move through the hot electrolyte membrane because of its high ionic conductivity at high temperatures. It is the current understanding that the mobility of such oxygen ions requires the presence of oxygen vacancies in the crystal lattice of the electrolyte. The existence of vacancies on the oxygen site gives rise to the high ionic conductivity of YSZ. Oxygen is transported by hopping through its vacancy sites (vacancy diffusion mechanism). The concentration of oxygen vacancies is determined by the concentration of the dopant [MSE5320 2010]. The chemical reaction at the anode is as follows:

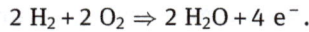

$$2\,H_2 + 2\,O_2 \Rightarrow 2\,H_2O + 4\,e^-.$$

The reaction product is 2 mol of water in the gas phase (vapor, steam) and 4 mol of electrons, which contribute to the Nernst voltage. The oxygen which is necessary for the anode reaction is delivered by the electrolyte (YSZ). The electrolyte in turn is receiving the oxygen via the porous cathode. The chemical reaction and conversion of the fuel take place at a region in the anode which is called "triple phase boundary." With this information, we can write down the Nernst equation with the "concentrations" of the reacting gases given in pressure p.

$$E = E^0 + \frac{RT}{2F}\ln\left(\frac{p^1_{H_2}p^{1/2}_{O_2}}{p^1_{H_2O}}\right)$$

6.4.4 Electronic structure and conductivity of SOFC cathode materials

The SOFC cathodes are typically made from LaSrMn-oxide which has an ABO_3 perovskite crystal structure [Jiang 2008, Maguire 2000] with rare earth metal A such as lanthanum, and a 3d metal B such as manganese. Under SOFC operation conditions, they are heated up and have an appreciably high electronic conductivity.

This is important for us to know: Metals are known for their high electronic conductivity. The conductivity of copper is 5.96×10^7 S/m at 20 °C. Manganese has at this temperature a conductivity of 2.07×10^6 S/cm; this is a factor of almost 30 smaller than for copper. The conductivity of MnO_2 ranges between 10^{-3} and 10^{-4} S/cm [Belanger 2008]. The oxidation of manganese to MnO_2 can lower its conductivity apparently by 10 orders of magnitude. $LaMnO_{3-\delta}$ may have 6.2 S/m, when we bring the Mn ions into a perovskite structure [Aruna 2000], but only when it is so sub-stoichiometric that it contains a considerable concentration of Mn^{4+}, and not only Mn^{3+}. These manganites have a high degree of spin polarization and can be considered as half-metals [Gunnarsson 2002].

Figure 6.8 (right panel) shows a sketch of a unit cell of $LaMnO_3$. Overlaid over the La^{3+} ions (large turquoise spheres) are Sr^{2+} ions (gray shells). The ionic radius of La^{3+} is 115 pm (10^{-12} m) and that of Sr^{2+} is 132 pm. The size of ions (ion radius) is a very important information for materials engineers which is listed in a paper by Shannon [Shannon 1976]. The Goldschmidt tolerance factor [Goldschmidt 1926] allows a prediction which ions fit together in a perovskite crystal lattice, for example. The sketch shows only the positions of the ions relative to the Cartesian coordinate axes and gives a rough symbolic indication of the size. However, the ions have atomic orbitals which do partially overlap and this can have great effects on the properties of the material [Kanamori 1959].

The manganese (Mn) ions are in blue color and the oxygen (O) ions are in pink color. The parent material $LaMnO^3$ has a low electric conductivity because of the poorly conducting Mn^{3+} ion. Substitution of La^{3+} by Sr^{2+} forces part of the Mn to become Mn^{4+}, which is the conducting ion in the LaSrMn-oxide. The conductivity of the cathodes is a problem of condensed matter physics and solid state chemistry. Not all fuel cell researchers are aware of this, and not all solid state scientists are aware of this. Some literature on the relevant topic of metal insulator transitions can be found here [Imada 1998, Park 1994, Toulemonde 1999, Zhou 2005].

The orbital overlap between O 2p orbitals and Mn 3d orbitals mediates charge transport throughout the crystal. The magnetic moments of the manganese and the spin direction of the oxygen influence the electron transfer along the Mn^{3+}–O–Mn^{4+} chains. There is a direct correlation between the electronic conductivity and ferromagnetic properties in $LaMnO_3$, for example [Zener 1951]. Charge transport by electron hopping along this chain is governed by the so-called Goodenough–Kanamori rules [Goodenough 2008].

The left panel in Figure 6.8, which you should read along with Table 6.1, is a schematic which shows how the 3d orbitals of Mn (long ellipses) overlap with the 2p orbitals of the oxygen (round). The electronic spin of the oxygen and the spins of the manganese ions determine whether electrons can hop from one manganese across the oxygen to the next manganese. Via the spin orientation, the magnetic properties affect the electronic transport properties of the material [Goodenough 1955].

I call the linkage of two cations with an anion typically the super exchange unit. Such units exist not only in metal oxides but also sulfides and so on. The "central" ion needs not be oxygen; it can be any other ligand molecule such as sulfur, phosphor and nitrogen. Such superexchange units exist also in the reaction centers in some metalloproteins such as the oxygen evolving complex in photosystem II (Mn bridged with oxygen) or in hydrogenases (iron and nickel bridged with sulfur and so on).

Substitution of ions by other elements is an important part of the tool box of the materials chemist. By doing so, you can for example force the oxidation states of one ion in a compound to increase of decrease. This can have consequences for other

Schematic electron-spin configurations	Mn – Mn separation	Transition temps.	Resistivity	Case
Ordered lattices				
Antiferromagnetic 4+ OR 3+ 4+ OR 3+	Smallest	$T_o > T_c$	High	1
Ferromagnetic 3+ OR 4+ 3+	Large	$T_o > T_c$	High	2
Paramagnetic 3+ 3+	Largest	$T_c \approx 0$	High	3
Disordered lattices				
Ferromagnetic 4+ 3+ 3+ 4+	Small	$T_c = T_o$	Low	4

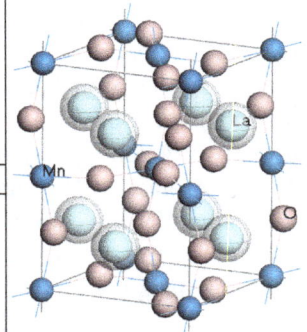

Figure 6.8: (Left) Schematic summary of orbital overlap of Mn 3d and O 2p orbitals and spin coupling. (Right) Crystallographic unit cell of SOFC cathode parent material $LaMnO_3$ with perovskite structure. The distance between Mn (dark blue) and O (pink) is moderated by the central atoms (cation) La^{3+} (light blue, small 115 pm) and their substitutes Sr^{2+} (light pink, large 132 pm).
Reprinted with permission from *Physical Review* and from J. B. Goodenough as follows: Goodenough JB, Theory of the Role of Covalence in the Perovskite-Type Manganites [La,M(II)]MnO$_3$, 100 (2), 564–573. Copyright (1955) by the American Physical Society" [Goodenough 1955].

Table 6.1: Semicovalent model for the magnetic coupling of manganese ions in the perovskite-type manganites [La,M(II)]MnO$_3$.

Low-energy, empty lattice orbitals

Ion	Outer electron configuration	Empty low energy orbitals	N
Mn^{4+}	d^2	Octahedral (d^2sp^3)	6
Mn^{3+}	d^4	Square (dsp^2)	4

N is the number of semicovalent bonds which can be formed by a manganese ion with its six neighboring oxygen ions. T_o and T are the transition temperatures for bond ordering and magnetic ordering, respectively. In the column for schematic electron-spin configurations, the cations, marked 4+ and 3+, have an empty orbital pointing toward the O2-p orbitals if they are joined by a dash.
Reprinted with permission from *Physical Review* and from J. B. Goodenough as follows: Goodenough JB, Theory of the Role of Covalence in the Perovskite-Type Manganites [La, M(II)] MnO$_3$, 100 (2), 564–573. Copyright (1955) by the American Physical Society [Goodenough 1955].

materials properties because these are typically linked. Figure 6.9 shows a conceptual phase diagram with regions for semicovalent exchange and double exchange as a function of the magnetization versus the relative amount of Mn^{4+} in lanthanum manganite substituted by a divalent ion.

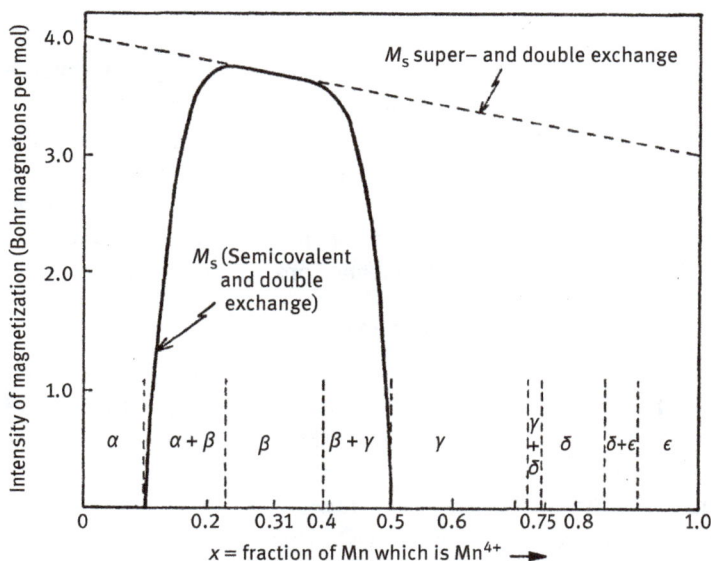

Figure 6.9: Predicted intensity of magnetization M and phase diagram for the system. $[La,M(II)]MnO_3$ according to covalent bond, semicovalent-exchange model.
Reprinted with permission from *Physical Review* and from J. B. Goodenough as follows: Goodenough JB, Theory of the Role of Covalence in the Perovskite-Type Manganites $[La, M(II)]MnO_3$, 100 (2), 564–573. Copyright (1955) by the American Physical Society" [Goodenough 1955].

The system is obviously rich in phases and it can depend delicate on the fraction of Mn^{4+} whether the compound is described by the superexchange model, double exchange model or semicovalent exchange model. The theory is not only important for SOFC anodes but also for lithium battery cathodes, for example. One important difference is however that SOFCs operate at very high temperatures which are typically considerably above the magnetic Curie points. It is an exercise for the advanced reader to speculate on the consequences of that for the operation of the SOFC. Is magnetism important at high temperatures? The reader is referred to the original and rich literature on this topic by Goodenough.

Hence, there are metal oxides or ceramics which have an appreciable conductivity necessary for use as electrodes in SOFC. This depends not only on the stoichiometry for the metal atoms on the A and B positions in the crystal. As I said already, it is also the oxygen content which plays a role. The synthesis and process parameters such as type of chemical precursors, gas concentrations and temperature in a furnace can play a decisive role for the later structure and function of the material.

6.4.5 The temperature-dependent conductivity of an SOFC cathode

A further way to increase the relative concentration of the Mn^{4+} ions is by replacing the La^{3+} ions on the A-site by Sr^{2+} ions. This stoichiometric trick forces some of the predominant Mn^{3+} ions to become Mn^{4+} ions. Like in any technology where components are being developed, materials engineers are constantly on the search for alternative materials in order to have better electrodes or electrolytes. One material is the iron perovskites where the B-site ion is Fe (iron) and not Mn (manganese). Researchers like to substitute and dope materials in order to enhance properties, such as

$$A'_xA''_{x-1}B'_yB''_{y-1}O_{3-\delta}$$

$LaFeO_3$ is an insulator. It is a yellow material (very large bandgap) with very poor conductivity. Knowing that oxygen is typically O^{2-} and lanthanum typically La^{3+}, we calculate that the oxidation state of iron must be Fe^{3+}. This is known to be a not well-conducting ion, such as like Fe_2O_3. When we substitute the La^{3+} partially with Sr^{2+}, we can force the iron to become oxidized from Fe^{3+} toward Fe^{4+}.

The electronic structure of iron Fe^0 is $3d^64s^2$. Oxidizing iron to Fe^{3+} takes away three electrons so that the electron configuration for Fe^{3+} is $3d^54s^0$. The 4s electrons are all gone and the five 3d electrons fill the shell half. We remember from chemistry class that the 3d shell can hold 10 electrons at large. Therefore, Fe^{3+} with $3d^5$ is a very stable configuration.

Removing one more electron to make Fe^{4+} with $3d^5$ turns out problematic. What actually happens is that the iron takes from the oxygen ion ligand an electron hole of the O 2p type which is written as L. The electron configuration of iron in $SrFeO_3$ reads then $3d^5L$. This material has a very high electronic conductivity almost like a metal. The charge carrier in this material is the electron hole L. Figure 6.10 shows the electric conductivity[6] of three different LaSrFe-oxide (LSF) samples as listed in Table 6.2.

Fuel cell engineers usually do not go that deep into solid state physics, but some of them acknowledge the importance of related fields in physics and chemistry for the use of materials in energy devices. The orbital overlap is often referred to as the exchange integral and can be computed with density functional theory methods. The same holds for materials in battery technology. Zuo and Vittoria, for

---

[6] Electric conductivity as it was measured by the 4-point method. This is likely the electronic conductivity. But no statement is made about a potential ionic contribution to the total electric conductivity.

Table 6.2: SOFC cathode model samples for 4-point DC conductivity measurements.

Sample no.	Stoichiometry	Microstructure
1/Blue	$(La_{0.8}Sr_{0.2})_{0.95}FeO_{3-\delta}$	Sintered powder slab
2/Black	$(La_{0.9}Sr_{0.1})_{1.00}FeO_{3-\delta}$	Sintered powder slab
3/Red	$(La_{0.8}Sr_{0.2})_{0.95}FeO_{3-\delta}$	Single crystal slab (you saw this one in the chapter on electroanalytical methods; it was a cylinder-shaped slab with four painted Pt rings)

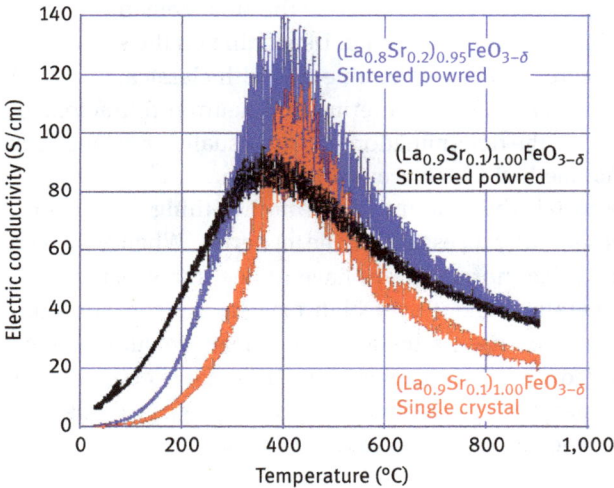

Figure 6.10: Variation of the electric conductivity of the LaSrFe-oxide model material for SOFC cathodes as a function of temperature and stoichiometry. The conductivity has a temperature maximum.

example, demonstrate the calculation of the exchange integrals for the manganese ferrites with spinel and inverse spinel structures. These structures are important in some battery cathodes (positive electrodes).

The exponential increase for low temperatures is due to the polaron conductivity activation process. The $1/T$ decay for high temperatures is an indication of metal-type conductivity.

The general characteristic of the conductivity is as follows. At ambient temperature, the conductivity is below 10 S/cm (specific conductivity). At around 400 °C, the conductivity has a maximum or around 100 S/cm. Before this maximum, the conductivity decreases not linear but exponential. This is an indication that the conductivity mechanism depends on polarons. After the maximum, the conductivity decreases with $1/T$ or $1/T^2$. The $1/T$ behavior is metal-type conductivity. The $1/T^2$

behavior would be an indication for a Fermi liquid. It is difficult to tell whether the exponent of decay α in $T^{-\alpha}$ is $\alpha = 1$ or 2, because the data points scatter so much.[7]

We notice another thing in the stoichiometry. In the end, I labeled the oxygen O with $O_{3-\delta}$, and not just O_3. Crystal structure analyses of metal oxides and many other compounds show that there exists always a concentration of oxygen vacancies in the lattice. This is because of fundamental thermodynamic conditions. Entropy would not allow normally that all lattice positions are perfectly occupied. Particularly at the surface of metal oxides, we find that there are more metal ions than the oxygen ions can compensate for. In the volume of the crystal is a similar situation, but not to that extent.

To account for the missing oxygen, we add the δ to the stoichiometry. It is not trivial to give an accurate estimate on the value for δ. Depending on the situation, it may be $\delta = 0.01–0.1$ or even more. You can determine this with classical analytical chemistry such as titration. Rutherford backscattering and neutron diffractometry can also be used. All methods for the determination of δ are usually very laborious. Therefore, you should use that method which is available to you.

One thing is to be able to tell the amount of δ. Another thing is to control the stoichiometry by synthesis and processing of the material. When you want to get a stoichiometric metal oxide, normally you have to heat the specimen in a furnace at an optimum temperature under very high oxygen gas concentration.

Sometimes, materials engineers need a metal oxide where the surface is not fully oxidized but partially reduced, for example for particular electrocatalytic properties [Hu 2016]. In that case, you may want to use a reducing gas atmosphere for synthesis and processing, such as H_2, Ar, N_2 or CO.

6.4.6 Chromium poisoning of cathodes

During SOFC operation, the cathode is exposed to air from the ambient environment. Therefore, we say that the cathode is operating under oxidizing conditions, because air contains 20% oxygen. Looking back at Figure 6.5 or 6.7, we see that the cathode layer is in direct contact with the steel interconnect plate. This plate is also the electric current collector to which the cathode gives the electrons. On the back side, the electrons are delivered to the anode.

---

7 When I worked on the high temperature properties of the iron perovskites 10 years ago, I was looking for a mathematical expression which would account for the exponential increase at the low temperature branch and the $1/T$ or $1/T^2$ high temperature branch. This is an exercise for the reader. Back then, I thought it would be a wild idea that the $1/T^2$ decay of the conductivity in a material like LSF could be a signature for a Fermi liquid. In Spring 2018, I became aware of a paper where a Fermi liquid was identified at very high temperature. My advice to the reader is when you get a novel idea, do not wait too long with publishing it.

The high operation of the SOFC makes that some of the chromium in the steel plates reacts with the oxygen in the air and maybe also with humidity in the air and forms a volatile chromium species which can enter the cathode.

The chromium ions can react with the LSM perovskite and then form a Cr-containing spinel phase which has a low electric conductivity. Over time, this accumulates to a real problem which is called chromium poisoning [Tsekouras 2014, 2015].

The poisoning of SOFC is typically studied by monitoring the electric signal, such as the Nernst potential (voltage) over the operation time. You record the conductivity of the fuel cell under "healthy conditions," and then you change the operation conditions to the worse by adding a malign component, a poison either to the fuel or to the air or to one of the components during synthesis and processing. And then, you compare the electric conductivity for the various conditions.

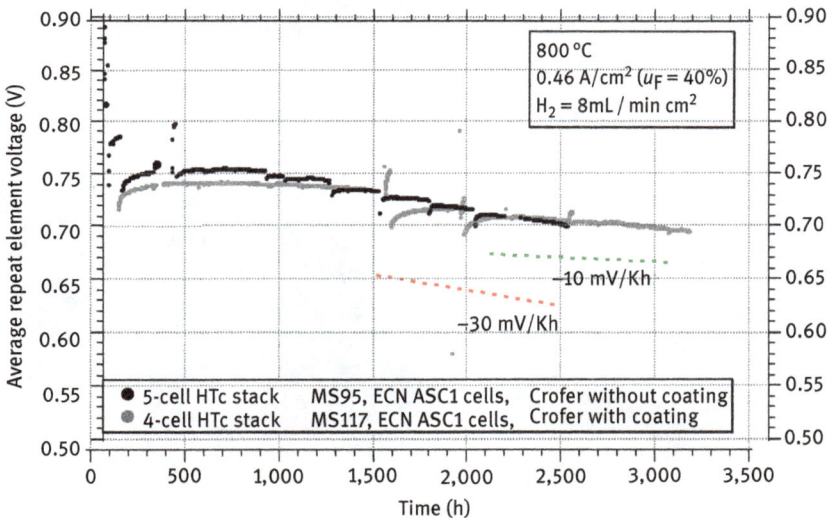

Figure 6.11: Durability tests run first- and second-generation HTceramix stacks (R-Design). The plot shows the average repeat element voltage, the only difference between the generations being the use of a protective coating on the metal interconnect air side. The dotted lines indicate the slope of the two curves [Steinberger-Wilckens 2007].

Figure 6.11 shows the transient of the voltage of an SOFC-averaged single repeating unit.[8] In the initial starting phase, the repeating unit shows a voltage of up to 0.9 V. Within 100 h of operation, it decreases to 0.75 V, which it maintains for almost 1,000 h. We then notice a gradual degradation of 0.05 V in the course of 1500 h, which is

---

[8] Note the different terminology in SOFC and battery technology. The SOFC community calls it "repeating unit," whereas the battery community calls it "cell" or "element."

almost 1 mV/day. It appears that a protective coating makes that the voltage is not further falling so much after 2500 hours of total operation.

Whether chromium evaporates from the metal interconnect and whether it interacts with any other component of the SOFC, such as the cathode, depends also on the temperature of the constituents. It is not so trivial to determine the temperature of an anode or cathode. We have therefore carried out calculations and simulations within in the EU FP7 project SOFC-Life [Steinberger-Wilckens 2011].

Figure 6.12 shows photos of the current collector face from a SOFC (left panel) and an anode face (right panel), the latter with traces from the anode. The current collector was disassembled from a SOFC after testing. Overlaid is the simulated temperature distribution in a SOFC under operation illustrated by a color code where red means hot and blue means cold. Variations in the temperature profile originate from the structuring of the metal interconnect which lays the path for the reactants and the combustions products. The letters A, B, C and D on the anode side (right panel) denote positions where materials samples were collected for synchrotron X-ray spectroscopy measurements.

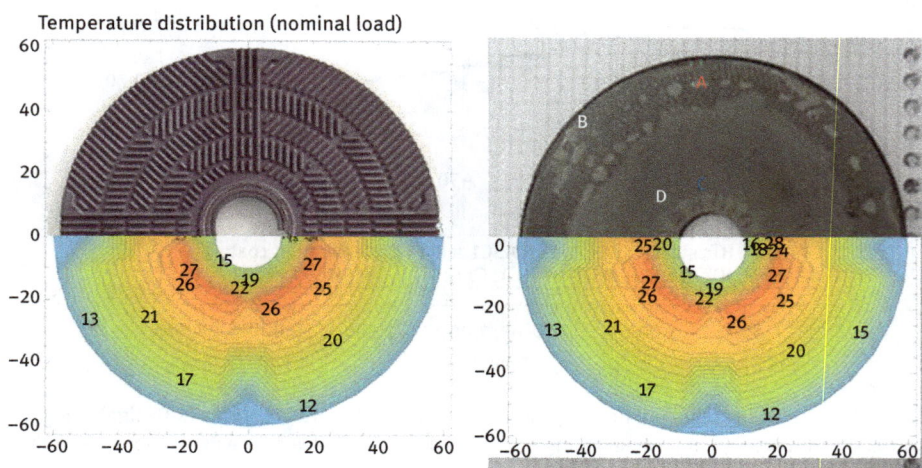

Figure 6.12: SOFC interconnect with traces from the anode material after disassembly. Overlaid is the calculated temperature distribution during operation. The calculation was performed by Markus Linder, ZHAW Winterthur.

6.4.7 Sulfur poisoning in SOFC anodes

One advantage of SOFC is that they can convert hydrocarbon fuels. Natural gas is a popular fuel. It contains sulfur impurities such as thiophene. Petrochemical industry has been researching long time for the desulfurization of fossil fuels including natural gas [Katsapov 2010].

Sulfur will react with the nickel catalyst layer in the SOFC anodes and form deleterious compounds, such as Ni_2S_3, for example. This is one form of anode degradation. We notice this first from the decrease of the Nernst voltage of the cells and stacks. When we disassemble the cells, we will also notice some structural degradation of the anodes. This effect is called sulfur poisoning.

We have investigated the interaction of the sulfur with the anode in a number of experiments in our Real-SOFC project [Steinberger-Wilckens 2007]. There had been speculations over which kind of compounds would be formed between the nickel in the anode and sulfur from the fuel under SOFC operation conditions. It was only when we carried out X-ray spectroscopy on several samples from our project partners that we learnt which kind of compounds formed, such as sulfate and even thiophene [Braun 2008a, Huggins 2008].

Figure 6.13 shows the voltage transient for a SOFC anode which was measured with X-ray spectroscopy at the same time when it was measured for its electrochemical properties. Such studies are called *operando* studies. The voltage is first 0.685 V. After 12 min, 0.25 ppm H_2S gas is introduced in the SOFC cell. Over a length of 10 min, the Nernst voltage is decreasing to 0.68 V. This is a small change but it is noticeable. After some time, H_2 is inserted as fuel. Then, the Nernst voltage is increasing again. In the next step, a higher concentration of H_2S is added, 2.5 ppm H_2S. Then, the sharp decrease of the voltage is more pronounced.

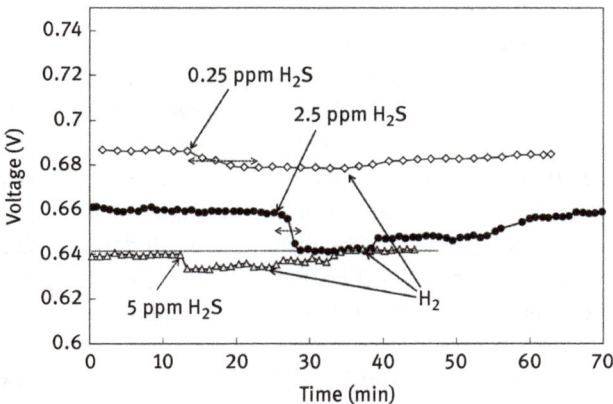

Figure 6.13: Evolution of voltage (Nernst voltage) of a model SOFC during exposure to H_2S [Nurk 2013]. Reprinted from Journal of Power Sources, 240, Nurk G, Huthwelker T, Braun A, Ludwig C, Lust E, Struis RPWJ, Redox dynamics of sulphur with Ni/GDC anode during SOFC operation at mid- and low-range temperatures: An operando S K-edge XANES study, 448–457, Copyright (2013), with permission from Elsevier.

Wood and biomass are valuable fuels which have been used for thousands of years in conventional combustion methods. You will later learn in this book that these fuels can be converted to generator gas (wood gas) and used in combustion engines. Our wood gas–SOFC project [Herle 2010] dealt with the tolerance of SOFC components toward potential harmful impurities which come in contact with SOFC components.

When you burn wood under ideal conditions, most of the matter will be converted to CO_2 and H_2O, and only the solid ash remains. Trace elements in the ash can be K, Na, P [Haga 2010], Cl, Mg, for example [Haga 2008]. It is possible that these trace elements can come with the generator gas into the fuel cell. We have built at Empa a rack which allows poisoning SOFC materials while their electric conductivity is being measured [Nurk 2011].

Figure 6.14 (left) shows the transient of the area serial resistance (ASR) of a SOFC anode when the vapor of an alkaline metal is guided over it. The resistance was determined with impedance spectroscopy, as is shown in the right panel in Figure 6.14. Note how the radius of the semicircle increases, and also note how its distance from the origin (0,0) in the Nyquist plot shifts toward larger real parts of the impedance. Upon adding the alkaline salt, the ASR increases by around 5%.

Why are we adding alkaline vapor? Well, there is an idea to run SOFC not only with fossil fuels but also with biomass gas from wood. Wood contains alkaline metals such as potassium. We therefore wanted to test the effect of such potassium on the SOFC operation and performance. What you need to do in the end is applying filters which take out the contaminants from the fuel before it enters the SOFC.

After the poisoning studies with electroanalytical methods, we carry out a so-called postmortem analysis. This term implies that the sample has died during the testing, which is not necessary the case. But the sample has been taken out of the reactor and is now being inspected for example with optical microscopy, electron microscopy, X-ray diffraction and so on. Such studies are necessary in order to understand the observations made with electroanalytical methods.

The left panel in Figure 6.15 shows an optical micrograph of the anode which was specifically prepared for sodium poisoning studies. The white or bright "stuff" is the 3YSZ electrolyte substrate (zirconia stabilized with 3% yttrium) and salt deposits. On top of the electrolyte is the anode layer from nickel-coated CGO. It appears that the operation of the SOFC has caused some delamination of the anode from the electrolyte.

The right panel in Figure 6.15 shows two energy dispersive X-ray analysis spectra recorded from different locations on the sample shown in the left panel in Figure 6.15. We can identify the signal from potassium (K) and this is indication that there could be interaction between K and electrode which caused the changes in the conductivity as seen in Figure 6.14.

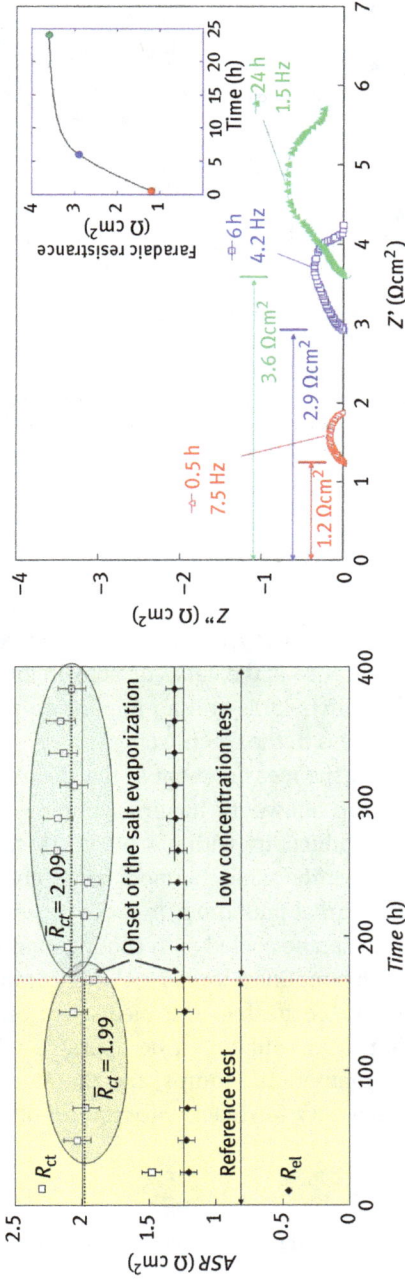

Figure 6.14: (Left) Transient for the area serial resistance of a SOFC anode when at around 180 h, some alkaline metal vapor is conducted over the anode. (Right) Three impedance spectra as Nyquist plots recorded from the anode poisoning study in [Nurk 2011]. Spectra were recorded at 0.5, 6 and 24 h. The distance of the high frequency intercept from the origin is the Faradaic resistance, plotted in the inset. Reprinted from Journal of Power Sources, 196 / 6, Nurk G, Holtappels P, Figi R, Wochele J, Wellinger M, Braun A, Graule T, A versatile salt evaporation reactor system for SOFC operando studies on anode contamination and degradation with impedance spectroscopy, 3134–3140, Copyright (2011), with permission from Elsevier, doi: 10.1016/j.jpowsour.2010.11.023.

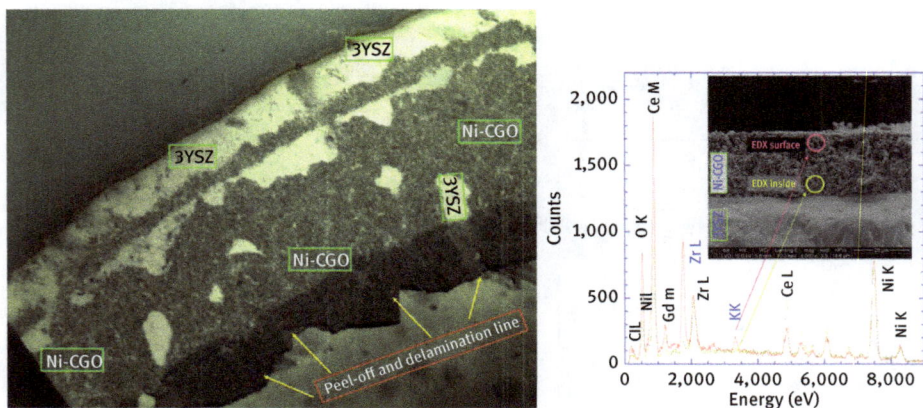

Figure 6.15: (Left) Optical micrograph of a SOFC electrode surface after operation under potassium vapor exposure. The white 3YSZ is the electrolyte layer. The Ni-CGO is the dark anode layer. White salt deposits are visible on the surface. (Right) EDAX spectrum of cross-section from broken electrode assembly shows which chemical elements are on the SOFC electrode assembly. Reprinted from Journal of Power Sources, 196 / 6, Nurk G, Holtappels P, Figi R, Wochele J, Wellinger M, Braun A, Graule T, A versatile salt evaporation reactor system for SOFC operando studies on anode contamination and degradation with impedance spectroscopy, 3134-3140, Copyright (2011), with permission from Elsevier, doi: 10.1016/j.jpowsour.2010.11.023.

6.4.8 Cathode conductivity experiments

The cathodes for SOFC are made from metal oxides with perovskite structure as already mentioned in the previous section. Their main purpose is the catalytic conversion of oxygen gas from the air. But their electronic conductivity is still an important issue because we do not want to have too much ohmic loss in the electric circuit.

Figure 6.16 shows a custom-built test stand for the measurement of the electronic conductivity of ceramic bars and slabs.[9] The setup allows for heating samples in a controlled gas atmosphere and recording their conductivity with a 4-point measurement. Four samples can be measured at a time under the same thermodynamic conditions (p, T) with controlled temperature T and gas partial pressure p. In order to protect the cell infrastructure from overheating, the setup can be cooled with cooling liquid.

On the left, there is an electric furnace which contains a large and wide ceramic open cylinder which can receive the samples to be heated and measured. Such furnaces come with a computer controlled hardware which can be operated with ease from a computer. It is possible to program temperature ramps, and thermocouples in the furnace make sure that there is necessary feedback information of the temperature in the ceramic tube.

---

9 Note that this is not a fuel cell. It is an experimental setup which allows for doing conductivity studies on ceramic monoliths.

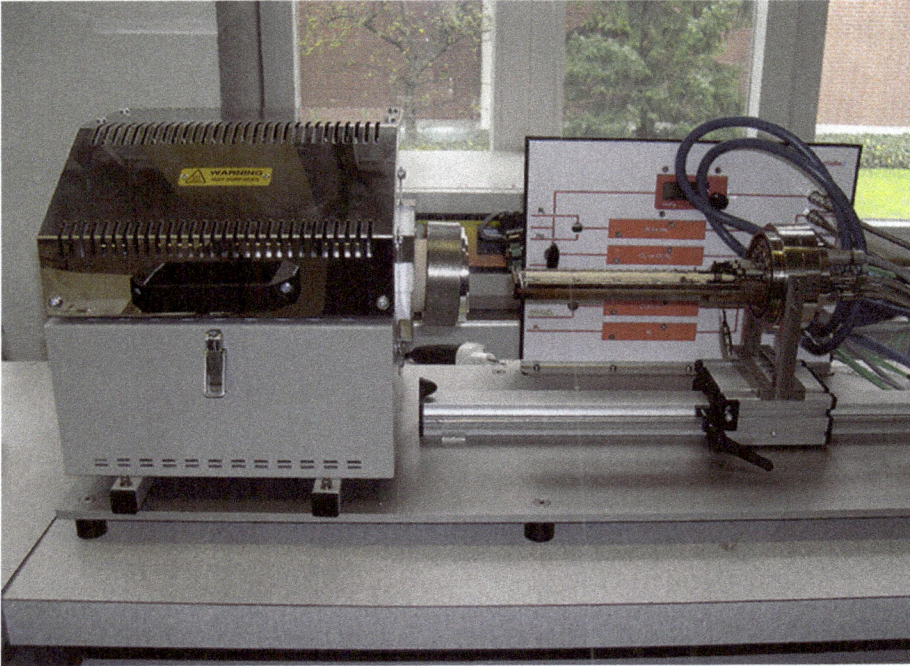

Figure 6.16: Conductivity measurement system at high temperatures under controlled gas atmosphere. The two wide blue plastic hoses can supply water or cooling liquid into the metal flange on the right side. The cooling effect at the sample position on the remote left side is warranted because the stainless steel is a good heat conductor.
Designed by Dr. Josef Sfeir (Empa) and built by Empa Machine Shop.

Figure 6.17 shows the setup with the furnace opened. You see the large diameter ceramic tube which is closed in the left end. The tube can contain the gas atmosphere which is brought into it via the gas feedthroughs in the metal flange on the very right. It is possible to move this flange and the sample stage on a rail to the left direction into the large ceramic tube and then close it with a huge ring with threads and thus have a hermetically sealed volume in which the four samples rest for the measurements.

In the middle between large ceramic cylinder and metal flange, you see a flat metal plate from stainless steel which hosts the sample and other supporting infrastructure, later of which should be protected from excessive heat by the external cooling. The plate is connected with a gastight flange which can be moved on the rack from right to left and thus inserts the plate into the ceramic tube in the furnace and closes the entire system. The flange is sealed with a rubber gasket so as to have a gas tight-controlled atmosphere in the ceramic tube. The white plate with red labels in the background is a control panel with gas mixture valves which are computer controlled. Figure 6.18 shows this plate in magnification.

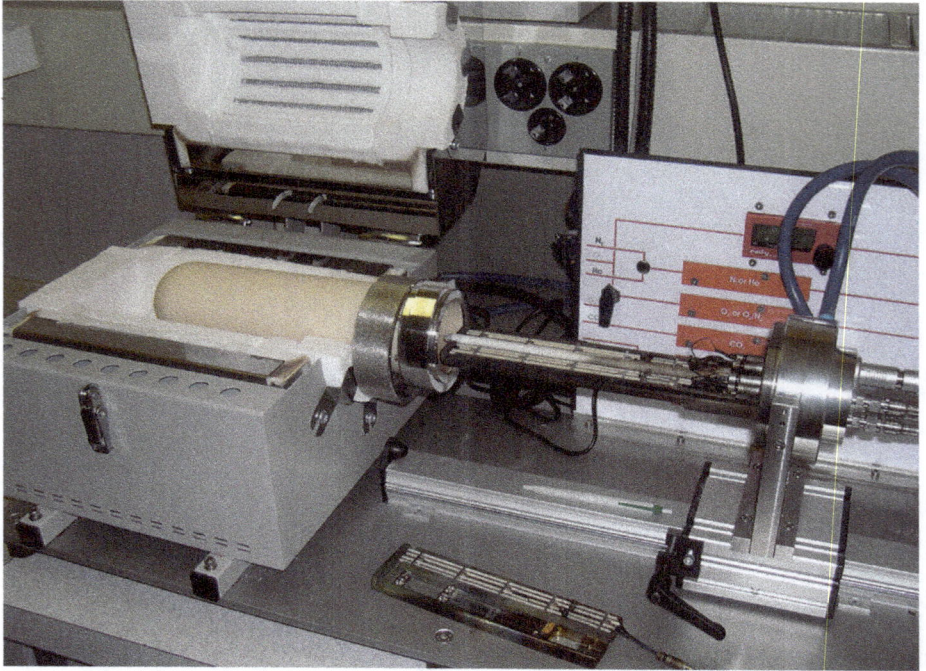

Figure 6.17: Conductivity measurement setup with the furnace top open. The furnace is insulted with thick white chamotte insulation. The yellowish closed cylinder in the furnace is a ZrO_2 tube to host the sample holder and the desired gas atmosphere. That tube has a glass-sealed metal ring with threads on it which will fit the flange shown on the right side of the photography. In the middle bottom is a separate stainless steel plate which can host two more samples.

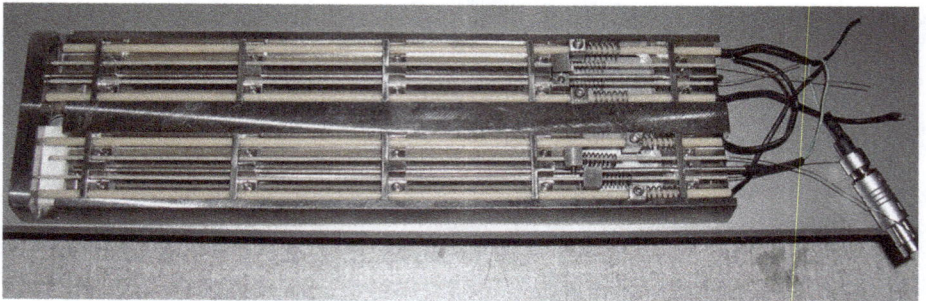

Figure 6.18: Stainless steel plate with two sets of spring-loaded ceramic tubes (2 × 4 tubes) which contain platinum wires for imposing electric current on two samples. The total number of cables and wires for this plate is 12: 2 × 2 wires for current and voltage for one sample; this makes eight wires for two samples. There are two wires for every thermocouple per sample – this makes four additional wires.

The stainless steel support plate has two terminals along the long axis which can host two ceramic plates of 2.5 cm length and 4 mm width each (see left side). Sample thicknesses can vary from the submillimeter range to 3 mm. Four metal springs located at the other side of the plate (right side) will press four very thin and long ceramic tubes onto the sample surface. As the ceramic sample will then be pressed on the end plate of the steel plate, it is necessary to put a thin insulating ceramic plate such as from ZrO_2 between this steel end plate and the ceramic sample.

The thin ceramic tubes contain platinum wires, the ends of which are in contact with the sample in a defined distance. The two outer ceramic tubes contain the platinum wires which impose the current on the sample in a well-defined distance. The two inner ceramic tubes contain the platinum wires which read the voltage over a well-defined distance. This allows applying the 4-point method on the samples. The other ends of the platinum wires are connected on the right side with plugs which can lead to a power supply and a meter, or to a potentiostat.

The somewhat thicker ceramic tube along the main axis of the steel plate hosts a platinum sponge at the left outermost end of the steel plate. This is an oxygen sensor which helps via the Nernst equation to determine the oxygen partial pressure in the vicinity of the samples as it is exactly located in between the samples.

At the high oxygen concentration side, the oxygen gas will become reduced at the platinum sponge:

$$O_2 + 4e^- \rightarrow 2O_{2-}$$

At the low oxygen concentration side, the oxygen ions will be oxidized:

$$2O_{2-} \rightarrow O_2 + 4e^-$$

When a constant current source is connected to the gas sensor rod, the oxygen ions generated at the platinum sponge can enter the ZrO_2 tube, which is a solid electrolyte at high temperatures. This process liberates an amount of oxygen gas at the sensor anode which is according to Faraday's law proportional to the current:

$$N = \frac{I \cdot t}{z \cdot F}$$

N is the number of moles oxygen which are transported through the ZrO_2, I is the current in ampere imposed on the sensor, t is the time over which the N moles are transported, F is Faraday's constant as 96487 C/mol and z is the chemical valence of the oxygen which we take as (2–). Basically, this works like an oxygen pump because with the electric current, you can produce gas from ions.

As we have two different concentrations c_1, c_2 of gas (or two different gas partial pressures p_1, p_2) across an electrolyte barrier, we have a thermodynamic potential difference and Nernst equation will apply so that we will sense a Nernst voltage V depending on the temperature and the logarithm of the ratio of the gas partial pressures:

$$V = \frac{k_B \cdot T}{e_0} \cdot \ln \frac{p_1}{p_2}$$

The constants are Boltzmann's constant k_B ($1.38064852 \times 10^{-23}$ J/K) and the electron charge e_0 ($1.6021766208 \times 10^{-19}$ C).

Figure 6.18 shows a second stainless steel plate which can be put on top of the first stainless steel plate. This one holds the second couple of ceramic samples. Close inspection shows that each of the two half compartments of the plate has a slot for a thermocouple which points very close to the sample position on the left side. Close inspection of the sample position as shown in Figure 6.19 shows us how crowded it is inside the measurement cell. Every sample has four wires as is necessary for the 4-point method.[10] Every sample has also one thermocouple in

---

[10] In several figures in this book, you saw ceramic slabs which are painted with platinum paste or silver paste (paste, paint, ink; there may be definitions for each of these three terms but what counts is that you can paint them with an artists' brush on the sample surface). Some researchers use gold paste. The pastes or inks can be purchased from various vendors. The inks and pastes are made from colloidal silver, gold, platinum which are immersed in some solvent. Solvent maybe not the correct term because nothing of the metal is dissolved. Those of you who work or worked with electron microscopy may have come across the silver paint. You can improve the contact between some electrode tip of a terminal and the sample by painting the silver over the sample. The paint will penetrate the porous surface and make some metallized grip to the sample. When the solvent, typically some organic, evaporates, a thin film of well-conducting metal layer makes a good current collector for electric measurement. Some paints require that the sample is heated in a drying furnace. When you work in the high temperature fuel cell field, you may use platinum paste which needs to dry in a drying furnace after you have painted it. After that, you bring the painted sample into a furnace and heat it to 950°C. This temperature will burn away all residual organics from the solvent. Sometimes, the paint becomes too thick over time and some colleague may add some arbitrary solvent for thinning the paint. Some other colleague may mistake this paint as silver paint and use it without heating it to such high temperature and will find that the metallized layer is not electrically conducting. Therefore, metal paint is not always the metal paint that you think you hold in your hands. Another aspect is the potential catalytic action of the metal species that you use in the paint. Platinum is a well-known catalyst and can do to your samples some things which you do not want to happen. You only want to have a metal layer for improved electric contact for analytical purposes, and not necessarily a catalyst layer on your sample. Therefore, read the instructions of the paint that you purchased, and if you find some used paint in the laboratory, make sure you know what paint it is. It also helps reading the literature from various fields where such paints and inks are used. This all will take you extra time, but it will make you a better and more experienced researcher and scientist.

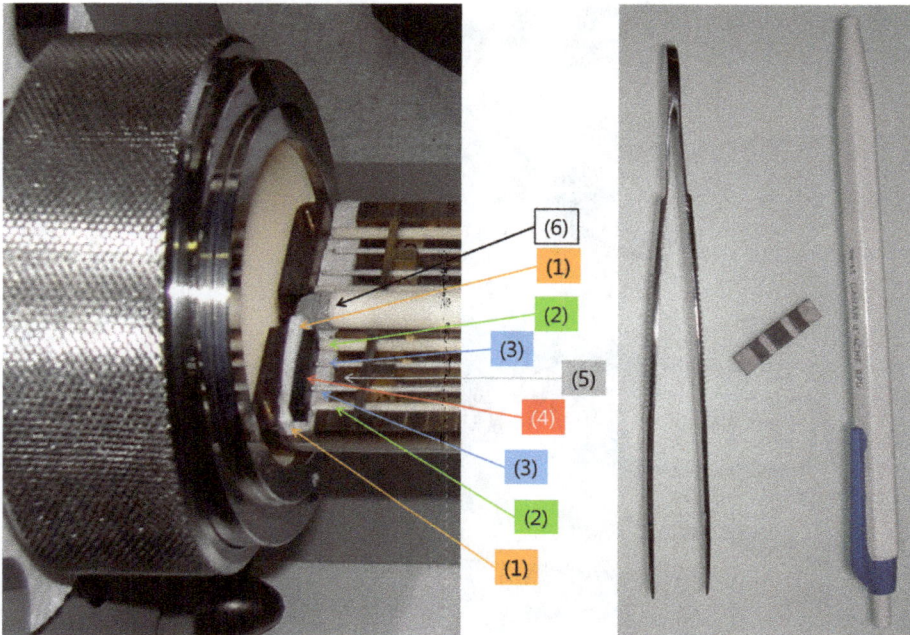

Figure 6.19: (Left) Situation of sample and sample holder near the furnace. (1) Electric insulation (white glass fiber fabric) behind and under sample. (2) Pt wire inside thick ZrO_2 tube for imposing external current on the sample. (3) Pt wire inside thin ZrO_2 tube for DC voltage reading. (4) Prismatic shape black SOFC cathode material sample. (5) Position for thermocouple. (6) Thick ZrO_2 gas sensor tube with Pt sponge. A second sample can be placed on the other half of the steel plate. (Right) Ceramic sample between a pen and a tweezer for size illustration. As this sample was not well conducting, its surface is painted with platinum paste for improved contacting. The four painted gray Pt compartments on the dark sample are in the same distance like the current and voltage Pt tips on the sample holder.

position. With four samples altogether, this makes 24 wires. The oxygen sensor has also two wires, which makes a total of 26 wires.

In order to handle the measurement, custom-built software was made at Empa based on LabView®. Figure 6.20 shows a typical data acquisition table with potentiostat and frequency response analyzer and desktop computer with data acquisition software.

Those who do the synthesis of materials in a chemistry laboratory may end up with a powder material, be it a perovskite or a spinel which appears suitable for further use in fuel cell or batter development. The synthetic chemist may now be finished and hand over the powder to the materials chemist or ceramic engineer, who is supposed to make electrodes from the powder.

For electric conductivity measurements, you typically press ceramic bars as the ones shown in Figure 6.21. For thermal conductivity measurements, you may

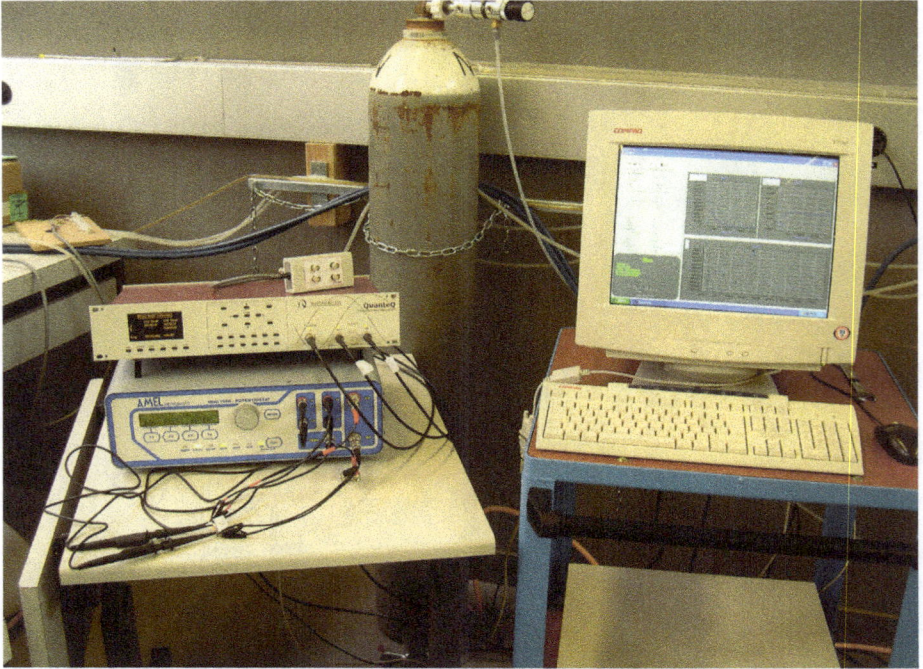

Figure 6.20: Typical data acquisition table with a potentiostat from AMEL, frequency response analyzer from QuanteQ, desktop computer with data acquisition software. The compressed oxygen bottle is connected with the flange at the measurement cell.

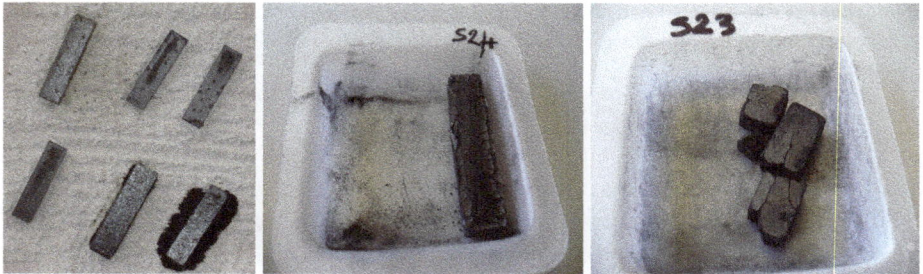

Figure 6.21: (Left) Ni-substituted LSF ceramic slabs with six different relative Ni concentrations; the mechanical integrity of the slabs is maintained. (Middle) A ceramic slab after sintering with very brittle surface, suggesting lack of mechanical integrity. (Right) A ceramic slab with a composition which did not survive the sintering process; it broke after recovery from the furnace.

have to press coin-type pellets. Then, you mix the powder with a small amount of binder material such as cellulose or some organic solvent which helps coagulate the metal oxide powder particles. Then, you fill the powder in a dye and compact it with a mechanical press. The pellet has not typically a mechanical integrity and you can load it in a furnace and heat it at some high temperature which is specific for the material. We learnt earlier in this chapter that the gas partial pressure under which the sample is exposed in the furnace is also an important process parameter.

When the processing is finished, you will find the pellet nicely sintered in the furnace. Then, you can pick it up with a tweezer and bring it to the 4-point DC measurement setup as shown in Figure 6.19. Researchers who spend more time on pellet making will eventually experience that some materials resist compaction and sintering. Figure 6.21 shows in the left panel a set of six Ni-substituted LSF samples. The pressed pellets were laid on an Al_2O_3 support and then sintered in air atmosphere in a muffle furnace. The two lower right bars have adversely interacted with the aluminum oxide underneath. It appears like some of the material in the bars has leaked out into the aluminum oxide. Likely, the LSF resisted the overdose of Ni.

In the middle panel, we see another ceramic slab in a plastic-weighing boat. This slab is labeled S24 and it could not be sintered the way as we had wished for. The surface is very brittle and this sample is likely very troubling for making 4-point DC measurements. The right panel shows another sample, labeled S23 in a plastic-weighing boat. This sample broke when we removed it from the furnace. It is not always possible to improve a material by endless (or pointless) doping. Materials often resist excessive substitution with foreign elements. It is therefore useful to consult the thermodynamic phase diagrams. They give a very good indication about which crystallographic phases can be considered possible or stable under particular conditions [West 1993].

Figure 6.22 shows the 4-point DC conductivity of $(La_{0.8}Sr_{0.2})_yFeO_{3-\delta}$ for various concentrations of the A-site occupancy y for temperatures from ambient to 900 °C. Overall, their specific conductivity ranges from 0 to 140 S/cm. The variation of conductivity with temperatures has also the same general shape: an exponential increase, then having a maximum conductivity and then decreasing again, but apparently not in an exponential manner. The sample with $y = 1$ has the ABO_3 stoichiometry and shall be considered the parent compound. It has maximum conductivity of 80 S/cm at 450 °C. From the stoichiometry formula $(La_{0.8}Sr_{0.2})_yFeO_{3-\delta}$, we can estimate that we have three oxygen atoms per formula unit with electric charge (−2). This makes an overall charge of (−6); 0.8 ions of La^{3+} make (2.4+) and 0.2 ions of Sr^{2+} make (0.4+). To have charge balance (0), we need the Fe ion to have the charge (3.2+). The valence of the Fe is therefore on average 3.2. Here is an exercise: How can an ion have an average valence of 3.2? We have to average the ionic charge by taking 20% Fe^{4+} and 80% Fe^{3+}: $0.2 \times 4 + 0.8 \times 3 = 0.8 + 2.4 = 3.2$. It is the Fe^{4+} ion which provides the conductivity in this material.

What do we *have* to do to increase the relative portion of Fe^{4+} in the sample? We must decrease the relative amount of the positive charge on the A-site ions which is La^{3+} and Sr^{2+}. To maintain charge balance with the (−6) from the three oxygen ions, the Fe has then to respond accordingly and obtain a relatively large oxidation

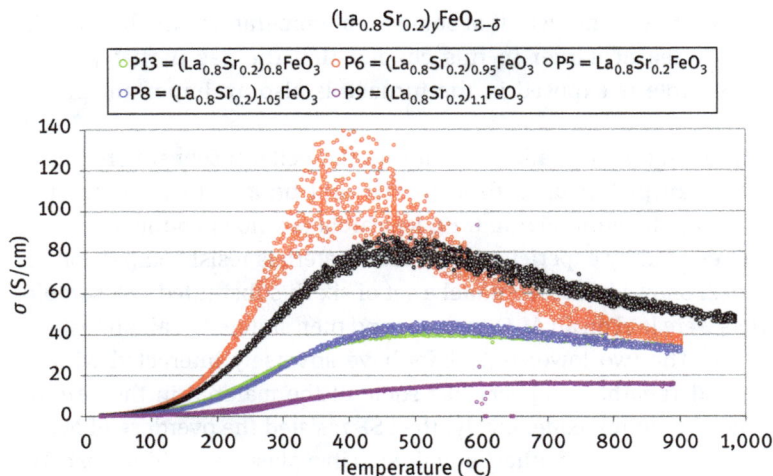

Figure 6.22: Electric conductivity of five samples of $(La_{0.8}Sr_{0.2})_yFeO_{3-\delta}$ with five different values for the stoichiometry parameter y (=0.8, 0.95, 1.0, 1.05, 1.10). The conductivity is plotted versus the temperature $T = 0.25–900$ °C in linear coordinates.

number than the (3.2+) that we had before. It is left to the reader of this chapter as an exercise to determine the relative amount of Fe^{4+} for the compound $(La_{0.8}Sr_{0.2})_{0.95}FeO_{3-\delta}$.

The conductivity for this compound $(La_{0.8}Sr_{0.2})_{0.95}FeO_{3-\delta}$ is the red data points in Figure 6.22. It has the highest conductivity with 140 S/cm, notwithstanding that the scattering of data is significant. The effect of manipulating the material by the self-doping on the A-site is very encouraging and we therefore further decrease the relative amount of La and Sr to arrive at the stoichiometry $(La_{0.8}Sr_{0.2})_{0.8}FeO_{3-\delta}$. However, we now see that the maximum conductivity for $y = 0.8$ (green data points) is not exceeding the previous one with $y = 0.95$. It is only 40 S/cm and thus even lower than the parent compound $y = 1$. When we substitute in the opposite direction and increase y to $y = 1.05$, we get a conductivity maximum slightly above 40 S/cm. When we further increase to $y = 1.1$, the conductivity maximum further decreases to below 20 S/cm.

It appears that not only the height of the maximum is changing, but also the position of the conductivity maximum on the temperature axis. We could try to determine this in our study. Furthermore, it appears also that the onset temperature of conductivity is changing with the composition. The data that we can extract from such seemingly simple conductivity measurements can have a somewhat complicated structure. When we restrict ourselves to the mere application as a SOFC cathode for this material, what do we want? Do we want a very high conductivity at a very high temperature, or would we be satisfied when the conductivity maximum is at a lower temperature? How wide should be the temperature range where the conductivity is very high?

When you transform the coordinates into an Arrhenius plot, you can determine an activation energy E_a from the slope of the conductivity curve. You see that the conductivities were measured in heating direction ($dT > 0$) and cooling direction ($dT < 0$). This is why, a small hysteresis is visible. This is because the material may have picked up oxygen from the air or leased oxygen into the air. This would change the δ in the stoichiometry. It is therefore important to disclose in which gas atmosphere at which partial pressure the conductivity was measured.

Close inspection of the Arrhenius plots in Figure 6.23 shows that there are small kinks in the conductivity curves. The red curve ($y = 0.95$) has such kink (or step) at around $1000/T = 3$ and $\ln(\sigma T) = 6$; the black curve ($y = 1.0$) has the step at around $1000/T = 2.7$ and $\ln(\sigma T) = 6.8$. The blue and green curve have the step at around $1000/T = 2.5$ and $\ln(\sigma T) = 7$. The purple curve has the step at around $1000/T = 2.4$ and $\ln(\sigma T) = 6$.

Such steps can happen for example when the sensitivity range of the meter which you use for measuring the conductivity is changing to a different sensitivity range. To rule this effect out (another exercise for the reader), you can change the size of the sample or the distance of the terminals which read the voltage in the 4-point measurement geometry. By doing so, you change the absolute resistivity of the sample portion that you probe – and thus avoid to pass through a resistivity range where the meter has to switch to the next sensitivity range.

$$(La_{0.8}Sr_{0.2})_yFeO_{3-\sigma}$$

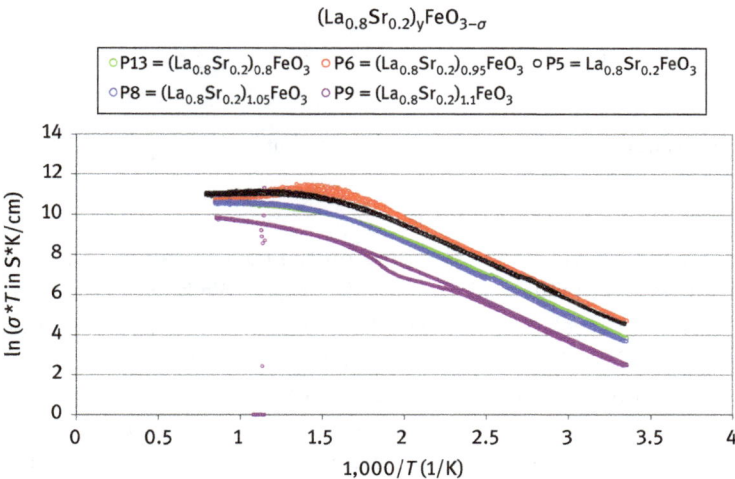

Figure 6.23: Electric conductivity of five samples of $(La_{0.8}Sr_{0.2})_yFeO_{3-\delta}$ with five different values for the stoichiometry parameter y ($y = 0.8, 0.95, 1.0, 1.05, 1.10$). The conductivity is plotted versus the reciprocal temperature $T = 0.25–900\ °C$ in a semilogarithmic plot (Arrhenius plot).

Another reason for such steps can be a structural transformation (phase transition) of the material with concomitant change of the resistivity. We have investigated this effect with valence band X-ray and electron spectroscopy and found correlations which point to structural changes in the material which can cause kinks in the conductivity data. Sometimes you also observe slight changes in the slope in the Arrhenius plot, which means the activation energy for conductivity processes may change [Braun 2008b, 2009b].

In Figure 6.24, you see two 4-point conductivity curves of two LSF samples which were substituted on the B-site with Ni and with Cu. Data were recorded in heating and in cooling direction, this is why we see the hysteresis. Depending on the oxygen concentration in the samples – which may vary during this kind "thermal cycling" – the conductivity of the electrodes will change accordingly.

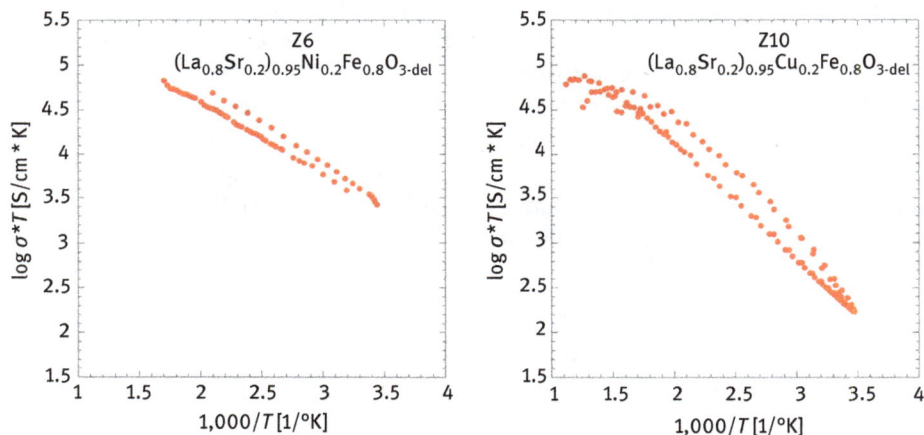

Figure 6.24: 4-Point DC conductivity of Ni-substituted and Cu-substituted iron perovskites as a function of temperature, measured in air.

Also, the thermal expansion will vary depending on the concentration or partial pressure of the gas in which the thermal cycling takes place. We have studied this effect in iron perovskites which were substituted with titanium and tantalum [Bayraktar 2008, Braun 2009a, 2012]. When you give these studies a deep thought, you will find that the term "thermal expansion" is not all justified because the driver for the expansion is a change in the chemical constitution, assisted by the annealing. Hence, it is a combination of chemical expansion and thermal expansion.

As we see that the electric conductivity will depend on the doping or substitution on the A-site and the B-site, and also on the temperature and on the gas partial pressure, we will end up with a large and complex matrix of transport data. We could easily introduce other parameters such as pellet compaction pressure, sintering time and sintering temperature.

You can summarize the conductivity data graphically in a contour plot such as shown in Figure 6.25. The activation energy (an exercise for the reader is how to determine the activation energy E_a from the Arrhenius plots shown in the previous graphs) is plotted as the stoichiometry parameter x for the relative strontium content (this is the A-site substitution in the ABO_3 perovskite) versus the stoichiometry parameter y for the relative nickel content (this is the B-site substitution in the ABO_3 perovskite), remember:

$$A'_x A''_{x-1} B'_y B''_{y-1} O_{3-\delta}$$

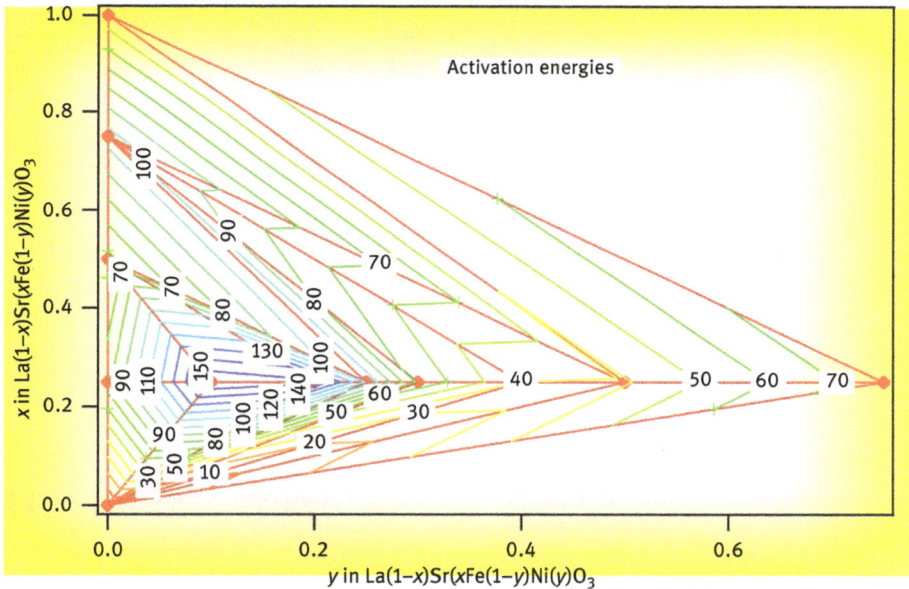

Figure 6.25: Contour plot for the activation energy of $La_{(1-x)}Sr_xFe_{(1-y)}Ni_yO_3$ as a function of stoichiometry parameters x and y.

This contour plot (or density plot) shows a clear maximum for $E_a = 150$ meV in the region $x = 0.25$ and $y = 0.1$. Here is the highest kinetic barrier for the polaron conductivity. For practical examples on conductivity in polaron systems, I recommend the papers by W.-H. Jung [Jung 2001a, 2006, 2007, 2000, 2001b, 2002]. It appears that there are some activation energy minima at $y = 0.4$ and $x = 0.25$, and from there toward the origin.

6.4.9 The triple-phase boundary

Until now, I implicitly assumed that the cathodes must have an electronic conductivity (only) and the electrolytes may have an ionic conductivity (only). This is a simplification which often holds, but not always. It turns out that some of the cathode materials may have an appreciable ionic conductivity under SOFC operation temperatures, in addition to the required electronic conductivity. Materials with both type of conductivities are called mixed ionic electronic conductors (MIECs).

An SOFC electrode assembly is sketched in Figure 6.26. We are looking here at three different materials. The support at the bottom is the solid electrolyte, for example YSZ. The electrolyte is a compact and "dense" ceramic which can conduct the oxygen vacancies [Zintl 1939] $V_O^{\bullet\bullet}$ which are the ionic (cation, positive charge) charge carriers. Coated on the electrolyte is the cathode layer. Let the cathode be MIEC with thickness L.

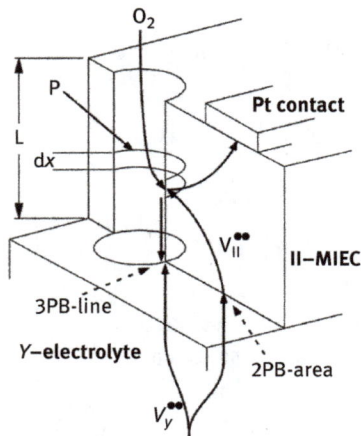

Figure 6.26: Cross-sectional schematic of the physical structure and chemical reactions occurring at a porous MIEC phase in contract with a solid electrolyte.
Reprinted with permission from *J. Electrochem. Soc.*, 150, 8, A1139–A1151 (2003). Copyright 2003, The Electrochemical Society. Coffey GW, Pederson LR, Rieke PC: Competition Between Bulk and Surface Pathways in Mixed Ionic Electronic Conducting Oxygen Electrodes. *Journal of the Electrochemical Society* 2003, 150. doi: 10.1149/1.1591758 [Coffey 2003].

The interface between electrode and cathode is a two-phase boundary (2PB). The aforementioned oxygen vacancies can travel across this 2PB. On top of the cathode is a current collector, for example, a platinum terminal. Unlike the electrolyte, the cathode (and the anode; not discussed here) must be porous.

A cylinder pore is shown in Figure 6.26; it extends from the top of the cathode to the electrolyte surface and has at least the length L. The perimeter of the pore at the contact area between electrolyte and cathode is the boundary of three different

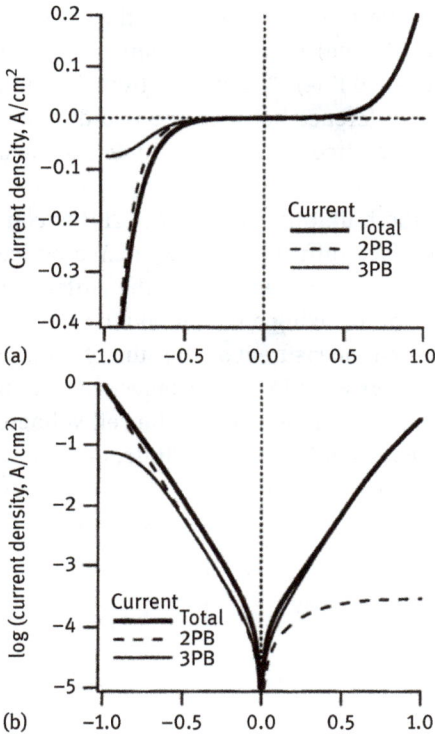

Figure 6.27: (a) *I*–*V* and (b) log *I*–*V* curves for the 2PB and 3PB pathways as well as the total. Results are for the standard conditions listed in Table 6.1.

Reprinted with permission from *J. Electrochem. Soc.*, 150, 8, A1139–A1151 (2003). Copyright 2003, The Electrochemical Society. Coffey GW, Pederson LR, Rieke PC: Competition Between Bulk and Surface Pathways in Mixed Ionic Electronic Conducting Oxygen Electrodes. *Journal of the Electrochemical Society* 2003, 150. doi: 10.1149/1.1591758 [Coffey 2003].

phases, that is, the three-phase boundary (3PB).[11] The phases are the electrolyte, the cathode and the gas phase (oxygen) in the pore.

The oxygen can travel to the electrolyte either over the surface of the cathode in the pore or through the bulk of the cathode to the electrolyte. The oxygen vacancies can come either from the electrolyte as $V_O^{\bullet\bullet}$ or from the MIEC as $V_{II}^{\bullet\bullet}$. They annihilate when they meet with the oxygen from the ambient.

Coffey et al. have simulated the electrode kinetics of the processes taking place as illustrated in this sketch in Figure 6.26 by using the Butler–Volmer equation and obtained two slightly different characteristics for the current originating from surface diffusion (via 3PB) and bulk diffusion (via 2PB).

---

11 The 3PB is a one-dimensional space, whereas the 2PB is a two-dimensional space.

The resulting $I-V$ curves are shown and compared in Figure 6.27. On the reducing branch, the current originating from the 3PB (solid line) is closer to 0 than the current based on bulk diffusion of oxygen (via 2PB, dashed line). The current from the bulk diffusion is strong in negative direction. The difference between the two different current modes becomes more pronounced when plotted on the logarithmic scale in a Tafel plot (Figure 6.27).

In oxidizing direction, the current from the bulk diffusion via the 2PB is considerably lower than the current from the surface diffusion of oxygen. We notice therefore that the current depends on whether it is conducted over the surface or through the bulk, and whether it is in reducing or oxidizing direction. But both types of current certainly take place together and we can measure it as the sum of both.

Finally, the influence of the oxygen partial pressure in cell becomes obvious in Figure 6.28 where the logaithm of current density is plotted versus the cell voltage. Notethat the oxygen partial pressure varies over six orders of magnitude, and more. The current density varies then over two orders of magnitudes.

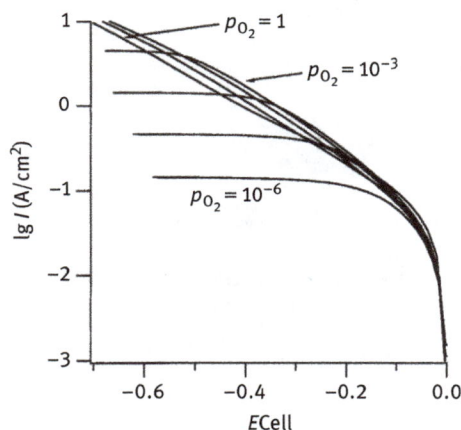

Figure 6.28: $I-V$ response for partial pressures of oxygen ranging from 1.0 to 10^{-6} atm. The parameters are as in Table 6.1 except $k_{ads} = 300$, $k_{rex} = 1$, $i_{0sPB} = 0.1$ and $g^{*}_{i03PB} = 0.1$. Reprinted with permission from *J. Electrochem. Soc.*, 150, 8, A1139–A1151 (2003). Copyright 2003, The Electrochemical Society. Coffey GW, Pederson LR, Rieke PC: Competition Between Bulk and Surface Pathways in Mixed Ionic Electronic Conducting Oxygen Electrodes. *Journal of the Electrochemical Society* 2003, 150. doi: 10.1149/1.1591758 [Coffey 2003].

6.5 Biofuel cells, or bio fuel cells?

The metabolism in human and animal bodies, actually in every living cell including the plant cells, bacteria, microbes and viruses, is the conversion of chemical energy into electromagnetic radiation and electric energy. A simple example is the conversion of glucose in our body, or the "burning of fat" when the hydrocarbon reserves in our body are converted into energy. While we say "burning," it does not mean the fat or glucose combust with a flame like in an open fire or explosion. The conversion is basically of electrochemical nature and we will come to that in a later chapter in this book.

It was therefore of interest to see whether biological systems such as living cells could be used, exploited by feeding them with a hydrocarbon and tap electric power out of them. I have to clarify here another misnomer, and this is the difference of the fuel against the difference of the conversion mechanism. The title in the review by Bullen [Bullen 2006] mentions "biofuel cell" but the biofuel is not what the fuel community would understand as biofuel. Are the biofuels derived from biological systems in contrast to, say, fossil fuels where a fossilization of biofuels has taken place? Biomass is a biofuel. Methanol produced from corn is a biofuel. In 2016, car manufacturer Nissan presented their e-Bio Fuel Cell system [Doi 2016]. Read carefully, because this Nissan car is powered by a SOFC which is fed by 100% ethanol or ethanol-blended water after it went through a reformer. Hence, the process of conversion is not biological, but the origin of the fuel is biological. The ethanol is for example produced from sugarcane.

When I talk here in this chapter of biofuel cell, then I mean a fuel cell where biological components carry out the electrochemical conversion of any fuel into electricity. The first such cell was reported by Potter over 100 years ago [Potter 1911]. In the review paper by Kannan et al. [Kannan 2009], you can see an electrochemical cell with a solid electrolyte membrane on which an anode is coated which contains hydrogenase as a catalyst, and on the other side of the electrolyte, a cathode is coated with laccase. The fuel can be glucose which is converted to protons H^+ which diffuse through the proton-conducting electrolyte membrane such as NAFION® [Mauritz 2004]. With the oxygen from the cathode, the protons will combine to water (H_2O). The anode reaction is therefore

$$\text{Glucose} \Rightarrow \text{Glucunolactone} + 2H^+ + 2e^-$$

And the cathode reaction is [Kannan 2009]

$$O_2 + 4\ H^+ + 4\ e^- \Rightarrow 2\ H_2O.$$

Lin et al. [Lin 2013] present what they call a photosynthetic microbial fuel cell which has deposited a *Spirulina platensis* biofilm and which produces a photovoltage with OCV of 0.49 V and a power of 10 mW/m^2. But this cell does not convert any fuel and should therefore be called bio-photoelectrochemical cell. The same holds for the cell presented by Kirchhofer et al. [Kirchhofer 2015] who assemble thylakoid anodes in an electrochemical cell along with a laccase cathode. No fuel is converted here.

The microbial systems presented in the review paper by Rosenbaum et al. [Rosenbaum 2010] do convert nutrients and CO_2 and air and produce electricity. They satisfy the definition of biofuel cells.

I will come in a later chapter in this book back to biological systems. Here at this point, I would like to conclude with a quote by Blanford [Blanford 2013] about the new business opportunities that arose from protein electrochemistry in the sensor technology:

> The results from a final-year undergraduate project led to an $876M sale of a spin-out company 19 years later: the 1977 communication from Mark Eddowes and Allen Hill seeded the rich field of protein electrochemistry, the technology that underpins commercial glucose biosensors.

It would be a great success when the biosensor business (signal processing) could mature into bio-energy storage and biopower conversion soon.

6.6 Direct methanol fuel cell

The direct methanol fuel cell (DMFC) uses methanol as fuel. Note that here the type of fuel is used to describe the kind of fuel cell. It is another type of proton exchange fuel cell, but it can also be run in an alkaline electrolyte. Their advantage is that they can be used at ambient temperature with a liquid fuel, methanol, which is diluted in water. A recent review about DMFC is found in Joghee (2015).

The electrochemical reaction at the anode reads

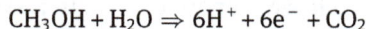

$$CH_3OH + H_2O \Rightarrow 6H^+ + 6e^- + CO_2$$

where the protons are formed and where carbon dioxide is produced as a reaction product. The protons react at the cathode with the oxygen from the air and the electrons produced at the anode. The reaction product at the cathode is water:

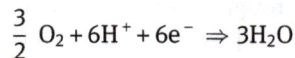

$$\frac{3}{2} O_2 + 6H^+ + 6e^- \Rightarrow 3H_2O$$

The overall reaction shows therefore the production of water and CO_2 which is

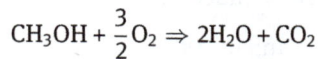

$$CH_3OH + \frac{3}{2}O_2 \Rightarrow 2H_2O + CO_2$$

These reaction schemes look very simple even when it is not the hydrogen which is being converted but a hydrocarbon molecule such as methanol. However, the electrochemical processes are mediated ad facilitated by electrocatalysts such as platinum, and there is spectroscopic evidence that chemical intermediates are being formed during conversion. For further reading about this, I recommend Cameron et al. [Cameron 1987].

It is possible to run a fuel cell with hydrogen obtained by reforming methanol. This is the indirect methanol fuel cell which requires a steam reformer [Han 2002, Jan 2000].

6.7 Phosphoric acid fuel cell

The PAFC was the first commercialized fuel cell [Jingang 2012]. It has its name from the phosphoric acid which is the liquid electrolyte. However, the liquid electrolyte is fixated in a polytetrafluorethylene, which is Teflon® fiber structure or silicon carbide. An X-ray imaging study on a model system is found in Eberhardt et al. [Eberhardt 2014].

The electrode reactions are the same like in conventional PEM-FC. At the anode, 2 mol of hydrogen gas is converted to four protons and four electrons. With 1 mol oxygen gas at the cathode, protons and oxygen will react to 2 mol of water. The electrodes are often carbon paper with dispersed electrocatalysts.

The fuel for the PAFC is hydrogen, but this needs not be high purity hydrogen. Traces of carbon monoxide, which for example is present in the hydrogen obtained from water gas, shift reaction – a rather low cost product in comparison to high purity hydrogen.

The oxidant can be pure oxygen or air. The operation temperature of the PAFC is from 130 to 200 °C, because the proton conductivity of the electrolyte has a favorable value at this temperature. The relatively high temperature along with the electrochemical corrosive environment makes that carbon and graphite are used as materials for electrode components.

As the vapor pressure of the H_3PO_4 phosphoric acid[12] is relatively low, it can be handled with ease in the PAFC. However, it will evaporate and in order to prevent evaporation of the electrolyte, a cooling trap is necessary which condensed the electrolyte and guides it back in the electrolyte container. The phosphoric acid electrode works as a proton conductor [Kreuer 1996]. The chemical reactions in the PAFC are catalyzed by platinum or some of its alloys. When a blend of Pt and Ru is used, the PAFC was more tolerant to CO poisoning. The necessary catalyst coverage for the anode and the cathode is in the order of 1/2 mg/cm^2 electrode area.

The PAFC has a very good dynamic behavior and was maybe for this reason employed for the First National Bank in Omaha, Nebraska. When there is a power failure in the IT center of a major bank, a power outage even for a very short time can

---

12 Phosphoric acid is an ingredient in Coca Cola by the way. The dilute phosphoric acid in the pop soda has a refreshing taste. Other soft drinks have citric acid or even acetic acid. I was brought up in a catholic home and catholic culture and learnt in my early juvenile age that Jesus Christ was fed by Roman soldiers during his crucification with a sponge soaked with vinegar, acetic acid. This story was told in a frightening way – such as like giving vinegar to a dying man was some kind of torture. However, water mixed with vinegar was common refreshment among the Roman soldiers and many other people 2000 years ago. My favorite refreshment during long experimental days in my diploma thesis work at KFA Jülich in summer 1995 (see my photo in Chapter 5) was a mixture of apple juice, sparkling water and vinegar. I mixed it by myself. However, concentrated acids are dangerous. Don't drink them!

Figure 6.29: PAFC Power Plant – To the First National Bank of Omaha just a few milliseconds without electricity can mean hours of headaches and millions of dollars of lost revenues. Four UTC fuel cells 200 kW power plants now serve as the bank's primary source of power. Heat from the fuel cell installation also provides energy for space heating, increasing the overall efficiency of the fuel cell system to more than 80%.
Reproduced with permission from The Electrochemical Society from Perry ML, Fuller TF: A Historical Perspective of Fuel Cell Technology in the 20th Century. *Journal of the Electrochemical Society* 2002, 149. doi: 10.1149/1.1488651 [Perry 2002].

cause loss of data which can amount in considerable loss of revenues. It is therefore mandatory for a bank to have an uninterrupted power supply which can react on the spot when there is a power outage.[13] Figure 6.29 shows you three of the four PAFC fuel cell systems at First National Bank [Perry 2002].

---

13 In the week of August 14, 2003, my colleagues from University of Kentucky and I had a synchrotron beamtime at the National Synchrotron Light Source in Brookhaven National Laboratory (BNL) at Long Island in New York State. We were doing STXM on soot. In the afternoon, suddenly, the light went out and the dim emergency lights went on. At some point, the loud vacuum pumps which provide for the vacuum in the synchrotron storage ring went out with a loud whistle tone which you know when the frequency of the rotation of the pump jet blades is slowing down. From the news in the radio, we learnt that there was a power outage, and after a while we knew it was a major power outage. I remember myself saying "we just lost Connecticut;" STXM pioneers Chris Jacobsen and Sue Wirick were with me at the beamline X1A at this time. I believe after 1 or 2 h of the first signs of outage it was announced that the vacuum in the storage ring had become so poor (the pressure too high) that even if there was an immediate return of electricity, it would take several days to get the storage ring working for user synchrotron beam time. This meant our beamtime was terminated by higher force!

What we learn is there are many alternatives of fuel cells and each system may have advantages and disadvantages. The good thing is that it seems we have choices.

The Westinghouse Electric Corporation Research Laboratories in Pittsburgh PA participated in the mid-1970s in a large project on the assessment of various alternatives for energy conversion for the NASA [Shah 1976, Walde 1976]. The reports on this project are available for free download from various websites in the United States. Figure 6.30 shows a comparison of the costs of four different fuel cell systems with a capacity of 900 MW h.

The PAFC power plant has total cost for 1 kW h electricity to 44.0 Mills (1 Mill is 0.1$ cent), therefore 4.4 cent/kW h. The capital cost is 1.4 cent, the fuel is hydrogen and costs 2 cent, and the operation and maintenance cost are 1.6 cent/kW h. The overall efficiency of the system is 29.3% – lower than any other system compared here.

The alkaline fuel cell provides electric power at the highest cost of 58.9 Mills/kW h, that is, 5.9 cent/kW h. This is because the capital costs and operation and maintenance costs are higher than in any of the other three fuel cell systems. There is little use in the slightly higher efficiency of 30.7% as compared to the PAFC system.

The MCFC ranges in the middle of all compared systems. The best solution appears to be the SOFC. It uses the cheapest fuel at a cost of 1.1 cent/kW h and has the lowest cost for operation and maintenance. The overall efficiency is more than 50% and allows (with a plant efficiency of 60%) for electric power at a cost of 4 cent/kW h.

And because of safety reasons, we were expected to leave the place. So, my senior and our PhD student went to our rental car and made a visit to the next supermarket for some food and refreshments. When we arrived there, supermarket staff was covering their fresh food and iced beverages and ice cream with metalized blankets in order to retard the warming. Needless to say, their lights were dim too. We purchased ice cream and fruit and produce and went for a late afternoon to the beach in the Hamptons. When we arrived back at BNL, we had to argue with security staff to get back to our guesthouse where we lived for our beamtime. They did not want to let us in because of safety reasons. We decided that we grab our stuff in the guesthouse and vacate the guesthouse, and then somehow move on. Our scheduled flights back to Kentucky were several days later. We – in the center of the chaos – were not fully aware of the chaos that we were in. We could not get quick flight back from Islip in Long Island or from La Guardia in New York back to Kentucky. Our senior decided to stay for the week to follow with his sister's family who lived in Queens NY. The PhD student and I were supposed to travel back to Kentucky with the rental car – 1250 km. The ride was so short I do not remember any details. I was the driver because the PhD student had no driver's license. Back home in Kentucky, I learnt that my father had called my wife when he, in Germany, learnt about the power outage [Minkel 2008] Minkel RJ: The 2003 Northeast Blackout–Five Years Later. *Scientific American* 2008. in the news. He thought I got stuck in an elevator in a Manhattan sky scraper because he knew I was working in New York. But instead being stuck in the sky scraper in New York, I had been walking barefoot in the sands of the Long Island beaches. Had Brookhaven National Laboratory had a UPS, a large uninterrupted power supply system such as the PAFC in Omaha, then we could have finished our experiments with success and without hassle. Then you would not have been reading this anecdote.

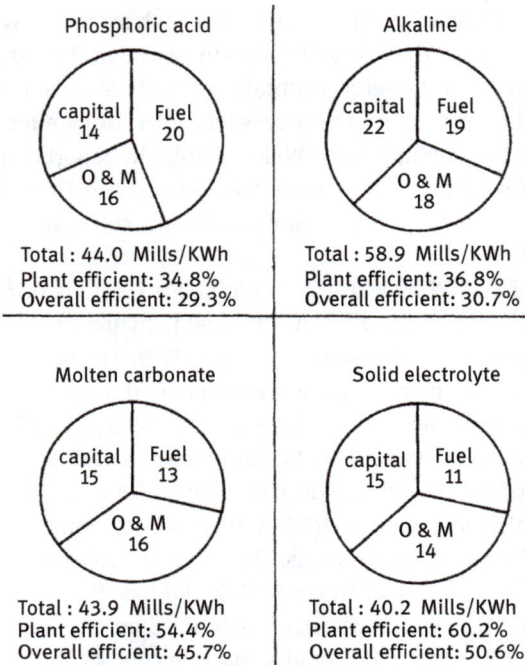

Phosphoric acid

capital 14 | Fuel 20
O & M 16

Total : 44.0 Mills/KWh
Plant efficient: 34.8%
Overall efficient: 29.3%

Alkaline

capital 22 | Fuel 19
O & M 18

Total : 58.9 Mills/KWh
Plant efficient: 36.8%
Overall efficient: 30.7%

Molten carbonate

capital 15 | Fuel 13
O & M 16

Total : 43.9 Mills/KWh
Plant efficient: 54.4%
Overall efficient: 45.7%

Solid electrolyte

capital 15 | Fuel 11
O & M 14

Total : 40.2 Mills/KWh
Plant efficient: 60.2%
Overall efficient: 50.6%

Figure 6.30: Breakdown of electricity costs for all 900-MW DC fuel cell power plants. The fuel gas is medium-Btu gas (OTF), and the useful life of all fuel cell subsystems is 10,000 h.
Reproduced from the Final Report by Walde CJ, Ruka RJ, Isenberg AO, Energy Conversion Alternatives Study (ECAS), Westinghouse Phase 1. Vol. X11: fuel cells. [Phosphoric acid, potassium hydroxide, molten carbonate and stabilized zirconia electrolytes are compared]. 1976. Westinghouse Research Labs. Pittsburgh, PA (USA) [Walde 1976].

Note that this comparison was made over 40 years ago. Numbers may look different today. The differences are not so huge; the lowest cost per kW h was 4 cent and the highest was 6 cent. Certainly, this is a difference of 50%. The lowest costs are not necessarily the main driver in the decision for a system.

7 Solid electrolytes

7.1 Some words about liquid electrolytes

We remember from middle school or maybe even elementary school that electrolytes are typically liquid. We used water and rock salt NaCl to do some simple electrolysis with battery connected to a simple electrochemical cell. There is no fundamental requirement which says an electrolyte has to be liquid. The requirement for an electrolyte is that it must be an ion conductor. Rock salt dissolved in water will dissociate in Na^+ and Cl^- ions and these are the electric (ionic, not electronic!) charge carriers which can close an electric circuit in an electrochemical cell.

But what is the purpose of the water in the electrolyte solution when the conducting species is the alkaline ion? If a liquid state was a necessary criterion for an electrolyte, why not take any other salt and melt it at high temperature, such as KCl?

Over 200 years ago, English Chemist Humphry Davy (1778–1829) conducted the first molten salt electrolysis with NaOH,[1] which was contained in a platinum dish. This is how he produced the alkaline metal sodium. Some salts can be melted at high temperature without any addition of any other solvent such as water. The mobile ions in the melt allow for electric current when a voltage is applied over two electrodes which are in contact with the molten salt. The process is realized in the Downs cell at the industrial scale [Downs 1924]. Lithium and also aluminum is produced with molten salt electrolysis. A book on the various analyses methods for molten electrolytes was published by Daněk [Daněk 2006].

I have already presented the molten carbonate fuel cell in Section 6.3. This is another application of the molten salt electrolytes.

We can return to the question above and ask what is the role of the water? Artemov et al. [Artemov 2015] posed this question recently and found that water by itself can be an electrolyte when fully dissociated in H_3O^+ and OH^-. They were able to mathematically model the ionic conductivity of pure water with existing physical and chemical principles. This contradicts to our experience because the electric resistivity of water is a measure for its purity: High purity water with no ions in it has a very high resistivity.

The conductivity of an electrolyte is related to the velocity of its charge carriers, specifically the time which the cations need to travel to the electrically charged cathode, and the anions need to travel to the charged anode. We must consider the electrolyte with its viscosity η and the ions with their charge ($z_i e$) and their radii r_i.

---

[1] When you burn wood, the white ash which remains after the combustion contains trace elements such as potassium and sodium. In old ages, the ash was used to make leach such as KOH.

https://doi.org/10.1515/9783110561838-007

When we apply an electric field E over the electrolyte, the ions will follow with the velocity v_i according to following relation (compare [Wagner 2012]):

$$v_i = \frac{z_i e}{6\pi\eta r_i} \times E$$

The viscosity coefficient η of the electrolyte will typically increase with increasing temperature. Therefore, the electrolyte conductivity will increase with increasing temperature. This is the reason why the molten carbonate fuel cell and phosphoric acid fuel cell operate at elevated temperatures. Note that a frozen aqueous electrolyte such as sulfuric acid may still have a considerable ionic conductivity.

Lithium batteries have typically organic electrolytes which protect the electrode from corrosion. An aqueous electrolyte would immediately react adverse with metal lithium, for example. The organic electrolyte is a polar solvent mixture such as dimethyl carbonate and ethylene carbonate with dissolved $LiPF_6$ crystals. Note that the ZEBRA battery has a molten salt positive electrode but the electrolyte (separator) is still solid. Ionic liquids have become interesting as electrolytes. An ionic liquid is a salt in a liquid state and has therefore inherent ionic character. They have often low vapor pressure and therefore do not evaporate so fast, which can be relevant for some battery and supercapacitor applications.

7.2 NAFION®, a polymer solid electrolyte

A well-known electrolyte in polymer form is NAFION®, which is a Teflon (tetra-fluoroethylene) backbone combined with perfluorovinyl ether groups terminated by sulfonate groups. Along the sulfonic acid (SO_3H) groups, protons can hop and thus work as charge carriers. NAFION® is therefore a proton conductor and ionic polymer (ionomer). The material is used as a thin foil with white color but you can obtain it also as a liquid which you can pour or brush for versatile use on electrode assemblies. The electrolyte membrane needs minimum humidity for proton transport. When it is used in devices such as fuel cells where heat can evolve, precaution must be taken to avoid that the membrane becomes too dry. In this case, the humidity from the ambient is not sufficient anymore and a water management is necessary.

There are various polymer electrolytes known in electrochemistry [Sequeira 2014], and some are used also in lithium ion polymer batteries [Long 2016]. But NAFION® has a special place in the materials science of organic electrolytes. Because of the popularity of NAFION®, many researchers from various scientific disciplines investigate the material. It is therefore not surprising that the actual physical processes which constitute the proton mobility in this material remain mysterious, notwithstanding that it is well studied as shown in the nice review by Mauritz and Moore [Mauritz 2004].

7.3 Solid electrolytes

Hundred years ago, Emil Baur worked at ETH Zürich on the controlled oxidation of carbon and organic materials, to some extent also on electrochemical energy conversion such as carbon elements [Tobler 1932].

In 1916, Baur and Treadwell mentioned in their patent (Deutsches Patent 325'783) a carbon element (Kohlekette) which had a solid electrolyte. It did not work so well, and therefore, Baur used a molten electrolyte for an element of the type as sketched below (find it in Tobler [Tobler 1932]):

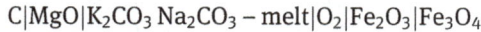

$$C|MgO|K_2CO_3\ Na_2CO_3 - melt|O_2|Fe_2O_3|Fe_3O_4$$

Note on the right side the air electrode with iron oxide as electrocatalyst, and on the left side the carbon anode. However, as Baur [Baur 1937] points out, all available molten electrolytes had to be discarded for the concept of the electrochemical cells he was working on, and he had to seek success with the solid electrolytes.

The solid electrolyte was a ceramic tube with 1.5 cm inner diameter and 25 cm^2 area inside which was ionically conducting. The tube was filled with carbon powder (grains), and the tube was inserted in a ceramic vessel that contained Fe_3O_4 powder ("Hammerschlag"). When the experimenters inserted this apparatus in a furnace with 1,000 °C, they measured an electromotive force of around 1 V.

You can imagine that oxygen from the air entered the Fe_3O_4 cathode and then diffused through the ceramic cylinder, the electrolyte and then reacted with the carbon inside the tube to form carbon dioxide. Baur and Preis mention that you can also use iron or any other metal powder instead of carbon.

In order to enhance the voltage and current density, Baur and Preis tested a whole range of ceramic materials and found that the so-called Nernstmasse, yttrium-substituted zirconia (YSZ) was the best electrolyte for their purpose. The Nernstmasse was already known for its use as ion-conducting filament in a patent by Walther Nernst from 1897.

Bauer and Preis made a techno-economic analysis for their solid oxide fuel cell and compared it against steam turbines and battery galvanic elements [Baur 1937]:

> Indessen braucht der ungünstige Ausblick auf die Investitur nicht abzuschrecken. Die keramische Industrie wird die benötigten Formstücke billig herzustellen wissen, wenn die Nachfrage einsetzt.
>
> Wir wollten nur zur Geltung bringen, dass mit tatsächlich zugänglichen Festleitern Brennstoffketten gebaut werden können, die in ihrem Raumbedarf sowohl neben Sammlerbatterien als auch neben Dampfkraftwerken bestehen dürften.

Wagner [Wagner 1943] explained the origin of the electric conductivity in yttrium-stabilized zirconia around 50 years after the material was developed. Specifically, the Y^{3+} fills a position of a Zr^{4+}, and because of the required charge neutrality, negative charge needs to compensate by creating oxygen vacancies. Because of the oxygen in the crystal lattice as the charge O^{2-}, it needs to such Y^{3+} cations to force one O^{2-} vacancy:

$$[Y'_{Zr}] = 2[V_O^{\bullet\bullet}]$$

Wagner concludes in his paper that exclusively anion conductors would be suitable for fuel cells and that a systematic analysis of YSZ with X-ray methods and further electrical methods would be desirable [Wagner 1943]:

> Für Brennstoffelemente mit festen Elektrolyten kommen daher ausschliesslich Anionenleiter in Betracht. Unter diesen Gesichtspunkten erscheint eine systematische Untersuchung der Mischkristallsysteme vom Typus der Nernststiftmasse (ZrO_2 + Y_2O_3) mit röntgenographischen und elektrischen Methoden wünschenswert.

Hund investigated later the crystal structure of YSZ and correlated it with its transport properties and confirmed Wagner's conductivity model [Hund 1951]. In particular, Hund investigated the phase diagram ZrO_2–Y_2O_3 and found that in addition to the pure phases ZrO_2 and Y_2O_3, an extended range of mixed crystals exists where the cation sites for Zr and Y are ordered, whereas the anion lattice has vacancies which are disordered. Wagner had already speculated that in the system ZrO_2–Y_2O_3, there would be "anomalous mixed crystals of the fluorite type."

There exist two possible atomic arrangements in the fluorite phases, that is,

1. A completed cation lattice where four cations are in the fluorite cube where Zr^{4+} is substituted by Y^{3+} and the ion lattice must have oxygen vacancies in order to warrant electric charge neutrality.
2. A complete oxygen ion lattice with 8 O^{2-} in the unit cell. When Zr^{4+} is substituted by Y^{3+}, then there must be additional cations be brought into the lattice, likely octahedral gaps, as interstitial cations, also to warrant charge neutrality.

Today, you can carry out a diffractometry phase analysis by using X-ray diffraction (XRD) with contrast variation (anomalous X-ray scattering) at synchrotron sources and with contrast variation in neutron diffraction. Still, this is a tedious work but it is necessary to be able to distinguish the two phases.

With the measured lattice constant a_w of the mixed phase from XRD and the known phase ratio x mol ZrO_2 and y mol Y_2O_3, you can calculate the mass density ρ_{calc} of the mixed phase. N_A is Avogadro's constant. Hund carried out standard XRD and measured also the density ρ_{exp} of the materials. You can calculate the density ρ_{exp} from the lattice constant and the stoichiometry x and y:

$$\rho_{calc} = \frac{xZrO_2 + yY_2O_3}{N_A \cdot a_w^3}$$

The actual phase (1) or (2) is the one whose calculated density ρ_{calc} is closest to the measured density ρ_{exp}. The discovery of the mechanism which allows ion conduction by vacancies required in the beginning of its discovery quite some creativity. With all the information that we have today, it is easy to follow various scenarios and accept them or reject them as models and theories. During my time as diploma thesis student in Jülich, diffusion by vacancy involvement was a matter of serious investigation. I can recommend the studies on the "Platzwechselmechanismus" by Thomas Flores [Flores 1997, Ibach 1996], who investigated surface diffusion of manganese atoms by vacancies on single crystals with computer simulations (I think with the first available IBM Pentium PC) and tunnel microscopy.

Today, zirconia is a commodity which you can purchase from the shelf in all kinds of shapes and stoichiometries and purities. Figure 7.1 (left) shows on the left two YSZ pellets with 1 cm diameter and 2 mm thickness, purchased from some vendor. The right panel in Figure 7.1 shows such pellets which on one side is coated with a black porous cathode layer from LaSrCrMn-oxide and a gold current collector.

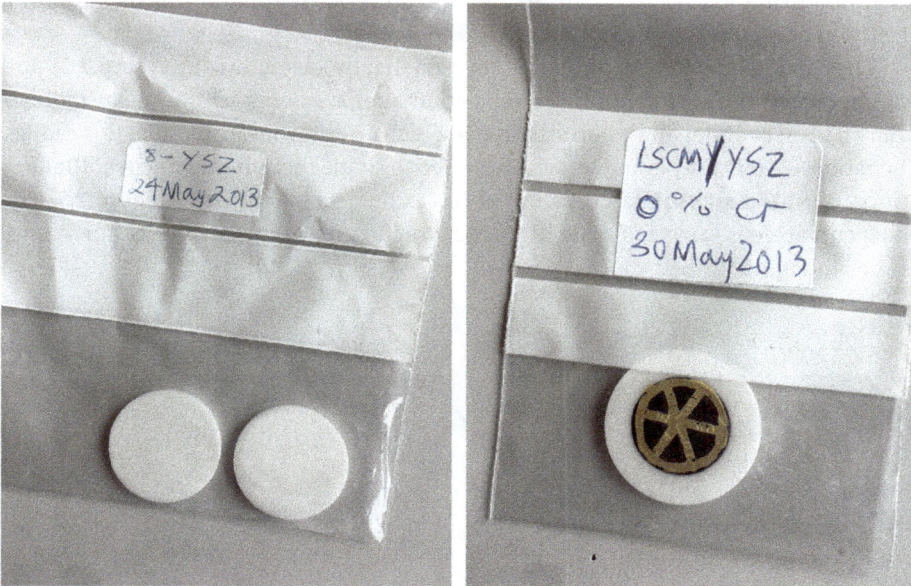

Figure 7.1: (Left) Two white pellets of 8% yttrium-stabilized zirconium oxide (8-YSZ), commercially available from some vendor. (Right) A YSZ pelet coated on the surface with a porous cathode layer from LaSrCrMn-oxide and a current collector layer from gold.

These pellets are compact and gas tight, as is required also for electrolyte applications. For some fuel cell concepts, YSZ is mixed with cathode materials and then we have a porous assembly. It is of interest to know the influence of porosity on the transport properties of such composite networks.

7.3.1 Influence of shape and porosity on the ionic conductivity

Aivazov and Domashnev [Aivazov 1968] found an empirical relationship between the thermal and electric conductivity λ and the porosity θ, which reads

$$\frac{\lambda}{\lambda_0} = \frac{1-\theta}{1+6\theta^2}.$$

Koh and Fortini [Koh 1973] expanded on this relationship, confirmed it (except for the constant 6, which may depend on geometrical specifics of the system under investigation) and found also that the Wiedemann–Franz law follows the influence of the porosity on the relationship between thermal and electric conductivities.

Yamahara et al. [Yamahara 2005] investigated the ionic conductivity of YSZ composites and took the geometry of bottle neck structures into account. They started out with a composite of YSZ coated by LSM and then leached out the LSM layer with an acidic treatment. The third component that they put in the electrolytes was scandium oxide (SYSZ). You can see the difference in morphology in the scanning electron micrographs in Figure 7.2.

They modeled such bottle necks mathematically, as shown in Figure 7.3, and parameterized the structure of the ionic conductor elements with a mean radius r from an ideal cylinder and a perturbation A which is radial deviation from this radius. The ratio $k = A/r$ is the parameter which is used to shape the cylinder into a bottleneck structure with a sinusoidal curvature of the radius, as exercised in Figure 7.3. The sinus curvature is an approximation to the structure observed in the electron micrographs in Figure 7.2(b).

Figure 7.2: (a) YSZ sintered at 1,475 K to a density of 3.70 g/cm³ (3.70 × 10³ kg/m³), corresponding to a fractional porosity of ~0.38 and (b) LSM–YSZ after LSM removal, sintered at 1,525 K to a YSZ density of 3.65 g/cm³ (3.65 × 103 kg/m³), corresponding to a fractional porosity of ~0.38. The LSM–YSZ shows evidence of considerable differential densification.
Reprinted from *Solid State Ionics*, 176, Yamahara K, Sholklapper TZ, Jacobson CP, Visco SJ, De Jonghe LC, Ionic conductivity of stabilized zirconia networks in composite SOFC electrodes, 1359–1364, Copyright (2005), with permission from Elsevier [Yamahara 2005].

You have to see the problem as follows. The specific resistance of a material is given as a material's constant ρ. But the resistivity depends on the geometry of the conductor. When it is a compact cylinder with length l and cross-section A, the resistivity R equals simply

$$R = \rho \times \frac{l}{A}$$

Remember the hydrostatic analogue which I presented in a previous chapter of this book – the flexible hose. When the water faucet, the metal pipe in the wall, delivers a particular constant water flow(or flux, like 20 L/min), you felt the pressure by the hose in your hand. Imagine now the hose is near the faucet as wide as the faucet, say with a 1.5 cm^2 diameter, and then 5 m further, you have attached another piece of hose, an extension of 5 m but thinner, like 1 cm^2 diameter. The pressure and resistance in the thin hose will be higher than the pressure and resistance in the wide hose.

You can make this more complex by attaching again a wide hose, and again a thin hose and so on. The structure sketched in Figure 7.3 resembles an alternating wide and thin hose (tube, pipe, conductor, etc.). It is therefore of general interest to know how the irregularity of a conductor with alternation from wide to thin affects the conductivity. Yamahara et al. have approximated such irregularity by a sinusoidal profile, maybe also for the ease of mathematical treatment.

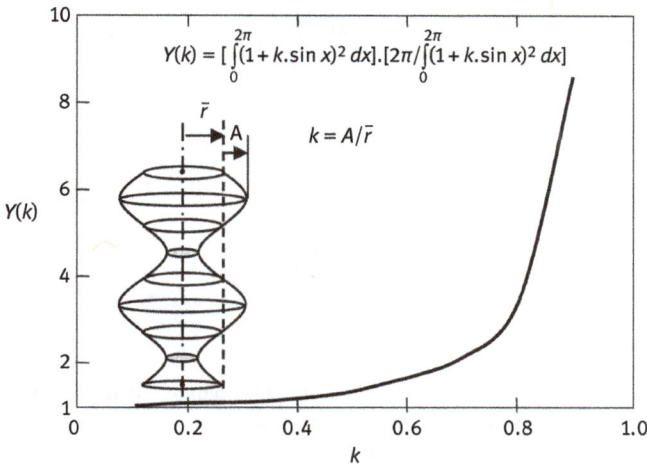

Figure 7.3: Relative increase in resistivity, $Y(k)$, of a sinusoidally perturbed column. A is the amplitude of the perturbation, while $\bar{r}$ is the mean radius.
Reprinted from *Solid State Ionics*, 176, Yamahara K, Sholklapper TZ, Jacobson CP, Visco SJ, De Jonghe LC, Ionic conductivity of stabilized zirconia networks in composite SOFC electrodes, 1359–1364, Copyright (2005), with permission from Elsevier [Yamahara 2005].

You see in Figure 7.3 and in the original paper in Yamahara et al. [Yamahara 2005] that the relative increase of the resistance $Y(k)$ is expressed as follows:

$$Y(k) = \left[\int_0^{2\pi} \left(\frac{1}{(1+k \cdot \sin x)^2} \right) dx \right] \cdot \left[\int_0^{2\pi} (1+k \cdot \sin x)^2 dx \right]$$

$Y(k)$ is therefore some kind of form factor for this particular shaped conductor. Figure 7.3 shows the graphical response of Y when k increases. Small k means small A and large r, basically a thick conductor with few wiggles only. $Y(k)$ is close to unity 1 and the resistivity can be taken like a cylinder with constant diameter. Large k means large A and small r, basically a thin conductor with large inhomogeneity in the radius. A normal cylinder does not reflect this geometry anymore, and the resistivity calculated from cylinder geometry needs a substantial correction.

The actual resistivity values, or better, conductivities of the electrolyte samples prior to and after leaching of the LSM layer are shown in Figure 7.4. The composite which contains the scandium oxide (SYSZ) is shown with the filled square and has the overall highest conductivity over the temperature range from 1100 to 1400 °C (0.02–0.08 S/cm). After leaching off the LSM component, the conductivity drops by one order of magnitude (open square symbols).

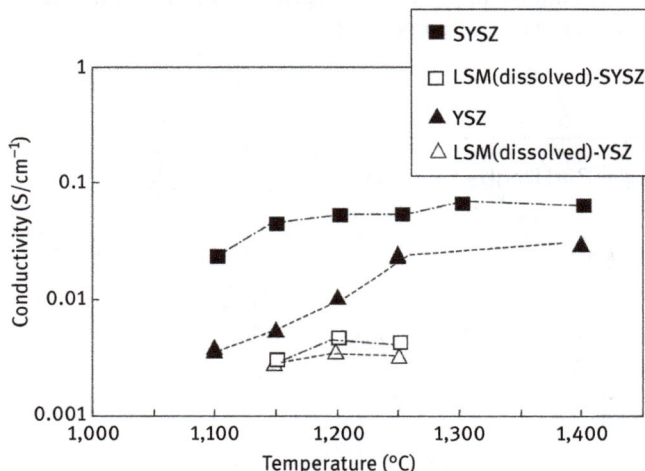

Figure 7.4: Effective oxygen-ion conductivity of SYSZ and YSZ (LSM-free) and of the SYSZ and YSZ network after acid leaching of the corresponding LSM–zirconia composites.
Reprinted from *Solid State Ionics*, 176, Yamahara K, Sholklapper TZ, Jacobson CP, Visco SJ, De Jonghe LC, Ionic conductivity of stabilized zirconia networks in composite SOFC electrodes, 1359–1364, Copyright (2005), with permission from Elsevier. [Yamahara 2005].

The conventional YSZ–LSM composite has a lower conductivity than the SYSZ–LSM, starting from 0.004 S/cm at 1100 °C and increasing to 0.02 at 1250 °C and then maintaining this value as a plateau. The corresponding sample with the LSM leached out has a conductivity slightly lower than the YSZ without LSM (open triangles).

We see how adding a cathode layer (LSM) and a scandium oxide phase can alter the conductivity considerably. As the addition or subtraction of such additional phases causes also changes in the morphology and backbone of the electrode or electrolyte, morphology should be taken into account. With a computational modeling, one can address this problem to some extent. The study by Yamahara et al. [Yamahara 2005] is a very good example how structure and transport properties are related.

Yamahara et al. have determined the conductivity in relation to the measured porosity. They could confirm for their sample the Koh–Fortini relationship: the conductivity decreases virtually like $1/\theta$ [Aivazov 1968, Koh 1973], where θ is the porosity. The correct functional relation is given above and in the original works.

7.3.2 Lithium ion conductivity in LiNbO₃ thin films

Electrolytes are not only used for energy storage and conversion applications. The applications can be manifold. Electrochromism is the effect that a material can change its color (Greek *chromos*, color) when it experiences an electric current or a potential. This effect can be used to change the transparency of windows. When the window is coated with an electrochromic layer, applying a voltage can make that the layer darkens and by doing so you can control how much light can enter a room.

The physical origin of this effect is that the optical properties of some materials can be changed when ions are inserted into the crystal lattice. This is not surprising because changing the situation of a crystal lattice by bringing extra ions in or removing ions will alter the electronic structure of the material and therefore also a number of physical or chemical properties, such as magnetic, electric, optical[2] and catalytic properties.

When you arrange this electrochromic material in an electrochemical cell, you can control by the choice of the electrolyte and counter electrode which ions are wanted to enter the crystal lattice of the material. This is essentially an intercalation process. Ozer and Lampert [Ozer 1995] prepared lithium niobate films with a sol–gel process, as shown in the flow diagram in Figure 7.5, and investigated the ion conductivity of the films for electrochromic applications.

They synthesize the films from two metalorganic Nb and Li precursors in solution which they mix and stir for 2 h. They spin coat the mixture on glass,

---

2 WO_3 is a yellow compound, but when you synthesize it, it may come out with a blue color and this is because of the oxygen deficiency of $WO_{3-\delta}$ which typically occurs unless you force the proper phase by synthesizing it at high oxygen pressure. Many other compounds change their color depending on the stoichiometry. Chromium oxides may be green or yellow depending on the oxidation state of the chromium. Iron cyan complexes may be red or yellow depending on the coordination of the iron ion.

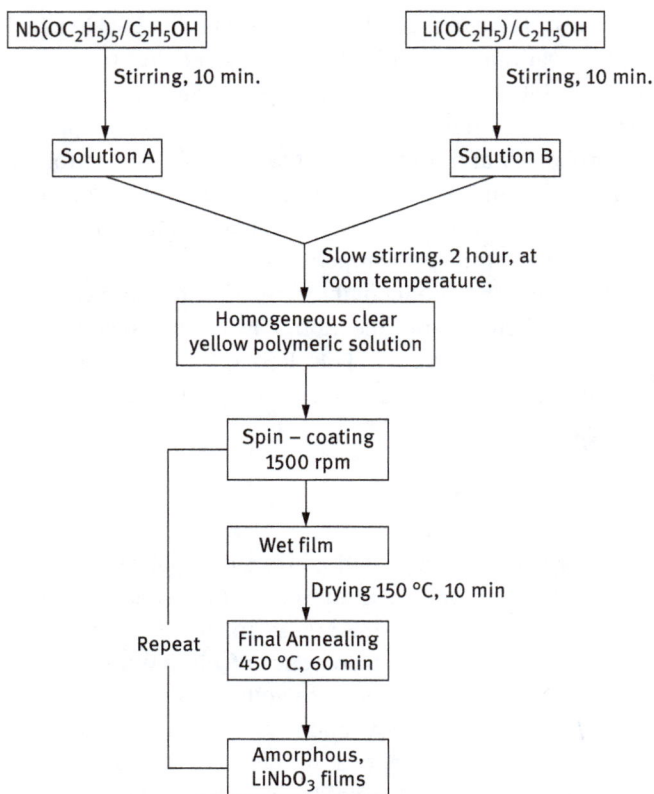

Figure 7.5: Process flowchart for sol–gel LiNbO₃ film preparation. [Ozer 1995] lithium niobate EIS. Reprinted from *Solar Energy Materials and Solar Cells*, 39, Ozer N, Lampert CM, Electrochemical lithium insertion in sol-gel deposited LiNbO₃ films, 367–375, Copyright (1995), with permission from Elsevier.

fused silica glass and ITO substrates and dried the wet film for 10 min at 150 °C and then annealed the films for 1 h at 450 °C. The spin coating was repeated when thicker films were needed.

They used anhydrous $LiClO_4$ crystals dissolved in propylene carbonate as electrolyte. Three cyclic voltammograms with scan speeds of 20, 40 and 80 mV/s are shown in Figure 7.6. The authors report that three to four cycles were necessary for a stable electrochemical behavior of the films. Shown are therefore the fifth cycles. The authors calculated the deposited and removed lithium via the transferred electric charges by Faraday's law and found the values to be roughly equal, which means virtually all inserted lithium on the cathodic scan will also be extracted on the anodic scan.

The Li^+ insertion and extraction process was highly reversible as demonstrated by over 1200 cycles. The anodic current and cathodic current differ by a factor of two

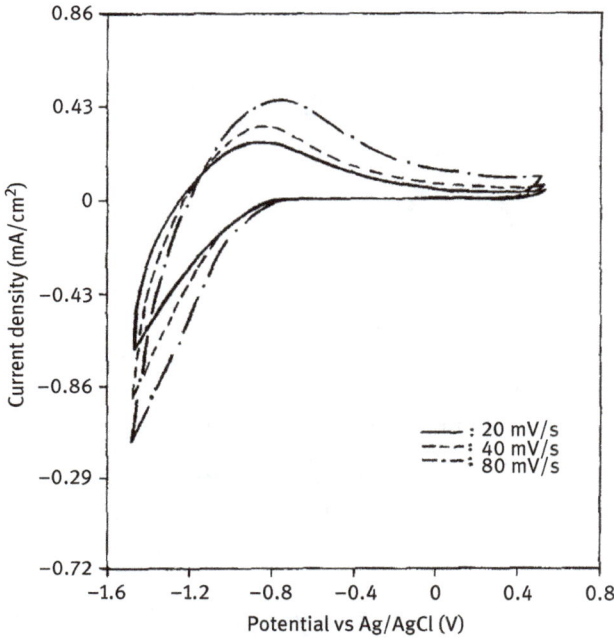

Figure 7.6: Cyclic voltammograms for Li⁺ intercalation/deintercalation in amorphous LiNbO₃/ITO/ glass electrodes in 1 M LiClO₄/PC. Different scan rates are shown.
Reprinted from *Solar Energy Materials and Solar Cells*, 39, Ozer N, Lampert CM, Electrochemical lithium insertion in sol-gel deposited LiNbO₃ films, 367–375, Copyright (1995), with permission from Elsevier [Ozer 1995].

which means the conductivity is changing during insertion and extraction of the lithium ions. This is not surprising because presence or absence of particular ions in a compound can alter the properties of a material significance, which may include the conductivity, as suspected here.

The conductivity is measured with impedance spectroscopy. A representative impedance spectrum, which was recorded from 100 mHz to 60 kHz, is shown in Figure 7.7. The spectrum occupies a range in the order of 1 kΩ and shows overall a large semicircle, notwithstanding that the axes are not drawn isometrical.

The spectrum is modeled with the circuit drawn in Figure 7.7, that is, the electrode resistance R_e (current collectors), the double layer capacitance C_d, the charge transfer resistance R_i (they call it ionic resistance [Ozer 1995]) and a geometric capacitance between the electrodes C_g.

Ozer and Lampert subjected the films to different heat treatment procedures so that the films changed their structure from amorphous to crystalline. The lithium conductivity ranged from 0.61 to 0.84 µS/cm. Films with higher crystallinity had the lower conductivity. Interestingly, the authors claim that the lithium conductivity

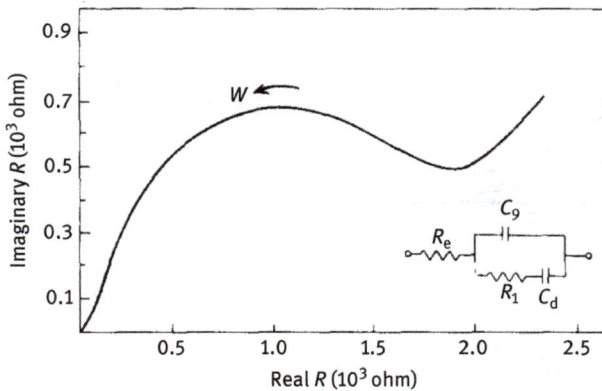

Figure 7.7: Typical impedance plot of a crystalline $LiNbO_3$ film (240 nm thick). Ionic conductivity is $u = 8.4 \times 10^{-7}$ S/cm. The inset shows the equivalent circuit used for analysis.
Reprinted from *Solar Energy Materials and Solar Cells*, 39, Ozer N, Lampert CM, Electrochemical lithium insertion in sol-gel deposited $LiNbO_3$ films, 367–375, Copyright (1995), with permission from Elsevier [Ozer 1995].

depends on the density of the films: denser films have lower conductivity. We have made the same observation with ceramic proton conductors.

7.4 Ceramic proton conductors

7.4.1 The barium zirconate and cerate proton conductors

YSZ is an excellent solid electrolyte and there has not been very much success with finding a better material in the past 100 years. One shortcoming however is that the performance of the material evolves only at temperatures of 600 °C and above. When we think of the SOFC, we have to consider how the other components in the fuel cell behave at such high temperature. The metallic interconnect will evaporate chromium which can poison the cathodes, for example.

The entire infrastructure of a SOFC can suffer from the high temperatures. It would therefore be worthwhile to have an electrolyte which works at lower temperatures. When you mix barium carbonate to the zirconia and maybe yttria, you obtain yttrium-substituted barium zirconate, a compound which crystallizes in the ABO_3 perovskite structure: $BaZr_{1-x}Y_xO_{3-\delta}$. The same holds with barium cerate $BaCe_{1-x}Y_xO_{3-\delta}$. Many materials have appreciable proton conductivity.[3]

---

3 Tantalum oxide (Ta_2O_5) is an important material for capacitors, which have a high dielectric breakdown voltage even at high temperatures. Those are very good for applications in automotive technology near hot components, particularly close to the combustion engine "under the hood." This

By substituting the Zr^{4+} or Ce^{4+} with a lower valence Y^{3+}, we force the crystal to form oxygen vacancies in order to main charge neutrality. When you expose this material to humidity, the water molecules H_2O will enter the material, and the oxygen ion of the H_2O will fill the oxygen vacancy. The two protons H^+ from H_2O will settle nearby and form O–H bonds according to following chemical reaction [Braun 2009a]:

$$H_2O(g) + V_{\ddot{O}} + O_O^x \Leftrightarrow 2OH_O^{\bullet}$$

In this condition, the protons are part of the crystal lattice and we can talk about a hydride: $H_{2\delta}BaZr_{1-x}Y_xO_{3-\delta}$. Before this stage, the material was only hydrated $BaZr_{1-x}Y_xO_{3-\delta}\cdot H_2O$. When we want to make a proton conductor from this material, we must liberate the protons from the crystal lattice so that they can diffuse "freely." It has turned out that heating the hydrated proton conductor will break the hydroxyl bonds and excite the crystal lattice to give the protons a kick.

With quasi-elastic neutron scattering (QENS), it was shown that protons in such ceramic proton conductors would rotate around the oxygen ions when the temperature was low [Hempelmann 1995, Karmonik 1995, Matzke 1996]. At elevated temperature, the protons can jump from one oxygen ion to an adjacent oxygen ion. This requires the breaking and new formation of hydroxyl bonds. The effect is that the protons act as electric charge carriers.

This is an important finding. There is not much gain for proton conductivity when the protons circle around the oxygen, basically bound on an idle trajectory by the orbital leash. But when the protons jump from one oxygen terminal to the next one, then we have a net charge transport.[4]

Ceramic proton conductors are not only relevant for solid oxide fuel cells. The proton-conducting electrolytes may have various applications, as shown in Figure 7.8

---

material can also be used as proton conductor in protonic devices [Ozer 1994a] Ozer N, He YX, Lampert CM: Ionic-Conductivity of Tantalum Oxide-Films Prepared by Sol-Gel Process for Electrochromic Devices. *Optical Materials Technology for Energy Efficiency and Solar Energy Conversion Xiii* 1994, 2255:456-466. Doi 10.1117/12.185388, [Ozer 1994b] Ozer N, He YX, Lampert CM: *Ionic conductivity of tantalum oxide-films prepared by sol-gel process for electrochromic devices.* Bellingham: SPIE - Int Soc Optical Engineering; 1994. doi: 10.1117/12.185388, [Ozer 1997] Ozer N, Lampert CM: Structural and optical properties of sol-gel deposited proton conducting Ta2O5 films. *Journal of Sol-Gel Science and Technology* 1997, 8:703-709. doi: 10.1023/a:1018396900214. Tantalum is a relatively rare material and there are not many countries in the world with large tantalum mining capacity. Similar holds for niobium. Because of its use in electronics industry, it is considered a strategic material for example in the United States. Until around 10 years ago, the Defense Logistics Agency (www.dla.mil) released a limited amount of tantalum every year for public sale.

4 I can give you here a somewhat simplistic analogue. You remember the hydrogen fuel cell car study in the beginning of this book. It appeared that the car was designed for use for short ranges in cities. When you have only one hydrogen gas station around, you can only operate around this single station. The fuel range may not be wide enough to make it to a very remote second hydrogen gas station. This is certainly frustrating and you cannot make long-range transport. The goal is to either extend the range of the vehicle or to make a denser hydrogen station network.

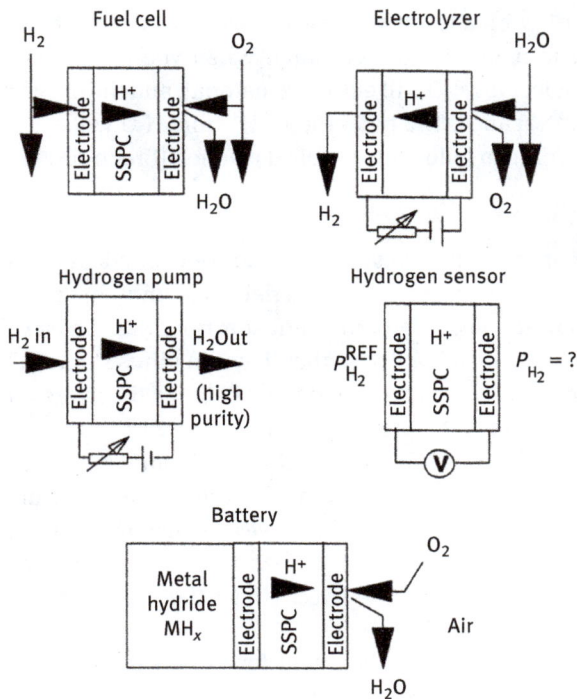

Figure 7.8: Various applications for solid state proton-conducting electrolytes. Reproduced from [Jewulski 1990].

[Jewulski 1990]. The high temperature electrolyzer (SOEC) may run with a proton-conducting electrolyte. Hydrogen pumps and hydrogen sensors could use proton-conducting electrolytes and the metal hydride batteries as well.

7.4.2 Mass gain of ceramic proton conductors upon protonation

The mass gain of the ceramic proton conductors upon humidification (hydration) is minute but it can be measured with a microbalance. You may wonder why one would measure the weight, the mass of such proton conductors but this was necessary for analytical reasons.

When we started our research on ceramic proton conductors, we were confused that we were apparently unable to load the samples with water vapor. After we had sintered the slabs in the furnace at some $T > 1500$ °C, we brought them into a tube furnace and exposed them to argon gas which was humidified. When we measured the impedance of the samples, they showed no change compared to the samples before humidification.

It took us a while to learn that the ceramic slabs would absorb the humidity from the ambient after they were removed from the sintering furnace. You are in the believe that it takes extra efforts to bring protons (water) into the crystal lattice but in fact the ceramics is prone to absorbing water already under ambient conditions. When you then attempt to load the ceramic with extra water, this will have no success because it is already fully adsorbed with water. In order to learn how much water your sample can absorb, you have to dry the sample first. The proper protocol was therefore as follows:

1. Sinter the pressed powder to a monolithic slab in a sinter furnace.
2. Remove the sintered slab from the sintering furnace and dry it in a furnace at 500 °C in dry argon, for example like in (Figure 7.9). Determine the weight (mass) of the sample on a microbalance – immediately after you removed it from the furnace.
3. Insert the sample in the furnace (Figure 7.9) and humidify it with water vapor, provided by humidified argon at 450 °C.

Figure 7.9: The tube furnace on the left contains several proton conductor slabs which are being fed with humidified argon gas. The argon comes from a pressurized bottle (not shown) via a thin white plastic hose into water [for quasi-elastic neutron scattering (QENS), we need H_2O; for neutron diffraction (ND), we need heavy water D_2O] a round bottom flask which is being heated in an electric glass wool heating mantle. The dry argon bubbles in the water which is heated and controlled to 60 °C. The water vapor escapes through the second hose on the top of the flask into the furnace where the samples pick up the vapor.

You may certainly wonder whether we are able to detect such small anticipated mass changes. Well, we had to monitor the presence of protons, of water to the best of our capacity with the means that we had available.

We determine the relative mass change of a ceramic slab to $\Delta m = 100 \times (m_{wet} - m_{dry})/m_{dry}$. The original data from the measurements are shown in Figure 7.10. The left column shows the mass of one slab after drying, $m_{dry} = 1.6416$ g, in the middle the wet mass of $m_{dry} = 1.6441$ g, and the mass after the QENS experiment was done, 1.6433 g. Obviously the sample lost some mass because of residence in the furnace, which is a plausible information.

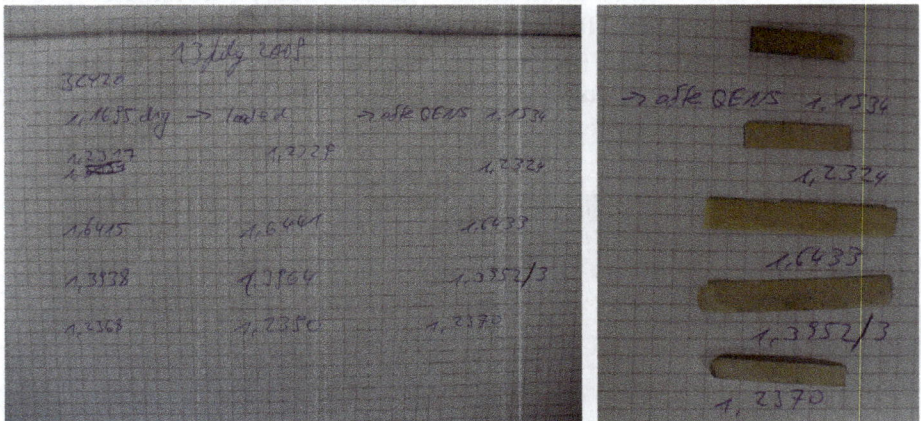

Figure 7.10: Five sintered ceramic slabs of nominal composition $BaCe_{0.8}Y_{0.2}O_3$ (BCY20) on a sheet of paper after they were measured at the Swiss Spallation Neutron Source SINQ. The numbers on the left show their mass after they were dried, m_{dry}, in the middle after they were humidified, m_{wet}, and the numerals on the right are their mass when the QENS experiment with exposure to vacuum and high temperature was finished. The sample on the top with the dark spots and freckles showed anomalous behavior in the impedance measurement and in the weight change. The other samples with almost no spots showed the largest gain in mass.

Figure 7.11 shows the set of sample after loading with water vapor. You can comprehend Figure 7.11 (photo) as follows. First, each dry sample is weighed on the microbalance and its mass written on the plastic bag in black ink and then a photo is made of all six plastic bags with samples inside or outside on the back plus the weight information. Printout the photo on paper for later use. The ceramic bars appear here in gray color.

When the samples are loaded with water vapor, they are weighed again on the microbalance. Now write down the new weight on the printout from the dry samples and put the samples on the position of each bag as shown in Figure 7.11. The samples appear now in yellow color and smaller because the photo is larger than the real size. In the middle in Figure 7.11, you see a large alumina boat which contained the ceramic slabs for heating in the furnace.

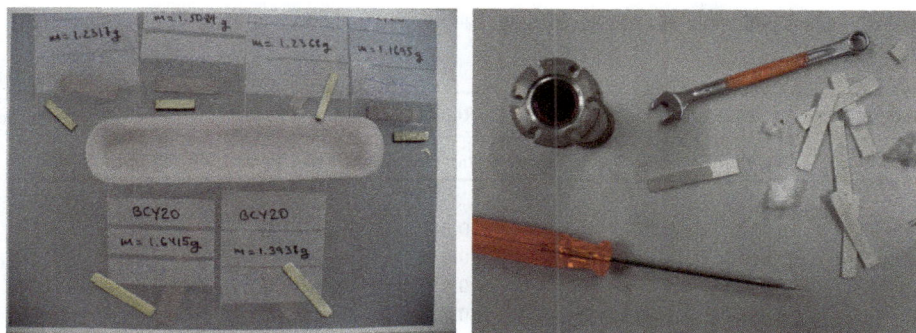

Figure 7.11: Sheet of DIN A4 paper (printout) with photos of six ceramic slabs (stoichiometry BCY20) and their weight after drying (black ink, large gray slabs) and the same slabs (small yellow slabs) after loading with water vapor (blue pen ink). The different size is not sample shrinking but magnification from the printout.

The right panel in Figure 7.11 shows seven such ceramic slabs, one of which is metallized with a platinum coating for electrical contacting. A screw driver and wrench are shown for size comparison. In the upper left, you see the top of a platinum cylinder which is later used for neutron scattering experiments. We will learn about this in the later section of this chapter.

From the mass changes during hydration, we are able to determine the concentration of the charge carriers in the proton conductor. We only need to weigh the relative mass change with the ratio of the molar mass of the water M_{H_2O} and the ceramic proton conductor, for example yttrium-substituted barium zirconate (BZY); M_{BZY10}:

$$[OH^-] = \frac{m_{hydr} - m_{dry}}{m_{dry}} \times \frac{M_{BZY10}}{0.5 M_{H_2O}} = 30.61 \times \frac{m_{hydr} - m_{dry}}{m_{dry}}.$$

This is a practical photographic way of keeping record because it allows sorting the samples without marking them specifically. The marker effect is on one hand given by the mass because the samples largely differ in weight. Irregular shape of samples helps also telling apart which sample is which sample. But a photo is a very good visual record.[5]

---

5 There may be good reasons for making many samples as identical as possible. But when absolute identity (same size, same color, same mass, shape, etc.) is not necessary, then you should not make extra efforts for something which is not necessary. When you need to make a parameter study on a dozen or several dozens of samples, let's say from a large glass pane, you can smash it with a hammer in a number of pieces and when they have roughly the same size, you can continue your experiments with them after you made photos of them for easier identification. Otherwise, you will need a glass cutter or glass pen and it is not always easy to mark them properly. It is also not always easy to make them perfectly equal. Then, don't waste your time on such useless task. Nowadays, many people have

Table 7.1: Mass of sample BCY20 ceramic slabs determined on July 13, 2009, at Empa with a microbalance.

Sample no.	Mass of slab		Rel. %	Mass of slab	Rel. %
BCY20	Dry; m_0	Loaded; m_l	$100 \times (m_l - m_0)/m_0$	After QENS; m_{QENS}	$100 \times (m_{QENS} - m_0)/m_0$
1	1.1659	–	–	1.1534	−1.0721
2	1.2317	1.2329	0.097	1.2324	0.057
3	1.6415	1.6441	0.158	1.6433	0.110
4	1.3938	1.3964	0.187	1.3952(3)	0.100
5	1.2368	1.2390	0.178	1.2370	0.016
6	1.5084	1.5094	0.066	–	–

Sample no. 6 was not included in the QENS measurement because of too low mass change (too low proton concentration).

Table 7.2: Mass of sample BCY20 ceramic slabs determined on 13 July 2009, at Empa with a microbalance.

Sample no.	Mass of slab		Rel. %	Mass of slab	Rel. %
BCY10	Dry; m_0	Loaded; m_l	$100 \times (m_l - m_0)/m_0$	After QENS; m_{QENS}	$100 \times (m_{QENS} - m_0)/m_0$
1	1.3142	1.3138	−0.030		
2	1.8527	1.8523	−0.022		
3	1.9773	1.9773	0.000		
4	1.9044	1.9040	−0.021		
5	2.4332	2.4309	−0.095		
6	0.9682	0.9682	0.000		
7	0.8572	0.8571	−0.012		

The mass and mass changes for sets of ceramic slabs with two different stoichiometries, BCY20 and BCY10, are listed in Tables 7.1 and 7.2.

We have not used the mass change for any quantitative analyses. Although when you know the mass gain during hydration, you should be able to tell how many protons are distributed in the crystal lattice. As the protons are the charge carriers, their concentration enters the equation for the conductivity, such as the Nernst-Einstein relation.

---

a smart phone with a camera. Ten years ago at a conference in Salzburg, Austria, I witnessed a traffic accident at a busy crossroads in the city center. The police officer stood in the accident and documented the scene with his mobile phone, back then not yet a smart phone. During your experiments make many photos, it will help you later because of GPS signatures, time stamps and so on. Today, as everything is digital, you need not worry about being economical with printing costs. It will also help the readers of your master thesis or doctoral thesis when you disclose photos from your experiments.

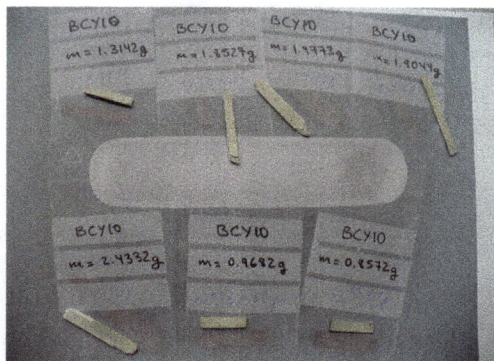

Figure 7.12: Sheet of DIN A4 paper (printout) with photos of six ceramic slabs (stoichiometry BCY10) and their weight after drying (black ink, large gray slabs) and the same slabs (small yellow slabs) after loading with water vapor (blue pen ink). The different size is not sample shrinking but magnification from the printout.

You remember that the conductivity is essentially the product of the charge carrier concentration and the charge carrier mobility.

We prepared the proton conductors by the conventional ceramic solid state reaction, where we mixed barium carbonate and cerium oxide (CeO_2), zirconium oxide (ZrO_2) and yttrium oxide (Y_2O_3) in the stoichiometric proportions, mixed it and ground it in a mortar in acetone. At times, we even mixed the powders in a ball mill. But be aware that ball milling may incur impurities in the powders. This is not withstanding that the powder precursors themselves may contain already impurities from ball milling at the manufacturer. You will only know about this if you do a thorough chemical analysis of precursors.

After drying the powder mixture, they were typically calcined for say 6–12 h at around 1000 °C. This condition would cause decomposition of the carbonate. We then would remove the material from the furnace, grind it again in a mortar and press it to pellets uniaxially at may be 50 kN. These pellets would be for 24 h at 1700 °C, resulting in white ceramic pellets suitable for further use in proton conductor experiments.

7.4.3 Thermogravimetric analysis (TGA) of hydration and proton loading

With a fine balance in a furnace, you can monitor mass changes during oxidation and reduction and adsorption and desorption processes while changing the temperature and atmospheric conditions of the sample. Since we measure the mass of the sample with a balance during temperature change, we call this thermogravimetric analysis (TGA). Such instruments are commercially available; we used a TGA Netzsch STA 409 CD.

The model that we should have in mind is the following: the ceramic powder is adsorbed with humidity from the ambient, and the water molecules from the humidity sit in pores in the powder particles, and some water molecules have entered the crystal lattice and have become dissociated; the protons from the H_2O are residing in the crystal lattice as interstitials and the oxygen ions from the water may have filled the oxygen vacancies which we have doped into the lattice by replacing tetravalent Zr^{4+} by trivalent Y^{3+}.

The TGA graph in Figure 7.13 looks confusing but we can handle its complexity. The green line shows the temperature ramp and goes with the abscissa on the right side from around 100 °C to over 900 °C. When you forget about the other colors, you notice that the temperature goes up to 900 °C and then down to around 100 °C, and then again up to 900 °C and down again.

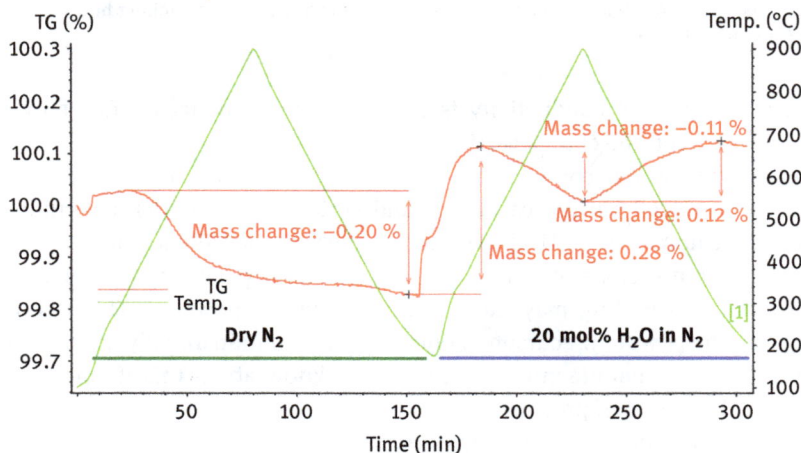

Figure 7.13: Mass change of 10% yttrium-substituted barium zirconate (BZY10) powder in dry conditions (left) and loading water to incorporate hydrogen (right). Sample synthesis, measurement and analysis were carried out by Dr. Alejandro Ovalle, Empa.

Then, we see the temperature ramp extends over 300 min, that is, 5 h. In the first 160 min, the temperature ramp is done in dry nitrogen (dry N_2, green thick horizontal bar). You have to handle and arrange this in the TGA instrument. You need a bottle of compressed nitrogen gas for this. After that, you repeat the temperature ramp in nitrogen loaded with 20 mol% water H_2O in the nitrogen (blue thick horizontal bar). This is the second phase of the experiment which extends over 150 min. So, we heat up and cool down in dry nitrogen, and then, we heat and cool in wet nitrogen.

The red line monitors the mass of the sample during the temperature and gas treatment as described before. For this, we look at the TG abscissa on the left, which begins with an arbitrary 100%, because this is where the experiment starts. In the first 5 min, the mass decreases from 100% to 99.98% and then increases in the following 5 min to 100.02%. This very small change occurs while the sample and sample compartment are being heated. During this time – we must watch now the green T-line and the abscissa on the right – the temperature increases from 100 to 160 °C.

Then over the next 20 min, the temperature increases linear with flat slope for the mass (100.03%) until it has reached 360 °C. Then, at 25–30 min, the mass drops relatively fast from 100.03% to 99.86% when the temperature is maximum 900 °C at 80 min. All potential water trapped in the BZY10 powder will be evaporated. Even during cooling down after t = 80 min, the mass slightly decreases. When the dry N_2 flow is stopped at t = 150 min, T has decreased to 230 °C. The mass at t = 150 min reads 99.83%. The mass had the maximum reading with 100.03% at around t = 25 min. The mass difference was therefore 0.20%. Now, we can consider the BZY10 powder "dry" and free of protons.

The data which we obtain from this sample can be considered reference data. Note however that this sample is a powder. We cannot make a 4-point DC conductivity measurement and we cannot make a dilatometry measurement because this here is a dispersed powder and not a monolithic pellet.

At around t = 160 min, we begin filling the sample compartment in the TGA system with humidified nitrogen while we start the next temperature ramp. The steep rise in the mass of the powder at t = 155 min is because of the adsorption of moisture (99.83–99.94%). At t = 160 min, there is a second onset of mass gain which finds a maximum at t = 185 min with a mass gain from dry condition of 0.28%. Here, the temperature is T = 450 °C.[6] With still increasing heating time and temperature, the mass is decreasing from slightly below 100.12% to 100.01%. This is a difference of 0.11%. Therefore beyond 450 °C, the BZY10 powder is again releasing water vapor.

Therefore, the best water loading (hydration procedure) temperature seems to be 450 °C because here the weight gain is 0.28%. We assume that the same temperature is the proper one for the proton loading of pressed pellets.

7.4.4 Conductivity of ceramic proton conductors

The conductivity of ceramic proton conductors is typically measured with electrochemical impedance spectroscopy (EIS). The proton conductors are pressed into pellets and sintered and then their structure is determined with XRD (phase analysis) and infrared or Raman spectroscopy (vibration properties). For these experiments, you scrape of some grains from the pellets. If you like, you prepare an extra pellet under

---

6 In order to follow my story step by step, it may be worthwhile that you copy Figure 7.7 and paste it in a graphics program and overlay it with a grid so that you can identify for every time the stamp in the TGA, also the mass in % and the temperature T. This is an exercise for the reader.

the same conditions in the furnace and scrape off material from the surface, and from inner regions, and then do XRD, IR and Raman on these. This is the way to get surface specific and bulk specific and intermediate region phase information.

Then, you coat the pellets at the bottom side and top side with platinum paste and burn it at 950 °C, similar to the ones shown in Figure 7.1. Then, you press a platinum mesh on these top and bottom faces and contact with them frequency response analyzer (FRA) and run the EIS. Preferably you do that with a dry and with a humidified pellet and see the difference in EIS spectra. When you can do the impedance measurement *in situ*, for example with a ProboStat™ from NorECs (Norwegian Electro Ceramics AS), then you have a better control over the relevant thermodynamic parameters T and p.

Figure 7.14 shows three impedance spectra from BZY10 which has been loaded with protons. The part spectrum shown in the left panel was recorded with the sample temperature being 300 °C (573 K). The open symbols (o) are data points recorded at frequencies from 1300 Hz to 3 MHz. When we increase the sample temperature, the impedance spectrum "shrinks" so that we can see semicircle extending from around 0 Ω cm at 3 MHz to 160,000 Ω cm at 0.1 Hz.

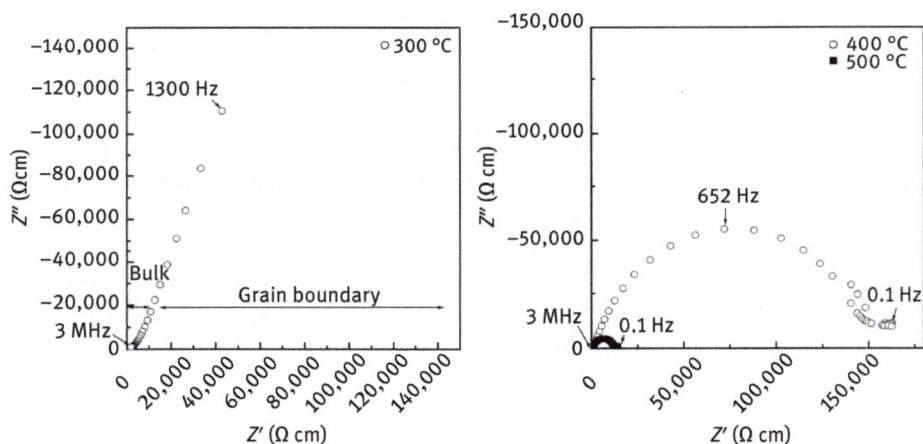

Figure 7.14: Nyquist plots of the impedance spectra of BZY10 for temperatures 773, 673 and 573 K. Reprinted with permission from Springer Nature: *Journal of Applied Electrochemistry*, 39:471–475, Proton diffusivity in the BaZr$_{0.9}$Y$_{0.1}$O$_{3-\delta}$ proton conductor, Braun A, Duval S, Ried P, Embs J, Juranyi F, Strässle T, Stimming U, Hempelmann R, Holtappels P, Graule T, (2009). [Braun 2009a].

With still increasing temperature, at 500 °C, the spectrum is dramatically shrinking, as you see from the position of the filled square symbols (■) which extends from around 0 Ω cm at 3 MHz to 12000 Ω cm at 0.1 Hz. We remember that the radius of the semicircle corresponds to the resistance. From the shrinking of the impedance spectra semicircles in the right panel, we now see that the spectrum in the left panel is actually only an incomplete large semicircle. The interpretation is that with increasing temperature, the proton conductivity is also increasing.

The impedance spectra of such ceramics are typically composed of three contributions, that is, the conductivity of the bulk material, the conductivity of the grain boundaries and the conductivity of the electrodes or current collectors attached to the ceramic. Grain boundaries are typically causing a resistivity toward ionic transport. Ceramic engineers who are familiar with ionic conductivity typically try to make ceramics with a small grain boundary portion. They will also avoid using dopants as sintering aids when these cause mediocre electric properties.

In Figure 7.15, you see three impedance spectra of hydrated $BaCe_{0.8}Y_{0.2}O_{3-\delta}$ (BCY20) which we have recorded at 125 °C. The inset in Figure 7.15 shows the change of the mass of specimen during temperature increase. The impedance spectra are modeled with an electric circuit made from three parallel circuits of a resistor and a capacitor. The grain boundary contribution is made from the parallel circuit of R_{GB} and C_{GB}. Hence, the electric properties are represented (and simplified!) into a parallel circuit from these two components. We do the same with the volume contribution R_{bulk} and C_{bulk}. The contribution from the two attached electrodes is summed up by just one set of $R_{electrodes}$ and $C_{electrodes}$.

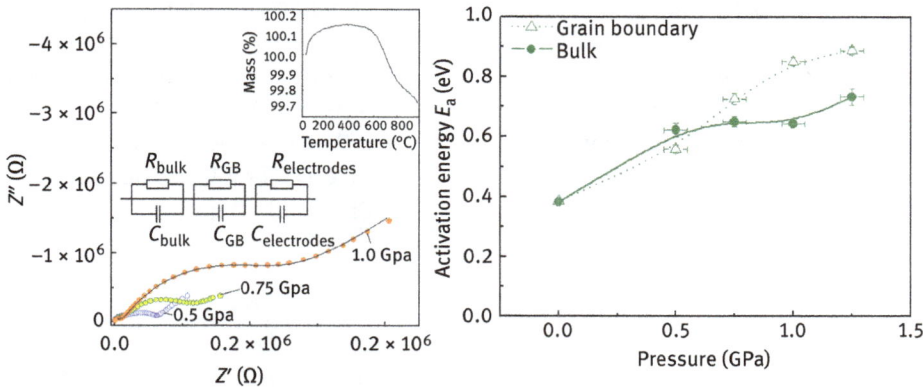

Figure 7.15: (Left) Nyquist plots recorded when the samples were at 125 °C under different pressures (0.5, 0.75 and 1.0 GPa) and one representative fit; inset: mass loss up to 1000 °C; the equivalent circuit used for the impedance spectra modeling. (Right) The activation energy of bulk and grain boundary conductivities.
Reprinted from *Journal of the European Ceramic Society*, 31, Chen QL, Braun A, Yoon S, Bagdassarov N, Graule T, Effect of lattice volume and compressive strain on the conductivity of BaCeY-oxide ceramic proton conductors, 2657–2661, Copyright (2011), with permission from Elsevier [Chen 2011b].

With these conventions clarified, we can inspect the spectra. The spectra were recorded at the Institute for Geosciences at Universität Frankfurt in Germany, who operates a setup for taking EIS spectra at high pressure (not gas pressure, but compaction of specimen by isostatic pressure) and high temperature. The spectrum with the blue data points was measured when the BCY20 was at 0.5 GPa pressure. With a tolerant eye, we can identify three semicircles, which are fitted according to the above model circuit. The spectrum with the yellow data points was measured when the same sample was further compacted to 0.75 GPa. We notice now that the spectrum is growing toward larger impedance on the real axis (Z') and imaginary axis (Z'').

The spectrum at 1 GPa extends even further. The tiny semicircle near the origin is the spectroscopic signature of the bulk conductivity. The bulk conductivity is therefore very high in comparison to the grain boundary conductivity and the electrode conductivity. At the lower pressures, this tiny semicircle on the left is virtually invisible, suggesting an even higher conductivity for the volume.

The deconvolution of the spectra into resistivities and capacitances yields the fit parameters which we can further evaluate. The reciprocal of R_{GB} and R_{bulk} yields the conductivities σ_{GB} and σ_{bulk}, which can be plotted versus the reciprocal temperature (you need a sufficiently wide series of temperature under which the impedance spectra are measured), and then yields from the slope in an Arrhenius plot the activation energy for the proton conductivity process. We have done this [Chen 2011b] and obtained the corresponding activation energy E_a.

The right panel in Figure 7.15 shows the thus obtained activation energies for varying pressure values under which the spectra were recorded. The activation energies for the bulk proton conductivity (filled circle) and the grain boundary (open triangle) are both increasing with increasing pressure. It would be worthwhile to make proton conductors under tensile strain and then see whether the activation energy would considerable decrease.[7] This would make a favorable proton conductor for operation at lower temperatures [Braun 2017c, Chen 2010, 2011b].

7.4.5 Determination of proton conductivity with neutron methods

We can determine the ionic conductivity at the molecular scale with QENS [Hempelmann 2000, Mashkina 2005]. This neutron method uses the broadening of the neutron peak when the neutron probes scatter on ions which are in motion. The neutron scattering cross-section is particularly large for light weight elements such as hydrogen and its isotopes, but also for lithium and oxygen, for example.

Therefore, when we investigate materials which contain hydrogen, we can anticipate a large scattering contrast. When the neutrons hit on ions in the crystal

---

[7] Looking at the green curves in Figure 7.15 (right panel), we have to extend these or extrapolate these to negative pressure; this would be tensile strain. It is possible to impose tensile strain on a material for example with epitaxy of ultrathin films.

lattice which are in motion, the elastic neutron scattering peak is accompanied by extra intensity at the flanks of the elastic peak. This extra intensity is a broadening of the neutron peak which originates from the diffusion of ions or other motion of ions in the crystal lattice. The width of this broadening is a measure for the diffusivity of the neutron scattering ions and can be converted to ion conductivity [Chudley 1961].

To carry out such experiment, we must bring the hydrated proton conductors in the neutron beam and heat the sample up during measurement. We also must assure that the high temperature does not cause evaporation of the humidity which is trapped in the ceramic sample. We therefore use a platinum cylinder (Pt has a very high transmission for neutrons; Pt allows the neutrons from the beamline to come to the sample and it will allow the scattered neutrons to pass to the neutron detector) which contains the ceramic slabs which I have shown in Figure 7.16.

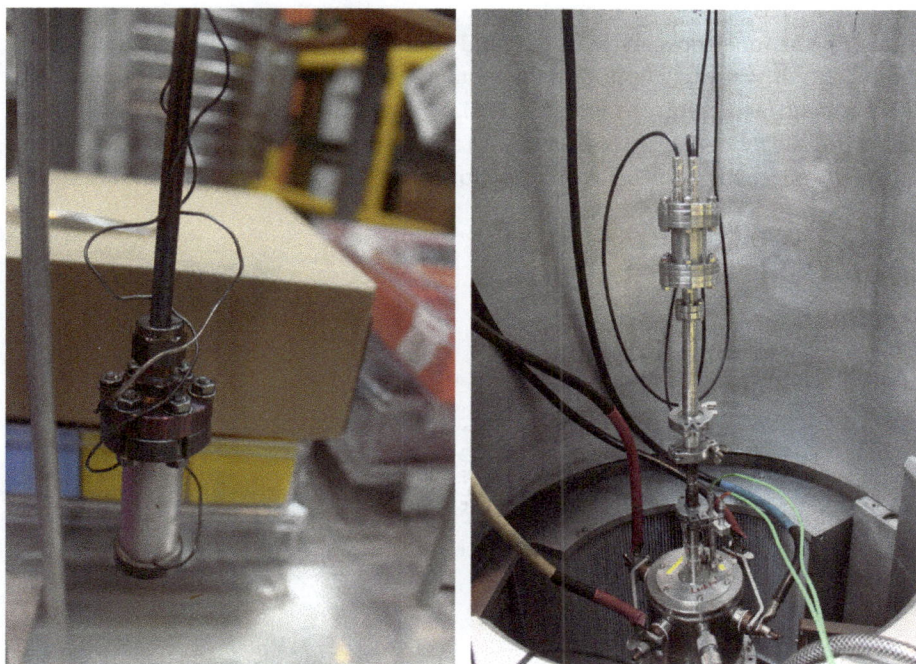

Figure 7.16: (Left) Platinum cylinder containing ceramic samples inside, two thermos couples attached at bottom and top for temperature registration. The metal tube on the top is connected with a pressure valve and contains the wires for electric to one metallized ceramic slab inside the cylinder. The other ends of the wires are connected with electric feedthroughs, which are connected with the sample cables to the frequency response analyzer (FRA). (Right) The platinum cylinder plus samples are inserted in the oven in the neutron measurement chamber. The lengthy metal tube arrangement above the furnace is a result of different mechanical and electrical standards for different vacuum flanges. The two black cables at the top of the flange yield to the FRA which is several meters away from the sample.

The Pt container has a pressure valve which must manage that the cell does not explode from evaporating humidity when inserted in the furnace in the neutron measurement chamber. One ceramic slab inside the Pt container is connected via electric feedthroughs to the FRA so that we can record impedance spectra while QENS spectra are being recorded. The pressure valve warrants that not all protons recombine with oxygen ions to water and become baked out and disappear from the ceramics.

Figure 7.17 shows one exemplary QENS spectrum (open symbols) recorded at 900 K. The green line is the elastic peak which contains no energy loss, and the blue line is the QENS broadening which is due to the disorder by moving protons during the experiment. The full width at half maximum (FWHM) Γ of this broad peak can be translated to proton diffusivity according to the Chudley–Elliott model [Chudley 1961]:

$$\Gamma(Q) = \frac{l}{\tau}\left(1 - \frac{\sin(Ql)}{(Ql)}\right),$$

where τ is the time between two jumps and l is the distance made by one jump, and Q is the neutron scattering vector (or the momentum transfer; it represents the reciprocal space and is therefore measured in 1/Å). You can run a least square fit to the QENS data with this model, or you can make a Taylor expansion of the above equation for small (Ql) and then obtain an approximation, which to calculate I leave as an exercise for the reader.

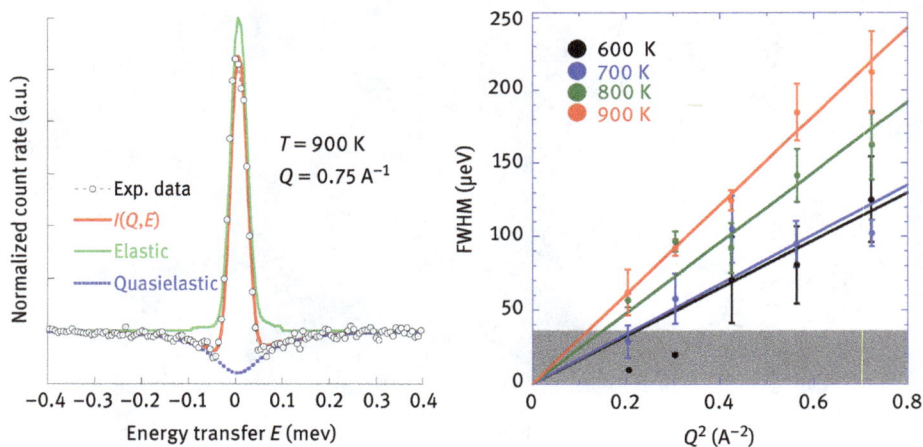

Figure 7.17: (Left) Neutron spectrum deconvoluted into elastic peak (green line) and quasi-elastic peak (blue). The spectrum was recorded from BZY10 at 900 K with neutrons with wave vector $Q = 0.75$ 1/Å. (Right) FWHM of QENS spectra from samples at four different temperatures plotted versus the square of the wave vector Q as first approximation of the Chudley–Elliott diffusion model.
Reprinted with permission from Springer Nature: *Journal of Applied Electrochemistry*, 39:471–475, Proton diffusivity in the BaZr$_{0.9}$Y$_{0.1}$O$_{3-\delta}$ proton conductor, Braun A, Duval S, Ried P, Embs J, Juranyi F, Strässle T, Stimming U, Hempelmann R, Holtappels P, Graule T, (2009) [Braun 2009a].

When you then plot the FWHM over Q^2, you may obtain a linear curve with a slope which is basically the diffusion constant D. The jump length, jump time and diffusion constant are related by the following expression:

$$l^2 = 6D\tau.$$

The right panel in Figure 7.17 shows such plot where the FWHM is plotted versus Q^2 for QENS spectra recorded at sample temperatures from 600 to 900 K. You notice that the slope increases with increasing sample temperature.

Using the Nernst–Einstein relation,

$$D = \sigma \times \frac{k_B T}{e^2 c_p}$$

we can transform from the diffusivity D to the conductivity σ. The constants are the Boltzmann constant k_B, the electron charge e and the proton charge carrier concentration c_p. With the thus determined conductivity, we can make Arrhenius plots and determine the activation energy and compare this with the activation energy determined from electroanalytical methods.

The advantage of using QENS is that we do not need any electric contacting for the sample. We even can measure powder samples with QENS. Macroscopic defects such as cracks in electrolyte layers will not be registered by QENS. QENS is therefore a very good method for the determination of the upper limit of ionic conductivity.

We have recorded with one exception all QENS data in our studies [Braun 2009b] at the Swiss Neutron Spallation Source (SINQ) at the Paul Scherrer Institute in Villigen, Switzerland. The FOCUS neutron beamline [Beck 2001, Janssen 1997, Janssen 2000, Juranyi 2003] has a time-of-flight (TOF) neutron spectrometer for that purpose.

Figure 7.18 shows the FOCUS neutron beamline in the SINQ experimental hall. The yellow metal fence keeps persons from accessing the area which is under control against exposure to neutrons and γ-radiation. The neutron experiment cannot be started unless the yellow door is closed and confirmation by the experimenter is acknowledged that no person is inside the yellow fence.

Neutrons arrive from the right side which shows a large diameter metal tube, the neutron guide tube. Thick walls from concrete and walls from thick lead bricks provide the general protection from exposure to radiation. The sample position is on the right side of the large white (polyethylene) surrounding of the spectrometer. The blue stairs allow to access the top of the spectrometer facility where additional equipment for the experiment can be installed. For our experiment, this was the FRA and supporting computers and other equipment.

Figure 7.19 (left) shows PhD student Qianli Chen [Chen 2012a] preparing the macro for the data acquisition of the EIS experiment, before the neutron experiment is started. EIS and QENS are done in parallel on the same samples. The PC is

Figure 7.18: FOCUS neutron beamline at the Swiss Spallation Neutron Source SINQ at PSI in Switzerland. At the top of the beamline, experimenters may put specific instrumentation as we did for out proton conductor studies, in particular the electrochemical data acquisition system. FOCUS is a joint venture of the Paul Scherrer Institute, Switzerland, and Universität des Saarlandes, Germany.

connected with the FRA. The right panel in Figure 7.19 shows the Solartron® FRA plus an extension kit which allows recording of impedance spectra from poorly conducting samples. This extension kit is necessary because the proton conductors have a very low conductivity at low temperatures. For a full assessment, it is therefore necessary to measure even under conditions which would not be relevant for a later application as proton conductor.

On top of the extension kit is the sample preparation holder which holds the platinum cell and the metal tubes and flanges and feedthroughs before they are inserted into the furnace at the FOCUS beamline. The cables from the FRA to the samples in the cell are several meters long. This is typically too long for conventional EIS experiments because long cables infer a large inductive signature in the EIS spectra which are not welcome during analysis of the spectra from the sample.

But there was no way around long cables in this experiment because the FRA could not be inserted into the furnace for the sake of shorter cables. In the end, it was necessary to record EIS from the cables shortcut with no sample, and EIS from the cables plus sample. The difference of both spectra would allow for information of the inductivity of the cables, which could be determined and accounted for later in the least square fit of the data for the samples.

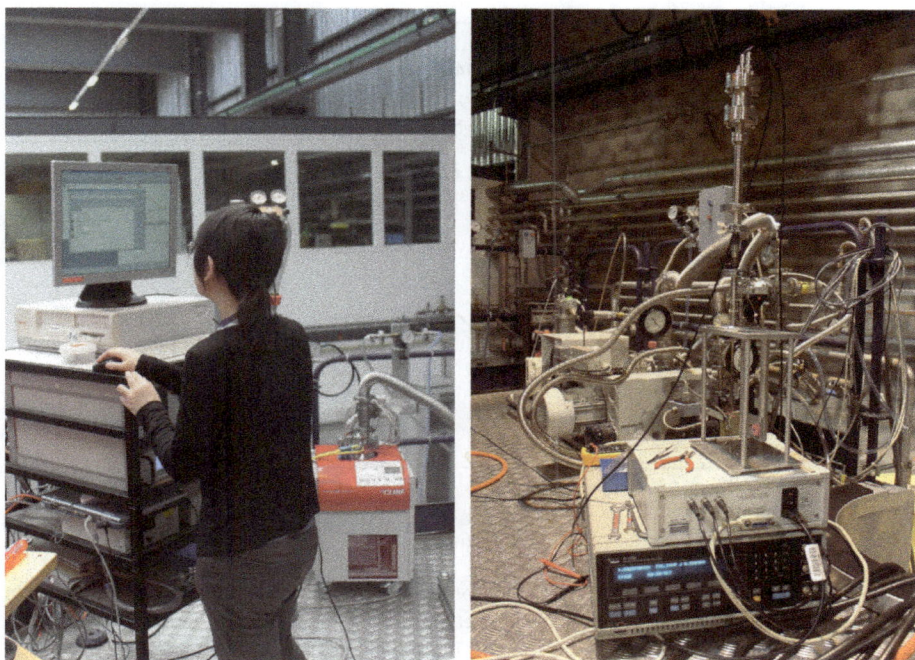

Figure 7.19: (Left) ETHZ/Empa PhD student Qianli Chen on top of the FOCUS beamline at PSI setting the macro for the EIS measurement. (Right) FRA from Solartron® (display with blue letters) and extension kit for low conductivity samples (white box on top) connected to the platinum sample cylinder with black cables; the cell is hooked up in the metal frame on top of the extension kit, waiting to be inserted into the furnace of the beamline.

An alternative to the TOF neutron spectrometer is the neutron spin-echo[8] spectrometer [Karlsson 2010]. We have explored this method at the RESEDA beamline at the FRM-II nuclear research reactor in Garching in Germany in December 2008. Back then, the beamline was not yet fully operational and thus the experiment had some pioneering character. As only one detector was available right after going operational, the statistical significance of the data points in Figure 7.20 is quite low.

The curvature of the spin-echo data from RESEDA is comparable with those found in Karlsson [Karlsson 2010]. Even when the statistics is poor, you should

8 I am grateful to my former supervisor Prof. Peter Böni, who suggested me on the occasion of one of the annual meetings of the Swiss Neutron Scattering Society in Villigen in 2006 or 2007 that I should apply for beamtime at his RESEDA beamline for my "proton-phonon coupling" experiments; these were his words. Maybe this combination of wording was the seed for our later discovery of the proton polaron [Braun 2017b] Braun A, Chen Q: Experimental neutron scattering evidence for proton polaron in hydrated metal oxide proton conductors. *Nature Communications* 2017, 8:15830. doi: 10.1038/ncomms15830.

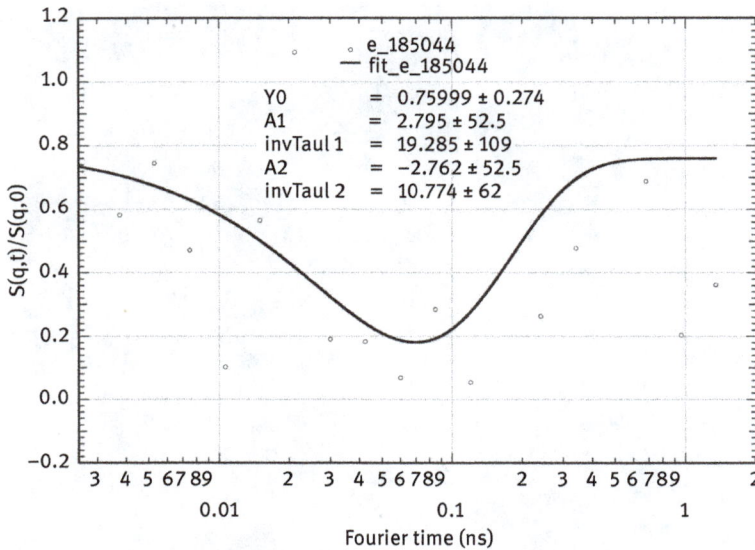

Figure 7.20: Neutron spin echo data of BZY10 recorded at the RESEDA neutron beamline at FRM-II in Garching, Germany in December 2008. The scatter of the data points is large because only one detector was available at the time of experiment. The solid line is a fit to exponentials with fit parameters τ_1 and τ_2 as jump times. The experiment was done together with Dr. Alejandro Ovalle, Empa. Preliminary data analysis by Dr. Wolfgang Häussler, FRM-II.

work with the data and "play with it." Another least square fit of the range with short times where the signal decays, the relation to which reads

$$I(Q, t) = \exp\left(\frac{t}{\tau_{(Q)}}\right),$$

yields a time constant (relaxation time) of $\tau = 82$ ns. With the relation $1/\tau = DQ^2$ (D is the self-diffusion coefficient), we find $D = 2.30 \times 10^{-5}$ cm^2/s. We obtained this very high diffusivity value from a sample which was heated to around 60 °C. This value is comparable to the QENS data which we obtain at high temperatures from 300 to 600 °C.

We know that such high diffusivities relate to the rotational mode of the protons around the oxygen in the crystal lattice. They have fairly low activation energies, but they do not provide any net charge transport. It is therefore not surprising that we observe such proton diffusion mode at low temperatures like 60 °C.

Maybe an interesting note is that the spin-echo data are plotted over the time domain. Impedance spectra are plotted over the frequency domain, which can be transformed to the time domain with a Fourier transformation. The QENS spectra are plotted versus the energy transfer, where the Chudley–Elliott model provides us with the time τ.

With the electroanalytical methods and neutron methods, we have investigated the charge carrier dynamics of the protons in proton conductors. We can apply these methods to all other types of proton conductors, and also to other kind of ionic charge carriers such as oxygen and oxygen vacancies, and lithium or sodium, for example. I have not covered this here in this book but the concept and approach are basically the same.

The methods are not exclusive. We have used also vibration spectroscopy, specifically Raman spectroscopy and learned about the phonon modes in BCY and BZY which are excited by raising the sample temperature [Chen 2011a]. Along with XRD, we measured the lattice spacing and changes thereof during changes of temperature and proton loading [Braun 2009c, Chen 2011a, 2012b, Duval 2007]. We carried out these experiments even under high pressure. With all different data combined, we found that the protons under proton conductivity conditions behave like polarons [Braun 2017b], a suggestion which we were led to by the theoretical work of Samgin [Samgin 2000].

The underlying chemical processes in the materials can be studied very well with X-ray spectroscopy methods [Chen 2013]. I have written about this extensively in my recent book [Braun 2017a] and will therefore refer to that one.

7.5 LiPON battery solid electrolytes

Solid state power sources include not only solid oxide fuel cells but also batteries and supercapacitors and hybrids thereof, as you can see in the movie in Meshcheryakov [Meshcheryakov 2017], where researchers from Russian technology startup company Comberry LLC demonstrate thin film energy storage devices, that is, hybrids of batteries and supercapacitors.

Batteries with solid electrolytes have three major advantages. They are safer because there is lesser chance for a thermal runaway. The performance is better because higher voltages are possible. And they have a much longer cycle life [Jones 2011]. This is why, solid state batteries have been of interest for many decades [Sequeira 1984].

There is a wide variety of solid electrolytes for solid state batteries [Dudney 2009]. Bates et al. [Bates 1992] have discovered in the early 1990s a compound with stoichiometry $Li_{3.3}PO_{3.9}N_{0.17}$ (LiPON) by sputtering of $LiPO_3$ in nitrogen, which has a lithium ion conductivity of 2 µS/cm. This is still a factor 1000 lower than conventional liquid lithium battery electrolytes.

7.6 Some words about ceramic insulators

In this book, there is large coverage about how to improve the conductivity of electrodes and electrolytes. But the control and optimization of the transport properties does not only mean maximize conductivity. Full control means also that you can close and lock some gates to make sure that electrons and ions or mass and heat go only the particular direction that you want them to go by your design of the device or system.

One such blockage is the electrolyte which has only ionic conductivity and no electronic conductivity. There may be situations where you want to have no conductivity at all. Then, we are talking about an insulator.

We were once approached by a company which produces plasma laser systems. The plasma is generated in ceramic tubes. It is important that these tubes are electrically insulating in order to prevent electric shortages near the plasma system. The manufacturer of the laser systems would purchase the ceramic tubes from some vendors and there were various vendors which would offer ceramic tubes that generally fit the requirement for the laser application. But as manufacturer, you must chose along the specifications and "quality" is one important criterion.

We received a set of different specimen from different tube suppliers and they were supposedly of different quality. It was obvious that some of them had few dark spots. The ones with the spots had the highest conductivity as measured with impedance spectroscopy. I do not recollect whether this was electronic or ionic conductivity. But the spots turned out to be the visual criterion that ruled them out for the laser application.

All tubes were from aluminum oxide (Al_2O_3), which is a very good insulator. But the defects associated with the spots ruined the material for that particular application. We never found out the underlying mechanism for the failure.

Another company produced X-ray cathode tubes which had ceramic components. The company reported that the X-ray tubes were only ready for operation in the X-ray diffractometer after they had been heated for quite some time. Without the heating, they would perform poor.

It is my speculation that the ceramic parts of the cathode tubes contained some humidity which would enter the monolith and the crystallites and then reside in there. When a very high voltage is applied, the water molecules in the ceramic may become dissociated (maybe they are already dissociated and the ceramic has protons as interstitials in the lattice) and then can work as protonic charge carriers; when you have a ceramic monolith as insulator, you do not want to have any charge carriers inside; otherwise, it is not insulating properly.

When you bake out the ceramic at several hundred centigrade, 500 °C and more, the humidity in the ceramic may evaporate and you are left in the end with a dry and properly insulating X-ray tube. The drying is then basically a forming or formation step prior to first use of the tube in the X-ray machine (diffractometer, spectrometer etc.).

The X-ray tube company said they had been aware of the problem for a long time and they had found a way to fix it (by heating), but they never understood what was really going on in the material. It is sometimes a matter of available and also qualified staff to work such problems out. It can take many years and millions of dollars before a problem is really understood.

A company which is living from revenues made by sale of products may chose in the end just for finding a pragmatic solution that for the clients and customers. The engineers and scientists personally often would be happy if they really had the time and opportunity to figure out the deeper sources of the problems. I can confirm that this is the case because I have worked in research laboratories in industry and also have collaborated often with such laboratories.

8 Photoelectrochemical cells

8.1 The oil crisis

I remember when I was a young boy, in 25 November 1973, that the government in Germany imposed a ban on driving cars on Sundays [Kusch 2013]. This Sunday ban had not a religious purpose or background. Had it?

On 6 October 1973, Egypt and Syria started a military attack on Israel which is nowadays known as the Yom Kippur War. Yom Kippur is a major Jewish holiday celebrated after the Jewish calendar in September or October. At the same time, the Organization of Petrol Exporting Countries (OPEC) curbed down the petroleum production to force western countries to curb down support for Israel in this war. On the 21 December 1975, an international terrorist team attacked the OPEC conference in Vienna and took several dozen of high ranking politicians as hostages. This attack had the purpose to coerce the OPEC (an organization whose members may have contrasting interests [Tippee 2014]) and manipulate the oil price, supposedly with interest by Libya leader Ghaddafi [Helm 2001].

As oil supply from the OPEC countries became limited, many nations faced an oil shortage and had to impose administrative measures on the consumption. A driving ban on Sunday would not hurt the economy so much because most workers who depended on commuting to work on their own car would need the cars on weekdays. Sunday, in catholic and protestant Germany, a holiday of rest would be the best day for imposing the ban. My cousins and I, because of our young age ignorant of the negative political causes and economic implications of such ban, were happy for the following reason. There was a steep road which invited for making fast rides downhill, but we never could enjoy this nice slope by bike or roller skates because it would cross a major national road. If we ever decided to make a fast run downhill on the bike, we would face the danger of getting hit by a car crossing our way at high speed. Now, with Sunday 25 November 1973, we kids had our free ride for the first time which we enjoyed the whole day. Not one single car was seen on the roads that day.

Why do I bring this history anecdote in this technical book?

Stephen Randolph, Director of the Office of the Historian at the United States Department of State, writes [Randolph 2017]:

> The 1973 Oil Embargo acutely strained a U.S. economy that had grown increasingly dependent on foreign oil.

The dependency on foreign oil had a negative impact on the US economy, and not only on the US economy. The economy worldwide had taken a shock. In 1973, we would not only become aware of this dependency, but we could actually feel this dependency. But a run for petroleum had already been in the second world war and in the years before [Hervey 1994].

https://doi.org/10.1515/9783110561838-008

When electrochemist Adam Heller [Barton 2014] received the ECS Gerischer Award in 2015 in Phoenix Arizona (see Figure 8.1), he made reference to "the first Arab oil crisis" in his speech. Heller works on electrochemical energy storage and conversion, including photoelectrochemistry and bioelectrochemistry.

Figure 8.1: Adam Heller at ECS Gerischer Award Ceremony at ECS Meeting in Phoenix 2015. Photo by Artur Braun. With kind permission from Adam Heller.

Heller's own scientific work appeared therefore affected by the oil crisis. You will notice when you make a thorough literature search that the works on solar energy conversion and artificial photosynthesis [Archer 1975] did spike in the mid-70s as a response to the oil crisis. You will see later in this book in chapter 9 on bioelectrochemistry where I show how generator gas technology was developed for the production of transportation fuel from wood or coal as alternatives to petroleum due to shortage in the 1930s and 1940s.

I used to work several years in Kentucky in a project which was administered by the Consortium for Fossil Fuel Liquefaction Science. Converting coal and organic waste such as plastic into liquid fuels was one of our objectives. South Africa was subject to several embargos due to its Apartheid politics. It had therefore no or little access to mineral oil and thus used the Fischer–Tropsch process for the conversion of its coal (South Africa is rich in coal) into oil. Access to energy in general and access to oil and transportation fuel in particular is essential for the welfare of the industrial countries, developing countries [Turner 1982] and advancing countries [Wakeford 2015]. Petroleum is the most important commodity on the globe and its frequent volatility in trading makes it also subject to speculation and manipulation on the stock market and finance market [Coleman 2012, Hammoudeh 1995]. Therefore, not only abundance but also lack of petroleum is a business opportunity.

The oil crisis is typically blamed on the Arab nations, but there are studies which point to more fundamental negative issues in world economy which may have caused the oil crisis in 1973 [Kitching 1983, Larouche 1995]. Ali Attiga, Director of the Organization of Arab Petroleum Exporting Countries (OAPEC), pledged for more cooperation between the western industrialized nations and the oil-exporting countries [Attiga 1987, Eden 1988]. A good example for this was the cooperation between the Soviet Union and Western Germany from engaging in what is called energy diplomacy [Bosch 2014], which in fact was trading oil and gas for high-tech goods and technology. Often, weaker countries become crushed in the geopolitics and in the fight for access to oil, and this can cause war [Austvik 1992], famine [Deng 2002]. National interests, national security interests are typically mentioned in cryptic terms such as geopolitics and geostrategy. In practice, it means sometimes someone is using force to access the required resources. Many countries with large oil reserves are troubled countries, like Venezuela, Nigeria, Persia, Libya and Iraq, to name a few. The Middle East plays an important role in this respect [Ross 2005]. Qatar wanted to deliver natural gas to Europe, and the necessary gas pipeline should cross Syrian territory. In 2009, Syria's President Assad decided that such pipeline could not be built in Syria. Russia, Syria's most important ally, delivers natural gas to Europe via pipeline. Natural gas from Qatar would be a competition in a European energy market which has a volume of hundreds of billions of dollars. In 2011, a civil war started in Syria with the aim to overthrow the Assad regime. Military from many countries stepped their boots on Syrian soil. Over 10% of the Syrian population got killed or injured in the war. Millions of Syrians sought refuge in other countries.

There is another civil war going on in Yemen now. Saudi Arabia invaded Yemen in 2015, leading a collation of several other countries who want to protect and stabilize the Arab peninsula by blocking Yemeni access to ports and airports with supplies. Saudi's strongest opponent, Iran – another country with huge petroleum reserves, is supporting Yemen. Currently, over 10 million people in Yemen are in danger of starvation. Diseases like cholera and diphtheria have been reported [Breul 2018]. When you have a look on the world map you will see that the countries near the major arterial maritime routes for petroleum, like Yemen, Somalia, Sudan, Eritrea and Ethiopia – have a history of distress and conflict. You may wonder why the small country of Djibouti in the Red Sea is host to many foreign military bases.

Bockris provides an interesting overview of the works around the 70s where the term hydrogen economy was coined [Bockris 2002, Chen 2017]. His colleagues came to the conclusion that sending hydrogen over a distance of at least 200 miles would be cheaper than sending electricity over the same distance.

What I want to point to is that the access to energy is very important for the welfare of nations. This is why many countries have given this topic highest priority. The United States have established specifically for that purpose a Department of Energy (DOE), which has a huge budget, major share of which is spent for applied science and basic science of everything related to energy. Particularly, the rich countries in the world depend on access to energy, primarily on fossil fuel from the OPEC countries which are concentrated in Arabia and the Middle East.

The majority of the conflicts in this region, including international terrorism, have their roots in the need for petroleum as energy source. A look on the world map shows that the maritime trade routes for oil and gas are associated with open or covert warfare. To the least until the 1970s, activity in terrorism would have benefits on the financial returns from oil business [Blomberg 2009].

8.2 The case for solar energy and solar fuels

In view of all that what was said in the previous section, we can understand that researchers, actually many researchers, were interested in finding alternative energy sources, particularly solar energy. Solar cells are based on photoelectrodes which provide a photovoltage from which we can draw a photocurrent. Such solar cells (photovoltaic cells) are electric power sources which we all know today. Around 20% of the world energy consumption is used in electric energy. So, theoretically with solar cells one could provide for 20% of the world energy consumption. The other 80% energy is consumed as all kinds of fuels such as fossil fuels, nuclear fuels and biomass. It would certainly be interesting to provide for these 80% of energy by solar energy. It is possible to use the electricity from photovoltaic (PV) cells for the electrolysis of water and thus produce hydrogen as fuel.

One alternative to this indirect combination of PV and electrolyzer technology would be to dip the photoelectrodes right in the water (or an aqueous electrolyte for the obvious reason of a necessity of ionic conductivity). Adam Heller was one of these many researchers who worked on the photolysis of water for solar hydrogen[1] production [Heller 1981].

---

[1] For those interested in nano science and solar hydrogen, I recommend the book by Lionel Vayssieres [Vayssieres 2010] Vayssieres L: *On Solar Hydrogen & Nanotechnology*. John Wiley & Sons (Asia) Pte Ltd: John Wiley & Sons (Asia) Pte Ltd; 2010. doi: 10.1002/9780470823996.

Utilizing of solar energy in photovoltaic cells was already an established scientific field in the 1970s, and when combined with an electrolyzer, one could produce "solar hydrogen fuel." But a new breakthrough made in 1972 by Akira Fujishima and Kenichi Honda showed how water could be splat (oxidized) with encouraging efficiency into H_2 and O_2 with a TiO_2 photoanode in an electrochemical cell [Fujishima 1972].

8.3 PEC water splitting cells

Fujishima and Honda demonstrated in 1972 [Fujishima 1972] how water could be "photolyzed" (photolysis) to hydrogen and oxygen by a titanium oxide (TiO_2) electrode when illuminated with ultraviolet radiation. When the photoelectrode is irradiated with photons, electrons are emitted from the atoms in the photoelectrodes. The underlying chemical reaction for water oxidation read as follows:

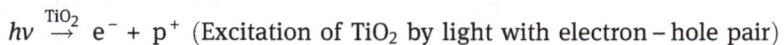

$$hv \stackrel{TiO_2}{\rightarrow} e^- + p^+ \text{ (Excitation of TiO}_2 \text{ by light with electron–hole pair)}$$

These photoelectrons e^- and their conjugated photoelectron holes p^+ (or h^+, depending on the notation), often referred to as electron hole pairs, can be used not only for photocurrent generation but also for chemical conversion of species at the photoelectrode surface. Four holes p^+ are necessary for the conversion of one formula unit of oxygen gas:

$$4\,p^+ + 2\,H_2O \rightarrow O_2 + 4\,H^+ \text{ (Water oxidation at TiO}_2 \text{ photoanode)}$$

The resulting protons H^+ are reduced at the cathode, which delivers the necessary electrons from the electric circuit of the photoelectrochemical cell:

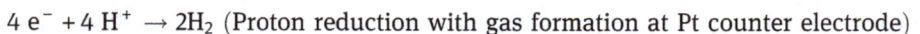

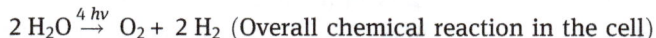

$$4\,e^- + 4\,H^+ \rightarrow 2H_2 \text{ (Proton reduction with gas formation at Pt counter electrode)}$$

$$2\,H_2O \stackrel{4\,hv}{\rightarrow} O_2 + 2\,H_2 \text{ (Overall chemical reaction in the cell)}$$

Fujishima and Honda suggest in their seminal work

> Although the possibility of water photolysis has been investigated by many workers, a useful method has only now been developed. Because water is transparent to visible light it cannot be decomposed directly, but only by radiation with wavelengths shorter than 190 nm [Coehn 1910].

In the years that followed Fujishima's and Honda's work, many researchers tried to find "better"[2] photoelectrode materials and also tried to better understand the

---

2 "better" sounds like a shallow term but it is commonly used and sometimes with great success. In the early 2000s, Kentucky Pizza company "Papa John's" used the slogan "Better Ingredients. Better Pizza. Papa John's" with great success, see for example [Fox 2017] Fox M. Blog. 6 November 2017.

underlying physics and chemistry of photoelectrochemical water splitting. There was also interest in developing different PEC concepts, as is demonstrated in the early review paper by Archer [Archer 1975].

For example, the first reaction equation above shows only the absorption of the photon by the TiO_2 and the production of the electron hole pair. The two subsequent equations show only the chemical reactions of the water and protons with holes and electrons. However, the role of the TiO_2 is not only that of a photon absorber, but also that of an electrocatalyst, on which intermediate compounds are formed, which cannot easily be identified and monitored. Electrocatalysis by itself is a complex field of science and sometimes overlooked by researchers who are looking for a better absorber material.

Titanium oxide is a well-studied and very stable photoelectrode material but it is hardly considered as a material for practical solar applications. TiO_2 has a relatively large energy band gap of around 3 eV (depending on the crystallographic phase rutile, anatase and brookite) and will therefore absorb only ultraviolet radiation. To understand this issue better, let us have a look at the solar spectrum. Figure 8.2 shows the solar spectral photon flux density (solar irradiation) measured on Earth.

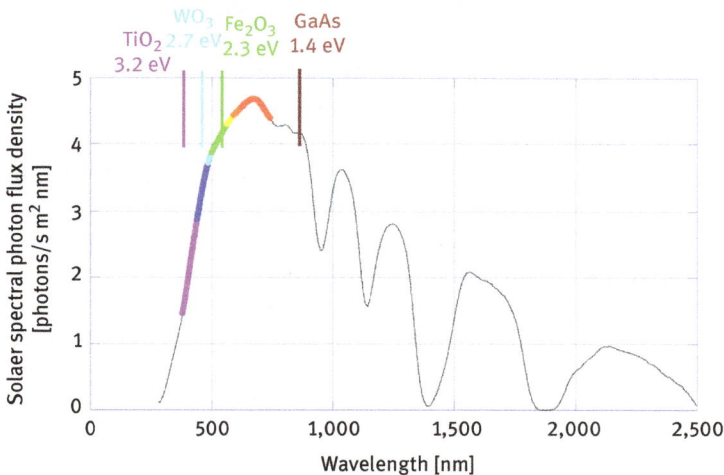

Figure 8.2: The solar spectral photon flux density as measured on Earth.

Papa John's – Better Ingredients, Better Pizza, Better Supply Chain Management Social Media for Business Performance. 2018. https://smbp.uwaterloo.ca/2017/11/papa-johns-better-ingredients-better-pizza-better-supply-chain-management/ . We certainly expect that better materials make better components, and better components make better devices, and better devices make better systems, and eventually a better life for all of us. While we do not specify in what respect we understand "better".

I have plotted the ultraviolet and visible part of the spectrum in the corresponding colors. The lower nm range is plotted in violet and represents the UV range. The relation E (eV) = $1240/\lambda$ (nm) allows to convert the wavelength scale to the energy scale. The band gap energy of TiO_2 (anatase) is around 3.2 eV and falls in the ultraviolet range. The photon flux density for this wavelength and energy range is relatively small ($1 \cdot 10^{18}$–$3 \cdot 10^{18}$), whereas materials like WO_3 and Fe_2O_3 have around double that value ($4 \cdot 10^{18}$–$4.5 \cdot 10^{18}$).

The fact that not so much ultraviolet radiation from the sun arrives here on Earth is the reason why TiO_2 is not a very good solar photoanode. Its photovoltaic efficiency is only 1–2%. We prefer therefore WO_3 and Fe_2O_3 with 5% and 18%, respectively. "Conventional" semiconductors like silicon and gallium arsenide have photovoltaic efficiencies of 25% and 28%, respectively. They are excellent absorbers; however, they are not stable in electrolytes (compare [Hardee 1977]). They will corrode and not survive the harsh electrochemical conditions for a long time.

The width of the bandgap is in general a measure for the strength of the chemical bonds [Grätzel 2001]. This dilemma has kept researchers from making great progress in PEC technology development. The good absorbers are not stable, and the stable materials are not good absorbers.

It is only since recently that researchers applied the simple idea of coating a chemically inert overlayer from aluminum oxide over the not so stable absorbers and then obtain a well-performing PEC photoelectrode. While this idea sounds trivial, its scientific background is not. Aluminum oxide is an insulator. Why would it permit electrons or holes or ions to pass?

To the best of my knowledge, no study has been published yet which explains this contradiction. What we know is that the very thin aluminum oxide film can hardly be made out as a crystallographic phase in X-ray diffraction. This is not only because it is very thin. Possibly this is an amorphous phase which, particularly as it is very thin, can allow the electron holes produced in the absorber electrode to tunnel through the layer into the electrolyte. This is only my speculation and it would require a number of dedicated analytical studies to clarify this question.

8.3.1 Semiconductor Photoelectrochemistry

Hardee and Bard [Hardee 1975] synthesized TiO_2 by chemical vapor deposition (CVD) and processed TiO_2 photoelectrodes under slightly reducing conditions (vacuum, high temperature) and thus obtained an n-type semiconductor which they compared with TiO_2 which was grown by anodization of a Ti metal substrate. In a follow-up work, they followed the same route for Fe_2O_3 [Hardee 1976].

We remember in the previous Chapter where we inserted a metal electrode in an electrolyte. We recall that the redox potential of the electrolyte should be considered constant, whereas the Fermi energy of the electrode can adjust to the level of the redox potential on the energy scale to obtain equilibrium of energy. As the electrode is a metal, the shift of the Fermi energy will be quite abrupt.

The situation is different when the electrode is a semiconductor such as GaAs or doped silicon or some seemingly simple metal oxide such as TiO_2, WO_3, or Fe_2O_3. These electrode materials do not have "free electrons" and therefore the Fermi level cannot shift like in a metal. The only way to match the redox potential is when the conduction band and valence band of the electrode "bend" toward the redox potential level, depending on which type of charge, positive or negative are accumulating at the surface on the electrode when in contact with the electrolyte.

When the semiconductor material (photoelectrode) absorbs photons of the right wavelength range, pairs of electrons and holes will be formed and the holes in a photoanode, such as hematite (α-Fe_2O_3), can migrate to the surface where they can oxidize water.

Kennedy and Frese hypothesized a second type of hole in hematite with a different structural or orbital origin because the observations made in their electrochemical and optical studies would not add up without such second electron hole [Kennedy 1977]. We were able to detect these two kinds of holes with *in situ* X-ray spectroscopy experiments [Braun 2012c, 2017] and relate one of them with surface capacitive states which one can measure with impedance spectroscopy [Bora 2013].

Such experiments with X-ray or neutrons are not trivial. Figure 8.3 shows a sketch how a microphotoelectrochemical cell is used for such experiments. The cell is a small PEEK® container on which a silicon wafer with a silicon nitride window (X-ray window) is glued. On the back side of the silicon nitride window, we have deposited a very thin hematite photoelectrode which is connected as working electrode with a wire to a potentiostat. The container is fed with KOH electrolyte via a peristaltic pump. A platinum counter electrode in the cell is connected with the potentiostat. A silver wire in the cell is the reference electrode and also connected with the potentiostat.

The X-rays can pass through the silicon nitride window and probe the hematite photoanode and also the electrolyte layer in the cell. When we switch on a solar simulator light and point the light beam on the cell, light will hit the hematite and make electron hole pairs.

When we apply via the potentiostat an external bias, we can lift the conduction band to the water redox potential. Then, the X-ray spectrum of the oxygen in the hematite will show two extra features in the X-ray absorption pre-edge which are shown in blue in the Figure 8.3 [Bora 2013, 2012c, 2016b]. These can be identified with structures found in the impedance spectra [Bora 2013].

Further reading on semiconductor photoelectrochemistry is the reviews by Lawrence "Laurie" M. Peter [Peter 2016, 1990] and the encyclopedia edited by Allen J. Bard [Bard 2002].

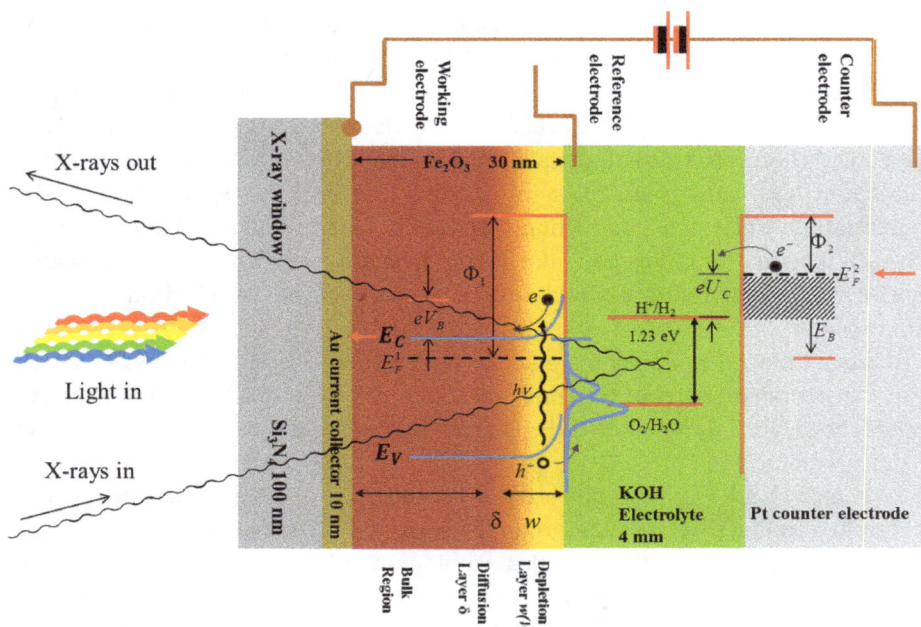

Figure 8.3: Schematic of a microphotoelectrochemical cell for *operando* and *in situ* X-ray spectroscopy measurements with energy levels from X-ray spectroscopy and electrochemistry [Braun 2012c, 2017].

8.3.2 Photoelectrochemical workstation

A typical photoelectrochemical workstation is shown in Figure 8.4. The computer controls the potentiostat, which is connected with three cables with the cappuccino cell. The electrochemistry data can then be viewed on the monitor, which is connected to the computer. The yellow arrow shows the optical path between light source (solar simulator) and sample (in the cell). There is a colored safety goggle which protects your eyes from the exposure to high intensity visible light. When no UV filter is used before the solar simulator, a darker safety goggle is necessary.

The solar simulator has a xenon bulb, the emission spectrum of which resembles quite well the solar spectrum. The solar simulator has two knobs which allow adjusting the bulb so that the focus of the light hits the desired position on the sample. There is a lens in front of the solar simulator which allows for adjusting the focus. Typically, there is also holder for a filter which you can place in front of the solar simulator. Often a UV filter is used. With a photodiode, which you place at the sample position, you can calibrate the light intensity on the sample position. Further literature on solar simulators and spectral calibration is found in following references [Emery 1986, Gueymard 2002].

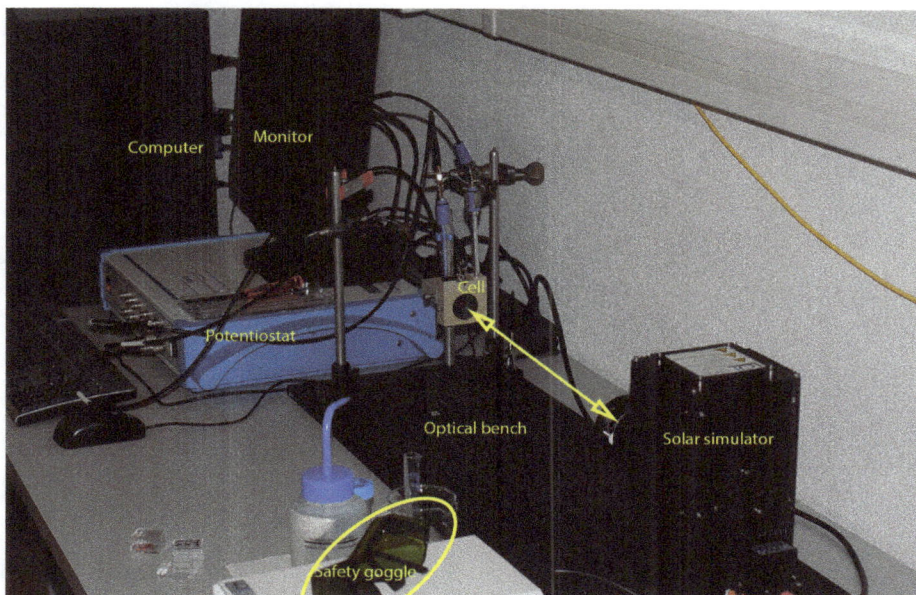

Figure 8.4: Photoelectrochemistry work station with solar simulator and cappuccino cell holder mounted on an optical bench.

The power supply for the solar simulator (in Figure 8.4 it is a white box under the green safety goggle [Baczynska 2012, Paul 2008, Williams 1972, Young 2000][3]) has an ignition button with which you start the solar simulator. Before that you have to select the power with a knob. Before you do the measurement, you should wait for maybe 30 min until the power output of the lamp is stable. It is advised that you shield the sample from the exposure to the light for that period, unless for any specific reason you chose to do so. Some manufacturers equip the solar simulators with an automatic shutter: the bulb is on but the light is not coming out until the shutter is opened. Some solar simulators allow inputting an electric signal which makes that the shutter opens and closes with a given transient pattern.

For example, you can program that the shutter is open for 3 s and then closes for three seconds and so on – automatically. By doing so, you can produce with ease a dark current spectrum and a light current spectrum in one single scan.

Figure 8.5 shows how the cappuccino cell is placed. The counter electrode and reference electrode are inserted into the electrolyte compartment. The photoelectrode (working electrode) is inside the cell and connected with a crocodile clamp, along

---

3 Protective eyewear such as safety goggles is mandatory in chemistry laboratories. When we work with solar simulators we need additional protection against the intense radiation from the xenon bulb.

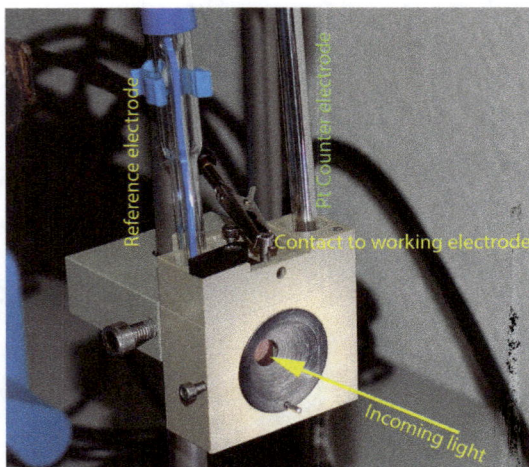

Figure 8.5: Spectrophotoelectrochemical cell (Cappuccino cell) held by a laboratory stative.

with the two other electrodes to the potentiostat. The cell has in the front a window with a transparent glass slide, over which a black aperture cone is placed. When only the solar simulator is providing light, the light beam will pass through the aperture and through the window, electrolyte and hit the photoelectrode. The aperture will cast a shadow everywhere else in the cell.

As the light shall shine perpendicular on the photoelectrode, no reference electrode of counter electrode shall be in the optical path. This makes a fundamental problem for the electrochemical experiment which in the ideal way would require that the counter electrode and working electrode (photoelectrode) are facing each other. Instead, they are standing side by side. Also, the reference electrode is quite far away from the photoelectrode. It is therefore necessary that we use electrolyte with a high conductivity so that the primary and secondary current distributions are still favorable for ions (compare [Newman 1967]). For the reactor design, the considerations on geometry are very important, see for example [Hankin 2017, Newman 1962].

Since we are using transparent glass slides as photoelectrode support, it is possible to turn the electrodes around and have the glass slide facing the light beam. The photoabsorber material is then hit by the back. You may then notice that the photocurrent density is higher than when you illuminate the absorber from the front.

8.3.3 Photocurrent spectroscopy

One purpose, maybe the principal purpose of photoelectrochemistry, is the conversion of solar energy into electric energy which can immediately be converted into chemical energy. This is why we study the photocurrent of materials which we consider relevant for photoelectrochemical cells (PEC).

Figure 8.6 shows the *I–V* curve of an iron oxide photoelectrode which was prepared by dip coating of an FTO-glass into an iron ion containing solution. I will discuss these curves and these data later in this section. For the synthesis, iron nitrate was dissolved in oleic acid and heated at 70 °C until a homogeneous solution was obtained. The solution was then heated for 90 min at 125 °C and then cooled to ambient temperature. 24 h later it was mixed with tetrahydrofuran and centrifuged, which caused a separation of iron oxide nanoparticles and a supernatant liquid [Bora 2011] – the aforementioned iron ion containing solution.

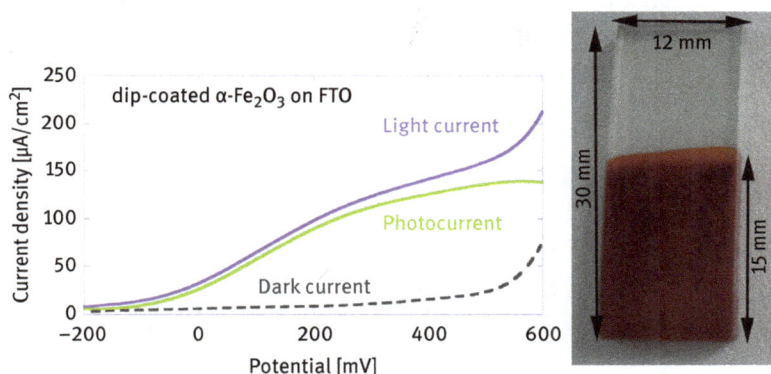

Figure 8.6: (Left) dark current density, light current density and photocurrent density of an α-Fe$_2$O3 photoanode versus the applied external potential with Ag/AgCl reference electrode. The photocurrent was obtained by subtraction of the dark current from light current density. (Right) FTO coated glass slide coated with a brownish layer of hematite. The entire assembly is a photoelectrode.

The FTO glass is dipped in this solution and then heated in a muffle furnace in air at 500 °C for 30 min, which gives a α-Fe$_2$O$_3$ film of around 100–150 mm thickness, depending on the solvent content of the supernatant liquid. The right panel in Figure 8.6 shows such sample of 3.5 cm length and 1.5 cm width. The brown red layer is the iron oxide film.

When you repeat the dipping procedure and heating procedure in the furnace, you will grow a thicker film. For your first photoelectrochemical studies, it may be just important that you get a sample which shows somehow a "photoeffect." But at some point, you may want to know how thick the film is. For the thickness determination, you may use a profilometer, which scratches with a needle a ditch into the film in one lateral direction, and with a scan of a probe tip it will determine a height profile perpendicular to that ditch. The ditch itself will then allow for the height or thickness of the film versus the substrate. Or you may use a stylus instrument which measures the height of the substrate (this is by definition height = 0) and then you move the sample a little and then you measure the height of substrate plus film. You may also break the sample and inspect the cross-section in an optical microscope or an electron microscope. When the film has optical quality, one may use optical ellipsometry or X-ray or neutron reflectometry for measuring the film thickness. The sensitivity and resolution limit of the aforementioned methods is different. The determination of film thicknesses is not always a trivial task. Sometimes, it requires some creativity by the researcher.

We have measured the thicknesses of our films with a profilometer because we had such available. Figure 8.7 shows the thicknesses of a number of hematite films, which were coated on FTO glass slides via the aforementioned synthesis method. It needs no keen eye to realize that the film thickness, which we measure in nanometer, scales linear with the number of dipped layers.

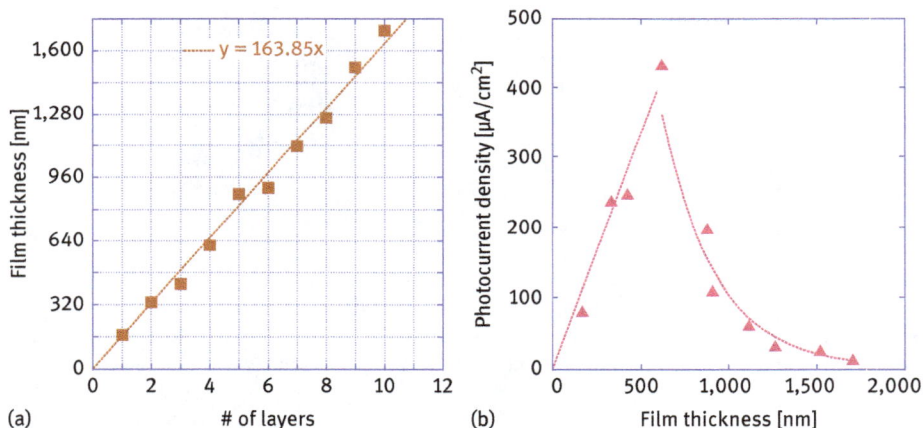

Figure 8.7: (Left) Increase of the film thickness with increasing number of dips for α-Fe_2O_3 after dip coating and firing in the furnace. (Right) Dependence of the photocurrent density from the absorber film thickness. Four layers warrant the maximum photocurrent density. More layers cause a decrease of the photocurrent density because the resistivity increases.

This is certainly what we expect – that the dips are adding arithmetically. However, unless we measure this we cannot be sure. I have laid a linear least square fit through the data points. I have also told the plot program (KaleidaGraph™) that the least square fit curve must go through the origin so that with 0 dip the film thickness is 0 ($y = a \cdot x + b$ with b=0.). This is a reasonable constraint. The fit program then returns $y = 163.85 \cdot x$. The slope of 163.85 (nm) is the average thickness for the film that results from one dip. The dip coating method is therefore quite reproducible.

For the better illustration, I have laid a grid in the plot with a thickness increment of 160 nm and a layer increment of 1. Now it is quite obvious that one dip has ~ 160 nm, two dips have ~320 nm and so on. The top data point for 10 dips yield 1710 nm, which would mean that one dip yields 1710/10 nm = 171 nm.

When you do not have access to a very sensitive thickness determination instrument such as a profilometer available and only a stylus, then you have to resort to the method of just adding up many layers (you must count them) and when the film is thick enough, measure it with the stylus and divide this thickness number by the number n of layers. Of course you cannot be sure that all layers have the equal thickness, but the method is already a very good help.

Let us now go back to Figure 8.6 (left panel), which shows the photocurrent density versus the potential (reference electrode Ag/AgCl). When the experiment is done in a room which is not illuminated, then we call the obtained current "dark current." The dark current is below 10 $\mu A/cm^2$ but increases noticeable at $V > 300$ mV. After 500 mV, the dark current increases considerably and this indicates that the water in the electrolyte is becoming oxidized.

When the solar simulator is switched off and the room light is switched off, no electron hole pairs will be produced in the absorber in the photoelectrode. The potentiostat will then record only the dark current, as shown in Figure 8.6. The dark current density is determined by the absolute current measured in Ampere (or μA) divided by the active electrode area which is exposed to the electrolyte in the cell.

We must be cautious when inserting the electrode in the electrolyte because we have to consider some geometric constraints. Let us therefore look in the schematic in Figure 8.8. It shows three arrangements how a photoelectrode (red color with area A) coated on an FTO slide (green color) is inserted in an electrolyte (dark blue) container (light blue). The left cell shows how the iron oxide layer is partially inserted into the electrolyte. The area which dips in the electrolyte and therefore forms an electrochemically active solid–liquid interface is $A' < A$. The dark current density is $j = I/A'$.

In the arrangement in the middle, the whole slide is dipped too deep into the electrolyte. The entire iron oxide electrode area A plus a part of the FTO substrate with area B dips into the electrolyte. When we connect this arrangement with the potentiostat, the total measured current, I, will have to be related to areas A and B. The iron oxide will likely produce a different current density than the FTO. The total measured current will then be $I = j_{\text{iron oxide}} \times A + j_{\text{FTO}} \times B$. We can avoid the hassle of singling this out by avoiding the mistake of dipping too deep into the electrolyte.

When we point the light source on the sample, only that part of the photoelectrode will produce a photocurrent which is exposed to the light. When the cell has an aperture that limits the area over which light will shine, then this is the illuminated area C. The photocurrent is then the product of photocurrent density and area C: $I_{\text{photo}} = j_{\text{photo}} \times C$. Note, however, that the iron oxide is largely dipped into the electrolyte. Therefore, the total measured current is the dark photocurrent over area A' and the photocurrent over area C: $I = j_{\text{iron oxide}} \times A + j_{\text{photo}} \times C$.

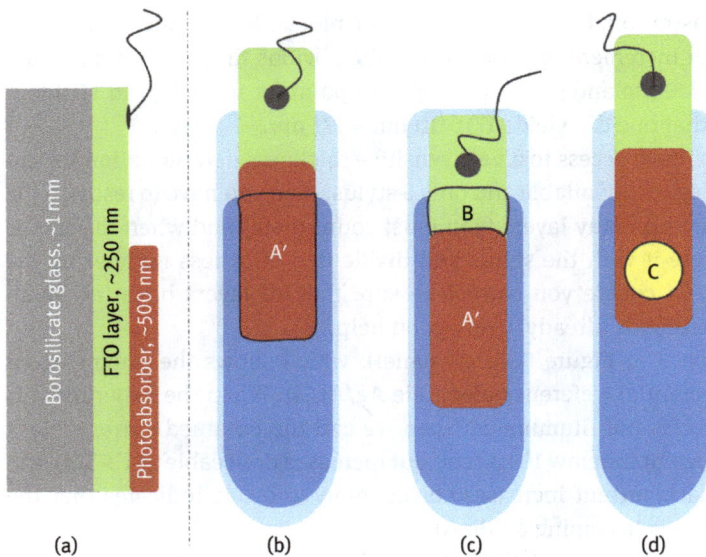

Figure 8.8: Geometrical arrangements of photoelectrode, with photoabsorber (red, with total area A) coated on FTO (green) inserted in electrolyte (dark blue) in a vial (light blue) in dark and illuminated conditions. For simplicity, reference electrode and counter electrode are not sketched. (a) Sagittal cross section view on the photoelectrode with borosilicate glass substrate (grey, ca 1 mm thick), transparent conducting oxide layer (FTO thickness can range from some 10 to some 100 nm), which is partially coated with an absorber layer (red, can be several hundred nanometers thick). The electric contact wire is soldered with indium metal on the TCO. (b) Frontal view on the photoelectrode partially dipped into the electrolyte; this is the proper situation. A' is the geometrical area of the photoelectrode in contact with the electrolyte. (c) Photoelectrode fully dipped into the electrolyte with active geometrical area A. Even the FTO is in contact with electrolyte, with active geometrical area B in contact with electrolyte. This arrangement is improper. (d) The photoelectrode is properly inserted in the electrolyte and the light beam hits only the photoabsorber through the electrolyte. The area C which is exposed to the light causes the photocurrent. The area A (not shown) adds the dark current to the total measured light current.

The current density that you measure from a cell like the one in Figure 8.5 when the photoelectrode is under illumination is therefore the light current, and not the photocurrent. For the determination of the photocurrent you will have to make a weighted subtraction of the dark current from the light current. This is shown in Figure 8.6. Note that the dark current is spiking at around 600 mV, and so is the light current. After the weighted subtraction, we obtain the photocurrent curve (green curve between black and violet curve) which runs relatively flat at high potentials.

Most researchers neglect this procedure of dark current subtraction because the photocurrent is often several orders of magnitude larger than the dark current. Then the contribution of the dark current to the light current is negligible and the result from a light current measurement is virtually the photocurrent.

It has become established in the PV research that efficiencies and other PV performance parameters are reported according to standards. Around 10 years ago, a similar need for standardization was identified by some of the increasing number of researchers in the PEC communities. I can recommend the paper written by Chen [Chen 2011] and published in the Journal of Materials Research.

Right now (Summer 2018) there is another initiative by some of the members of the PEC communities to update these standards and include the device and reactor fabrication.

8.3.4 Electronic defect states

In Section 4.6 I wrote about the relation of transport properties and structure. Defects are important inhomogeneities in materials. While the term defect has an entirely negative connotation, defects[4] can play also a benign role in structure and in transport. Therefore, in the analytical sciences like here, we should beforehand be neutral about the role of defects.

A defect-free photoelectrode or photoabsorber would be a single crystal. In electrochemistry however, we prefer porous electrodes which have a high internal surface area and allow for a large electrode area to be in contact with the electrolyte. This way you can make from 1 cm^2 geometrical electrode area 1000 m^2 active surface area for electrochemical reaction, for example. The pores and grain boundaries are two-dimensional defects.

When photons hit a semiconductor, electron hole pairs are formed and these can be considered as an exciton and a defect. The electrons and holes may diffuse through the semiconductor and may arrive for example at a point defect and then annihilate. This effect is not welcome, and such defects would therefore be malign.

The term defect can be understood as a symmetry breaking. The surface of a material is the archetype of symmetry breaking in materials science. All atoms (ions) in the volume inside the material have saturated chemical bonds and orbitals, but the atoms which constitute the surface of the material may not have the necessary bonding partners which the ions in the volume have. Igor Tamm [Tamm 1932a, b] pioneered the theory of localized electronic states which are formed at surfaces from such two-dimensional defect [Cole 1997]. Such surface states may influence the electron

---

4 Defects are not only possibly benign for electronic or magnetic and optical properties. Also mechanical properties can benefit from the presence of defects, for example dislocations which are relevant for Orowan strengthening or hardening [Queyreau 2010] Queyreau S, Monnet G, Devincre B: Orowan strengthening and forest hardening superposition examined by dislocation dynamics simulations. *Acta Materialia* 2010, 58:5586-5595. doi: 10.1016/j.actamat.2010.06.028. It is regrettable that many researchers who work in materials science have a negative prejudice about defects because of ignorance about some benign action of defects.

transport and hole transport between the electrolyte and the electrode. When we monitor the current and potential in an I/V curve, we basically should be able to sense such electronic surface states and other relevant states.

You can see early works from Tomkiewicz and Tomkiewicz et al. who studied, for example, TiO_2 and CuInSe surfaces and developed a method for experimental determination of surface states [Shen 1986, Tomkiewicz 1980a, b, 1980c, Ullman 1980].

I am showing here a study from the PhD thesis of Yelin Hu [Braun 2011b, Hu 2016a] where iron oxide photoelectrodes had been processed with an oxygen plasma for up to 20 min [Hu 2016b]. Depending on the plasma processing time, the photocurrent would worsen. Figure 8.9 shows the photocurrent density of iron oxide photoelectrodes which were subject to 2.5, 5, 10 and 20 min oxygen plasma treatment (all solid lines). The dashed lines are the corresponding dark currents. The black curve has the highest photocurrent density of around 0.6 mA/cm^2, followed by the red curve with lower photocurrent density (2.5 min plasma treatment).

You see from the photocurrent data that it is not so trivial to compare them. The photocurrent onset potential may change and also the photocurrent at any other potential may be different from another sample. What we do here then is to select two arbitrary potentials, 1.23 and 1.43 V and compare the photocurrent density versus the oxygen plasma treatment time. Now it is obvious how the photocurrent is decreasing with plasma treatment time t.

A scientific more meaningful approach for comparing such dissimilar curves is by mathematically modeling them with respect to physical and chemical parameters. One early mathematical model for the photocurrent in semiconductors was developed by Gärtner [Gärtner 1959]. One important relation in this model that we should keep in mind when considering the distribution of charge carriers in polarized electrodes is the depletion layer width.

The width w of the charge carrier depletion or accumulation region scales square root with the bias potential V and depends on the flat potential V_{fb} and a material-specific constant w_0 (the "width constant"):

$$w = w_0 \sqrt{V - V_{fb}}$$

A mathematical model that describes the photoelectrode inserted in an electrode was provided by Butler [Butler 1977].

The concentration of the electron holes h is a function of the bias potential V and the relative position x in the film measured versus the top surface of the electrode. L_h is the hole diffusion length and a is the optical absorption coefficient.

$$h = h_0 - (h_0 + A \cdot e^{-aw}) \cdot \frac{e^{w-x}}{L_h} + A \cdot e^{-ax}$$

The photocurrent is the result of holes in depletion layer and holes in the bulk [Butler 1977]. We have investigated the electron holes with X-ray spectroscopy and found

(a)

(b)

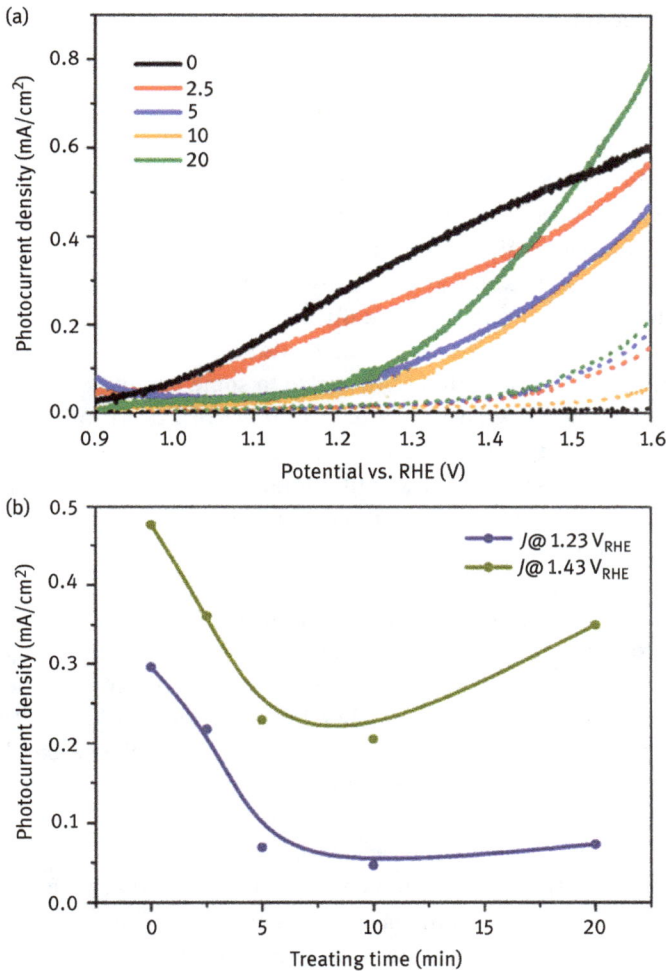

Figure 8.9: Current–voltage (*J–V*) characteristics of hematite electrodes in 1 M KOH following oxygen plasma treatment for indicated periods (min) in dark (dotted lines) and under one sun illumination (solid lines); scan rate = 10 mV/s. (b) Variation of photocurrent density with oxygen plasma treatment time under 1.23 and 1.43 V vs RHE applied bias.

Reprinted with permission from Hu Y, Boudoire F, Hermann-Geppert I, Bogdanoff P, Tsekouras G, Mun BS, Fortunato G, Graetzel M, Braun A: Molecular Origin and Electrochemical Influence of Capacitive Surface States on Iron Oxide Photoanodes. *The Journal of Physical Chemistry C* 2016, 120:3250-3258. Copyright (2016) American Chemical Society.

that there are two hole states in the valence band. The low energy peak originates from O2p electron holes, whereas the higher energy peak is due to charge transfer from Fe3d to the upper Hubbard band [Braun 2012c]. These two kind of holes had been predicted or hypothesized 30 years before [Kennedy 1978].

The photocurrent can be fitted with the relation

$$J = q\, \varphi_0 \left(1 - \frac{e^{-\alpha W_0 \sqrt{(V - V_{fb})}}}{1 + \alpha\, L_h} \right),$$

where q is the electron charge and φ is the potential. The photocurrent curves can therefore be compared with respect to the parameters absorption coefficient α, hole diffusion length L_h, flat band potential V_{fb} and width constant w_0.

It is my experience that metal oxides are typically under-stoichiometric after synthesis. Our iron oxide photoelectrode may actually be not α-Fe_2O_3 but α-$Fe_2O_{3-\delta}$ and δ accounts for the missing oxygen ions at the iron oxide surface. Therefore, the iron ions are not only in the oxidation state Fe^{3+} but also to some extent Fe^{2+}. The Fe^{2+} is the electrocatalytic driver of the water oxidation. However, anodization of the photoelectrode takes place during PEC operation and this will convert the Fe^{2+} to Fe^{3+}, which can be determined with resonant photoemission spectroscopy [Gajda-Schrantz 2013]. When we apply the external bias to the PEC reactor to bring it in hydrogen production condition, we must be aware that we are anodizing, i.e. oxidizing the anode (photoanode).

We can therefore expect that anode material becomes oxidized. This sounds surprising because the iron oxide photoanode (hematite, α-Fe_2O_3) is already fully oxidized. The iron is in the oxidation state Fe^{3+}, but we just learnt that the iron may be throughout in Fe^{2+} oxidation state at the surface. Therefore, Fe^{2+} will become Fe^{3+}. Moreover, it turned out that under such anodizing conditions, the valence band spectra shift toward the Fermi energy and this implies that we are doping holes into the electrode [Braun 2012b].

It is similar with tungsten oxide but there it is well visible; the stoichiometric WO_3 is yellow but the as-synthesized samples have a blue color and have oxygen vacancies, not only at the surface but throughout the material, hence $WO_{3-\delta}$. This material has metal type conductivity and therefore will be the right photo absorber for PEC applications. You have to oxidize the $WO_{3-\delta}$ under the right processing conditions to obtain WO_3, but it turns out that this is not so trivial a task [Braun 2011a, 2012a].

When we oxidize the surface with a plasma treatment, the performance obviously becomes worse. This might be related with the oxidizing of the electro-catalytic Fe^{2+} to Fe^{3+}.

The impedance spectra of the iron oxide photoanodes (pristine, and in an oxygen plasma treated for 20 min) are shown in Figure 8.10. The two larger spectra were obtained from the pristine electrode at DC bias potentials of 1 V (green spectrum) and 1.25 V (red spectrum) versus reversible hydrogen electrode (RHE) reference. The spectra were fitted with the Randles circuit which is shown in Figure 8.10(b). The photoanode bulk is represented by a capacity C_{bulk}. $R_{trapping}$ is a resistance due to the annihilation of conduction band electrons with valence band holes at surface states. In series with $R_{trapping}$, there is a parallel circuit from a trap capacity C_{trap} and a charge transfer

(a)

(b)

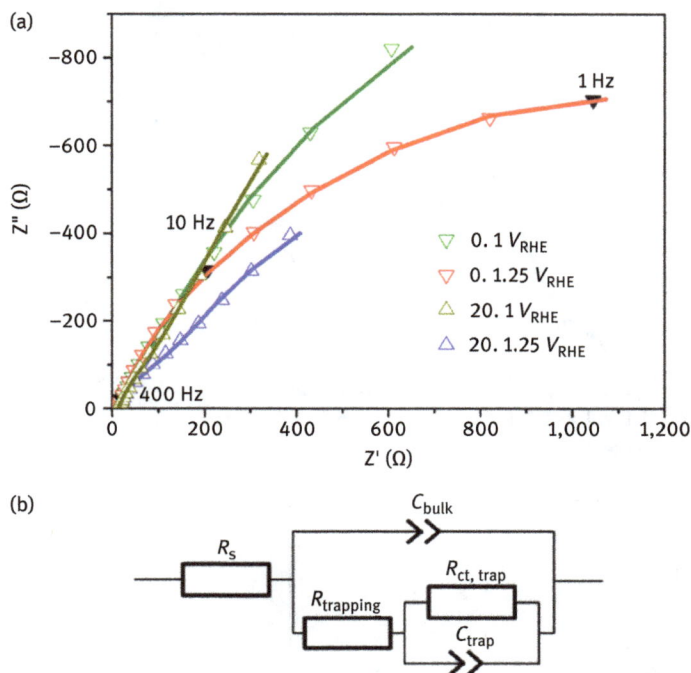

Figure 8.10: (a) Nyquist plots of untreated and 20 min oxygen plasma-treated hematite electrodes under 1 sun illumination and indicated applied dc bias in 1 M KOH. Fitting results indicated as solid lines and selected ac perturbation frequencies labeled as filled black symbols. (b) Equivalent circuit used to interpret spectra.
Reprinted with permission from Hu Y, Boudoire F, Hermann-Geppert I, Bogdanoff P, Tsekouras G, Mun BS, Fortunato G, Graetzel M, Braun A: Molecular Origin and Electrochemical Influence of Capacitive Surface States on Iron Oxide Photoanodes. *The Journal of Physical Chemistry C* 2016, 120:3250-3258. Copyright (2016) American Chemical Society.

resistivity $R_{ct,trap}$. C_{trap} is a trap chemical capacitance which mediates the charging and discharging of surface states and thus the current generation. The corresponding charge transfer resistance $R_{ct,trap}$ – we should better name it charge transfer conductivity – is the transfer of the holes from surface states to the adsorbed water. We ignore here the series resistance R_S.

This Randles circuit model was fitted to the impedance spectra and the trap state capacitance C_{trap} plotted versus the bias potential V, which was applied to the working electrode when the impedance spectra were recorded. The trap state capacitance of the pristine hematite electrode (was not subject to oxygen plasma) has over the potential range from 1.1 to 1.5 V a large peak with maximum of 0.004 F/cm².

The electrodes treated in the plasma have also increased intensity in this potential range, specifically at 1.43 V, but the intensity (0.0001–0.0003 F/cm²) is by a factor >10 smaller than from the pristine electrode, as shown in the magnification in Figure 8.11(b). At this potential, we obviously have an

(a)

(b)

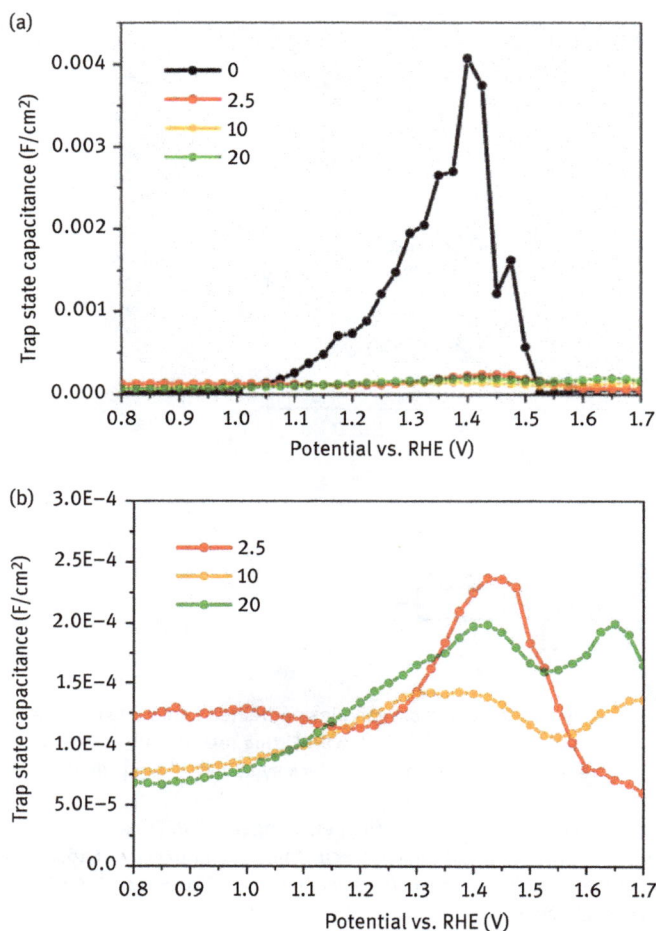

Figure 8.11: (a) Trapped surface state capacitance of hematite electrode vs applied potential in 1 M KOH under 1 sun illumination with oxygen plasma treatment times (min) indicated. (b) Trapped surface state capacitance of hematite electrode vs. applied potential in 1 M KOH under 1 sun illumination with oxygen plasma treatment times (min).
Reprinted with permission from Hu Y., Boudoire F, Hermann-Geppert I, Bogdanoff P, Tsekouras G, Mun BS, Fortunato G, Graetzel M, Braun A: Molecular Origin and Electrochemical Influence of Capacitive Surface States on Iron Oxide Photoanodes. *The Journal of Physical Chemistry C* 2016, 120:3250-3258. Copyright (2016) American Chemical Society.

electronic defect state which causes the high photocurrent. The close inspection of the magnified plot shows that there is more than one peak in the spectra of plasma-treated samples.

This approach is comparable to the Mott–Schottky plot where we run impedance spectra across a range of potentials. The surface state capacitance is then plotted as $1/C^2$ versus the potential and we obtain the charge carrier sign, concentration and the flat band potential V_{fb}.

Every peak is the electric signature of a chemical reaction or a similar scenario between orbitals and electronic states. We have investigated the electrodes with resonant X-ray spectroscopy to get element-specific and orbital-specific information which we can link with the transport properties and photoelectrochemical properties [Hu 2016a, b]. You can read details in [Braun 2017].

I have shown so far commercial batteries, fuel cells, electrolyzers and double-layer capacitors, but you will likely not find a water-splitting cell that you can buy commercially as a consumer. As of yet there exists no technology for hydrogen fuel production based on PEC.[5]

8.4 Dye sensitized solar cells (DSSC)

8.4.1 Sensitization of semiconductors with dye molecules

The only commercialized photoelectrochemical cells that I know are the dye sensitized solar cells (DSSC). It is possible to make some semiconductors, when they are not absorbing particular wavelength ranges, susceptible for these wavelengths when appropriate organic dye molecules are adsorbed on the semiconductor surface. You may then observe in the semiconductor photoelectrode an additional photo effect in that spectral range which originates and which is attributable to the dye molecule. This effect is called spectral sensitization[6] [Gerischer 1968].

It has always been considered as problematic that TiO_2 cannot absorb photons from the visible wavelength range. You can read this from the solar spectrum in Figure 8.2. One idea was therefore to add chromophores or dyes on the TiO_2 which would make the substrate sensitive to visible light.

A very early study on such sensitization can be found by Allisson [Allisson 1930a, b], who used chlorophyll to sensitize photochemical reactions in the group of electrochemist Emil Baur at ETH Zürich, whom we know from the research nonsolid electrolytes and the carbon element and zirconia electrolyte.

Works on sensitization of semiconductors were carried out on ZnO, another important metal oxide semiconductor [Dudkowski 1967]. Dudkowski et al. made

---

5 I would be pleased if someone sends me an email after this book has been published and tells me I did bad research for the book and there actually was a manufacturer and commercial supplier of PEC reactors. The time for this is mature now, I believe. Please send your email to Artur.Braun@alumni.ethz.ch.

6 I am not sure whether there is a clear-cut definition what sensitizer and sensitization means. I believe it relates to the optical absorption of materials, but it may also sometimes be used where the term electrocatalyst would be more appropriate.

"photoconductive couples" where dyes such as methylene blue and eosin, were aggregated on ZnO. The photocurrent at visible wavelength varied with the square-root of the light intensity. The build-up and the decay of the photocurrent had exponential curvatures [Dudkowski 1967].

Helmut Tributsch and colleagues [Ellmer 2008] received (personal communication with the book author at the [Braun 2016a] meeting 2015) the ZnO crystals from Gerhardt Heiland [Ibach 2005] in Aachen. Heiland is known for his works on the electric conductivity of surfaces and depletion regions which he studied on ZnO [Heiland 1954, 1955, 1957, 1961].

The dye can transfer upon excitation by visible light an electron to the conduction band of the semiconductor and then relaxes back in its ground state by the oxidation of an adjacent water molecule. An example for this reaction is given by a ruthenium dye complex (observe the abbreviation; ruthenium bipyridine, $Ru(bipy)_3$) covalently bond on TiO_2 [Anderson 1979, Creutz 1975].

When the dye is excited with light with wavelength below 560 nm (this is a photon energy larger than 2.214 eV), the molecule gets into a charge transfer excited state which is in principal able to reduce H_2O to H_2. The sensitizer molecule Ru $(bipy)_3^{2+}$ will then be oxidized to $Ru(bipy)_3^{3+}$. It is fundamentally bad when the sensitizer dye is "discharged" that way unless there is a process which can return the dye to its original state.

Such process is indeed possible by the assistance of hydroxyl according to following reaction which can regenerate the starting complex [Creutz 1975]:

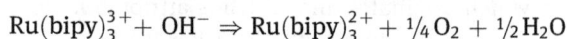

$$Ru(bipy)_3^{3+} + OH^- \Rightarrow Ru(bipy)_3^{2+} + \tfrac{1}{4}O_2 + \tfrac{1}{2}H_2O$$

It is therefore possible to use this system for PEC water splitting.

There are also water splitting systems where the Ru is not used for the spectral sensitization but as electrocatalyst. We have investigated such system [Toth 2016], that is, $[Ru(1)_3][PF_6]_2/Co_4POM$ (1 = 4,4'-bis(^{n}nonyl)-2,2'-bipyridine), which was originally designed by Besson et al. [Besson 2010]. We made drop-casted films and Langmuir–Blodgett films of these molecules on FTO substrates.

One thing is what sensitizer or electrocatalyst molecule you have. Another thing is how it is deposited on the electrode. An excellent molecule can show poor performance when the processing of the electrode is not optimized. Mistakes can be made at every stage of the development of molecules, materials, components, devices and systems.

We chose two different methods for electrocatalyst coating, that is, drop-casting (DC) and Langmuir–Blodgett (LB) films. Figure 8.12 shows the evolution of the oxygen gas over time when the DC bias is applied in the electrochemical cell shown in Figure 8.13. The amount of evolved oxygen is measured with a gas chromatograph (GC).

We see four data curves in Figure 8.12. LB stands for the film produced by Langmuir–Blodgett technology; DC stands for the drop-casted film. CO_4POM is just

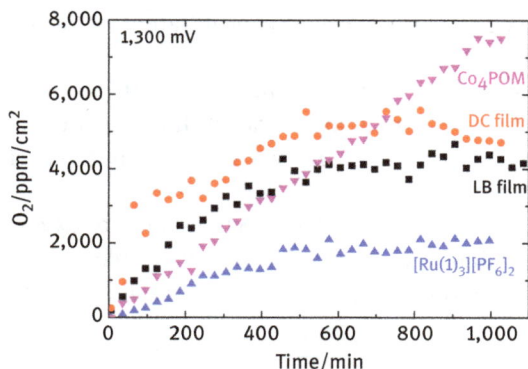

Figure 8.12: O_2 gas evolution vs. time at +1.3 V bias for LB and DC films of the $[Ru(1)_3][PF_6]_2/Co_4POM$ system, and separate DC Co_4POM and $[Ru(1)_3][PF_6]_2$ films [Toth 2016]. A self-assembled, multi-component water oxidation device.
R. Tóth, R. M. Walliser, N. S. Murray, D. K. Bora, A. Braun, G. Fortunato, C. E. Housecroft and E. C. Constable, *Chem. Commun.*, 2016, **52**, 2940. **DOI:** 10.1039/C5CC09556E. Published by The Royal Society of Chemistry.

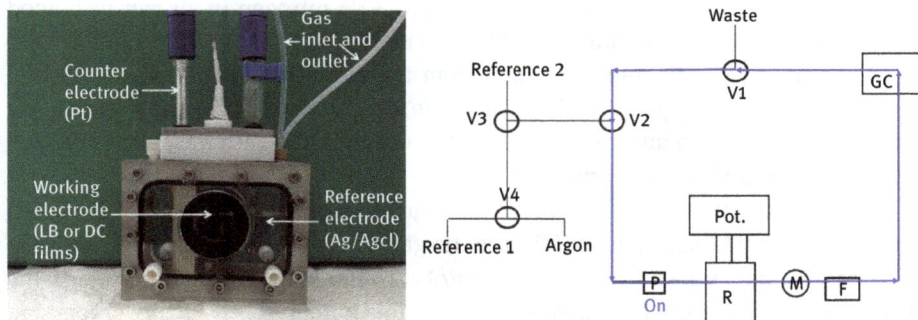

Figure 8.13: (left) Photograph of the electrochemical reactor and (right) scheme of the experimental setup for online, recirculated O2 measurement, where P: pump pumps the evolved gas and the Ar carrier gas through M: manometer, F: flowmeter, R: electrochemical reactor and GC: gas chromatograph. V: valves close/open the lines to the O_2 reference gases and for Ar carrier gas to flush the system. Pot.: potentiostat is connected to the electrochemical reactor. The cell (reactor) was designed by Krisztina Schrantz (Szeged University and Empa) and Rita Toth (Empa) [Toth 2016]. A self-assembled, multicomponent water oxidation device.
R. Tóth, R. M. Walliser, N. S. Murray, D. K. Bora, A. Braun, G. Fortunato, C. E. Housecroft and E. C. Constable, *Chem. Commun.*, 2016, **52**, 2940. **DOI:** 10.1039/C5CC09556E. Published by The Royal Society of Chemistry.

one component of the electrocatalyst molecule, and $[Ru(1)_3][PF_6]_2$ is also just the other component of the molecule.

The isolated ruthenium sensitizer molecule produces the lowest amount of oxygen over 1000 min of +1.3 V bias. This is the accumulated amount of 2000 ppm/cm^2 electrode area. There is a steep increase of oxygen concentration for the first 250 min but then a plateau is approached.

The LB film produces oxygen with a steeper slope until 450 min and then approaches a plateau of 4000 ppm/ cm^2 electrode area. The DC film has an even higher concentration of oxygen when compared with the two aforementioned specimens with a plateau of over 5000 ppm/cm$_2$.

Interestingly, the Co$_4$POM alone shows a steady steep accumulation of oxygen over 1000 min, outperforming the two films and the Ru sensitizer molecule. We can directly read the oxygen production rate from this plot which is 8000 ppm/cm^2 per 1000 min, or 8 ppm/cm^2 min.

Note that we have here an electrochemical cell which is connected to a potentiostat which provides the necessary bias potential. The electrochemical cell is sealed in this case to warrant that the evolved gases, the hydrogen and oxygen, can be guided to the gas chromatograph.

There are not so many research groups, still at this time, who have the capability of measuring the evolved gases in PEC cells, in addition to the dark currents and photo currents. But it is not so difficult to make a gas tight cell and connect it with a GC. Even when there is a leak and air enters the system, this can be used as an advantage because the ratio of ~20% oxygen to 80% nitrogen in air can be a good quantitative marker for calibration of the system.

We can also use the electric signal from the potentiostat, that is, the current density and convert it via Faraday's law into evolved hydrogen and oxygen. We certainly have done so and can compare the thus obtained gas concentrations with the gas concentrations measured with the GC.

We compare the oxygen evolution in Figure 8.14. The data points, filled red squares (DC) and black squares (LB), are the gas concentrations measured with the GC over 600 min under +1.3 V versus Ag/AgCl reference. The monitoring of the current density during the oxygen evolution at this potential is basically chronoamperometry [Toth 2016]. The integral over this curve gives the transferred electric charge, which enters the Faraday's law.

The black and red solid lines in Figure 8.14 denote the oxygen concentration obtained via Faraday's law. It is an exercise for the student to convert the electroanalytical data into gas volume, and the gas volume data into the current density.[7] We had a similar relation using Faraday's law in the Chapter on batteries, where the charging current was integrated to obtain the charge in mAh/g for a lithium ion battery cathode (positive electrode).

---

[7] The paper in [Toth 2016] Toth R, Walliser RM, Murray NS, Bora DK, Braun A, Fortunato G, Housecroft CE, Constable EC: A self-assembled, multicomponent water oxidation device. *Chemical Communications* 2016, 52:2940-2943. doi: 10.1039/c5cc09556e. is open access and can be downloaded for free. In the paper and in the supporting information, you can find the chronoamperometry data.

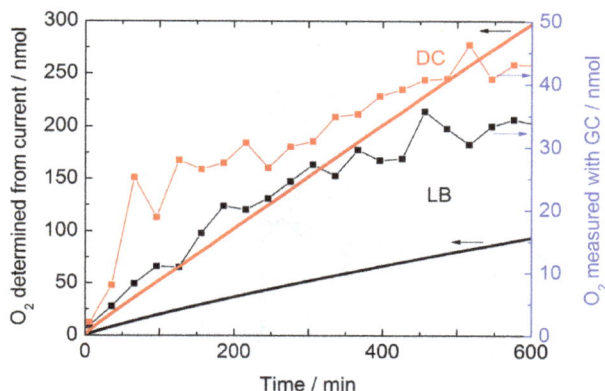

Figure 8.14: Comparison of the evolved O_2 from the LB (closed symbols) and DC (open symbols) films measured by gas chromatography (GC) and calculated from the current density data using Faraday's Law [Toth 2016]. A self-assembled, multicomponent water oxidation device.
R. Tóth, R. M. Walliser, N. S. Murray, D. K. Bora, A. Braun, G. Fortunato, C. E. Housecroft and E. C. Constable, *Chem. Commun.*, 2016, **52**, 2940. **DOI:** 10.1039/C5CC09556E. Published by The Royal Society of Chemistry.

8.4.2 The Grätzel cell

In 1991, O'Regan and Grätzel [Kalyanasundaram 2008] came up with an improved DSSC [O'Regan 1991] which was based on low-cost materials with medium purity but a relatively high efficiency of over 7% in simulated solar light and over 12% efficiency in diffuse sunlight. Their paper has been cited around 20000 times since.

Figure 8.15 shows the concept how the DSSC works. O'Regan and Grätzel used a ruthenium dye complex (again – observe the abbreviations; $RuL_2(\mu(CN)Ru(CN)L_2')_2$, **1**, where L is 2,2' bipyridine-4,4'-dicarboxylic acid and L' is 2,2'-bipyridine), which is adsorbed on a 10–mm-thick highly porous TiO_2 photoelectrode and exposed to iodine containing organic electrolyte (0.5 M tetrapropylammonium iodide and 0.04 M iodine in a mixture of ethylene carbonate with acetonitrile).

In this condition, the sensitizer is at the energetic ground state S (1) and becomes lifted by the absorption of a photon $h\nu$ to the excited state $S\star$ (2). The sensitizer will then pass on one of its electrons e^- to the semiconductor photoanode (3). Because of the loss of the electron, the sensitizer is in the oxidized state S^+.

The redox couple at charge state R^- in the electrolyte will pass one electron to the oxidized sensitizer molecule (5) and thus regenerate it to its original energy state S. The redox couple is thus charged from R^- to R and will receive another electron through the electrolyte in contact with the counter electrode (cathode).

The potential difference of the Fermi level of the illuminated photoanode TiO_2 and the redox potential (Nernst potential) of the redox couple R/R^- is equal to the photovoltage ΔV (4), which can power the external load. This kind of DSSC has been further developed and commercialized since several years ago [Baxter 2012].

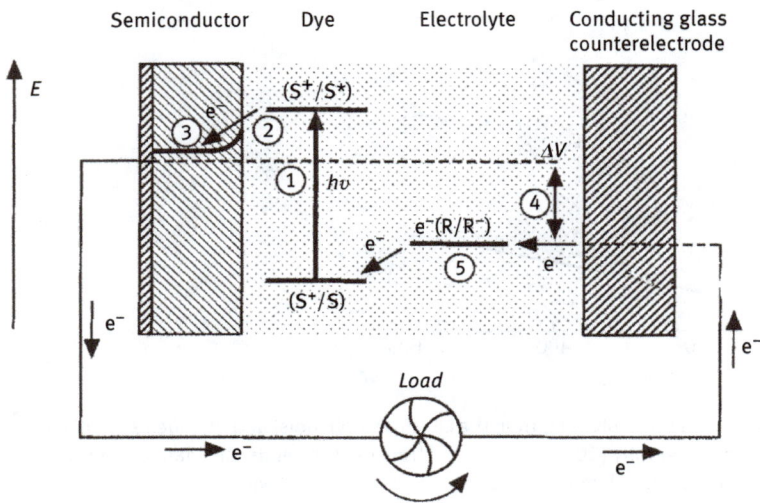

Figure 8.15: Schematic representation of the principle of the dye-sensitized photovoltaic cell to indicate the electron energy level in the different phases. The cell voltage observed under illumination corresponds to the difference, ΔV, between the quasi-Fermi level of TiO$_2$ under illumination and the electrochemical potential of the electrolyte. The latter is equal to the Nernst potential of the redox couple (R/R$^-$) used to mediate charge transfer between the electrodes. S, sensitizer; S*, electronically excited sensitizer; S$^+$, oxidized sensitizer.
Reprinted by permission from Springer/Nature **Nature** 353:737-740, A low-cost, high-efficiency solar cell based on dye-sensitized colloidal TiO2 films, O'Regan B, Grätzel M, (1991). [O'Regan 1991].

The I/V characteristic of the DSSC is shown in Figure 8.16 for two different illumination powers. At large, they have the same shape. The I/V curve in the top panel (a) was obtained when the cell was illuminated with 83 W/m^2. The DSSC delivers a photocurrent of up to 1.3 mA/cm^2 for the low voltages and rapidly decreases beyond the working point, which we identify as 0.5 V. The current at $V > 0.65$ V is 0. The determination of the electric power from current and voltage is an exercise for the reader.

The I/V curve of the cell in Figure 8.16(b) was obtained at 750 W/cm^2. This is a factor nine larger than for the case (1). Yet, the current density is not larger than under "dim" illumination. This shows that there would be no use when we concentrated the sunlight from a large reflector are and then direct it on a DSSC. Diffuse light appears to be sufficient for operation of the DSSC.

The photocurrent density under strong is even lightly lower than 1.2 mA/cm^2, which signifies that it is in this case not worth to look for places in the world which have particularly intense sunshine for DSSC application. You can use this DSSC also for indoor applications where light enters the building through windows. I do remember that the Swiss Embassy in Seoul, Republic of Korea, had one large

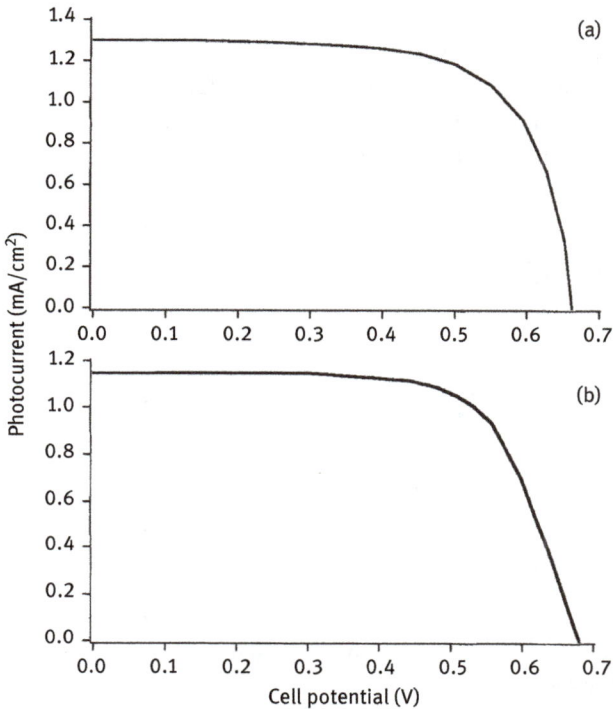

Figure 8.16: Photocurrent–voltage characteristics of a cell, based on a colloidal TiO_2 film sensitized by 1; the film, supported on a conducting glass sheet, was used in a sandwich-type configuration. The size of the dye-coated TiO_2 photoanode was 0.5 cm². The counter electrode, consisting of conducting glass coated with a few monolayers of platinum, was placed directly on top of the working electrode. A thin layer of redox electrolyte is attracted into the intra-electrode space through capillary forces. The cell was exposed to simulated sunlight with AM1.5 spectral distribution. (a) Light intensity 83 W/m², electrolyte: 0.5 M tetrapropylammonium iodide +0.04 M iodine in a mixture of ethylene carbonate (80% by volume) with acetonitrile. Fill factor was 0.76; surface area 0.5 cm² (before multiplication by roughness factor). Conversion efficiency was 7.9%. (b) Light intensity 750 W/m, electrolyte: 0.5 M tetrapropylammonium iodide, 0.02 M KI × 0.04 M I_2 in the same solvent. Fill factor was 0.684; conversion was 7.12%.
Reprinted by permission from Springer/Nature. Nature 353:737-740, A low-cost, high-efficiency solar cell based on dye-sensitized colloidal TiO2 films, O'Regan B, Grätzel M, (1991). [O'Regan 1991].

electronic clock mounted at the wall and the front face of the clock looked like an ancient Asian painting but this painting was in fact a DSSC that powered the clock.

Unlike PV cells, DSSC are to some extent translucent and it is possible to tune their color so that the DSSC can at the same time also serve as decoration. The Convention Center at EPF Lausanne has one side in the grand reception hall decorated with around one dozen large, long DSSC with mixed shades of red and yellow [Gast 2014].

The Grätzel cell [O'Regan 1991] is a very good example how one monolithic device such as the silicon photovoltaic cell can be substituted by a more complex multi-component system where costs and application range favor the latter one.

Figure 8.17 compares the normalized incident photon to current efficiency (IPCE) of the bare TiO_2 photoanode (A) with the photoanode coated with one layer of ruthenium dye molecule (B) versus the wavelength of the incident light.

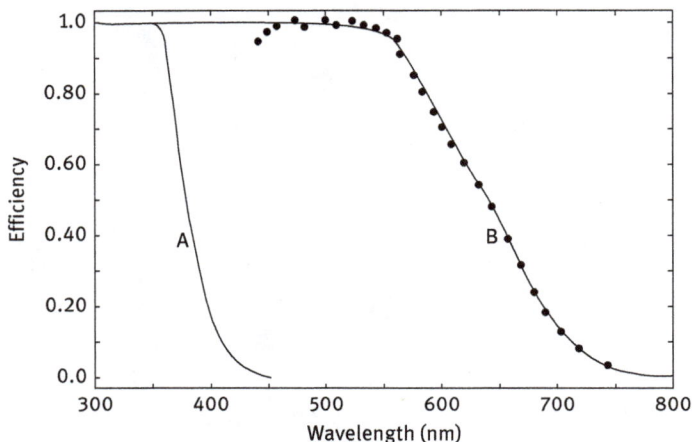

Figure 8.17: Absorption and photocurrent action spectra of TiO_2 films supported on conducting glass. A, absorption efficiency of the bare TiO2 film corrected for conducting glass background; B, absorption efficiency of the same film coated with a monolayer of 1; full circles, monochromatic current yield at short circuit as a function of excitation wavelength. Yield is corrected for 15% loss of incident photons through light absorption and scattering by the conducting glass support. Reprinted by permission from Springer/Nature. Nature 353:737-740, A low-cost, high-efficiency solar cell based on dye-sensitized colloidal TiO2 films, O'Regan B, Grätzel M, (1991). [O'Regan 1991].

Only for photons with wavelengths shorter than 450 nm does TiO_2 have a noticeable efficiency. Photons with a larger wavelength just do not have the necessary energy $E = h\nu$ to overcome the band gap energy of 3.1 eV (400 nm). However, adding the dye to the electrode shifts the IPCE curve considerably to larger wavelengths[8] by around 250 nm!

This means that the layer of dye molecules helps collect the photon energy from quanta which have wavelengths of 600 up to 700 nm, which corresponds to band gap

---

[8] We remember that larger wavelengths correspond to lower energies, and vice versa. It is an exercise for the reader to find and write down the relationship between wavelength λ and energy E, and frequency ν.

energies[9] of 2.1 down to 1.77 eV. The trick with applying dyes helps therefore collecting hitherto unutilized photon energy.

This type of solar cell is overall cheaper than a conventional silicon-based solar cell because TiO_2 costs less than silicon when processed for solar cells. The ruthenium dye however is a costly component. It is not only the cost of the ruthenium as element but also the synthesis of the organic dye which adds to the cost of the DSSC. It is therefore not surprising that researchers tried to find a yet cheaper alternative to ruthenium.

8.5 Excursion to noble metal mining and refining

When we consider for example the Grätzel cells, we realize that Ruthenium is an essential component for the function of the cell. Ru is the central element in the dye which absorbs the visible light. This effect was called sensitization. We make the TiO_2 sensitive to visible light. Ruthenium is a noble metal and we mentioned it already in a previous Chapter when we dealt with supercapacitors. Ruthenium oxide is a great supercap electrode material and has a very high pseudo-capacity. Platinum is a very important electrocatalysts used in PEM fuel cells. Iridium is a very good oxygen evolution electrocatalyst. These noble metals are relatively rare and quite expensive. In the volcano curve in electrocatalysis [Parsons 2011], these noble metals typically are constituting the maximum exchange current when plotted over the free energy of adsorption [Quaino 2014].

In 2013, I visited South Africa on a diplomatic mission. I used this opportunity to pay an unsolicited visit to University of Pretoria and met there by coincidence researchers who were interested in making a project on artificial photosynthesis with me. When my colleagues from South Africa and I discussed the project, I bragged about how good it was that we were making photoelectrodes from low-cost and abundant 3d metals, and not be the expensive and rare noble metals such as platinum, palladium, ruthenium and iridium. We were lucky and received funding which turned into a very interesting project with a lot of capacity building [Braun 2014].

My South Africa colleagues were not overly happy over my (ignorant) remark that we should avoid noble metals. They replied that South Africa was a major exporter of such noble metals. These noble metals were basically piled up in the backyard of Pretoria. So, while many researchers are trying to find low cost and good solutions for materials, researchers in countries who are rich in particular resources like shown

---

9 I have not been entirely correct here. Molecules do not have energy bands and thus also no bandgap. Molecules have highest occupied and lowest unoccupied molecular orbitals (HOMO, LUMO).

here want their material to be a solution for the materials problems. Therefore, I had to open up for the noble metals.

Figure 8.18 shows the periodic table of elements as it is printed on glass bricks. My hand is stretched out over the 44th element in the fifth row, this is, Ruthenium Ru. It is just below the third element in the fourth row, Iron Fe. The other "interesting" noble metals such as palladium, platinum and iridium are in the vicinity as well.

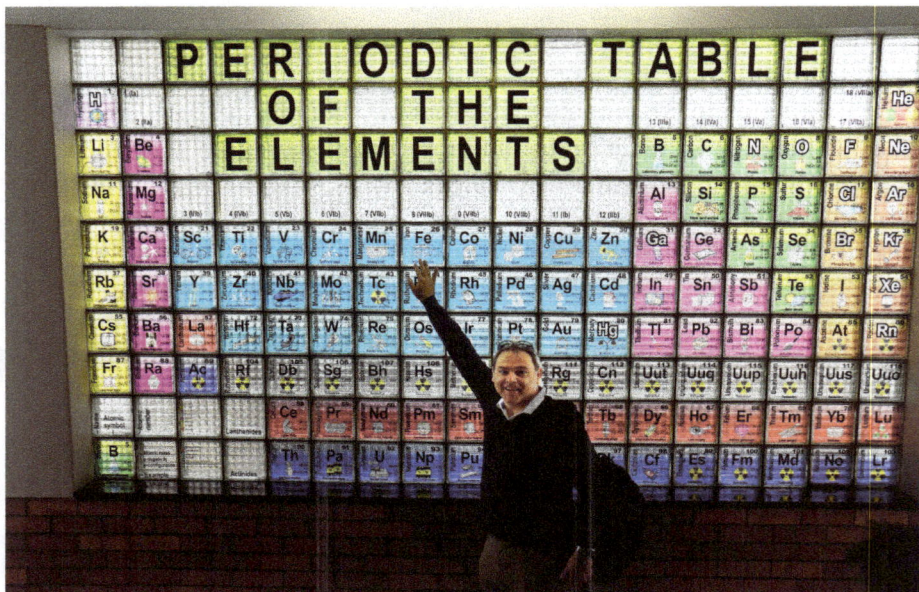

Figure 8.18: Periodic Table of Elements printed on glass blocks in the chemistry building at University of Pretoria. I am holding my hand over element number 44, Ruthenium in the fifth row.
Photo by Florent Boudoire.

Major platinum mining companies in South Africa are listed in Table 8.1. On one of my trips my colleagues and I went to the Pilanesberg Safari Resort, which is half way between Pretoria, the capital of South Africa, and Gaborone, the capital of Botswana. On the way to Pilanesberg along the R556 road I saw numerous piles of gravel and mining facilities. Figure 8.19 shows such facility near Maroelakop in the Gauteng province. In the beginning, I did not know what I was seeing. So I took photos with my smartphone. Nowadays, it is easy with geo-tagging of smartphone photos and Internet search machines to find out about these facilities. In the background of the photo, you see the EPL (EPL – Lonmins Eastern Platinum) Concentrator facility of the Lonmin plc mining company. Lonmin plc was founded in 1909 and is registered in

Table 8.1: The ten largest platinum producing companies and some of their key data in 2014.

Rank	Name	Year founded	Number of employees	Sales	Headquarter	Produced kg
1	Anglo American	1917	133900	27 Bio $	London	53580
2	Impala Platinum Holdings	1973	55000	37.6 Bio $	Illovo, ZA	33450
3	Norilsk Nickel	1993	83600	8.5 Bio $	Moscow	18626
4	Lonmin plc	1909	25000	965 Mio $	London	12360
5	Stillwater Mining	1992			Littleton CO, USA	8080
6	Northam Platinum Ltd.					6832
7	Aquarius Platinum					5220
8	Vale SA					5160
9	Glencore plc				Zug, Switzerland	2580
10	Asahi Holdings					1270

The data are approximate only and mentioned for a rough comparison and overview.

Figure 8.19: People in South Africa near Maroelakop and Segwaelane at R556, 80 km west of Pretoria. In the rear of the photo you see an ore concentrator (EPL Concentrator) facility of the Lonmin mining company.

the United Kingdom. In 2014, its revenues was 965 million US$ and an operating income of 52 million US$. Lonmin has around 25000 employees.

For comparison, Swiss mining company Xstrata (founded 1926 as Südelektra AG, a company which in South America produced energy facilities) had around 37000 employees and a 31 billion US$. Xstrata was bought several years ago by Glencore, which is a global materials trading company and has 156000 employees and a sales volume 170 billion US$.

World platinum production rank 2014

This facility is right next to the Marikana mine, which was in 2012 the scene for a major miners' strike which turned violent; in the subsequent clash with the police, 40 deadly casualties emerged, 36 of which were mine workers [Alexander 2013, 2016, Anonymous 2012, Bernard 2016, Holmes 2015, Naicker 2016].

In the foreground in Figure 8.19, you see a dozen local people who apparently left the bus or who are waiting for the bus. All over South Africa, you can see small shuttle buses who serve the rural locations like shown here. Some people in remote areas have to walk long distance before they reach a road which is served by such shuttles.

Living conditions for the predominantly black community are not overly good. Mining companies must cut costs to be competitive on the world market for platinum group metals (PGM). Labor costs in the mining stage are factors in the value chain of PGM processing [Ndlovu 2014]. And most of the mining workers are black people. They happen to be in a position where cutting labor costs can have severe impact in the value chain. After the standoff between miners, mining management the police, the world turned its attention on the mine in Marikana. The labor in mines is very hard and very dangerous and the salaries do not allow for creation of wealth for the families of the miners.

Lonmin promised to take better care of the people in the neighborhood near the mines because this is where they recruit the miners. While improvements were made by Lonmin following the years of the violent strike, Lonmin apparently fell behind its original promise as was recently outlined in a report by amnesty international [Amnesty 2016].

The flow diagram in Figure 8.20 gives a schematic overview of the various technical steps and factory facilities which are necessary for the production of noble metals. The ore (rock, stones, gravel, sand, other solid raw material) is transported from the mine (mined surface or underground) to the concentrator. At Lonmin mines, the concentration of the ore is 4.6 g per ton material. The largest trucks admitted on German roads may weigh 38 tons. Their payload is typically 24 tons. Such a truck full of mined material would then carry 110.4 g ore containing platinum metals (0.0000046).

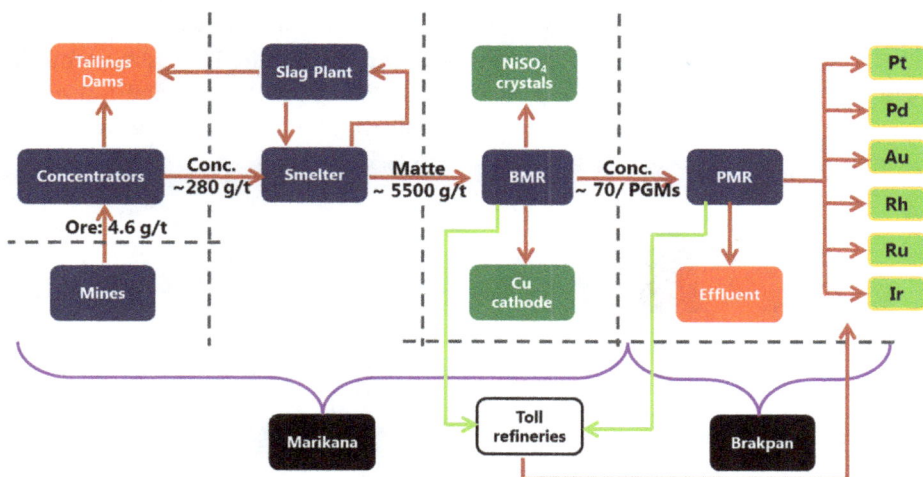

Figure 8.20: Flow diagram for noble metal ore mining and processing at Lonmin. [Turner Jones 2011]. (PGM = platin group metals). https://www.thebalance.com/the-10-biggest-platinum-producers-2014-2339735.

The concentrators use flotation methods to separate the ore from the other solid material. This is a concentration by a factor 60. The truckload of material that comes from the concentrator would then contain 6.7 kg platinum ore. This material will be heated in a smelter. The metal-enriched product from this smelting process is called "matte" and has a platinum metal (and other noble metals) of 5.5 kg/ton. This is another concentration by a factor of 20. A truckload of matte contains then 132 kg metal materials of commercial interest for the company.

The BMR base materials refinery separates the nickel and copper and cobalt sulfates in an electrochemical process. Copper will be deposited as metal on an electrode by cathodic deposition. The electrolyte will contain nickel sulfate which will be crystallized. The rest of the material, except the platinum, group will be given to special refineries which process the material for a service fee. One product from the electrochemical process is the platinum metal refinery/is the platinum metal group (PMG). The waste water will be collected in artificial lakes in the city of Brakpan.

8.6 Solar cells on flexible substrates

Electrochemical converters come typically as rigid devices with a clear-cut size and geometry. There may however be applications where a tolerant geometry would be desirable. This is for example relevant for wearable devices. Creativity has virtually no limit. It is therefore not surprising that designers worked in a DSSC into a back pack. For this it was necessary to have a flexible DSSC.

DSSC can be fabricated on flexible substrates as we have shown in 2011 [Seong 2016, Tinguely 2011]. This opens new manufacturing horizons for DSSC as sketched in Figure 8.21, the roll-to-roll process (R2R). The substrate for R2R is a flexible foil which is rolled up in a coil. The end of the foil is then rolled off the coil and guided through a bath which contains a coating solution.

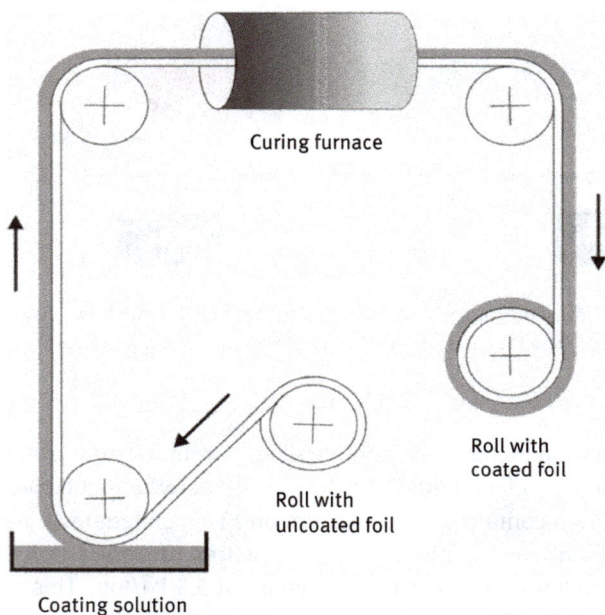

Figure 8.21: Simplified schema of the roll-to-roll process used for coating of TiO$_2$ on a PET-ITO foil. The arrows indicate the flow direction.
Reproduced from Tinguely J-C, Solarska R, Braun A, Graule T: Low-temperature roll-to-roll coating procedure of dye-sensitized solar cell photoelectrodes on flexible polymer-based substrates. *Semiconductor Science and Technology* 2011, 26:045007. doi: 10.1088/0268-1242/26/4/045007. "© IOP Publishing. Reproduced with permission. All rights reserved.

The surface of the foil must be in a condition that the coating solution will have the proper adhesion. The coated foil is then further guided into a furnace for curing of the coating material. Behind the furnace the foil is again rolled up and ready for further processing elsewhere in the manufacturing process.

The R2R coating is well established in many production processes in various industries. A basic problem is that the foils for R2R are from plastic materials like hydrocarbons which cannot sustain very high temperature without damage. In the extreme case, they would just burn away. The plastic foil used in this study was polyethylene (PET), which was coated with an indium tin oxide (ITO) layer. Table 8.2 lists the various substrates used for this master thesis, which was carried out in an industrial setting with Swiss company [Tinguely 2009].

Table 8.2: Specifications of various conductive substrates used in the master thesis by Jean-Claude Tinguely [Tinguely 2009].

Substrate	Abbreviation	Specific resistance [Ω/m^2]	Manufacturer
ITO-coated Glass	Glass	4–8	Delta Tech. (USA)
ITO-coated PET	ITO	≤10	Delta Tech. (USA)
ITO-coated PET	KTI14	14	Kintec (Chi)
ITO-coated PET	ITO80	60–90	CPF Films (UK)
Baytron P-coated PET	Bayt	>100	Celfa AG (CH)
ITO-coated PET	ITO300	250–350	CPF Films (UK)

When you want to sinter TiO_2 nanoparticles to make a compact highly porous film for DSSC, this cannot be done on such plastic foils because the sinter temperature can be around 500 °C. It was therefore necessary to find an alternative solution to bond the nanoparticles for the DSSC film together. Our master student decided to use the block copolymer PLURONIC®, which is made from poly ethylene oxide and poly propylene oxide blocks [Tinguely 2009, 2011].

The TiO_2 particulates used in this study were the well-known P25 (see, e.g., [Long 2006]), which has a particle size of around 25 nm. It is of interest to know how much of the PLURONIC® can be volatized during thermal treatment. Figure 8.22 shows the thermogravimetry analysis (TGA) data for three TiO_2 – PLURONIC® coatings which had received different drying treatment.

One sample was dried in the laboratory at 25 °C. There is a massive decrease of mass at 200 °C to below 90% of the original mass, and with increasing temperature the mass decreases further to below 80%. The dip at 600 °C comes from the change of the purging gas nitrogen to oxygen.

When the sample was dried at 150° in the laboratory furnace, the mass decreases to 90% at $T = 250$ °C and with increasing temperature the mass decrease further to below 90% at $T > 300$ °C. Treating the foil in the R2R furnace at 120–140 °C yields samples which show only little mass decrease upon annealing in the TGA. This shows that the R2R coating is a good process for making TiO_2-coated ITO films for DSSC.

With the coating method shown to be successful, it was now possible to prepare a real DSSC device. It was therefore necessary to prepare a solution which contained the necessary dye. The dye N-719 (di-tetrabutylammonium cis-bis(isothiocyanato)bis (2,2′-bipyridyl-4,4′-dicarboxylato)ruthenium(II)) was used in our study [Tinguely 2009, Tinguely 2011]; the dye was dissolved in water-free ethanol and stirred until no solid was left. The molarity was 0.3 mM. The TiO_2 samples were dipped into the solution so that the dye would attach, and then the sample was removed and then dried. The electrolyte was iodine triiodine dissolved in acetonitrile.

The development of suitable dyes is a field of research of its own, specifically coordination chemistry. Figure 8.23 gives an overview of the molecule structure of a

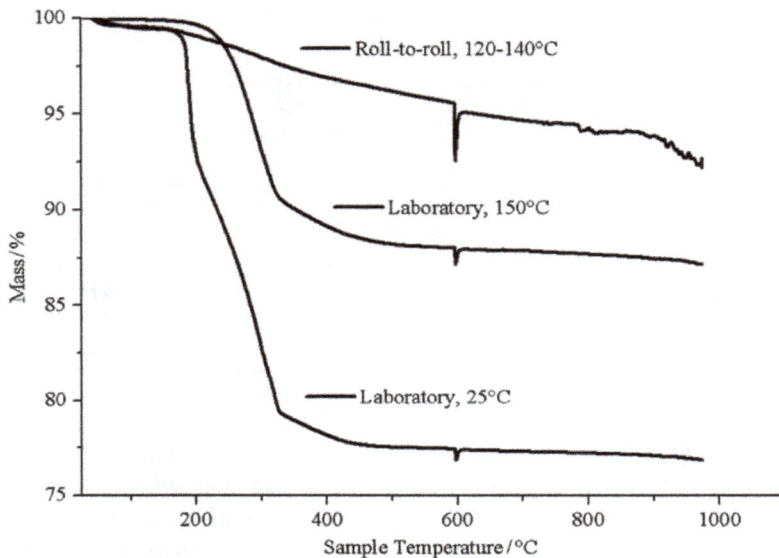

Figure 8.22: Thermogravimetric data of TiO$_2$ coatings showing remaining Pluronic contents after different drying conditions. Laboratory coating, drying at room temperature: no Pluronic elimination (about 25% concentration). Laboratory coating, drying at 150 °C: about 10% left. Roll-to-roll coating, 120–140 °C: no remaining Pluronic detected. At 600 °C, the nitrogen gas was replaced by oxygen gas. Reproduced from Tinguely J-C, Solarska R, Braun A, Graule T: Low-temperature roll-to-roll coating procedure of dye-sensitized solar cell photoelectrodes on flexible polymer-based substrates. *Semiconductor Science and Technology* 2011, 26:045007. doi: 10.1088/0268-1242/26/4/045007. © IOP Publishing. Reproduced with permission. All rights reserved.

number of dyes developed in one single synthesis study [Klein 2004]. Ruthenium is the central ion. Characteristic for such dyes is also the pyridine rings which contain one nitrogen ion, and specifically in this study the thiocyanate groups (SCN).

The DSSC were then actually prepared from R2R-coated foils. The counter electrode was a glass slide coated with ITO and then sputtered with a thin platinum layer for samples (DSSC) R1 and R4 in Figure 8.24; this is the rigid counter electrode. Samples R2 and R3 had countered electrodes where the PET foil coated with ITO and a sputtered platinum layer was used; this was the flexible counter electrode.

The TiO$_2$-PLURONIC® mixture of the photoelectrodes of DSSC R1 and R2 had been prepared under neutral pH 7, whereas R3 and R4 were prepared under alkaline pH 10.5. We read from Figure 8.24 that the current densities vary between 0.5 and 2.3 mA/cm^2 depending on how the photoelectrodes and the counter electrodes were prepared.

Imagine you have a foldable DSSC which occupies only a small volume, like a small camping tent, but when you need electric power you can spread it over the meadow with an active light absorbing area of 5 m^2. Or you can design the entire camping tent as a dye sensitized solar cell. It will protect you from the weather like a

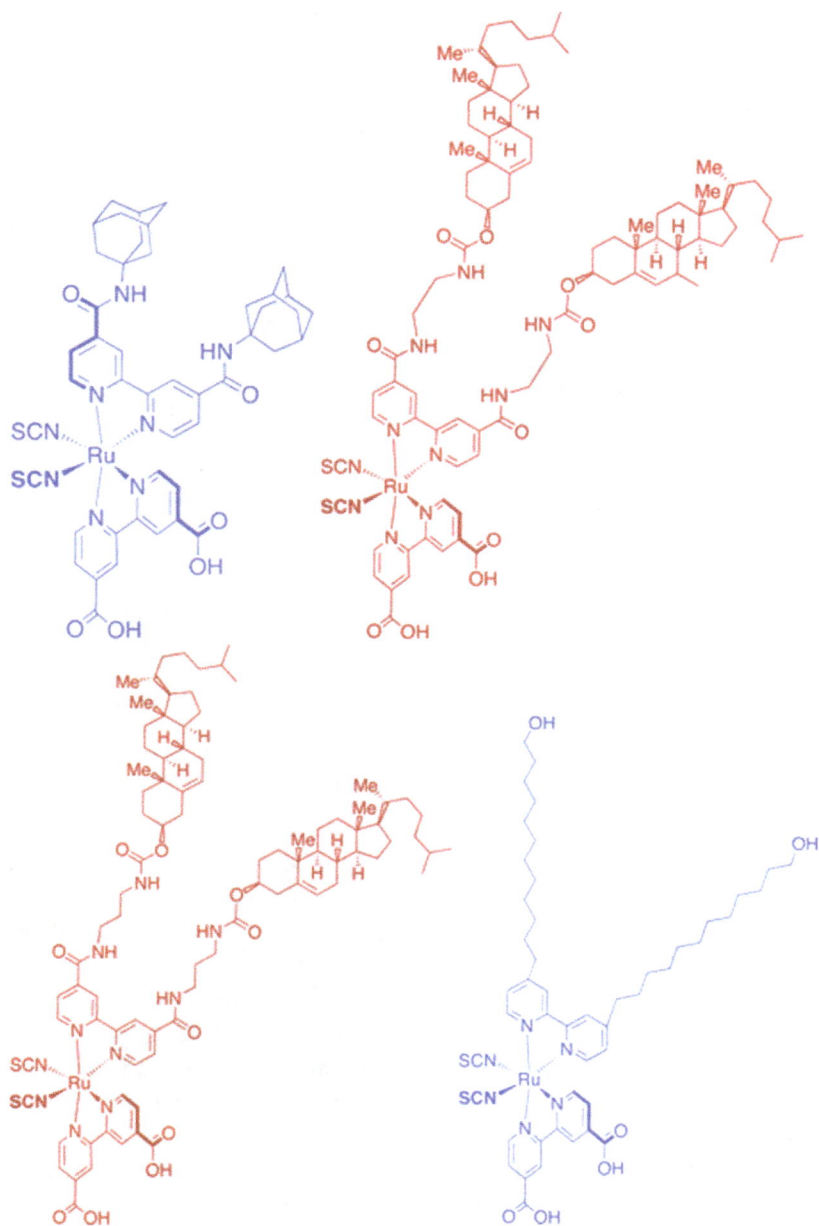

Figure 8.23: Structure of some ruthenium complexes for DSSC studies or applications.
Reprinted with permission from Klein C, Nazeeruddin MK, Di Censo D, Liska P, Gratzel M: Amphiphilic ruthenium sensitizers and their applications in dye-sensitized solar cells. *Inorg Chem* 2004, 43:4216-4226. doi: 10.1021/ic049906m. Copyright (2004) American Chemical Society. [Klein 2004].

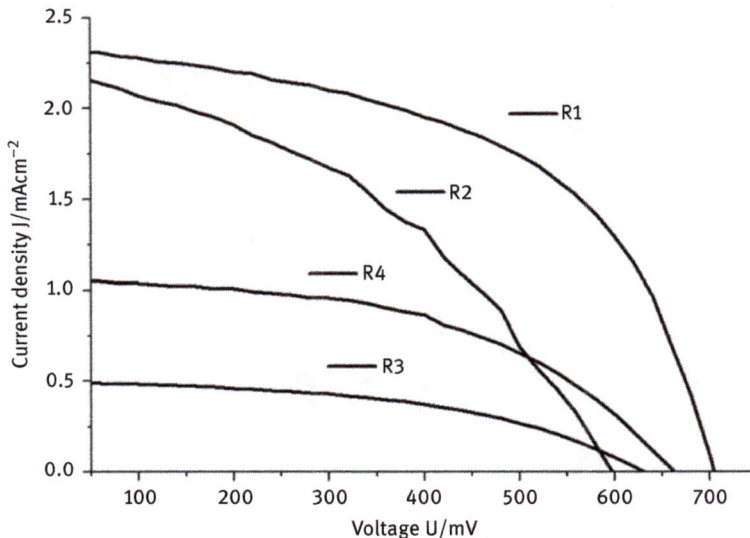

Figure 8.24: *I–V* characteristics of the DSSC using roll-to-roll coated photoelectrodes for four different set-ups, all 1:3 Pluronic: TiO_2. R1 and R4: Pt-sputtered ITO-glass carrier counter electrode, R2 and R3: Pt-sputtered ITO-PET carrier counter electrode. R1 and R2: pH 7, R3 and R4: pH 10.5 adjusted with NH_3.

Reproduced from Tinguely J-C, Solarska R, Braun A, Graule T: Low-temperature roll-to-roll coating procedure of dye-sensitized solar cell photoelectrodes on flexible polymer-based substrates. *Semiconductor Science and Technology* 2011, 26:045007. doi: 10.1088/0268-1242/26/4/045007. © IOP Publishing. Reproduced with permission. All rights reserved.

tent but will also deliver you electric power. The only problem might be at night when there is no sunshine or daylight and the DSSC is therefore not working.

This problem has already been taken care of on the small scale. Figure 8.25 shows a backpack which I received as a farewell present from one of the PhD students in my research group. The backpack has a small (few 100 cm² area) DSSC woven into the back side. The DSSC is quite stiff, but flexible enough to be tolerant to volume changes in the backpack. It will not break during movements or stress, and it appears quite shock resistant. I do not know whether it is advised to use it during heavy rain or high humidity levels.

But the DSSC works and it stores the electricity in a small battery which is nowadays called power bank. The battery is the red rectangle shape container with black and white cable attached. The black cable comes from inside in the backpack which is connected with the DSSC. The white USB cable is connected with my smartphone to charge it.

The more light the DSSC is exposed to, the more of the four small lighting elements on the black disk on the red battery will be shining. During a very sunny day outside, all four elements would lighten up and supposedly the battery would be

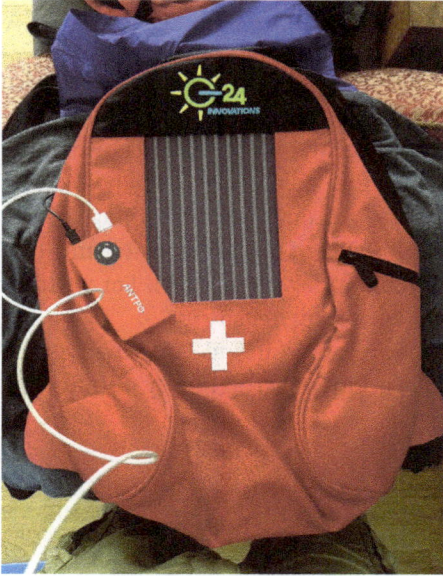

Figure 8.25: This backpack has a DSSC (rectangular lined panel between green-yellow sun and white cross) integrated which charges the small red power bank (lithium ion battery) via the black cable coming from the inside of the backpack. The white USB cable is connected to my iPhone with which I took the photo in April 2015.

charged as best as possible. You can think of a similar solution for the aforementioned DSSC blanket or DSSC camping tent. With a sufficiently large battery you can store the energy for later use.

We could certainly put the blanket over a car while it is driving in the sunshine or daylight, and we even can integrate the DSSC for automotive applications on the rooftop of the car or on a boat. This is nothing special because we remember that the PV panels were used in satellite and spaceship applications already long time ago.

The wrapping artists Christo and Jeanne-Claude [Kalfatovic 2002] wrapped the German Reichstag in Berlin 1994 [Verhüllter Reichstag – Projekt für Berlin 1994]with cloth and tissue [Jodidio 2001, Singh 2012]. This is a huge building with a large area exposed to daylight, which could serve for holding an equally large DSSC blanket. If a sufficiently large and flexible DSSC was available, one could wrap such building today with the DSSC...

8.7 Some diagnostics mathematics on DSSC

When we look at the $I–V$ curves in Figure 8.24, we wonder how we best can compare them. We need to define a way of characterizing a curve which is not "properly shaped." We remember we had a similar problem with the fitting of the photocurrent in a PEC electrode.

Figure 8.26 displays an I–V curve (red solid line) of a hypothetical DSSC. Two tangents (black dotted lines) coming from the high current side and the high voltage side meet at what we call the point of maximum power MP. We recall that power is the product of current and voltage, and the I–V curve therefore encloses the power. The current of MP, I_{MP}, is the horizontal line which goes through MP. The voltage of MP, V_{MP}, is the vertical line which goes through the MP. When no current is flowing, that is, $I = 0$, the measured voltage is the open circuit voltage V_{OC} (OCV). For virtually absent voltage during short circuit, the measured current density is I_{SC}.

The efficiency of the solar cell η is its figure of merit and is defined as the ratio of the power input P_{in} by the sunlight and the electric power output P_{out}:

$$\eta = \frac{P_{out}}{P_{in}} = \frac{V_{MP} \cdot I_{MP}}{P_{in}}.$$

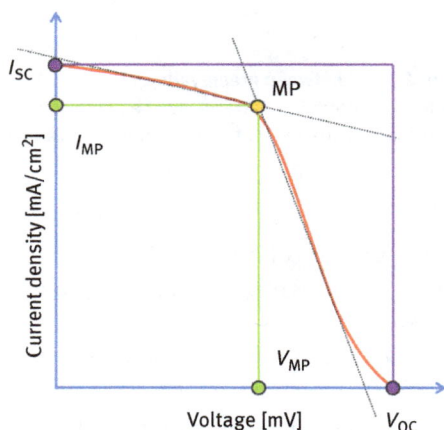

Figure 8.26: Characteristic parameters of a DSSC with red I–V curve. MP = point of maximum power. V_{MP} = voltage of maximum power. I_{MP} = current density at maximum power. I_{SC} = current density at short circuit. V_{OC} = open circuit voltage.

An "ideal" DSSC should have a rectangular characteristic where the rectangle is span by I_{SC} and V_{OC}. We see from Figure 8.24 that this is not the case; the I–V curves are bending away downward and inward, and thus the actual MP is not in the upper right corner of the aforementioned rectangle with power P_T. The ratio of the rectangle areas defined by the points MP and PT is called fill factor ff:

$$ff = \frac{V_{MP} \cdot I_{MP}}{V_{OC} \cdot I_{SC}}$$

8.8 Some remarks on design

The term *design* has not a very strong scientific connotation as it is more known to art then to science. However, there is nothing wrong when you as scientist spend some extra thoughts how a process works, how a device works. Often you have to sketch a physical, chemical, biological, economic, or social system to bring an idea, a concept on the paper or on the chalkboard, either for your own reflection over the concept or for explaining it to other people. Even if it is merely theory and you just want to explain mathematical relations.

When you want to make an experiment you maybe have to design your own apparatus for a measurement. When you are an engineer, you also start from scratch and develop this into a machine or apparatus that works. Or you take someone else's machine and make an improvement on this part or that component. When your school offers a design course or design class, it may be worthwhile to give it a try. Design is not necessary an unscientific activity. Form and function (Le Corbusier) are related with each other. We realize this in the organization of cells in the plants and in the skeleton of humans and animals, and in the layout of factories.

The science and technology of materials for energy storage and conversion in batteries, capacitors, fuel cells, solar cells and electrolyzers has been field of great interest for researchers for hundreds of years. The comprehension and control of the transport properties of such often electrochemical energy systems is essential for the function and integrity of the devices.

The transport properties in this context, as I mentioned already in a previous Chapter, include the (i) electric charge transport, (ii) the mass transport, the (iii) heat and thermal transport and (iv) the optical transport. The charge carriers (i) include the electrons, electron holes, ions and more complex things such as polarons, for example. The mass transport in (ii) includes fluids such as liquids and gases, and as these can be of ionic nature, this mass transport can be coupled with electric transport. When we are dealing with devices, Joule heat may involve and its control may be very critical for operation and integrity; this is a very complex topic of its own typically handled by the mechanical engineers. And whenever we deal with solar cells and photo electrochemical cells, the optical transport is considered.

All these transport properties depend on the "structure". Structure, however, is a diffuse term. A microscopist has a different understanding of structure than a neutron scatterer, to name only these two. In my very own and admittedly poor definition, structure is something multidimensional which starts at the atomic scale and extends to the design of components and architecture of systems.

I have written these lines as an invited paper for the XRM2018 Conference in Saskatoon, Canada [Braun 2018]. As (1) this conference was about microscopy and (2) I am not a microscopist, I had to span the arc from my own position and background to the scope of the conference. If we really want to do a good job as researchers, be it in microscopy, be it in scattering, or any other scientific method, we do not only make a measurement but we also look at the potential consequences that arise from the data. Our work as researchers is therefore to a very large extent of analytical nature.

When we go back and look into the architecture of systems, we can rest conceptually on what architects do, and this is giving material a shape, a *form* which brings about a particular *function*. The relationship between form and function is well known to designers and architects. "Form and Function" was first coined by artist Horatio Greenough [Greenough 1957] in the nineteenth Century, as Editor Harold E. Small explains:

> "Greenough was three generations ahead of his time. ... It was Greenough, not Whitman, who first protested against meaningless ornamentation. It was Greenough, not Ruskin, who first expressed the idea that the buildings and art of a people express their morality. It was Greenough, not Le Corbusier, who first said that buildings designed primarily for use "may be called machines." It was Greenough, not Louis Sullivan, who first enunciated the principle that, in architecture, form must follow function."

The aforementioned generalizations made by Greenough are noteworthy for what we researchers do in our field. Let me therefore identify the term "form" with structure, which mandates the function. In our structure analyses, we seek to identify those objects which are components that assemble up to "machines," as we know from architect Le Corbusier. Objects with function are the organs which make machines.

It was the Bauhaus half a century later whose philosophy rejected ornaments so that form was left with function only. This reductionist philosophy is widely essential to our scientific method, particularly in physics which historically aims at reducing everything into systems with high symmetry. Unintentionally, we mistake that we end up with less and lesser function, because function is a result of complexity and not simplicity, as Physics Nobel Laureate Anderson [Anderson 1972] explains in his paper on interdisciplinarity and complexity.

It is possibly because of the complexity of our vision apparatus that we prefer the visuals of microscopy data over almost any other form of data. Let me therefore begin with an example from biology where the "Bauplan" of the photosynthetic apparatus – an energy storage and conversion device – has been elucidated progressively by microscopy methods. The naked eye sees structures in the leaf of plants which become visible as cells in the spyglass and microscope. We can even identify the chloroplasts with an optical microscope, provided we do the necessary "sample prep." It requires further "sample prep" when we want to resolve the thylakoids in the chloroplasts with an electron microscope. I will come to biological systems later in this book.

Considerable improvement was made with the combination of electron microscopy with energy dispersive X-ray (EDX) spectroscopy, because we gain now images with high spatial resolution and element specificity. The electron energy loss spectroscopy (EELS), which is nowadays available with transmission electron microscopes

(TEM), permits to some extent the resolution of the electronic structure (or molecular structure) of materials with resolution virtually at the atomic scale.

As we zoom further into the photosynthetic apparatus, we find the thylakoids with the thylakoid membrane and the photosystems, which include the light harvesting complexes and the reaction centers. I have not come across a microscopy visualization of these yet, but they may exist in some publication. The metalloproteins such as oxygenase and hydrogenase have metal centers which are bridged with ligand groups, that is, Mn-O-O-Mn and Fe-S-S-Fe/Ni. Their molecular scale structure cannot be resolved with microscopy. Not yet.

This is why protein crystallography (PX) was developed – with much success – by taking advantage of the anomalous dispersion of X-rays. Multianomalous diffraction (MAD) phasing brings about the element specificity with high accuracy in protein crystallography [Karle 1980] and in crystallography in general.

Not all relevant issues at the molecular can be addressed with diffraction. The charge transfer across the aforementioned Mn-O-Mn and Fe-S-Fe superexchange units in proteins follows the Goodenough-Kanamori rules [Goodenough 2008], the same way like the corresponding structures in lithium ion battery positive electrodes and solid oxide fuel cell cathodes. The spin of the ions can determine whether electrons can hop across the unit, or not. This is why K-β spectroscopy or L-edge NEXAFS spectroscopy, methods very sensitive to the spin structure, are used for analyses of such systems. But here no atomic resolution is available yet. Notwithstanding that methods like X-PEEM and STXM allow for high-quality electronic and molecular structure determination with spatial resolution down to 30 nm.

The bond length, spin state, oxidation state and hybridization effects determine to a large extent the electric transport, which can be elucidated with the aforementioned hard and soft (and tender) X-ray spectroscopy methods. But problems still persist. For example, the hydrogenase cofactors are accompanied by an iron-sulfur cluster, which spectroscopically cannot be distinguished from the iron-sulfur cofactor. This is why typically the nickel hydrogenases and not the iron hydrogenases are studied with X-ray spectroscopy.

If there was a chance to probe the hydrogenase locally with X-ray spectroscopy, we would maybe be able to tell the difference in operation of a nickel hydrogenase from a sulfur hydrogenase, if there was any difference. Local probes are therefore necessary for the further discrimination of structures in a larger assembly of materials with different structure and morphology, such as electrode assemblies.

There are cases where we can probe deeper into complex structures by taking advantage of the penetration depth of X-rays and electrons. X-ray interferometry, reflectometry and gracing-angle methods allow for probing flat and laminar structures with depth sensitivity in the nanometer range [Nemsak 2014]. The penetration depth of X-rays allows therefore to some extent for *in situ* and, more interesting, for *operando* experiments where the materials in a component or complete device is probed nondestructive with X-rays while at the same time, for example, electroanalytical methods can be employed.

This approach can permit a parameterization of device settings and corresponding structure changes which in the ideal case can be mathematically modeled to the extent that the relationship of "form and function" is fully recovered at the quantitative level.

Not only the atomic scale but also the mesoscale can be relevant for function. In electrochemistry, this can be for example nanoparticle electrocatalysts or the high internal surface area of porous electrodes. Small angle scattering (SAS) is the method of choice for probing this size range and correlate it for example with the ionic conductivity. An example from photosynthesis is the conformational changes in the *bacteriorhodopsin* proton pump: small angle scattering with neutrons maps the conformational changes at the mesoscale, and quasi-elastic neutron scattering produces the proton conductivity [Braun 2015], but with no real space images.

We are therefore still left in the position that we have to "imagine" the structure of devices and systems based on their function, unless we can map all details at all relevant scales in the real space. Once we have mapped out all components, we can think about new designs. This can be for example the adding of a histidine tag to a light harvesting protein to better bind the protein to a semiconductor electrode, for example [Braun 2012d, Faccio 2015a, Faccio 2015b, Ihssen 2014, Schrantz 2017].

Design is a serious activity in the industrial development chain of a product, including also the manufacturing. Therefore, design is also a serious profession. I have therefore tried on several occasions to involve industry designers in electrochemical energy projects.

To give you an impression how a designer looks at scientific advancements and their use in materials and products I can recommend the TED talk by industry designer Prof. Mareike Gast [Gast 2017], TEDxUniHalle.

9 Electricity in nature

9.1 The electricity between air and earth

Looking at the old historic experiments which are associated with the name of Benjamin Franklin, we find a conjecture between the air and electricity. In the eighteenth century, people became interested in electricity because of Franklin's interest in this field [Priestley 1775]. Franklin, an illustrious character, suggested for example the lightning rod. Some people believed they could use the electricity to influence the growth of the plants. Soon the "electrophysiology" became an essential part in the research on electric phenomena. However, plant physiology and medical sciences went different routes right from the beginning. The former used electric (dis-)charges in the atmosphere, whereas the latter focused on the Faradaic currents in living organisms.

From our twenty-first century perspective, the experiments, studies, views, imagination and perception of the researchers of the eighteenth and nineteenth centuries may look naïve. But as science is an ongoing quest [Feynman 1955], researchers back then were probably no less ignorant than researchers of the twentieth and twenty-first centuries. Even today it would sound weird to some scientists who study condensed matter or physical chemistry when they learn about other researchers who study phenomena outside in nature which are not directly related to the manufacturing of new technological devices. Wilson is an example of a researcher who was capable of doing both.

Irrespective of the living systems, plants or animals, there is always an electric current between the negatively charged ground and the positively charged atmosphere. This is why Wilson coined it atmospheric electricity [Wilson 1900]. Physics Nobel Prize Laureate Charles Thomson Rees Wilson (developer of the Wilson cloud chamber for measuring radiation) determined a current of $2.2 \cdot 10^{-16}$ $\mu A/cm^2$ with an electrometer [Wilson 1900, 1901, 1902a, b, 1903, 1904a, b, 1906a, b, 1908, 1909, 1911, 1921, 1925].

In areas with noticeable vegetation, the mediator for this exchange of electric charge is the phytosphere—the layer of plants. Here you can see in this photo (Figure 9.1) not just unspecified scenery of nature. You see the blue sky with the air. You imagine the brown soil in the field underneath the plants. And in between you see grass, corn plants, crops, the trees, the woods and the forest. The vegetation looks actually like a film with different layers of vegetation—with a total thickness of 10–20 m. Look how "homogeneous" this film is when we look at it from remote distance.

9.2 Some words on reproducibility and stability of biological systems

In my scientific career which extends now over 25 years, I often heard colleagues saying that biological systems are not to be preferred over inorganic ones because the

https://doi.org/10.1515/9783110561838-009

Figure 9.1: (left) Photographs of corn plantation and wheat field with a forest in the background. (right) Far view of a corn plantation between soil plot, a grass field and forest in the background. You can interpret forest, grain plantation and grass field as layers or films on a substrate. The soil is the empty substrate, when we ignore the roots of the plants and the microbial cultures in the soil.

former would be less "reproducible" than the latter ones (metalloproteins instead of inorganic semiconductors for example). What does reproducibility in this respect mean? Or what do my colleagues mean by reproducibility?

My colleagues meant when you prepare a sample of algae or biofilms or microbes, then their size distribution or the scattering of other data would be quite large. They would not be all equal to some desired extent, at least not as much as for example silicon wafers or microchips that you could produce en masse in a semiconductor factory. Anyway, the crop field in Section 9.1 shows around several 10000 corn plants with virtually the same height, all aligned in the same direction.

I could show you also photos from sunflower plants which extend as a uniform layer of a large field. There is no lesser uniformity in these macrolayers than in the uniformity of semiconductor films, metal films or metal oxide films in the micro region and nanoscale. Therefore, I could not really agree with my colleagues' argument that one hardly can do proper science with biological samples.

Figure 9.2 shows the size distribution of seed grains from morning glory flowers (*Convolvulaceae* family) along with a flower, which I grow in my garden. When the colorful blossoms fade away in early fall, they leave you with pockets of seeds, which I collect and plant again the following year. I took a handful of dried seeds and put them on a microbalance in our lab and measured their mass one by one because I was interested how much their mass would scatter, statistically.

While it was a tedious task to put every single seed (180 in total) with a tweezer on the balance and also take it back, it helped me becoming convinced that they virtually looked all alike in shape except in size and volume. The bottom panel in Figure 9.2 shows the frequency n (in German: Häufigkeit), this is the number of grains which fall in a particular mass range.

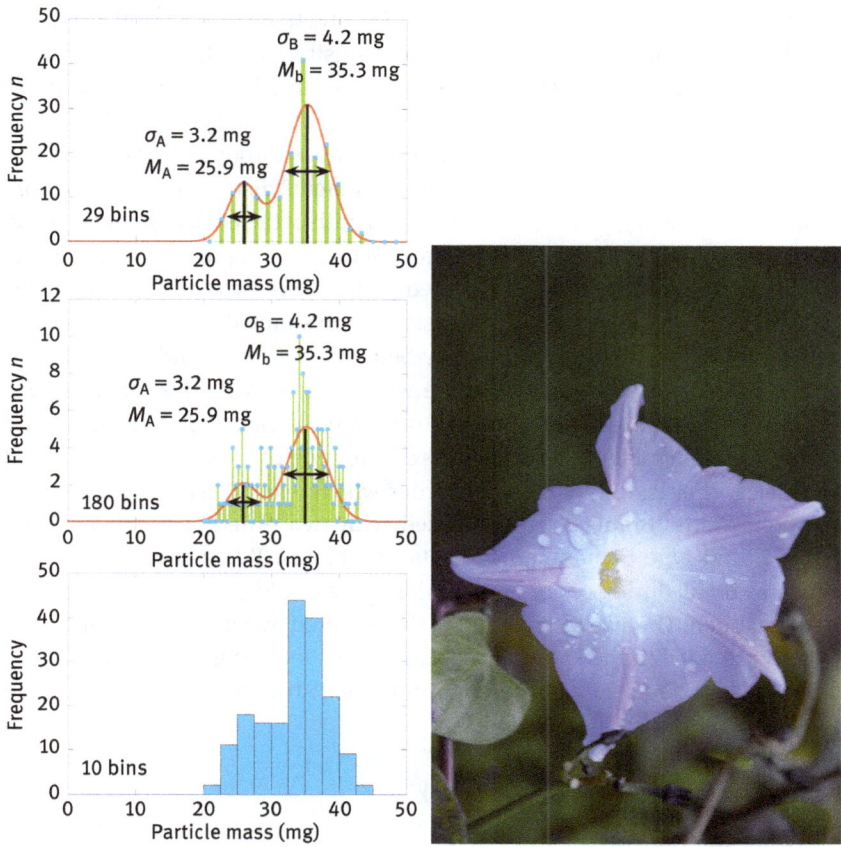

Figure 9.2: Statistical distribution of the mass of 180 seed particles from morning glory flowers (plus one flower on the right) with a clear maximum at around 35 mg. The bottom panel shows a simple bar plot histogram with 10 bins, suggesting a relevant size from 20 to 45 mg per seed grain. Finer resolution with 180 bins is shown in the middle panel. Top panel shows distribution in 29 bins. Bimodal mass distribution is obvious from all three representations.

Figure the following: you have now a table where every seed grain is listed with its mass, the first one with 36.2 mg, the second one with 27.3 mg and the last one with 41.2 mg. You now have to sort the 180 data points and begin with the lowest in mass (22.095 mg) and end with the highest in mass (42.765 mg), or the other way around if you like. To better organize the dataset, you may decide to section the span of 22.095–42.765 mg into 10 equally large compartments, drawers, boxes, bins. (1) 20.000–22.500 mg; (2) 22.500–25.000 mg; ...; (10) 42.500–45.000 mg.

Totally, 44 of the 180 seeds fall in the sixth size range bin of 32.500–35.000 mg, and 40 seeds fall in the next larger mass bin of 35.000–37.500 mg. Therefore, this mass range around 35 mg constitutes a very strong mode. Also the third bin of around 26 mg is statistically prominent.

The next you can do is widen the numbers of bins and try to assess all 180 particles more individually than in just 10 bins. This is demonstrated in the middle panel in Figure 9.2. The data points make now a more spiky impression but a keen eye identifies two modes in the same mass area as seen before with 10 bins. The red solid line in the middle panel is a convolution of two Gaussian distributions through these data points.

With 29 bins, we arrive at a pretty well outlined bi-modal distribution with maxima for the peaks at $M_A = 25.9$ mg and $M_B = 35.3$ mg. The width of the peak is 3.2 and 4.3 mg, respectively. I cannot draw further conclusions from this result other than that there is a bimodal distribution for the mass of the seed grains. When you work in materials science, you may occasionally come across systems with a bimodal (size) distribution.

What I want to show here is the analogy between processes and conditions in nature and in man-made technology. People working with statistical distributions will often notice, empirically that the properties of objects, such as size and mass, are distributed in two modes. Figure 9.3 shows the distribution of the size of particles and mass of particles, that is, glassy carbon powder which was investigated for super-capacitor applications [Braun 2002]. I worked with two different types of glassy carbon: type K and type G monolithic sheets and plates. When these plates and sheets somehow break during or after production, they would be ground into powder and then sold as such. The size distribution of the powered was determined with light scattering and provided in the materials specification sheet. It turned out that these powders had a bi-modal[1] size distribution and mass distribution. Particle sizes are

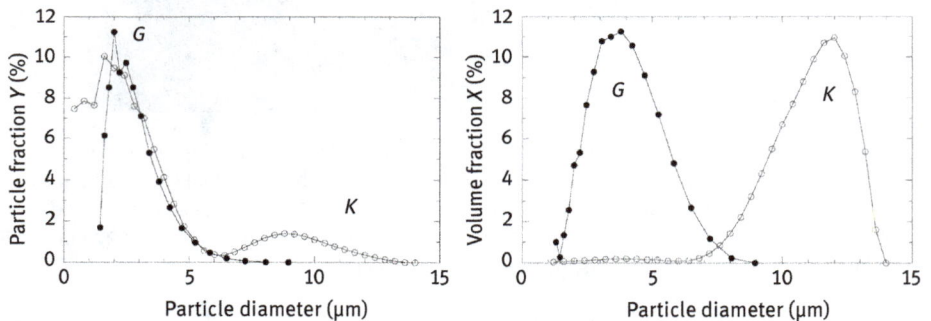

Figure 9.3: Particle fraction distribution and mass fraction distribution of glassy carbon powder type K and type G, as obtained from the manufacturer [Braun 2002]. The left panel shows how many % of the number of particles have a particular diameter. Both *K* and *G* have a maximum at around 2 mm. Only *K* has an additional maximum at around 9 mm. The right panel shows how many % of the volume of the particles have a particular diameter. As the volume scales with the third power of the radius, we notice a considerable shift in the profile fractions between both representations.

---

1 The "mode" is what statistically shows up significant in a spectrum. Basically, it is a peak in a spectrum. Mode is also the German and French word (the French fashion magazine *Dépêche Mode*) for fashion. When many people suddenly are wearing the same, then this is a mode in the spectrum of how people dress.

also important when we make electrodes for lithium batteries. When we consider for example the experiment illustrated in Braun [Braun 2015], there it was necessary to have some knowledge on the size of the spinel particles because the different probing depth of different spectroscopy methods could yield different information and insight on the processes taking place during lithium intercalation and battery function. It may be important to differently weigh the spectroscopic signals with respect to the relative abundance of larger and smaller particles.

I want to briefly exercise how we can determine a mean radius from a size distribution. We form the weighted average of the radius by integrating the product of the size distribution $P(R)$ with the size parameter R over the entire size range; theoretically this size range extends from 0 to infinity. Sometimes such calculations extend from $-\infty$ to $+\infty$, but in the problem we are addressing here there are no negative numbers and also no negative infinity because we have no negative particle radius.

Practically, this integration runs from the smallest found particle size to the largest found particle size. When looking back at the example with the flower seeds, we can do the same with the mass, for which the lowest was 22.2 mg and the largest was 42.8 mg, for example. Then, we integrate the size distribution $P(R)$ in the same range, as shown in the below equation. The ratio of both integrals is then the mean radius $<R>$, the first statistical moment of R:

$$<R> = \frac{\int_0^\infty P(R) \cdot R \, dR}{\int_0^\infty P(R) dR}$$

When we know the density ρ of the particles, that is, the ratio of mass and volume $\rho = m/V$, then we can convert the mass into volume. The volume distribution X to particle distribution Y is obtained from

$$X_i = \frac{V_i}{V} = \frac{N_i \cdot (4/3)\pi R_i^3}{V}$$

where V is the volume of all N particles in a specific quantity of powder, N_i of them having radius R_i, where V_i is therefore the fraction of the volume which arises from the particles with radius R_i. Therefore,

$$Y_i = \frac{N_i}{N} = \frac{V}{N} \frac{X_i}{(4/3)\pi R_i^3}$$

and, of course,

$$V = \sum_i V_i$$

We see that only the quantities X_i and R_i are required for the calculation of the distribution $P(R)$ [Braun 1999].

There may be systems with a bimodal distribution which can be rationalized, particularly in biology where evolution and population dynamics may favor development of an additional mode, such as for the Darwin finches with two different beak sizes at Galapagos Islands [Hendry 2009]. Patterns of human communication were also shown to have a bimodal characteristics [Wu 2010]. But I am drifting too much away from electrochemical energy storage and conversion already.

Before I end with this section, I want to make the readers aware that such universal principles, which the careful observer will inevitably find here and there, may be frequently termed mysterious by one group and immediately rejected as unscientific mystification.

An example for such dispute is found in a communication about Arnold Sommerfeld [Benz 1972] and a particular Dr Engesser, an Engineer. Arnold Sommerfeld had given a lecture to physicians and nature researchers in Kassel on September 24, 1903, where he said with reference to Castigliano's minimum principle

„Die Natur bevorzugt diejenige Wahl, bei welcher sie mit der geringsten Formänderungsarbeit auskommt".

"Nature prefers the case with the least deformation of shape"

In a Letter to the Editor [Engesser 1904], Engesser wrote that such finding should not be mistaken by a natural mysticism of science. Rather, it would be a consequence of a narrow set of mathematical conditions which would yield such universal principle.

No matter how we term these findings. They present a regularity which attracts our attention, and Engesser provides a simple mathematical rational for such behavior in nature.[2]

9.3 Electric properties of the earth subsurface

We learnt in the previous chapters how materials for electrodes and solid electrolytes have electronic and ionic and mixed electric conductivities. We remember how materials with spinel structure and olivine structure were used for lithium ion battery cathodes [Kulka 2015, Kulka 2016]. And we remember how zirconium oxide is used as oxygen ion conducting electrolyte membrane material for SOFC electrolytes [Biswas 2013, Yokokawa 2011], and barium zirconate as proton conducting electrolyte. We

---

2 Zipf's law is another such observation which can be expressed mathematically, see for example [Newman 2005] Newman MEJ: Power laws, Pareto distributions and Zipf's law. *Contemporary Physics* 2005, 46:323-351.doi: 10.1080/00107510500052444.

had put a proton conductor under compressive strain to learn about the pressure-dependent conductivity [Braun 2017, Chen 2011a, Chen 2010, Chen 2011b, Chen 2012].

The purpose for such high pressure experiments is not far-fetched. Strained phases should have higher proton conductivity. Compressed phases are relevant for geoscientists who study the phases in the earth core and earth mantle, for instance olivine under high pressure (my colleague Frank E. Huggins did his PhD thesis on this matter with Mössbauer spectroscopy [Huggins 1975]). Spinel and olivine are phases which are present in the earth mantle. Their electric (electronic and ionic) and thermal conductivities are of interest for geoscientists. An exhaustive review about these phases for example in the oceanic asthenosphere is available in [Katsura 2017]. Many soil and clay and minerals are used as materials in technology, such as diatomaceous earth materials [Saltas 2013].

Because the Earth has an electric conductivity, we can anticipate that electric currents may flow through the Earth. Such currents are typically called telluric currents. They can be magnetically induced by the rotation of the Earth and influenced by the solar bursts, which can extend to the Earth surface [Caglar 2000]. Such currents can be modeled by physical principles and mathematically evaluated [Mosnier 1985].

We are basically looking at a gigantic electric circuit, and its electric transport properties may depend not only on the microscopic structure of the matter but also on macroscopic inhomogeneities such as earthquake faults. In return, one can assess to some extent such geological inhomogeneities with electrical measurements [Madden 1996]. Imagine a pipeline made from metal tubes of say 500 km length buried in the soil. With the presence of telluric currents, we can expect that the pipeline is a huge electric conductor [Vanyan 2002], which will interact with the soil as it has electrolytic properties. It is therefore not surprising that we can measure a Nernst voltage along this gigantic wire and observe also corrosion[3] processes [Martin 1993] and try to counter such induced corrosion by cathodic protection [Martin 1994].

Now that we have learnt that the Earth materials can have similar properties like batteries and electrolytes, are we surprised when some researcher talk about Earth batteries? An Earth battery [Khan 2008] is nothing less, nothing else than a galvanic couple such as a zinc plate and a copper plate which are inserted in the soil and then produce a Nernst voltage, principally not different from the lemon battery or the Daniell element. The use of the soil is that we need not purchase an extra electrolyte and container. The danger is that we can contaminate the soil with metal ions. Copper and zinc ions may be environmentally harmful. As many other elements can be harmful as well.

---

3 Corrosion certainly takes place also at the device scale. Imagine a photoelectrode which you coat for better electric contacting with silver or tin and then mount it on a stainless steel frame. Depending on the electrochemical potentials of the involved metals, the electrons and holes generated without intention in the photoelectrode can trigger corrosive reactions at the interface of the metal layers. We observed this recently in a PEC reactor which was not even operated once.

9.4 Natural electric field

When we walk on ground we do not assume that there is any electricity we are walking on. Only during thunderstorms and lightning, we are reminded of the naturally occurring electricity. Yet, even in the absence of lightning, the surface area of the Earth is negatively charged with around 500000 Coulombs. The radius of the Earth is on average $R = 6371$ km; the geometric surface area of the Earth is then $S = 4 \pi R^2 = 510$ million km^2. The electric charge per m^2 is therefore quite small, just less than 1 millicoulomb.

Against the ionosphere above us, which is positively charged, the potential difference is 300,000 V. As the air mass between Earth and ionosphere is not completely insulating, there is always a flow of electricity, which amounts to a total current of around 1350 A. The current density is thus 2.6 µA/km^2. With Ohm's law $R = U/I$, this yields a resistance between "Earth and Air" of 222 Ohm. With such high voltage and strong current over the entire Earth, the power is $P = U*I = 405$ MW. But when we normalize it to m^2, the power is a mere 0.8 µW/m^2. It seems pointless to even think about tapping this small amount of energy (power), given that the solar radiation delivers 50–1000 W/m^2; this is a factor of 1 billion.

When there is an electric current between ionosphere and Earth, why is it that the charge becomes not exhausted? What was just described looked like a giant electric capacitor made from the Earth surface and the ionosphere. When there is a current, the capacitor must become discharged, right? Well, there is an enormous constant supply of new charge carriers coming from our Sun, which has done so for the past four billion years and will do so for the next four billion years to come.

We have made already some simple geometric calculations in this Section. Here is another one: What is the capacity C of this giant capacitor: $C = Q/U = 500000$ Coulomb/300000 V = 5/3 Farad = 1.67 Farad. More on this topic can be found in the book by Coroniti and Hughes [Coroniti 1969]. The field vector **B** of the magnetic field of the Earth and the plasma belt at some height over ground may interact in a way that the scalar product **E·B** = 0 produces a phenomenon of charged rings. This is where the terms electrosphere and magnetosphere become a meaning.

9.5 Electricity in the atmospheres

At sea level, the atmospheric electric conductivity is quite low in the range of 10^{-14} S/m. For example, measurements in the equatorial Indian Ocean and Arab Sea in 1991 with relative humidity of 70–80% yielded 2.3×10^{-14} S/m. Interestingly, at the Somali current, when the relative humidity was 80–90%, the conductivity was only 1.1×10^{-14} S/m [Kamra 1997]. Kamra et al. explain this reverse relationship between conductivity and air humidity by the attachment of specific ions to aerosols [Kamra 1997].

The presence of aerosols can have influence on the conductivity in the atmosphere. Therefore, the conductivity can be used for the assessment of air quality [Kamsali 2011]. This method is similar to the measuring of the purity of water. When you purchase high purity water, it may be specified with its conductivity.

The gradient of the electric potential over the city of Delhi in winter 1968 [Srivastava 1972] is shown in Figure 9.4. Season, weather conditions and the presence of aerosols in the air (not all aerosols and particulates in the air should be termed air pollution) can have an influence on the electric conductivity of the atmosphere. Figure 9.4 shows how the electric potential gradient increases strong from below 40 V/m to nearly 150 V/m around 500 of meters above surface, while the temperature is increasing accordingly from around +10 °C to +20 °C. Temperature T is shown as the right curve in the figure, and the V/m as the curve on the left. The temperature scale is shown underneath the V/m scale. Observe how the temperature decreases above 500 from +20 to −60 °C at around 15 km height. Above 500 m, the potential gradient decreases sharply from 150 to 20 V/m. This range of sharp change of potential gradient is known as exchange layer and is sometimes accompanied by the observed temperature inversion:

All meteorological events are accompanied by characteristic changes of the electrical parameters. [Srivastava 1972]

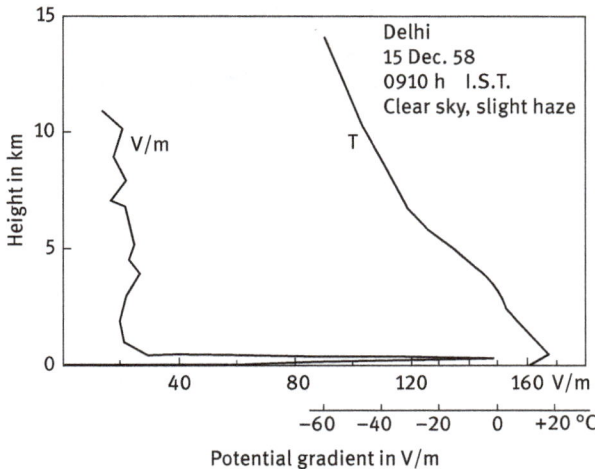

Figure 9.4: Variations of potential gradient with height over Delhi in winter. (1ST: Indian Standard Time= GMT+ 5 h 30 min).
Reprinted with permission from Springer Nature, *Pure and Applied Geophysics PAGEOPH*, 100:81–93, Electrical conductivity and potential gradient measurements in the free atmosphere over India Srivastava GP, Huddar BB, Mani A, (1972) [Srivastava 1972].

The electric conductivity in the height up to approximately 50 km from the Earth surface originates predominantly from the ions which are produced by ionizing radiation. This ionizing radiation is the cosmic radiation from outer space and in the lower regions up to 1 km level also by the natural radioactivity of the Earth. At very high altitudes of 60 km and above, it is not ions but "free" electrons, which are the major contribution to this conductivity and electric currents [Rakov 2007].

The Earth has a magnetic core which produces a magnetic field, the poles of which coincide in good approximation with the rotational axis of the Earth. The magnetic field lines thus truncate the North pole with maximum field strength, span around the Earth and close the magnetic loop in the South pole where the field lines condensate again to maximum density of magnetic flux.

The incoming charged particles from the sun, mostly electrons and protons and some helium, and the ions formed by the cosmic radiation which ionizes the molecules in the atmosphere are interacting with the aforementioned magnetic field lines and experience the so-called Lorentz force F, which is the vector product of the magnetic field B and the velocity v of the particles with electric charge q: $F = q\,(\mathbf{v} \times \mathbf{B})$.

This Lorentz force causes the electrons and protons from the sun to accumulate around the magnetic field lines, where they hit on oxygen and nitrogen molecules from the atmosphere so these become ionized. When the ions recombine again with their opposite charge, they emit light which can be seen as the spectacular aurora borealis provided the concentration of particles is high enough and provided there is sufficient dark background (polar night). This is typically observed as green polar light in the Arctic and Antarctica. Figure 9.5 shows such green light bands at the dark night sky (23:00 pm) in the city of Tromso in the Norway Arctic.

The electric charge density of the higher spheres above the Earth surface is still today a matter of scientific and technological interest. The HAARP (High Frequency Active Auroral Research Program) project makes it frequently in the public, where it has a negative connotation [Bailey 1997, Busch 1997, Gordon 1997] because of its geophysical use in defense and military projects [Cole 2005]:

> The program's facility is a high power transmitter located in Alaska, capable of broadcasting powerful VLF radio waves into the Earth's ionosphere. These waves propagate along the Earth's magnetic field lines to the system's geomagnetic conjugate point situated nominally 600 miles south of New Zealand in the southern Pacific Ocean. By studying the radio signals at this point, the VLF Group seeks to discover how energetic particles in the planet's radiation belts interact with very low frequency electromagnetic waves.

Writes Milikh: "It is well known that strong electron heating by a powerful HF-facility can lead to the formation of electron and ion density perturbations that stretch along the magnetic field line. Those density perturbations can serve as ducts for extremely low frequency (ELF) waves, both of natural and artificial origin" [Milikh 2010, Milikh 2008].

Figure 9.5: Green polar light (*Aurora Borealis*) in the Arctic City of Tromso in Norway. Photo Artur Braun. Artistic enhancement by Philipp Rogenmoser, Empa.

It appears that it is possible to manipulate the ionosphere by this "high power transmitter" [Cole 2005, Pedersen 2015] and enhance charge densities [Fallen 2011] and measure currents [Papadopoulos 2011] literally across the globe. And "Disturbances" can be created with HAARP facility [Bernhardt 2016]. In summary, it appears such HAARP can be used like an enabler or even like a transistor which allows controlling the stream of cosmic radiation with a relative small energy input. Relative small refers here to the huge energy coming from the cosmos, notwithstanding that HAARP is consuming a lot of energy with reference to the global scale.

With this knowledge on the electrical properties of Earth and atmosphere, it is not surprising that scientists speculate about similar situations on other planets [Certini 2009, Langlais 2010] such as on the Mars [Farrell 2001], Uranus [Melin 2011] or in outer space.

9.6 Excursion to energy storage in wood

This book is about electrochemical energy storage and conversion. Is there any other energy storage than by electrochemistry? For sure there is. But why do we store energy? What is the purpose of energy storage? When fire was discovered by man supposedly in Africa half a million years ago, he took control of an energy source which could provide light in the dark and heat in the cold. This energy source was wood or other pieces of dry plants and combustible material. In later ages, coal and other fossil fuels such as petroleum, mineral oil and natural gas were used as fuels for all kind of purposes. Later, sperm oil from whales and fish oil was used for lamps and lighting.

Nowadays fossil fuels are used for transportation purposes, in addition to heating and cooking. Heating includes also the huge amount of industrial processes in factories. There is consensus today that oil, gas and coal are of biological origin; this is why they are called fossil fuels. There are however theories which say that these could also be of inorganic origin. There exist chemical and physical processes deep in the earth which make that carbon and hydrogen react to hydrocarbons at high pressure and high temperature.

The formation of oil is in this model of genesis not made by decomposition of bio-organic material, but by synthesis of its elemental constituents [Kundt 2014]. It is not so easy to experimentally verify such theory, but colleagues working in high pressure research [Kolesnikov 2009] were able to experimentally simulate the scenario which would allow for the production of hydrocarbons heavier than methane CH_4 from methane as the precursor. Note that this is a theory as much as the broadly established view is a theory.

Professor Wolfgang Eberhardt made an ironic remark in his talk in 2013 at Empa [Beni 2013] that the invention of fire was a "bad discovery" when we consider it as a cause of manmade global warming. As the Carnot machines have a low thermodynamic efficiency when compared with electrochemical processes, fire as the source of the explosions in such machines is thus a less desirable process.

The biological photosynthesis processes which take place in plants that build the wood can also be interpreted within Carnot theory [Jennings 2005, Meszéna 1999]. We certainly can write down an energetic balance and budget for biological systems. Some biological processes such as protein folding can certainly be interpreted along the classical lines of the Carnot theory, see for example [Shibata 1998]. Wood is a widely used fuel particularly in rural areas and in underdeveloped regions and countries. Fire wood is harvested by cutting trees in forests with chain saws, operated manually by local wood farmers and loggers. The work is heavy labor and dangerous. The trees are cut in short pieces of around 1 m length and then split with a splitting hammer. Then the wood logs are piled up as shown in Figure 9.6. The location is near the mountain top of the Hulftegg at the border of the Swiss cantons of Zürich and Sankt Gallen.

The wood logger has used a machine to bundle the split wood logs with tough metal strips and piled them up near the unpaved road where it can dry for one year. The wood is exposed to the weather and will dry. To prevent direct rain from the top to moisten the wood, plastic sheets are used as cover. The owner of the wood may come from time to time to pick up bundles on a trailer which is pulled by a tractor. The owner may sell the bundles to local dealers, who transport the wood to their barns

Figure 9.6: Beech wood piled up for drying in the nature at the Hulftegg Pass at the border of Kanton Zürich and Kanton Sankt Gallen, Switzerland. Plastic and metal sheets are covering the piles to protect them from excessive rain. The piles are clamped together by metal bandages for easy piling and pick up with heavy equipment.

where it can further dry. This is for example shown in Figure 9.7 which displays the back side of a large barn near a Swiss village; the barn is used as a weather cover for the wood piled up vertically so that it can further dry. This way the wood will lose most of its moisture and remains typically with residual moisture of 20%, which makes it ready for burning.

The farmer who owns this wood may cut it further with a saw to logs of 30 or 40 cm length and then sell small amounts of wood to local home owners who need the wood to feed their residential stoves. It is quite common in Switzerland that home owners provide their tenants with an extra stove for the nicer atmosphere in the living room, notwithstanding that the majority of homes are equipped with a modern central underfloor heating system.

This central heating is traditionally fueled with heating oil or natural gas. In rural areas with no connection to natural gas pipelines, the central heating may be fueled with the very wood logs shown in Figures 9.6 and 9.7. Figure 9.8 (left) shows a stove which is fed with dry wood. A pot with water is put on the top of the stove. This fire will last for a couple of hours. The exhaust from the wood combustion is led to the

Figure 9.7: Wood piles at a barn in Illnau, Switzerland. The logs from split beech wood have a length of around 80 cm. They dry for 2–3 years outside in the ambient and are then ready for burning.

chimney of the home which is connected with the back of the stove through a metal stove pipe. When all wood is burnt and the fire is extinct, ash will remain at the bottom of the stove.[4] It will be collected with a shovel and small broom and then disposed. The right photo in Figure 9.8 shows a sign "Wir heizen mit Holz!" (we are heating with wood), which was mounted at the outer wall of a rural home close to the Schnebelhorn mountain in Switzerland, near the Sunegg. At this remote place, it is common to use the local wood for heating.

---

4 I am coming here back to an earlier remark I made in this book about whether we can use water as an energy source. We cannot. The opposite is the case. Hydrogen and oxygen gas can be chemically reacted as an open flame and then release heat, or in combustion engine and produce mechanical work and heat or reacted in an electrochemical cell and the energy is the electromotive force, that is, electricity. The reaction product is then water. Water in so far has no energy anymore, except for its heat that it has at temperature $T > 0$ K. Here when you burn wood, you end up with the ash and with CO_2 and H_2O as reaction products. Their energy is gone already. We cannot pull more energy out of it. Compare water to a relaxed expander. It has no energy anymore unless you expand it to hydrogen gas and oxygen gas. I hope this picture helps with the understanding.

Figure 9.8: (left) An iron stove in a residential home. Wooden logs are burnt in the stove and release the heat in the room. The flow of air into the stove can be optimized so that only minimum soot is produced. A pipe at the backside of the stove leads smoke and carbon dioxide through the wall to the rooftop of the home into the environment. The ash from the burnt wood is collected in a tray at the bottom of the stove and can be emptied regularly. New wood logs are fed to the stove by opening the front door. With proper adjustment of the air flow in the stove, the logs can burn for several hours and provide a homogeneous flow of heat in the room. (right) We are heating with wood! Secured quality." A slogan from the Swiss wood energy association, posted at a home in a remote rural area where wood is a major supply of energy for the residential homes.

The wood shown in these figures is usually taken from beech trees because beech is a very good sort of trees for this purpose. Table 9.1 lists 16 sorts of trees which are often used as fire wood. The list shows the caloric heating value of the wood. Beech is leading with a 2200 kWh per stacked cubic meter (stcm). The stacked cubic meter is the most practical metric for the loggers who work in the forest and sell the wood. Note that the wooden logs are not totally uniform in shape. When you pile them up, there is some empty space in between these logs. This is ok because these spaces facilitate the drying of the wood. Air can come in and moisture can come out.

Here follow some rules of thumb: 1 stcm of fresh cut wood typically weighs 800 kg and has on average a caloric value of 1500 kWh for the pile or 1.9 kWh/kg. After drying in the ambient environment, the weight shrinks from 800 kg to around 420 kg while the caloric value is increasing from 1500 to 1800 kWh. Per kilogram this is an increase to 4.3 kWh/kg. Every 10% increase in moisture causes a 9% decrease in caloric value. This

Table 9.1: Energy content of various sorts of wood and trees in decreasing order.

Wood type	Energy content kWh/stcm	Energy content (kWh/kg)	Logging age/year	Maximum height/m	Maximum age/year
White/Red beech	2200	4.2	120–160	35	200–300
Ash	2100	4.2	100–140	30	250–300
European beech	2100	4.2	60–100	20	150
Oak	2100	4.2	180–300	25	500–800
Robinia	2100	4.1			
Birch	1900	4.3	60–80	25	100–120
Elm	1900	4.1	124–140	30	400–500
Maple	1900	4.1	100–120	20	150
Douglas fir	1700	4.4	60–100	55	400–700
Larch	1700	4.4	100–140	30	200–400
Pine	1700	4.4	80–140	36	200–300
Spruce	1600	4.4	80–120	40	200–300
Alder	1500	4.1	60–80	25	100–120
Fir	1500	4.4	90–130	40	500–600
Poplar	1400	4.2	30–50	25	100–150
Willow	1400	4.1			

Stacked cubic meter (stcm) is the volume unit of wood logs piled as shown in Figures 9.6 and 9.7 [Krajnc 2015]. Data based on the assumption that the wood is dry from exposure to ambient environment and thus would contain around 20% moisture. Data taken from http://www.energie.ch/heizwerte-von-holz and http://www.wald-prinz.de/umtriebszeit-wie-lange-benotigt-ein-baum-bis-zur-hiebsreife/3697.

is because water has a relatively large heat capacity. Much of the produced heat from burning moist wood will be wasted to evaporate the water. The presence of water in the wood will also influence the combustion chemistry towards formation of carboxylic functional groups which could be dangerous when they accumulate as creosote in the chimney of your home, which eventually may light up in fire [Braun 2008].

While the caloric value per stcm ranges from 2200 to 1400 kWh/stcm (50% difference), the normalized value per kg ranges from 4.4 to 4.1 kWh/kg (less than 10% change). The caloric value of these different sorts of wood is therefore quite similar. For comparison, coal has a caloric value of a 8 ± 1 kWh/kg, and heating oil has a caloric value of 12 kWh/kg.

These nice monolithic wooden logs are actually quite expensive wood when you have to heat your entire home from it. When you experiment with wood burning, you will notice that pine wood burns much faster than oak wood, for example.

Wood is also used for other purposes, for example for the manufacture of furniture and the like. It would be more worthwhile to let the trees grow for the use of furniture wood, rather than burning it as fire wood. Oak is very heavy and stable, whereas pine wood is quite lightweight, when you compare the furniture. This manifests also in the 50% difference in the caloric value per stcm of the different sorts of wood. The right columns in Table 9.1 show the maximum height and the maximum age of the trees.

It takes at least two generations, that is, over 50 years before a tree is ready to be cut at full size. Figure 9.9 shows a photo of a pile of wood in the forest which is waiting for further processing. The cross-section where the wood was cut with a chain saw shows characteristic rings which give account of the age of the tree. I have taken the tree in the middle as an example where I can demonstrate with arrows in color what the rings mean.

Figure 9.9: Cross section of cut trees showing the annual rings of growth.

The tree in the middle is probably an oak or beech wood. The dark spot in the middle is the center. The tree has grown concentric but not isometric. The north side of the tree has not faced the sun directly and therefore has thinner rings which add to a smaller radius of this part. The south side was exposed to the sun and has thicker rings and thus an overall thicker portion of wood on that side. The rings have dark and bright shading. The bright rings are from growth in the spring, as exemplified by the two thin blue lines. The dark rings are from growth in the autumn, as exemplified by the two thin yellow lines. The individual rings are indicated along the two arrows with green thin bows. The tree has around 70 rings and would be around 35 years old.

However, the natural maximum possible age of trees can range from 100 to 1000 years. We have to pause now over the fact that a tree grows over a 100 years, whereas it only needs 1 h to burn it in a wildfire, for example.

A very large oak tree, when you burn it as fire wood, can warm your home for one entire year. When you estimate your own lifetime to 80 years, then it needs 80 such trees in your lifetime to keep yourself warm. When you live in a family of four, be it together with your sibling and your parents when you are young and together with your spouse and your children and grandchildren when you are old, then you may average the wood-for-heat consumption to only 20 such huge trees in a family of four, for example.

At full age, such huge trees require a lot of space, like 180 m² per tree, as I have sketched in Figure 9.10. A rule of thumb says that a tree needs as much projected area above surface like in the subsurface underground. The total area occupied by 20 such trees necessary for a family of four is thus 3600 m² which is a square of 60 m length. This space is required and occupied for your own family's energy needs for having it warm throughout the year. As it is believed that the global population counts now seven billion people, the necessary area for all of them are 25 million km².

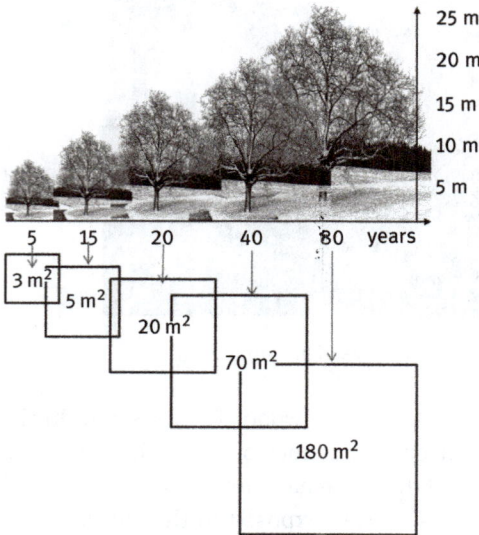

Figure 9.10: Required area of a tree subsurface. A five-year-old tree would have a height around 5 m and require around 3 m² area underneath the surface. Inspired by http://www.baumpfleger.at/image_3/platz.jpg.

This is around the territory of Russia and Brazil, which can supply the world population with fire wood.

$$3600 \text{ m}^2 * 7 * 10^9 = 3.6 \ 10^3 * 7 * 10^9 \approx 3.6 * 7 * 10^{12} \text{ m}^2 \approx 25 * 10^{12} \text{ m}^2 = 25 * 10^6 \text{ km}^2.$$

Germany has a population of around 80 million people, which, according to my calculation, would require almost 300000 km² area for wood. The territory of Germany counts 357000 km². We are obviously running into a problem when we want to heat the homes for all Germans with firewood. There is no space left for anything else. Plus: where would all the other energy come from that we need for our daily lives, for our mobility, for our industrial activity and economy?

$$3600 \text{ m}^2 \times 80 \times 10^6 = 3.6 \times 80 \times 10^9 \text{ m}^2 = 3.6 \times 8 \times 10^{10} \text{ m}^2 = 28.8 \times 10^{6+4} \text{ m}^2 = 28.8 \times 10^4 \text{ km}^2 = 3 \times 10^5 \text{ km}^2$$

Firewood is normally not grown the way I just described. When you plant a forest, small trees which are just one year old are planted in rows with a distance of around 1 m from the next small tree and row. As the trees are growing, they require more space and some of the trees are taken out, particularly those who are not doing so well. These can be used as firewood. With more and more years to follow, the remaining trees claim more space and continuously trees will be cut out and used for firewood or for other purposes. This means while the forest is growing, it will remain to be cultivated and firewood can be taken out before the trees have reached the maximum height and volume.

For the distinct production of fire wood, you may have to grow the trees for maybe 3–10 years only; this is way before they have reached the logging age for furniture. In the last couple of years, a new technology and industry evolved which uses scratch wood, shreds it into powder and then presses it into small pellets. The pellets have a mass and volume which makes them fluidizable: they can be handled to some extent like a gas or a liquid.

The wood pellets are typically made from timber waste and wood waste. The pellet diameter is 6–25 mm and length is 3–50 mm. Water content is typically lower than 15%. Note that the wood pellets are not produced from the aforementioned long wooden logs shown in Figures 9.6 and 9.7. The long logs are cut by the wood farmers later into pieces of around 30 cm which are then sold on the local market "at the porch" to residential home owners. In rural areas in Europe and in the United States, many home owners have open fire places or wood stoves which give a particular atmosphere in the living rooms. This causes however air pollution which can be detected with the naked eye as wood smoke in valleys and also with air sampling methods [Braun 2008].

Over 20% of the world energy consumption is electric energy. The other 80% are all kind of fuels, this is the fossil fuel, biomass and nuclear fuels. Biomass includes the wood. While many people in third world countries use fire wood and other biomass for coking and heating, technologically high developed countries like Switzerland, Austria, Italy, France, Germany and the United States have a part of their population which uses wood for heating purposes. The wood is the energy storage material.

There is the saying that wood can warm your body up three times: when you cut the wood in the forest, you will warm up because of the work you do; when you transport and process the wood at home, you warm up again; and when you finally burn the wood, you take advantage of the heat from wood combustion. The energy content of wood varies from type to type. It makes also a difference whether the wood is dry or damp. In damp wood, some of the energy from burning gets lost because of the evaporation of the water stored in the wood. The energy ranges between 4 and 4.5 kWh/kg. For comparison, coal can have 7.8–9.8 kWh/kg. Heating oil has 12 kWh/kg. Note that trees will regrow (over a period of around 20–25 years), whereas coal and heating oil will be irreplaceably lost—unless the theory explained in Kolesnikov (2009) and Kundt (2014) is the correct one.

Figure 9.11 shows a small truck which delivers wood pellets to a residential home. Supposedly, the pellets have a caloric value of 5 kWh/kg. The cylindrical container on the truck is reminiscent of containers for liquids and gases. Indeed, the pellets are fluidizable and can be handled to some extent like a liquid or a gas. Figure 9.12 shows the rear end of the truck and two flexible hose pipes which are about to be connected to the truck's pellet release pump at the rear end. The hoses are connected with the underground pellet storage room in front of the home, which was opened up for the delivery. Figure 9.13 shows the two hoses in close view connected into the basement of the home which contains a large pellet fuel container for the central heating. There are two hoses necessary for the delivery: one pumps the pellets into the basement and

Figure 9.11: Truck delivering small pellets from compressed wood waste to a residential Minergie home near Zürich, Switzerland. The pellets are contained in the cylindrical vessel.

Figure 9.12: Two hoses right before connection to the pellet delivery truck, one of which blows the pellets into the pellet storage container in the basement of the residential home. The second hose sucks air from this storage container in the basement so as to remove wood dust.

the other sucks air from the container in the basement to allow for the necessary circulation of air and pellets.

The right panel in Figure 9.13 shows how the delivery company advertises that their products, the wood pellets, are a CO_2-neutral fuel. This claim is not farfetched because the wood is from trees which have consumed CO_2 which was contained in the atmosphere. Certainly, combusting the pellets in the burner in the basement will again produce CO_2 which is released in the atmosphere. But this CO_2 will be consumed again by plants during photosynthesis, as is indicated by the dim blue circular arrow (metabolic cycle) on the back of the pellet container on the truck. Water H_2O and carbon dioxide CO_2 are the main ingredients for the production of sugars and starch and cellulose, and fresh oxygen O_2 in the photosynthetic process, which is powered by solar energy. After the combustion of the wood, we will find white ash in the burner which contains potassium K, sodium Na, magnesium Mg, calcium Ca, phosphorus P, sulfur S, silica Si and chlorine Cl. These are some of the necessary

Figure 9.13: (Left) Access inlet to the pellet storage container in the basement of the residential home. (right) Back side of the wood pellet container of the truck. The timber and wood industry advertise that wood pellets are a CO_2 neutral and renewable fuel.

trace elements which are needed to build up the molecules of the photosynthetic apparatus.

9.7 Generator gas: fuel gas produced from wood

We see that the energy density ratio between oil (12 kWh/kg) and wood (~ 4 kWh/kg) is around 3. It is possible to convert wood with a reformer into combustible gas and use the gas as a fuel for transportation or for other use. This reformation process was a substitute for fossil fuels which were not available for example in war times in Germany when the country was cut from foreign oil supplies. Three kilogram of wood could thus replace one liter of benzene in so-called "producer gas vehicles" http:// www.petrolmaps.co.uk/german30.htm. In Germany, this producer gas was called Generatorgas [Reed 1979]. German Generatorkraft A.G.[5] was a specifically established company which organized the wood gasification [Flachowsky 2017].

---

5 "Erwerb und Vertrieb von festen Generatorstoffen aller Art, vornehmlich Holz, Kohle und Torf, sowie der Betrieb aller Geschäfte, die der Förderung des Generatorwesens und der für den Generator erforderlichen Kraftstoffe dienen. Gegründet am 29.6.1940, handelsgerichtlich eingetragen am 21.1.1941. Bis 11.7.1941 lautete die Firma: Generatorkraft AG für feste Kraftstoffe. Bis 30.11.1942: Generatorkraft AG für Tankholz und andere Generatorkraftstoffe, danach: Generatorkraft AG. Aufsichtsrat (1943): Staatssekretär Günther Schulze-Finitz (Reichsministerium für Bewaffnung und

Countries rich in wood and forestry do have some basic interest in wood gas and generator gas. After all, wood is a fuel resource. Sweden was interested in wood gas during the Second World War and funded a major study on wood gas technology [Reed 1979]. When the United States became interested in wood gas, they decided to translate the Swedish report into English, rather than performing their own study [Reed 1979].

Generator gar is produced by the gasification of a solid fuel (wood, coal, char, . . .) at high temperatures. We have a complete gasification when we burn the fuel in excess oxygen; but this is not our goal. The aim is not the combustion of the fuel, but the reforming of the solid fuel to gas fuel. When instead we have surplus of the solid fuel component, then the carbon dioxide and water vapor pass over a glowing layer of coal, such as char coal so that the two gases become chemically reduced to hydrogen H_2 and carbon monoxide CO, which both are combustible gases, that is, the gaseous fuels that we wish for. The extent of the reduction depends on the process parameters and is at large governed by the water-gas shift reaction

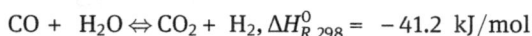

$$CO + H_2O \Leftrightarrow CO_2 + H_2, \Delta H_{R\ 298}^0 = -41.2 \ kJ/mol$$

and the Boudouard equilibrium.

$$CO_2 + C \Leftrightarrow 2CO, \Delta H = +172.5 \ kJ/mol$$

The solid fuel is composed of carbon, hydrogen and oxygen, and also trace elements which we may find in ash after the fuel is burnt. Complete combustion with excess oxygen from air forms carbon dioxide from the carbon in the fuel, and water from the hydrogen in the fuel, typically as water vapor (steam). The oxygen in the fuel will, of course, be a part of the combustion products, and therefore the amount of oxygen needed for complete combustion is decreased. The following chemical reaction formulae describe this burning:

$$C + O_2 = CO_2$$

$$H_2 + \frac{1}{2} O_2 = H_2O$$

A heat quantity of 241.1 kJ (57.6 kcal) results from the burning of 1 mol, this is, 2.016 g, hydrogen into water vapor.

Munition), Berlin, Vorsitzer; Fabrikant Carl F. W. Borgward, Bremen; Professor Dr. Karl Hettlage, Berlin; Dr. Heinrich Machemer (Reichsministerium für Bewaffnung und Munition), Berlin; Hugo Stinnes, Mülheim (Ruhr) u.v.a. 1950 Berliner Wertpapierbereinigung, 1954 verlagert nach Frankfurt/Main, 1955 aufgelöst, 1958 nach Abwicklung erloschen." (Quelle: Peus Nachf.) http://www.schoene-aktien.de/wertpapier-KU01018.html

Generator gas is composed of carbon monoxide, hydrogen and methane as the major combustible gases. There are also some amounts of other hydrocarbons and also tar vapor. These are generated as pyrolysis products.

The most important reactions that can take place in the reduction zone and between the gases formed are listed in Table 9.2. The heat quantities are measured in kJ (kcal) per mole, where the combustion of the carbon has been assumed to yield 4019 kJ (96.0 kcal). A positive (plus) sign means that heat is generated in the reaction. A negative (minus) sign means that the reaction requires heat.

Table 9.2: Chemical reactions and necessary energy (+) and available energy (−) for producer gas processes.

Chemical reaction	kJ/mol	kcal/mol
$C + CO_2 = 2\ CO$	−164.9	−39.4
$C + H_2 = = CO + H_2$	−122.6	−29.3
$CO_2 + H_2 = CO + H_2O$	−42.3	−10.1
$C + 2\ H_2 = CH_4$	+83.3	19.9
$CO + 3\ H_2 = CH_4 + H_2O$	+205.9	49.2

Reproduced from the translated Swedish wood gas report [Reed 1979].

Generator gas is produced when oxygen from the air is guided over glowing wood or coal under lean combustion conditions. This means that the oxygen does not react with the carbon and hydrogen from the wood to CO_2 and water H_2O. Rather, it will form CO, H_2 and CH_4, which are combustible gas fuels. Generator gas was not only used for mobility applications. For example, a glass factory [Denk 1920a, b] in Brazil was partially operated with generator gas over 100 years ago [Dralle 1915a, b]. Even nowadays generator gas is of interest for instance for using it for stoves [Mukunda 2010].

Wood is certainly not the only energy source for making fuel gas. Coal can be used and also petroleum oil [Nations 1979]. An example of a large gas factory which used residual fuel oil as energy source—almost 100 years ago—is explained in [Pike 1929]. Hot water steam and the oil (a low quality residual which hitherto could not be used as transportation petroleum) are then sued to produce what is called city gas. According to the chemical analysis of the city gas produced in the test plant (Table 9.3, [Pike 1929]), almost one-third of the weight is methane and one-fourth of the weight is carbon monoxide. Hydrogen, ethane and benzol are produced to 6 weight % each as well. And there is almost 20% of carbon dioxide. Pike and West give a very nice quantitative socioeconomic explanation about which processes take place and which budget applies and how much a household would have to pay for the particular energy products.

Table 9.3: Analysis of city gas produced in a test plant.

Molecule	Volume per cent	Weight per cent
H_2	45.9	6.30
CH_4	26.2	28.75
CO	13.2	25.35
C_2H_4	3.2	6.14
C_6H_6	1.2	6.42
CO_2	6.5	19.60
O_2	0.5	1.10
N_2	3.3	6.34
	100.0	100.0

Reproduced from Pike and West [Pike 1929]. Reprinted (adapted) with permission from Pike RD, West GH: Thermal characteristics and heat balance of a large oil-gas generator. Industrial and Engineering Chemistry 1929, 21:104–109. Copyright (2017) American Chemical Society.

Generator gas, producer gas, city gas and wood gas are certainly not only useful for burning it to heat a home. We learnt already that it was used to power combustion engines for vehicles in the Second World War.

Figure 9.14 shows a contemporary motor bike which is equipped with a wood gas producer; this is the metal container mounted on the back of the motor bike. It

Figure 9.14: Yamaha motor bike equipped with a wood gas reactor.
Photo by courtesy of Markus Schmid, Moto Sport Schweiz.
https://www.motosport.ch/media/motosport/archivfiles/32965_wallpaper_9.jpg.

contains a wood reactor which is heated with wood fire and which will pyrolyze the wood and produce hydrocarbon gas: wood gas. The wood gas is led to the combustion engine (Yamaha) which will run the motor bike.

We can certainly use the wood gas or producer gas also for electrochemical conversion. This is possible in solid oxide fuel cells (SOFC), which operate at temperatures from 600 to 1000 °C. These SOFC then typically need a reformer, which prepare the fuel for the use in SOFC. I have not yet come across an application of this type, but it should be technically possible. Some 10 years ago we worked in a project where biomass gas was considered as fuel for SOFC [Herle 2010].

The wood gas would enter an SOFC and then produce a considerable amount of heat and also the electromotive force necessary to drive the electric motor. This would be an analogous use without the need for clean hydrogen and a PEM FC. Nissan has been involved in SOFC research (anodes, electrolytes, catalysts) for several years [Nabae 2005, Nabae 2008, Sumi 2011, Yamanaka 2007, Zuo 2006] and has recently announced [Doi 2016] it will develop an electric vehicle which is propelled with biofuel.

10 Electricity and biology

10.1 Electrocultures

The trees, the plants have water pipelines inside which is a huge extended network ranging from the outer leaves through the branches to the roots down in the soil. This water is not just water but an electrolyte carrying ions from minerals. Therefore, the phytosphere is electrically – ionically – well connected with the ground. Therefore, the electric potential of the phytosphere is by and large the potential of the ground. The negative ions travel from the bottom through the plant to the upper leaves. So, here we have a vertical current.

Some researchers were therefore interested in how the growth of plants would depend on the electric properties of the environment. Some of them hoped that the output and performance of agriculture industry could benefit from the research on electrocultures [Oswald 1933].

During the preparation of the EU Flagship project SUNRISE [Aro 2017], I have been made aware that in the Netherlands, a country which is well known for its agriculture technology and business, some farmers feed industrial CO_2 into their greenhouses in order to boost the growth of their crops. The farmers pay for this CO_2 – not as CO_2 tax, but as a CO_2 price like for any other good that you would pay for.

Figure 10.1 shows a small greenhouse in a village in Switzerland. These greenhouses operate throughout the year, and they contain heating stoves so that no freezing occurs on the plants during winter. I do not know whether the resulting CO_2 is kept or led into the greenhouse over the plants so that they grow better. But I learnt lately that some CO_2 produced in factories in the Netherlands such as in Rotterdam would be sold to local greenhouse farmers who would use it for that very purpose.

Karl Selim Lemström observed during his research expedition in the arctic that the plants grow actually very well during the short growing periods during the year. Lemström, an expert on the *aurora borealis*, wondered whether the electric fields near the poles could be a factor that would enhance growth of the crops and flowers. He became therefore interested in the effect of electric fields on horticulture and agriculture [Lemström 1904]. As Lemström had tried to make a synthetic *aurora borealis* in his laboratory, it is no surprise that he also tried to reproduce the naturally occurring electric fields in nature and subject plants to them.

The idea is to enhance these electric fields and thus enhance the growth of the plants. Breslauer explains in his report to the Zeitschrift für Elektrochemie how he produced electricity in a hutch and let it via a set of parallel wires in some height over the crops. This was one electric pole of the setup. The soil and grounds was the opposite pole. He therefore believed that the electric field between ground and wires would enhance the growth of the crops under the wires [Breslauer 1910].

https://doi.org/10.1515/9783110561838-010

Figure 10.1: Greenhouse for flower growth in a village in Switzerland.
Photo from 26 December 2017, Artur Braun.

Breslauer refers to a study of Kähler who found that the 250 rainfalls in the year 1908 provided an electric charge of 2×10^{-9} A h/m^2 for the ground. The electric current of charge provided by the experiment in the hutch would be 1000–10000 times higher than that in nature, but he would provide the actual outcome of the crops at a later time.

The electrocultures[1] were a seriously discussed matter in the beginning of the twentieth century, as can be seen from the 21st General Meeting of the German Bunsen Society in 1914 [Löb 1914], where Professor Löb gave a general description of the problems, chances and current views of electroculture. No lesser than Fritz Haber had worked on this topic, but the last comment on the transactions reported in Löb [Löb 1914] was given by Haber, which reads quite reluctant:

> Einen nützlichen Einfluss durch noch weitere Verminderung hervorzubringen, ist uns aber nicht geglückt. Die Möglichkeit einer solchen nützlichen Wirkung bei bestimmten unbekannten

---

1 Briggs et al. give a definition for electrocultures in their Bulletin [Briggs 1926] Briggs LJ, Campbell AB, Heald RH, Flint LH. Electroculture. United States Department of Agriculture – Department Bulletin. 1379. The term "electroculture" as used in this bulletin refers to practices designed to increase the growth and yield of crops through electrical treatment, such as the maintenance of an electric charge on a network over the plants or an electric current through the soil in which the plants are growing.

Bedingungen lässt sich auf Grund der vorliegenden negativen Resultate freilich nicht unbedingt in Abrede stellen, aber man wird gut tun, solche nützlichen Wirkungen nicht anzunehmen, bis positive Daten für ihr Vorhandensein beigebracht werden.

It seems the excitement over the electrocultures did not produce great results. But still, many years later, electrocultures were a serious scientific topic and a topic of potential economic interest. For example, Chree calculated the distribution of the electric potential on the ground which was set by the aforementioned parallel wires and presented this to the Physical Society of London [Chree 1921]. And no lesser than Sir Edward Victor Appleton, who would 25 years later be awarded with the Nobel Prize in Physics for proving the existence of the ionosphere, commented on Chree's derivation that Maxwell had come up with a similar solution for a similar problem long time ago [Chree 1921].

The United States Department of Agriculture presented in its Bulletin report in 1926 experimental results from the Office of Biophysical Investigations of the Bureau of Plant Industry, which were negative [Briggs 1926]. Briggs et al. end their report with a mixture of positive and negative outcomes. In 1933, Oswald carried out his doctoral thesis on the theory of electrocultures [Oswald 1933]. The last entry on "electrocultures" that I could find in www.webofknowledge.com was from 1976 with title "Electroculture Cuts Food Costs" [Storey 1976].

10.2 "Wires in bugs"

When I worked in Berkeley in the years 1999–2001, I knew one father from our children's soccer team who was known for his ingenious experiments on the brain physiology of insects. I learnt that he, a Berkeley Professor, was known for "putting wires in the heads of insects" [Borst 1983, 1984, 2014, Egelhaaf 1993]. This certainly sounds spectacular, and I am still wondering what sounds more weird, wiring insects in the 1990s or early 2000s, or wiring frogs 200 years before that.

How do you make such measurements with small animals such as insects? One way is to stimulate the insects with some odor and then measure with a position sensitive detector where they will move.

> The fly's reaction to the stimulus is registered by a "locomotion recorder" [Buchner 1976]. In this device, a fly is glued with its head and thorax to a metal hook, but is free to walk on a red styrofoam ball carrying black dots. The movement of the ball is recorded photoelectrically as forward and turning counts (for details see Buchner [Buchner 1976]). The experiment is conducted in total darkness. Data are stored on a punch tape for subsequent computer evaluation [Borst 1982].

The experimental setup is sketched in Figure 10.2. The drosophila fly is kept under control as it is glued to a wire hook (this system was developed by Buchner in the early 1970s during his PhD thesis [Buchner 1976]). Therefore, it cannot fly away. But the wire is flexible enough so that the fly can walk on a ball from red painted

Figure 10.2: Drosophila on the red styrofoam ball with black dots painted with the two capillaries of the olfactometer directed toward their antennae.
Journal of Comparative Physiology·A (1982) 147: 479–484, Alexander Borst and Martin Heisenberg, (© Springer-Verlag 1982) with permission of Springer.

Styrofoam. To be able to assign any motion on the red ball, black dots are painted on it. Any movement in sufficient increment will generate an optical contrast which is detected by a photoelectrical sensor and can be recorded manually or automatically.

Next, the fly is exposed to an olfactometer. This is a device which produces odor at particular concentration and guides it with a tube in a particular direction so that a "nose" would smell it. Parallel to the first tube is a second tube which guides either clean air with no odor or some other odor so that there is an odor contrast in the outlet of the two olfactometer tubes. An "intelligent" nose would now be able to detect the difference and make a choice whether to walk to one particular odor tube or to avoid a particular odor tube.

The olfactometer used in their study is quite complex and outlined in Figure 10.3. A thermal conductivity cell is connected with four glass bottles (B) which can contain four different types of odor molecules. We basically have a "battery" of odor molecule bottles.

Gas from the thermal conductivity cell can move through each of these four bottles and carry the odor molecules to the two ends of the pipes at which the fly will be positioned. Six valves (V) allow to precisely controlling the selection among the four odors. Valve V1 allows choosing either bottle 1 or bottle 2. Valve 3 allows to guide the selected odor either to further in the olfactometer or to outside. The same holds for the second set of bottles, B3 and B4, with valves V2 and V4. Valves V5 and V6 allow to switch the odor from the left hose to the right hose, and vice versa.

We can mix two different odors from bottle B1 and B2 to a particular ratio by taking out the odor from B1 over a time interval $t1$ (which may be a long time and thus with high concentration) and from bottle B2 for a somewhat shorter time interval $t2$ (accordingly lower concentration than from B1). This is illustrated in Figure 10.4 with the time profile over $t1$ and $t2$ and the corresponding lengths of the small horizontal tubes. In analogy to B1, B2 and $t1$, $t2$, we can proceed with the two other bottles B2, B4 with $t3$, $t4$.

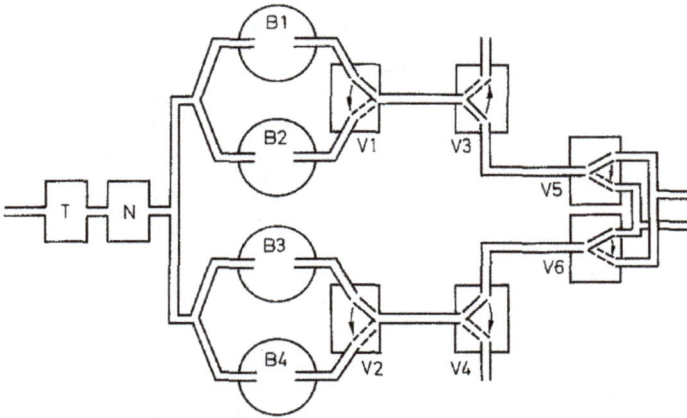

Figure 10.3: Olfactometer consisting of thermal conductivity cell (T), needle valve (N), four glass bottles (B) and six air valves (V). All connections are made from glass. Air flows from left to right. *Journal of Comparative Physiology·A* (1982) 147: 479–484, Alexander Borst and Martin Heisenberg, (© Springer-Verlag 1982) With permission of Springer.

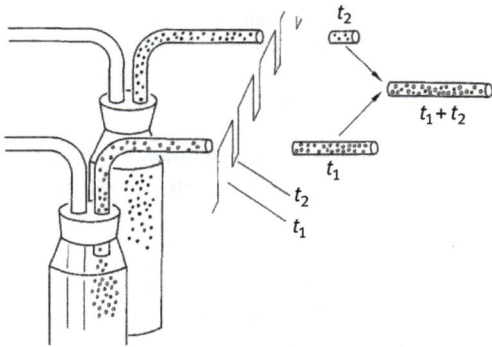

Figure 10.4: Scheme to illustrate the operating principle of the olfactometer. Certain concentrations are provided by adjusting the "take-out" times *t*1 and *t*2 from the respective bottles (R. Wolf, unpublished). The frequency of the duty cycle is $f = 1$ cps (V1 and V2 of Figure 10.3). Mixing is assured by the comparatively large cross-sections of the connecting tubes. *Journal of Comparative Physiology·A* (1982) 147: 479–484, Alexander Borst and Martin Heisenberg, (© Springer-Verlag 1982) with permission of Springer.

The schematic for the technical circuit is shown in Figure 10.5. The experiment, planned and carried out in the early 1980s, is largely computer controlled. A motor presses the syringe for the odor bottles. Two pumps press air with control by an electronic flow meter along with a thermal conductivity cell, a needle valve and a regulator connected to both. These set the odor condition in front of the fly, which has a recording electrode connected with the antenna. The recording electrode signal is

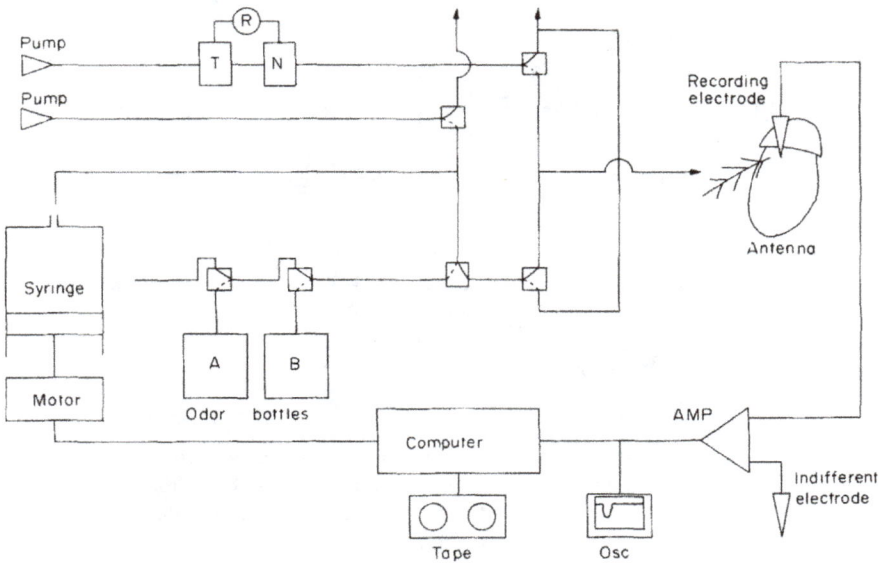

Figure 10.5: Experimental setup showing the motor-driven syringe with the odor bottles (A and B lower left), air pumps with electronic flow meter (upper left) and recording electrode on the proximal funiculus (upper right). Indifferent electrode (lower right). Amplifier, computer with tape and oscilloscope to register EAG response. Three way air valves (D) are shown in the stimulation mode of the olfactometer (follow solid line from the syringe to the fly's antenna). T = Thermal conductivity cell, N = needle valve. R = regulator.
Reprinted from J. Insect Physiol. Vol. 30, No. 6. 1984, Alexander Borst, Identification of different chemoreceptors by electroantennogram-recording, 507–510, Copyright (1984), with permission from Elsevier.

amplified and sends the electric signal to oscillator and the computer. The data were recorded with a magnet tape. With this setup, we can monitor and record the movement of the fly depending on the excitation with a particular odor.

A calibration curve for the olfactometer is shown in Figure 10.6. The electric response is given in relative units from 0 to 100 over the relative concentration for arbitrary mixtures from two components in percentage. The two chosen odors are 3-octanol and 4-methylcyclohexanol. The solid line is the response by a gas detector for the 3-octanol. With increasing concentration, the response is increasing, too. Note that the axes are in double logarithmic scale. The dashed line is the response for the 4-methylcyclohexanol. Here, the response is overall slightly lower than for the previous odor. The mixture of both gases gives the dotted curve which has the highest overall response over the entire concentration range.

The "response versus concentration" profile is called electroantennogram . This is demonstrated from an actual drosophila fly in Figure 10.7. The upper left inset shows two different signals versus the time axis. The upper transient is the chemical stimulus via the odor bottles which is provided to the animal for 5 s. The reaction of the animal is

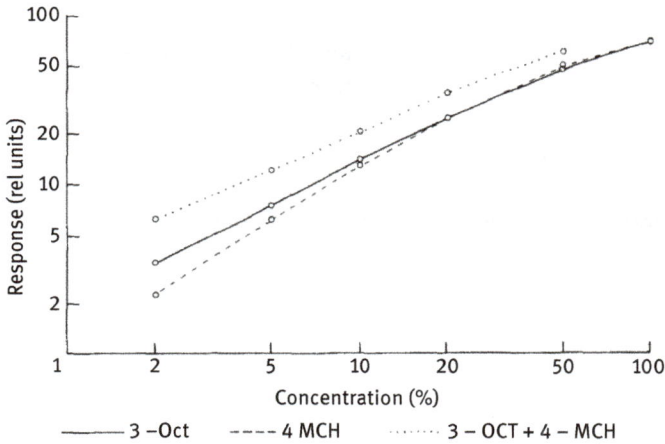

Figure 10.6: Calibration of the olfactometer with a gas detector showing double-logarithmic dose–response characteristics.
Reprinted from J. Insect Physiol. Vol. 30, No. 6. 1984, Alexander Borst, Identification of Different Chemoreceptors by Electroantennogram-Recording, 507–510, Copyright (1984), with permission from Elsevier.

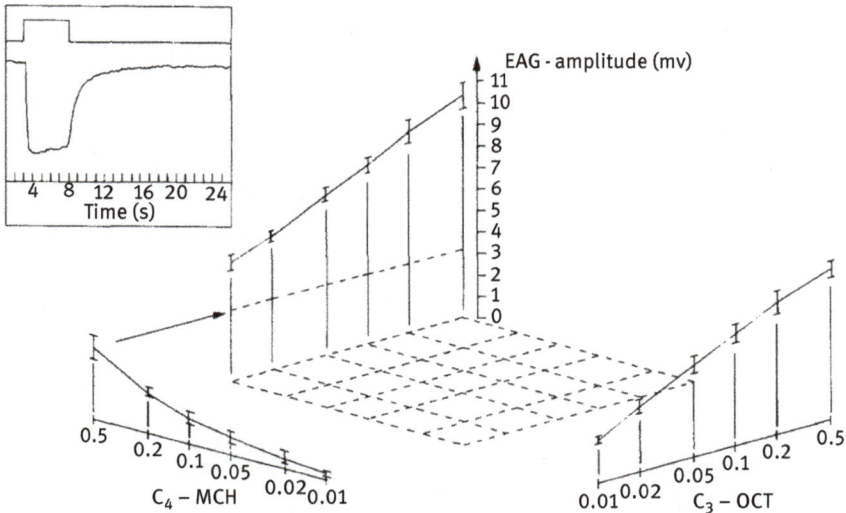

Figure 10.7: EAG-response to 3-octanol, (1–50%, right), to 4-methylcyclohexanol (variation of relative concentration from 1–50 (left) and to 50% 4-methylcyclohexanol plus 3-octanol (1–50%, middle rear). Note that the latter is very similar to the 3-octanol dose–response curve super-imposed on the response to 50% 4-methylcyclohexanol. The inset shows the typical time course of an EAG response (lower trace) to a chemical stimulus (upper trace).
Reprinted from J. Insect Physiol. Vol. 30, No. 6. 1984, Alexander Borst, Identification of Different Chemoreceptors by Electroantennogram-Recording, 507–510, Copyright (2017), with permission from Elsevier.

shown in the lower curve which at large overlaps with the stimulus curve. Close inspection shows that the reaction of the animal is delayed by around 0.2 s.

The amplitude of the electric response, in mV, from the animal is then plotted versus the concentrations of the two different odors in a three-dimensional plot, see Figure 10.7. You realize that this is not an electrochemical experiment. It is not even a purely electrical experiment. It is an indirect approach for linking the reaction of an animal to a chemical stimulus and the electric signal is only a help, an assistant for creating the link. It is not a direct measure for any physiological process in the animal. Notwithstanding that the conclusion drawn from such experiments is groundbreaking.

10.3 Early and historic studies on animal electricity

Since the invention of the Leiden jar over 270 years ago [Von Kleist 1745], it was known that people who work with the Leiden jar may experience an electric shock which triggers reaction in humans, such as muscle contractions which humans cannot immediately control. A true researcher with a natural curiosity is certainly interested in such phenomena.

Over 230 years ago, when electricity was becoming an extensively studied though not yet completely understood field, one of these true researchers with intrinsic curiosity, Italian physician and researcher Aloysii ("Luigi") Galvani (09 September 1737 in Bologna, Italia; 04 December 1798 Bologna), experimented with animals and animal parts which he subjected to the new phenomenon of electricity.

Galvani published his work in 1791 as "Comments on the electric forces of muscles in motion" on 72 pages including tables and illustrations [Galvani 1791]. We know already that back then, it was common in Europe to publish scientific work in ancient Latin language. English translations were made and published in the second half of the twentieth century and can be found in Green [Green 1953] and Moment [Moment 1955].

Figure 10.8 shows what Galvani describes as "Tab. 1" (lat. Tabula), which is Table 10.1. Galvani uses in his book four such tables which literally show tables, desks on which experimental equipment is spread out for presentation. It is left up to our own imagination whether Galvani uses the laboratory table, the laboratory desk as the main location of experimental activity, or whether the graphical representation in the book is for didactic purposes. As for the latter, I know at least one graduate text book [Ibach 1988] where the author used "Tafeln" (tables in the meaning of a black board, not in the meaning of a written list of data) as didactic tool for the representation of experiments and concepts. Tafel is the German word for blackboard and chalk board.

What does Galvani show us on his table, on his lab desk? Galvani adds on page 56 in his book [Galvani 1791] a list with written explanations (see Figure 10.9) of the tools and specimen that he used, which I have listed and translated in Table 10.1.

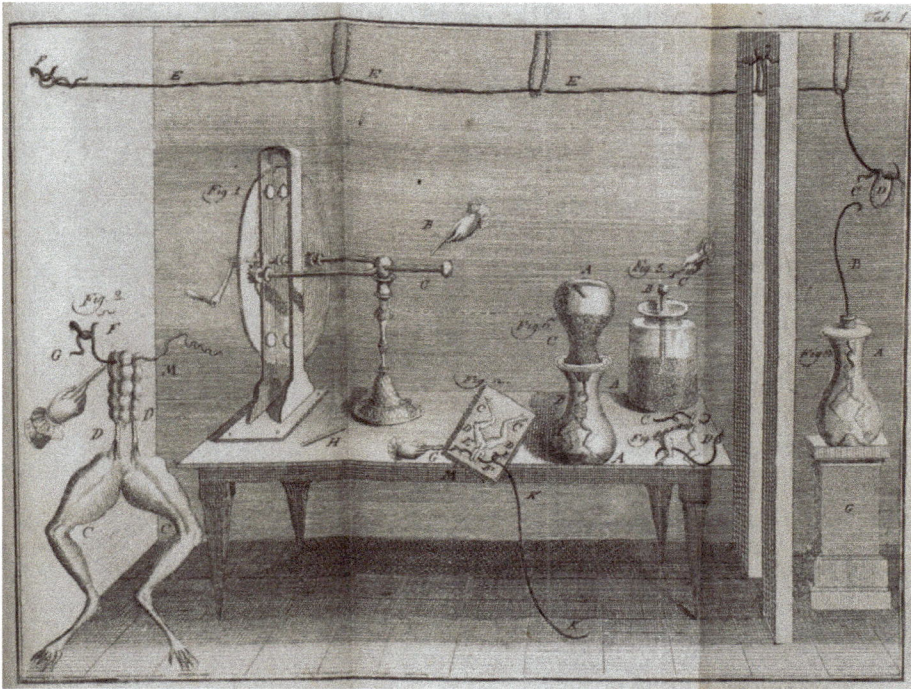

Figure 10.8: Illustration of experiment on frog legs by Aloisi Galvani [Galvani 1791] Table 1 on page 59 in Galvani's book [Galvani 1791], showing six figures (Figs. 1–6) with experimental setup. Descriptions of the instruments are shown in Figure 10.18 in ancient Latin, with English translation given in Table 10.2.

Galvani begins in his list on page 56 with Fig. Ω what he calls *Rana ad experimentum praeparata*: Frog prepared for experiment. He then lists the legs (C C) of the frog, the nerves (D D) which he has set free by dissection, a metal wire (F) which pierces through the openings in the spinal cord across the spinal marrow, then an iron slab (G) in a virtual hand and the spinal cord (M). This is not placed on the table but is attached to a hook on the wall of the experimenter's room, see lower left corner in our Figure 10.8 labeled "Fig. 2 Ω."

Galvani continues then in his list with "Fig. 1" which he summarizes as "*Machina electrica*," that is, the electricity producer, which is placed on the left side of the desk. It is made from the rotatable disk (A) and an electric conductor (C) which conduct the electricity from the disk away to the handle of C which a hand can touch with an iron slab (B), at which sparks will form when getting in contact with C. Upon close inspection, we notice that the two contacts on the disk have brushes which collect the electric charge from the disk which is likely an insulating material. This was just one of Galvani's sources for electricity.

Table 10.1: English translation of the list of instruments, items and samples displayed in Figs. 1–6 on page 56 of Galvani's book [Galvani 1791].

	Frog prepared for the experiment
Fig. Ω	
CC	Legs
DD	Sacral nerves which stimulate the lower legs
F	Metal wire that pierces through the openings in the spinal cord across the spinal marrow
G	Cylinder from iron
M	Spinal chord
Fig. 1	Electrical machine
A	Disk
B	Iron cylinder, from which sparks are extracted
C	Conductor
Fig. 2	
C C	Legs
D D	Internal femoral nerve
E	Iron wire pulled through a F guide
G	Iron cylinder in contact with the iron wire so that sparks from the conducting machine can be extracted
H	Glass cylinder which is connected with the iron wire E so that sparks can be predicted
K K	Conductor of the nerves
Fig. 3	
A	Glass bowl in which the frog is enclosed
B	Iron wire which is hooked up to the frog
C	Outer end of the iron wire which is connected with the iron wire B
D	Iron loop
E E E	Very long iron wire connected with iron wire B
F	Iron hook attached to iron wire E
Fig. 4	
C	Nerve conductor
D	Muscle conductor
Fig. 5	Leiden jar
A	Very small spheres (bullets) contained in the Leiden jar
B	Conductor in the jar
C	Hand of the person who draws the sparks from the conductor B
Fig. 6	
A	Jar turned upside down, which contains bullets
B	Matching jars which contain the animate and the bullets which represent the conductivity of the muscles

Galvani, a physician and researcher, was on the search for the source of life, the location of life within the animal body (compare also [Fritzsch 2014, Oschman 2009]). So, he dissected fish, bird, frogs and other small animals and studied them. He carried out most of his studies around 1780. In November of 1780, he learnt when one of his assistants touched a nerve of the frog with his scalpel, the muscles of the legs contracted as if the frog had a cramp. Another assistant, who aided with the

56

Tab. 1.

F*Ig.* Ω Rana ad experimentum præparata.

C C Crura.

D D Nervi facri, qui in crurales nervos abeunt, quos crurales internos placet appellare.

F Filum metallicum, quod per foramina fpinæ dorfi
trajectum fpinalem medullam perforat.

G Cylindrus ferreus.

M Spina dorfi.

Fig. 1. Machina electrica.

A Difcus.

B Cylindrus ferreus, quo fcintilla extorquetur.

C Conductor.

Fig. 2.

C C Crura.

D D Crurales nervi interni.

E Filum ferreum per medullam F trajectum.

G Ferreus cylindrus, quo tangitur filum ferreum, dum
fcintilla e conductore machinæ extorquetur.

H Cylindrus vitreus, quo tangitur filum ferreum E, dum
fcintilla elicitur.

K K Nervorum conductor.

Fig. 3.

A Phiala vitrea, intra quam præparata rana eft conclufa.

B Filum ferreum cum unco ranæ conjunctum.

C Extremitas fufpenfi fili ferrei, cui adnectitur filum
ferreum B.

D Laqueus fericus.

E E E Filum ferreum, quod conjunctum cum filo ferreo B
efficit nervorum conductorem, eumque longiffimum.

F Uncus ferreus, cui adnectitur filum ferreum E.

Fig. 4.

C Nervorum conductor.

D Mufculorum conductor.

Fig. 5. Leidenfis phiala.

A Minuti globuli venatorii intra phialam contenti.

B Conductor phialæ.

C Manus ejus, qui fcintillam e conductore B extor-
quet.

Fig.

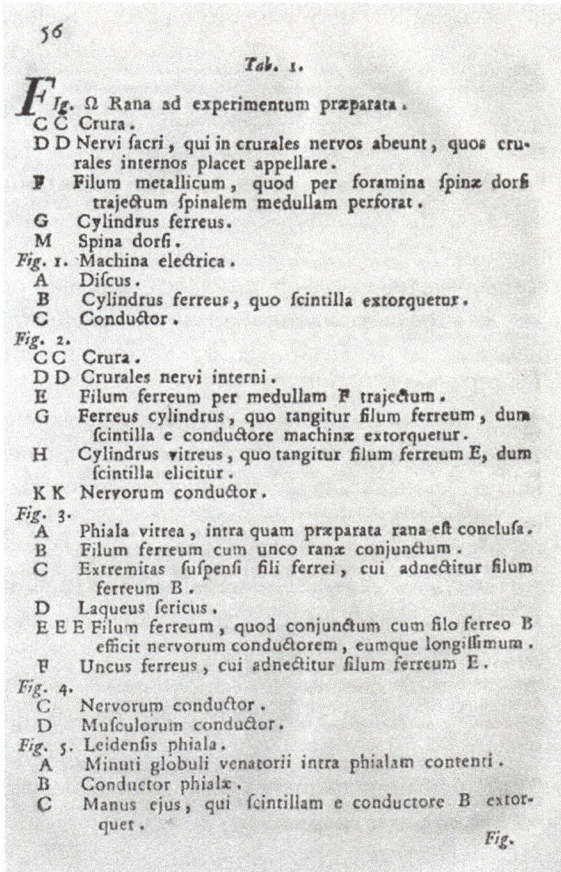

Figure 10.9: Page 56 in Galvani's book, listing the experimental tools and specimen as shown in the table in the illustration from page 59. The English translation is provided in Table 10.1.

electrical experiments, mentioned that he only witnessed the muscle contraction in the frog when at about the same time, the electricity generator produced sparks.

Does this mean that it was actually Galvani's assistants and not Galvani himself who made the first discovery of animal electricity? I think it does. Note that Galvani's Commentaries count as single author publication. His name is found today in the literature, and not the names of his assistants. At the least, Galvani refers in his commentaries to his assistants and thus partially warrants scientific credit to them. Nowadays, it is the rule, though not necessarily common, to share credit officially in publications as shared authorship or with an appropriate acknowledgment. This is notwithstanding that scientific fraud of all kinds can happen today.

Anyway, Galvani speculates that unintentional touching of nerves does not cause any muscle contraction with the frogs. But when there is a spark from his

Machina electrica, the muscles will react. He therefore plans to make a targeted, systematic and well-designed study in order to verify a causal relationship between muscle action and electric nerve stimulation.

Galvani and his assistants also observed that it made a difference whether the scalpel was held with the hand of the experimenter at the horn handle or at the metal shaft. Back then, it was already known that metals were electric conductors but materials such as horn and bone were electric insulators. Galvani wrote therefore that the electric fluid, which would somehow act inside the frog, would not participate in the experiment unless the scalpel was touched by hand at the metal shaft or any other metal piece in direct contact with the metal part of the scalpel. The electric fluid in this context is what we know today as the electrolyte, the ionic conductor (again, I have to refer to Oschman here [Oschman 2009]).

Now that the role of the scalpel was clarified, Galvani refined the study by employing an electrically insulating glass bowl and an electrically conducting metal cylinder. They found that a conducting metal container was necessary to generate the muscle contraction, along with the spark generation. Galvani thus had the proof that a conducting material needed to touch the nerves in order to cause the muscle contraction.

I refer the reader to Galvani's original work [Galvani 1791] and finish this section with two of his conclusions which are (1) that a closed electric circuit was necessary to stimulate the muscles electrically and (2) that it was the nerves and not the muscles which needed to be excited electrically to induce the muscle contraction.

Galvani's work was pioneering and groundbreaking because it established the causal relationship between electricity and muscle contraction, muscle action [Piccolino 1998]. It was thus possible to stimulate animate matter with external electric signals. Carlo Matteucci, an Italian Physicist, was inspired by Galvani's work and continued this field of research on animal electricity.

Meanwhile, electrical engineering and technology had progressed, which allowed Matteucci to carry out electrical experiments with better control and higher precision than his predecessors had been able to. This enabled Matteucci for example to design an instrument which could detect small currents by looking at the muscle stimulation. He discovered that injured tissue would produce electric currents during the healing and recovery process, the so-called injury currents [Matteucci 1850]. Matteucci became thus known as a founder of bioelectricity and electrophysiology [Bresadola 2011, Schiff 1876].

Figure 10.10 is taken from one of the papers Matteucci reported in 1850 to the Royal Society of London [Matteucci 1850]. Dissected frogs are fastened on two parallel glass tubes. He put on the legs of these fastened frogs (and thighs, and articulation of claws) what he calls legs of "highly sensitive galvanoscopic frogs."

Wade provides an interesting historical [Wade 2011] on how electricity was used to stimulate living systems. When Alessandro Volta had built his pile of Voltaic cells (Volta pile), he made simple experiments which are not different from those which we children did when we got a hold of batteries at early age. We would take two wires

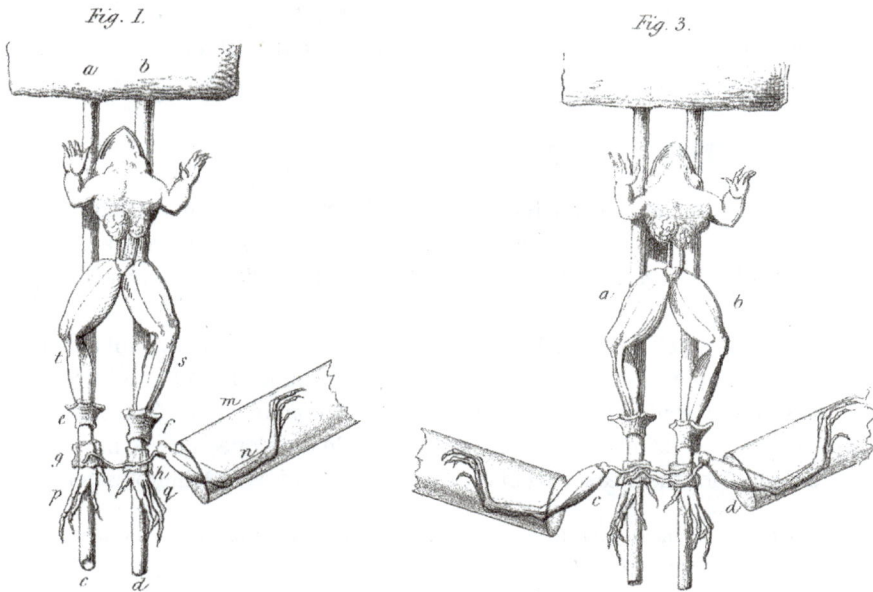

Figure 10.10: Details on electric experiments with dissected frogs by Matteucci [Matteucci 1850].

and contact them with the two poles of the battery on one end of the wire, and then we hold the other two ends against our tongue, and then we would taste a sour feeling on our tongue. Volta held the other end of one wire then not only at the tongue, but at the eye, nose and ear and thus experienced other effects of sensitization, as Wade lines out in his graphical review paper, but a substantial credit is given to Volta for providing a reliable and easy to use electric power source for the next generation of researchers [Wade 2011].

It is rather common in science than uncommon that someone who makes a new tool or instrument is also the first one to use it in some field, as Volta did when studying electrically triggered physiological effects. It was then Galvani who took advantage of readily available electric power sources such as the Leiden jar and Volta pile and focus on the animal electricity. Interested in progress and in the evolution and further development of their science, Volta and Galvani exchanged letters in the years of their research on biological systems [Galvani 1793].

We notice that over the centuries, the size of the objects which researchers studied became smaller and smaller. Was it in the beginning the large animals like dogs, cats, grogs and mice, and certainly also humans, researchers went on to look into the smaller animals also because microscopy methods were available. Nowadays, we can look into single biological cells and even in their subunits which are as small as the DNA, for example. At some point, the microscopic visual methods are not fruitful anymore and we need to employ other methods where our eye, even when supported by microscopes, cannot follow the "form-and-function" relationship anymore. See more on this topic in Chapter 7 in my book [Braun 2017a].

10.4 Electrophysiology in the nineteenth century

Some eminent electrochemists such as Nernst and Helmholtz have worked in electrophysiology. To the most of us, Helmholtz is not known as a bio-electrochemist, but in 1842, Helmholtz proved in his doctoral thesis that nerve fibers are based on the ganglion cells.

As a military physician in Potsdam near Berlin, in 1846, Helmholtz built a laboratory and wrote a paper on the metabolism in muscle action. In 1849 at Königsberg/Prussia, as a professor of physiology and pathology, Helmholtz laid the focus of his scientific work on the vision and hearing apparatus. He developed an eye mirror and built an apparatus for the measuring of the nerve speed in frogs [Helmholtz 1850]:

> I found that there is a measurable time between the electric current stimulus on the hip tissue of a frog and the propagation of the current in the nerve of the leg. In large frogs with nerves of 50 to 60 millimeter length, this time was 0.0014 to 0.0020 of a second. The frogs were kept at 2–6° Celsius and measured at 11–15°Celsius.

Also, Walther Nernst was interested in the transport of nerves and published over several years relevant papers on the matter [Nernst 1904, Nernst 1908a, b]. Inspired by Nernst's work on diffusion potentials [Nernst 1889] which arise as a function of concentration and temperature, Julius Bernstein [Seyfarth 2006] treated nerves and muscles as (electric) concentration circuits and contributed thus to the foundations of membrane theory. Bernstein was also the first one to correctly describe which we know today as the action potential [Schuetze 1983], an electrophysiological quantity which is important for the communication among cells in life.

I can close this section with the conclusion, not only my conclusion, that the organs of animals and humans have an electric nature, the exploration and discovery of which have led to the development of batteries. But still, these organs remain interesting for their own potential irrespective of parallel technological advances like the batteries. This is nicely explained in the comment of Fritzsch [Fritzsch 2014] in Science with the title "Electric Organs – history and potential." Many of the electric characteristics though are not fully understood yet. And, whenever a field is entered which cannot be understood and explained with the readily available rational models and tools, those who enter are likely considered "esoteric."

Says Bertrand Piccard, aerospace pioneer, in his ZeitgeistMinds presentation in 2013:

> "Good afternoon to everyone. Let me start by a question. Who are we? You know, in the entrance of this conference hall is written: "Everyone is looking for new things all the

time." And actually I'm not sure it's true. I believe that maybe here in this room we are all looking for new things. We are curious. And actually **we find our balance into the unknown**. But so many people in life don't trust life at all. So many people are afraid of the unknown. Afraid of the doubts. Afraid of the question marks. So what do they do? They try to find completely other tools than curiosity. They try to find control, power, speed. Because this helps them to fight against the doubt and the question marks. This helps them to fight against the uncertainty, against the "changeants." Against everything that can threaten their comfort zone. So what I love so much in ballooning actually is the fact that when you fly a balloon you learn exactly the other things, exactly the opposite, exactly the contrary. You learn to have – no power. Because you have no engine. You learn to have no control, because you're pushed by the wind…."

10.5 The "electric branch" of analytical psychology

With the improvement of technical apparatus, it became possible to study the reactions of humans which were not only considered a purely physiological event but also events of a psychological nature. The measuring of electric transients with respect to human behavior and reaction was therefore groundbreaking (Schmidgen [Schmidgen 2004] calls it the speed of thoughts and emotions), for which for example Bernstein was a pioneer too with the invention of the rheotome [Bernstein 1912, Seyfarth 2006].

Otto Veraguth, a neurologist at University of Zürich, was one of the researchers who conducted electric studies with living humans [Veraguth 1907]. He used a very sensitive galvanometer, sketched in Figure 10.11 which he connected with the skin of the persons, and with an electric power source of low voltage.

The electric resistance of the skin, actually of the entire human body, is measured by running an electric current from the battery (5) through it via the electrodes (6) and (7). The signal from the human is measured via the galvanometer (4) which forms with the resistors (1) and (3) a Wheatstone Bridge. This allows centering the mirror of the galvanometer. The shunt resistance (2) parallel to the galvanometer (4) dampens the oscillations of the galvanometer.

Veraguth connected human test persons with the electrodes to the battery and the galvanometer. When the persons sat at rest and fully connected, he could tune with the resistors (1) and (3) the galvanometer and bring it to "0" amplitude reference position, which was considered the reference position for the skin resistance. Over time, this resistance was slightly decreasing with a very flat slope.

Then, Veraguth pinched quickly a needle in the skin of the head of the person without the person anticipating such action from the doctor. Immediately, there was linear increase in the resistance for around 1.5 s, which was followed by an abrupt and steep increase in the resistance for about the same magnitude. Within another 3 s, the resistance was assuming a plateau value. The overall shape of the resistivity curve was therefore a slight linear decrease, a quick linear increase and a sharp

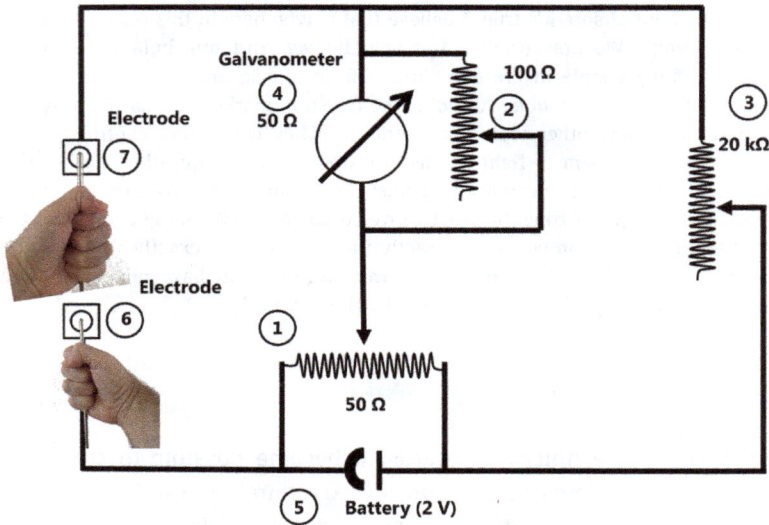

Figure 10.11: Electric circuit of the galvanometer setup for measuring the psycho-galvanic reflex phenomenon by Veraguth (1907).
Adapted from Meier (1994).

sigmoidal increase, followed by a plateau. This experiment was reproducible for many other human beings [Veraguth 1907].

Other researchers, physicians, neurologists and psychologists used this analytical method as well, which supposedly was first used by Professor Tarchanoff [Peterson 1907b]. Carl G. Jung [Bash 1945], the world famous founder of analytical psychology, conducted similar studies at the Zürich laboratory [Ricksher 1907].

What confidence can we have in such electrical measurements? Peterson considered this electric signal as an indicator for emotions [Peterson 1907a]; the lie detector is based on the electrophysiology of this galvanometer [Bunn 2012]. Is it possible to enter the subconsciousness of the human mind empirically with electric apparatus [Meier 1994]?

The experiments are carefully planned and the physicians try to recruit as many test persons as possible in order to warrant a certain statistical significance for their experimental data. The persons are excited with a needle, which is a painful sensation which has various reactions from the human body: pain, muscle contractions, maybe an accelerated heart beat frequency, maybe sweating of the skin. The signal, which is extracted, is only of electrical nature with relatively simple characteristics. How can such simple signal reflect the complexity of the human body reaction under pain?

When we conduct a 4-point study on a metal slab or a slab of a sintered metal oxide or some other ceramic material, then we are dealing with a comparably simple system. Likely, we have a single phase system, if we checked carefully with X-ray diffraction or with neutron diffraction. We have measured the geometrical dimensions of the slab: length and diameter. We have determined its specific weight with a pycnometer and therefore know about the porosity of the slab. We can control which current we impose on the slab from its outer terminals, and we know precisely the distance of the terminals in between where we read the voltage drop. These data allow us to determine the specific resistivity of the material from which the slab is made.

We can make slabs at different sintering temperatures and then find out how the resistance differs with sintering temperature. We can put the slab in a hot furnace or in a refrigerator (thermostat, cryostat), put a thermometer next to the slab and then determine the resistivity as a function of temperature of exposure. We can change the composition of the material of the slab during synthesis and then get the resistivity as a function of composition. Finally, we can expose the slab to various gases while we measure the resistivity and then get the resistivity as a function of gas exposure. When we keep the gas concentration constant but monitor the resistivity over time, we will be able to determine the bulk gas diffusion constant for the slab.

We can also polarize the slab and determine the electric current as a function of applied potential magnitude and direction, with the very same parameter change as I just listed above. With these experiments, we can learn already a lot about the electronic structure of the material and how its transport properties change upon circumstances. This is only because the model that we implicitly and correctly assume for the slab is a crystallographic lattice with atomic orbitals of cations and anions that overlap and help transfer electrons across the lattice. We are dealing with a body which has a functional and structural homogeneity over its entire size. The human body is not structurally and functionally homogeneous. Therefore, studies on human bodies and animal bodies are way more complex and difficult. The same holds for plants and microbes, even for single cells and their subunits.

During my time in California, I worked with a protein spectroscopy group. One of the usual systems they worked on was hydrogenase, a metalloprotein which contains a substantial amount of iron. As iron is very prone to oxidation, dealing with hydrogenase is a delicate act which requires working in a chemically reducing environment to prevent the hydrogenase from denaturation. This difficulty keeps many researchers away from working with hydrogenase.

Hydrogenase is typically used as purified crystal or it is dissolved in some solution. I heard that the problems with hydrogenase become somewhat less severe when it is adsorbed on an electrode. This makes sense to me because you can control with an electrode in an electrochemical cell potentially unwanted electric charges which may damage your sample otherwise. This is the reason why an electrochemical cell can be of great assistance when making experiments on samples with ionizing radiation where radiation damages may occur.

I owe this insight to the head of the aforementioned protein spectroscopy group, Stephen P. Cramer, who made a brief remark to me about this possibility in the year 2001 when we discussed spectroelectrochemical cells. Walther Nernst made similar remarks exactly 100 years before [Nernst 1901]. The attachment of the hydrogenase to the electrode or current collector is an electric grounding. Oschman refers in his paper "Charge transfer in the living matrix" to the benefits of barefoot walking, for example, in relation to charge transfer between human body and ground, where the electric charges may result from inflammation, wounds, oxidative burst and so on [Oschman 2009].

10.6 The biological cell

In my presentations on bio-electrodes, I typically show a figure which was sketched by Melvin Calvin [Calvin 1960] for a paper which he delivered in 1960 at the McCollum-Pratt Symposium on Light and Life. It shows the sample assembly on a glass support with interdigital electrodes over which he had sublimed chlorophyll from shredded spinach (Figure 10.12). This arrangement allowed for conductivity measurements on chlorophyll under various illumination conditions.

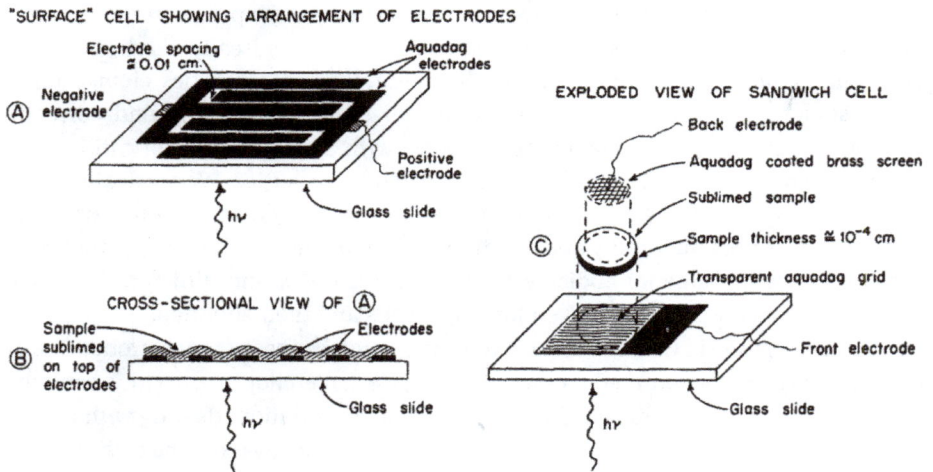

Figure 10.12: Diagram of sample cells (conductivity).
Reprinted from Tollin G, Kearns DR, Calvin M: Electrical Properties of Organic Solids. I. Kinetics and Mechanism of Conductivity of Metal-Free Phthalocyanine. *The Journal of Chemical Physics* 1960, 32:1013–1019. doi: 10.1063/1.1730843, with the permission of AIP Publishing [Tollin 1960]. See [Calvin 1960].

Said Calvin in Japan in his talk titled "Petroleum Plantations" [Calvin 1978]:

> We should not have to grow plants in order to catch the sunshine. But in order to do that, we must understand how it is done by the plant. If we understand exactly how the plant does it, then we have a possibility of doing it without the plant—synthetically and artificially.

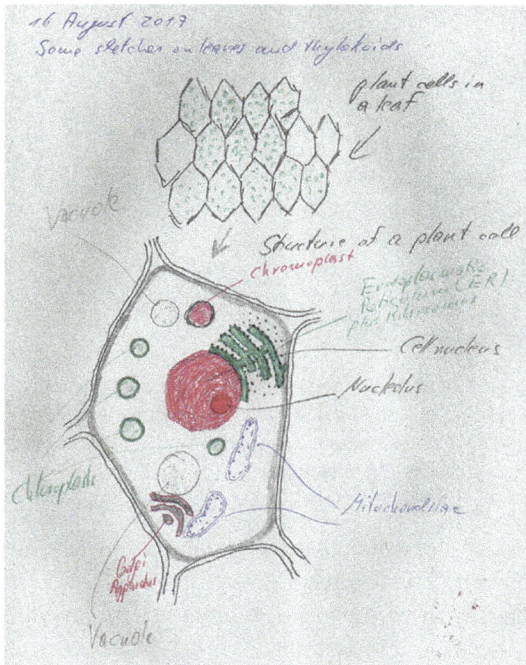

Figure 10.13: The construction of a plant cell and its components. The leaf is built from hexagonal units which are the cells. The cells contain nucleus, chloropoasts and other components as explained in the text.

We should therefore be interested how spinach cells or plant cells in general are built up. Figure 10.13 shows a sketch of a plant leaf and its substructures. Many of us have learnt maybe already in elementary school how plant leaves look under a microscope (optical microscope). Leaves are typically sectioned by a large number of plant cells which look like hexagonal units like sketched in the upper portion in Figure 10.13.

Every plant cell has a nucleus and a number of chloroplasts, chromoplasts, vacuoles, mitochondria, Golgi apparatus, endoplasmatic reticulae and ribosomes. The plant cell is therefore already a very heterogeneous "unit" built from many different components. Compared to a battery or fuel cell or solar cell, the plant cell can be compared with an industrial complex.

The chloroplast has a size of around 1–5 μm and is shown in the left portion in Figure 10.14. The chloroplast is contained by one outer membrane and one inner membrane, and both membranes contain, naturally, the intermembrane space. The chloroplast is filled with stroma liquid and contains also DNA, starch balls and ribosomes.

When you look at the chloroplasts with an electron microscope, you will see lamellae stacked and piled together. This is probably the first strong visual impression

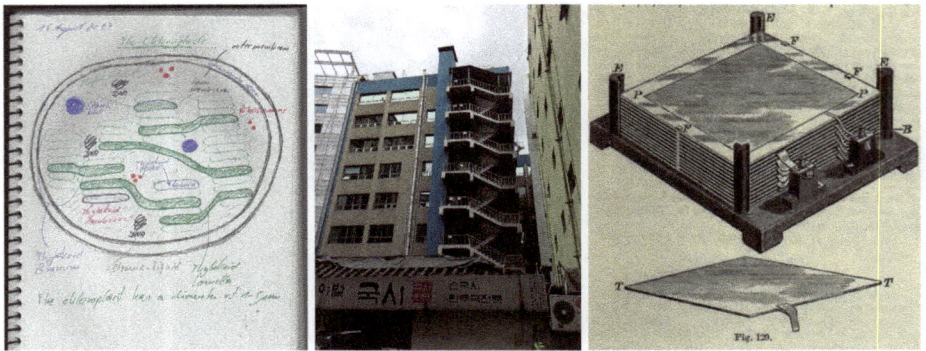

Figure 10.14: (Left) Schematic of the chloroplast and its components. Observe how the thylakoid grana are connected with thylakoid lamella to other thylakoid grana. (Middle) Apartment building. The neighboring apartment floors are connected via outside stairways. (Right) Sketch of a capacitor stack from Ayrton (1891) resembling with connectors thylakoid lamellae.

which one gets in electron micrographs, see for example the seminal work by Rabinowitch and Govindjee in Scientific American [Rabinowitch 1965].

One such pile or stack of lamellae is called thylakoid granum. The individual lamella or disk is called thylakoid. The thylakoid is built from a lumen which is confined by the thylakoid membrane. Some thylakoid grana are connected with neighboring thylakoid grana via long thylakoids which are called lamella. The thylakoid lamellae are reminiscent of staircases, stairways in apartment blocks, which are connecting outside the floors one by one. When I came across some buildings in Seoul, Korea, I saw stairways which reminded me of these thylakoid lamellae, as shown in the middle panel in Figure 10.14.

The geometry and architecture of the grana in the thylakoids is not coincidental. It was recently found that the diameter of the grana and the number of membrane layers per granum become decreased under illumination, whereas the concentration, that is, the number of the grana per single chloroplast, becomes increased. Because of this arrangement, there is a larger contact area between the grana and the stromal lamellae, as Wood and coauthors point out in their very recent paper in Nature Plants [Wood 2018].

This geometrical rearrangement has consequences for the electric transport in the chloroplasts. There are linear electron transfer pathways and cyclic electron transfer pathways which need to be properly balanced for efficient and long-lasting photosynthesis. When the grana are smaller, the distance between the electron shuttles, plastoquinone and plastocyanin is shorter and this helps promote diffusion of these shuttles and thus linear electron transfer. It is an exercise for the reader to sketch this and map the relevant components and their mutual distances from each other to explain the "form and function."

In contrast, larger grana increase the partition of granal and stromal lamellae plastoquinone pools. This augments the efficiency of the circular electric transfer and warrants some photoprotection by what is called non-photochemical quenching [Wood 2018].

Figure 129 in the old book of Ayrton [Ayrton 1891] shows a capacitor stack, where all capacitor electrodes are connected with leads to a terminal. These leads look not much different than stairways outside a multistorey building which connects the floors with each other, as shown in the middle panel in Figure 10.14. We learn that geometrical conditions that we apply to energy conversion and storage devices may hold also for systems in photosynthesis (I recall the speech of Calvin in Nagoya 1978 [Calvin 1978]).

Let us now zoom deeper into the system. The thylakoid is built up like shown in Figure 10.15 from a paper coauthored by my colleague Eva-Mari Aro at Turku, who is a consortium partner in our SUNRISE flagship project [Aro 2017]. We now have to wonder what does the term thylakoid "membrane" mean? The inner membrane and outer membrane are actually shells from parallelly arranged lipid molecules of around 5 nm length (the red dots with the two legs opposing the next pair of legs with red dot).

This lipid layer prevents electrons and ions and matter from passing through the thylakoid either into the lumen or out to the stroma. It is basically a firm border. The lipid layer blocks any transport. This lipid layer border is interrupted by a number of functionalities such as photosystem I (PS I), photosystem II (PS II) and cytochrome. Shown in Figure 10.16 is for example the PS II.

The processes taking place in the thylakoid membrane are very complex and only the structure by itself is already complex. It is a very good exercise for the reader to sit down and sketch all components and try to work out step by step how the form and function principle applied to photosynthesis, beginning from the absorption of photons, then production of electric charge carriers, then the transfer of charges including ions which participate in the redox processes which build up the NADPH and the ATP. This is an example for all processes in biological cells.

10.7 Coating electrodes with biological components

The actual experiments in artificial photosynthesis can be very filigree and require sample preparation at the molecular scale, or they can be clumsier. You can take dyes which are based on the juice of red bead from your garden or a shredded spinach cocktail and start experimenting with that one. For beginners, this may be a good exercise to start working with biological systems. You will notice that they are very prone to degradation and need much care.

Or you can restrict yourself to particular proteins which you want to coat on electrodes and purchase these and hope they will sustain long enough during your experiment. In the end, you will learn that you have to acquire some minimum laboratory equipment for photosynthesis research and also the necessary experimental skills. When you have a background in conventional electrochemistry (as opposed to

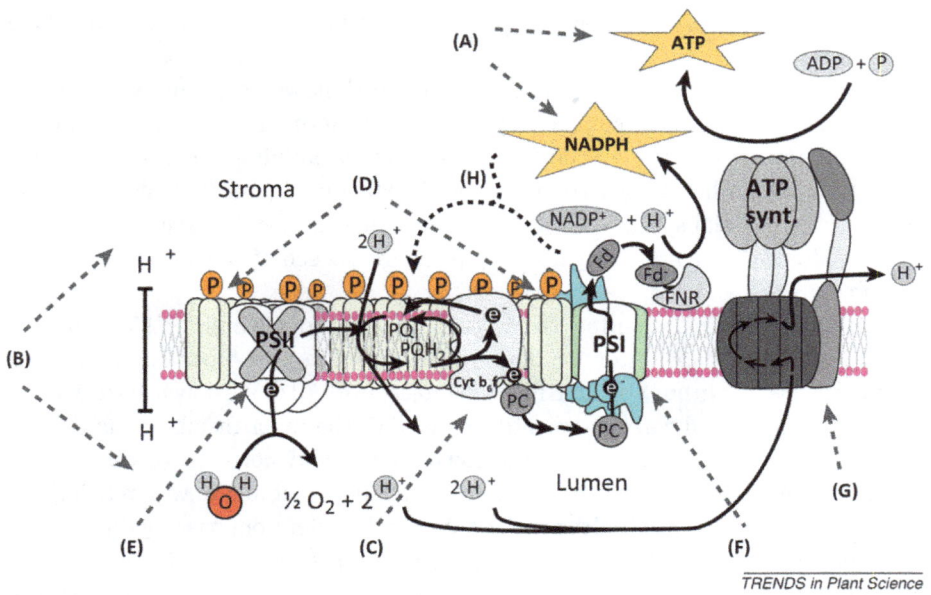

TRENDS in Plant Science

Figure 10.15: Regulation of photosynthetic light reactions. The efficiency of photosynthetic light reactions is regulated according to the light intensity and the metabolic state of the chloroplast to fulfill the needs of plant metabolism and to avoid photooxidative damage of the photosynthetic machinery. This is possible only via coordinated cooperation of all photosynthetic protein complexes (LHCII, PSII, PSI, Cyt b6f and ATP synthase enzymes) to convert light energy into NADPH and ATP. (A) Photosynthetic light reactions are responsible for conversion of light energy into NADPH and ATP. This requires tightly coordinated cooperation of the photosynthetic protein complexes. (B) The proton gradient between the thylakoid lumen and stroma (DpH) is dependent on (1) the accumulation of protons in the lumen from the water-splitting activity of PSII and from the electron transfer via Cyt b6f and (2) the rate of proton efflux from the lumen (i.e. activity of the ATP synthase in releasing the DpH). (C) The Cyt b6f complex couples the electron transfer to proton transfer. This not only enhances the generation of DpH but also allows the control of electron transfer according to the DpH. The higher the DpH, the slower the electron transfer from PSII to PSI. (D) The thylakoid membrane is rich in LHCII complexes serving to capture the energy for PSII and PSI. STN7 kinase-dependent LHCII phosphor-ylation enhances the energy transfer to PSI. The STN7 kinase senses the redox state of the PQ pool, enabling balanced excitation of PSII and PSI via the common LHCII matrix. The PSBS protein in the antenna system senses the DpH, allowing LHCII to dissipate the excess excitation energy as heat. The energy transfer efficiency from LHCII to the photosystems is enhanced by a decrease in DpH and reduced by an increase in DpH. (E) The D1 protein of PSII has a high turnover rate; when photodamage of PSII exceeds the rate of repair, the overall PSII activity decreases. Although the rate of PSII damage is linearly dependent on the intensity of light, the recovery rate is a dynamically regulated process that enables plants to tune their PSII activity to the level of energy requirements. (F) Although PSI is a long-living enzyme, electron donation to PSI needs to be regulated according to the capacity of the PSI electron acceptors. Excess electrons fed to PSI induce irreversible photodamage of PSI. However, when the electron donation to molecular oxygen is prevented, PSI is an extremely robust enzyme capable of safely dissipating all the excitation energy distributed to PSI from LHCII and PSII. (G) ATP synthase is one key factor in the regulation of photosynthetic light reaction. It can sense the metabolic state of the chloroplast and regulate the thylakoid DpH accordingly, which in turn regulates the excitation energy

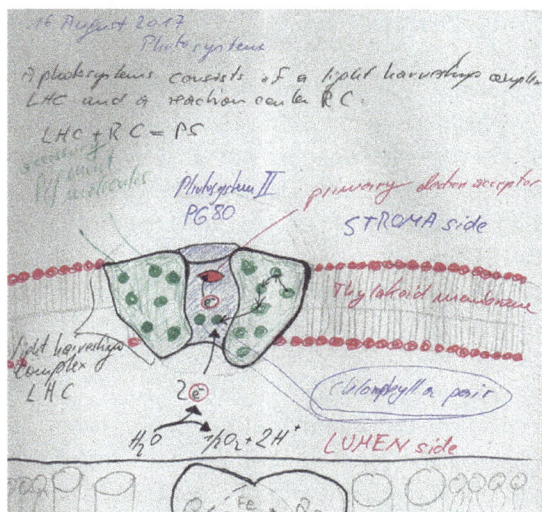

Figure 10.16: Sketch of a fraction of a thylakoid membrane with bilayer lipid membrane, and a photosystem II (PS II) with light harvesting complex (LHC) and reaction center (RC).

high temperature electrochemistry, which is not compatible with life biological systems), dealing with aqueous electrolytes and cuvettes, you may find it easy to accommodate with the biological world.

When you have the opportunity to work with a bio-laboratory which prepares your samples, that can save you a lot of work and disappointment but one negative aspect can be that you are not going to learn so much because you are not involved so much in the biological work. The more you are actively involved in all parts of the research project, the more authority you can demonstrate when you talk about your work at a conference.

What also counts is that you observe carefully and discriminate properly what you are doing and observing. I mentioned in a previous section how hydrogenase may behave different depending whether it is in solution or whether it is adsorbed. Rabinowitch et al. [Rabinowitch 1954] reported at the XIII International Congress on Pure and Applied Chemistry that how the spectroscopic signature of chlorophyll in solution was different from chlorophyll in the cell.

Figure 10.15: (continued) transfer from LHCII to PSII and PSI, and the electron transfer from PSII to PSI via Cyt b6f. The molecular mechanism that determines how the proton gradient is regulated according to the light intensity is not known but is dependent on PGR5. (H) PSII-independent electron flow to PQ from PSI electron acceptors functioning in cyclic electron transfer and in chlororespiration. Abbreviations: Cyt b6f, Cytochrome b6f complex; DpH, electrochemical gradient across thylakoid membrane; LHC, light harvesting complex; PGR5, proton gradient regulation 5; PQ, plastoquinone; PS, photosystem; PSBS, photosystem II subunit S; STN7, state transition 7.
Reprinted from *Trends in Plant Science*, 19, Tikkanen M, Aro EM, Integrative regulatory network of plant thylakoid energy transduction, 19–17, Copyright (2014), with permission from Elsevier. [Tikkanen 2014]

10.7.1 Coating porous photoelectrodes with proteins

We began our first work on biological systems by looking around what other researchers were doing. I came across a paper which was very inspiring because it combined hydrogenase with TiO_2 [Reisner 2009]. The paper fell right in the topic of a symposium which my colleagues and I were organizing [Braun 2009a] for the Materials Research Society Spring Meeting 2009 in San Francisco, so we invited it as late-breaking paper.

As for our own research work, our PhD student chose to decorate α-Fe_2O_3 with the light harvesting protein C-phycocyanin, which turned out to be a fruitful combination [Bora 2012b]. One way of preparing an electrode for photoelectrochemical studies with protein layers is shown in Figure 10.17. The fluorine doped tin oxide (FTO) glass slide is first coated with the absorber layer, in this case iron oxide.

The left and right side of the absorber layer are soldered with an indium layer and then a pair of wires is soldered on it. This assembly is coated with an epoxy resin layer in order to provide strong mechanical contact with the wires on the indium and in order to prepare a defined area which is later exposed to the electrolyte. This area is now coated with a solution which contains the protein molecules, such as phycocyanin. Note that the two wires provide only one pole, that is, the pole of the working electrode.

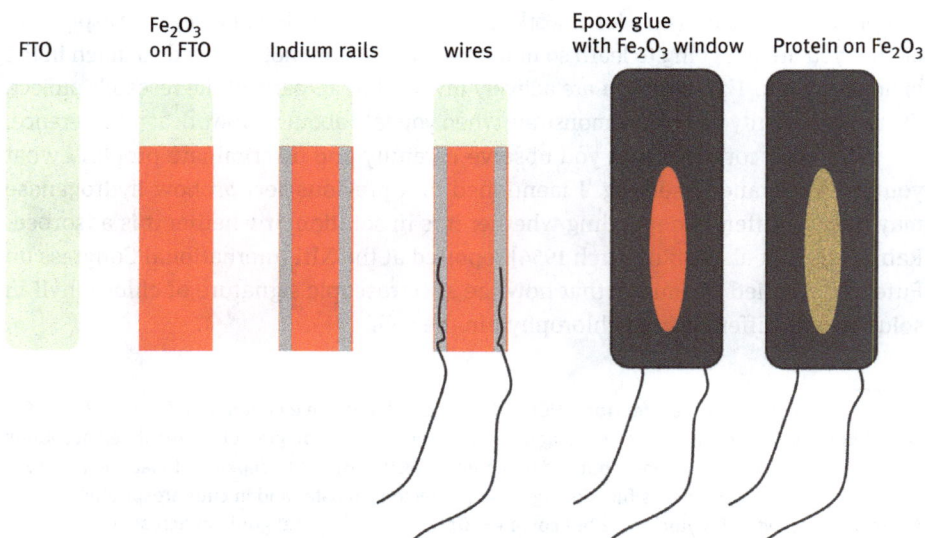

Figure 10.17: Steps for the preparation of an electrode coated with a protein layer. Note that the two wires are not meant to constitute two different poles. They belong to one pole only and will both be connected to the working electrode slot in the potentiostat.

10.7.2 Protein films drop casted on single crystal metal oxides

The substrate for this film is an iron oxide single crystal, cut to a particular crystallographic surface index [*hkl*] and polished to optical quality, as demonstrated in Figure 10.18. This single crystal has a size of 5 × 5 mm area and 0.5 mm thickness. This single crystal is glued with silver paste on a metal sample holder for good electric contact. Then, a layer of Lacomit varnish is coated over the single crystal boundary in order to prevent electrolyte from getting in contact with the silver underneath.

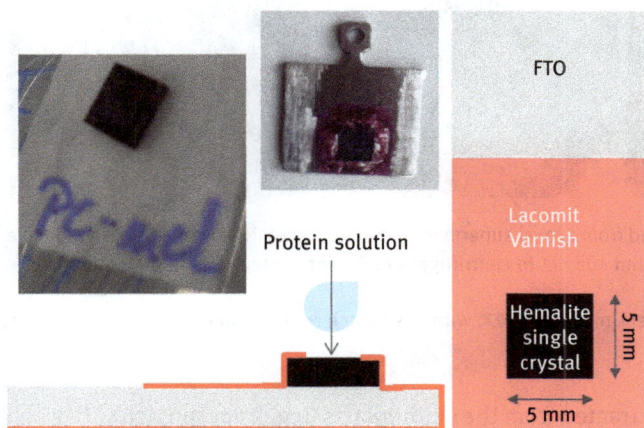

Figure 10.18: Procedure how single crystal substrate is prepared for receiving a functional component such as phycocyanin on its surface.

Otherwise, this silver would add its signature to the cyclic voltammogram or impedance spectrum. We want to rule out such signatures because they only complify our measurement and do not add any relevant information. Figure 10.18 shows a sketch how this is done also on an FTO glass substrate. Then, a drop of the solution which contains phycocyanin is cast on the single crystal surface. This sample can also be used for X-ray photoelectron spectroscopy (XPS) and photoemission spectroscopy experiments at a synchrotron beamline [Faccio 2015a].

10.7.3 Thylakoid films deposited on gold electrodes

Thylakoids are obtained by shredding spinach in a blender. The green mass is then exposed to a salt solution which will explode the chloroplasts by osmotic shock. You will need a centrifuge in order to separate the chloroplasts from the other components of the spinach leaves. And you will need an ultracentrifuge in order to separate the thylakoids from the other components in the chloroplasts. Figure 10.19 shows three visual impressions during the thylakoid extraction protocol, recorded during my sabbatical at Yonsei University in Seoul, Korea [Braun 2017c].

Figure 10.19: Spinach purchased from a local supermarket (Seoul, Republic of Korea) is shredded in a kitchen blender, then filtered and subject to centrifugation. Essential steps for the extraction of thylakoids from plant leaves.
Photos by Artur Braun at Yonsei University 2017, with assistance from group members of Professor Ryu, WonHyoung.

Once the thylakoids are extracted from the chloroplasts (e.g. from spinach), they can be deposited on solid supports and even on electrodes. Figure 10.20 (left image) shows thylakoids deposited on seven electrodes. The electrodes are made from a

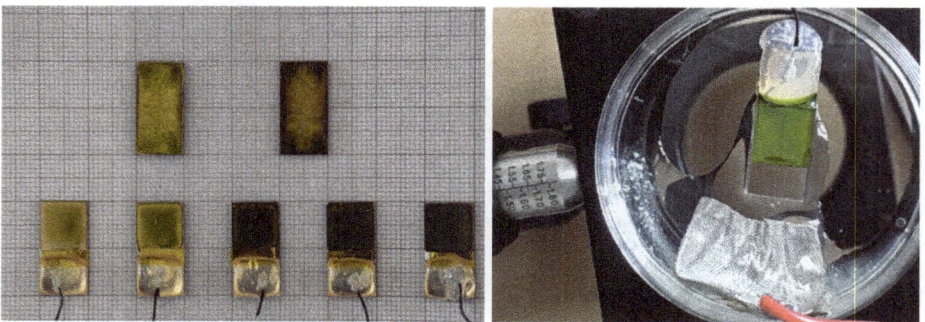

Figure 10.20: Photographs of seven thylakoid films deposited on gold electrodes. Millimeter paper allows for more accurate determination of the electrode area. The films were deposited from aqueous sugar solutions which contained different concentrations of thylakoid. By this way, different concentrations of thylakoids were obtained on the films. The lower left film has 10 µg/ml and the lower right film has 50 µg/ml. In between, we have 20, 30 and 40.
Photo by Artur Braun at Yonsei University [Braun 2017c].

solid float glass slide which is coated with a titanium adhesion layer on which a thin gold layer is deposited. This gold layer serves as current collector.

Then, a solution of thylakoids and sugar is coated on the gold and refried under ambient conditions. If these are to be used as electrodes, they need first to be contacted with a wire. For this, a thin wire is contacted with silver paint on the gold layer. For mechanical stability of this contact, it is necessary to put rubber glue over the electric contact area. This rubber glue warrants also that no electrolyte will get in contact with the wire or with the silver paint.

It is important to avoid contact between electrolyte and any other electronically conducting materials. The reasons for this are that any additional solid–electrolyte interface may give rise to redox reactions which could contaminate your experiment. This may for example show up as an additional current wave in the cyclic voltammogram. A cyclic voltammogram is therefore a good and sometimes necessary test for the cleanliness of your system.[2]

I have arranged the electrodes on millimeter paper for you to demonstrate that their base area is 2 × 1 cm. The two upper samples are made for use in UV–vis absorption spectroscopy in reflection mode. They have the right size to fit in a conventional UV–vis cuvette. The five lower samples were prepared as follows.

The solutions that were dropped on the electrodes had accurately known thylakoid concentrations such as 10 mg/L. They were dropped with a micropipette on the electrode. Thus by the drop size setting and the number of drops, the amount (mass) of thylakoids was determined per sample. The higher the concentration of the thylakoids is, the larger the absorption of the light is during the photoelectrochemical experiment. This is what we can reasonably expect according to the Lambert–Beer law. It remains to be seen to which extent the thylakoids can produce or moderate a photocurrent.

Part of the electrochemical setup for the experiment is shown in the right panel in Figure 10.20. Such thylakoid film electrode is in a plastic Petri dish and fixed with a cable to the potentiostat. The platinum mesh is the counter electrode. About 400 mV DC bias is applied in the phosphate buffer saline (PBS) electrolyte. The light source is mounted on top, a light emitting diode (LED) with a particular wavelength range. I used blue, green orange and red wavelengths and the measured currents are shown in the plot in Figure 10.21.

In order to allow also for "dark exposure," I had placed the entire setup under a cardboard box which just happened to suit the size of the setup. The setup is run by two synchronized potentiostats which operate in a master-slave mode. One potentiostat drives the color LED. The other potentiostat records the current of the working electrode.

---

2 The chapter/section on single crystal hematite photoelectrodes shows a CV with a very sharp peak. That peak originates from the oxidation of silver paint which was exposed to KOH electrolyte. The electrolyte had crept under the lacomite varnish and reached the silver paste contact. Consequently, the silver participated in the electrochemical reaction of the system.

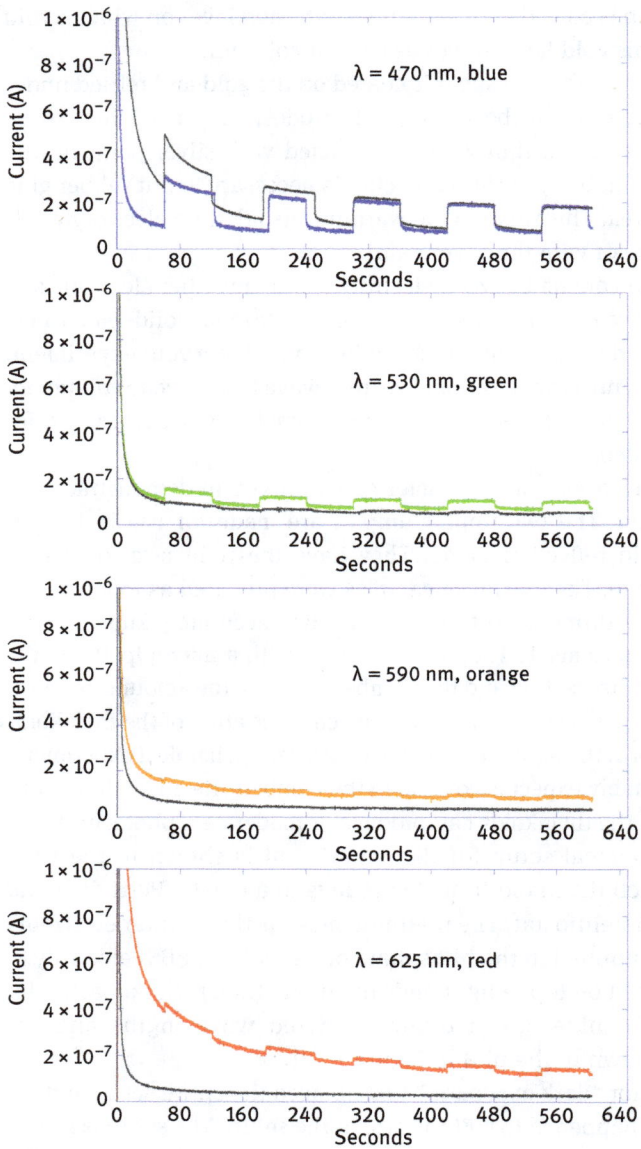

Figure 10.21: Current transients of thylakoid film from spinach recorded with light of four different wavelengths, 625 nm, 590 nm, 530 nm and 470 nm. The dark transient at the bottom is from the bare gold film coated on the glass substrate. The zig-zag profile comes from the automatic switch on, switch off of the LED by the programed potentiostat.
Data unpublished and recorded in the Lab of Prof. W. Ryu at Yonsei University, Dept. of Mechanical engineering, Seoul, Korea [Braun 2017c].

The current transient at the bottom in Figure 10.21 was obtained with the red LED. The potentiostat assembly was such programmed that the LED would switch on and off in a period of 1 min. The smaller current in a sequence is then obtained when the LED is switched off, and the high current is observed when the LED is switched off. The dark curve at the bottom was recorded from the same set of gold-coated substrates but with no thylakoid film on it. I observed that also the bare gold electrode shows a photocurrent which is particularly strong under the blue light.

From the different intensity of the light currents, you can see how sensitive the thylakoid film electric response depends on the wavelength of the light under which it is stimulated. When you take the maximum light current and compare it with the absorption spectrum of chlorophyll, you will notice that they scale quite well with the absorption of the chlorophyll.

10.8 Biofilms on photoelectrodes

10.8.1 How to get biofilms

The Swiss Federal Institutes of Technology (Swiss-FIT) constitutes the two schools in Zürich (École polytechnique fédérale de Zurich, ETHZ) and Lausanne (Eidgenössische Technische Hochschule Lausanne, EPFL), and the four research laboratories, that is, the Paul Scherrer Institut (PSI), the Eidgenössische Materialprüfungs- und Forschungsanstalt (Empa), the Eidgenössische Anstalt für Wasserversorgung, Abwasserreinigung und Gewässerschutz (Eawag) and the Eidgenössische Forschungsanstalt für Wald, Schnee und Landschaft (WSL).

These six entities are governed by the ETH Council and funded by the general budget from the Swiss Federation. Often, international guests who visit me at Empa ask me what is the difference between ETH and Empa. And then, I typically explain what I just wrote above.

Empa and Eawag happen to share the same campus in Dübendorf. Dübendorf is in walking distance to Zürich. Some seven or eight years ago when it became dark in fall or winter and when I looked out of the window in my office, I saw some Hg bulbs lighting in a large garage-type building just across the road. This building belonged to Eawag and the fact that light was on there at night made me curious. I figured there must be something going on with plant growth and water.

Eventually, I walked over, went into the building and found myself somebody[3] whom I could ask about these lights. These are the long slim Hg vapor or neon-type

---

[3] When you are shy, then you likely will not be able to make yourself such situation for collaboration. I once met at a conference a lady who had gotten into serious problems with finishing her PhD thesis. Her supervisor could not supervise her with all necessary instrumental and intellectual support, despite supervisor's good will. I believe supervision had changed during the thesis as well. I asked

bulbs which you see often in large buildings on the ceiling. The lights were illuminating a testing rig where algae would grow in something that looked like a gutter tube through which water – water from the nearby creek – "Kriesbach" was flowing. The gentleman who showed me the testing rig referred me to a lady at Eawag who was overseeing this experiment.

I found this very exciting because algae do make photosynthesis and many researchers are working on it. Also, I knew that Eawag possessed what is called in German "Algothek" [Braun 2008c], a collection of various living algal cultures which could be used for reproduction of algae when you take a sample thereof. I looked up that colleague at Eawag and eventually we organized a small informal project together. She was interested in nano-ecotoxicology and wanted to study the malign interaction of microbes with potentially malign nanoparticles. We could supply her with our iron oxide nanoparticles which our PhD student would produce basically as a waste product [Bora 2012a] when he made hematite photoanodes [Bora 2011a, b, 2013].

In return, she would grow for us algal cultures which would produce biofilms, first on microscopy glass slides and later on iron oxide photoelectrodes on FTO glass [Braun 2015, Burzan 2016]. A biofilm is not something biological which is deposited as film on some substrate. A biofilm is grown by microbes who settle and colonize on some surface. The biofilm is the comfortable environment made by the microbes.

There are some algae who produce biofilms (like *Anabaena spirulina*), and there are other algae who do not (like *Arthrospira spirulina*). In one case, we studied such *Anabaena* sp. biofilm which was grown on a hematite photoelectrode with a novel method of XPS. The objective of that study was to measure the valence band spectrum of the interface between biofilm and iron oxide while under electrochemical DC bias, while under light and while under water vapor. These are electrophysiological conditions which would also hold when such colony of algae, potentially genetically modified and harnessed on an electrode, would produce hydrogen in a PEC reactor (of the modified BIQ-house type) or electric charges as solar cells.

---

her why she did not contact those researchers who had the possibility of making the necessary measurements for her with their equipment. And, the PhD candidate replied to me that it was not possible for her to write emails to strangers and ask for help. You decide for yourself how you handle such situation. On the occasion of the MRS Spring Meeting 2015 in San Francisco, my fellow Meeting Co-Chairs and I [Braun 2014b] Braun A, Fan HY, Haenen K, Stanciu L, Theil JA: Braun, Fan, Haenen, Stanciu, and Theil to chair 2015 MRS Spring Meeting. *Mrs Bulletin* 2014b, 39:740-741.doi: 10.1557/mrs.2014.183. had the duty and privilege to take the major speaker of the conference, which is typically attended by 5000 researchers worldwide, out for lunch. He was an imminent North Western University Professor and gave us the advice "you have to be sociable for being more successful in science." As we agreed with him, I am sharing this advice with the readers of this book.

10.8.2 *In situ* photoelectrochemical biofilm studies at the synchrotron

XPS studies are typically done under ultrahigh vacuum and this is not a condition under which algal biofilms can flourish. The synchrotron end station with beamline 9.3.2 at the Advanced Lightsource (ALS) in Berkeley (Figure 10.22) however operated an instrument which would allow for electrochemical experiments even when the pressure in the chamber was 1000 mTorr, for example from water vapor.

Figure 10.22: (Left) Access road (Cyclotron Rd.) to Lawrence Berkeley National Laboratory (LBNL) in Berkeley, California. (Right) The dome of the Advanced Light Source at LBNL. The structure is still from the 1930 when Ernest Orlando Lawrence built the large cyclotron.

The PhD students in my group [Yelin Hu (EPFL Prof. Grätzel) and Florent Boudoire (Uni Basel Prof. Constable)] could mount the biofilm in the XPS chamber and connect it with the potentiostat. The only remaining problem was that the sample position in the chamber could not really be easily illuminated with a lamp. The chamber has many windows but nonesuch where one could point a lamp directly on the sample. Photoelectrochemical experiments were therefore difficult. I needed a strong large lamp, a strong lighting which could illuminate the whole chamber from all around the windows of the chamber.

Eventually – near the stock room outside behind the synchrotron, I found a number of large lighting. They were stored there near the gas cylinders (Figure 10.23) and waiting to be used for the ALS Users Meeting next day. The synchrotrons have a Users Meeting once a year with oral presentations, poster sessions, merchant exhibitions and lunch and dinner. At the ALS, in sunny California, these events are outside, and for this, they need large lighting at night. In need of such lamp, I took them inside to the end station, plugged them in and had around the experiment station exactly the illumination I was looking for (Figure 10.23, see also Chapter 7 in Braun [Braun 2017a]).

Figure 10.23: (Left) Mounted halogen lamps near the gas bottle storage, ready to be used for the ALS Users Meeting 2013 in Berkeley. (Right) Mounted halogen lamps dispatched to beamline 9.3.2 at the ALS for illumination of the first ambient pressure XPS measurement on a living algal biofilm photoelectrode under DC polarization.
Published in Braun (2015).

10.9 Thylakoid membrane electrochemistry

10.9.1 Extraction of electrons from photosynthetic cells

We have seen in the beginning how we can use the work of photosynthesis by cutting trees and burning them to get heat, and harvesting plants and eating them, converting wood by reforming into fuel gas which can propel vehicles such as cars, trucks, motorbikes.

Photosynthetic organisms are electrochemical systems, but they hardly interact with their outer environment in the sense that they could deliver electric current for us. Rather, they produce starch and keep it in their chloroplasts, as we have seen in Figure 10.14. It is therefore of interest to explore whether such opportunities to extract electric charge or ions theoretically exist and how to test them in experiment.

We have seen in the early studies of Galvani [Galvani 1791] how frogs were connected with metallic conductors, and Becker's experiments on salamander [Becker 1961] and other animals were not very much different from Matteucci's [Matteucci 1850] studies.

In Figure 10.24, we see how photosynthetic cells (*Chlamydomonas reinhardtii*) are punctured with a tip electrode and then subject to electrochemical measurements. Such experiments require micro machining of electrodes to make them small enough so that one can truncate the cells at the right spot.

For such experiment to be possible, you have to micromachine a tip with a tip diameter size in the sub-micrometer range, and a micromanipulator to be able to point the tip and fix the cell, not much different from trying to run a needle with your left hand into a balloon in your right hand.

This approach has been known for quite a long time, though not yet at the nanoscale like is done today. The method is known as voltage–patch clamp in biology

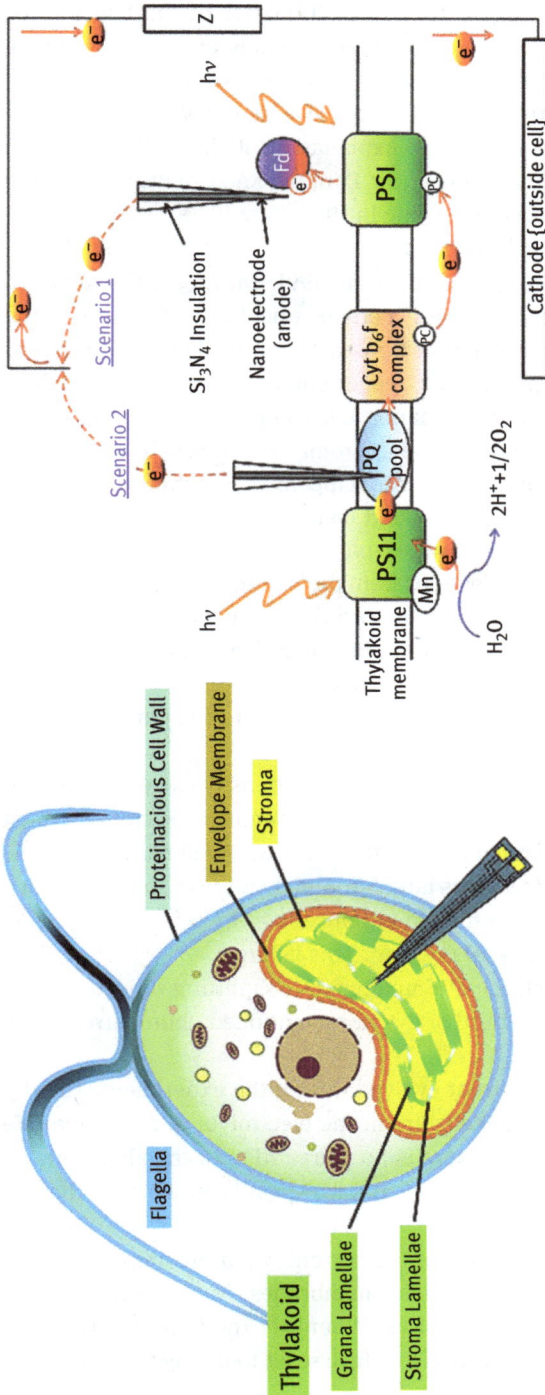

Figure 10.24: Graphical abstract.
Reprinted with permission from W. Ryu, S.-J Bai, J.-S. Park, Z. Huang, J. Moseley, T. Fabian, R. J. Fasching, A. R. Grossman, F. B. Prinz, Direct Extraction of Photosynthetic Electrons from Single Algal Cells by Nanoprobing System, Nano Letters, (2010) 10:1137–1143. Copyright (2010) American Chemical Society.

and was for example used by Andrew F. Huxley (half-brother of the famous and infamous Aldous Leonard Huxley) and colleagues for the study of nerve currents or action potentials of frogs or squids [Birks 1960, Hodgkin 1939, 1945, 1946, 1949, 1952a, 1952b, Huxley 1950] for which Andrew Fielding Huxley eventually was awarded with the Nobel Prize in Medicine in 1963; read the explanation of the "voltage clamp" technique in his Nobel Lecture [Huxley 1963]. Sir Bernard Katz, multiple coauthor with Huxley, received the Nobel Prize in Medicine in 1970 [Katz 1970].

Neher and Sakmann used the patch clamp method [Neher 1992] for their studies on muscles and received the Nobel Prize in Medicine in 1991. Drilling with wires and electrodes has become therefore a successful method, and those who use it and develop it further stand in the tradition of eminent researchers like Galvani and Nernst.

Today, scientists are trying to probe even deeper and enter single cells such as shown in Figure 10.24 [Ryu 2010]. With the nanotube, you pinch through the algal cell and through the envelope membrane into the Stroma and access the thylakoids.

We wish then for out of two specific scenarios to happen: The electrode (the anode) shall hit on the plastoquinone pool in the thylakoid membrane and then extract electrons which are produced by PS II during the water splitting. Or the electrode hits on the ferredoxin complex which has received electrons from the photosystem (PS I).

With this approach, the colleagues from Stanford University and Yonsei University succeeded to "steal" a current of 1.2 pA from the algae *C. reinhardtii* [Ryu 2010], see Figure 10.25. This sounds like a very small current but we have to understand that this current results from only one small single cell only. Per m^2, the current is 6000 mA or 6 A.

It is an exercise for the reader to carry out the current density calculation based on the size and number of algae that fit on 1 m^2. As 1 m^2 equals 10^5 cm^2, we must divide the current density by this factor and arrive thus at a current density of 0.6 mA/cm^2. This was in the year 2010, but few years later, around 20 pA was obtained with a nano-patterned electrode support, which is an enhancement by a factor of over 15 [Kim 2016], that is, a current density of 10 mA/cm^2. This is large!

Our best metal oxide photoelectrode at Empa for water splitting ranged around 3.3 mA/cm^2 [Wang 2016], – this is one-third of the aforementioned photocurrent from the living algal cells.

Figure 10.25 compares how the current evolved during the experiment. The top left image shows the current when the cells are in the electrolyte (a biocompatible tris–acetate phosphate medium). When there is no penetration of the algal cells by the nanostructured pillars, the current profile is flat irrespective of whether there is illumination or not (light on, light off).

The top right panel shows the variation of the current when the nanostructured electrodes are penetrated into the chloroplast membranes. My colleagues have switched on the illumination in intervals of 10 s. Whenever the light is turned on, the current rises from 0 pA to 1 pA or above. The algae react to the light on stimulus with a corresponding photocurrent.

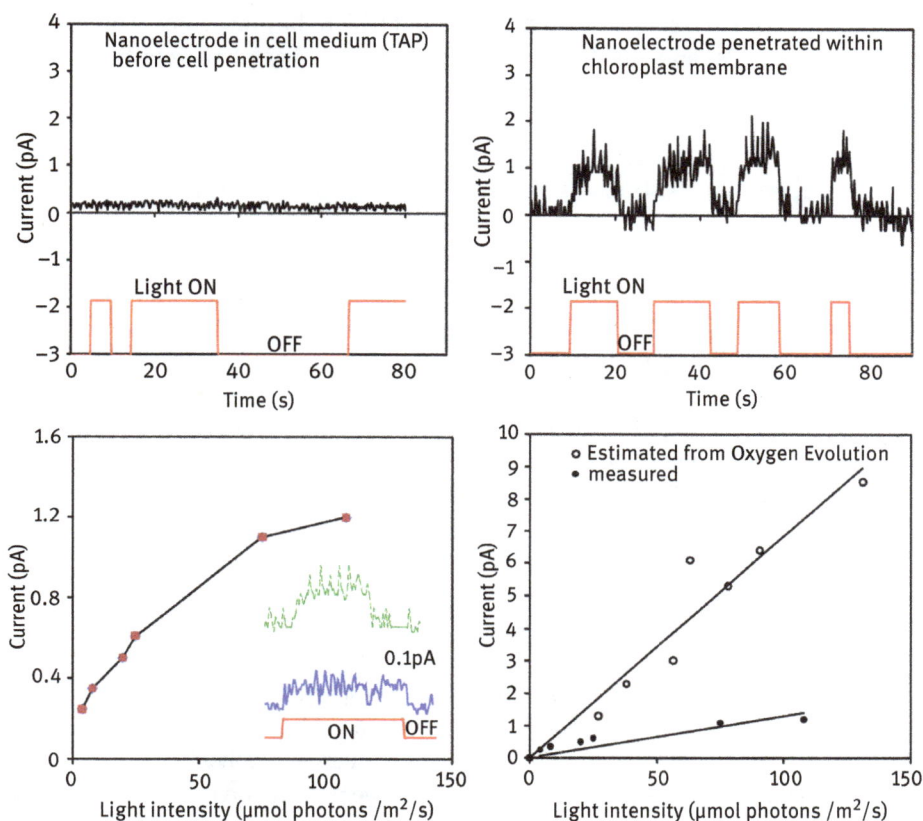

Figure 10.25: Light-dependent oxidation currents from a single *Chlamydomonas* cell. An Au nanoe-lectrode was biased at 200 mV with respect to reference electrode (Au), and PET was initiated by exposing the cells to light of 108 μmol photon/m^2/s (halogen lamp). (a) Au nanoelectrode was placed in the medium outside of the cells; no light-dependent signal was detected. (b) Au nanoelectrode was inserted into the chloroplasts of a cell; light-dependent oxidation currents were detected. (c) Dependency of current on light intensities is shown in the range of 4–108 μmol photon/m^2/s. Examples of the current signals observed (right, top and bottom for 108 μmol photon/m^2/s and 4 μmol photon/m^2/s, respectively) are shown. (d) Comparison of photosynthetic currents measured by the nanoelectrode relative to the current estimated from O$_2$ evolution measurements. Reprinted with permission from W. Ryu, S.-J Bai, J.-S. Park, Z. Huang, J. Moseley, T. Fabian, R. J. Fasching, A. R. Grossman, F. B. Prinz, Direct Extraction of Photosynthetic Electrons from Single Algal Cells by Nanoprobing System, Nano Letters, (2010) 10:1137–1143. Copyright (2010) American Chemical Society.

My colleagues have varied the light intensity for the stimulation of the algae and their photocurrent and found a profile as shown in the lower left portion of Figure 10.25. It looks like a square root function but it is likely that it can be modeled well with a logarithmic function [Titien 1970].

Finally, Ryu et al. compare the photocurrent which they measured with electro-analytical methods, and which they calculated via Faraday's law from the evolved oxygen, see lower right panel in Figure 10.25.

10.10 Solar cell assembly from bilayer lipid membrane and nanoparticles

A different concept is shown in a paper by Tien and Ottova [Tien 1999], where a self-assembled bilayer lipid membrane (BLM) is used as separator between two semiconductor electrodes and then used as semiconductor septum electrochemical photo-voltaic cell.

It has been discovered in the late 1960s that BLM can produce an electromotive force (EMF, E.M.F., e.m.f., emf) upon illumination [Tien 1968, Titien 1970] with an open circuit potential E_{OCV} that depends on the light intensity I as

$$E_{OCV} = A \log \left(1 + \frac{I}{B} \right)$$

with two constants A and B that depend on the BLM [Titien 1970].

Figure 10.26 shows the concept where a LBM serves as the ultrathin barrier or matrix and separates an n-type semiconductor nanoparticle from a p-type semiconductor nanoparticle. Electrochemical reductions and oxidations of species take place at their surfaces; the entire system is in contact with an electrolyte.

Note that in this biomimetic[4] design, the charge production does not take place in a biological component but in an inorganic semiconductor. It is entirely open at this time how future artificial photosynthesis systems will be built. They could be entirely inorganic and not much reminiscent of biological systems. They could be at large biological and just organized in a way that the biological components are biological at the microscopic scale only, or they could be hybrid systems. Future will show how it will look like. Different systems may coexist. The Otto engine and the Diesel engine coexist on the road as well, for example now alongside with electric vehicles.

10.11 Other nano-biohybrid systems

The bioelectrochemical cells mentioned in this chapter produced electric current but no chemical work. Many researchers are however interested in making systems which can produce fuels from artificial photosynthesis [Rozhkova 2012].

---

4 There is no "bio" in biomimetic, by the way.

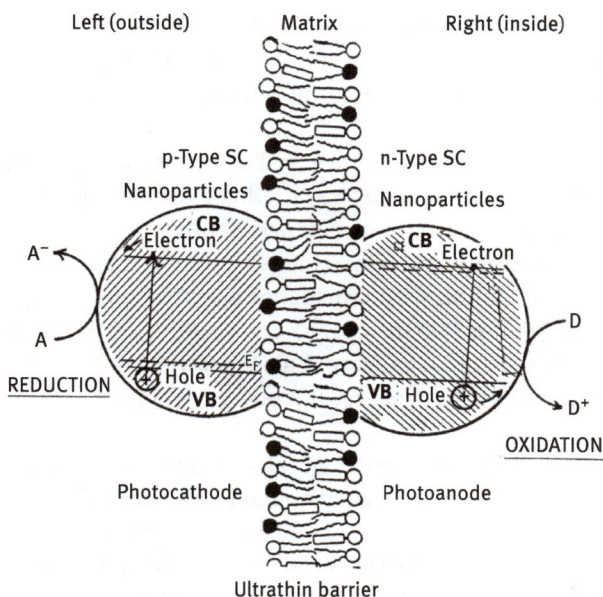

Figure 10.26: SC-SEP cell modeled on natural photosynthetic thylakoid membrane. Shown schematically is the basic light-induced process of the SC-SEP cell. In the center is a matrix (an ultrathin barrier) separating two aqueous solutions, onto which nanoparticles of semiconductor (SC) origins are attached. A SC, as typified by CdSe, is characterized by filled valence band (VB) and empty conduction band (CB). The energetic separation of these bands is termed bandgap. Light excitation results in charge separation into discrete electrons and holes. These entities being the most powerful reducing (electron) and oxidizing (hole) agents, respectively, initiate a reduction reaction on one side of the matrix and oxidation on the other side as indicated. SC nanoparticles will be used in this novel type of electrochemical photovoltaic solar cell.
Adapted and modified from H.T. Tien, S.P. Verma, Nature 277 (1970) 1233. Reprinted from Colloids and Surfaces A: Physicochemical and Engineering Aspects 149 (1–3), H. Ti Tien, Angelica L. Ottova, From self-assembled bilayer lipid membranes (BLMs) to supported BLMs on metal and gel substrates to practical applications 217–233, Copyright (1999), with permission from Elsevier. [Tien 1999].

My colleague Elena Rozhkova at Argonne Lab, along with her colleagues there, and at Northwestern University have functionalized TiO_2[5] nanoparticles with *bacteriorhodopsin* (bR) [Balasubramanian 2013, Wang 2014, 2017]. bR is known in nature as a proton pump which gets its pump energy by absorbing light of particular wavelengths in a cycle, as I explain in a chapter in the book [Chan 2016], and which is further explained in the paper by Renger and Renger [Renger 2008].

The proton pumping is there not an electric process, but a mechanical one which can to some extent be compared to conformational changes which are quite common

---

5 What the *drosophila* fly is to the biologist is TiO_2 to photocatalysis.

in large molecules like proteins. Pure electroanalytical methods are therefore not sufficient anymore to understand the structural origin of the charge carrier dynamics, that is, proton dynamics which takes place in these bR aggregates during light-stimulated and light-driven operation.

My colleague Jörg Pieper, now at University of Tartu in Estonia, is an expert on neutron scattering and has used quasi-elastic neutron scattering (QENS) for looking into this issue, together with his colleagues in Darmstadt and Berlin.

We learned about this method already in connection with the ceramic proton conductors, which we used in combination with impedance spectroscopy at high temperature and high pressure [Braun 2009b, 2017b, Chen, 2010, 2012, 2013, Holdsworth 2010]. We all are generally quite eager and also creative in designing novel neutron experiments for example for "solving the energy problem" [Braun 2014a].

Pieper et al. used a sophisticated combination of QENS and laser spectroscopy which they call laser neutron pump probe, which in my opinion was a groundbreaking experiment [Pieper 2008]. They used the *Halibacterium salinarum* whose purple membrane contains also the bR. It is necessary to synchronize the laser light pulse of the right wavelength for the stimulation of the bR with the period during which the QENS spectra are recorded.

The hybrid system which was built by the researchers in Illinois [Balasubramanian 2013] was made from TiO_2 coated with Pt and bR. It was tested in a neutral pH 7 solution which needed a so-called sacrificial electron donor (methanol). The H_2 production rate of this system was around 1.5 molecules of H_2 per second per bR molecule.

10.12 Enhancement of photocurrent by light harvesting proteins

10.12.1 Genetically engineered algae produce phycocyanin with his-tag

We have functionalized iron oxide photoelectrodes with light harvesting proteins in order to enhance the photocurrent and hydrogen production. Phycocyanin is such a protein which can enhance the photocurrent when attached to hematite, for example [Bora 2012b, 2013]. In strongly alkaline electrolyte, the protein will denaturate, but we found that the chromophores of phycocyanin remain operational.

One interest was in how to better attach the proteins on the electrode. The keyword "covalent attachment" is important in this context but we had for some time a more mechanical approach in mind. For example, it should be possible to genetically engineer those algae who produce phycocyanin in a way that they would grow the phycocyanin a so-called histidine-tag (his-tag) which would make that the phycocyanin produced from such genetically modified algae bonds preferentially with the metal ions of the semiconductor photoelectrode [Faccio 2015a, b, Ihssen 2014].

10.12.2 Cross-linking of phycocyanin with the organic semiconductor melanin

Another approach was to make a lateral cross-linking of phycocyanin by using melanin, which is an organic semiconductor. I have extensively written about this in my previous book [Braun 2017a] because there was a debate whether melanin can be considered a semiconductor, or not.

The purpose of using a polymerization of the melanin with the phycocyanin was the immobilization of the latter on the metal oxide electrode. The melanin was synthesized by tyrosinase as enzymatic cross-linking. Two alternative synthesis methods were carried out and explored as schematized in Figure 10.27. The photoelectrodes prepared this way [Schrantz 2017] were measured for their photocurrent and their hydrogen production in PBS as electrolyte.

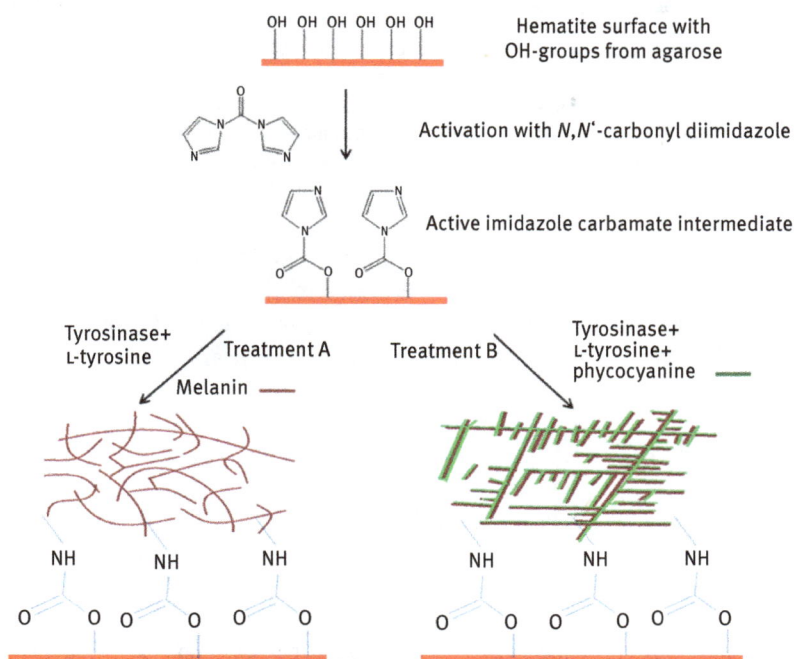

Figure 10.27: Scheme of the reaction steps for hematite photoanode functionalization. Reprinted from *Catalysis Today*, 284, Schrantz K, Wyss PP, Ihssen J, Toth R, Bora DK, Vitol EA, Rozhkova EA, Pieles U, Thony-Meyer L, Braun A, Hematite photoanode co-functionalized with self-assembling melanin and C-phycocyanin for solar water splitting at neutral pH, 44–51, Copyright (2017), with permission from Elsevier. [Schrantz 2017].

Part of the experimental setup is shown in Figure 10.28. We used a Teflon®-based sealed photoelectrochemical cell which was connected with a gas chromatograph so that we could measure the H_2 in parallel with the photocurrent.

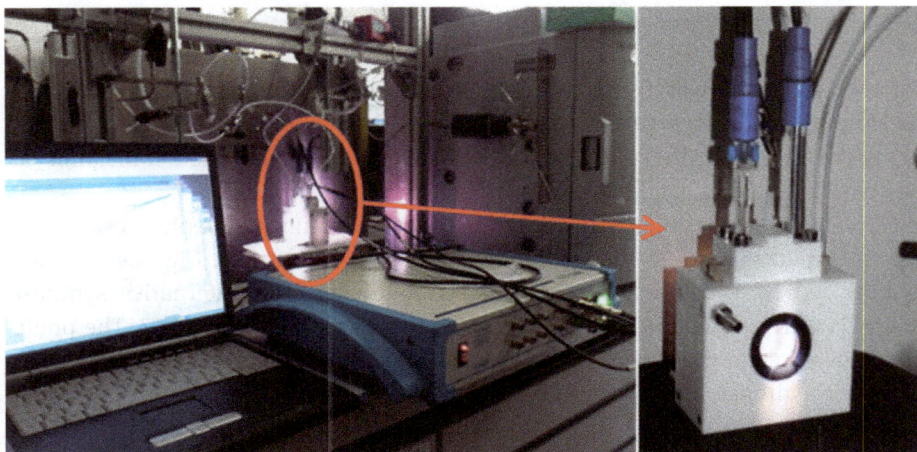

Figure 10.28: (Left) PEC test station for solar H_2 generation and detection, with the PEC cell illuminated and in operation; (right) PEC cell for *in situ* H_2 generation and online GC analysis. Reprinted from *Catalysis Today*, 284, Schrantz K, Wyss PP, Ihssen J, Toth R, Bora DK, Vitol EA, Rozhkova EA, Pieles U, Thony-Meyer L, Braun A, Hematite photoanode co-functionalized with self-assembling melanin and C-phycocyanin for solar water splitting at neutral pH, 44–51, Copyright (2017), with permission from Elsevier. [Schrantz 2017].

The experimental data were quite encouraging. The production of the hydrogen is shown in the left panel in Figure 10.29. Upon biasing the bio-electrode, hydrogen was produced virtually linear in time. During 120 min of operation, the H_2 production increased cumulative to almost 1375 ppm when only the iron oxide (hematite)

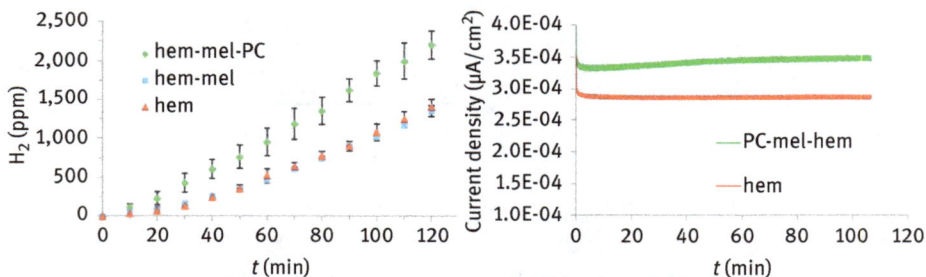

Figure 10.29: (left) H_2 evolution on pristine hematite (▲), melanin coated hematite (■) and PC-melanin-coated hematite (◆) photoanodes during a chronoamperometric measurement at 1000 mV in 0.05 M PBS. (right) Chronoamperometry at 1000 mV in 0.05 M PBS on pristine hematite (red line), and PC-melanin-coated hematite (green line) photoanodes, measured while following the H_2 evolution by GC measurement. Reprinted from Catalysis Today, 284, Schrantz K, Wyss PP, Ihssen J, Toth R, Bora DK, Vitol EA, Rozhkova EA, Pieles U, Thony-Meyer L, Braun A, Hematite photoanode co-functionalized with self-assembling melanin and C-phycocyanin for solar water splitting at neutral pH, 44-51, Copyright (2017), with permission from Elsevier. [Schrantz 2017].

photoelectrode was operated or such photoelectrode coated only with melanin. But when the phycocyanin was coated on the electrode with enzymatic cross-polymerization, the cumulated H_2 rose to 2,250 ppm; this is an increase of 50%.

The H_2 production rates were thus 687 ppm/h for the iron oxide electrode and 1,120 ppm/h for the phycocyanin-coated electrode cross-linked via melanin.

10.13 Impedance spectra of bilayer membranes with rhodopsin

It is of general interest to know how charge transfer takes place across bilayer membranes. We remember that the membranes in photosynthesis are lipid bilayers which are interrupted by PS II, PS I and the like, such as bR. Figure 10.30 shows a simple model of such membrane which is prepared by detergent dialysis.

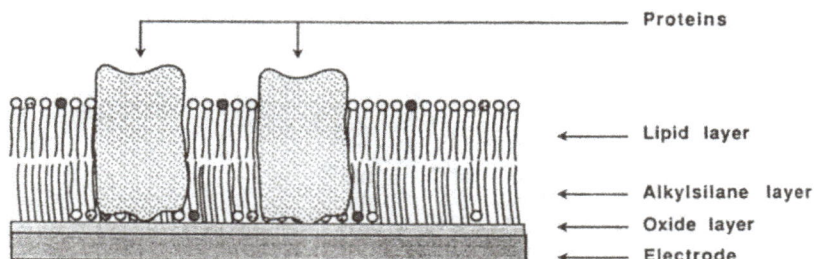

Figure 10.30: Model of a single surface-bound membrane formed by detergent dialysis on an alkylsilanated electrode surface.
Reprinted with permission from Li, J. G.; Downer, N. W.; Smith, H. G., Evaluation of Surface-Bound Membranes with Electrochemical Impedance Spectroscopy. *Biomembrane Electrochemistry* **1994,** *235*, 491–510. Copyright (1994) American Chemical Society. [Li 1991, 1994].

Specifically, the electrode assembly is a metal current collector with a conducting metal oxide and an alkylsilane (octadecyldichlorosilane [OTS]) layer. Li et al. [Li 1991, 1994] have coated on top of this the visual receptor protein rhodopsin and a lipid layer.

Li et al. employed impedance spectroscopy on this biomembrane structure for the assessment of its electrical properties. Figure 10.31 shows three impedance spectra for different stages of electrode design and assembly. The first spectrum was recorded from the platinum oxide (PtO) electrode, the next spectrum from the electrode coated with the OTS layer and the third spectrum with the bilayer membrane on top which contained the rhodopsin (Rh).

The spectra are not presented in the Nyquist plot but as the Bode plots with the log modulus of Z and with the phase angle versus the frequency – in logarithmic scale. The spectrum of the bare PtO electrode (open symbols) has the overall lowest impedance for the entire frequency range. The impedance increases significantly for all frequencies, when the silane layer is added to the electrode. There is no further significant increase of the impedance for the very low and very high frequencies. But in the range 100–10,000 Hz, there is a bump in the spectrum of the electrode with the Rh – bilayer membrane on top, basically a diffuse resonance.

The phase angle which is shown in the bottom panel in Figure 10.31 shows resonances which can be modeled according to the electric circuits which Li et al. have suggested in Figure 10.32 (see their original manuscript, Figure 9).

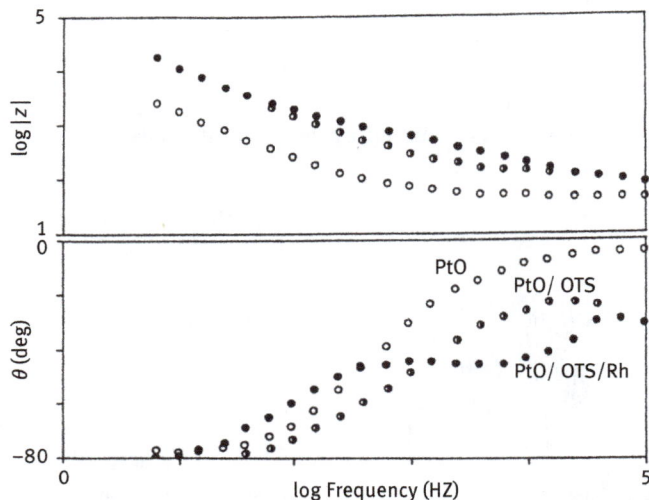

Figure 10.31: Bode plots for the PtO, PtO-OTS and PtO-OTS-Rh electrodes measured at 0.4 V in 0.1-M KCl containing 20 mM HEPES, pH 6.98.
Reprinted with permission from Li, J. G.; Downer, N. W.; Smith, H. G., Evaluation of Surface-Bound Membranes with Electrochemical Impedance Spectroscopy. *Biomembrane Electrochemistry* **1994,** *235,* 491–510. Copyright (1994) American Chemical Society. [Li 1991, 1994].

They begin with a circuit (Figure 10.32a) which they base on the actual physical design and common characteristics of an electrode inserted in an electrolyte: a parallel circuit for the oxide surface, a parallel circuit for the electrochemical double layer, and a somewhat more complex parallel circuit for the bilayer membrane which has a porosity and thus a double layer inside its structure as well. Not shown here, the experimental spectra can be compared quite well with the simulated spectra based on this circuit.

The high frequency signature can be simulated very well with a simplified circuit (Figure 10.32b) which has no redox contributions but only capacities. This is not surprising because chemical reactions are typically slow when compared to electro-static double layer charging processes. Therefore, the reactions are typically resolved at the low frequencies.

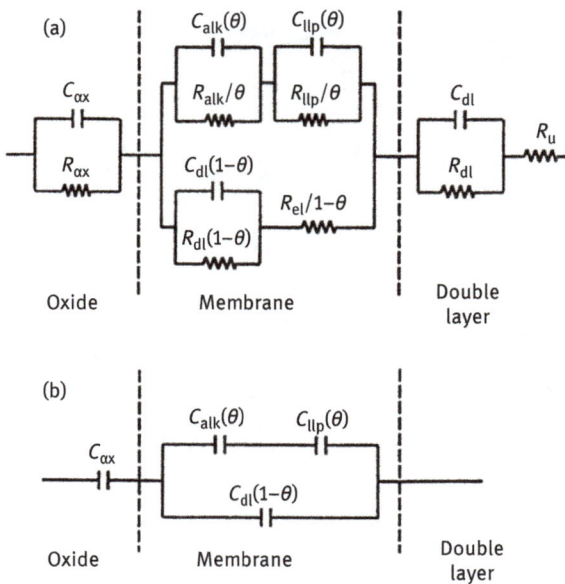

Figure 10.32: (a) Proposed equivalent circuit for surface-bound membrane electrode interface. (b) Simplified equivalent circuit valid at higher frequency region [Li 1991, 1994].
Reprinted with permission from Li, J. G.; Downer, N. W.; Smith, H. G., Evaluation of Surface-Bound Membranes with Electrochemical Impedance Spectroscopy. *Biomembrane Electrochemistry* **1994,** *235,* 491–510. Copyright (1994) American Chemical Society.

Li et al. [Li 1991, 1994] conclude from the impedance spectra that the surface structure is one complete single-membrane bilayer with coverage of 97% which has a long-term stability [Li 1994].

10.14 Impedance spectra of electrodes coated with proteins and with biofilms

We can picture the surface on electrode coated with a protein layer like as shown in the schematic in Figure 10.33. The charge transport processes and potential chemical reactions occurring in an electrochemical cell or system in general can be represented by resistors and capacitors like already shown in a previous chapter.

The photoanode (working electrode, hematite) and the counter electrode are connected to appropriate current collectors; the working electrode is coated with a protein layer plus we have an electrolyte. Inscribed in the physical representation is the electric Randles [Randles 1947, 1952a, b, c] circuit.

When we measure the impedance of an iron oxide photoanode in KOH as a function of the DC bias potential, we notice that the overall impedance decreases

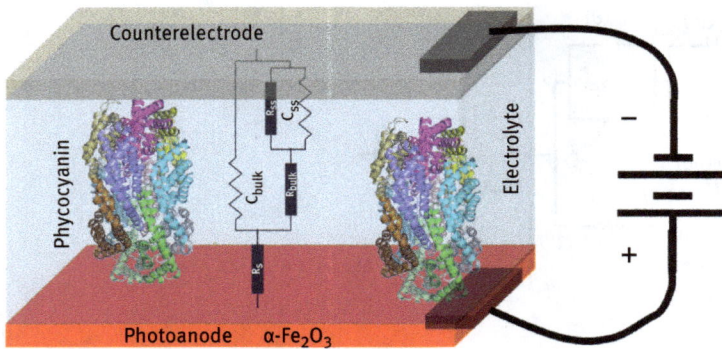

Figure 10.33: Schematic representation of an iron oxide photoelectrode with attached protein molecules (here CPC, Protein Database ID: 4FOT according to [Marx 2013]) along with an electric circuit representing the bioelectrochemical, electrochemical and solid state processes taking place during water splitting.
Reproduced from Braun [Braun 2015] with permission from Wiley.

with increasing bias, as shown in Figure 10.34. This effect is becoming stronger when the photoanode becomes illuminated. For the low DC bias of 0 mV and 200 mV, the difference in the effect is rather small because the spectra extend on the imaginary axis from 4000 Ω to 7000 Ω. At those potentials where we observe in the cyclic voltammograms or I/V curves the onset of a photocurrent, the radii of the semicircles in the impedance spectra decrease significantly. This is particularly obvious for the spectra recorded at 600 mV and 800 mV bias , which show two semicircles with small

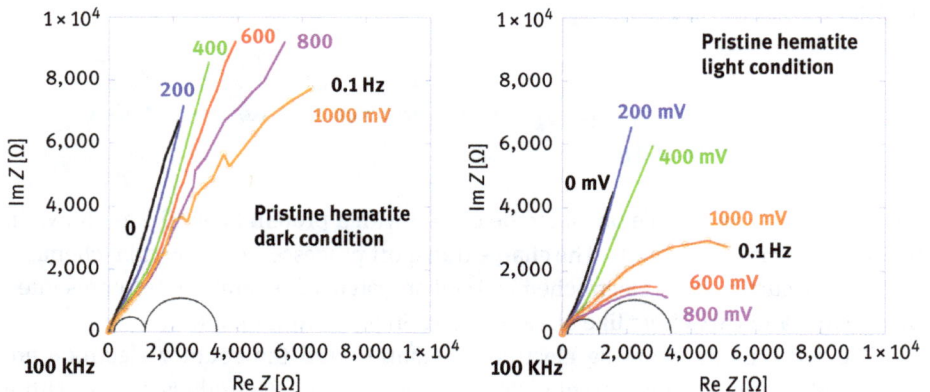

Figure 10.34: Impedance spectra of a pristine hematite photoanode in KOH for DC bias from 0 mV to 1000 mV versus Ag/AgCl reference. The left set of spectra was recorded in dark condition. The right set of spectra was recorded in light condition with 1.5 AM solar simulator. The frequency range is 0.1 Hz to 100 kHz.

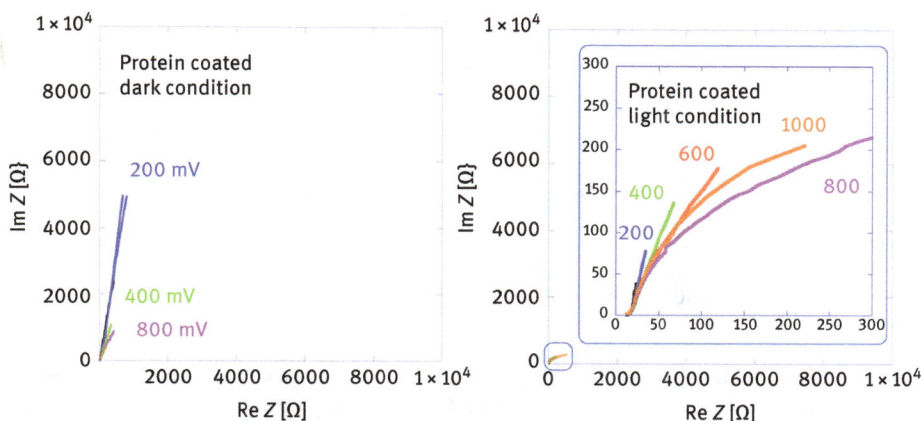

Figure 10.35: Sample caption impedance spectra of a phycocyanin-coated hematite photoanode in KOH for DC bias from 0 mV to 1,000 mV versus Ag/AgCl reference. The left set of spectra was recorded in dark condition. The right set of spectra was recorded in light condition with 1.5 AM solar simulator. The frequency range is 0.1 Hz to 100 kHz.

radius of 1000 Ω and 400 Ω, respectively, as demonstrated in the right panel in Figure 10.34.

Figure 10.35 shows impedance spectra from iron oxide which was coated with phycocyanin. The first observation is that the impedance spectra measured in dark extend to 5000 Ω only where the pristine hematite had an impedance spectrum range that reached almost 10000 Ω. The coating of the inorganic electrode with a bioorganic layer of proteins appears to lower the impedance.

I would like to recall that the coating of the carbon electrodes with conducting polymers like polypyrrole and polyaniline, as we have seen in the case of the microbial fuel cell [Zou 2009], led to higher fuel cell currents. An organic coating can therefore be of great assistance when it comes to lowering the impedance of electrode assemblies.

The impedance of the protein coated electrode in the dark at higher DC bias values becomes so small that the axis window of 10000 Ω is not suited anymore for presenting the data properly. The lower right panel in Figure 10.35 shows the impedance spectra of the same sample when recorded under illumination. The small blue rectangle shows that the spectra near the origin hardly can be made out.

I have therefore made an inset with a magnification of the range of interest which fits in a window of 300 Ω only. The spectra in the inset show nicely how the impedance of the protein coated-electrode decreases considerably and systemically with increasing bias when under illumination.

Simply measuring the radius of the semicircles allows already for a rough estimation of the charge transfer resistance R_{CT}. The values for R_{CT} are plotted in Figure 10.36 versus the bias potential. Under 0 mV bias in the dark, the phycocyanin-coated electrode has a very high charge transfer resistance, and it is decreasing

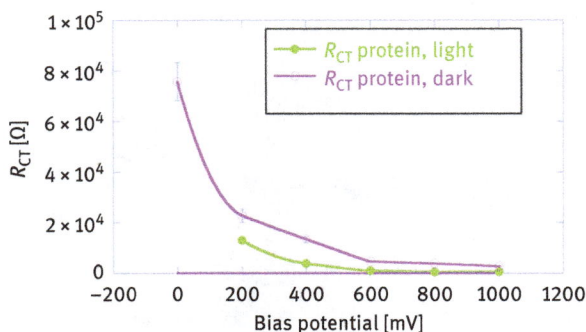

Figure 10.36: Charge transfer resistance R_{CT} of phycocyanin-coated electrode in dark and in light condition obtained from impedance spectroscopy. Electrolyte was 0.05 m phosphate buffer saline. Illumination with 1.5 AM solar simulator. Electrode was iron oxide. Impedance recorded from 0.1 Hz to 100 kHz. Reference electrode was Ag/AgCl.

rapidly when the bias potential approaches the water splitting potential of iron oxide. But the illuminated phycocyanin electrode has around half of the resistance once the light is switched on.

The semicircles in the spectra can certainly be fitted to a Randles circuit so that the electric parameters can be quantitatively determined. Table 10.2 lists the fit parameters.

Table 10.2: Fit parameters for the impedance spectra of pristine hematite and protein (C-phycocyanin)-coated hematite in dark and illuminated condition.

Electrode		R_S [Ω]	R_{tr+} [Ω]	$R_{ct,tr+}$ [Ω]	CPE1T ($\pm$)	CPE1P (+)	CPE2–T(+)	CPE2–P(X)
Pure α-Fe$_2$O$_3$	Dark	11.5	9.81	2377	7.1E − 06	0.97	4.24E − 05	0.815
	Light	11.6	10.3	1271	5.1E − 06	1	3.83E − 05	0.831
Phycocyanin	Dark	14.8	7.66	4070	5.0E − 05	0.86	2.38E − 05	0.832
on α-Fe$_2$O$_3$	Light	13.9	5.99	443	7.3E − 05	0.82	2.98E − 05	0.8

The fit model is based on the electric circuit shown in Figure 10.37. Least square fits were computed with ZView.

10.15 Microbial fuel cells

A maybe better known application of bioelectrochemical cells is in fuel cells. The fuel can be some organic waste and microbial actors in the bio fuel cells convert fuel and oxygen to electric power; see the review article by Bullen et al. [Bullen 2006] and Santoro et al. [Santoro 2017], and on miniaturized enzymatic fuel cell systems the review in Song et al. [Song 2011].

An example for a microbial fuel cell is demonstrated in Zou et al. [Zou 2009], see Figure 10.38, and in Zou et al. [Zou 2010]. The researchers used low-cost Nalgene®-based laboratory plastic container which served as the fuel cell reaction vessel. The

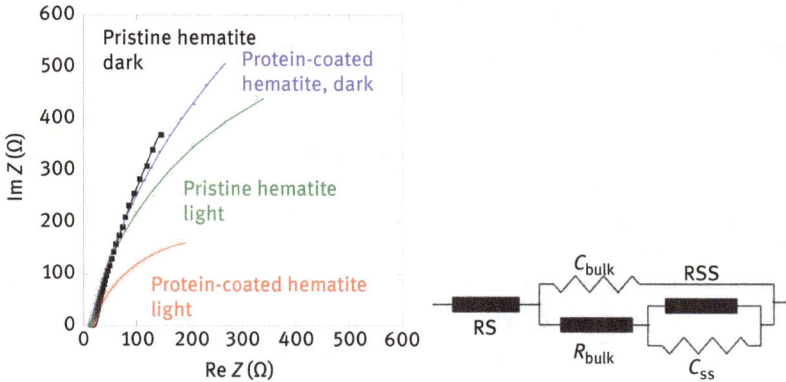

Figure 10.37: Impedance spectra from pristine hematite photoelectrode and protein-coated hematite photoelectrode in dark and light condition, along with the least square fit (solid lines through data points) obtained from modeling against the circuit shown in the right panel.

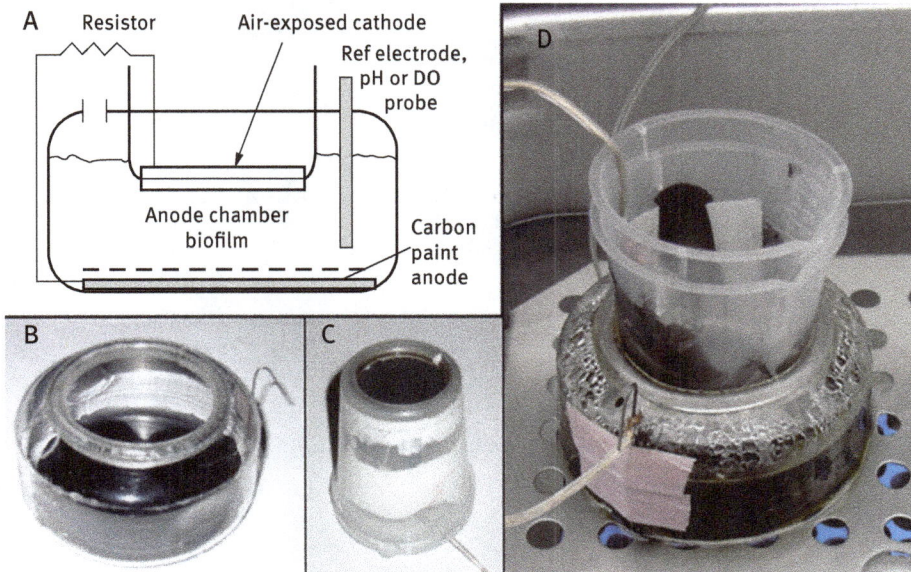

Figure 10.38: Single-chamber PMFC: (A) schematic diagram, (B) anode chamber with the bottom surface painted with carbon black, (C) cathode and (D) biofilm PMFC under operation. [Color figure can be seen in the online version of this article, available at www.interscience.wiley.com.] Reproduced from Zou Y, Pisciotta J, Billmyre RB, Baskakov IV: Photosynthetic microbial fuel cells with positive light response. *Biotechnol Bioeng* 2009, 104:939–946. doi: 10.1002/bit.22466 with kind permission from Wiley.

bottom of that vessel was coated with electrically conductive carbon paint (four layers). When we return in this book to the section about Melvin Calvin's interdigital aquadag electrodes [Acheson 1910], we see that carbon paint can be quite useful in bioelectrochemistry. This carbon anode layer was further functionalized with electroactive polymers (polypyrrole and polyaniline) because this made higher currents.

The cathode was a platinum-coated carbon cloth and carbon paste and several polytetrafluorethylen (PTFE) diffusion layers; this was basically an air electrode similar to those known from the zinc air batteries.

Then, a biofilm from the species *Synechocystis PCC-6803* was grown and later transferred into the Nalgene® vessel with a modified mineral medium as electrolyte and subjected to 12 h day/night cycles under which it produced an emf in the range of 0.1 V under light. The maximum power density was 1.3 mW/m^2.

11 Land use and power plants

I have been presenting my research for many years at the Materials Research Society (MRS) Spring Meetings in San Francisco, California. I happened to be one of the Co-Chairs for the Meeting in 2015 [Braun 2014b] – the last Spring Meeting in San Francisco.

Since 2016, the MRS holds their Spring Meetings in Phoenix, Arizona. The map in Figure 11.1 shows via the red dots my flight route back from PHX to SFO. The flight goes from metropolitan Phoenix over the Mojave Desert and mountains to the Pacific shore over the wetlands and marshes in the Silicon Valley and San Francisco Bay Area.

The map was prepared with iPhoto, an Apple Inc. software.

When I have a window seat in the plane, I usually look out of the window and scan and investigate the land and sea on the ground. When I have a seat on the aisle, I rather write a manuscript.

The small city of Needles, CA (aerial view in Figure 11.2) is at the border to Nevada and Arizona in the Mojave desert, in the population area of the native Mojave tribes called "Âka Makhav" (Makhav = Mojave), which means "the people living alongside the river."[1] The river here is the Colorado River.

The area is mostly desert land with average temperatures in June–August above 40°C and in the extreme case even above 50°C. The average precipitation per year is 117 mm water. The living conditions in this area are extremely hard. Some remarks on the living conditions of settlers and native American tribes in the harsh desert environment[2] can be found in the book of Richard E. Lingenfelter "Death Valley & The Amargosa" [Lingenfelter 1986].

But along the river is sufficient water for farming and agriculture, as you will notice from the dark circles and squares which are crops artificially watered through-out the year. Also, the low humidity in the Mojave desert would still be high enough to run a PEC reactor – not with liquid water but with the low humidity – and thus generate hydrogen fuel in the desert with solar energy.

All across the desert lands in the United States, you can see these green little patches where local farming is possible for the community. Around 50 km northwest of

---

1 The reader may wonder why I bring up here the native Indian tribes but I consider this book not only a technical one. That said, I would like to mention that on my trip from San Francisco in California to Santa Fe in New Mexico in 2014, which I made by car through the deserts, I came along the Coconino Forest in Arizona. At one place on the road, I purchased some crafts from Native Americans, probably members of the Navajo Nation. One of them lectured me on the electroplating they used for making the metal jewellery. Another one lectured me on the spread of the Sanskrit languages over the globe and on the linguistic origin of the first names and family names of the people from the Baltic States, particularly Lithuania.
2 Countries like Australia, Canada, Norway and USA are easily blamed for using too much energy, but, fair enough, some of their population is living in cold or hot desert land where you need air conditioning and central heating for living. This certainly costs a lot of energy.

https://doi.org/10.1515/9783110561838-011

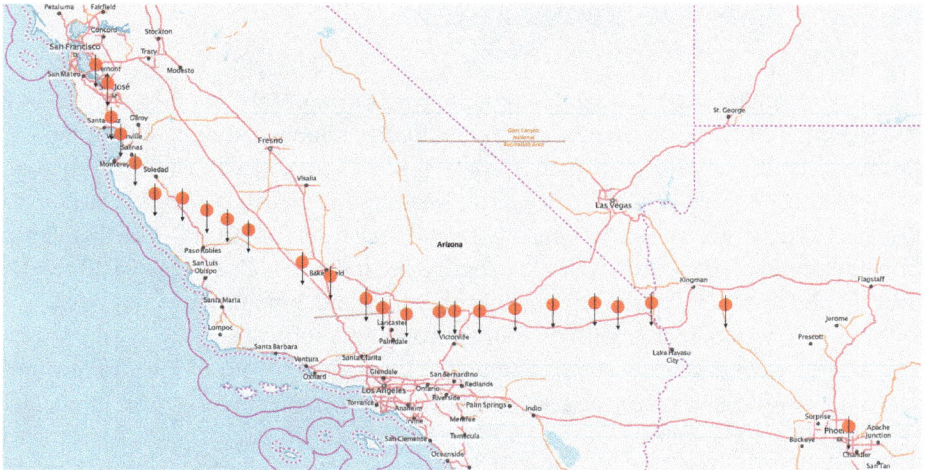

Figure 11.1: Map with red dots showing flight route from Phoenix to San Francisco, United Airlines. The GPS coordinates are stored in the images taken with a smart phone.

Figure 11.2: (Left) The small city of Needles (5,000 population) in California, at the Colorado River. Dark circles and rectangles are from artificially watered crops and plant fields. The circles can have a diameter of almost 1 km. (Right) Water irrigation machine from 10 segments allowing for circular watering of a crop field.
Photo taken near Chandler by Artur Braun, April 9, 2018.

Needles is the city of Ivanpah, where >170000 heliostats (>340000) mirrors (2.5 × 3.5 m each; they need to be c leaned regularly [Alon 2014]) follow the sun trajectory and focus the solar light on three central towers where water is heated to $T > 500$ °C so that steam turbines can produce electricity. The power plant operates since 2014 and can provide electricity to 140000 homes [Stillings 2016].

Figure 11.3 shows from aerial view three very shiny ellipses which reflect the sunlight to the observer. I took the photo in Spring 2017 on my flight back from Phoenix to San Francisco. A pilot who happened to sit next to me told me it was a new solar power plant in the desert (Ivanpah Solar Electric Generating System). The height of one tower is 140 m and the most remote mirror is in 1400 m distance. With a total reflective area of 800000 m^2, [3] the power of one of the three blocks is 126 MW electricity [Polo 2017]. All three towers together provide 377 MW net electric energy from an area of 14 km^2 [Overton 2014].

Figure 11.3: Ivanpah Solar Electric Generating System, ISEGS (Mojave Desert) with three large solar fields and towers reflecting the sunlight into the sky, incidentally captured by the photographer during a United Airlines flight from Phoenix AZ to San Francisco CA.

The design and construction of such towers is subject to ongoing research and development work in order to reduce cost and also risk for accidents during

---

3 I calculated a few years ago that our simple hematite photoelectrode – without any materials optimization such as doping with silicon or adding a co-pi catalyst, with a 1000,000 m^2 area PEC area when it received 12 h sunshine daily for one full year, would be able to fill one Boeing 747 ("Jumbo Jet") per year.

construction [Peterseim 2016]. Also, the heliostats are constantly being improved [Ortega 2015].

While we may feel happy about an additional power plant that does not use fossil fuels and does not produce CO_2, we are made aware of some negative repercussions that come with concentrating solar light so intense that it can heat up water vapor to over 500 °C. Birds may be caught in the focusing light and then basically become fried, or boiled in the air, but scientists are working toward understanding the causes of such accidents and also find mitigation measures [Ho 2016]. The light reflected from the towers is so intense that pilots and copilots have worried about the safety [Ho 2015] of flights because the glare may disturb the pilots.

Also, we may think that the desert land in this sparsely populated area is of no further interest than let's say for such a remote power plant, some people believe it should be rather preserved for recreational purposes and as wildlife habitat. Forced place making has therefore become a social, civic and political issue [Moore 2016].

We believe that environmental activists would welcome when coal power plants are shut down and instead wind farms and solar farms are installed. But this is not necessarily the case. The Beacon Solar Plant in near Death Valley was eventually moved from CA to NV because of litigation from a group of environmentalists and labor union (compare also [Rubin 2015]).

When you fly around 350 km further west over the Mojave Desert, you will come to Willow Springs and Edwards Air Force Base, see Figure 11.4. The area is still desert but there is some population. You notice four large circular crop fields and a few miles away westward a large solar PV power plant. Not visible in this photo, many more PV power plants stretch over almost 20 mi to the west where the mountains begin. In the north before the mountains, you see many small irregular arranges lines of white pins – these are wind turbines which belong to several different wind power companies. The land in the desert is very cheap and the electric power can be brought to the metropoles in the west or to cities such as Las Vegas and Phoenix. This is not withstanding that the USA operate 99 commercial nuclear reactors with a 100000 MW power. In 2016, they provided 800 TW h electricity, which was 20% of the nation's electricity generation.

Note that the electric energy generated by all aforementioned centralized solar and wind power plants is only available in real time of generation and cannot be stored without further storage technologies. This, however, is a secondary problem. The most important step is the first one: the power production.

Figure 11.5 shows the tip of the San Francisco South Bay between Palo Alto and Fremont, where the Silicon Valley starts. This is a densely populated place, confined between mountains and ocean, with accordingly high energy consumption but very little energy production. The climate there is very mild, the region is very rich in wildlife, despite the overpopulation of 7 million people. To the native tribes of the Ohlone [Hoikkala 1996, Jackson 1998, Johnson 1997, Johnson 1996], the environment provided likely better living conditions than the desert to the Mojave tribes. Before

Figure 11.4: Area between Mojave and Willow Springs, 20 mi west of Edwards Airforce Base in California. Round dark objects denote plantations. Prismatic dark objects are solar panel (PV) power plants. White lines in the center-left region are wind parks and contain hundreds of smaller wind turbines.

the arrival of the Spanish invaders, the population of the Ohlone Indian tribes in this area counted 10000–20000 in the year 1770. It rapidly decreased to below 1000 by the year 1850. The Spanish Missions were built along the Pacific coast and were connected via the El Camino Real from San Bruno in Baja California to Sonoma in Northern California.

The Bay Area has about 15000 km^2 greenbelt, two-thirds of which is used for farming and ranching (this corresponds a 100×100 km large square). We recall that the space used for the three solar towers in Ivanpah CA was 14 km^2; it can provide electricity for 140000 homes.

A similar marshland near the coast is shown in Figure 11.6, an aerial view over a larger neighborhood of Rotterdam in the Netherlands. The area is densely populated and also used for agriculture and for fishing industries and for tourism. In short, this place is extensively used. Even the air space of the region is extensively used as the nearby Amsterdam Airport Schiphol is the third largest in Europe in terms of passenger capacity use.

At large, the Dutch have reclaimed the land from the North Sea several hundred years ago. Rotterdam has operated for many hundred years a large sea port which

Figure 11.5: Wet lands in the San Francisco South Bay with Dumbarton Bridge from Palo Alto to Fremont and marsh lands (Don Edwards San Francisco Bay National Wildlife Refuge, California). Photo by Artur Braun, 2017.

was essential for creating wealth in northern Europe. Today, the port of Rotterdam is one of the largest in the world. For Europe, it is the most important port for oil trade, which totals to 100 million tons crude oil per year. The total amount of goods is likely 450 million tons per year.

Air Liquide has major operations in Rotterdam. The hydrogen gas which I pumped in Empa's Hyundai ix35 Fuel Cell car on my trips to the Netherlands was provided at the Air Liquide H_2 station in Rhoon near Rotterdam. The hydrogen is sent to there via pipeline from the port, where Air Liquide has a large factory. Air Liquide has 1700 km pipeline in the Netherlands but the pipeline network extends even to Le Havre in France.

In Rotterdam, there is a connection to the major European river Rhein (Rhine) which allows for shipping goods to the major Dutch, German and Swiss industry

Figure 11.6: Aerial view over Rotterdam area, the Netherlands. Shown is agricultural area close to urban residential and industrial area, intersected by waterways.
Photo by Artur Braun 2017.

zones such as Ruhrgebiet (Duisburg, steel), Cologne, Ludwigshafen (chemistry such as BASF), Strassbourg (France) and Kehl, Basel (chemistry). Another important river is the Maas (French Meuse) which connects the international port in Rotterdam with the Belgian steel industry. An extensive network of artificial channels connects many rivers in Europa indirectly with Rotterdam.

A seaport is a gate to the world. But there exist bottle necks and choke points which pose a risk for international trade. The risks around countries regionalizes the world oil market, and shipping chokepoints regionalize the world oil market [Kaufmann 2014]. Figure 11.7 shows a world map [Shibasaki 2017] where the focus is on the Suez Canal, which is a major choke point for international ship and maritime traffic. The choke points are not only for the oil trade but also for the food trade [Wellesley 2017]. The world is sectioned in that map in 23 regions and the regions depend on the canals which are choke points for the trade.

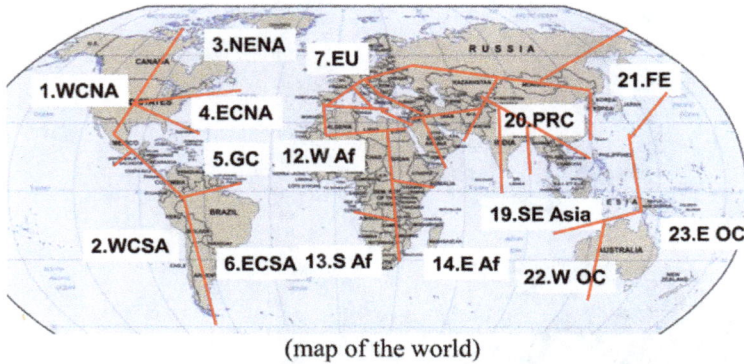

(map of the world)

(detailed map around the Suez Canal)

Figure 11.7: World division for the Suez Canal estimation.
Reprinted from *Research in Transportation Business and Management*, 25, Shibasaki R, Azuma T, Yoshida T, Teranishi H, Abe M, Global route choice and its modeling of dry bulk carriers based on vessel movement database: Focusing on the Suez Canal, 51–65, Copyright (2017), with permission from Elsevier. [Shibasaki 2017].

In order to plan for the transportation of goods over the oceans, you have to know also alternative routes in case a canal is blocked. The canals are typically not blocked but this is also because of international precautions such as diplomacy and military presence. You may use search engines such as http://www.swisscows.ch and find maritime choke points for example at https://www.worldoiltraders.com/sea-ports/ and then get an idea why international conflict zones and regions are always centered around the same spots on the world map.

We are therefore looking at a transport problem at the global scale in principal not much different from the transport problems in electrochemical energy converters (including biological ones), where electrons and holes and ions and polarons somehow have to fit through membranes and electrodes and electrolytes. George Friedman

[Friedman 2011] and to a lesser extent Zbigniew Brzezinski [Brzezinski 1997, Brzeziński 2015] explain in their books very well how geography and geology – and this is the structure – dictates which opportunities arise for the people in which countries.

Transportation by ship is very cheap, transport over land is much more expensive, but a railway connection can help defray transportation costs when compared to transport with trucks, for example. The initiative of China to build a new intercontinental Silk Road railway, the Eurasia Land Bridge, is a new opportunity for trade between Asia and Europa. The train will go 8000 km from Chengdu in China to Rotterdam. Further opportunities for maritime trade arise when the arctic ice melts and the Bering Strait will open up.

11.1 Coal power plants

A major supply of electric energy comes from coal power plants. The coal is burnt and used to heat steam which will power turbines which then will power electric generators. The burning of the coal causes conversion from chemical energy to thermal energy which is converted in mechanical energy as hot water steam which in the end is converted in electric energy. Countries which are rich in coal resources want to use that resource for their own power supply. And, they may sell their coal to other countries.

Figure 11.8 shows the Cholla[4] coal power plant in Arizona between Holbrook and Joseph City. This 995 MW plant is located at Highway 40, the scenic Route 66 in the Arizona Desert between Petrified Forest National Park and Coconino National Forest. The plant was commissioned in 1962 and closed down in 2016. It received the necessary coal from the McKinley mine which is around 180 km away near Gallup in New Mexico. The mine is located in Navajo County and provided high quality bituminous coal by surface mining.

McKinley mine [Sourcewatch 2012] was owned by the Chevron Corporation and produced from 1983 to 2009 almost 150 million tons of coal which it sold to Cholla power plant and other coal power plants in New Mexico Arizona. Figure 11.9 shows the annual production of coal from 1983 to 2009. At peak times, the mine produced 8 million tons per year. McKinley mine used to have around 214 employees, 86% of which were recruited from the native Navajo tribes. McKinley mined out in 2009 and was closed [Donovan 2009]. In its last year, the production was still over 2 million tons. This would be enough for the Cholla power plant, but Arizona Power Supply, the utility company which operated Cholla Plant, had found already a different supplier of coal.

The coal used to be transported from the mine to the power plant by railway which follows the southwest chief route from Chicago to Los Angeles. The trains along this rail can have over 100 wagons which are pulled by several locomotives. I have counted a train with 130 coal wagons, each of which can load around 100 t of coal. One train can

---

4 Cholla is a long-branched cactus in the south-west regions of USA.

Figure 11.8: Cholla coal power plant at Highway 40 (Route 66) between Joseph City and Holbrook, Arizona, in April 30, 2014. The power plant was closed in 2016. Photo Artur Braun.

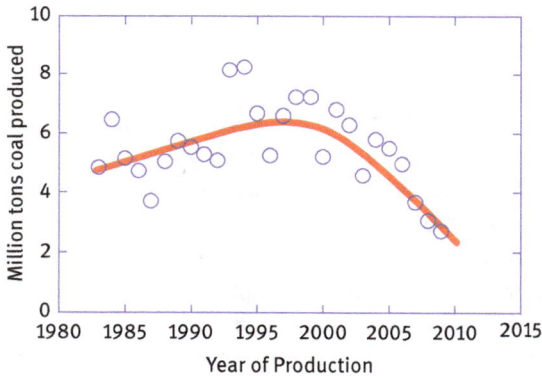

Figure 11.9: Production of coal at Chevron Corporation's McKinley mine near Gallup, New Mexico, which was closed in 2009. Data from Coaldiver (2010).

thus deliver over 10000 t. When every day one such train arrives at Cholla, a total of 3.6 million tons of coal can be converted to electricity in 1 year. In 2009, Cholla purchased from McKinley 2.2 million tons of coal. This means on average every day during the week Monday to Friday one such train unloaded its coal at Cholla power plant.

Coal power plants and nuclear power plants need large amounts of water for steam generation and cooling. The cooling water for the Cholla power plant was stored in the artificial Cholla Lake. The lake was also used for recreational purposes and for commercial fish production. To prevent growth of algae in the cooling system of Cholla plant, chlorine had to be added to the water which in turn could have adverse impact on the fish growth and recreational purpose of the lake. Consequently, the Cholla Lake was closed to the public in 2002 or 2003.

In 2011 Chevron Corporation announced, it would withdraw from the coal business; "Chevron Mining" spokeswoman Margaret Lejuste stated [Lejuste 2011]:

"New coal technologies are too far off to make staying in coal a good strategy".

When you look up the power plant in a map, you will find that it occupies a territory of around 375 ha (1 ha = 10000 m^2, 100 ha = 1 km^2), including the cooling water lake and the railway tracks and the coal pile. The McKinley mine is around 2400 ha wide, and around one-third of its annual coal production of 6 million tons per year was sold to Cholla. We can therefore say that around 800 ha of the mine is necessary to provide the coal for the power plant. In total, around 1200 ha is occupied for the delivery of 1000 MW electric power. We can therefore say that the production of 1 MW coal power requires a land use of around 1 ha. Such estimation is for example important when we want to assess the necessary area for production of energy from alternative sources such as wind power plants or solar power plants [Aro 2017]. The molar weight of carbon is 12 g/mol. The molar weight of CO2 is 44 g/mol. With 2 million tons of coal burnt per year at Cholla, $2 \times 44/12 = 7.33$ million tons of CO2 would be produced.

11.2 Wind power plants (wind parks)

Around 2003, 10 years before Chevron Corporation started pulling out from the coal business, rancher Bill Elkins from nearby Snowflake AZ became interested in exploring the opportunities for wind power on his land. Elkins looked into the local electric power grid and concluded that it was feasible to connect a wind power plant to the electric grid. His interest, together with the efforts of developer John Gaglioti and scientists from Northern Arizona University, amounted into a wind power plant in its first phase with 30 wind turbines that produce 63 MW electricity (Dry Lake Wind Power Project) [Wadsack 2010], which is financed by Iberdrola Renovables, a Spanish company, for $100 million [Renewables 2018].

We need 15 of such 63 MW wind power plants when we want to replace the 1,000 MW coal power plant. One result is that we do not produce carbon dioxide. We also do not need any cooling water. Switching from coal to wind eliminates therefore a considerable amount of CO$_2$ and saves around 3 million m^3 water per year.

The operation of a coal mine can require discharge of waste material into the environment which is subject to approval by the government. The reader can have a

Figure 11.10: Ancient wind mill between Oegstgeest and Voorhout in the Netherlands. Photo Artur Braun 2018.

look at what procedures are necessary with the government based on the public available permit for Chevron issued by the EPA [Nn0029386 2017].

Wind power plants are not an invention of the twentieth century. Wind has been recognized as much as power source as water power long time ago. An early account of the everlasting and never-ending power wind mills is found in the story of Don Quixote by Miguel de Cervantes in 1605 [De Cervantes 1605].[5]

Figure 11.10 shows an ancient wind mill in the Netherlands. The wind mill would transform the mechanical power from the blades to mechanical power of some tools,

---

5 In June 2018, I participated in the 5th Summer School on Photobiology, organized by University of Padua and held in Brixen (Bressanone), the German speaking part in Italy in Süd-Tirol. I used this opportunity to go there by car, with hydrogen. Half an hour away from Brixen is Bozen (Bolzano), there is a H_2 station. It was a challenge to go from Zürich to there across the Alps with a FCEV. But the real challenge was to go from Bozen to Venice in Italy – and back to Bozen to the H_2 station. I tried it and I succeeded to make the >400 km to Venezia and back. At night at the conference dinner, I briefed my colleagues at the dinner table on my record. The gentleman next to me was wondering whether there were H_2 stations in Spain, his home country. I looked it up in the internet with my smartphone (https://www.netinform.de/H2/H2Stations/H2Stations.aspx?Continent=EU&StationID=-1) and found a station in the city of Albacete in Spain. His accompanying wife told me Albacete was his hometown. Albacete is in the region of La Mancha, which we know from the adventures of Don Quixote fighting wind mills. There are actually now wind parks in La Mancha. We may wonder whether the hydrogen at the H_2 station in Albacete is made from electricity from the wind park.

such as grinding stones when the wind mill was operated as a grain mill. Back then, no electric power was involved and no chemical energy was involved. Every power transmission was purely mechanical. This includes also water power.

Since about 100 years, the wind power plants are based on electricity generators which deliver directly electric power. Today, the electricity can be delivered right into the electric grid. Many communities decide to use part of their land for some wind power plant, operated by their public utility companies.

Meanwhile, developers think of not using costly land for the development of wind power plants, but using the low cost seashores instead. Figure 11.11 shows the wind turbines of the Gunfleet Sands offshore wind farm before the British coast at Clacton-on-Sea. It has 48 turbines with 3.6 MW each and two new turbines with 6 MW each. This is a total power of around 185 MW. Six of such wind farms add up to a total power of 1000 MW and thus can replace the aforementioned Cholla coal power plant.

Figure 11.11: Off-shore wind farm Gunfleet Sands with 50 turbines at the English coast of Clacton-on-Sea, Essex. There is 6 × 7 turbines plus 3 × 2 turbines of 3.6 MW each, and two new 6 MW turbines. Photo Artur Braun 2017.

The area for this plant is roughly 20 km^2 or 2000 ha. Six times this area is necessary to accomplish the 1000 MW power, which is 12000 ac. This is 10 times the area which is occupied by the Cholla coal power plant and the one-third share of the McKinley coal mine. When we replace all 3.6 MW turbines by the 6 MW turbines, we need only 7200 ac for the 1000 MW power. This is still six times the area which we needed for the coal power plant solution. But notice that the offshore wind farm is in the ocean and not on land.

Figure 11.12 shows the Thorntonbank windfarm, which is even 30 km offshore deep in the North Sea, near Brugge in Belgium. This is a 325 MW plant with 54 turbines with 6 MW each.

Figure 11.12: Thorntonbank off-shore windfarm in the North Sea 30 km away from Brugge, Belgium.

11.3 The biosphere

The place in Earth where life exists is called the biosphere. There are some regions with extreme conditions where living is very hard, such as in the deserts [Warren-Rhodes 2013]. At the microscopic level, there may be organisms which have adapted to the harsh living conditions, these are called extremophiles[6] [Baqué 2013]. These can be fungi or algae which can sustain extreme temperatures, or very high

---

6 Extremophiles: "Organisms which require extreme physico-chemical conditions for their optimum growth and proliferation. Extremophilic microorganisms are e.g. thermophiles or psychrophiles, halophiles, alkalophiles or acidophiles, osmophiles and barophiles, based on their growth at extremes of temperature, salt concentration, pH, osmolarity or pressure, respectively." [McNaught 2014] McNaught AD, Wilkinson A: IUPAC. Compendium of Chemical Terminology. In *IUPAC Compendium of Chemical Terminology – The Gold Book* (Nic M, Jirat J, Kosata B, Jenkins A eds.),

concentrations of antimicrobial elements (*Amycolatopsis tucumanensis* [Alvarez 2017, Braun 2017, Costa 2013, Davila Costa 2011, Davila Costa 2012] is a strain which is found near copper mines in Argentina and can deal well with antifungal copper), or extreme electromagnetic radiation (my colleagues at SCK CEN nuclear research center in Belgium have worked on algal bioreactors which would be used on space missions for the production of food; ARTEMIS project with International Space Station).

My international colleagues and I have tried on several occasions to get the funding for studying the use of extremophiles in PEC reactor bio-electrodes [Braun 2016a, Braun 2016b]. Their use should be basically applicable in bio fuel cells as well [Bullen 2006].

You can think now how it would be possible to establish life and settle a human population on the Moon or Mars. At some point, you want to grow there your food and do not depend on supplies shipped to there from Earth. For this, you have to establish a new and independent biosphere. When you have an energy supply such as by the sun, you should be able to establish a circular metabolism, a circular economy where you can regenerate the human waste into food. You need minerals, CO_2 and water, and energy to force the metabolic cycle. Natural photosynthesis maybe would work.

Around 30 years ago, researchers have tried to build a new biosphere in the Sonoran desert near Tucson, Arizona. They called the project Biosphere 2 and let eight researchers stay for 2 years in a specially constructed building complex (Figure 11.13) which would provide from the outside only energy, but all other supplies and food and oxygen and water were inside the building [Allen 1991, Avise 1994, Cohen 1996, Cohn 2002, Galda 1992, Jiang 1997, Macilwain 1994, Odum 1993, Walford 2002, Watson 1993].

The main building is a huge green house near the city of Oracle AZ which is basically the green lung of Biosphere 2. In addition, there are hermetically sealed and connected buildings with laboratories, sleeping rooms, living rooms and social rooms and kitchen which allowed the crew of eight, men and women, to work and live and research.

The Biosphere 2 project has received from a number of people ("the press") quite negative opinions. Some even say the project was a failure.[7] This of course is big

---

2nd ed. (the "Gold Book") edition. Oxford: Blackwell Scientific Publications; 2014. doi: 10.1351/goldbook.

7 A good example for similar ignorance of pseudo-scientists is the well-known Michelson–Morley experiment, which some of them called a flop or a failure [Michelson Livingston 1987] Michelson Livingston D. Michelson-Morley: The Great Failure. The Scientist. 13 July 1987. https://www.the-scientist.com/books-etc-/michelson-morley-the-great-failure-63642. The two experiments of Michelson in Potsdam 1881 and Morley in Cleveland in 1887 was solid proof that the established ether theory, a theory of classical physics, was not valid. Their discovery, irrespective their own

Figure 11.13: Biosphere 2 building complex in Oracle, Arizona. The large hermetically closed green-house with glass windows hosts a tropical rainforest, savannah and desert climate regions in addition to a large water basin. The white buildings are connected and served as working and living areas for eight humans. The half-dome is the energy supply center.
Photo with fisheye lens by Artur Braun 2018.

nonsense from people who have little understanding of science, research and devel-opment. Biosphere 2 was the first major attempt to lock up a group of humans for a long time in a sealed environment without external support. It was predictable that such experiment will bring about numerous problems, both at the technical level and also at the social level. That was essential part of the mission. The future work is then to work these issues out in follow-up missions.

Figure 11.14 shows the inside view of the big green house which hosts inside also a large water basin. Another part of the greenhouse contains a section where the microclimate is like the rainforest, see Figure 11.15. The team of eight had to do their own farming for living. At some point, the food supply from the Biosphere 2 was so lean that the team needed to be reduced from eight to seven humans. Imagine this mission had been carried out on Mars where you cannot simply send one person out through the door back to normal civilization with all supplied that we need and that we know, including suffi-cient food and water.

Basically, low food supply means people would fight over food. Historically, this is certainly nothing new for mankind. In extreme situations, even canni-balism can occur, as can happen during war, or when an airplane crashes in the Andes (the survivors of Uruguayan Air Force Flight 571 in the year 1972), for example.

The humans in Biosphere 2 consumed not only food and water, they also con-sumed oxygen and produced CO_2. Over time, the CO_2 concentration rose too high. When you have a steel construction, it may pick up a lot of oxygen from the ambient

---

potential personal disappointment about the outcome, settled one of the most important scientific questions of the nineteenth century and thus set one of the most important foundations for the soon to come birth of relativity theory in the early twentieth century.

Figure 11.14: Inside the large greenhouse of Biosphere 2 is a large water reservoir. Photo Artur Braun 2018.

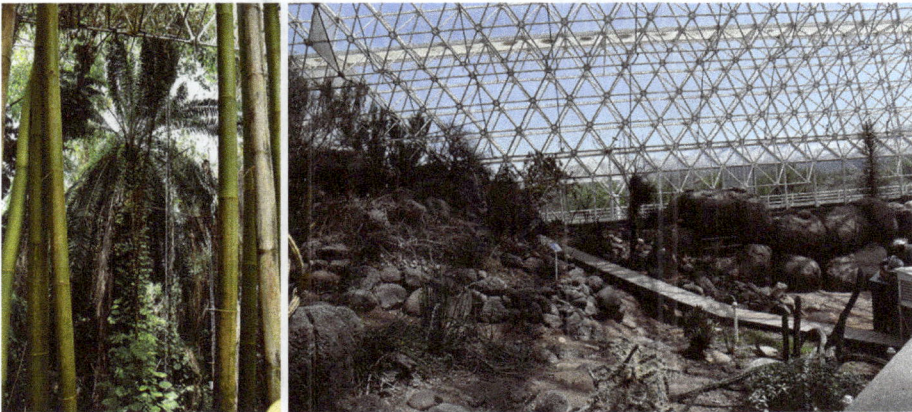

Figure 11.15: (Left) One section of the large greenhouse in Biosphere 2 has a tropic rain forest microclimate. (Right) One section in the large greenhouse of Biosphere 2 has a desert microclimate. Photo Artur Braun 2018.

inside and form rust. This oxygen is then not available anymore for the humans. And when the CO_2 concentration becomes too high, some plants may not function anymore the way as "scheduled."

Insects, ants, bugs and rodents may populate spaces in Biosphere 2 and cause more than nuisance. Living together in a team of eight over long time can cause stress. Somebody may cheat when it is about sharing the common food and so on. All this needs to be explored and in my opinion, the Biosphere 2 mission did a great job.

It is therefore necessary to carry out further such missions to learn everything about the occurring problems, including the social, cultural, psychological and anthropological issues. While this sounds like life in jungle, the eight researchers were actually living in a comfortable high-tech environment with single rooms and, for example, a well-equipped modern kitchen which is shown in Figure 11.16. One goal however was to always have enough self-grown food in that kitchen.

Figure 11.16: The kitchen for the crew in Biosphere 2 as it was used from 1991 to 1993. Photo Artur Braun 2018.

I should note for the fairness that the electric power for Biosphere 2 was provided from an external lifeline. This power was necessary because the Biosphere 2 complex has a huge infrastructure for acclimatization in its basement: pumps and pipes and heat exchangers and so on. The electricity did not come from any renewable sources but from the electric grid. Today, the campus around Biosphere 2 has a number of small renewable energy projects and installations and also a charging station for electric vehicles.

In so far, the living of the crew of eight was not entirely independent. Hence, there is a lot do for making a true independent next Biosphere 2. It is possible to power Biosphere 2 entirely with PV, for example. But this would require entirely new installations which would occupy a large sun collector area.

Biosphere 2 contains also modern laboratories, some of which are still used today for the training of high school students and college students. During my last visit to Biosphere 2 in April 2018, I visited one of the labs and spoke with a marine biologist whose prime interest was cyanobacteria. Figure 11.17 shows a snapshot of the "Marine Science Laboratory." To a large extent, Biosphere 2 looks like a showcase for the general public.

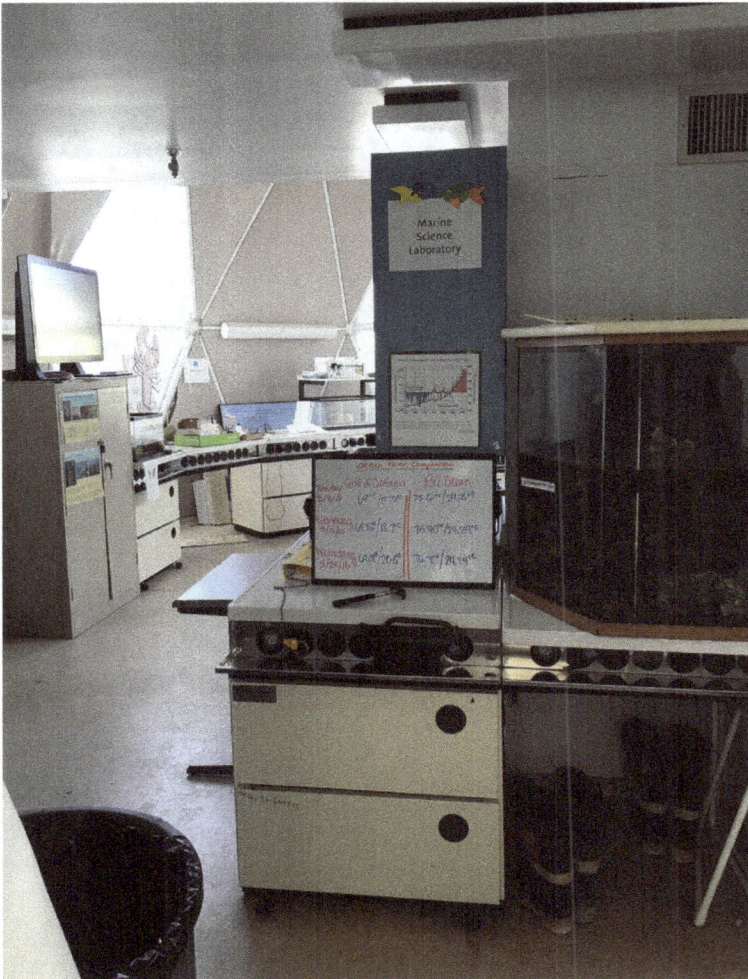

Figure 11.17: One of the laboratory rooms ("Marine Science Laboratory") in Biosphere 2 which today are used for student projects.
Photo by Artur Braun 2016.

Since several years, the Biosphere 2 is a training center, and education and conference center and also some kind of museum. Biosphere 2 has changed ownership several times during its existence and it is funded not by tax payers' money but by donations and by entrance fees and paid services such as catering and conference venue services.

Biosphere 2 is open to visitors almost every day during the year and provides also guided tours. Figure 11.18 shows a container where tomatoes are grown under controlled conditions pretty much like in a normal greenhouse. For live in a hermetically sealed environment such as in outer space or on Moon and Mars or in some hot or cold desert on Earth, or under sea, you have to control the environment and make sure that plants do not get diseases.

Figure 11.18: Container in Biosphere 2 for controlled growth of plants and crops. Photo by Artur Braun 2018.

In our readily existing Biosphere 1, our Earth, we can handle everything pretty well because we can move to different farming sites when one site is for whatever reason corrupted. A drought in one region in the world may be devastating, also an invasion by insects or an epidemic by plants diseases or on cattle farming – the whole subcontinent is usually not terminally affected by this. Biosphere 1 is still self-organized and it has been working well for billions of years.

We are not yet able to live self-sustained in a spaceship on an orbit around Earth. We have to bring food and water up there in order to survive. And, we have to protect ourselves against cosmic radiation. However, it appears that we can handle already the energy demands by using solar PV technology in space.

But even on Earth, we cannot be self-sustained yet as was demonstrated with Biosphere 2. The energy lifeline was necessary in order to power the environment stability. When you book a guided tour at Biosphere 2, the guide will bring you also to the basement of the complex where you can see the engineered facilities for air conditioning and so on. All this is not rocket science but rather conventional civil engineering.

Figure 11.19 shows two impressions how it looks like underneath Biosphere 2. The left photo in Figure 11.19 shows the "lung" of the complex which is under the half-dome structures in Figure 11.13. On the right in Figure 11.19, you see one of the narrow hallways in the basement with pipes and tubes which are part of the climate management system.

Figure 11.19: Environment and climate management infrastructure in the basement of Biosphere 2. (Left) The mechanical "lung" and (right) one of the basement hallways with tubing and piping infrastructure for climate management in Biosphere 2.
Photo by Artur Braun 2018.

A fully self-sustained and independent synthetic Biosphere is a highly complex project where the supply of energy is only one but essential part. It does not necessary have to be solar energy, but when it is electric, then we have a large number of degrees of freedom how to use them in the kind of technology we can master today.

The management of water and gases like oxygen, nitrogen, carbon dioxide and their absorption and transpiration in a photosynthetic cycle are engineering projects where agriculture and farming technology become relevant. Electrochemical technology is not yet ready for application in this biological context but the potential for using electrochemical methods for sensors and energy conversion and storage is definitely there and advised. My colleagues and I are working toward establishing a circular economy [Aro 2017] which could be one important part for a general strategy for a synthetic Biosphere.

It will depend on further progress in bioelectrochemistry and molecular agriculture where and whether we can expand artificial photosynthesis from a mere future energy conversion technology to a food production technology. From the perspective of humans on Earth, this development is maybe not yet necessary. But observations of technology and architecture trends such as urban farming lead me to think that future farming may look entirely different from what we are used to know as farming historically and today.

In parallel to this, researchers may try again to design new Biosphere projects. Once we can master new synthetic Biospheres, we can expand our civilization to outer space and other planets and solar systems. I believe this idea is not far-fetched. Space exploration is an ongoing science and technology, and space colonization will eventually follow as a practical application.

12 Electrochemical engineering and reactor design

12.1 Safety and security issues in electrochemical reactors

Electrochemical reactors are important for the industrial processing such as aluminum electrolysis, electrometallurgy, chlor-alkaline electrolysis and water electrolysis. In the laboratory, you may be looking at an electrochemical cell. At the factory or large industrial complex, you are looking at electrochemical reactors.

Industrial scale reactors are usually large. This holds for example for the huge electrolysis plants or small fuel cell and battery plants which we learned about in the previous chapters. In general, the danger for the general public and wildlife and for investments scales usually with the size of the reactor. Precautions for working with small or tiny lab scale cells include wearing personal protective clothing such as goggles, lab coat, and gloves and boots. These are also necessary when working with larger scale reactors, but there are additional safety measures required because of the size of the reactors.

A chemical reactor can pose a great danger to the general public particularly when it is located in a densely populated area such as in chemical factories like in Seveso [Fortunati 2004, Hay 1976, Sambeth 1983, 2004] or in Bhopal [Gonsalves 2010] or in petroleum refineries and related factories [Anonymous 1985, Berenblut 1985, Sissell 1998, 2003]. There can be many causes for accidents. For example, a tsunami can cause a flood which can cause a failure in a nuclear reactor and thus a major secondary nuclear catastrophe, such as in Fukushima in 2011 [Fujii 2011, Iino 2018, Nasu 2015, Ramana 2015]. And, there is also the risk of sabotage and terror acts in various industries [Bajpai 2007, 2008, Early 2013, Friedrich 2008, Helberg 1947, Vamanu 2014, Weisman 1987].

The main purpose of a company and firm is the gain of a profit from the manufacture and sales of a product or a service. It is therefore important to make a high quality product at low cost. The yield is also important. The design of the reactor (chemical, electrochemical, bio etc.) can have a large influence on the yield (the quantity) of the product and the quality of the product. You also want to make the reactor safe so that no costly damages can occur, and you want to avoid human casualties. The companies and their engineers and technicians who design and built reactors bear therefore a very high responsibility.

12.2 Role of mathematics in electrochemical engineering

The design optimization and operation of electrochemical reactors belong to chemical engineering. A substantial part of chemical engineering is the mastering of the necessary mathematics for the computation of chemical reactions and yield of products, for example. I can recommend here for example the works of Charles

https://doi.org/10.1515/9783110561838-012

Tobias and Norbert Ibl, and the textbook on Electrochemical Systems by John Newman and Karen Thomas-Alyea [Newman 2004].

For example, when you do water electrolysis for hydrogen production, you are interested in pure hydrogen with no traces of impurities. Maybe even water can be considered an impurity and you want to produce the hydrogen as dry as possible. You want to use as little energy as possible and therefore minimize electric losses in the reactor. Therefore, you use electrodes with a high electric conductivity.

Or consider an electroplating company which puts zinc layers on auto parts. When you carry out a simple (a seemingly simple) electroplating experiment in a custom built cell, you might realize that the zinc is not homogeneously thick coated on the template. This holds particular when the template, the auto part has an irregular shape. When it is curved, some region of the part is further away from the counter electrode than some other region of the part.

As the ionic resistance of the electrolyte depends also on the length of the electrolyte path, the most remote regions from the counter electrode may experience larger ionic losses and therefore less zinc is deposited over these regions. Those regions of the auto part which are very close to the counter electrode may be experiencing a faster deposition of zinc. In the end, the curved and irregular auto part is not homogeneously coated.

If you are interested in improving the homogeneity of the zinc coating, you maybe need to adjust the counter electrode shape so that the electrolyte path between working electrode (auto part) and counter electrode is identical for the entire part. When you have a flat square auto part to plate, you can take a flat square counter electrode of the same size and bring both electrodes in very close and parallel distance to each other. Then, the current distribution between the two electrodes is very homogeneous.

When you do not have a same size and same shape counter electrode, you may want to try it with just a rod-like counter electrode. The ionic current path is then of radial geometry. Depending how you place both electrodes to each other, the middle of the auto part may be coated thick and the regions to the left and to the right, which are further away from the rod counter electrode, have a thinner coating.

Another use of mathematical tools is the "Plausibilitätsbetrachtung," the plausibility check. You sketch a rough idea about what you want to build and calculate how much energy you can theoretically gain from the sun as primary energy source and then multiply this with the efficiencies from absorbers, catalysts and other components. For example, Cai et al. have calculated the design of an energy system which has to fit in an unmanned submarine [Cai 2010]. Or you can estimate how much reactors and components will cost and how much the cost will decrease in mass production. This can amount to a full techno-economic analysis. A very good example for such is found in the review paper by Helmut Tributsch [Tributsch 2008].

12.3 Parallel electrode plates in car starter battery

A very naïve design is the parallel alignment of two prismatic electrodes which are facing each other. This geometry is also very intuitive and easy to study from the mathematical perspective. We know this problem from the plate capacitor where the electric field lines point perpendicular to one electrode surface to the opposite electrode. The adding of electrode couples separated by electrode layers and membranes is basically the same like making a pile or a stack.

Figure 12.1 shows the manufacturing sequence of a VARTA® lead acid accumulator (VARTA® Silver Dynamic AGM) which is well known for its use in the auto starter batteries, which I have shown in Chapter 4 of this book. The manufacturing of the positive electrode begins with the making of a metal frame (mesh) current collector which is manufactured by a metal casting method. The electrochemical active electrode material is a paste which is bonded throughout and over that metal frame. This positive electrode comes then in a folder which is made from a microporous membrane separator material. The plate is basically put in a bag.

The negative electrode is also made from a paste of electrochemical active material which is bonded throughout a metal frame mesh. But this mesh is not made with metal casting method but with a stretch method. The precursor of the mesh is forcefully stretched and this is how the "grippy" structure evolves.

Eight positive electrodes and eight negative electrodes are piled up as a bundle in interdigital manner. The separators are already there with the positive electrodes. The electrodes with the same type polarity are electrically connected with cell connectors now so that the pile constitutes now one unit with eight times the area of one single electrode. This is here the purpose of stacking: using area more efficient.

Meanwhile, plastic containers large enough for one whole battery are made by injection molding. Each container is divided into six compartments which get holes punched through the separation compartments. The holes are necessary to receive the cell connectors when the cell bundles are inserted in one such compartment. The cell connector is reaching through the holes into the next neighboring compartment which will also receive such cell bundle and so on until the six battery compartments are filled and connected with cell bundles.

The cell connectors are then welded together electrically in order to maintain a firm mechanic and electric contact. The battery is now closed with an injection-molded plastic cap, which has two electric poles and six holes for battery acid filling. This top plate is then plastic welded with the battery container.

Now, the battery is being filled with the battery acid. When the liquid level is sufficient, the battery is being electrically charged; this is called formation charging. This warrants that the battery electrodes are brought in the right electrochemical condition (formation; conditioning). The acid holes are now being closed with a plastic welding process.

Figure 12.1: Geometry, design and manufacturing steps of the VARTA car starter battery. JC-8054_Fertigungsprozess_Silver_Dynamic_AGM_A1 Reproduced with permission from Johnson Controls Autobatterie GmbH & Co.KGaA.

In the last steps, the battery is being cleaned, then checked with a 300-A high current test and also mechanically tested. Then, the labels come on the battery including instruction information, and the poles become protected with a cover.

The same principle is realized also in PEM fuel cell stacks, where the electrodes (bipolar plates) are typically prismatic. The small SOFC stacks and also our previously mentioned supercap stacks have round flat cylinder-shaped disks as electrodes. The same holds for the membranes and separators.

The zinc coal elements have a different geometry. A cylinder beaker from zinc is also the mechanical container for the entire element. In the center of the beaker, in line with the central axis of the beaker is the solid carbon cylinder stick. This is a radial or azimuthal geometry. The electric field lines are pointing from the cylinder radial to the coal stick.

The "Swiss roll" is a sandwich of flexible electrodes and separators or membranes which you stack together like one electrode foil, such as carbon cloth or a coated aluminum foil on the bottom, one flexible separator on top of the electrode and then another flexible electrode with the same size and geometry on the top. Now, you roll the three components together to a coil, a roll. To the best of my knowledge, this concept was first presented by Robertson et al. [Robertson 1975].

12.4 Primary and secondary current distribution

When you want to plate a rod-like cylinder homogeneously, you maybe bend a large sheet of metal to a wide hollow cylinder as counter electrode, with height H and radius R, and place the rod with length $L = H$ in the axial center of the cylinder. All points on the rod which you want to plate are then in equal distance from the cylinder counter electrode.

The extent of homogeneity and inhomogeneity of plating can also depend on the concentration and thus conductivity of the electrolyte. When the electrolyte conductivity is very high, you may be somewhat more relaxed about geometrical adjustments of your electrode geometry. When the conductivity is very poor, you want to bring the part as close as possible to the counter electrode.

The geometric shape and placement of the electrodes are therefore important for an efficient and homogeneous plating process. The first key word here is the (ionic) current distribution. I have mentioned in the various previous chapters the current density j, which is the total measured current I related to the total exposed electrode area A in the electrolyte.

j certainly is only a geometrically averaged value. If we were able to measure the current density at one single specific point x on the electrode or on any arbitrary (actually, all points) position, then we could introduce the relative local current density $j_x/j = f(x,y,z) = f(\mathbf{r})$.

When there is no significant overpotential involved in the electrochemical reaction, we call the above expression the primary current distribution. When the concentration overpotential is negligible but the reaction overpotential is significant, than we will measure the secondary current distribution. For a list on terminology in electrochemical engineering, see Ibl [Ibl 1982].

When you consider the slope (or the differential) of the characteristic of the overpotential and the current, $d\eta/dj$, and multiply it with the conductivity κ of the electrolyte, you obtain the global electrochemical characteristic of your system except for a characteristic length l of your electrochemical reactor, specifically the electrode. You obtain then a dimensionless number called Wagner number

$$Wa = \frac{\kappa}{l}\frac{d\eta}{dj},$$

which is a measure for the secondary current distribution [McNaught 2014]. The electrochemical cells which researchers are typically using in their laboratories are not optimized for an optimum current distribution.

12.5 Solar hydrogen reactor concepts

12.5.1 The EPFL cappuccino cell

The electrochemical cells in laboratories are typically made from glassware including quartz glass, for example. Sometimes Plexiglas® (polycarbonate) is a good choice as well. Glass can be very inert and another advantage is that you can peek into the region where the reaction is taking place. You can see whether the electrolyte changes the color or whether the electrode changes the color during reaction. You can see whether the electrode has the right depth into the electrolyte and many more things. These cells are very useful for very precise inspection of conditions and reactions.

Here, I am presenting a cell which was designed for high throughput screening of photoelectrodes for PEC applications (Figures 12.2 and 12.3). It was originally designed by the group of Professor Grätzel at EPFL Laboratory for Photonics and Interfaces, and in the event of a large European project on PEC (NanoPEC, [Augustynski 2009]) with many partners worldwide, it was decided that all partners should use this cell design so as to be able to compare all data from measurements from different project partners with ease. We at Empa then volunteered to have our machine shop build several dozens of these cells and distribute them – for the cost of manufacturing only – to research groups on all continents.[1]

---

1 We actually lent out one of the cappuccino cells to Mareike Gast on the occasion of the Designers' Open Festival in Leipzig in October 2012 [Hartmann 2012] Hartmann J, Neubert A: Designers' Open. In

The advantage of this cell is that glass slide PEC electrode specimen can be clamped in between two titanium leads in a holder and fixed with screws. The holder can easy be inserted in what is now being called "cappuccino cell." The cell body and sample holder are machined from costly but corrosion-resistant PEEK® plastic (PEEK – polyether ether ketone).

The change of samples in this cell is quick and easy. This is why, the cappuccino cell is being used for rapid screening of electrodes. The name cappuccino cell originates maybe from the fact that the walls of the cell are thick and that the volume inside for the liquid is rather small. The milk coffee color of the PEEK® fits to the cappuccino description plus there is a "cap" (cap -> cappuccino, something on the top), a PEEK® lead which can close the cell. The small electrolyte volume may be a disadvantage because of cleanliness. A larger electrolyte volume may be easier to distribute and dilute impurities than a small one.

The black plastic cap which you can push over the aluminum conus has an aperture with exactly 1 cm area. This was a requirement set by a smart designer because it makes determination of the light current density easy:

$$\pi R^2 = 1\,\text{cm}^2$$

so that the radius R equals

$$R = \sqrt{\frac{1}{\pi}}\,\text{cm} = 5.6419\,\text{mm}$$

Therefore, the current which you read with the potentiostat (say e.g. 31.25 mA) from the sample in an illuminated area of just 1 cm^2 yields a current density of 31.25 mA/cm^2. This is a practical trick for that you need not make any extra size corrections for the area of the sample. Note, however, that this holds only for the light current. The dark current does not worry with the aperture of the cap. The dark current is obtained from that portion of the sample which is exposed to the electrolyte. This must always be kept in mind.

Some senior researcher in our NANOPEC project [Augustynski 2009] worried that the light could cast a corona on the sample behind the aperture so that the light exposed area would be actually somewhat larger than 1 cm^2. In order to check for this effect, we made a set of black caps where the aperture was smaller than 1 cm^2, with the smallest aperture having 1 mm radius only. We then took one hematite photo-electrode and measured its light currents when we had the different apertures attached. It is an exercise for the reader to calculate the influence of the corona on the light current. You do not know how broad the corona is. You have the radius of the smallest aperture (1) given as R_1, and the corona has a width of say ΔR so that the

Das Festival für Design Leipzig (GmbH DO ed. Leipzig: Designers' Open GmbH; 2012. Frau Gast, an industry designer with a design studio in Frankfurt am Main, was one of the curators of that design festival and exhibited the cell for us. She also showed a poster with the new Teflon cell.

Figure 12.2: Spectrophotoelectrochemical analysis cell from PEEK® plastic designed by EPFL, the so-called cappuccino cell. (Top left) Cell body from front view. The aluminum ring is fixed with four metal screws on a quartz window and a rubber gasket for sealing. The black aperture plastic cap is pushed on the aluminum disk and defines the entrance window for the light from the solar simulator. The diameter of the hole in the black cap is 5.64 mm, corresponding to 1 cm² area. The small piece on the top is the sample holder which is made from two PEEK® pieces which contain titanium 90° bent, between which the electrode is clamped by tightening the two visible white plastic screws. With two metal screws on the top of the clamp, we make electric contact with the FTO glass of the electrode. A crocodile clamp can be connected with one of the metal screws. (Top right) View into the cappuccino cell from the top. Two large holes on the left and right side of the cell allow for inserting glass tubes of a reference electrode and a counter Pt electrode into the electrolyte. The dovetail guide shape on the back is used to fix the cell to a stative. (Lower left) View from the top where the sample holder piece is now inserted into the cell. The glass slide would be fixed now between the two titanium metal leads. (Bottom right) An extra piece of PEEK® is put over the sample holder in order to close the cell. This feature has never been used in our experiments.

totally illuminated area is $A_1^{\text{light}} = \pi(R_1 + \Delta R)^2$. The expected light current with the next larger aperture (2) is then $A_2^{\text{light}} = \pi(R_2 + \Delta R)^2$ and we expect that the width of the corona ΔR is the same for all apertures. We carried out the systematic study and found that the width of the corona cast by the aperture must be negligibly small, at least for the sample which we used. Therefore, no corona correction was necessary for the determination of the light current densities.

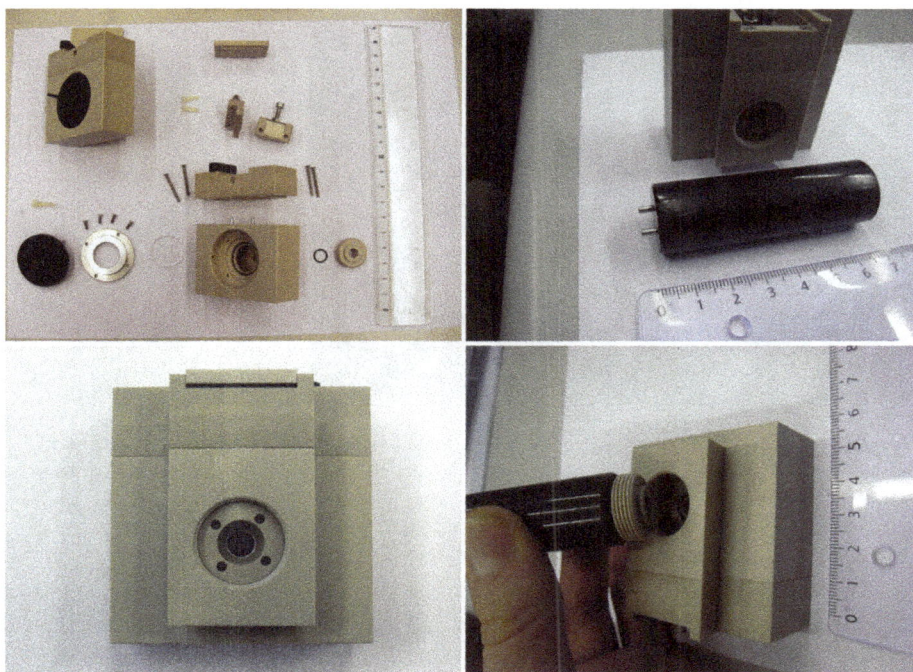

Figure 12.3: (Top left) Assembled and a disassembled cappuccino cell with all components spread out, alongside with a ruler in centimeter scale. (Top right) cappuccino cell from rear view, plus a wooden tool with two metal pins, which helps remove the back screw, as shown in the (bottom right) image. The back screw from PEEK® pushes a glass slide on a small black rubber gasket for sealing the cell on the back. The back screw has a hole (bottom left) which warrants a free optical path throughout the entire cell so that the photoelectrochemical experiment can be accompanied by a parallel UV–vis experiment on the same sample, for instance.

12.6 PEC reactor for sealed GC measurements

As part of my equipment grant [Braun 2008d] at Empa, we purchased also a gas chromatograph (GC). At some point, we felt that we should connect the PEC cell to the GC so that we could determine the mass of the evolved gas during water splitting, in parallel to determining the photocurrents.

For this, you need to design a PEC cell which is sealed and gas tight. It was Asst.-Prof. Dr. Krisztina Schrantz and Dr. Rita Toth in my research group who designed, along with the design department at Empa, an extra PEC cell specifically designed for gas evolution experiments with the GC. The machine shop made one such cell and it is shown in Figure 12.4. This time, the cell was made from Teflon (polytetrafluorethylene, PTFE), as you can see from the white color.

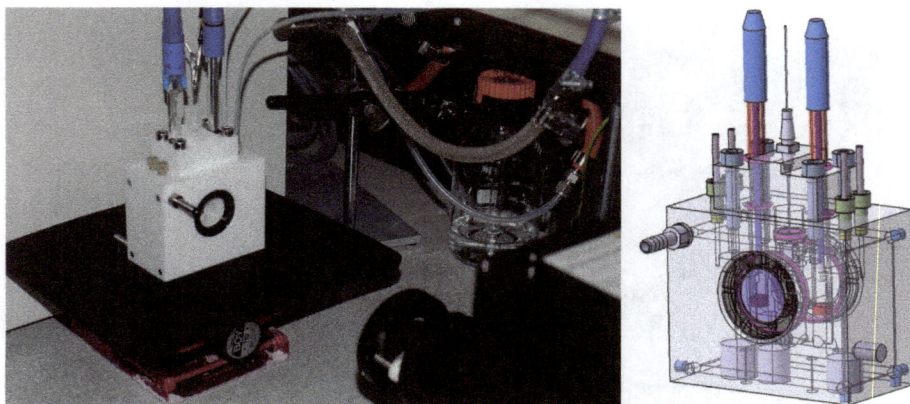

Figure 12.4: Hermetically sealed photoelectrochemical cell from Teflon® for using with gas chromatograph. (Right) Design of the hermetically sealed photoelectrochemical cell to be connected with a gas chromatograph.

The two glass tubes with reference electrode and counter electrode can be pushed tightly through two holes in the Teflon® body and it is gas tight. The sample holder too is gas tight and has the working electrode sample underneath, with an electric feedthrough through the holder to allow for electric contact with the potentiostat.

The cell has inlets for the GC carrier gas and outlets for the gas. The cell can be heated and cooled with a liquid from a thermostat. Two metal pipes, one at the front side and one at the rear side, can be connected with tubing or hoses to a potentiostat and the temperature of the cell can be controlled.

The cell is in figure in front of the solar simulator. When line shines on the photoelectrode inside the cell, filled with electrolyte, the potentiostat will record the light current as a function of the bias potential supplied by the potentiostat and controlled by the reference electrode. When there is oxygen and hydrogen gas evolved, the carrier gas from outside will transport the evolved gas in the cell through separate tubing to the GC, which is separately computer controlled and which will record the amount of the evolved oxygen at the photoanode, for example.

The right image in Figure 12.4 shows a sketch of the cell from the design department, where all features become literally transparent for the interested observer. Teflon has many advantages as a material for cells, but one disadvantage is its porosity. The material may absorb some kinds of molecules which over time may surface and mix to the electrolyte. I have never tested this but I was told this could happen. So if you use a Teflon cell and see in your cyclic voltammogram signatures, potentially they might come from some alcohol that you used to clean the cell before experiment, and which was trapped in the Teflon material and slowly is picked up by the electrolyte. During my PhD thesis work, however, I never

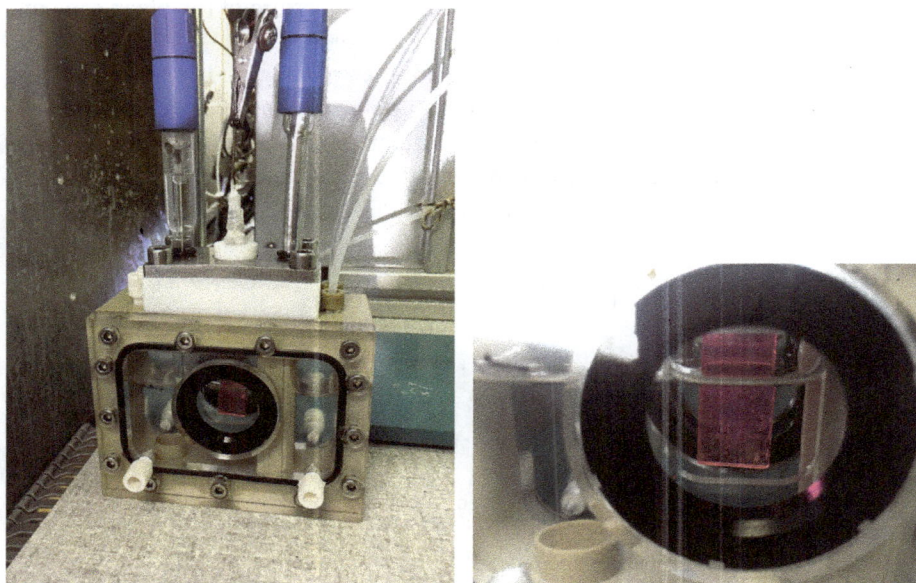

Figure 12.5: Photoelectrochemical cell from PEEK® plastic body and polycarbonate front and end plates for use. In the left image, you see how the electrode is not fully dipped in the electrolyte. This is ok because the electrode is coated with Lacomite® varnish and only a particular area in the electrode center is not coated with the varnish and thus in direct contact with electrolyte.

observed this but did not pay attention either. If this effect is really there, it may be very small.

A next generation PEC cell for use with GC was made with PEEK® plastic body (Figure 12.5), and with a transparent polycarbonate back panel and front panel. It is a great advantage when the cell allows for as much as possible peeking into it so that you can observe bubble formation or color changes or anything that you may find noteworthy for the interpretation of the data curves that you see in the monitor from the potentiostat and from the chromatograph.

You should check the entire experiment for integrity. When there is one cable not properly attached, you may see an I/V which you cannot interpret properly and you may come up with a wild interpretation although the only important thing is that the counter electrode was not properly connected or that the working electrode had no contact with the current collector of the sample holder. Or the working electrode did not dip deep enough in the electrolyte, or the working electrode fell off into the cell and there was no electric circuit closed at all. With a transparent cell, you can spot many mistakes easier.

One disadvantage that brings the transparency of the cell in Figure 12.5 is that a large light beam (large cross section) shines over the entire sample inside the cell. In the cappuccino cell, only a well-defined area by the aperture of the black cap cone warrants that the sample is exposed to light by that area.

Typical chromatograms for H_2 detection are shown in Figure 12.6. Hydrogen was forced into the system with 50 ppm and 100 ppm concentration as reference. A silicon-doped hematite photoelectrode was operated for half an hour in one of our PEC GC cells and the evolved hydrogen measured with the GC. As you can see, the cumulated hydrogen volume increases over the PEC operation time.

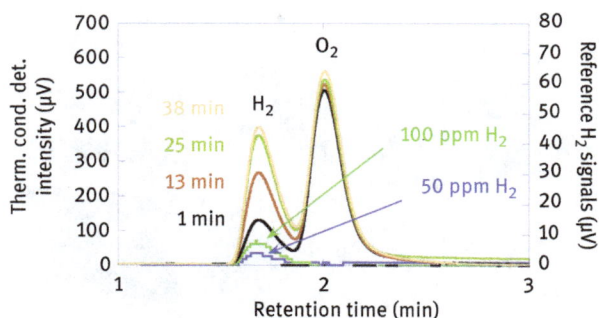

Figure 12.6: H_2 chromatograms recorded during 30 min of H_2 generation from 1% Si-doped α-Fe_2O_3 with sub-monolayer of cobalt cocatalyst [Bora 2013].

It certainly took us some efforts to set up this combined system with potentiostat and GC and the necessary cells. But I was surprised over the quite small number of research groups in the world that actually had such systems in their labs. I heard complaints about leakages in the pipes and tubing and hoses. It is certainly not always easy to make out the spot where the leak occurs, but leak detection is a common task in laboratory practice.

I have been trained in ultrahigh vacuum physics [Braun 1996] and I know therefore that there is no absolute vacuum possible. Leaks exist always. Important is that you are aware of the leaks and that you somehow can control them. We used in our GC system the minute leak from air in the system the ratio of ~20% oxygen over ~80% nitrogen. This gas mixture in air is a marker which you can use for calibration of the system. While this approach is of course not 100% accurate, it gives you still gas concentrations in your PEC experiment which can complement the electrocatalytic data very well.

In the end, via Faraday's law, the electric current should be balanced by the evolved gases. When you want to find out how much of the current or electric charge at which potential is used for oxidizing the water and how much is lost by anodizing the photoelectrode or other components in the system, you hardly can solve this task with the potentiostat alone.

12.7 Large scale PEC reactors

12.7.1 Starting from scratch: Design of a primitive structural model

In October 2012, I was invited to participate in the AMPEA Workshop on Artificial Photosynthesis [Holzwarth 2012] at one of the Max Planck Institutes in Germany (Max Planck Institute für Chemische Energiekonversion in Mülheim an der Ruhr). Much, actually most of the time, in this workshop was on the chemistry of photosynthesis, photocatalysis and electrocatalysis, photovoltaics, physical processes and so on: all the stuff which we physicists and chemists do like. One of the more eminent participants cautioned us that now were the time to build devices and focus not anymore only on theoretical questions and issues.

When I was in the train on my way back from Mülheim a. d. Ruhr to Switzerland, I remembered what the Professor from Uppsala University had said. During the train ride, I was deliberating how to proceed with my solar fuel research. I became convinced that I should build a large PEC reactor. I will continue to illustrate in this section how the making of such reactor evolved. While I am writing this section and search for evidence, I find in my files in my PC that the first plan for the reactor was made on October 13, 2012, one day before I left for the workshop in Germany.

How do you plan for such project? You simply start thinking, sit down and start writing and sketching. I did it literally on the back of an envelope, as you can see in Figure 12.7. At home, I found an old CD (compact disk) holder from IKEA for which I had no further use. It had a size of somewhat larger than 10 cm × 10 cm and therefore was as large as some FTO panes that were available from solar cell suppliers.[2] It was made from a not too hard plastic material which I could cut with a scalpel. I was therefore simply thinking on how I could make from this and some other cheap material near my hands something that looked like a PEC reactor. At this early and premature stage of development, it was not necessary to make a functional device.

What I wanted was simply a structural device, a design which should show to other people how the later device could look like. As I am writing these lines here in late February 2018, I am looking forward to attending the 88th Geneva International Motor Show (compare [Wood 1993]) on February 8, 2018. I will go there with our Hyundai ix35 Fuel Cell car. I have been joking sometimes about the difference between structural models and functional models with reference to the Automobilsalon. We see the beautiful new car models in the showroom – but we never see them driving there, right? Maybe they even have no engine. What counts at this stage – on stage – is that the cars are very good looking. This holds for many things which are being engineered.

---

2 The size 10 cm × 10 cm is very suitable for making devices because many components in the solar cell industry have that size.

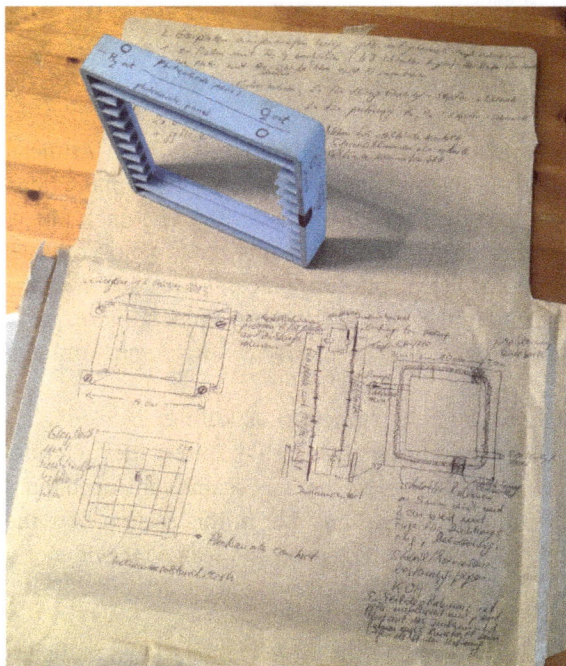

Figure 12.7: First thoughts written down "on the backside of an envelope" for what is necessary to build a simple PEC reactor model, along with sketches on how to add components to the PEC reactor frame, shown as the blue plastic CD holder with cutout center and pencil marks on where to drill holes for gas inlet and outlet.

Now as I am writing this, I remember a story which Richard Feynman told at the Caltech commencement address in June 1974. He refers to the so-called Cargo Cult on the Melanesian Islands of Vanuatu[3] [Jahan 2016, Meuzelaar 2015] around the Guru John Frum [Crowley 1996, Guiart 1956, Macclancy 2007, Mondragon 2009, Raffaele 2006, Tabani 2010, Wood 1976], where ignorant indigenous Islanders would build

---

3 Vanuatu is a group of 83 small islands which are scattered over a stripe of 1300 km in the South Pacific, almost 2000 km northeast of Australian Brisbane. Vanuatu has almost 300,000 citizens and was decolonized from British and French New Hebrides in 1980. In the United Nations Human Development Index, Vanuatu ranks with 0.597 on the 134th place, which is a medium human development level. During my appointment with the Consortium for Fossil Fuel Liquefaction Science (CFFLS) at University of Kentucky, I collaborated with Dr. Henk Meuzelaar, a spectroscopy expert from University of Utah who authored numerous books on spectroscopy and also on carbon. Henk along with his dear wife Nelleke, both are Dutch, from the Netherlands are the founders of the Project MARC – Medical Assistance to Remote Communities in Vanuatu. During our project meetings, Henk explained me the harsh living conditions of the indigenous people in Vanuatu. On the more remote Islands, people may suffer from malnutrition and poor medical treatment.

landing stripes and wooden models of airplanes in order to invite the Gods back on the Island who used to bring food and utilities on the Island long time ago – when the U.S. military supplied the Island. As the Islanders were not engaging in any kind of scientific thinking, they were unable to learn, to understand, to gain insight that their wooden planes (structural models) would not result in any aircraft or flight (functional models), and thus in no supplies [Feynman 1974].

When you read my previous book on X-ray spectroscopy and electrochemistry [Braun 2017], you will come across a section where I present some L-edge spectra of a structural model of hydrogenase [Bora 2013] which I had received from my colleague Prof. Xile Hu at EPFL. The molecule that you see in Obrist (2009) is such structural model. When I discussed with Prof. Hu his research on hydrogenase, he made me aware that his approach was to first mimic the structure of components of the photosynthetic apparatus before he would eventually proceed – and succeed – with presenting or synthesizing an actual functional model which would be able to produce hydrogen or contribute to the production of hydrogen. Writing this section here, I quickly looked up his updated publication record and found that he recently published results on an actual functional hydrogenase model which he had synthesized [Xu 2016].

I think I have now explained well enough what it means to make a model and this I will continue with the PEC reactor model. At some point, I decided to not call it anymore PEC cell, but PEC reactor.[4] Figure 12.7 shows the blue CD holder from which I have cut out the center part. What is then left is something like a frame which can hold two glass panes. The two glass panes do not have to be functional photoelectrodes. It is enough when you take simple float glass and give it a reddish furnishing with red paint (Figure 12.8), if red is the color that you have in mind for the absorber material. Note – what counts at this point is only the color, not the function.

We are now basically already scaling up the components of a PEC reactor from the small laboratory scale experiments shown before in the chapter on photoelectrochemical cells. We therefore need larger equipment, for example furnaces which can contain the large glass panes.

Figure 12.9 shows such panel in a box furnace. You can see the red color of the hot glowing heating elements. The 10 cm × 10 cm wide panel fits nicely in the furnace. When we succeed to stack the panels with some technical aids, that is, piling them up, then we can produce several electrodes at the same time. We can even use wider

---

4 Here is my advice to the readers who are passionate about the MINT subjects – Mathematics, Informatics, Natural Sciences, Technology. It is very important, once you get out of the lab and speak with people who are not from your field, what language you use. The very successful colleagues whom I know are also very skilled in the language arts. Language helps you to communicate your ideas with the outside world. This concerns the written language and the spoken language, and also the graphical language. When you are a gifted writer or speaker, you can turn this in science into an asset.

Figure 12.8: A wide glass pane of 10 cm by 10 cm coated with an iron-ion containing precursor material.

Figure 12.9: By courtesy of Asst.-Prof. Krisztina Schrantz (Empa and Szeged University) and B.Sc. student Florian Häusler (Empa and TU Freiberg).

panes to make larger units, but the maximum size in this box furnace is 20 cm × 20 × cm. We therefore ordered also a package of FTO glasses of that large size.

The distribution of the temperature in such large furnace is less homogeneous than in a small tube furnace where you have smaller samples (does this remind us of the primary and secondary current distribution? Anyway, this again is an engineering problem). Scaling up can require often changing various parameters, which have to be optimized again before good results are obtained. The right photo in Figure 12.9 shows a 10 cm × 10-cm large glass pane with iron oxide absorber layer, and the top of that layer is making flakes which are peeling off.

The design of the PEC reactor was continued and finished by PhD student Florent Boudoire (Empa and Univ. Basel). He met with engineers of Empa's design department and thought about how to assemble components, explained which purpose each component had and under which conditions they should operate. It is very important that you sit together with the designer vis-à-vis and explain what the whole idea is about.

Designers (I think the proper German word is "Konstrukteur"; you may also call them engineers) which you would meet in an office sitting in front of a PC at a research lab have typically been trained for many years in a machine shop before they decided to become designers. They have hands-on skills and practical experience with the needs of their very various clients and customers. They have also great experience with materials, be it polymer, metal, wood, glass, ceramics and they know what these materials can do and what not. Design engineers know about their corrosion resistance, temperature stability and mechanical toughness. They know low-cost solutions and expensive solutions. They know solutions that must last several years and solutions which need no long durability. So, I advise everybody who wants to build a reactor to seek discussion with and counsel from a design technician or engineer.

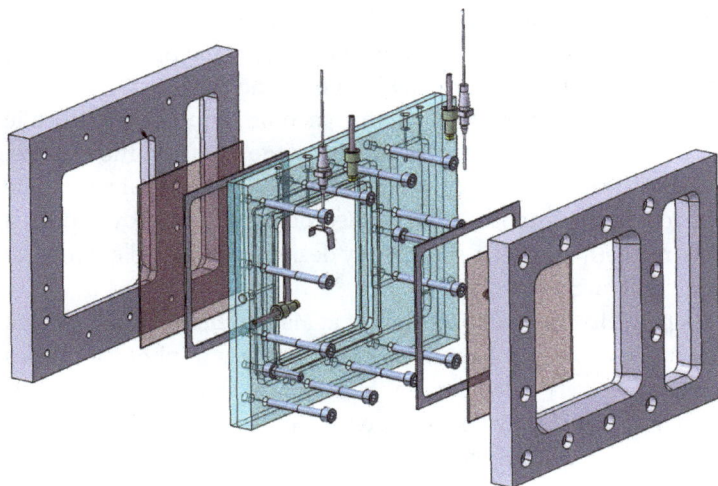

Figure 12.10: Basic structure of the PEC reactor prepared by computer-aided design from the construction shop at Empa (Heinz Altorfer, Florent Boudoire).

The reactor, as displayed in Figure 12.10, has as the two major structural components two massive frame plates from stainless steel (gray). With bolts, they clamp together one iron oxide photoelectrode coated on FTO glass of 10 cm × 10 cm size (reddish). The glass is pressed on a center frame from Plexiglas® (green), with a quadratic flat rubber gasket in between them. On the other side, the frame is closed with a second glass plate and gasket. The two glass plates and the frame present the electrolyte container. The counter electrode is in an extra compartment in the Plexiglas® frame, which has some rigid separator against the electrolyte volume where the oxygen is produced. Two horizontal holes in the Plexiglas® frame allow for replacing electrolyte, and two holes on top frame allow for collection of oxygen and hydrogen. Two smaller holes are feedthroughs for the electric wires for biasing the cell at photoanode and counter electrode, which is a long platinum wire. The complete and assembled PEC reactor is displayed in Figure 12.11.

Note that nothing in this reactor is optimized yet for function. The only requirement was that it must be able to produce hydrogen – even in minute amounts. Optimization of components comes later. The two steel plates are too massive for future use, for example, but for experimenting that it is better, the reactor has a heavy weight so the wind cannot blow it away. The counter electrode, when in future still platinum, will make the reactor an extremely costly device. So, we will find a low-cost alternative for this. The photoelectrodes have a current density which maybe not worthwhile to mention; but we chose a material with a formulation which we have in the shelf already. Note that we use no polymer separator membrane in this design. The gaskets which we use are flat and on flat support. For an ideal sealing, we would have the machine shop make groves as advised in the industrial standards for sealing and gaskets. The second glass pane can be also chosen to be a light absorber. We can use this prototype and make our first experience with it while other researchers work on optimizing these components in their lab. Once they have come up with a better solution, we can implement the better components in our reactor design.

While the reactor prototype is being made in the machine shop, it is worthwhile to experiment already with the large electrodes that we made in the muffle furnace. Providing a proper electric contact is one of the many technical issues that need to be solved. The cappuccino cell is designed that the electric contact is provided by a simple mechanical clamp method. This is specifically designed and good enough for fast diagnostic screening of many samples.

When I was on sabbatical at University of Hawaii at Manoa [Braun 2010], I learnt they used a different method. They soldered the end boundary of photoelectrodes with indium metal. Indium is very soft, more so than the conventional tin which is used for soldering. It requires soldering temperatures lower than for tin. This is good to prevent heat damage to the electrodes. You purchase a coil of indium and use a conventional soldering tool for distributing a tiny amount of In over the entire area which you want to coat. Then, you coat a metal wire with In. In the third step, you solder the wire and the electrode together which will provide an excellent electric contact. Because In is so soft, the contact is mechanically not stable. The wire will fall off from the electrode unless you put extra glue such as epoxy resin over it.

Figure 12.12 shows the large iron oxide photoanode with a rim of In[5] for better electric contact in the reactor shown in Figure 12.11. It was not necessary to provide

---

5 You may have only Ag silver paint at hand and think this would also be a good material for electric contacting. We had such a case where a researcher chose silver, but over time the silver

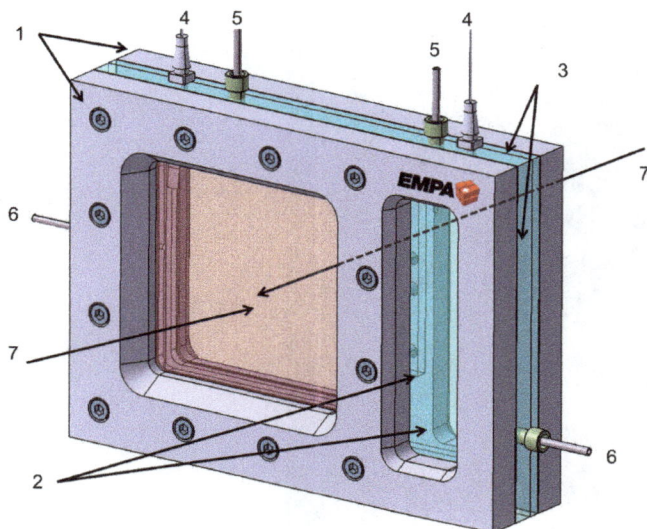

Figure 12.11: CAD sketch of the complete PEC reactor as designed by Heinz Altorfer and Florent Boudoire: (1) 12 metal bolts keep the frames together and apply the pressure on the reactor for sealing and electric contact, where necessary; (2) separate compartment in the electrolyte volume for holding the Pt wire as counter electrode; (3) Plexiglas® frame which holds the electrolyte volume; (4) electric feedthrough for the Pt counter electrode; (5) small exhaust pipes for the collection of the evolved hydrogen gas (right) and oxygen gas (left); (6) small hoses for inserting and extracting electrolyte and (7) two glass panes which close the electrolyte volume, one of which is the photoanode.

an extra wire contact with this large electrode because it will be pressed on a current collector when the PEC reactor is assembled. The larger the electrodes, the easier they break. They may break already in the furnace when the absorber material is being sintered on the panes. The proper control of the process parameters is therefore very important for production of vital components such as photoelectrodes. And certainly, a very important question is: what are the process parameters? These have to be identified as well. You would say that the temperature is the most important parameter. You want to make sure that the temperature is homogeneous around the electrode. The same holds for the atmosphere around the electrode; the gas concentration is the second process parameter which needs to be controlled and optimized.

Other parameters may be the choice of substrate, because the glass panes with the transparent conducting oxide coating may have different resistivities. The scaling

---

changed its color to brown. Silver likes to react with molecules from the environment and forms, for example, AgS. Therefore, I would always choose indium or tin over silver for the aforementioned purposes.

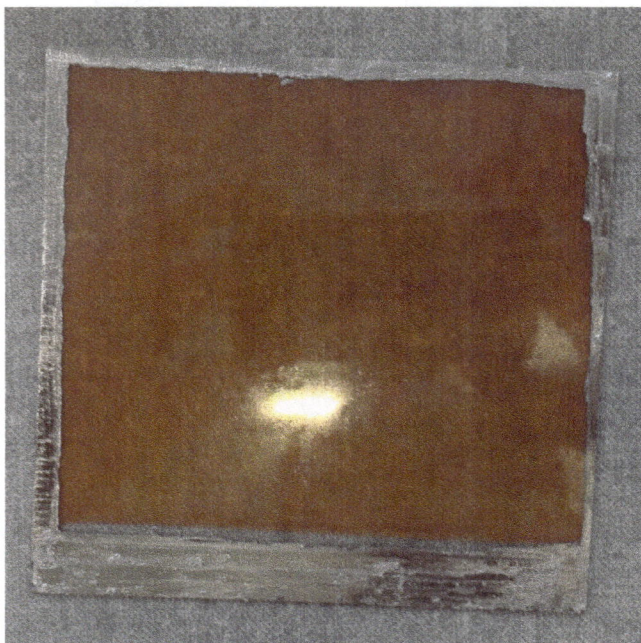

Figure 12.12: Large iron oxide photoanode from α-Fe₂O₃ coated on FTO glass, 10 cm × 10 cm size. The boundary is soldered with metal indium as current collector.

up of a system from the laboratory to a device, even a small one like shown here, is very time consuming and often comes with disappointment. Behind every product can be a long and painful way of development.

A few sentences ago I mentioned the easy breaking of large electrodes. In Figure 12.13, you see the larger part of such broken electrode connected to a power supply and dipped in an electrolyte container. This is a "quick and dirty" experiment which shall serve to find out the general behavior of a large electrode when biased in the dark and under light. A simple 1 L glass beaker is filled with 1 M NaOH as electrolyte. The filling level is set that just the iron oxide coating of the electrode is soaked in electrolyte.

A small crocodile clamp has contact to the upper FTO rim of the electrode and is connected with the power supply which provides the electric bias. Another crocodile clamp holds a platinum sheet which is dipped in the electrolyte. A large clamp holds the crocodile clamp tight to the glass wall so that it does not drop in the electrolyte. The two red cables lead to the power supply and one multimeter. Position and distance of both electrodes are fixed, but neither controlled nor optimized.

The beaker is in a tray so that electrolyte spill will be recovered. As we are using strongly caustic NaOH, we must wear safety goggles [Paul 2008, Williams 1972, Young 2000] to protect our eyes from injuries. The beaker is lifted on a stone in

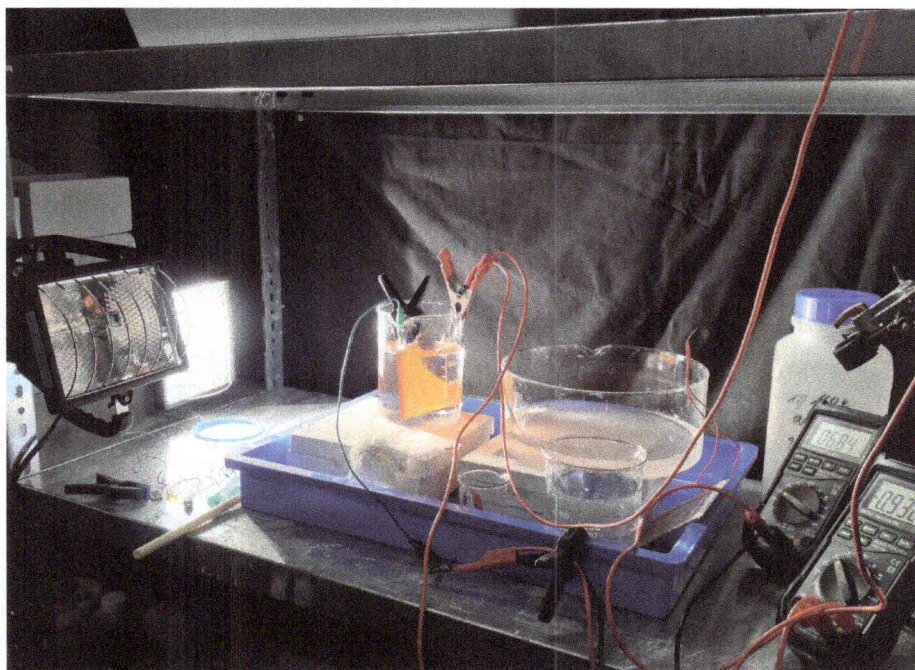

Figure 12.13: Large piece of iron oxide photoanode dipped in 1 M NaOH and connected to power supply for quick electrochemical assessment. One halogen lamp is switched off in order to allow for taking the photo.

order to raise its level to the focus of the halogen lamp, which is used to illuminate the electrode and thus do a photoelectrochemical experiment. When you pay close attention to the photo, you can retrace which cables come from the power supply to the photoanode and counter electrode, and which go to the multimeter for voltage reading and which go to the multimeter for the current reading. This is left as a practical exercise for the reader.

With one large electrode illuminated by two halogen lamps, the photocurrent was around 2–3 mA at around 1.6–1.7 V bias. As this seemed not overly high, I tried to improve the electric contact between clamp and iron oxide by soldering the afore-mentioned indium layer. Then, I used two clamps on the left and right side of the large electrode as I was expecting a better current distribution. Still, the photocurrent for this setup was 2–3 mA. Poor electric contacting could not be the reason for the low photocurrent. You would certainly divide the photocurrent by the total exposed electrode area. Note that we have here two different kinds of exposure.

First, the exposure is with respect to the electrolyte. We have to dip the electrode fully into the electrolyte without exposing the indium layer or FTO to the electrolyte. Current that we measure shall only originate from the iron oxide exposed to electrolyte. Second, the exposure is with respect to the light source. When there are electrode areas in the shadow, they may be exposed to electrolyte and thus contribute to the dark current, but as they are getting little light, their contribution cannot be fully considered for a photocurrent. When these questions of exposure are correctly answered, then it makes sense to divide the total current by the exposed area.

I have to make a remark on the illumination for this experiment. As we have now large area photoelectrodes, we need also a large area light source. The light source has to be proportionate with the sample that is to be illuminated. The photoelectrochemical cells which I showed in one of the previous chapters were very small, with an active photoelectrode area of around 1 cm^2 only. For this, we could easily use a solar light simulator, which works with a xenon lamp of 500 W, and the light is focused on the sample with an optical lens. This is relatively expensive equipment and costs in the same range like a potentiostat.

Figure 12.14 shows such 500 W halogen tube in the rail of a local supermarket. The two lamps to that kind of tube are shown on the left side in Figure 12.13. For the large electrode setup shown here, I used several low-cost halogen lamps which may cost around 10 or 20 Swiss Francs each. They have bulbs with a power of 500 W, but they are not focused on a small area. Certainly, this light source is not calibrated to 1.5 air mass intensity, but I feel this is not necessary here. Calibration is a fine tuning which can come at a later stage. I also want to remark that I used no potentiostat but a low-cost power supply and two low-cost multimeters. The setup shown here is very simple and costs little money. Meanwhile, the expensive and calibrated equipment can be used for the smaller samples in the other laboratory, where materials development or in-depth analysis is required.

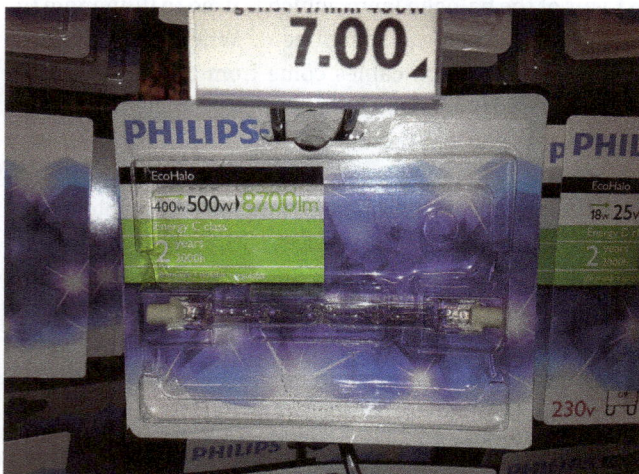

Figure 12.14: Low-cost halogen light tube with 500 W and 8,700 lumen light output used for the illumination of the large area (10 cm × 10 cm) photoelectrode.

A slightly improved setup for recording dark current and photocurrent data from large electrodes is shown in Figure 12.15. Instead of the cylindrical narrow beaker, I chose a wide prismatic glass container where I could fix the electrode easier. I actually purchased several of these large glass containers from IKEA when I found these by coincidence during shopping. My colleagues at University of Hawaii at Manoa had used such shaped glass containers as infrared radiation absorbers for long-term measurements.

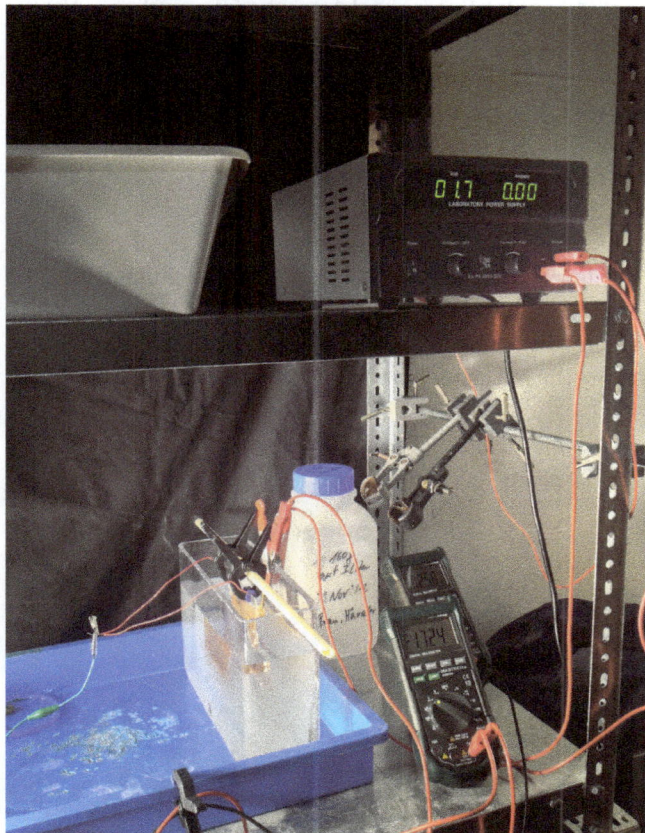

Figure 12.15: Large piece of iron oxide photoanode dipped into a container with KOH electrolyte and connected to power supply set to 1.7 V DC bias. The voltage reading in the multimeter is 1.724 V and the current reading in the second multimeter is 2.58 mA.

A large volume of water will absorb the heat from the halogen lamp but will allow the visible portion to pass through. Such trick is necessary. Otherwise, the photoelectrodes will become too warm in the container, whereas we typically require performance data at ambient temperature – for convenience. I do not know whether the community has already agreed on a particular standard temperature at which the performance data should be reported. On the occasion of the first MRS Symposium on photoelectrochemistry and photocatalysis, which I organized with eminent colleagues [Braun 2009], my colleagues wrote a paper about the rapid development of PEC materials and components [Chen 2011]. In this paper, it was agreed which technical standards should be used when reporting performance data.

Currently (January–June 2018), the DoE PEC Working Group is looking at further benchmarking procedures. This working group is a loose and informal assembly of researchers mostly from the USA but also Europa and Asia who frequently meet at the MRS Meeting or ECS Meetings.

The top of the rack shows the small power supply, with a voltage reading of 1.7 V. The current reading on the power supply is 0.00 because the current that I set with the knob – I did set a current! – is smaller than the range of the reading. The two multimeter on the lower rack have a higher sensitivity and show 1.724 V and 2.58 mA, while illuminated with the halogen lamps.

Figure 12.16 shows photocurrent data obtained from a small 2.5 cm^2 hematite photoanode recorded in the large scale setup explained in this chapter. So, the electrolyte container and the light source were certainly over-dimensionalized. But when you measure small electrodes in small containers and large electrodes in large containers, you cannot compare the influence of the container size because this is not set constant.

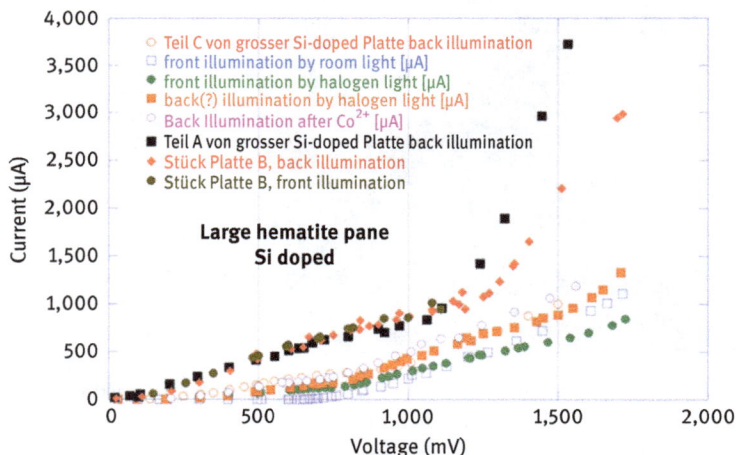

Figure 12.16: A series of light current data from differently large hematite electrode broken pieces measured under front and back illumination with halogen light or dim room light.

The data were recorded manually – by reading and writing down the current and voltage values on the multimeter. The blue curve was obtained when the hematite surface was pointed to the light source. The green data points (there are only around 10 data points in total) were obtained when the sample was turned around and the substrate bottom was pointing to the light source. Illumination of the hematite photoanode from the back side yields a higher photocurrent.

It may occur that you break one of the large electrode panes, for example when you want to mount them in a PEC reactor prototype. You still can use the pieces of the broken electrode for the following reason. Most experimental data in literature are taken from samples with area of around 1 cm^2. When you make a 100-cm^2 electrode, you expect that the photocurrent density and the hydrogen production are by a factor 100 higher.

This however is not necessary the case. Typically, you cannot easily scale up a component in size. Many steps are necessary in order to maintain the high performance by increasing the device size. But before we can fix this issue, we have to measure actually electrodes of different size. When you make larger samples, then you have to make larger containers, larger cells where you can measure the large samples.

Sometimes you cannot simply homogeneously expand the size of a component or a device and this is where unforeseen or unwanted but necessary differences in design can alter the performance. Figure 12.17 shows the broken pieces of a large iron oxide photoanode. Important is the geometric area of the electrodes, but as these were broken in irregular shapes, the size determination is made via a mass determination by using a balance. The mass and size of the pieces are listed in Table 12.1.

One 100 cm^2 large pane has a mass of 26.66 g, and the broken piece with mass M has the area $A = M/26.66 \times 100$ cm^2. When we connect such small broken piece as electrode in the reactor, we can measure the dark current and light current and see whether the current density is larger or smaller than the current density of a small 1 cm^2 photoelectrode or a large 100 cm^2 photoelectrode.

Table 12.1: Mass and area of the electrode pieces, broken from a 10 cm × 10-cm large hematite photoanode. The mass was determined with a balance. The area was determined from the ratio of 26.66 g/100 cm^2 for one plate, assuming proportionality between mass and area.

Mass [g]	Pane area [cm^2]	Hematite-coated area [cm^2]
0.99	3.73	3.73 − 1.3 = 2.43
5.97	22.49	22.49 − 3.8 = 18.69
19.59	73.79	73.79 − 5 = 68.79
26.55 (Sum of parts)	100	
26.52 (Reconstructed)	100	
26.66 (Other plate)	100	

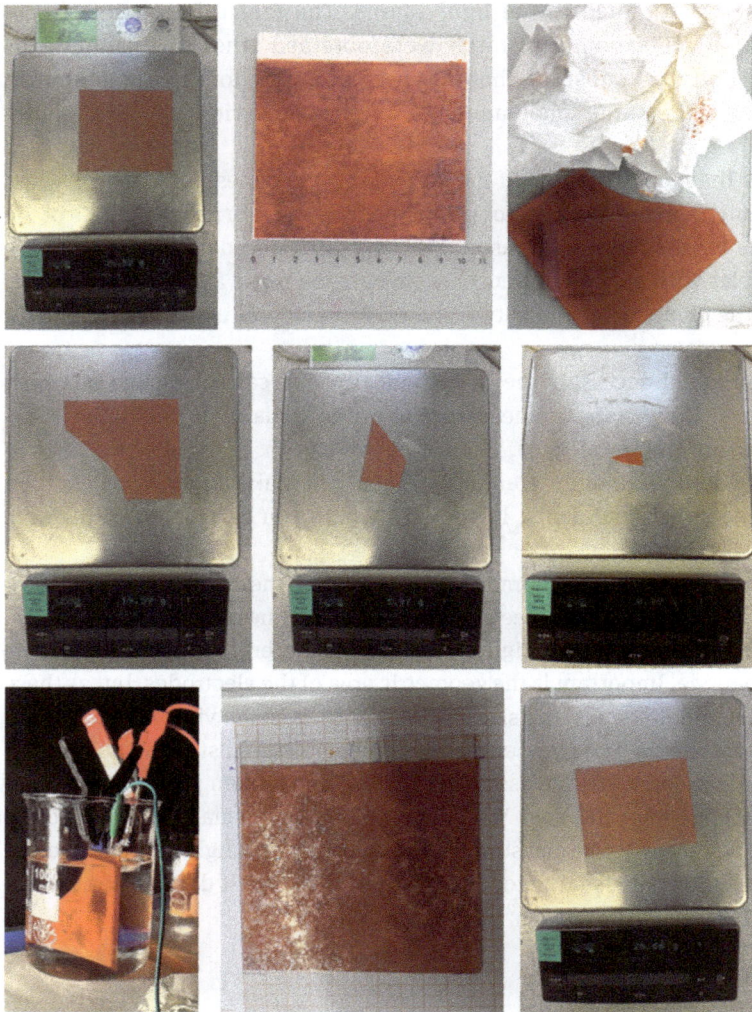

Figure 12.17: (a) Large hematite photoelectrode on balance showing 26.52 g mass. (b) Large electrode in comparison with ruler, revealing 10 cm × 10 cm electrode area. Note the dim transparent stripe on the top which is not coated with the red hematite. (c) Electrode broken in three pieces. The paper towel was used to wipe away peeled off material from the hematite surface. (d) Larger piece of the broken electrode on balance showing 19.59 g. (e) Middle size piece with 5.97 g mass. (f) Smallest piece with 0.99 g mass. Note that the three pieces are not uniform and their area cannot be measured directly with a ruler. (g) Largest piece with 19.59 g inserted in electrolyte; compare with Figure 12.13. (h) The three broken pieces reassembled together on a millimeter paper. Note that the millimeter paper is not calibrated; 10.5 cm on the paper is 10 cm in real. (i) A second hematite photoelectrode with 26.66 g total mass.

12.7.2 A functional model of a PEC reactor

When the machine shop had manufactured all parts for the PEC reactor, the PhD students in my research group had assembled the reactor and mounted it in the basement in a room where one could make long-term experiments with low supervision (Figure 12.18). You can use such room for running reactors for several days and monitor from time to time the operation. You can use a timer which switches the halogen lamps on for a couple of hours or for half a day, so that the reactor has a simulated daylight and night period. This is how it would operate in real life anyway.

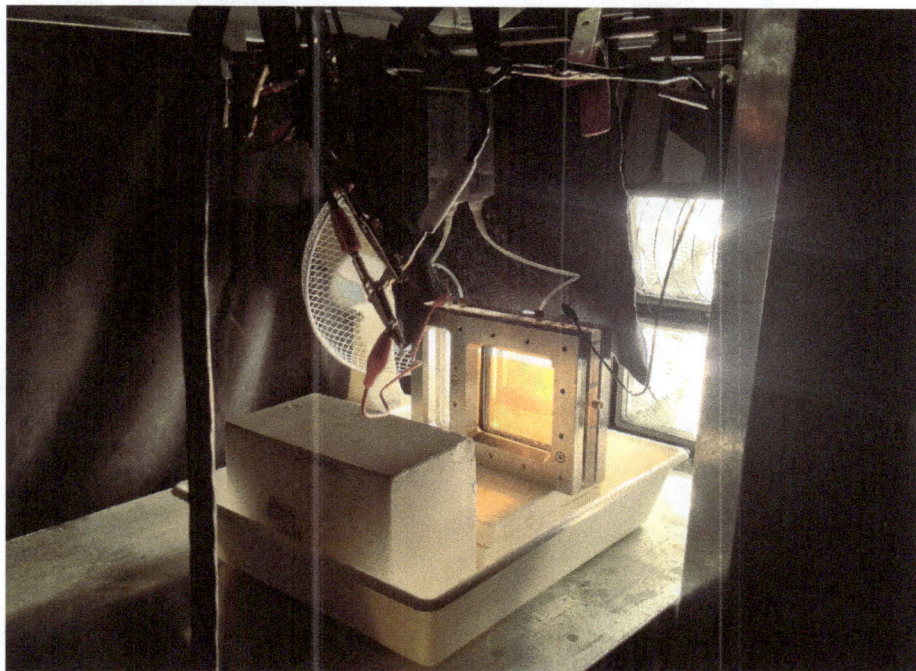

Figure 12.18: PEC reactor with large 10 cm × 10 cm photoanode from iron oxide with KOH electrolyte in operation condition. The necessary DC bias is provided by a power supply via the red and blue cables. The two thin white tubing guide the produced H_2 and O_2 gas into two sealed plastic syringes which are filled with water; the evolving gas volume will repel the water. Four 500 W halogen lamps provide the illumination to produce the photocurrent – plus an enormous undesired heat. The electric fan is used in order to cool the PEC reactor.

When you think of using PEC reactors with bio-organic components such as algae, cyanobacteria, biofilms, then you may have to consider their natural biorhythm which is synchronized with the diurnal and circadian rhythms and periods of life on Earth.

After we managed to run the PEC reactor for the first time with success down in the basement, we figured out how to run it without external power supply out in the sunshine. Since we still needed for the hematite photoanode an external bias, we used several small PV panels which turned out to be powerful enough to lift the conduction band of the photoanode above the redox potential of water.

Figure 12.19 demonstrates how the 10 cm × 10 cm large PEC reactor is mounted in a plastic tray outside in the sunshine before a patch of grass. Three small PV panels with 500 mV each are connected in series to provide 1.5 V DC bias.

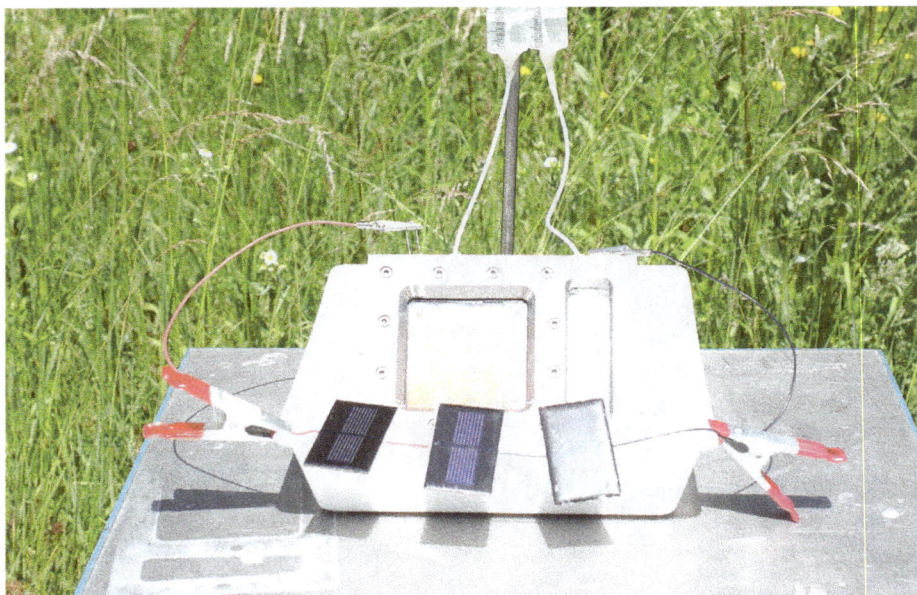

Figure 12.19: PEC reactor outside in 90°F (32.2 °C) sunny day, biased by three small silicon PV cells. The evolving H_2 and O_2 are collected by two sealed plastic syringes filled with water. Experiment done by doctoral student Florent Boudoire and Physiklaborant apprentice Nadja Rutz. Photo by Artur Braun.

12.7.3 More advanced structures

Now that one PEC reactor was shown to be working under real conditions outside in the nature, it was necessary to think about making larger reactors. We have seen in the previous chapters how large solar farms and wind farms and hydropower plants produce (convert, store) gigantic amounts of energy. When we want to run equal with PEC technology, we need to occupy huge areas, many square kilometers in order to collect the necessary sunlight or daylight.

I mentioned previously that we had purchased also FTO glasses with 20 cm length; these would quadruple the photoelectrode area. Another approach is to settle with 10 cm × 10 cm electrode area as the largest reasonable unit because this size is canonical in PV technology and in other industries. A 1-m^2 large photocatalytic hydrogen reactor was developed by Schröder et al. at TU Berlin [Schröder 2014].

We then have to stack these reactor units together and thus enlarge the area of the reactor system. We have then to think how to connect the reactors because they will require electrolyte feeding once extensive gas production would deplete the liquid level in the reactors. We also have to decide how the DC bias is to be arranged.

The result was a setup where four PEC reactors are arranged in a 2 × 2 platform as shown in Figure 12.20. The entire setup is still massive but the purpose of this construction is the study of the operability. Supplying the electrolyte, collecting the gases, providing mechanical stirring and agitation of the electrolyte to prevent diffusion barriers at electrode surfaces, management of temperature and many other issues need to be studied on this setup. Eventually, these need to be calculated as well.

When electrochemical reactions take place at electrode, typically a concentration gradient evolves which can pose a diffusion barrier for reaction partners. This amounts to an additional resistance which you want to avoid when you

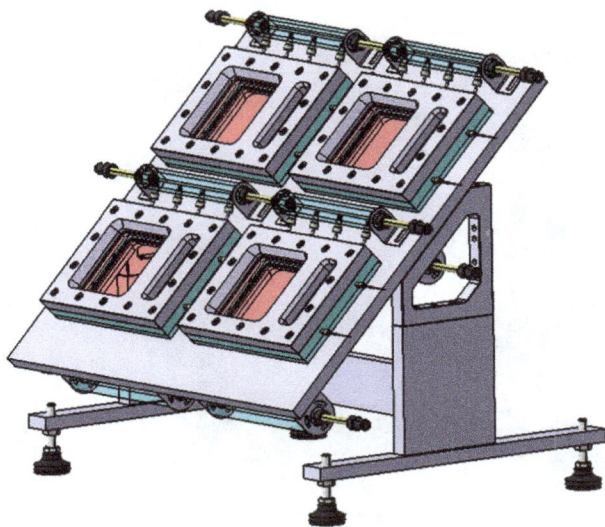

Figure 12.20: Four single-cell PEC reactors arranged in a 2 × 2 panel.
Design and construction by Dr. Minkyu Son, and Erich Heiniger, Daniel Rechenmacher and Ardian Salihu, all Empa.

want to increase the yield of the reactor. You can destroy this barrier by stirring or by agitation with a gas bubble flow [Ibl 1971, Sigrist 1979]. Do you use a pump for this stirring, which will cost extra electric energy? Or is it possible to use the natural convection when the reactor is exposed to warm sunlight?

Will high pump pressure possibly break the glasses in the reactor, and the caustic electrolyte will spill? Yes, it will.

12.8 Energy harvesting with window facades

When you consider the cappuccino cell from EPFL and the two PEC GC cells which Krisztina Schrantz and Rita Toth developed at Empa, you see how the concept of prismatic window-type cell was continued. It is therefore not surprising that the large reactor with the 10 cm × 10-cm size electrodes looks like a somewhat complex window with a liquid inside. In some way, this resembles also the window facades which were made for the famous BIQ house in Hamburg, Germany [Kolarevic 2015], as you see in the photo in Figure 12.21.

Figure 12.21: Arup, Strategic Science Consult (SSC) and Splitterwerk Architects, BIQ House, Hamburg, 2013. This first algae-powered building is covered with over 100 bio-reactive panels. Reproduced with permission from Wiley, Kolarevic B: Actualising (Overlooked) Material Capacities. *Architectural Design* 2015, 85:128–133, and Joannes Arlt, and laif. [Kolarevic 2015].

These facades contain algae which can be fed with CO_2 and thus provide biomass [Wolff 2015] which can be used as primary energy source by gasification [Elsayed 2014], for example. It is very encouraging that architects and civil engineers and companies who produce utilities for residential homes pick up the idea of having a "wet" energy source direct mounted at the homes.

It certainly would be interesting to have such window facades not filled with algal water, but with algae or biofilms attached to electrodes where hydrogen or hydrocarbons could be produced electrochemically. You then would not harvest the algae and turn them into combustible biomass. Rather, the algae would be working for you as colonies and produce a fuel without being sacrificed as biomass. This concept is outlined in the perspective paper in Braun (2015).

As my colleague from University of Uppsala has put it with some humor: When you want to get milk from the cow, you cannot press the whole cow and then expect the milk to come out. This will only kill the cow. You have to press and squeeze at the right spots of the cow to get her milk. You hopefully understand with a smirk in your face the analogy.

A whole range of algal photobioreactor concepts is shown in the book chapter of Koller [Koller 2015]; the reader may realize in same examples in there similarities with PEC reactors and fuel cells designs, including those with parallel plates. At the engineering level, sometimes seemingly disparate fields of technology become sometimes unified again. This is possible for example when a bionics professor becomes interested in artificial photosynthesis.

In the 1980s, bionics professor Ingo Rechenberg experimented with algae in the Sahara desert [Rechenberg 1994]. He used for example *Chlamydomonas oblonga* (green algae), which excrete carbohydrates and produce oxygen. The carbohydrates are being decomposed by *Rhodobacter capsulatus* (purple bacteria) into H_2 and CO_2. The carbon dioxide is fed back in the bioreactor. In a separate bioreactor, *Nostoc muscorum* (blue algae), water is being split [Rechenberg 1998].

For the scaling up toward large scale reactors, Rechenberg invented the heliomites photobioreactor, which has a cone geometry by winding up a very long hose of up to 620 m length with 6 cm diameter filled with water and algae for biomass, or with purple bacteria who will grow and produce hydrogen gas. One such cone would be 5 m tall and stand on a circular base area of 4.5 m diameter, see Figure 12.22. Rechenberg had presented them on an art exhibition in Berlin 1986, "Heliomites in the Sahara" [Eroglu 2008, Rechenberg 1998]. A 100 of such heliomites would find place on a 60 × 60-m square area.

It appears that this design inspired Matthes et al. to build a photobioreactor which they called GICON® photobioreactor, who based their idea on the shape of a "Christmas tree PBR" [Matthes 2015], see Figure 12.23. They look very similar to Rechenberg's heliomites.

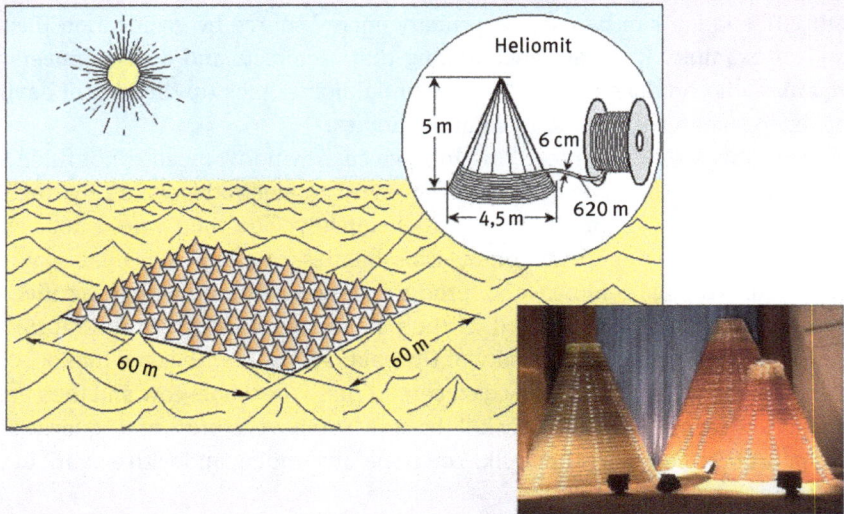

Figure 12.22: Concept of heliomite photobioreactor in the Sahara with 100 heliomites adding to 100 kW power hydrogen source.
Reprinted with kind permission from Ingo Rechenberg, Morocco (07 July 2018). The reactor has a cone shape with the cone top removed, as shown in the photo on the lower right taken at the "Heliomiten in der Sahara" art exhibition in 1986, where the reactors produced under flood light for several weeks.

Figure 12.23: Microalgae platform with four photobioreactor units (located at Anhalt University of Applied Sciences, Koethen, Germany).
Reprinted by permission from Springer Science+Business Media Dordrecht 2014, *Journal of Applied Phycology* 2015, 27:1755–1762. doi: 10.1007/s10811-014-0502-4, Reliable production of microalgae biomass using a novel microalgae platform, Matthes S, Matschke M, Cotta F, Grossmann J, Griehl C, (2014).

12.9 An artificial photosynthesis carpet rolled out in the desert

A somewhat different approach is pursued by my colleague Dr. Heinz Frei at Berkeley Lab, and his close colleagues. He has been thinking of membranes for a quite a while but the membranes are built from highly functional aligned nanotubes [Kim 2016].

Figure 12.24 shows a section of such membrane, a sketch thereof, where count-less short tubes are aligned like piece of turf or a carpet. In these tubes, which are the functional smallest units of the membrane, water is split into oxygen and water using solar energy (hv), similar to photosystem II in nature.

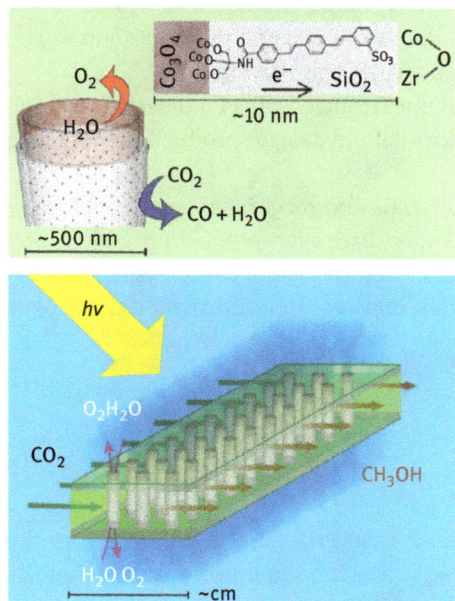

Figure 12.24: Functionalized Co oxide–silica core–shell nanotubes as complete artificial photosyn-thetic units in the form of a macroscale array for CO_2 reduction by H_2O under membrane separation. Top: Sketch of an individual core–shell nanotube showing spaces for CO_2 reduction and H_2O oxidation catalysis separated by an ultrathin silica membrane. Inset: Expanded view of the Co_3O_4–SiO_2 core–shell nanotube wall with silica-embedded oligo-para-(phenylenevinylene) molecular wires and a heterobinuclear light absorber. Bottom: Array of nanotubes with separation of oxygen evolu-tion space inside the tubes from CO_2 reduction space between the tubes.
Reprinted with permission from the American Chemical Society. Fabrication of Core–Shell Nanotube Array for Artificial Photosynthesis Featuring an Ultrathin Composite Separation Membrane, Eran Edri, Shaul Aloni, Heinz Frei, ACS Nano 12, 533-541. Copyright (2018) American Chemical Society.

In addition, the greenhouse gas CO_2 from the atmosphere or from concentrated sources can diffuse into this arrangement of tubes and react with the readily pro-duced hydrogen or protons to hydrocarbon molecules.

The structure of the nanotubes is shown in the right panel in Figure 12.24. The inner tube is a Co_3O_4 water oxidation catalyst which is surrounded by an outer tube from SiO_2. This outer tube is entranched by molecular wires which transport electric charge from the light absorber ZrOCo to the water oxidation catalyst.

The reduction of the diffusing CO_2 will not interfere with the water oxidation. The highly structured and compartmentalized nanotube design warrants that no chemical and electrical shortcuts take place. The architecture is therefore reminiscent of natural photosynthesis where we have too a well-organized management of flow of electrons, holes, protons and gases. We are looking here at countless reactors at the nanoscale.

They are able to produce these membranes in the size of 1 in at this time. The idea is to make huge carpet-like membranes of m^2 and larger and then roll them out in the sun. In theory, it should be possible to cover many square kilometers of desert land in the Mojave Desert and use the humidity in the dry desert as the water source. The concentration of the CO_2 though might be somewhat lean, though.

In his review paper [Tributsch 2008], Helmut Tributsch calls for massive research in a few innovative directions, before photovoltaic hydrogen production would be able to become reality.

Pinaud et al. have compared various materials and reactor concepts (photoelectrochemical, photocatalytic) for centralized solar hydrogen production and carried out a techno-economic analysis [Pinaud 2013]. They arrive at a cost of delivery at the dispenser of 2–4\$ per kilogram H_2, contingent however that future material targets can be met.

12.10 Test cell for separator measurements under pressure and temperature

At some point during my PhD thesis, it became necessary to look deeper into the properties of separators for supercaps and batteries. The separator warrants that the two electrodes in an electrochemical cell do not have direct electronic contact with each other. This is why a separator for that purpose cannot be made from metal. Aluminum foil therefore would be not useful because it makes directly an electric shortcut. The separators are typically membranes like foils which you can soak with an electrolyte. A fully blocking separator would not provide the necessary ionic conductivity. A sheet of paper may already do a good service as a separator. Other materials are glass fiber fabric or polymer foils.

The separators have to have some minimum porosity that the electrolyte can conduct ions from one electrode to the other electrode. The separators must be mechanically stable so that grains or spikes in the electrodes do not puncture through the separator and thus make an electric shortcut. Sometimes this can happen, for example when during battery operation or fuel cell operation, grains in the electrodes or electrocatalysts undergo recrystallization and reshaping.

The separators need also some chemical stability so that they do not become corroded when in contact with electrolyte or under electrochemical potentials. The separators are therefore materials which too need to be optimized for electrochemical engineering and technology. The chemical industry is making lots of efforts to optimize separators for the battery and fuel cell business.

At some point, we were therefore exploring separators that were commercially available. They were tested for compatibility with sulfuric acid electrolyte, for example.

Some were too thin and electric shortcut was therefore possible when rough electrode surfaces could truncate the separator. Some were too thick and they had therefore a large electric resistance. And others were just too expensive per square meter.

One task was the determination of the electric conductivity of the separator. For this, the separator material was soaked with an electrolyte and then an impedance spectrum was recorded. For this, I developed some test cells where we could easily measure the separators. One such cell is shown in Figure 12.25. Two glassy carbon rods serve as electrodes, between which a separator can be placed. The carbon rods are pressed into KelF® plastic cylinders and connected from the back with a brass screw. This "electrode" is put in a Teflon® cylinder which has somewhat less than double the length of the aforementioned glassy carbon electrode plus KelF® cylinder and brass screw.

Then, you place a separator and a drop of electrolyte on it and insert the second electrode, press both electrodes together to provide firm electric or ionic contact, connect the electrodes with cables to the impedance analyzer and run impedance spectra. I found in my studies that the impedance spectra will change depending on whether I squeeze the two electrodes very firmly together or not. It was therefore necessary to improve this cell to be able to apply a well-defined and well-quantified pressure to make the results as reproducible as possible.

12.10.1 Conductivity of separators for various thicknesses

You can now insert one separator between the electrodes and measure the impedance and thus determine the resistance of the entire setup. When you add a second and third separator, the resistance will increase accordingly. There may be contact resistances from the cell and the electrodes but the resistance by the separators should increase proportional to the number of separators added. This is indeed the case as is shown in Figure 12.27 for five separators that were consecutively added in the cell.

The resistance was determined from impedance values obtained at 10 kHz (filled symbols in Figure 12.13) and 100 kHz (open symbols), respectively. The nominal thickness of the separator was 50 μm. For both frequency points, the data points for the resistance show a linear trend, as expected. The specific conductivity for this particular kind of separator was 0.42 S/cm and 0.53 S/cm. For this experiment, no pressure was applied and only the nominal thickness of 50 μm per separator was assumed.

When you make experiments with a cell as designed as in Figure 12.26, you will find out that the conductivity of the separator varies depending how firm you squeeze the two electrodes and thus the separator. You have to see the separator like a thin sponge with many channels inside which can be soaked with electrolyte. The electrolyte conducts the ions and warrants that the electrochemical cell works properly. When you squeeze the separator too much, the pore channels can be clocked (blocked) and the ions do not flow anymore with ease. This manifests in an increase

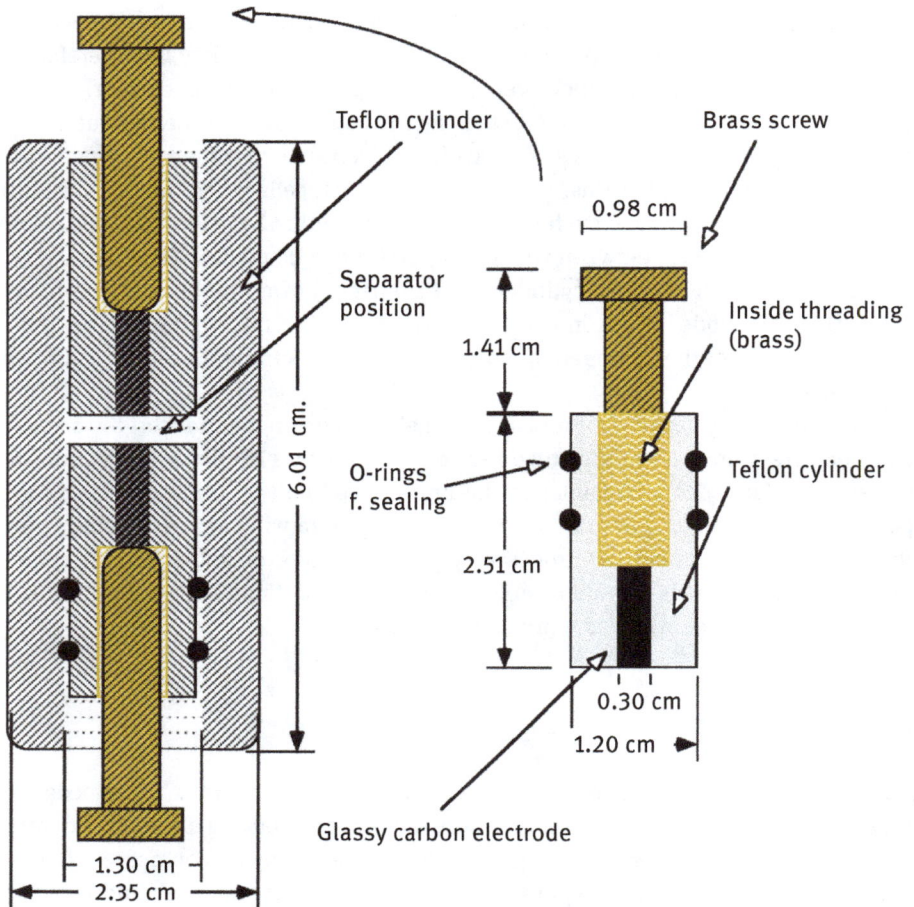

Canvas: Zelle 1 / Arthur Braun 23.9.96

Figure 12.25: Electrochemical cell for the conductivity measurement of separator membranes under various pressures and temperature (manufactured by Christian Marmy, Paul Scherrer Institut). Two glassy carbon electrodes with cylinder shape are pressed into two KelF® cylinders and electrically contacted on the back with two brass screws. The lower carbon electrode in KelF® is inserted in a Teflon® cylinder with the carbon electrode facing up. One or several separators (glass fiber, polymer, cellulose etc.) can be laid over the glassy carbon electrode and soaked with a drop of electrolyte. The second electrode in KelF® is turned upside down and inserted into the Teflon® cylinder. This electrochemical cell is then inserted into a Teflon® socket which holds the entire cell arrangement. The brass screw of the upper electrode component has a plate on top, on which heave weights can be placed. Thus, the weight will act on the separators in between the glassy carbon electrodes. The two brass screws are electrically connected with an impedance analyzer.

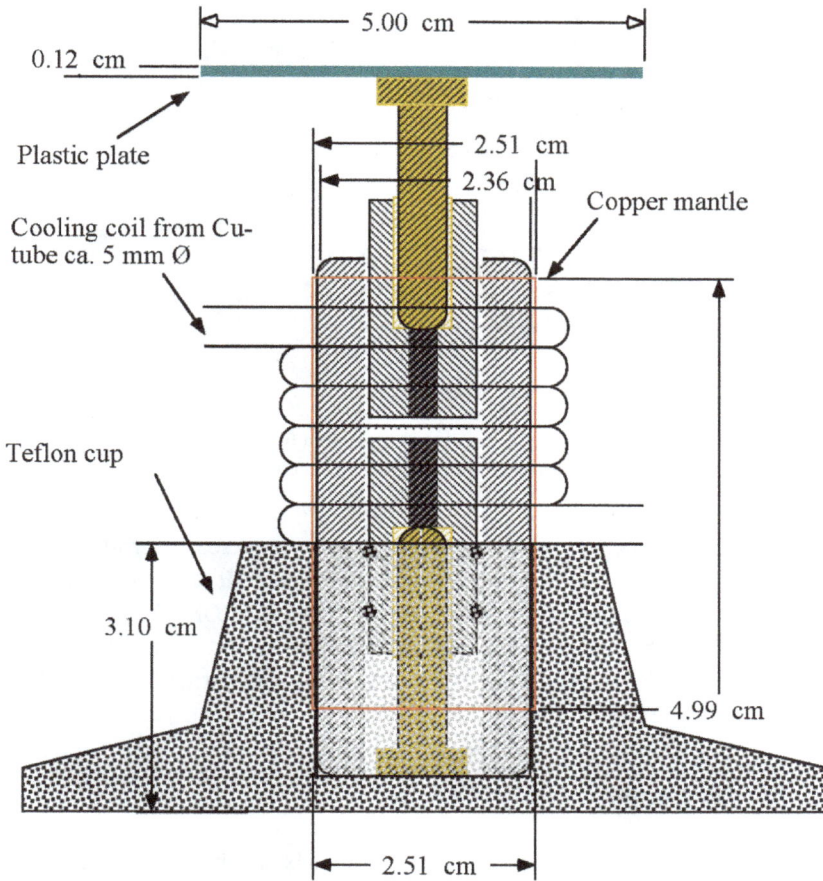

Canvas: Zelle 3 / Artur Braun 23.9.96

Figure 12.26: Electrochemical cell for the conductivity measurement of separator membranes under various pressures and temperature. Two glassy carbon electrodes with cylinder shape are pressed into two KelF® cylinders and electrically contacted on the back with two brass screws. The lower carbon electrode in KelF® is inserted in a Teflon® cylinder with the carbon electrode facing up. One or several separators (glass fiber, polymer, cellulose etc.) can be laid over the glassy carbon electrode and soaked with a drop of electrolyte. The second electrode in KelF® is turned upside down and inserted into the Teflon® cylinder. This electrochemical cell is then inserted into a Teflon® socket which holds the entire cell arrangement. The brass screw of the upper electrode component has a plate on top, on which heave weights can be placed. Thus, the weight will act on the separators in between the glassy carbon electrodes. The two brass screws are electrically connected with an impedance analyzer.

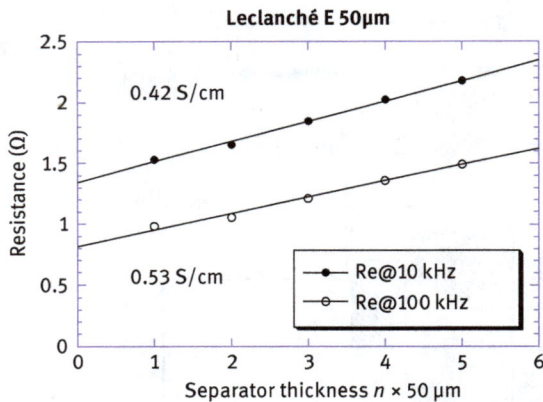

Figure 12.27: Dependence of the resistance of separators as a function of number of stacked separators. The resistivity was determined with impedance spectroscopy from the real part axis at 10 kHz and at 100 kHz.

of the resistance. It is better here to say it manifests in a decrease of the conductivity. The conductivity of the dry separator is extremely poor. It is basically insulating, unless you have an ion conduction separator such as a NAFION® foil. But this is here not the case. Only the liquid electrolyte phase conducts the ionic current. You can mathematically model the conductivity based on the geometrical considerations of an ionic path through the separator which connects both electrodes.

12.10.2 Conductivity of the electrolyte

The ionic conductivity of the separator membrane is basically the conductivity of the electrolyte volume which is in contact with both electrodes on either side of the electrochemical cell. It is therefore necessary to measure the conductivity of the electrolyte without any separator membrane involved. This can be practically accomplished when you use the cell shown in Figures 12.25 and 12.26 and put a gasket between the two electrodes. When you know the inner open diameter of the gasket and the thickness of the gasket, then you know the width and thickness of the electrolyte path. Now, you can apply the same simple geometrical relationship between specific resistivity and resistance, or specific conductivity and conductance:

$$R = \rho \cdot \frac{l}{A}$$

with the gasket thickness l and the inner open diameter A. When you now determine the ionic resistance R, you can calculate the specific ionic resistivity ρ, and the specific conductivity is $\sigma = 1/\rho$ in S/cm, for example.

Figure 12.28 shows the impedance spectrum of a 100-µm thin ionic path of 3 M sulfuric acid.

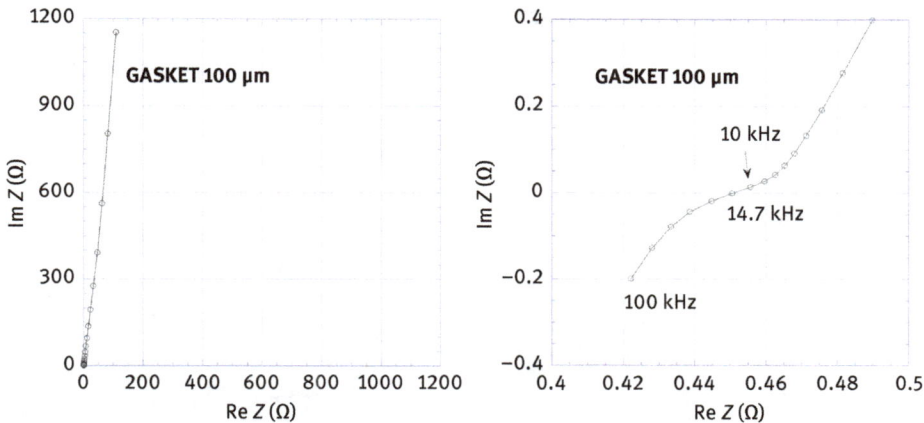

Figure 12.28: Impedance spectrum recorded from the cell shown in Figures 12.25 and 12.26 with 3 M sulfuric acid and the electrodes separated by a 100-μm thick gasket. About 3 M sulfuric acid has a conductivity of around 0.8 S/cm.

12.10.3 Pressure dependence of the thickness of a separator

You may notice this when you squeeze the Swagelok® cells too much when you assemble a lithium battery, as I have shown in Chapter 3. A similar effect happened when we squeezed the ceramic proton conductor in a specifically designed high pressure cell [Chen 2012, Holdsworth 2010].

Figure 12.29 shows the results of a study where I measured the thickness of a separator. I used a Stylus apparatus for that purpose. I had put the separator between two microscopy glass slides and then measured the total thickness. Measuring the two glass slides without the separator in-between them then yields the necessary reference thickness, the difference of which provides the thickness of the separator. I then put various weights on the glass slides so as to press the separator in between them. The initial thickness was determined as 180 μm. The nominal thickness is 102 μm (BSB20).

I have put one dozen different weights on the separator, and it is clear from Figure 12.29 how the thickness gradually decreases with a $1/d$ profile. When no weight was applied, the thickness read around 180 μm. I do remember that the thickness reading for some fibrous separators was complicated, possibly because small fibers were sticking out and these yielded ill-defined results. Flipping up the stylus pin several times and let it drop on the glass slides cured this problem, and consistent and reproducible thickness readings were obtained.

It appears that 80 μm thickness is obtained when a very heavy weight is put on the separator without destroying its microstructure. The red solid line is a least square fit of the arbitrary chosen function

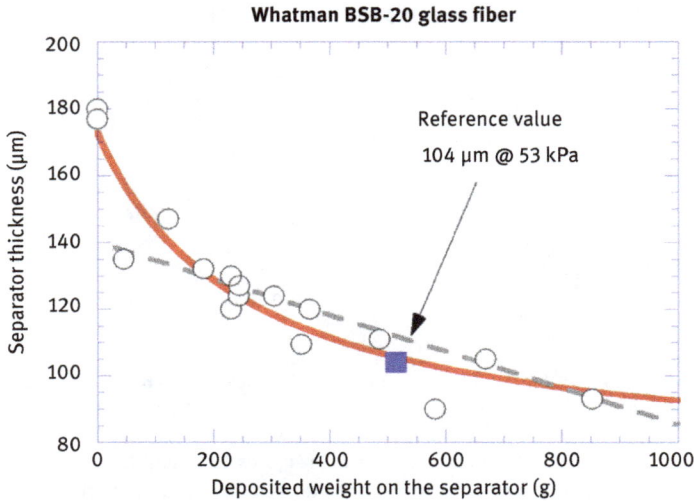

Figure 12.29: Dependency of the separator thickness in micrometers from the pressure put on it by weights in Gramm. The red solid line is the least square fit to a simple mathematical model shown above. The dashed line is a forced linear least square fit which obviously is not everywhere representative to the data trend. The blue square data point is a reference value provided by the manufacturer.

$$d_{(m)} = d_0 - d_p \cdot \left(\frac{m - m_a}{m - m_b} \right)$$

to the thickness data which obviously represents the trend of the data points pretty well.

The filled blue square is a data point that was provided from elsewhere as a reference value, which said at 53 kPa pressure, the thickness was 104 μm. The least square fit happens to go through this reference data point and this adds confidence that the fit curve suits the purpose quite well for the engineering of devices; at some point, the engineers want to know how thick the separator in the device is. The nominal thickness does not necessarily count because it can become squeezed together during device manufacturing. I leave it up to the reader as an exercise to calculate mass and pressure and transform the data in Figure 12.29 on a pressure axis, rather than on a mass axis.

The fit function is an arbitrary one. There should certainly be a functional relationship between pressure and thickness which can be based on first physical principles. While it is beyond the scope of this book to get into the details of the mechanical properties, more specifically, the elastic properties of the separator membranes, we can lend one or two more thoughts on the matter. The fit parameters are listed in Table 12.2. You see that the errors shown in the right column are huge, notwithstanding that the fit curve reproduces the trend quite well.

Table 12.2: Least square fit parameters to the red solid fit line in Figure 12.29, reproducing the decrease of the thickness of separator BSB-20 with increasing weight put on the separator.

Least square fit parameters for separator pressing		
$d_{(m)} = d_0 - d_p \times$ $(m - m_a)/(m + m_b)$	Value	Error
d_0	164.63 µm	1.4336e + 8
d_p	92.76 µm	1.4336e + 8
m_a	22.774 g	4.3454e + 8
m_b	258.39 g	110.19
X^2	1006.3	NA
R	0.9455	NA

When you keep a keen eye on the data points in Figure 12.29 and remember that at low pressure the determination of the thickness was problematic, then you may find that the data points at zero mass, which is zero pressure, are outliners. On the other hand, they are very close to the nominal thickness of 200 µm as supplied by the manufacturer. When we ignore these data points near zero pressure, we find actually a linear trend which can be fitted also quite well with a linear function, as I have done with the weak gray dotted straight line in Figure 12.29. I have ignored for this linear fit those data points which are significantly below the gray line.

It seems that the linear fit would make most sense from the physical point of view because it reminds us of the elasticity of materials and linear force laws which we know as Hooke's law: $F = kx$. The length of a spring coil changes when we pull or push it, and the change of length is proportional to the exerted force. The equation for the red fit curve which I suggested seems less qualified from this aspect because I could not supply a physical model for the observed behavior. Regression curves can be a practical help for solving engineering tasks, but we cannot simply derive physical models from a well-fitting regression curve [Webster 1997].

12.10.4 Selection of separators and membranes

A representative list of a number of separator materials is given in Table 12.3.

12.10.5 Temperature dependence of the conductivity

We have read in the beginning of this book in Chapter 1 how the power of a battery can change with the temperature. At some point during the development of a battery

Table 12.3: Technical data, that is, thickness, conductivity and porosity of a number of membranes which could be used as separators in electrochemical applications.

Type	$d_{nom.}$ [µm]	$d_{exp.}$ [µm]	Conductivity o [S/cm]	Porosity $p1.5, p_2$ [%]
Fluoropore	–	17	0.265	48.2; 57.5
GoreTex	30	–	0.151	33.2; 43.4
Celgard 3401	25	19	0.059	17.9; 27.2
Celgard 3501	25	19	0.1	25.3; 35.3
Leclanché C	–	80	0.133	30.7; 40.8
Leclanché C'	–	50	0.29	51.2; 60.2
Leclanché A	–	61	0.213	41.8; 51.6
Leclanché B	–	45	0.185	38; 48
Whatman cyclopore	15	–	0.01	5.8; 11.5
DBSB 30	140	110	0.48	71.2; 77.3
DBS 30	140	120	0.54	76.8; 81.9
Whatman cyclopore	9	–	0.052	16.4; 25.4
BSB 20	104	95	0.921	109.7; 107.3

or fuel cell or any device, it may be worthwhile how temperature alters the behavior of the device, its components and materials. This is certainly of practical importance for a consumer. I remember after my iPhone had become old and when I was walking for a long time outside in the cold in the winter, it would switch off very soon, indicating that the battery would empty very fast, way faster than during normal operation inside a room.

I have shown in Figure 12.26 that the test cell had a coil from copper tubing which allowed flushing a cooling liquid or hot water around the cell body. This allowed for cooling and heating the cell and the samples inside. For this, a thermostat had to be attached to the copper tubes. Figure 12.30 shows impedance spectra of a separator between the two glassy carbon electrodes in the cell when a DC bias of 0.4 V and 0.9 V was applied.

The purple spectra occupy the widest range in the Nyquist plot and reflect therefore the largest impedance of the separator between the glassy carbon electrodes. The spectrum recorded under 0.4 V bias shows a nice Warburg-type linear behavior at the very high frequencies and then a capacitive behavior with a constant phase element. When the bias is increased to 0.9 V, the imaginary part of the impedance becomes lower and the data points at low frequencies appear to obey a semicircle, suggesting that a charge transfer takes place at higher bias of 0.9 V. Note that this was the low-temperature behavior at 4–6 °C.

When we increase the temperature to 58 °C, the impedance spectra shrink considerably as we can see from the two red spectra in Figure 12.30. This is direct

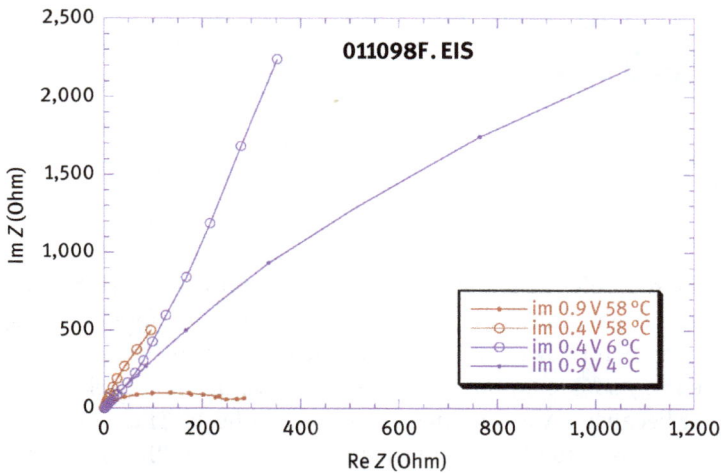

Figure 12.30: Impedance spectra of a sample recorded at 0.4 V and 0.9 V bias at low temperature (4 °C and 6 °C) and at high temperature (58 °C). The shortage of the branch and radii shows that increased temperatures lower the impedance and increase the conductivity.

proof that the conductivity of the system increases when the temperature is raised. This is not to be mixed up with the increase of the electromotive force when the temperature is raised. There exists a linear relationship between EMF and T.

These impedance spectra do not give immediately account of the conductivity of the separator temperature dependency. We would have to deconvolute the spectra and assign to electrodes, electrolyte and separator individual components. But we do see the strong effect of temperature on the conductivity of the whole system, which is already an interesting observation from the pure technical point of view.

This holds not only for aqueous electrolytes and also not only for inorganic systems, see for example [Barthel 1979]. Electrolytes in biological membranes show also a temperature dependency [Kuyucak 1994].

13 Reaching for the inner of the sun – by nuclear fusion

13.1 Reaching for the inner of the sun

13.1.1 How fossil are fossil fuels?

The primary energy sources referred to in this book were either the well-established fossil fuels (coal, natural gas, mineral oil) or the so-called renewable energies, particularly solar power, wind power and hydropower. When we include biomass as a still important energy source on the globe, we can trace this one back immediately to solar power because it was produced by photosynthesis. Then, even the fossil fuels can be traced back to photosynthesis products made millions of years ago.

Until recently, I was believed that all live on Earth originated from photosynthetic life. The beginning of all life took place at the surface of the Earth. Then, when I visited the Smithsonian Institute in Washington D.C. in 2016, I saw an educational movie which showed how some form of life was created in darkness in the deep sea. The energy necessary for starting and maintaining life comes from chemical compounds in the water and deep sea minerals or from the heat in there. Hence, the paradigm

> Without light there can be no life, so let there be hν! [Tien 2000]

that light as energy source is necessary for the creation is not correct. There are at least few examples which show alternatives can work as well.

I did not mention yet the thermal energy which is produced and delivered from within the center of our own planet, mostly because of natural nuclear reactions in the Earth core. There is indication that "fossil" hydrocarbons can be formed under conditions which persist deep in our Earth between crust and core[1] [Kolesnikov 2009, Kundt 2014], which implies they are not necessarily of fossil origin and thus not of solar origin. Certainly, these findings can mount further to highly controversial theories which are not necessarily supported by the established communities. However, whether a theory is right or not cannot depend on the consent in a community. Can it?

---

[1] That study was conducted at the Geophysical Research Laboratory of the Carnegie Institution of Science in Washington D.C., which I happened to visit in 2014, where Alexander Goncharov was my host. They have a high pressure research laboratory with facilities which can be interesting not only for geoscientists but also for materials scientists.

https://doi.org/10.1515/9783110561838-013

13.1.2 Some remarks on the "scientificness" of science

From the scientific perspective, there is nothing wrong with contesting a theory or hypothesis. Great philosophers have spent thoughts on the essence of right and wrong, true and false, or truth, Wahrheit, in general [Heidegger 1976]. Nietzsche suggested the introduction of a range of gray levels between right and wrong, rather than insisting on the two opposing extreme. It is interesting that this beginning of relativism in philosophy started at about the same time when the classical physics was challenged by new discoveries for which the new ideas of quantum physics and wave mechanics needed to be developed. Since, a new field of science evolved which dealt with the validity and validability of scientific discovery and judgment, one of their most prominent protagonists is Karl Popper:

> I have taught for more than 38 years, that all observations are theory-impregnated, and that their main function is to check and refute, rather than to prove, our theories. [Popper 2009]

The position of this school of thought has been challenged by Paul Feyerabend, who argues that there is no scientific metric with ultimate authority based on which a theory can be verified or falsified [Feyerabend 1964, 1975, 1984, 2005]. What we can take home from these disputes is the value of the doubt. This, I believe, is the common ground despite all controversy. Being critical is actually the essence of scientific and academic work and life, when we can believe the words by Physics Nobel Prize Laureate Richard Feynman who elaborated on the freedom to doubt in a speech at Caltech in 1955 [Feynman 1955]:

> I would now like to turn to a third value that science has. It is a little more indirect, but not much. The scientist has a lot of experience with ignorance and doubt and uncertainty, and this experience is of very great importance, I think. When a scientist doesn't know the answer to a problem, he is ignorant. When he has a hunch as to what the result is, he is uncertain. And when he is pretty darn sure of what the result is going to be, he is in some doubt. We have found it of paramount importance that in order to progress we must recognize the ignorance and leave room for doubt. Scientific knowledge is a body of statements of varying degrees of certainty – some most unsure, some nearly sure, none *absolutely* certain.

> Now, we scientists are used to this, and we take it for granted that it is perfectly consistent to be unsure – that it is possible to live and not know. But I don't know whether everyone realizes that this is true. Our freedom to doubt was born of a struggle against authority in the early days of science. It was a very deep and strong struggle. Permit us to question – to doubt, that's all – not to be sure. And I think it is important that we do not forget the importance of this struggle and thus perhaps lose what we have gained. Here lies a responsibility to society.

Let me therefore, now that we heard Feynman's word, continue and conclude this book on another hypothesis which is not shared by the general electrochemistry community, that is, the cold fusion[2] hypothesis [Simon 2002]. I mentioned briefly in a recent commentary paper [Chen 2017] how "unimaginably, astronomically long times" warrant that physical processes with extremely small probability such as proton tunneling eventually manifest in the proton–proton reaction where two hydrogen nuclei approach each other so close that they melt, fuse into a deuterium nucleus and further react to a helium nucleus. Subsequent chain reactions cause radiative energy release in the MeV range [Adelberger 2011, Bertulani 2016]. This is one of the nuclear fusion reactions, based on which the sun delivers us its energy on Earth. There is an interest in being able to simulate this reaction on Earth and thus tap the strongest forces of nature for a virtually never-ending power supply here on Earth.

As of yet, cold fusion has not been a success and has not become a success. But as Paul Feyerabend has said, "it will become clear that there is only one principle that can be defended under all circumstances and in all stages of human development. It is the principle: anything goes" [Feyerabend 1976].

> My intention is not to replace one set of general rules by another such set: my intention is, rather, to convince the reader that all methodologies, even the most obvious ones, have their limits. The best way to show this is to demonstrate the limits and even the irrationality of some rules which she, or he, is likely to regard as basic. In the case that induction (including induction by falsification) this means demonstrating how well the counterinductive procedure can be supported by argument. [Feyerabend 1976]

Progress in one direction is not possible without deleting the possibility of progress in the other direction [Feyerabend 1979]. While this quote bears some exaggeration, the core of the message is correct. Here is a fundamental dilemma in science and also in technology. It is unlikely that two alternatives in technology will gain equal support. Is it reasonable when a society invests much tax payer money in the development of fossil fuel combustion technology, way more than in the development of renewable energy technologies? Would it be nonsense when we invest more resources in the discovery of novel nuclear technologies such as cold fusion, although it has been judged a "flop"?

13.1.3 The power of the sun

Planet Earth and the sun were formed 4.5 billion years ago. By gravitation, sun keeps our Earth on its trajectory. The time which Earth needs for one cycle around the sun – this is what we call 1 year. The energy from our sun has created life on Earth. After 1 year, the cycle of life repeats. Its four seasons have major impact on life in nature. The Earth rotates around its own axis. This warrants that every point on Earth is facing the sun once per cycle and thus becomes illuminated. In the changes from night to day, from winter to summer, sun is the almighty and reliable constant upon which life depends.

---

2 Today, low energy nuclear reactions (LENR) is used instead of cold fusion.

Per every second, the sun emits 20000 times more energy from its 6000°C hot surface than was used for the entire industrialization of mankind. Two-hundred years of modern industrial revolution required just 200 h of sunshine. One full month of solar power was sufficient to build our modern civilization. The amount of energy which we receive from our sun, by human imagination, is unmeasurable and inexhaustible (Figure 13.1).

Figure 13.1: Sunrise over Death Valley, California, one of the hottest spots on Earth [Kubecka 2001]. The enormous amount of solar energy originates from nuclear fusion of hydrogen isotopes. Since 4 billion years, the sun has delivered this energy to Earth. The hydrogen pool of the sun will be exhausted in the next 4 billion years to come [Braun 2015b].
Photo by Artur Braun.

Hundred million miles away from the sun, in safe distance, Earth receives in 1 h enough energy to provide for mankind's energy demand for one entire year. Sun has completed now half of its lifetime. Half of its fuel – hydrogen – has already burnt to helium by nuclear fusion. There is more hydrogen left for us – for another 5 billion years.

The emission spectrum of the sun resembles to some extent the theoretical spectrum of the black body, which was derived by Max Planck from fundamental statistical ideas [Planck 1901]. The distribution of the electromagnetic radiation with frequencies v of a black body with temperature T is described by the following relation:

$$U_v^0(v, T)\mathrm{d}v = \frac{8\pi h v^3}{c^3} \frac{1}{e^{\left(\frac{hv}{kT}\right)} - 1}\mathrm{d}v.$$

The black body radiation law has a functional relationship between temperature of the black body and its emission spectrum; the well-known laws from Rayleigh and Jeans, Stefan–Boltzmann and Wien can be derived from it. From the position of the maximum of the emitted intensity, we can derive that temperature. As we cannot peek so easily into the core of the sun, the determined temperature represents the conditions on the sun surface.

13.2 Energy from the nuclear forces

13.2.1 The nuclear fusion reactions

As electrochemists, chemists and condensed matter physicists, we are typically dealing with electron binding energies which arise from the relations between the atoms or ions and within molecules and condensed matter. They range in the order of 1 eV and below. We pay no attention to the nucleus of the atom, with the exception that its relatively heavy mass as compared to the mass of the electrons plays a role in the total mass of the matter we are dealing with.

The nuclear power however originates from reaction among the particles that constitute the nuclei, and these are protons and neutrons. Nuclear science is a field of its own and beyond the scope of this book. But we need to deal with some very simple nuclear model here in order to understand the working principle and scope of fusion and cold fusion. This is notwithstanding that nuclear science is still a field that further develops. We restrict ourselves here on the picture that the nucleus of the atom is built from protons with mass 1 and charge +1, and neutrons with mass 1 and 0 charge. The chemical element is defined by the number of protons it has in its nucleus. Elements may have varying number of neutrons, which is schematized on a nuclide chart.

Let us consider two protons which shall be fused together so as to build a helium nucleus. Protons are positively charged, and their Coulomb interaction (Coulomb force) will make the protons repel each other. But there is another force acting as well, and this is the nuclear force. The nuclear force is an attractive force and has a very short range and follows a different law than Coulomb's law. The nuclear force is also stronger than the Coulomb force. Both forces are acting on the proton. When an additional force will come into play and push the two protons beyond the minimum of the nuclear force, which is achieved at around 1.3 fm (femtometer, 10^{-15} m), the protons can fuse and form a new nucleus such as deuterium ^{2}H, which contains one proton and one neutron.[3] This situation is illustrated in the proton–proton chain (cycle) in Figure 13.2. Consider a pool of protons ^{1}H which somehow we can force to approach each other beyond the aforementioned minimum of the potential located at around 1.3 fm.

This process seems simple here in the book, but in reality, it requires a tunneling process which is extremely rare and it will occur in the sun only once every 1.4×10^{10} years. This is the reason why the sun burns such a long time and is not yet exhausted. The reaction product is a deuterium nucleus ^{2}H (which is built from a proton and a neutron), one positron (p^+) and one electron neutrino ν_e. The energy released from this process step is 0.42 MeV per proton. The neutrino carries away 0.267 MeV of this energy. Since neutrinos hardly interact with any matter, this energy is carried away from the sun and basically lost. The positron (p^+) will immediately annihilate with an electron e^- and release 1.022 MeV in the form of two γ quanta. The deuterium ^{2}H will find further protons from the pool and then make fusion to a helium nucleus ^{3}He and release a γ quant of 5.493 MeV. With 1.4 s, the lifetime of the deuterium is very short. The helium nuclei ^{3}He can now make fusion toward a heavy ^{4}He nucleus with release of two further protons which will add to the proton pool. The released energy is 12.86 MeV. This nuclear reaction takes place at temperatures between 10 million and 14 million Kelvin. There are two alternative and competing proton–proton processes which form boron, lithium and beryllium as reaction products, in addition to ^{4}He. These have lesser probability and will be ignored here. The energy released from the chain is 26.196 MeV. I leave it as an exercise for the reader to figure out from Figure 13.2 how this value is obtained by addition. It is also interesting to estimate how much

---

3 Here in this chapter, we are dealing therefore with isotopes. Consider the periodic table of elements, which organizes all chemical elements in order of rows and columns. Consider not the first element, hydrogen (H). It contains one proton in the nucleus and one electron in its orbit. When you add one proton in the nucleus, you will not have hydrogen anymore, but helium (He) (an additional electron in the shell is required to make the helium atom neutral), which is in the next right column in the periodic table. If instead you are adding not another proton to the hydrogen but a neutron, the mass of the new hydrogen will be doubled and in so far be as heavy as helium, but it is still hydrogen; no additional electron is required in the shell for charge neutrality. This new element will therefore not switch position in the table of elements. It will stay on the first position and is called heavy hydrogen or deuterium. Because it stays in the same position in the table of elements, it is called an "isotope." To note the double mass, the mass number 2 is added to the symbol of hydrogen: ^{2}H. For better distinction, it is sometimes called ^{2}D. There exists another isotope of hydrogen, tritium which has one proton and two neutrons in the nucleus: ^{3}H or ^{3}T.

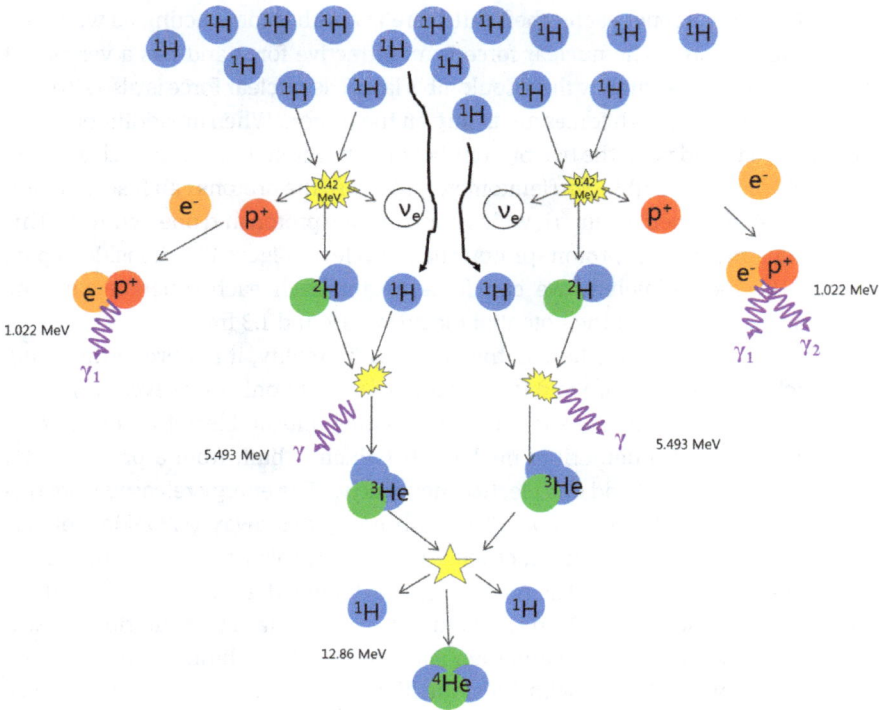

Figure 13.2: The nuclear reactions ("hydrogen burner") of the proton–proton chain toward ^{4}He production.

energy is released when we begin the chain with 1 mol of nuclei. This compares then with 1 mol of hydrogen converted in a fuel cell, and with 1 mol of carbon representative of coal (fossil fuel) or wood (biomass). This too is left as an exercise for the reader. The huge amount of energy released from nuclear fusion is the strong motivation for research and technology in this field.

To the physicist, nuclear fusion is a scattering problem like any other mechanical problem that deals with collisions of objects (Figure 13.2) with an elastic and inelastic contribution. The scattering cross-section of the nuclei and the kind of interaction determined whether a nuclear reaction such as nuclear fusion can take place. Scientists were quite early able to determine by simple calculations how likely it was that a proton would enter an atom when it was in a proton cloud with a kinetic energy distribution according to Boltzmann statistics [Atkinson 1929, Gamow 1938]. The reader is referred to literature about nuclear physics, specifically [Adelberger 1998, 2011]. What we need to remember here is that the conditions for a nuclear fusion are very extreme and not all possible so easy on Earth.

As soon as scientists (e.g. see [Bethe 1939, Gamow 1938, Von Weizsäcker 1937, 1938]) had figured which nuclear reaction would cause fusion of nuclei with the corresponding release of huge amounts of energy, they thought about how to harness these nuclear reactions in order to make them useful for mankind on Earth. It is certainly cynical that the first application of nuclear fusion was realized in the hydrogen bomb.

Today, it is established which of the various nuclear reactions release the most energy in the sun and in the universe [Siegel 2017]. In a hydrogen bomb, deuterium ^2D and tritium ^3T are brought at extremely high temperature and pressure in a condition where they can react, fuse and release energy as electromagnetic radiation to an amount which corresponds to their mass loss of 0.3%. This extreme condition is realized by the ignition of a nuclear bomb or by the activity inside the sun and in other stars.

13.2.2 The TOKAMAK

Engineers have tried to simulate these high temperatures and pressures in a controlled environment, which is provided by high temperature and strong magnetic fields in a fusion reactor. This reactor keeps a plasma at temperatures of 150 million ° C. This is high enough to provide the thermal environment for the proton–proton chain, for example.

Well-known Russian Physicists Igor Tamm (who established the term of surface states [Tamm 1932 a, b]) and his illustrious doctoral student, later dissident and Nobel Peace Prize Laureate [Sakharov 1975] Andrej Sacharow [Gorelik 2013] were among the first who figured that one could harness such plasma in a "**то**роидальная **ка**мера в **ма**гнитных **ка**тушках," a toroidal chamber with magnetic coils. The acronym built from the Russian term adds up to TOKAMAK, and this kind of fusion reactor is called TOKAMAK. The first reports on the TOKAMAK design appeared in the early 1960s [Matveev 1961]. In the mid-to-late 1960s, TOKAMAKs were considered for nuclear reactions [Artsimov 1969, Gashev 1965, Holcomb 1969, Hubert 1969]. Several national and multinational projects aimed and aim at demonstrating a controlled nuclear fusion for future energy production. Energy production for mankind by nuclear fusion in TOKAMAK is not being considered a realistic solution for the midterm. Maybe we must wait for another 50–100 years for this to become realized. It is technically just too difficult at this time to control a very hot proton and deuterium plasma for that purpose.

When Fleischmann and Pons [Fleischmann 1989b], and Hawkins, whose name was omitted in the original paper [Fleischmann 1989a], discovered that nuclear fusion could be forced when protons and deuterons were in or on palladium electrodes, the excitement and response by the scientific community were big. It seemed now that nuclear fusion was possible not only under extremely hot conditions but also under cold conditions. Cold fusion: the "mother load" of energy. Fleischmann writes later that he had figured this concept long before, but he had to conduct this project as a "hidden agenda" [Fleischmann 2006a].

13.3 The cold fusion

13.3.1 Electrochemists go nuclear

Thirty years ago, electrochemists Fleischmann and Pons discovered that their electrochemical cells, operated under conventional ambient conditions ($T \sim 300$ K, $p \sim 1$ bar), were slightly warming up when they were experimenting with electrocatalysis on noble metal electrodes in deuterated electrolytes [Fleischmann 1989a, b]. The amount of heat released from the cell was small but noticeable and reproducible. They found no other explanation for this observation than that of the nuclear fusion of protons and deuterons, mediated by the electrochemical environment. This implies a very strong claim that nuclear fusion was possible without the harsh, hot conditions that you have in sun or in a TOKAMAK.

Consequently, their discovery, published by 10 April 1989 as a "Preliminary Note," was called "cold fusion." This sounds unbelievable. The scientific community was therefore caught between excitement, skepticism and doubt, and ultimately by rejection. When a great discovery has been reasonably claimed, soon there will be followers, critical or not, who try to reproduce the claimed results. It is interesting how some researchers put great efforts in disproving claims made by other researchers. By 10 May 1989 (1 month after the first cold fusion paper appeared), *Physical Review Letters* received a theoretical work from Sun and Tomanek [Sun 1989], who calculated with density functional theory (DFT) the structure of a hypothetical PdD_2 crystal and compared it with Pd and PdD structures. Even at very high loading of Pd with deuterium, the distance between the deuterium nuclei would be so far apart that a fusion of the deuterium nuclei was "very improbable":

Our results show that the intramolecular distance $d(H_2)$ is expanded to 0.94 Å in the Pd lattice even at very high H concentrations in the hypothetical crystal PdH_2. At this large distance, cold nuclear fusion is even less probable than in the deuterium gas phase.

This sounds like an early and final verdict against the possibility of cold fusion. Gittus and Bockris were less negative and sought for possible theoretical explanations for the hypothesized cold fusion, which they published in May 1989 in *Nature* [Gittus 1989]. The aforementioned DFT study does not take into account potential dynamic effects when the lattice is properly excited. We learnt in this book that the protons in ceramic proton conductors can move along with the thermal excited lattice vibrations and thus behave as proton polaron [Braun 2017]. Resonant excitation of crystal lattices with infrared radiation can cause dramatic enhancement of proton transport [Samgin 2014, Spahr 2010]. It certainly would be interesting to see how such experimental conditions could affect the old experiment by Pons and Fleischmann. They also argue that the protons and deuterons could act as oscillators and delocalized species in the crystal lattice.

13.3.2 The experiment by Pons and Fleischmann

Pons and Fleischmann worked with specifically designed electrolysis cell which was integrated in a calorimeter Dewar which is sketched in Figure 13.3 [Fleischmann 1993].

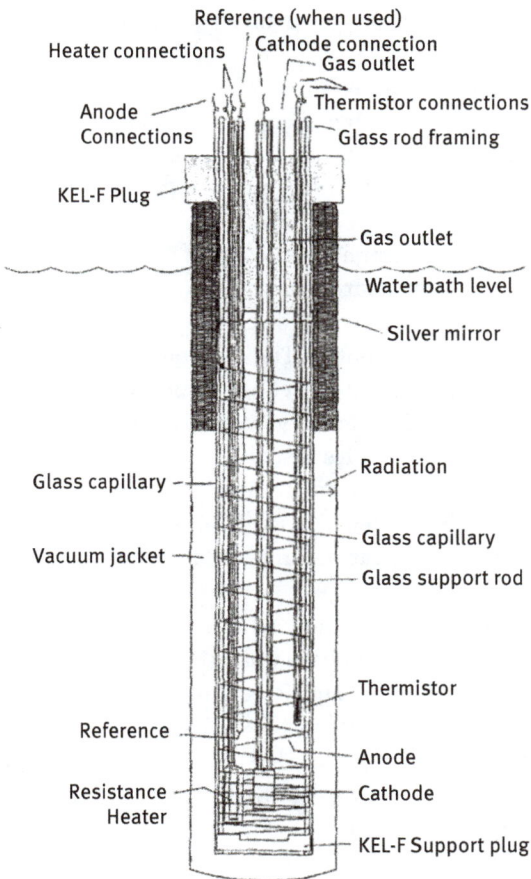

Figure 13.3: Schematic diagram of the single compartment vacuum Dewar open calorimeter cells used in this work.
Reprinted from Physics Letters A, 176, Fleischmann M, Pons S, Calorimetry of the Pd-D2O System – from Simplicity Via Complications to Simplicity, 118–129., Copyright (1993), with permission from Elsevier. [Fleischmann 1993].

They felt that a calorimeter was necessary for their study in order to use the suspected Joule heat evolution in the cell as a proof for nuclear fusion. Some reader may wonder why such a delicate analytical instrument like a calorimeter is used, which is designed to detect minute amounts of heat in a reaction, whereas cold fusion should produce so much heat that the experimenter would notice it immediately on their own body or maybe with a thermometer.

However, I am here reminded of my own research work, specifically hydrogen production by photo electrochemical water oxidation. In the beginning of my projects, I felt we would need a gas chromatograph for the precise quantification of the produced hydrogen and oxygen. Only later when we were working with quite successful metal oxide semiconductor photoanodes, we could see the gas bubbles evolve and catch them with simple glassware and measure the evolved gas volume with a simple ruler – for every bystander to see.

I guess that Fleischmann and Pons too had wished they would be able to produce such pretty obvious results to the general public, but they were bound to use the calorimeter. Moreover, they and their followers, for example Melvin Miles [Fleischmann 2006b, Miles 1990a, b, 1994, 2000, 2001], became experts in calorimetry and the device technology.

The cathode was a thick palladium sheet of 2 mm thickness and 8 cm length and 8 cm width. The palladium was surrounded by a large platinum-counter electrode. These electrodes were inserted into a large Dewar so that the electrochemical experiment could take place under precise temperature control. The electrolyte was a 0.1-M solution of LiOD dissolved in 99.5% D_2O with 0.5% H_2O. Let us go one step back and think what this means. Lithium hydroxide (LiOH) can be solved in water (H_2O) and this could be a conventional aqueous electrolyte. Both substances deliver the protons $^1H^+$ in the electrolyte. Fleischmann and Pons however wanted to deuterate their system; they wanted to have $^2H^+$ (this can be written also as $^2D^+$) instead of $^1H^+$. We recall that the deuteron contains one proton and one neutron and has thus the mass number 2.

When you look back in the section on ceramic proton conductors, you will read that we hydrated these ceramic slabs with vapor either from water (H_2O) or from heavy water (D_2O). There may be various reasons for doing so, but it was always for analytical purposes. You may for example see an isotope effect in the diffusion constants of protons $^1H^+$ or deuterons $^2H^+$. In my experiments, I protonated the ceramic slabs with water H_2O when we did quasi-elastic neutron scattering (QENS) [Braun 2009a, b, Chen 2013, 2012, Holdsworth 2010]. When we did ND, I deuterated it by using heavy water D_2O [Braun 2009c]. ^1H and ^2D have different coherent and incoherent scattering cross-sections for neutrons. Proper choice of H or D for scattering methods where we are interested in the coherent of incoherent response yields optimized results.

Pons and Fleischmann calculate that the pressure necessary to push deuterons together would be in excess of 10^{29} bar. They use the term "astronomically high" with respect to such high pressure. Fleischmann and Pons write they "compressed" the D^+

ions ($^2H^+$) from the electrolyte into the palladium cathode by using a galvanostatic method with a moderate current density of 1.6 mA/cm². This means they applied a negative potential to the cathode so that the D^+ ions would adsorb at its surface and potentially enter the palladium crystal lattice and become intercalated. This is the same principle that is used in the lithium intercalation battery.

The electrolysis steps are as follows [Fleischmann 1989b]: The heavy water molecule D_2O is oxidized at the cathode (palladium), and the resulting deuterons are adsorbed at the electrode surface. Instead of the corresponding hydroxyl group OH^-, we have now a deuteroxyl group OD^-:

$$D_2O + e^- \rightarrow D_{ads} + OD^-$$

With more heavy water available from the electrolyte, more deuterons will be produced which will bind to deuterium gas molecules at the electrode surface.

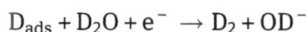

$$D_{ads} + D_2O + e^- \rightarrow D_2 + OD^-$$

Deuterons adsorbed at the electrode surface will diffuse into the crystal lattice of the palladium electrode

$$D_{ads} \rightarrow D_{lattice}$$

$$D_{ads} + D_{ads} \rightarrow D_2$$

Fleischmann had the opinion that the anticipated cold fusion was a bulk effect [Fleischmann 1994]. They therefore used electrodes of various geometry and size, that is, rods, sheets and cubes of palladium. They carefully monitored the evolution of the Joule heat, and it is interesting to note that they used current densities up to 512 mA/cm². Size of electrodes and evolved heat is listed in Table 13.1. I have not found any study which looked into the change of the molecular and electronic structure of the palladium before, during and after deuteration.

By doubling the diameter of the palladium rods (0.1–0.2–0.4 cm), the excess rate of heating increased from 0.0075 W by a factor of 4.8, and then again by a factor of 4.25 to finally 0.153 W, when the current density was 8 mA/cm². The increase by roughly a factor 4 corresponds to the parabolic increase of the mass of the Pd cylinder due to doubling the diameter. When the experimenters increased the size of the cylinders and the current density, the excess specific heating rate increased however disproportionate, as you can read from the right column in Table 13.1.

I have multiplied for the reader in Table 13.1 the values for the excess specific rate of heating at 8 mA/cm² by the factors 8 and 64 to check how much these values differ from the actually measured rates. The bulky rod with 0.4 cm diameter produces almost three times more heat when it is heavily loaded with deuterons due to the very high current density. Pons and Fleischmann attribute this considerable gain in heat due to the cold fusion of the deuterons in the palladium, which they "compressed" galvanostatically into the palladium.

Table 13.1: Generation of excess enthalpy in Pd cathodes as a function of current density and electrode size.

Electrode type	Dimensions (cm)	Current density (mA/cm^2)	Excess rate of heating (W)		Excess specific rate of heating (W/cm^3)
Rods	0.1 × 10	8	0.0075	0.095	=0.095
		64	0.079	1.01	>0.76 = 8 × 0.095
		512[a]	0.654[a]	8.33	>6.08 = 64 × 0.095
	0.2 × 10	8	0.036	0.115	=0.115
		64	0.493	1.57	>0.92 = 8 × 0.115
		512[a]	3.02[a]	9.61	>7.36 = 64 × 0.115
	0.4 × 10	8	0.153	0.122	=0.122
		64	1.751	1.39	>0.976
		512[a]	26.8[a]	21.4	>7.808
Sheet	0.2 × 8 × 8	0.8	0	0	
		1.2	0.027	0.0021	
		1.6	0.079	0.0061	
Cube	1 × 1 × 1	125	WARNING! IGNITION? See text		
		250			

[a]Measured on electrodes of length 1.25 cm and rescaled to 10 cm.
Reprinted from *Journal of Electroanalytical Chemistry and Interfacial Electrochemistry*, 261, Fleischmann M, Pons S, Electrochemically induced nuclear fusion of deuterium, 301–308. Copyright (1989), with permission from Elsevier. [Fleischmann 1989b].

The large palladium sheet shows an enhancement by the factor 3 (0.0021 -> 0.0061) when the current density is increased by 33% from a low 1.2 mA/cm^2 to a still low 1.6 mA/cm^2. This is a huge effect. But the catastrophic effect sets on when the experiments used the bulky palladium cube of 1 cm × 1 cm × 1 cm size: The experimenters issue in their paper the warning that ignition sets in at 125 mA/cm^2 with the palladium in cube geometry [Fleischmann 1989b].

The cube geometry is apparently successful, which was already acknowledged by Werner Heisenberg when he worked in World War II on nuclear fission projects with Uranium in Germany [Mayer 2015]:

> At the Kaiser Wilhelm Institute for Physics in Berlin, Werner Heisenberg experimented with uranium plates, whereas Kurt Diebner was working with uranium cubes at the German Army Ordnance Office. Heisenberg later acknowledged the superiority of the cube design.

The primary energy which is released from the fusion of deuterium and protons comes in the form of electromagnetic waves with extremely short wavelengths and very high energy in the MeV range. These are the γ-rays which cause severe harm to human body. The collisions of the produced particles such as deuterium will cause a warming up of the entire matter which will raise the temperature. This is the excess

Table 13.2: Generation of excess enthalpy in Pd rod cathodes expressed as a percentage of break-even values. All percentages are based on $^2D + {^2}D$ reactions, that is, no projection to $^2D + {^3}T$ reactions.

Electrode type	Dimensions (cm)	Current density (mA/cm^2)	Excess heating (% of break-even)		
			a	b	c
Rods	0.1 × 10	8	23	12	60
		64	19	11	79
		512	5	5	81
	0.2 × 10	8	62	27	286
		64	46	29	247
		512	14	11	189
	0.4 × 10	8	111	53	1224
		64	66	45	438
		512	59	48	839

[a]Percentage of break-even based on Joule heat supplied to cell and anode reaction $4OD^- \rightarrow 2D_2O + O_2 + 4e^-$. [b]Percentage of break-even based on total energy supplied to cell and anode reaction $4OD^- \rightarrow 2D_2O + O_2 + 4e^-$. [c]Percentage of break-even based on total energy supplied to cell and for an electrode reaction $D_2 + 2OD^- \rightarrow 2D_2O + 4e^-$ with a cell potential of 0.5 V.
Reprinted from *Journal of Electroanalytical Chemistry and Interfacial Electrochemistry*, 261, Fleischmann M, Pons S, Electrochemically induced nuclear fusion of deuterium, 301–308. Copyright (1989), with permission from Elsevier. [Fleischmann 1989b].

heat measured during cold fusion. With the right geometry of the palladium, they would be able to name a figure of merit for their reactor which could deliver an enormous amount of energy way beyond rthe break-even point, see Table 13.2. Pons and Fleischmann also claimed that they detected γ-rays and provided the evidence for that [Fleischmann 1989c].

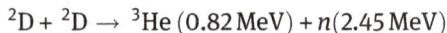

$$^2D + {^2}D \rightarrow {^3}T\,(1.01\,\text{MeV}) + {^1}H\,(3.02\,\text{MeV})$$

$$^2D + {^2}D \rightarrow {^3}He\,(0.82\,\text{MeV}) + n(2.45\,\text{MeV})$$

13.3.3 Other electrochemists aid to help

Soon after the claim by Pons and Fleischmann was published, Lin et al. in the Bockris group at Texas A&M University submitted as a preliminary note a speculative explanation about the potential mechanisms that lead to cold fusion observed by Fleischmann and colleagues [Lin 1990], which is sketched in Figure 13.4. Lin et al. speculated that during the extended electrochemical treatment of palladium, dendrites would grow with sharp tips which would allow for extraordinary large electric fields (A), over which a dielectric breakdown of the water molecules could take place, along with formation of a fluctuating gas volume

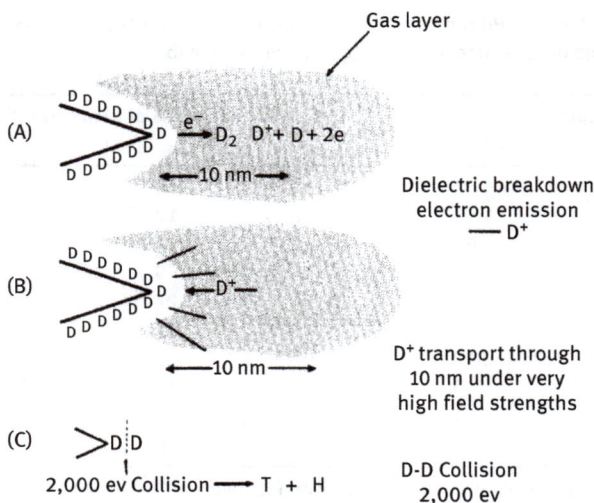

Figure 13.4: Model for the cold fusion.
Reprinted from *Journal of Electroanalytical Chemistry and Interfacial Electrochemistry*, 280, Lin GH, Kainthla RC, Packham NJC, Bockris JOM, Electrochemical fusion: a mechanism speculation, 207–211, Copyright (1990), with permission from Elsevier. [Lin 1990].

over the dendrite tip (B). Then, the strong electric field could accelerate one deuteron from the gas phase to a deuteron present at the tip with a collision of a very high kinetic energy of 2000 eV (C).

The aforementioned process of nuclear tunneling, which we know is a statistically extremely rare event, was determined by Gamow to be probable as [Gamow 1928]

$$G = \exp\left\{ -\pi e_0^2 \left(\frac{M_D 4 \, \pi^2}{h^2 E} \right)^{1/2} \right\}$$

with M_D being the rest mass of the deuteron and E the energy which the deuteron has when it transits from the electrolyte solution into the electrode (as shown in Figure 13.4(C)). Li et al. consider that deuterium gas can evolve electrochemically at the palladium surface until it is fully covered, depending on the current density i. The rate of collision between two deuterium atoms is then the current density divided by Faraday's constant, i/F, in mol/cm^2/s.

Lin et al. speculate now further and write if there is a fraction Γ of the electrode surface with *abnormally* high field strength, as suggested in Figure 13.4, then the energy at this location could be equal with the energy necessary for a D–D collision, and then the rate for the penetration (this is the fusion rate f) of the Gamow barrier would read

$$f = \Gamma(i/F) \cdot \exp\left(- \frac{2\pi^2 \epsilon^2}{h} \frac{M^{1/2}}{E^{1/2}} \right)$$

Lin et al. have calculated the fusion rate (f) for various energies (E) to be achieved at the tip of the dendrite, see for example Table 13.3. $E = 4000$ eV is the lowest energy where the agreement of their model with the experiment would be achieved [Lin 1990].

Figure 13.5 shows data by Fleischmann and Pons [Fleischmann 1993] recorded from calorimetry experiments. The temperature of the electrolysis cells is monitored for an entire week. In the beginning, the temperature of the cell is 38.7°C, and it is increasing homogeneously 39.7°C during 1 day. The water bath temperature was set to 30°C in an experiment room which had ambient temperature of 21°C. The increase of the temperature was accompanied by a decrease of the cell potential from 5.0 V to 4.9 V over the entire week.

Table 13.3: Relation between required energy and deuteron fusion rate, f. G is the probability for tunneling through Gamow barrier. E is the energy to be achieved at the dendrite tip due to electric field enhancement.

E (eV)	G	f ($i = 1$ A/cm^2)	$f\Gamma$ ($i = 1$ A/cm^2)
1000	2.63×10^{-14}	1.64×10^5	1.64×10^2
1500	8.18×10^{-12}	5.11×10^7	5.11×10^4
2000	2.50×10^{-10}	1.56×10^9	1.56×10^6
2500	2.59×10^{-9}	1.61×10^{10}	1.61×10^7
3000	1.44×10^{-8}	9.00×10^{10}	9.00×10^7
3500	5.52×10^{-8}	3.45×10^{11}	3.45×10^8
4000	1.62×10^{-7}	1.01×10^{12}	1.01×10^9

Reprinted from *Journal of Electroanalytical Chemistry and Interfacial Electrochemistry*, 280, Lin GH, Kainthla RC, Packham NJC, Bockris JOM, Electrochemical fusion: a mechanism speculation, 207–211, Copyright (1990), with permission from Elsevier. [Lin 1990].

At some point, the electrolyte would start boiling and the electrolyte would rapidly evaporate. In order to monitor the time properly where this would happen, they used a video camera and recorded time lapse images so that they could inspect later the condition of all cells in the photos and assign the state of cell to the time from the time stamp. One photograph from this time lap series is shown in Figure 13.6.

Also, other researchers commented on the possible mechanism for the reported cold fusion. Around 5 years later after their discovery, Fleischmann and Pons, and Preparata speculate on possible theories for the cold fusion [Fleischmann 1994]. Today, research on cold fusion goes by low energy nuclear reactions. A small group of researchers is active in the field, said Ludwik Kowalski in a commentary [Kowalski 2010] in *Physics Today*:

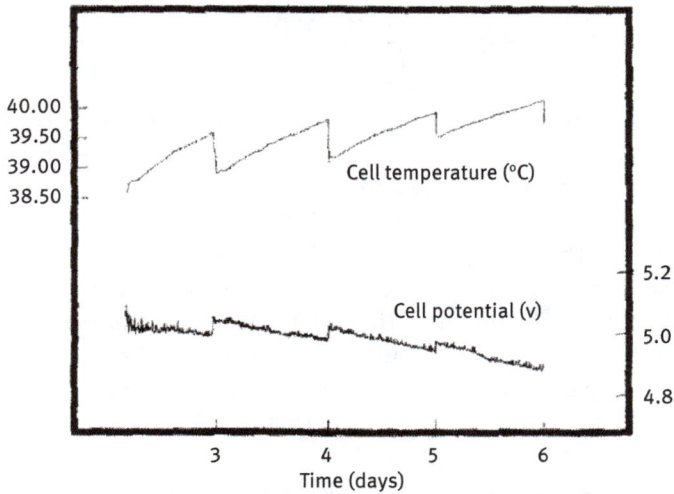

Figure 13.5: Cell temperature (upper) and cell potential (lower) versus time since the cell was started for the electrolysis of D_2O in 0.6 M Li_2SO_4 solution at pH 10 at a palladium rod cathode (0.4 × 1.25 cm). The cell current was 400 mA, the water bath temperature was 30.00°C and the room temperature was 21°C. The rate of excess enthalpy generation at the end of each day was 0.045 W (day 3), 0.066 W (day 4), 0.086 W (day 5) and 0.115 W (day 6). The accumulation of excess enthalpy for this period was on the order of 26 kJ (1.5 MJ/(mol Pd)).
Reprinted from Physics Letters A, 176, Fleischmann M, Pons S, Calorimetry of the Pd-D_2O System – from Simplicity Via Complications to Simplicity, 118–129, Copyright (1993), with permission from Elsevier. [Fleischmann 1993].

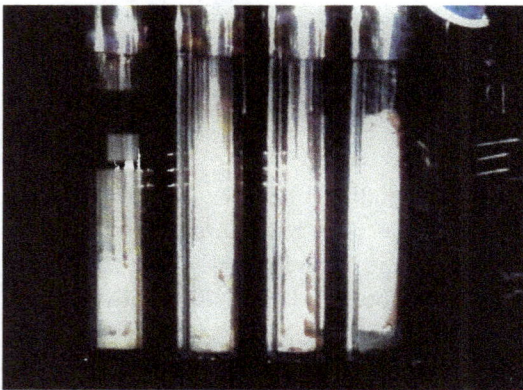

Figure 13.6: Still of video recordings of the cells described in Figure 13.3 showing the last cell during the final boiling period, the other cells having boiled dry.
Reprinted from Physics Letters A, 176, Fleischmann M, Pons S, Calorimetry of the Pd-D_2O System – from Simplicity Via Complications to Simplicity, 118–129, Copyright (1993), with permission from Elsevier. [Fleischmann 1993].

I believe that reports made by recognized scientists should be taken seriously, even when their results conflict with what is expected.

13.4 About reproducibility of experiments: some personal remarks

During the literature search for my diploma thesis on magnetic films in the library at KFA Jülich, I became aware how published data on magnetic materials scattered in the early stages of a new research field. Different groups published different results although nominally the systems that they worked on were very similar or identical. With increasing activity and increasing progress in that field, as the years went by, the scattering of data points became less. It appeared like the data points were asymptotically merging together over time to one solid conclusion.

I believe the reason for this was that the different research groups had, for example, not the same vacuum conditions for their experiment. Possibly their evaporation sources for film growth were not well calibrated. Small technical peculiarities can have a significant influence on the outcome of experiments. A very low concentration of CO_2 in the vacuum chamber may be still enough carbon in the system so that carbon impurities react with an iron film and then form steel, which can have considerable consequences for the properties of this iron film. When these researchers get in contact by mutual visits in their laboratories or attendance of conferences, they may learn about the specifics of the experiments of the other groups; specifics which have not been disclosed in publications because they were considered not worthwhile to mention or were not considered relevant by the research leader of the study, or the author of the manuscript, or a reviewer or an editor. When a peer reviewer suggests that you skip some technical detail in the paper for that reason, you perhaps follow the suggestion in order to get your paper accepted for publication. You maybe do so, rather than arguing with a reviewer and thus jeopardizing acceptance of your paper. You may even strongly disagre and be ready to argue and fight, but the academic degree of one or more of your coauthors of the paper depends on the rapid acceptance of the paper. And as a responsible principal investigator, you may find yourself in a situation of several conflicting interests. Are you going to fight over a seemingly ridiculous technical detail, the consequences of which you cannot reasonably foresee at the time of submission or peer review, and thus delaying graduation of a good student?

And when after several years the data points are less scattering, when they are densifying toward one solid value, are therefore all the other previous researchers "wrong"? The efforts of the very first pioneers were necessary in order to get the entire research field launched and started. Certainly, those who followed make it better. But the first one is the first one. You cannot make it to the top step of a ladder without using the lower steps. This is an essential insight from research and development.

Economic thinking would like to cut away these lower steps in order to make research more efficient.

While working in one of the many laboratories that I have been to in my scientific career, I was with an undergraduate student in our lab whose task was to reproduce the results of an eminent battery professor from another university. The student tried everything he could for about 1 year but was unable to reproduce the results. I guess one reason for that was that not all details were disclosed in the paper. And I am not blaming anyone for this.

When I was getting my first experiences with lithium ion batteries, I necessarily got in contact with alkaline metals. In order to foresee the potential dangers when working with lithium, I cut some small pieces of the shiny metal and kept it in plastic weighing boats in open atmosphere on my desk. At some point, one piece of lithium would eventually turn white. At some other occasion when I had again lithium, I kept a piece on my desk and that would eventually turn black. Obviously, the lithium metal had reacted with the gases from the atmosphere and changed its phase from metal to some lithium compound. I did some literature study and found that the white powder would be lithium hydroxide LiOH and the black powder lithium nitrite Li_3N. Now as I am writing this, I regret that I had not done an x-ray diffractogramm (XRD) on the samples back then to actually check for the crystal phases by myself. But I have no doubt that I was seeing LiOH, likely after the laboratory atmosphere must have been quite humid so that the H_2O from humidity would react with the lithium to LiOH. On the drier days in the laboratory, the lithium would favorably react with the nitrogen N_2, of which the atmosphere has almost 80%. This was from the same batch of lithium and it was kept in the same laboratory in the same location. The atmosphere in the lab would make the difference on the material.

For the preparation of my sabbatical in Hawaii, I made sputter targets from iron oxide and tungsten oxide. For the tungsten oxide sintering, I followed the recipe which I received from my colleague in Hawaii. When I opened the furnace on a Monday morning in our lab in Switzerland, only the iron oxide sputter target was there. The tungsten oxide sputter target disappeared. I never really figured out why the WO_3 had evaporated in the furnace. But I learnt that WO_3, like many other oxides of heavy elements, is volatile in relatively low temperatures – unlike their metals. But why would the WO_3 disappear in Switzerland but not in Hawaii? Maybe because University of Hawaii at Manoa is close to sea level (less than 100 m above) and Dübendorf in Switzerland at 440 m above sea level. I leave it as an exercise to the reader to calculate the pressure difference (more than 50 mbar) which arises from the height different of the two different locations, and whether such pressure difference could influence the temperature at which the tungsten oxide evaporates in a furnace over 24 hours.

Melvin Miles argues that those researchers who failed to reproduce the findings of Pons and Fleischmann and the findings of Miles did not wait sufficiently long for the experiments to become successful. It would require several weeks before the claimed effect could be observed, provided the electrode material was good one [Miles 1998].

My PhD thesis centers on the thermal gas phase oxidation of carbon. This process occurred practically at around 450°C, which is a high temperature. This temperature was sufficient to give the carbon a very high internal surface area [Braun 1999]. The same effect could be obtained electrochemically by anodization of the carbon electrode. The potential necessary for this effect was around 2 V versus Ag/AgCl reference [Sullivan 1997, 2000]. Back then, it appeared to me that such low electrochemical potential was quite disproportionate to the temperature of 450 °C. Around 2 V is what a small battery can accomplish. Can that be equal to a high temperature like 450 °C? Are electrochemical methods so powerful? Obviously they are.

Let me recall the double layer, over the thickness of which we have an incredibly high electrical field of 1 V/Å. Are fields as strong as this capable of forcing deuterons together so that they can fuse? No, they are not. If so, then we would have observed catastrophic reactions already long time ago.

13.5 When the sun sets

The anthropologic relevance of the sun is deeply rooted in its practical importance for mankind, its simple utility in human daily life. Sun bears strong symbolism in ancient and contemporary and popular cultures. The daylight has become the center of attention to some of my colleagues and this is why we founded, as a spin-off from the VELUX-Foundation, the Daylight Academy [Norton 2017]. Daylight is very important for the well-being of humans. And daylight is offered by the sun. There are countless songs and music featuring the sun literally.

I have started this chapter with a photo of the sunrise over Death Valley. I am ending this book with a photo of the sunset in Figure 13.7. There you are looking at the sunset over Mount Tamalpais in Mill Valley, California, the native home of the Coast Miwok People, and Indian tribe. To the Miwok People, Mount Tamalpais was the bride of the sun god. The rays of the sun setting over the Pacific Ocean at Point Reyes were the paths for Miwok souls into eternity. Meanwhile, sun maintains reliably its eternal cycle of sunrise and sunset for us. As night dawns over the San Francisco Bay Area, people in Australia and Eastern Asia are preparing for the new day. Night and winter are the dark phases to sun's cycle. Since creation of life billions of years ago, nature has coped with night and with winter by energy storage. Scientists, engineers and technologists are working now worldwide to make sure that in future you have all the energy you need – every time you need it – with solar energy storage by artificial photosynthesis.

We believe that we will run out of fossil fuels in the next few hundreds of years. After that, we have to get our energy from the sun as renewable energy, or from nuclear power plants. Maybe geothermal energy will help us considerably. But can solar energy and its derivatives hydropower and wind power be considered renewable energy, when the cycles of day and night, summer and winter will fade away as the sun, our hydrogen star, will have entirely turned into helium in the next 4–5 billion years?

Figure 13.7: Sunset over Mount Tamalpais, Marin County, California.
Photo recorded by Artur Braun at Lawrence Berkeley National Laboratory. [Braun 2015a].

If for any unforeseen reason and instance we cannot get our energy anymore from the sun, we should be prepared for this by having preserved enough fossil fuels and by having developed a viable nuclear energy technology, preferably nuclear fusion, as featured in this last chapter of this book. But irrespective of all bad scenarios, it is now worthwhile to tap the sun as long it is there.

Did you know? In only 1 hour, we receive from our sun the same amount of energy which we use in one entire year.

Appendix Resistor Color Code

Color	1. Ring 1. Numeral	2. Ring 2. Numeral	3. Ring Multiplier	4. Ring Tolerance
none				20 %
silver	–		0.01	10 %
gold	–		0.1	5 %
black	–	0	1	
brown	1	1	10	1 %
red	2	2	100	2 %
orange	3	3	1000	
yellow	4	4	10000	
green	5	5	100000	0.5 %
blue	6	6	1000000	0.25 %
violet	7	7	10000000	0.1 %
grey	8	8	100000000	0.05
white	9	9	1000000000	

Farbe	1. Ring 1. Numeral	2. Ring 2. Numeral	3. Ring 3. Numeral	4. Ring Multiplier	5. Ring Tolerance	6. Ring Temp.- Coefficient
silver				0.01	10 %	
gold				0.1	5 %	
black	–	0	0	1		200 ppm/K
brown	1	1	1	10	1 %	100 ppm/K
red	2	2	2	100	2 %	50 ppm/K
orange	3	3	3	1000		15 ppm/K
yellow	4	4	4	10000		25 ppm/K
green	5	5	5	100000	0.5 %	
blue	6	6	6	1000000	0.25 %	10 ppm/K
violet	7	7	7	10000000	0.1 %	5 ppm/K
grey	8	8	8	100000000	0.05 %	
white	9	9	9	1000000000		

https://doi.org/10.1515/9783110561838-014

Bibliography

[Acevedo-Pena 2017] Acevedo-Pena P, Baray-Calderon A, Hu HL, Gonzalez I, Ugalde-Saldivar VM: Measurements of HOMO-LUMO levels of poly(3-hexylthiophene) thin films by a simple electrochemical method. *Journal of Solid State Electrochemistry* 2017, 21:2407–2414. doi: 10.1007/s10008-017-3587-2.

[Acheson 1910] Acheson EG: A Pathfinder: Discovery, Invention and Industry; How the World Came to Have Aquadag and Oildag; Also Carborundum, Artificial Graphite. New York: The Press Scrap Book; 1910.

[ACS 1999] ACS: The chemical revolution of Antoine-Laurent Lavoisier – international historic chemical landmark. 1999. ACS. http://www.acs.org/content/acs/en/education/whatischemistry/landmarks/lavoisier.html

[Adamson 1973] Adamson A: *A Textbook of Physical Chemistry*. 2nd edn. Burlington: Elsevier Science ; 1973.

[Adelberger 1998] Adelberger EG: Solar fusion cross sections. *Review of Modern Physics* 1998, 70:1265–1291. doi: 10.1103/RevModPhys.70.1265.

[Adelberger 2011] Adelberger EG, García A, Robertson RGH, Snover KA, Balantekin AB, Heeger K, Ramsey-Musolf MJ, Bemmerer D, Junghans A, Bertulani CA, Chen JW, Costantini H, Prati P, Couder M, Uberseder E, Wiescher M, Cyburt R, Davids B, Freedman SJ, Gai M, Gazit D, Gialanella L, Imbriani G, Greife U, Hass M, Haxton WC, Itahashi T, Kubodera K, Langanke K, Leitner D, Leitner M, Vetter P, Winslow L, Marcucci LE, Motobayashi T, Mukhamedzhanov A, Tribble RE, Nollett KM, Nunes FM, Park TS, Parker PD, Schiavilla R, Simpson EC, Spitaleri C, Strieder F, Trautvetter HP, Suemmerer K, Typel S: Solar fusion cross sections. II.
The p p chain and CNO cycles. *Reviews of Modern Physics* 2011, 83:195–245. doi: 10.1103/RevModPhys.83.195.

[Agnew 2005] Agnew GD, Bozzolo M, Moritz RR, Berenyi S, SOFC Asme: *The Design and Integration of the Rolls-Royce Fuel Cell Systems 1 MW*. New York: Amer Soc Mechanical Engineers; 2005.

[Aguiar 2008] Aguiar P, Brett DJL, Brandon NP: Solid oxide fuel cell/gas turbine hybrid system analysis for high-altitude long-endurance unmanned aerial vehicles. *International Journal of Hydrogen Energy* 2008, 33: 7214–7223.doi: 10.1016/j.ijhydene.2008.09.012.

[Ahn 2005] Ahn SY, Jeon US, Ahn BK: *Development of the BOP Bench System for PEM Fuel Cell Vehicles at Hyundai MOBIS*. 2005.

[Ahn 2006] Ahn BK, Lim TW, IEEE: *Fuel Cell Vehicle Development at Hyundai-Kia Motors*. 2006.

[Aivazov 1968] Aivazov MI, Domashnev IA: Influence of porosity on the conductivity of hot-pressed titanium-nitride specimens. *Soviet Powder Metallurgy and Metal Ceramics* 1968, 7:708–710. doi: 10.1007/BF00773737.

[Alexander 2013] Alexander P: Marikana, turning point in South African history. *Review of African Political Economy* 2013, 40:605–619. doi: 10.1080/03056244.2013.860893.

[Alexander 2016] Alexander P: Marikana commission of inquiry: From narratives towards history. *Journal of Southern African Studies* 2016, 42:815–839. doi: 10.1080/03057070.2016.1223477.

[Alexandrov 2010] Alexandrov AS, Devreese JT: *Advances in Polaron Physics*. Berlin Heidelberg: Springer; 2010. doi: 10.1007/978-3-642-01896-1_3.

[Alfaro 1998] Alfaro P, Castro S: The zinc refinery of IMMSA in San Luis Potosi, Mexico. *Zinc and Lead Processing* 1998:71–83.

[Ali 2016] Ali M, Zhou F, Chen K, Kotzur C, Xiao C, Bourgeois L, Zhang X, Macfarlane DR: Nanostructured photoelectrochemical solar cell for nitrogen reduction using plasmon-enhanced black silicon. *Nature Communications* 2016, 7:11335. doi: 10.1038/ncomms11335.

[Allen 1991] Allen JL: *Biosphere 2: The Human Experiment*. London: Penguin Books; 1991.

https://doi.org/10.1515/9783110561838-015

[Bard (Editor) 2002] Allen J. Bard MSE , Stuart Licht (Editors): *Encyclopedia of Electrochemistry, Volume 6, Semiconductor Electrodes and Photoelectrochemistry*. Weinheim: Wiley, VCH; 2002.

[Allisson 1930a] Allisson F: Oxydo-Reduktionen mit Chlorophyll und anderen Sensibilatoren – ETH E-Collection. *Theses*. ETH Zürich: Chemie; 1930a.

[Allisson 1930b] Allisson F: Oxydo-Reduktionen mit Chlorophyll und anderen Sensibilatoren. *Helvetica Chimica Acta* 1930b, 13:788–805. doi: 10.1002/hlca.19300130509.

[Alon 2014] Alon L, Ravikovich G, Mandelbrod M, Eilat U, Schop Z, Tamari D: *Computer-Based Management of Mirror-Washing in Utility-Scale Solar Thermal Plants*. New York: Amer Soc Mechanical Engineers; 2014.

[Alvarez 2017] Alvarez A, Saez JM, Davila Costa JS, Colin VL, Fuentes MS, Cuozzo SA, Benimeli CS, Polti MA, Amoroso MJ: Actinobacteria: Current research and perspectives for bioremediation of pesticides and heavy metals. *Chemosphere* 2017, 166:41–62. doi: 10.1016/j.chemosphere.2016.09.070.

[Amnesty 2016] Amnesty: *Smoke And Mirrors: Lonmin's Failure to Address Housing Conditions at Marikana, South Africa. 2016*. London: . Amnesty International. AFR 53/4552/2016; 11 August 2016.

[Amphlett 1997] Amphlett JC, De Oliveira EH, Mann RF, Roberge PR, Rodrigues A, Salvador JP: Dynamic interaction of a proton exchange membrane fuel cell and a lead-acid battery. *Journal of Power Sources* 1997, 65:173–178. doi: 10.1016/s0378-7753(97)02472-5.

[Andaloro 2011] Andaloro L, Sergi F Napoli G, Dispenza G, Ferraro M, Antonucci V: *Modelling of a Range Extender Power Train for a City Bus*. 2011.

[Anderson 1972] Anderson PW: More is different. *Science* 1972, 177:393–396. doi: 10.1126/science.177.4047.393.

[Anderson 1979] Anderson S, Constable EC, Dare-Edwards MP, Goodenough JB, Hamnett A, Seddon KR, Wright RD: Chemical modification of a titanium (IV) oxide electrode to give stable dye sensitisation without a supersensitiser. *Nature* 1979, 280:571–573. doi: 10.1038/280571a0.

[Anonymous 1985] Anonymous: Pemex – the forgotten disaster. *Chemical Engineer-London* 1985:16–17.

[Anonymous 2012] Anonymous: Peace sought following violence at Lonmin's Marikana mine. *E&Mj-Engineering and Mining Journal* 2012, 213:6–7.

[Antonucci 2015] Antonucci V, Brunaccini G, De Pascale A, Ferraro M, Melino F, Orlandini V, Sergi F: Integration of mu-SOFC generator and ZEBRA batteries for domestic application and comparison with other mu-CHP technologies. In *Clean, Efficient and Affordable Energy for a Sustainable Future. Volume 75*. Edited by Yan J, Shamim T, Chou SK, Li H; 2015: 999–1004: *Energy Procedia*. doi: 10.1016/j.egypro.2015.07.335.

[Archer 1975] Arche r MD: Electrochemical aspects of solar energy conversion. *Journal of Applied Electrochemistry* 1975, 5:17–38.

[Archives 1959] Archives A: American Chemical Society Official Reports 136th National Meeting. *Chemical & Engineering News* 1959, 37:64–83. doi: 10.1021/cen-v037n044.p064.

[Arlt 2014] Arlt T, Schroder D, Krewer U, Manke I: In operando monitoring of the state of charge and species distribution in zinc air batteries using X-ray tomography and model-based simulations. *Physical Chemistry Chemical Physics* 2014, 16:22273–22280. doi: 10.1039/c4cp02878c.

[Arnold 2004] Arnold VI: Das Doppelschichtpotential. In *Vorlesungen über partielle Differentialgleichungen, Springer-Lehrbuch, Springer*. Berlin, Heidelberg. 2004: 107–1170. doi: 10.07/3-540-35031-4_10.

[Aro 2017] Aro E-M, Vincent A, Azzolini R, Barbieri A, Baumann S, Bercegol H, Braun A, Campus P, Chandezon F, De Groot H, De La Pena V, Durrant J, Faber C, Fleischer M, Fresno F, Gaertner T, Hammarstrom L, Joanna K, Antoni L, Lopez L, Lorena T, Loreto F, Meier M, Noble A, Tondelli L,

Van Der Velden E: SUNRISE – Solar energy for a circular economy. In *H2020 CSA Project Grand Proposal for EU Flagship Initiative*, SEP-210522421. Leiden, The Netherlands; 2017.

[Arriaga 2007] Arriaga LG, Martinez W, Cano U, Blud H: Direct coupling of a solar-hydrogen system in Mexico. *International Journal of Hydrogen Energy* 2007, 32:2247–2252. doi: 10.1016/ j.ijhydene.2006.10.067.

[Artemov 2015] Artemov VG, Volkov AA, Sysoev NN, Volkov AA: Conductivity of aqueous HCl, NaOH and NaCl solutions: Is water just a substrate? *EPL (Europhysics Letters)* 2015, 109. doi: 10.1209/0295-5075/109/26002.

[Artsimov 1969] Artsimov L, Bobrovsk G, Gorbunov EP, Ivanov DP, Kirillov VD, Kuznetso.Ei, Mirnov SV, Petrov MP, Razumova KA, Strelkov VS, Shcheglo D: Experiments in Tokamak Devices. In *THIRD INTERNATIONAL CONFERENCE ON PLASMA PHYSICS AND CONTROLLED NUCLEAR FUSION RESEARCH HELD BY THE INTERNATIONAL ATOMIC ENERGY AGENCY AT NOVOSIBIRSK, 1–7 Aug 1968*; Novosibirsk, Soviet Union. INTERNATIONAL ATOMIC ENERGY AGENCY, Vienna; 1969: 17.

[Aruna 2000] Aruna ST, Muthuraman M, Patil KC: Studies on combustion synthesized LaMnO3-LaCoO3 solid solutions. *Materials Research Bulletin* 2000, 35:289–296. doi: 10.1016/ S0025-5408(00)00212-9.

[Ashcroft 1976] Ashcroft NW, Mermin N: *Solid State Physics*. Holt, Rinehart and Winston, New York: Cengage Learning, Inc; 1976.

[Ashley 2012] Ashley S. Fuel-cell vehicles from Hyundai coming soon. *SAE International*. 27 Nov 2012. http://articles.sae.org/11518/

[Atchison 1986] Atchison F, Fischer WE, Pepin M, Takeda Y, Tschalaer C, Furrer A: The Spallation-Neutron-Source Project SINQ. *Physica B & C* 1986, 136:97–99. doi: 10.1016/ s0378-4363(86)80029-8.

[Atchison 1989] Atchison F, Bucher W, Hochli A, Horvath I, Nordstrom L: *The D2 Cold-Neutron Source for SINQ*. Bristol: IOP Publishing Ltd; 1989.

[Atkins 1997] Atkins P: *Physical Chemistry*. 6th edn. New York: W.H. Freeman and Company; 1997.

[Atkins 2010] Atkins P: *Inorganic Chemistry*. 5th edn. New York: W. H. Freeman and Company; 2010.

[Atkinson 1929] Atkinson RDE, Houtermans FG: Zur Frage der Aufbaumöglichkeit der Elemente in Sternen. *Zeitschrift fr Physik* 1929, 54:656–665. doi: 10.1007/BF01341595.

[Attiga 1987] Attiga AA: *Arabs and the Oil Crisis, 1973–86*. Kuwait: Organization of Arab Petroleum Exporting Countries (OAPEC); 1987.

[Augustynski 2008] Augustynski J, Braun A, Grätzel M, Kuznetsov A, Mendes A, Meda L, Rothschild A, Van De Krol R, Weidenkaff A: NANOPEC (Nanostructured Photoelectrodes for Energy Conversion). *Europe: European Commission* 2008..

[Augustynski 2009] Augustynski J, Braun A, Grätzel M, Kuznetsov AM, Mendes A, Meda L, Rothschild A, Van De Krol R, Weidenkaff A: 2009. NANOPEC – nanostructured photoelectrodes for energy conversion. European Commission. Europe. EUR 2 699 909. 3 years. finished. https:// cordis.europa.eu/project/rcn/89410_en.html

[Austvik 1992] Austvik OG: The oil price war – oil and the conflict in the Persian Gulf area. *Internasjonal Politikk* 1992, 50:277–289.

[Avise 1994] Avise JC: THE REAL MESSAGE FROM BIOSPHERE-2. *Conservation Biology* 1994, 8:327–329. doi: 10.1046/j.1523-1739.1994.08020327.x.

[Aylward 2008] Aylward G, Findlay T: *SI Chemical Data*. Australia: John Wiley & Sons; 2008.

[Ayrton 1891] Ayrton WE: *Practical Electricity: A Laboratory and Lecture Course*. 5th edn. London, Paris & Melbourne: Cassell & Company, Limited; 1891.

[Baczynska 2012] Baczynska K, Price L: Efficacy and ocular safety of bright light therapy lamps. *Lighting Research and Technology* 2012, 45:40–51. doi: 10.1177/1477153512443062.

[Bailey 1997] Bailey PG, Worthington NC: History and applications of HAARP technologies: The high frequency active auroral research program. *Iecec-97 – Proceedings of the Thirty-Second Intersociety Energy Conversion Engineering Conference, Vols 1–4* 1997:1317–1322. doi: 10.1109/Iecec.1997.661959.

[Bajpai 2007] Bajpai S, Gupta JP: Terror-proofing chemical process industries. *Process Safety and Environmental Protection* 2007, 85:559–565. doi: 10.1205/psep06046.

[Bajpai 2008] Bajpai S, Gupta JP, Uddin M: *Security Risk Management Programme for Chemical Process Industries: Indian Perspective.* Beijing: Science Press Beijing; 2008.

[Bakst 2014] Bakst A: DFATR-PEMFC The Probable AIP Structure of Submarine SMX® Ocean (concept) Part 1. Bakst Engineering 2014.

[Balasubramanian 2013] Balasubramanian S, Wang P, Schaller RD, Rajh T, Rozhkova EA: High-performance bioassisted nanophotocatalyst for hydrogen production. *Nano Letters* 2013, 13:3365–3371. doi: 10.1021/nl4016655.

[Bannister 1992] Bannister M: *Science and Technology of Zirconia V.* 1st edn. Boca Raton: CRC Press; 1992.

[Baqué 2013] Baqué M, De Vera J-P, Rettberg P, Billi D: The BOSS and BIOMEX space experiments on the EXPOSE-R2 mission: Endurance of the desert cyanobacterium Chroococcidiopsis under simulated space vacuum, Martian atmosphere, UVC radiation and temperature extremes. *Acta Astronautica* 2013, 91:180–186. doi: 10.1016/j.actaastro.2013.05.015.

[Barcia 2002] Barcia OE, D'elia E, Frateur I, Mattos OR, Pébère N, Tribollet B: Application of the impedance model of de Levie for the characterization of porous electrodes. *Electrochimica Acta* 2002, 47:2109–2116. doi: 10.1016/s0013-4686(02)00081-6.

[Bard 1985] Bard AJ, Parsons R, Jordan J: *Standard Potentials in Aqueous Solution.* 1st edn: New York, Basel: CRC Press; 1985. ISBN: 978-0824772918.

[Bard 2001] Bard AJ, Faulkner CJ: *Electrochemical Methods. Fundamentals and Applications.* 2nd edn. Hoboken, NJ: John Wiley and Sons Inc; 2001.

[Barlow 1999] Barlow S, Bunting HE, Ringham C, Green JC, Bublitz GU, Boxer SG, Perry JW, Marder SR: Studies of the electronic structure of metallocene-based second-order nonlinear optical dyes. *Journal of the American Chemical Society* 1999, 121:3715–3723. doi: 10.1021/ja9830896.

[Barsoukov 2005] Barsoukov E, Macdonald JR: *Impedance Spectroscopy.* Hoboken, NJ: John Wiley & Sons, Inc.; 2005. doi: 10.1002/0471716243.

[Barsoukov 2005] Barsoukov E, Macdonald JR: *Impedance Spectroscopy: Theory, Experiment, and Applications,* 2nd ed.; John Wiley & Sons, Inc. 2005. doi: 10.1002/0471716243

[Barthel 1979] Barthel J, Wachter R, Gores HJ: Temperature dependence of conductance of electrolytes in nonaqueous solutions. In *Modern Aspects of Electrochemistry.* Series Editor Conway BE. Boston, MA: Springer; 1979: 1–79. doi: 10.1007/978-1-4615-7455-2_1.

[Bartolozzi 1989] Bartolozzi M: Development of redox flow batteries – a historical bibliography. *Journal of Power Sources* 1989, 27:219–234. doi: 10.1016/0378-7753(89)80037-0.

[Bartolozzi 1994] Bartolozzi M, Braccini G, Marconi PF, Bonvini S: Recovery of zinc and manganese from spent batteries. *Journal of Power Sources* 1994, 48:389–392. doi: 10.1016/0378-7753(94)80035-9.

[Barton 2014] Barton SC, Minteer S: In recognition of adam heller and his enduring contributions to electrochemistry. *Journal of the Electrochemical Society* 2014, 161. doi: 10.1149/2.0201413jes.

[Bärtsch 1999] Bärtsch M, Braun A, Schnyder B, Kotz R, Haas O: Bipolar glassy carbon electrochemical double-layer capacitor: 100,000 cycles demonstrated. *Journal of New Materials for Electrochemical Systems* 1999, 2:273–277.

[Bash 1945] Bash KW: Zum 70. Geburtstag von C. G. Jung. *Experientia* 1945, 1:126–126. doi: 10.1007/BF02152976.

[Bates 1992] Bates JB, Dudney NJ, Gruzalski GR, Zuhr RA, Choudhury A, Luck CF, Robertson JD: Electrical properties of amorphous lithium electrolyte thin films. *Solid State Ionics* 1992, 53–56:647–654. doi: 10.1016/0167-2738(92)90442-R.

[Batteriesinternational 2016] Battery Pioneer Johan Coetzer: Molten salt and zebras... http://www.batteriesinternational.com/battery-pioneers-johan-coetzer/

[Batterybro 2015] Batterybro. Blog. 5 July 2018 . Batteries for electric airplanes: Solar Impulse II – flying across the Pacific. Battery bro. 2018. https://batterybro.com/blogs/18650-wholesale-battery-reviews/31850627-batteries-for-electric-airplanes-solar-impulse-ii-flying-across-the-pacific

[Bauer 1998a] Bauer GS: Operation and development of the new spallation neutron source SINQ at the Paul Scherrer Institute. *Nuclear Instruments & Methods in Physics Research Section B-Beam Interactions with Materials and Atoms* 1998a, 139:65–71. doi: 10.1016/s0168-583x(97)00956-7.

[Bauer 1998b] Bauer GS, Fischer WE, Rohrer U, Schryber U: *Commissioning of the 1 MW Spallation Neutron Source SINQ.* New York: IEEE; 1998b.

[Bauer 1999] Bauer GS, European Nucl SOC: SINQ as a versatile alternative neutron source. *Rrfm'99: 3rd International Topical Meeting on Research Reactor Fuel Management* 1999:1–9.

[Baur 1937] Baur E, Preis H: Über Brennstoffketten mit Festleitern. *Zeitschrift Fur Elektrochemie Und Angewandte Physikalische Chemie* 1937, 43:727–732. doi: 10.1002/bbpc.19370430903.

[Baxter 2012] Baxter JB: Commercialization of dye sensitized solar cells: Present status and future research needs to improve efficiency, stability, and manufacturing. *Journal of Vacuum Science & Technology A* 2012, 30. doi: Artn 02080110.1116/1.3676433.

[Bayraktar 2008] Bayraktar D: La0.5Sr0.5Fe1-yMyO3-δδδδ (M = Ti, Ta) Perovskite oxides for oxygen separation membranes ÉCOLE POLYTECHNIQUE FÉDÉRALE DE LAUSANNE LA FACULTE SCIENCES ET TECHNIQUES DE L'INGÉNIEUR 2008.

[Beaty 2013] Beaty HW, Fink DG : Calculation of voltage regulation and I2R loss. In *Standard Handbook for Electrical Engineers*, 16th ed. McGraw Hill Professional, Access Engineering; 2013.

[Beck 2001] Beck C, Janssen S, Gross B, Hempelmann R: Neutron time-of-flight spectrometer focus at SINQ: Results from nanocrystalline matter studies. *Scripta Materialia* 2001, 44:2309–2313. doi: 10.1016/s1359-6462(01)00894-6.

[Becker 1956] Becker EW: Die Herstellung von schwerem Wasser. *Angewandte Chemie* 1956, 68:6–13. doi: 10.1002/ange.19560680103.

[Becker 1961] Becker RO: Search for evidence of axial current flow in peripheral nerves of salamander. *Science* 1961, 134:101–102. doi: 10.1126/science.134.3472.101.

[Beladi 2011] Beladi H, Marjit S, Weiher K : An analysis of the demand for skill in a growing economy. *Economic Modelling* 2011, 28: 1471–1474. doi: 10.1016/j.econmod.2011.02.032.

[Belanger 2008] Belanger D, Brousse T, Long JW: Manganese oxides: Battery materials make the leap to electrochemical capacitors. *Interface* 2008, 17:4.

[Beni 2013] Beni A, Braun A, Huthwelker T, Van Bokhoven JA: Meeting report: Exploratory workshop on soft x-rays and electrochemical energy storage and converters. *Synchrotron Radiation News* 2013, 26:36–38. doi: 10.1080/08940886.2013.832590.

[Benz 1972] Benz U: Der akademische Lehrer Arnold Sommerfeld. *Physik Journal* 1972, 28:292–299. doi: 10.1002/phbl.19720280702.

[Berejnov 2012] Berejnov V, Martin Z, West M, Kundu S, Bessarabov D, Stumper J, Susac D, Hitchcock AP: Probing platinum degradation in polymer electrolyte membrane fuel cells by synchrotron X-ray microscopy. *Phys Chem Chem Phys* 2012, 14:4835–4843. doi: 10.1039/c2cp40338b.

[Berenblut 1985] Berenblut BJ, Whitehouse SM, Phillips HCD, Salter RE: PEMEX IN RETROSPECT – LESSONS IN RISK ASSESSMENT AND LOSS PREVENTION. *Chemical Engineer-London* 1985:17–21.

[Bernard 2016] Bernard T: "They Came There as Workers": Voice, dialogicality and identity construction in textual representations of the 2012 Marikana miner's strike. *Stellenbosch Papers in Linguistics Plus* 2016, 49:145–165. doi: 10.5842/49-0-662.

[Bernhardt 2016] Bernhardt PA, Siefring CL, Briczinski SJ, Mccarrick M, Michell RG: Large ionospheric disturbances produced by the HAARP HF facility. *Radio Science* 2016, 51:1081–1093. doi: 10.1002/2015rs005883.

[Bernstein 1912] Bernstein J: *Elektrobiologie – Die Lehre von den Elektrischen Vorgängen im Organismus auf Moderner Grundlage Dargestellt.* Braunschweig, Germany: Friedr. Vieweg & Sohn; 1912. doi: 10.1007/978.3.663.01627.4.

[Bertulani 2016] Bertulani CA, Kajino T: Frontiers in nuclear astrophysics. *Progress in Particle and Nuclear Physics* 2016, 89:56–100. doi: 10.1016/j.ppnp.2016.04.001.

[Besson 2010] Besson C, Huang Z, Geletii YV, Lense S, Hardcastle KI, Musaev DG, Lian T, Proust A, Hill CL: Cs(9)[(gamma-PW(10)O(36))(2)Ru(4)O(5)(OH)(H(2)O)(4)], a new all-inorganic, soluble catalyst for the efficient visible-light-driven oxidation of water. *Chem Commun (Camb)* 2010, 46:2784–2786. doi: 10.1039/b926064a.

[Bethe 1939] Bethe HA: Energy production in stars. *Physical Review* 1939, 55:434–456. doi: 10.1103/PhysRev.55.434.

[Beuys 1980] Beuys J, Schreiber H. Interview. 1980. Lebensläufe. 27 Januar 1980. https://archive.org/details/JosephBeuysInterviewLebenslaufe

[BfE 2016] BfE: Schweizerische Elektrizitätsstatistik 2016. 2016. Bundesamt für Energie BFE. Bern. Energie BF..

[Bhadra 2015] Bhadra S, Hertzberg BJ, Hsieh AG, Croft M, Gallaway JW, Van Tassell BJ, Chamoun M, Erdonmez C, Zhong Z, Sholklapper T, Steingart DA: The relationship between coefficient of restitution and state of charge of zinc alkaline primary LR6 batteries. *Journal of Materials Chemistry A* 2015, 3:9395–9400. doi: 10.1039/c5ta01576f.

[Bicer 2017] Bicer Y, Dincer I: Performance assessment of electrochemical ammonia synthesis using photoelectrochemically produced hydrogen. *International Journal of Energy Research* 2017, 41:1987–2000 doi: . 10.1002/er.3756.

[Biedcharreton 1993] Biedcharreton B: Closed-loop recycling of lead-acid-batteries. *Journal of Power Sources* 1993, 42:331–334. doi: 10.1016/0378-7753(93)80162-I.

[Biello 2008] Biello D: Inside the solar-hydrogen house: No more power bills – Ever. *Scientific American* 2008.

[Birks 1960] Birks R, Huxley HE, Katz B: The fine structure of the neuromuscular junction of the frog. *Journal of Physiology* 1960, 150:134–144. doi: 10.1113/jphysiol.1960.sp006378.

[Biswas 2013] Biswas M, Sadanala KC: Electrolyte materials for solid oxide fuel cell. *Journal of Powder Metallurgy & Mining* 2013, 02. doi: 10.4172/2168-9806.1000117.

[Blanford 2013] Blanford CF: The birth of protein electrochemistry. *Chem Commun (Camb)* 2013, 49:11130–11132. doi: 10.1039/c3cc46060f.

[Blomberg 2009] Blomberg B, Hess G, Jackson JH: Terrorism and the returns to oil. *Economics & Politics* 2009, 21:409–432. doi: 10.1111/j.1468-0343.2009.00357.x.

[Bocklisch 2015] Bocklisch T: Hybrid energy storage systems for renewable energy applications. *Energy Procedia* 2015, 73:103–111. doi: 10.1016/j.egypro.2015.07.582.

[Bockris 1963] Bockris JO, Devanathan MaV, Muller K: On the structure of charged interfaces. *Proceedings of the Royal Society A: Mathematical, Physical and Engineering Sciences* 1963, 274:55–79. doi: 10.1098/rspa.1963.0114.

[Bockris 2000] Bockris JOM, Reddy AKN, Gamboa-Aldeco M: *Modern Electrochemistry 2A. Fundamentals of Electrodics.* 2nd edn. New York, London: Kluwer Academic/Plenum Publishers; 2000.

[Bockris 2002] Bockris J: The origin of ideas on a Hydrogen Economy and its solution to the decay of the environment. *International Journal of Hydrogen Energy* 2002, 27: 731–740. doi: 10.1016/s0360-3199(01)00154-9.

[Body 1991] Body PE, Inglis G, Dolan PR, Mulcahy DE: Environmental lead – a review. *Critical Reviews in Environmental Control* 1991, 20:299–310.

[Boillat 2017] Boillat P, Lehmann EH, Trtik P, Cochet M: Neutron imaging of fuel cells – Recent trends and future prospects. *Current Opinion in Electrochemistry* 2017, 5:3–10. doi: 10.1016/j.coelec.2017.07.012.

[Bökman 1992] Bökman F, Bohman O, Siegbahn HOG: Electric double layers at solution surfaces studied by means of electron spectroscopy: A comparison of potassium octanoate in formamide and ethylene glycol solution. *Chemical Physics Letters* 1992, 189:414–419. doi: 10.1016/0009-2614(92)85224-x.

[Bolling 2012] Bolling AK, Totlandsdal AI, Sallsten G, Braun A, Westerholm R, Bergvall C, Boman J, Dahlman HJ, Sehlstedt M, Cassee F, Sandstrom T, Schwarze PE, Herseth JI: Wood smoke particles from different combustion phases induce similar pro-inflammatory effects in a co-culture of monocyte and pneumocyte cell lines. *Part Fibre Toxicol* 2012, 9:45 doi: 10.1186/1743-8977-9-45.

[Boman 1970] Boman CE: Refinement of crystal structure of ruthenium dioxide. *Acta Chemica Scandinavica* 1970, 24:116. doi: 10.3891/acta.chem.scand.24-0116.

[Bones 1987] Bones RJ, Coetzer J, Galloway RC, Teagle DA: A sodium/iron(II) chloride cell with a beta alumina electrolyte. *Journal of The Electrochemical Society* 1987, 134:2379. doi: 10.1149/1.2100207.

[Bones 1989] Bones RJ, Teagle DA, Brooker SD, Cullen FL: Development of a Ni,NiCl[sub 2] Positive electrode for a liquid sodium (ZEBRA) battery cell. *Journal of The Electrochemical Society* 1989, 136:1274–1277. doi: 10.1149/1.2096905.

[Bonhoeffer 1943] Bonhoeffer KF: The memoirs of Walther Nernst. *Naturwissenschaften* 1943, 31:257–257.

[Bönninghausen 2017] Bönninghausen D. Schweiz: Esoro erhält Zulassung für Brennstoffzellen-Lkw. Electrive. 3 Juni 2017. https://www.electrive.net/2017/06/03/schweiz-esoro-erhaelt-zulassung-fuer-brennstoffzellewn-lkw/

[Bonzel 1977] Bonzel HP: The role of surface science experiments in understanding heterogeneous catalysis. *Surface Science* 1977, 68:236–258. doi: 10.1016/0039-6028(77)90209-6.

[Boonpong 2010] Boonpong R, Worayingyong A, Arunchaiya M, Wongchaisuwat A: Effect of LaCoO(3) additive on the electrochemical behavior of zinc anode in alkaline solution. *Optoelectronic Materials, Pts 1and 2* 2010, 663–665:596-599. doi: 10.4028/www.scientific.net/MSF.663-665.596.

[Bora 2011] Bora DK, Braun A, Erat S, Ariffin AK, LöHnert R, Sivula K, TöPfer JR, GräTzel M, Manzke R, Graule T, Constable EC: Evolution of an oxygen near-edge X-ray absorption fine structure transition in the upper hubbard band in α-Fe2O3 upon electrochemical oxidation. *The Journal of Physical Chemistry C* 2011, 115:5619–5625. doi: 10.1021/jp108230r.

[Bora 2011a] Bora DK, Braun A, Erat S, Ariffin AK, Löhnert R, Sivula K, Töpfer JR, Grätzel M, Manzke R, Graule T, Constable EC: Evolution of an oxygen near-edge x-ray absorption fine structure transition in the upper hubbard band in α-Fe2O3upon electrochemical oxidation. *The Journal of Physical Chemistry C* 2011a, 115:5619–5625. doi: 10.1021/jp108230r.

[Bora 2011b] Bora DK, Braun A, Erni R, Fortunato G, Graule T, Constable EC: Hydrothermal treatment of a hematite film leads to highly oriented faceted nanostructures with enhanced photocurrents. *Chemistry of Materials* 2011b, 23:2051–2061. doi: 10.1021/cm102826n.

[Bora 2012a] Bora DK, Braun A, Erat S, Safonova O, Graule T, Constable EC: Evolution of structural properties of iron oxide nano particles during temperature treatment from 250°C–900°C:

X-ray diffraction and Fe K-shell pre-edge X-ray absorption study. *Current Applied Physics* 2012a, 12:817–825. doi: 10.1016/j.cap.2011.11.013.

[Bora 2012b] Bora DK, Rozhkova EA, Schrantz K, Wyss PP, Braun A, Graule T, Constable EC: Functionalization of nanostructured hematite thin-film electrodes with the light-harvesting membrane protein C-Phycocyanin yields an enhanced photocurrent. *Advanced Functional Materials* 2012b, 22:490–502. doi: 10.1002/adfm.201101830.

[Bora 2013] Bora DK, Braun A, Constable EC: "In rust we trust". Hematite – the prospective inorganic backbone for artificial photosynthesis. *Energy & Environmental Science* 2013, 6:407–425. doi: 10.1039/c2ee23668k.

[Bora 2013] Bora DK, Hu YL, Thiess S, Erat S, Feng XF, Mukherjee S, Fortunato G, Gaillard N, Toth R, Gajda-Schrantz K, Drube W, Gratzel M, Guo JH, Zhu JF, Constable EC, Sarma DD, Wang HX, Braun A: Between photocatalysis and photosynthesis: Synchrotron spectroscopy methods on molecules and materials for solar hydrogen generation. *Journal of Electron Spectroscopy and Related Phenomena* 2013, 190:93–105. doi: 10.1016/j.elspec.2012.11.009.

[Borst 1982] Borst A, Heisenberg M: Osmotropotaxis in Drosophila-Melanogaster. *Journal of Comparative Physiology* 1982, 147:479–484.

[Borst 1983] Borst A: Computation of olfactory signals in Drosophila-Melanogaster. *Journal of Comparative Physiology* 1983, 152:373–383.

[Borst 1984] Borst A: Identification of different chemoreceptors by electroantennogram-recording. *Journal of Insect Physiology* 1984, 30:507–510. doi: 10.1016/0022-1910(84)90032-5.

[Borst 2014] Borst A: Fly visual course control: Behaviour, algorithms and circuits. *Nature Reviews Neuroscience* 2014, 15:590–599. doi: 10.1038/nrn3799.

[Bosch 2014] Bosch F: Energy Diplomacy: West Germany, the Soviet Union and the oil crises of the 1970s. *Historical Social Research-Historische Sozialforschung* 2014, 39:165–185.

[Bottone 1889] Bottone SR: *Electric Bells and All About Them – A Practical Book for Practical Men*. London, Paternoster Square, E.C.: LWHITTAKER & CO.; 1889.

[Boukamp 2004] Boukamp BA: Impedance spectroscopy, strength and limitations. *Technisches Messen* 2004, 71:454–459. doi: 10.1524/teme.71.9.454.42758.

[Bowman 1984] Bowman CL, Baglioni A: Application of the Goldman-Hodgkin-Katz current equation to membrane current-voltage data. *Journal of Theoretical Biology* 1984, 108:1–29. doi: 10.1016/s0022-5193(84)80165-4.

[Bragg 1915a] Bragg WH: The distribution of the electrons in atoms. *Nature* 1915a, 95:344–344. doi: 10.1038/095344a0.

[Bragg 1915b] Bragg WH: Bakerian lecture: X-rays and crystal structure. *Philosophical Transactions of the Royal Society A: Mathematical, Physical and Engineering Sciences* 1915b, 215:253–274. doi: 10.1098/rsta.1915.0009.

[Bragg 1922] Bragg WL: *Nobel Lecture: The Diffraction of X-Rays by Crystals*. Stockholm; 1922,. Elsevier Publishing Company, Amsterdam, 1967.

[Braidwood 1981] Braidwood S: 1981, 6 Products that reflect the age we're in + SONY Walkman designed by Sumita, K. *Design* 1981: 26–26.

[Braun 1996] Braun A: Korrelation von Struktur, Magnetismus und Morphologie ultradünner Ni-Filme auf Cu3Au(100). *Diplomarbeit*. Rheinisch-Westfälische Technische Hochschule Aachen, Mathematisch-Naturwissenschaftliche Fakultät; 1996.

[Braun 1997] Braun A, Feldmann B, Wuttig M: Strain-induced perpendicular magnetic anisotropy in ultrathin Ni films on Cu3Au(0 0 1). *Journal of Magnetism and Magnetic Materials* 1997, 171: 16–28. doi: 10.1016/S0304-8853(97)00010-3.

[Braun 1999] Braun A: Development and characterization of glassy carbon electrodes for a bipolar electrochemical double layer capacitor. *Doctoral Thesis*. ETH Zürich, Chemistry; 1999.

[Braun 1999a] Braun A: Development and characterization of glassy carbon electrodes for a bipolar electrochemical double layer capacitor. *Doctoral Thesis.* ETH Zürich, Chemistry; 1999a.

[Braun 1999b] Braun A, Bärtsch M, Schnyder B, Kötz R, Haas O, Haubold HG, Goerigk G: X-ray scattering and adsorption studies of thermally oxidized glassy carbon. *Journal of Non-Crystalline Solids* 1999b, 260:1–14. doi: 10.1016/s0022-3093(99)00571-2.

[Braun 2000] Braun A, Bärtsch M, Schnyder B, Kötz R: A model for the film growth in samples with two moving reaction frontiers – an application and extension of the unreacted-core model. *Chemical Engineering Science* 2000, 55:5273–5282. doi: 10.1016/s0009-2509(00)00143-3.

[Braun 2001] Braun A, Seifert S, Thiyagarajan P, Cramer S, Cairns E: In situ anomalous small angle X-ray scattering and absorption on an operating rechargeable lithium ion battery cell. *Electrochemistry Communications* 2001, 3:136–141. doi: 10.1016/S1388-2481(01)00121-7.

[Braun 2002] Braun A, Bärtsch M, Schnyder B, Kötz R, Haas O, Wokaun A: Evolution of BET internal surface area in glassy carbon powder during thermal oxidation. *Carbon* 2002, 40:375–382. doi: 10.1016/s0008-6223(01)00114-2.

[Braun 2002] Braun A, Wang H, Bergmann U, Tucker MC, Gu W, Cramer SP, Cairns EJ: Origin of chemical shift of manganese in lithium battery electrode materials – a comparison of hard and soft X-ray techniques. *Journal of Power Sources* 2002, 112:231–235. doi: 10.1016/s0378-7753(02)00367-1.

[Braun 2003] Braun A, Shrout S, Fowlks AC, Osaisai BA, Seifert S, Granlund E, Cairns EJ: Electrochemical in situ reaction cell for X-ray scattering, diffraction and spectroscopy. *Journal of Synchrotron Radiation* 2003, 10:320–325. doi: 10.1107/s090904950300709x.

[Braun 2003a] Braun A, Bärtsch M, Merlo O, Schnyder B, Schaffner B, Kötz R, Haas O, Wokaun A: Exponential growth of electrochemical double layer capacitance in glassy carbon during thermal oxidation. *Carbon* 2003a, 41:759–765.doi: 10.1016/s0008-6223(02)00390-1.

[Braun 2003b] Braun A: Conversion of thickness data of thin films with variable lattice parameter from monolayers to angstroms: An application of the epitaxial bain path. *Surface Review and Letters* 2003b, 10:889–894. doi: 10.1142/s0218625x03005761.

[Braun 2003c] Braun A, Shrout S, Fowlks AC, Osaisai BA, Seifert S, Granlund E, Cairns EJ: Electrochemical in situ reaction cell for X-ray scattering, diffraction and spectroscopy. *Journal Synchrotron Radiation* 2003c, 10:320–325. doi: 10.1107/s090904950300709x.

[Braun 2003d] Braun A, Wokaun A, Hermanns HG: Analytical solution to a growth problem with two moving boundaries. *Applied Mathematical Modelling* 2003d, 27:47–52. doi: 10.1016/s0307-904x(02)00085-9.

[Braun 2004a] Braun A, Kohlbrecher J, Bärtsch M, Schnyder B, Kötz R, Haas O, Wokaun A: Small-angle neutron scattering and cyclic voltammetry study on electrochemically oxidized and reduced pyrolytic carbon. *Electrochimica Acta* 2004, 49:1105–1112. doi: 10.1016/j.electacta.2004a.10.022.

[Braun 2004b] Braun A, Kohlbrecher J, Bärtsch M, Schnyder B, Kötz R, Haas O, Wokaun A: Small-angle neutron scattering and cyclic voltammetry study on electrochemically oxidized and reduced pyrolytic carbon. *Electrochimica Acta* 2004b, 49:1105–1112. doi: 10.1016/j.electacta.2003.10.022.

[Braun 2005] Braun A: Carbon speciation in airborne particulate matter with C (1s) NEXAFS spectroscopy. *Journal of Environmental Monitoring* 2005, 7: 1059–1065. doi: 10.1039/b508910g.

[Braun 2005] Braun A: Supercapacitors: Novel High Octane – High Protein Elements in Electric Energy Storage. Aurora, CO: Energy Central; 2005.

[Braun 2006] Braun A, Holtappels P, Vogt U, Soltmann C. Investigations on the electronic, ionic, and thermal conductivity of nickel and cobalt substituted lathanum strontium ferrous oxides.

57th Annual Meeting of the International Society of Electrochemistry Edinburgh. Aug 27–Sep 1, 2006.

[Braun 2006] Braun A: Comment on "Studies on nanoporous glassy carbon as a new electrochemical capacitor material [Y. Wen, G. Cao, Y. Yang, J. Power Sources 148 (2005) 121–128]. *Journal of Power Sources* 2006, ?160:762–763. doi: 10.1016/j.jpowsour.2006.03.083.

[Braun 2008a] Braun A, Graule TJ: 2008a. Fundamental aspects of photocatalysis and photoelectrochemistry/Basic research instrumentation for functional characterization swiss national science foundation. 206021–121306. Dübendorf. 96227 CHF. 96227 CHF. 12 Months. finished. http://p3.snf.ch/project-121306

[Braun 2008b] Braun A, Huggins FE, Kubatova A, Wirick S, Maricq MM, Mun BS, Mcdonald JD, Kelly KE, Shah N, Huffman GP: Toward distinguishing woodsmoke and diesel exhaust in ambient particulate matter. *Environ Sci Technol* 2008b, 42:374–380. doi: 10.1021/es071260k.

[Braun 2008c] Braun ABB: Die Auswirkungen des Antibiotikums Ciprofloxacin auf das Cyanobakterium Microcystis aeruginosa. *Maturitätsarbeit*. Kantonsschule Glattal, and Eawag, Fachbereich Biologie; 2008c.

[Braun 2008d] Braun A, Janousch M, Sfeir J, Kiviaho J, Noponen M, Huggins FE, Smith MJ, Steinberger-Wilckens R, Holtappels P, Graule T: Molecular speciation of sulfur in solid oxide fuel cell anodes with X-ray absorption spectroscopy. *Journal of Power Sources* 2008d, 183:564–570. doi: 10.1016/j.jpowsour.2008.05.048.

[Braun 2008e] Braun A, Richter J, Harvey AS, Erat S, Infortuna A, Frei A, Pomjakushina E, Mun BS, Holtappels P, Vogt U, Conder K, Gauckler LJ, Graule T: Electron hole–phonon interaction, correlation of structure, and conductivity in single crystal La0.9Sr0.1FeO3–δ. *Applied Physics Letters* 2008e, 93:262103. doi: 10.1063/1.3049614.

[Braun 2009a] Braun A, Alivisatos AP, Figgemeier E, Je J, Turner JA. Call for papers – MRS symposium S: Materials in photocatalysis and photoelectrochemistry for environmental applications and H2 generation. Materials Research Society Spring Meeting. San Francisco, CA. 4 June 2008a. Materials Research Society.

[Braun 2009b] Braun A, Bayraktar D, Erat S, Harvey AS, Beckel D, Purton JA, Holtappels P, Gauckler LJ, Graule T: Pre-edges in oxygen (1s) x-ray absorption spectra: A spectral indicator for electron hole depletion and transport blocking in iron perovskites. *Applied Physics Letters* 2009b, 94:202102. doi: 10.1063/1.3122926.

[Braun 2009c] Braun A, Alivisatos AP, Figgemeier E, Je J, Turner JA. Call for Papers – MRS Symposium S: Materials in photocatalysis and photoelectrochemistry for environmental applications and H2 generation. Materials Research Society Spring Meeting. San Francisco, CA. 4 June 2009c. Materials Research Society.

[Braun 2009d] Braun A, Duval S, Ried P, Embs J, Juranyi F, Strässle T, Stimming U, Hempelmann R, Holtappels P, Graule T: Proton diffusivity in the BaZr0.9Y0.1O3–δ proton conductor. *Journal of Applied Electrochemistry* 2009d, 39:471–475. doi: 10.1007/s10800-008-9667-3.

[Braun 2009e] Braun A, Embs JP, Strässle T: 2009e. Effect of lattice volume and imperfections on the proton-phonon coupling in proton conducting lanthanide transition metal oxides: High pressure and high temperature neutron and impedance studies. Swiss National Science Foundation. Dübendorf, Zürich, Villigen. 201'242.00 Swiss Francs. 01.09.2009–31.08.2012. finished. http://p3.snf.ch/project-124812

[Braun 2009f] Braun A, Zhang X, Sun Y, Müller U, Liu Z, Erat S, Ari M, Grimmer H, Mao SS, Graule T: Correlation of high temperature x-ray photoemission valence band spectra and conductivity in strained LaSrFeNi oxide on SrTiO[sub 3](110). *Applied Physics Letters* 2009f, 95:022107. doi: 10.1063/1.3174916.

[Braun 2009g] Braun A, Ovalle A, Pomjakushin V, Cervellino A, Erat S, Stolte WC, Graule T: Yttrium and hydrogen superstructure and correlation of lattice expansion and proton conductivity in the

BaZr[sub 0.9]Y[sub 0.1]O[sub 2.95] proton conductor. Applied Physics Letters 2009g, 95:224103. doi: 10.1063/1.3268454.

[Braun 2009h] Braun A, Ovalle A, Pomjakushin V, Cervellino A, Erat S, Stolte WC, Graule T: Yttrium and hydrogen superstructure and correlation of lattice expansion and proton conductivity in the BaZr[sub 0.9]Y[sub 0.1]O[sub 2.95] proton conductor. *Applied Physics Letters* 2009h, 95:224103. doi: 10.1063/1.3268454.

[Braun 2010] Braun A, Nicolas G: 2010. Oxide heterointerfaces in assemblies for photoelectrochemical applications. Swiss National Science Foundation. 133944. Honolulu, HI. 11380 CHF. 11380 CHF. 01.11.2010–31.01.2011. finished. http://p3.snf.ch/Project-133944

[Braun 2011] Braun A, Seifert S, Ilavsky J: Highly porous activated glassy carbon film sandwich structure for electrochemical energy storage in ultracapacitor applications: Study of the porous film structure and gradient. *Journal of Materials Research* 2011, 25:1532–1540. doi: 10.1557/jmr.2010.0197.

[Braun 2011a] Braun A, Erat S, Zhang X, Chen Q, Huang T-W, Aksoy F, LöHnert R, Liu Z, Mao SS, Graule T: Surface and bulk oxygen vacancy defect states near the Fermi level in 125 nm WO3−δ/TiO2(110) films: A resonant valence band photoemission spectroscopy study. *The Journal of Physical Chemistry C* 2011a, 115:16411–16417. doi: 10.1021/jp202375h.

[Braun 2011b] Braun A, Grätzel M: 2011b. Defects in the bulk and on surfaces and interfaces of metal oxides with photoelectrochemical properties: In-situ photoelectrochemical and resonant x-ray and electron spectroscopy studies. Swiss National Science Foundation. 200021–132126 Dübendorf and Lausanne, Switzerland. 273'582 218'761.00. 48 months. finished. http://p3.snf.ch/Project-132126

[Braun 2012] Braun A, Erat S, Bayraktar D, Harvey A, Graule T: Electronic origin of conductivity changes and isothermal expansion of Ta- and Ti-substituted La1/2Sr1/2Fe-oxide in oxidative and reducing atmosphere. *Chemistry of Materials* 2012, 24:1529–1535. doi: 10.1021/cm300423m.

[Braun 2012a] Braun A, Aksoy Akgul F, Chen Q, Erat S, Huang T-W, Jabeen N, Liu Z, Mun BS, Mao SS, Zhang X: Observation of substrate orientation-dependent oxygen defect filling in thin WO3−δ/TiO2Pulsed laser-deposited films with in situ XPS at high oxygen pressure and temperature. *Chemistry of Materials* 2012a, 24:3473–3480. doi: 10.1021/cm301829y.

[Braun 2012b] Braun A, Erat S, Bayraktar D, Harvey A, Graule T: Electronic origin of conductivity changes and isothermal expansion of Ta- and Ti-substituted La1/2Sr1/2Fe-Oxide in oxidative and reducing atmosphere. *Chemistry of Materials* 2012b, 24:1529–1535. doi: 10.1021/cm300423m.

[Braun 2012c] Braun A, Chen Q, Flak D, Fortunato G, Gajda-Schrantz K, Graetzel M, Graule T, Guo J, Huang T-W, Liu Z, Popelo AV, Sivula K, Wadati H, Wyss PP, Zhang L, Zhu J: Iron resonant photoemission spectroscopy on anodized hematite points to electron hole doping during anodization. *Chemphyschem* 2012c, 13:2937–2944. doi: 10.1002/cphc.201200074.

[Braun 2012d] Braun A, Sivula K, Bora DK, Zhu JF, Zhang L, Gratzel M, Guo JH, Constable EC: Direct observation of two electron holes in a hematite photoanode during photoelectrochemical water splitting. *Journal of Physical Chemistry C* 2012d, 116:16870–16875. doi: 10.1021/jp304254k.

[Braun 2012e] Braun A, Sivula K, Bora DK, Zhu JF, Zhang L, Gratzel M, Guo JH, Constable EC: Direct observation of two electron holes in a hematite photoanode during photoelectrochemical water splitting. *Journal of Physical Chemistry C* 2012e, 116:16870–16875. doi: 10.1021/jp304254k.

[Braun 2012f] Braun AI, Julian: 2012f. Biomimetic photoelectrochemical cells for solar hydrogen generation (BioPEC). VELUX stiftung. 790. Dübendorf and Sankt Gallen. 230000. 230000. 1. Sept 2012–31. Aug 2014. finished.

[Braun 2014] Braun A, Fan HY, Haenen K, Stanciu L, Theil JA: Braun, Fan, Haenen, Stanciu, and Theil to chair 2015 MRS Spring Meeting. *MRS Bulletin* 2014, 39:740–741. doi: 10.1557/mrs.2014.183.

[Braun 2014] Braun A, Rohwer ER, Ozoemena KI, Constable EC, Diale MM, Roduner E: 2014. Production of liquid solar fuels from CO2 and water: Using renewable energy resources. Swiss South African Joint Research Programme (SSAJRP) Swiss National Science Foundation and NRF – National Research Foundation (RSA). IZLSZ2-149031 Dübendorf and Basel, Switzerland, and Pretoria, South Africa. 229'950 CHF. 224'358 CHF. 48 Months. finished. http://p3.snf.ch/Project-149031

[Braun 2014a] Braun A, Embs JP, Remhof A: 2013 ESS science symposium: Neutrons for future energy strategies. *Neutron News* 2014a, 25:6–7. doi: 10.1080/10448632.2014.870430.

[Braun 2014b] Braun A, Fan HY, Haenen K, Stanciu L, Theil JA: Braun, Fan, Haenen, Stanciu, and Theil to chair 2015 MRS Spring Meeting. *Mrs Bulletin* 2014b, 39:740–741. doi: 10.1557/mrs.2014.183.

[Braun 2015] Braun A, Boudoire F, Bora DK, Faccio G, Hu Y, Kroll A, Mun BS, Wilson ST: Biological components and bioelectronic interfaces of water splitting photoelectrodes for solar hydrogen production. *Chemistry* 2015, 21:4188–4199. doi: 10.1002/chem.201405123.

[Braun 2015] Braun A, Nordlund D, Song SW, Huang TW, Sokaras D, Liu XS, Yang WL, Weng TC, Liu Z: Hard X-rays in–soft X-rays out: An operando piggyback view deep into a charging lithium ion battery with X-ray Raman spectroscopy. *Journal of Electron Spectroscopy and Related Phenomena* 2015, 200:257–263. doi: 10.1016/j.elspec.2015.03.005.

[Braun 2015] Braun A: X-ray and neutron spectroscopy methods for photosynthesis and photoelectrochemistry. In *Topics in Current Chemistry*. Edited by Chan CK, Tüysüz H. Switzerland, Cham: Springer; 2015. doi: 10.1007/128_2015_650.

[Braun 2015a] Braun A: Sunset over Mount Tamalpais, Marin County, California. Edited by Braun A. Zürich, Switzerland: Youtube; 2015a.

[Braun 2015b] Braun A: Shine on you crazy diamond: Sunrise at Mesquite Dunes, Death Valley, 15 April 2015. Edited by Braun A. Zürich, Switzerland: Youtube; 2015b.

[Braun 2016] Braun A, Mun BS, Housecroft CE, Constable EC: 2016. Molecular and physical aspects of dye sensitization of photoelectrodes with copper-based sensitizer molecules. Swiss National Science Foundation/Korean-Swiss Science and Technology Programme (KSSTP) IZKSZ2_162232 Dübendorf, Gwangju, Basel. 250000 CHF. 36 Months. current.

[Braun 2016] Braun A: Structural characterization techniques: Advances and applications in clean energy. In *Structure and Transport Properties in Ceramic Fuel Cells (SOFC), Components and Materials*. Edited by Malavasi L. Pan Stanford Publishing; 2016.

[Braun 2016a] Braun A, Alvarez A, Davila Costa JS: 2016a. Charge transfer between inorganic and bio-organic copper systems for solar fuel cells Swiss National Science Foundation, CONICET, MINCyT. IZSAZ2-173405 Tucuman (Argentina), Dübendorf (Switzerland). 223,975 Swiss Francs. 0. 3 years. finished.

[Braun 2016b] Braun A, Diale M, Huthwelker T, Van Bokhoven JA: International exploratory workshop on catalysis, photoelectrochemistry, and x-ray spectroscopy for renewable energy. *Synchrotron Radiation News* 2016b, 29:14–16. doi: 10.1080/08940886.2016.1124677.

[Braun 2016c] Braun A, Cowan DA, Makhalanyane TP, Diale MM, Ponomarev A, Reutemann J, Janssen P, Aguey-Zinsou K-F, Grossmann G: 2016c. Algal solar hydrogen fuel reactor for renewable logistics in the Antarctic. EPFL. ID19. Antarctica. 200,000 Euro. 0 Euro. 12 months. finished.

[Braun 2016d] Braun A, Hu Y, Boudoire F, Bora DK, Sarma DD, Grätzel M, Eggleston CM: The electronic, chemical and electrocatalytic processes and intermediates on iron oxide surfaces during photoelectrochemical water splitting. *Catalysis Today* 2016d, 260:72–81. doi: 10.1016/j.cattod.2015.07.024.

[Braun 2017a] Braun A, Chen Q: Experimental neutron scattering evidence for proton polaron in hydrated metal oxide proton conductors. *Nature Communications* 2017, 8:15830. doi: 10.1038/ncomms15830.

[Braun 2017b] Braun A, Davila Costa JS, Alvarez A. Photoelectrode for solar hydrogen production with copper bacteria. 20th Topical Meeting of the International Society of Electrochemistry. Buenos Aires. 19–22 March 2017. International Society of Electrochemistry.

[Braun 2017c] Braun A, Diale M, Malherbe JB, Braun M: Introduction. *Journal of Materials Research* 2017c, 32: 3921–3923. doi: 10.1557/jmr.2017.426.

[Braun 2017] Braun A: X-ray Studies on Electrochemical Systems – Synchrotron Methods for Energy Materials. Berlin/Boston: Walter De Gruyter GmbH; 2017. ISBN 978-3-11-042796-7.

[Braun 2017d] Braun A: In situ photoelectron spectroscopy. In *Encyclopedia of Interfacial Chemistry: Surface Science and Electrochemistry*. Edited by Wandelt K. Amsterdam: Elsevier; 2017d. doi: 10.1016/B978-0-12-409547-2.13291-3.

[Braun 2017e] Braun A, Ryu W: 2017c. Electrochemistry of the thylakoid membrane – metal oxide electrode assemblies Swiss National Science Foundation. IZK0Z2-175249. Yonsei University, Seoul, Korea. 10100 CHF. 10100 CHF. 3 months. archived. http://p3.snf.ch/Project-175249

[Braun 2018] Braun A: "Thou Shalt Not Make Unto Thee Any Graven Image": Some remarks on x-ray scattering and materials science. In *XRM2018; Saskatoon, CA*. Edited by Urquhart SG. Cambridge, UK: Cambridge University Press; 2018.

[Bresadola 2011] Bresadola M: Carlo Matteucci and the legacy of Luigi Galvani. *Archives Italiennes de Biologie* 2011, 149 (Suppl.):3–9.

[Breslauer 1910] Breslauer M: Some numbers on the electroculture of energy and currents. *Zeitschrift Fur Elektrochemie Und Angewandte Physikalische Chemie* 1910, 16:557–559. doi: 10.1002/bbpc.19100161408.

[Breul 2018] Jemen: Reisewarnung. https://www.auswaertiges-amt.de/de/aussenpolitik/laender/jemen-node/jemensicherheit/202260

[Briggs 1926] Briggs LJ, Campbell AB, Heald RH, Flint LH. Electroculture. United States Department of Agriculture – Department Bulletin, 1926. 1379.

[Brousse 2015] Brousse T, Belanger D, Long JW: To Be or Not To Be Pseudocapacitive? *Journal of The Electrochemical Society* 2015, 162:A5185–A5189. doi: 10.1149/2.0201505jes.

[Brunauer 1938] Brunauer S, Emmett PH, Teller E: Adsorption of gases in multimolecular layers. *Journal of the American Chemical Society* 1938, 60:309–319. doi: 10.1021/ja01269a023.

[Brzezinski 1997] Brzezinski Z: *The Grand Chessboard: American Primacy and Its Geostrategic Imperatives*. 1st edn. New York: BasicBooks; 1997.

[Brzeziński 2015] Brzeziński Z: *Die einzige Weltmacht: Amerikas Strategie der Vorherrschaft*. Kopp Verlag; 2015.

[Büchi 2014] Büchi FN, Hofer M, Peter C, Cabalzar UD, Bernard J, Hannesen U, Schmidt TJ, Closset A, Dietrich P: Towards re-electrification of hydrogen obtained from the power-to-gas process by highly efficient H2/O2 polymer electrolyte fuel cells. *RSC Advances* 2014, 4: 56139–56146. doi: 10.1039/c4ra11868e.

[Büchi 2018] Büchi FN: Personal Communication. Braun A. 2018. Bei Brennstoffzellen(-Systemen) sinkt der Wirkungsgrad mit zunehmender Last. Bei Verbrennungsmotoren ist das umgekehrt.

[Buchner 1976] Buchner E: Elementary movement detectors in an insect visual system. *Biological Cybernetics* 1976, 24:85–101. doi: 10.1007/BF00360648.

[Buchner 2002] Patent. Buchner P, Nölscher C, Von Rittmar H, Waidhas M.Stapel aus Brennstoffzellen mit Flüssigkeitskühlung und Verfahren zur Kühlung eines BZ-Stapels.2002.Office EP.EP19990930997.http://google.com/patents/EP1086502B1?cl=de

[Buckely 2012] Buckely D: *Kraftwerk: Publikation*. Omnibus Press, 2012.

[Bujalski 2008] Bujalski W: Solid Oxide Fuel Cells (SOFC). 2008.

[Bullen 2006] Bullen RA, Arnot TC, Lakeman JB, Walsh FC: Biofuel cells and their development. *Biosens Bioelectron* 2006, 21:2015–2045. doi: 10.1016/j.bios.2006.01.030.

[Bunker 2015] Bunker J: Outreach report, project title: Wind energy R&D park and storage system for innovation in grid integration 2015. Office of Energy Research and Development Natural Resources Canada (Weican) WEIOC. 31 March 2015.

[Bunn 2012] Bunn GC: Truth Machine: A social history of the lie detector. *Truth Machine: A Social History of the Lie Detector* 2012:1–246.

[Bunsen 1841] Bunsen R: Ueber die Anwendung der Kohle zu Volta'schen Batterien. *Annalen der Physik und Chemie* 1841, 130:417–430. doi: 10.1002/andp.18411301109.

[Burzan 2016] Burzan N, Kroll A, Diale MM, Housecroft CE, Braun A: Algae mediated conductivity of hematite bio-hybrid photoelectrode. Empa. Swiss Federal Laboratories for Materials Science and Technology; University of Basel; Eawag. Swiss Federal Institute of Aquatic Science and Technology; University of Pretoria; 2016.

[Busch 1997] Busch L: Atmospheric physics – Ionosphere research lab sparks fears in Alaska. *Science* 1997, 275:1060–1061. doi: 10.1126/science.275.5303.1060.

[Butler 1977] Butler MA: Photoelectrolysis and physical properties of the semiconducting electrode WO2. *Journal of Applied Physics* 1977, 48:1914–1920. doi: 10.1063/1.323948.

[Caglar 2000] Caglar I, Eryildiz C: Monitoring of the Telluric Currents Originated by Solar Related Atmospheric Events in Northwestern Turkey. 2000. Astronomical Society of the Pacific Conference Series, Orem, Utah. Vol. 205, pp. 208. Editors Livingston, W.; Özgüç, A. ISBN 978-1-58381-541-0.

[Cai 2009] Cai ZP, Zhou HB, Li WS, Huang QM, Liang Y, Xiao XH, Chen JQ: Surface treatment method of cathodic current-collector and its effect on performances of batteries. *Rare Metal Materials and Engineering* 2009, 38:1676–1680.

[Cai 2010] Cai Q, Brett DJL, Browning D, Brandon NP: A sizing-design methodology for hybrid fuel cell power systems and its application to an unmanned underwater vehicle. *Journal of Power Sources* 2010, 195:6559–6569. doi: 10.1016/j.jpowsour.2010.04.078.

[Cai 2014] Cai G, Fung KY, Ng KM, Wibowo C: Process development for the recycle of spent lithium ion batteries by chemical precipitation. *Industrial & Engineering Chemistry Research* 2014, 53:18245–18259. doi: 10.1021/ie5025326.

[Calvin 1960] Calvin M. Some photochemical and photophysical reactions of chlorophyll and its relatives. McCollum-Pratt Symposium on Light and Life. Johns Hopkins University, Baltimore. 11 April 1960; 20 May 2008. University of California <Go to WoS>://WOS:A1972N338500004

[Calvin 1978] Calvin M: Petroleum plantations. Agricultural Chemical Society of Japan. Nagoya, Japan. 1 April 1978. Berkeley. http://escholarship.org/uc/item/6xp5p246

[Cameron 1987] Cameron DS, Hards GA, Harrison B, Potter RJ: Direct methanol fuel cells – recent developments in the search for improved performance. *Platinum Metals Rev* 1987, 31.

[Capasso 2014] Capasso C, Veneri O: Experimental analysis of a zebra battery based propulsion system for urban bus under dynamic conditions. In *International Conference on Applied Energy Icae2014*. Volume 61. Edited by Yan J, Lee DJ, Chou SK, Desideri U, Li H; 2014: 1138–1141: *Energy Procedia*. doi: 10.1016/j.egypro.2014.11.1040.

[Capasso 2015a] Capasso C, Sepe V, Veneri O, Montanari M, Poletti L, IEEE: Experimentation with a ZEBRA plus LDLC based hybrid storage system for urban means of transport. *2015 International Conference on Electrical Systems for Aircraft, Railway, Ship Propulsion and Road Vehicles (Esars)* 2015a.

[Capasso 2015b] Capasso C, Veneri O: Laboratory bench to test ZEBRA battery plus super-capacitor based propulsion systems for urban electric transportation. In *Clean, Efficient and Affordable*

Energy for a Sustainable Future. Volume 75. Edited by Yan J, Shamim T, Chou SK, Li H; 2015b: 1956–1961: *Energy Procedia.* doi: 10.1016/j.egypro.2015.07.235.

[Carmo 2013] Carmo M, Fritz DL, Merge J, Stolten D: A comprehensive review on PEM water electrolysis. *International Journal of Hydrogen Energy* 2013, 38:4901–4934. doi: 10.1016/j.ijhydene.2013.01.151.

[Carter 2001] Carter M: BMW group presents first car with petrol fuel cell for on board electricity supply. *Carpages.* 20 February 2001. http://www.carpages.co.uk/bmw/bmw_sofc_fuel_cell_20_02_01.asp

[Casanova 1998] Casanova A: A consortium approach to commercialized Westinghouse solid oxide fuel cell technology. *Journal of Power Sources* 1998, 71: 65–70. doi: 10.1016/s0378-7753(97)02757-2.

[Certini 2009] Certini G, Scalenghe R, Amundson R: A view of extraterrestrial soils. *European Journal of Soil Science* 2009, 60:1078–1092. doi: 10.1111/j.1365-2389.2009.01173.x.

[Chan 2016] Chan CK, Tuysuz H, Braun A, Ranjan C, La Mantia F, Miller BK, Zhang LX, Crozier PA, Haber JA, Gregoire JM, Park HS, Batchellor AS, Trotochaud L, Boettcher SW: Advanced and in situ analytical methods for solar fuel materials. In *Solar Energy for Fuels. Volume* 371. Edited by Tuysuz H, Chan CK. Cham: Springer Int Publishing Ag; 2016: 253–324: *Topics in Current Chemistry-Series.* doi: 10.1007/128_2015_650.

[Chandra 1984] Chandra S, Singh DP, Srivastava PC, Sahu SN: Electrodeposited semiconducting molybdenum selenide films. II. Optical, electrical, electrochemical and photoelectrochemical solar cell studies. Journal of Physics D: Applied Physics 1984, 17:2125–2138. doi: 10.1088/0022-3727/17/10/023.

[Chandra 1987] Chandra S, Khare N: Electro-deposited gallium arsenide film: II. Electrochemical and photoelectrochemical solar cell studies. *Semiconductor Science and Technology* 1987, 2: 220–225. doi: 10.1088/0268-1242/2/4/004.

[Chang 2011] Chang Y, Braun A, Deangelis A, Kaneshiro J, Gaillard N: Effect of thermal treatment on the crystallographic, surface energetics, and photoelectrochemical properties of reactively cosputtered copper tungstate for water splitting. *The Journal of Physical Chemistry C* 2011, 115:25490–25495. doi: 10.1021/jp207341v.

[Changyen 1992] Changyen I, Emrit C, Hoseinrahaman A: Incidence of severe lead-poisoning in children in Trinidad resulting from battery recycling operations. *Abstracts of Papers of the American Chemical Society* 1992, 204:17-Envr.

[Chen 2004] Chen H, Wang JM, Zheng Y, Zhang JQ, Cao CN: Effects of pH value of reaction solution on structure and electrochemical performance of calcium-containing active material of secondary zinc electrodes. *Transactions of Nonferrous Metals Society of China* 2004, 14:406–411.

[Chen 2010] Chen HI, Chiu YW, Hsu YK, Li WF, Chen YC, Chuang HY: The association of metallothionein-4 gene polymorphism and renal function in long-term lead-exposed workers. *Biological Trace Element Research* 2010, 137:55–62. doi: 10.1007/s12011-009-8564-x.

[Chen 2010] Chen QL, Braun A, Ovalle A, Savaniu CD, Graule T, Bagdassarov N: Hydrostatic pressure decreases the proton mobility in the hydrated $BaZr0.9Y0.1O3$ proton conductor. *Applied Physics Letters* 2010, 97:041902. doi: 10.1063/1.3464162.

[Chen 2011] Chen Z, Jaramillo TF, Deutsch TG, Kleiman-Shwarsctein A, Forman AJ, Gaillard N, Garland R, Takanabe K, Heske C, Sunkara M, Mcfarland EW, Domen K, Miller EL, Turner JA, Dinh HN: Accelerating materials development for photoelectrochemical hydrogen production: Standards for methods, definitions, and reporting protocols. *Journal of Materials Research* 2011, 25:3–16. doi: 10.1557/jmr.2010.0020.

[Chen 2011] Chen Z, Jaramillo TF, Deutsch TG, Kleiman-Shwarsctein A, Forman AJ, Gaillard N, Garland R, Takanabe K, Heske C, Sunkara M, Mcfarland EW, Domen K, Miller EL, Turner JA, Dinh HN: Accelerating materials development for photoelectrochemical hydrogen production:

Standards for methods, definitions, and reporting protocols. *Journal of Materials Research* 2011, 25:3–16. doi: 10.1557/jmr.2010.0020.

[Chen 2011] ChenQL, Braun A, Yoon S, Bagdassarov N, Graule T: Effect of lattice volume and compressive strain on the conductivity of BaCeY-oxide ceramic proton conductors. *Journal of the European Ceramic Society* 2011, 31:2657–2661. doi: 10.1016/j.jeurceramsoc.2011.02.014.

[Chen 2011] Chen Q, Braun A, Ovalle A, Savaniu C-D, Graule T, Bagdassarov N: Protons in lattice confinement: Static pressure on the Y-substituted, hydrated BaZrO3 ceramic proton conductor decreases proton mobility. arXiv:1106.1091 [cond-mat.str-el]. 2011 .doi: 10.1063/1.3464162.

[Chen 2011] Chen QL, Huang TW, Baldini M, Hushur A, Pomjakushin V, Clark S, Mao WL, Manghnani MH, Braun A, Graule T: Effect of compressive strain on the Raman modes of the dry and hydrated BaCe0.8Y0.2O3 proton conductor. *Journal of Physical Chemistry C* 2011, 115:24021–24027. doi: 10.1021/jp208525j.

[Chen 2012] Chen QL, Holdsworth S, Embs J, Pomjakushin V, Frick B, Braun A: High-temperature high pressure cell for neutron-scattering studies. *High Pressure Research* 2012, 32:471–481. doi: 10.1080/08957959.2012.725729.

[Chen 2012a] Chen Q: Effects of Pressure on the Proton-Phonon Coupling in Metal Oxides with Perovskite Structure. ETH Zürich, Physics; 2012a.

[Chen 2013] Chen LY, Hou Y, Kang JL, Hirata A, Fujita T, Chen MW: Toward the theoretical capacitance of RuO2 reinforced by highly conductive nanoporous gold. *Advanced Energy Materials* 2013, 3:851–856.doi: 10.1002/aenm.201300024.

[Chen 2013] Chen Q, Banyte J, Zhang X, Embs JP, Braun A: Proton diffusivity in spark plasma sintered BaCe0.8Y0.2O3−δ: In-situ combination of quasi-elastic neutron scattering and impedance spectroscopy. *Solid State Ionics* 2013, 252:2–6. doi: 10.1016/j.ssi.2013.05.009.

[Chen 2013] Chen QL, El Gabaly F, Akgul FA, Liu Z, Mun BS, Yamaguchi S, Braun A: Observation of oxygen vacancy filling under water vapor in ceramic proton conductors in situ with ambient pressure XPS. *Chemistry of Materials* 2013, 25:4690–4696. doi: 10.1021/cm401977p.

[Chen 2015] Chen K, Liu SS, Ai N, Koyama M, Jiang SP: Why solid oxide cells can be reversibly operated in solid oxide electrolysis cell and fuel cell modes? *Phys Chem Chem Phys* 2015, 17:31308–31315. doi: 10.1039/c5cp05065k.

[Chen 2017] Chen Q, Braun A: Protons and the hydrogen economy. *MRS Energy & Sustainability* 2017, 4. doi: 10.1557/mre.2017.16.

[Cheng 2013] Cheng JN, Liu DL, Yang L, Han JJ: Preparation and study of zinc-air batteries with carbon-supported amorphous MnO2 catalyst. In *Applied Energy Technology, Pts 1 and 2. Volume 724–725.* Edited by Wang A, Che LK, Dong R, Zhao G.. Stafa-Zurich: Trans Tech Publications Ltd; 2013: 813–817: *Advanced Materials Research.* doi: 10.4028/www.scientific.net/AMR.724-725.813.

[Chree 1921] Chree C: An electroculture problem. *Proceedings of the Physical Society of London* 1921, 33:377-387. doi: 10.1088/1478-7814/33/1/335.

[Christen 2000] Christen T, Carlen MW: Theory of ragone plots. *Journal of Power Sources* 2000, 91:210–216. doi: 10.1016/s0378-7753(00)00474-2.

[Chudley 1961] Chudley CT, Elliott RJ: Neutron scattering from a liquid on a jump diffusion Model. *Proceedings of the Physical Society of London* 1961, 77:353-361. doi: 10.1088/0370-1328/77/2/319.

[Clark 1872] Clark L: On a voltaic standard of electromotive force. *Proceedings of the Royal Society of London* 1872, 20. doi: 10.1098/rspl.1871.0089.

[Clements 2001] Clements W: A boxed set of virtual Marilyn. *The Globe and Mail.* May 26, 2001. https://beta.theglobeandmail.com/arts/a-boxed-set-of-virtual-marilyn/article1338231/?ref=http://www.theglobeandmail.com&

[Coaldiver 2010] . McKinley http://coaldiver.org/mine/MCKINLEY

[Coehn 1910] Coehn A: Studien über photochemische Gleichgewichte. IV. Das Lichtgleichgewicht Knallgas-Wasserdampf. *Berichte der deutschen chemischen Gesellschaft* 1910, 43:880–884. doi: 10.1002/cber.191004301149.

[Coetzer 2000] Coetzer J, Sudworth J: Out of Africa – the story of the Zebra battery. Beta Research & Development Ltd; 2000.

[Coffey 2003] Coffey GW, Pederson LR, Rieke PC: Competition between bulk and surface pathways in mixed ionic electronic conducting oxygen electrodes. *Journal of the Electrochemical Society* 2003, 150. doi: 10.1149/1.1591758.

[Cohen 1996] Cohen JET, David: Biosphere 2 and biodiversity: The lessons so far. *Science* 1996, 274:1150–1151.

[Cohn 1884] Cohn E: Ueber die Gültigkeit des Ohm"schen Gesetzes für Electrolyte. *Annalen Der Physik* 1884, 257:646–672. doi: 10.1002/andp.18832570407.

[Cohn 2002] Cohn JP: Biosphere 2: Turning an experiment into a research station. *Bioscience* 2002, 52:218–223.

[Cole 1997] Cole MW: The surface state electron. In *Two-Dimensional Electron Systems Physics and Chemistry of Materials with Low-Dimensional Structures. Volume* 19. Edited by EYA. Dordrecht: Springer; 1997. doi: 10.1007/978-94-015-1286-2_1.

[Cole 2005] Cole R, Reddell N, Inan U, Kery S, Cappellini J, Smit P, Greider G: From Alaska to the South Pacific in one-hop. *Proceedings of OCEANS 2005 MTS/IEEE New York.* IEEE; 2005: 917–922. doi: 10.1109/OCEANS.2005.1639872.

[Coleman 2012] Coleman L: Explaining crude oil prices using fundamental measures. *Energy Policy* 2012, 40:318–324. doi: 10.1016/j.enpol.2011.10.012.

[Collection 2012] Collection M: Carbon black in the form of inhalable dust [MAK Value Documentation, 2002]. In *The MAK Collection for Occupational Health and Safety.* Edited by Andrea Hartwig. Weinheim: Wiley VCH; 2012. doi: 10.1002/3527600418.mb133386e001.

[Collivignarelli 1986] Collivignarelli C, Riganti V, Urbini G: Battery lead recycling and environmental-pollution hazards. *Conservation & Recycling* 1986, 9:111-125. doi: 10.1016/ 0361-3658(86)90138-4.

[Colthorpe 2015] Colthorpe A: 'Physics meets art' at Germany's 100% renewable village grid battery. *Energy Storage News.* 23 Sep 2015. https://www.energy-storage.news/news/ physics-meets-art-at-germanys-100-renewable-village-grid-battery

[Comsol 2018] 1D Isothermal Zinc-Silver Oxide Battery. https://www.comsol.com/model/1d-isothermal-zinc-silver-oxide-battery-17229

[Connelly 1996] Connelly NG, Geiger WE: Chemical redox agents for organometallic chemistry. *Chemical Reviews* 1996, 96:877–910. doi: 10.1021/cr940053x.

[Constable 2009] Constable EC, Hernandez Redondo A, Housecroft CE, Neuburger M, Schaffner S: Copper(I) complexes of 6,6'-disubstituted 2,2'-bipyridine dicarboxylic acids: new complexes for incorporation into copper-based dye sensitized solar cells (DSCs). *Dalton Transactions* 2009:6634–6644. doi: 10.1039/b901346f.

[Conway 1962] Conway BE, Gileadi E: Kinetic theory of pseudo-capacitance and electrode reactions at appreciable surface coverage. *Transactions of the Faraday Society* 1962, 58. doi: 10.1039/tf9625802493.

[Conway 1991] Conway BE: Transition from "Supercapacitor" to "Battery" Behavior in electrochemical energy storage. *Journal of the Electrochemical Society* 1991, 138. doi: 10.1149/1.2085829.

[Conway 1999] Conway BE: *Electrochemical Supercapacitors.* Springer; 1999. doi: 10.1007/ 978-1-4757-3058-6.

[Coroniti 1969] Coroniti SC, Hughes J: *Planetary Electrodynamics.* New York, London, Paris: Gordon and Breach Science Publishers; 1969.

[Costa 2013] Costa JSD, Kothe E, Amoroso MJ, Abate CM: Overview of copper resistance and oxidative stress response in amycolatopsis tucumanensis, a useful strain for bioremediation. *Actinobacteria: Application in Bioremediation and Production of Industrial Enzymes* 2013:74–86. doi: BOOK_DOI 10.1201/b14776.

[Cotton 1999] Cotton FA, Wilkinson G, Murillo CA, Bochmann M: *Advanced Inorganic Chemistry.* 6th edn. New York: Wiley-Interscience; 1999.

[Cottrell 1903] Cottrell FG: The cut off current in galvanic polarisation, considered as a diffusion problem. *Zeitschrift Fur Physikalische Chemie-Stochiometrie Und Verwandtschaftslehre* 1903, 42:385–431.

[Courtney 2014] Oxidation Reduction Chemistry of the Elements. http://www.wou.edu/las/physci/ch412/redox.htm

[Creutz 1975] Creutz C, Sutin N: Reaction of tris(bipyridine)ruthenium(III) with hydroxide and its application in a solar energy storage system. *Proceedings of the National Academy of Sciences* 1975, 72:2858–2862. doi: 10.1073/pnas.72.8.2858.

[Crowley 1996] Crowley DJ, Crowley ML: Religion and politics in the John Frum Festival, Tanna Island, Vanuatu + Melanesian folklore, South Pacific. *Journal of Folklore Research* 1996, 33:155–164.

[Crozier 2014]. Crozier R. Australia's first fuel cell bicycle. UNSW Sydney Newsroom. 6 Sep 2014. https://newsroom.unsw.edu.au/news/science-technology/australia%E2%80%99s-first-fuel-cell-bicycle

[Crusto 2008] Crusto MF: Enslaved constitution: Obstructing the freedom to travel. *University of Pittsburgh Law Review* 2008, 70:233–275.

[Cui 2017] Cui BC, Zhang JH, Liu SZ, Liu XJ, Xiang W, Liu LF, Xin HY, Lefler MJ, Licht S: Electrochemical synthesis of ammonia directly from N-2 and water over iron-based catalysts supported on activated carbon. *Green Chemistry* 2017, 19:298–304. doi: 10.1039/c6gc02386j.

[Daimler 1994] Daimler: Daimlerag. Film or Broadcast. E-MOBIL „Rügen" (Rohschnitt). 1994. Mercedesstraße 137, 70327 Stuttgart, Deutschland. https://mercedes-benz-publicarchive.com/marsClassic/de/instance/video/E-MOBIL-Ruegen-Rohschnitt.xhtml?oid=49509

[Damaskin 2011] Damaskin BB, Petrii OA: Historical development of theories of the electrochemical double layer. *Journal of Solid State Electrochemistry* 2011, 15:1317–1334. doi: 10.1007/s10008-011-1294-y.

[Daněk 2006] Daněk V: *Physico-chemical Analysis of Molten Electrolytes.* Elsevier; 2006. doi: 10.1016/B978-0-444-52116-3.X5000-4.

[Daniell 1839] Daniell JF: Fifth letter on voltaic combinations, with some account of the effects of a large constant battery. *Philosophical Transactions of the Royal Society of London* 1839, 129:89–95.

[Darling 2016] Darling R, Gallagher K, Xie W, Su L, Brushett F: Transport property requirements for flow battery separators. *Journal of The Electrochemical Society* 2016, 163:A5029–A5040. doi: 10.1149/2.0051601jes.

[Darowicki 2004] Darowicki K, Ślepski P: Instantaneous electrochemical impedance spectroscopy of electrode reactions. *Electrochimica Acta* 2004, 49:763–772. doi: 10.1016/j.electacta.2003.09.030.

[David 1992] David J, Cadmium A: Technical and economical aspects of nickel-cadmium battery recycling. *Nickel-Cadmium Battery Update 92 – Conference Report: Munich, October 1992*–1992:108–112.

[David 1995] David J: Nickel-cadmium battery recycling evolution in Europe. *Journal of Power Sources* 1995, 57:71–73. doi: 10.1016/0378-7753(95)02244-9.

[Davies 2007] Davies KL, Moore RA: Unmanned underwater vehicle fuel cell energy/power system technology assessment. *IEEE Journal of Oceanic Engineering* 2007, 32: 365–372. doi: 10.1109/Joe.2007.893690.

[Davila Costa 2011] Davila Costa JS, Albarracin VH, Abate CM: Responses of environmental Amycolatopsis strains to copper stress. *Ecotoxicol Environ Saf* 2011, 74:2020–2028. doi: 10.1016/j.ecoenv.2011.06.017.

[Davila Costa 2012] Davila Costa JS, Kothe E, Abate CM, Amoroso MJ: Unraveling the Amycolatopsis tucumanensis copper-resistome. *Biometals* 2012, 25:905–917. doi: 10.1007/s10534-012-9557-3.

[De Cervantes 1605] De Cervantes M: El ingenioso hidalgo Don Quixote de la Mancha. 1605.

[De Levie 1963] De Levie R: On porous electrodes in electrolyte solutions. *Electrochimica Acta* 1963, 8:751–780. doi: 10.1016/0013-4686(63)80042-0.

[De Lorenzo 2014] De Lorenzo G, Andaloro L, Sergi F, Napoli G, FerraroM, Antonucci V: Numerical simulation model for the preliminary design of hybrid electric city bus power train with polymer electrolyte fuel cell. *International Journal of Hydrogen Energy* 2014, 39: 12934–12947. doi: 10.1016/j.ijhydene.2014.05.135.

[Deiss 1997] Deiss E, Wokaun A, Barras J-L, Daul C, Dufek P: Average voltage, energy density, and specific energy of lithium-ion batteries calculation based on first principles. *Journal of the Electrochemical Society* 1997, 144. doi: 10.1149/1.1838105

[Deiss 2001] Deiss E: Modeling of the charge–discharge dynamics of lithium manganese oxide electrodes for lithium-ion batteries. *Electrochimica Acta* 2001, 46. doi: 10.1016/S0013-4686 (01)00703-4.

[Delahay 1951] Delahay P, Pourbaix M, Van Rysselberghe P: Potential-pH diagram of lead and its applications to the study of lead corrosion and to the lead storage battery. *Journal of The Electrochemical Society* 1951, 98. doi: 10.1149/1.2778106.

[Deleebeeck 2017] Deleebeeck L, Gil V, Ippolito D, Campana R, Hansen KK, Holtappels P: Direct coal oxidation in modified solid oxide fuel cells. *Journal of the Electrochemical Society* 2017, 164:F333–F337. doi:10.1149/2.0961704jes.

[Delnick 1993] Delnick FM: Carbon supercapacitors. In *Symposium on Science of Advanced Batteries*, Nov 1993 Cleveland OH: U.S. Dept. of Energy, DOE; 1993.

[Deng 2002] Deng LB: The Sudan famine of 1998 – unfolding of the global dimension. *Ids Bulletin-Institute of Development Studies* 2002, 33:28+.

[Denison 1957] Denison IA, Pauli WJ, Snyder GR. Treating zinc plates of zinc-silver oxide battery 1957. Patent Office USP.US3368925A.

[Denk 1920a] Denk FJ: Producer gas as fuel for the glass industry. *Journal of the American Ceramic Society* 1920a, 3:94–113. doi: 10.1111/j.1151-2916.1920.tb17269.x.

[Denk 1920b] Denk FJ: Producer gas as fuel for the glass industry. *Journal of the American Ceramic Society* 1920b, 3:918–921. doi: 10.1111/j.1151-2916.1920.tb17861.x.

[Dennison 2015] Dennison CR, Agar E, Akuzum B, Kumbur EC: Enhancing mass transport in redox flow batteries by tailoring flow field and electrode design. *Journal of the Electrochemical Society* 2015, 163:A5163–A5169. doi: 10.1149/2.0231601jes.

[Depolarisator 1998] Depolarisator: Depolarisator. In *Lexikon der Physik*. Heidelberg: Spektrum Akademischer Verlag; 1998.

[Deshpande 2014] Deshpande RP: *Ultracapacitors*. 1st edn. New Delhi: McGraw Hill Education (India) Private Limited; 2014.

[Dessalegne 2001] Dessalegne M: *Replacement Migration: Is It a Solution to Declining and Ageing Populations ?* New York: Population Division, Department of Economic and Social Affairs, United Nations Secretariat; 2001.

[Dirkse 1964] Dirkse TP. Silver migration study. 1964. Air Force Aero Propulsion Laboratory Research and Technology Division, Air Force Systems Command, Wright-Patterson Air Force Base, Ohio Defense Documentation Center (DDC) (formerly ASTIA), Cameron Station, Bldg. 5, 5010 Duke Stree.t, Alexandria, Virginia, 22314. Office of Technical Services USDOC, Washington, DC 25, D.C. AFAPL-TR-64-144.

[Dispenza 2012] Dispenza G, Andaloro L, Sergi F, Napoli G, Ferraro M, Antonucci V: Modeling of a range extender power train for a city bus. In *Fuel Cell Seminar 2011. Volume* 42. Edited by Williams M; 2012: 201–208: *ECS Transactions*. doi: 10.1149/1.4705496.

[Doi 2016] Doi K, Kamijo M: Nissan intelligent power – e-bio fuel-cell system. Nissan Motor Co. Ltd.; 2016. https://newsroom.nissan-global.com/releases/160614-01-e

[Doi 2016] Doi K. Nissan announces development of the world's first SOFC-powered vehicle system that runs on bio-ethanol electric power. 2016. http://nissannews.com/en-US/nissan/usa/releases/nissan-announces-development-of-the-world-s-first-sofc-powered-vehicle-system-that-runs-on-bio-ethanol-electric-power?la=1

[Dolcourt 1981] Dolcourt JL, Finch C, Coleman GD, Klimas AJ, Milar CR: Hazard of lead exposure in the home from recycled automobile storage batteries. *Pediatrics* 1981, 68:225–230.

[Dolezalek 1901] Dolezalek F, Gahl R: On the resistance of lead accumulators and its distribution on the two electrodes. *Zeitschrift fur Elektrochemie* 1901, 7:0437–0441.

[Dolezalek 1904] Dolezalek F, Ende CLV: *The Theory of the Lead Accumulator (Storage Battery)*. New York: J. Wiley & Sons; London, Chapman & Hall, limited; 1904.

[Dollard 1992] Dollard WJ: Solid oxide fuel cell developments at Westinghouse. *Journal of Power Sources* 1992, 37: 133–139. doi: 10.1016/0378-7753(92)80070-R.

[Dong 2015] Dong YR, Kaku H, Hanafusa K, Moriuchi K, Shigematsu T: A novel titanium/manganese redox flow battery. *ECS Transactions* 2015, 69:59–67. doi: 10.1149/06918.0059ecst.

[Dong 2017] Dong YR, Kawagoe Y, Itou K, Kaku H, Hanafusa K, Moriuchi K, Shigematsu T: Improved performance of Ti/Mn redox flow battery by thermally treated carbon paper electrodes. In *Batteries and Energy Technology Joint General Session. Volume* 75. Edited by Doeff M, Manivannan M, Jow R, Kalra V, Liu G. Pennington: Electrochemical Soc Inc; 2017: 27–35: *ECS Transactions*. doi: 10.1149/07518.0027ecst.

[Donovan 2009] Donovan B: McKinley Mine to cease operations in December. Navajo Times. 24 Sep 2009. http://www.navajotimes.com/business/2009/0909/092409mine.php

[Downs 1924] Downs JC. Electrolytic process and cell. 1924. Patent Office USP.US 1501756 A.

[Dralle 1915a] Dralle R: The Carmita glass factory in Rio de Janeiro with generator gas- and oil firing. *Zeitschrift Des Vereines Deutscher Ingenieure* 1915a, 59:724–730.

[Dralle 1915b] Dralle R: Glass factory in Rio de Janeiro with generator-gas- and oil firing. *Zeitschrift Des Vereines Deutscher Ingenieure* 1915b, 59:697–704.

[Dudkowski 1967] Dudkowski SJ, Kepka AG, Grossweiner LI: Spectral sensitization of ZnO thin films with organic dyes. *Journal of Physics and Chemistry of Solids* 1967, 28:485. doi: 10.1016/0022-3697(67)90318-6.

[Dudney 2009] Dudney NJ: Glass and Ceramic Electrolytes for Lithium and Lithium-Ion Batteries. In *Lithium Batteries – Science and Technology*. Edited by Nazri G-A, Pistoia G. Boston, MA: Springer; 2009 10.1007/978-0-387-92675-9_20.

[Dufo-Lopez 2007] Dufo-Lopez R, Bernal-Agustin JL, Contreras J: Optimization of control strategies for stand-alone renewable energy systems with hydrogen storage. *Renewable Energy* 2007, 32: 1102–1126. doi: 10.1016/j.renene.2006.04.013.

[Dustmann 2004] Dustmann CH: Advances in ZEBRA batteries. *Journal of Power Sources* 2004, 127:85–92. doi: 10.1016/j.jpowsour.2003.09.039.

[Duval 2007] Duval S, Holtappels P, Vogt U, Pomjakushina E, Conder K, Stimming U, Graule T: Electrical conductivity of the proton conductor $BaZr0.9Y0.1O3-\delta$ obtained by high temperature annealing. *Solid State Ionics* 2007, 178:1437-1441.doi: 10.1016/j.ssi.2007.08.006.

[Early 2013] Early BR, Fuhrmann M, Li Q: Atoms for terror? Nuclear programs and non-catastrophic nuclear and radiological terrorism. *British Journal of Political Science* 2013, 43:915–936. doi: 10.1017/s000712341200066x.

[Eberhardt 2014] Eberhardt SH, Marone F, Stampanoni M, Buchi FN, Schmidt TJ: Quantifying phosphoric acid in high-temperature polymer electrolyte fuel cell components by X-ray tomographic microscopy. *Journal of Synchrotron Radiation* 2014, 21:1319–1326. doi: 10.1107/ S1600577514016348.

[Eberle 2012] Eberle U, Müller B, Von Helmolt R: Fuel cell electric vehicles and hydrogen infrastructure: Status 2012. *Energy & Environmental Science* 2012, 5: 8780. doi: 10.1039/c2ee22596d.

[ECS 1997] ECS: ECS and ISE meet in Paris: 1997 joint international meeting program, le Palais de Congrès, August 31–September 5, 1997. *The Electrochemical Society Interface* 1997, 6.

[ECS 2013] ECS: ECS Classics: Vladimir Sergeevich Bagotsky, scientist and teacher, 1920–2012. *Interface Magazine* 2013, 22:28–31. doi: 10.1149/2.F01131if.

[Eden 1988] Eden R: The arabs and the oil crisis 1973–1986 – Attiga, AA. *Energy Policy* 1988, 16: 330–330.

[Editors 1960] Editors: The chemical world today. *Industrial & Engineering Chemistry* 1960, 52: 29A-30A. doi: 10.1021/i650604a718.

[Egelhaaf 1993] Egelhaaf M, Borst A, Warzecha AK, Flecks S, Wildemann A: NEURAL CIRCUIT TUNING FLY VISUAL NEURONS TO MOTION OF SMALL OBJECTS .2. INPUT ORGANIZATION OF INHIBITORY CIRCUIT ELEMENTS REVEALED BY ELECTROPHYSIOLOGICAL AND OPTICAL-RECORDING TECHNIQUES. *Journal of Neurophysiology* 1993, 69:340–351.

[Ehl 1954] Ehl RG, Ihde AJ: Faraday's electrochemical laws and the determination of equivalent weights. *Journal of Chemical Education* 1954, 31:226. doi: 10.1021/ed031p226.

[Einstein 1905] Einstein A: Generation and conversion of light with regard to a heuristic point of view. *Annalen Der Physik* 1905, 17:132–148.

[Ekinci 2014] Ekinci M, Ceylan E, Cagatay HH, Keles S, Altinkaynak H, Kartal B, Koban Y, Huseyinoglu N: Occupational exposure to lead decreases macular, choroidal, and retinal nerve fiber layer thickness in industrial battery workers. *Current Eye Research* 2014, 39:853–858. doi: 10.3109/02713683.2013.877934.

[Eklund 1983] Eklund G, Von Krusenstierna O: Storage and transportation of merchant hydrogen. International Journal of Hydrogen Energy 1983, 8: 463–470. doi: 10.1016/ 0360-3199(83)90168-4.

[El Nashar 1982] El Nashar MIM: Nitrogen fertilizer industry in Egypt with some details on Helwan fertilizer plant. technical conference on Åmaonia fertilizer technology for promotion of economic co-operation among developing countries. Beijing, People's Republic of China. 13–28 March 1982. United Nations Industrial Development Organization.

[Ellmer 2008] Ellmer K, Fiechter S: Laudatio for professor helmut tributsch. Physica Status Solidi (b) 2008, 245:1743–1744. doi: 10.1002/pssb.200879549.

[El-Sayed 2010] El-Sayed AR, Mohran HS, El-Lateef HMA: Effect of minor nickel alloying with zinc on the electrochemical and corrosion behavior of zinc in alkaline solution. *Journal of Power Sources* 2010, 195:6924–6936. doi: 10.1016/j.jpowsour.2010.03.071.

[Elsayed 2014] Elsayed S, Boukis N, Kerner M, Hindersin S, Patzelt D: Gasification of Microalgae in Supercritical Water: Experimental Results from the Gasification of the Algal Species Scenedesmus Obliquus Cultivated in the BIQ-House in Hamburg Florence: Eta-Florence; 2014.

[Emery 1986] Emery KA: Solar simulators and I–V measurement methods. *Solar Cells* 1986, 18: 251–260. doi: 10.1016/0379-6787(86)90124-9.

[Emin 1982] Emin D: Small polarons. *Physics Today* 1982, 35:34–40. doi: 10.1063/1.2938044.

[Engesser 1904] Engesser: Zuschriften an die Redaktion. *Zeitschrift Des Vereines Deutscher Ingenieure* 1904, 48:1.

[Ensling 2006] Ensling D: Photoelektronenspektroskopische Untersuchung der elektronischen Struktur dünner Lithiumkobaltoxidschichten. *Doctoral Thesis*. Technische Universität Darmstadt, Material- und Geowissenschaften; 2006.

[Ensling 2010] Ensling D, Thissen A, Laubach S, Schmidt PC, Jaegermann W: Electronic structure ofLiCoO2thin films: A combined photoemission spectroscopy and density functional theory study. *Physical Review B* 2010, 82. doi: 10.1103/PhysRevB.82.195431.

[EPCOS 2013] EPCOS: *Aluminium Electrolytic Capacitors Handbook 2013*. München: EPCOS AG; 2013. https://en.tdk.eu/tdk-en/180390/tech-library/publications/capacitors

[Eroglu 2008] Eroglu I, Tabanoğlu A, Gündüz U, Eroğlu E, Yücel M: Hydrogen production by Rhodobacter sphaeroides O.U.001 in a flat plate solar bioreactor. *International Journal of Hydrogen Energy* 2008, 33:531–541. doi: 10.1016/j.ijhydene.2007.09.025.

[Ertl 2015] Ertl G: Walther Nernst and the development of physical chemistry. *Angewandte Chemie-International Edition* 2015, 54:5827–5834. doi: 10.1002/ange.201408793.

[Esoro 2016] Esoro: Factsheet Lastwagen. (Lastwagen F ed.). Fällanden, Schweiz: ESORO; 2016.

[Euler 1980] Euler K-J: Sinsteden – Plante – Tudor: Zur Geschichte des Bleiakkumulatirs. In *Geschichte der Elektrotechnik 13 Gespeicherte Energie Geschichte der elektrochemischen Energiespeicher. Volume 13*. Berlin. Offenbach: Vde-Verlag GmBH; 1980.[Jäger K (Series Editor): *Geschichte der Elektrotechnik 3 Gespeicherte Energie Geschichte der elektrochemischen Energiespeicher*].doi..

[Evans 2009] EvansAR, Bieberle-Hütter A, Galinski H, Rupp JLM, Ryll T, Scherrer B, Tölke R, Gauckler LJ: Micro-solid oxide fuel cells: Status, challenges, and chances. *Monatshefte für Chemie – Chemical Monthly* 2009, 140: 975–983. doi: 10.1007/s00706-009-0107-9.

[Evans 2017] Evans M. Lethal gases from leaking battery most likely to have sunk Argentine submarine San Juan. The Times. 23 November 2017. https://www.thetimes.co.uk/article/lethal-gases-from-leaking-battery-most-likely-to-have-sunk-argentine-submarine-san-juan-52pvpfw20

[Ewald 1962] Ewald PP: Laue's discovery of x-ray diffraction by crystals. In *Fifty Years of X-Ray Diffraction*. Edited by Ewald PP. Boston MA: Springer US; 1962: 733. Ewald PP (Series Editor). doi: 10.1007/978-1-4615-9961-6.

[Faccio 2015a] Faccio G, Gajda-Schrantz K, Ihssen J, Boudoire F, Hu Y, Mun BS, Bora DK, Thöny-Meyer L, Braun A: Charge transfer between photosynthetic proteins and hematite in bio-hybrid photoelectrodes for solar water splitting cells. *Nano Convergence* 2015a, 2:1–11. doi: 10.1186/s40580-014-0040-4.

[Faccio 2015b] Faccio G, Schrantz K, Thoeny-Meyer L, Braun A, Ihssen J: Engineering of proteins to develop biomimetic hematite-based biohybrid materials. *Protein Science* 2015b, 24: 184–185.

[Fallen 2011] Fallen CT, Secan JA, Watkins BJ: In-situ measurements of topside ionosphere electron density enhancements during an HF-modification experiment. *Geophysical Research Letters* 2011, 38:4.doi: 10.1029/2011gl046887.

[Fang 2015] Fang Z, Smith R. L. Jr., Qi X (Eds.) *Production of Hydrogen from Renewable Resources*. Heidelberg, New York, London: Springer Dordrecht; 2015.

[Faraday 1833] Faraday M: Experimental Researches in Electricity. Third Series. *Philosophical Transaction of the Royal Society of London* 1833, 123:33–54. doi: 10.1098/rstl.1833.0006.

[Faraday 1834] Faraday M: Experimental researches in electricity. Seventh series. *Philosophical Transaction of the Royal Society London A* 1834, 124:77–122. doi: 10.1098/rstl.1834.0008.

[Farrell 2001] Farrell WM, Desch MD: Is there a Martian atmospheric electric circuit? *Journal of Geophysical Research-Planets* 2001, 106:7591–7595. doi: 10.1029/2000je001271.

[Ferraro 2011] Ferraro M, Napoli G, Sergi F, Brunaccini G, Dispenza G, Antonucci V: SOFC/ZEBRA Hybrid System. In *Clean Technology 2011: Bioenergy, Renewables, Storage, Grid, Waste and Sustainability*. Edited by Lauon M, Romanowicz B. Austin, TX: CTSI; 2011: d143–146.

[Ferreira 2002]Ferreira AL, Lifshutz N, Choi WM. Gel-forming battery separator. 2002. Google Patent. http://www.google.ch/patents/WO2002007237A2?cl=en

[Ferreira 2006] Ferreira AL, Lifshutz N, Choi WM: Gel-forming battery separator. Google Patents; 2006.

[Feyerabend 1964] Feyerabend PK: A NOTE ON THE PROBLEM OF INDUCTION. *Journal of Philosophy* 1964, 61:349–353. doi: 10.2307/2023161.

[Feyerabend 1975] Feyerabend P: Science – myth and its role in society. *Inquiry-an Interdisciplinary Journal of Philosophy* 1975, 18:167–181. doi: 10.1080/00201747508601758.

[Feyerabend 1976] Feyerabend P: *Wider den Methodenzwang: Skizze einer anarchistischen Erkenntnistheorie.* Frankfurt: Suhrkamp; 1976.

[Feyerabend 1979] Feyerabend P: *Erkenntnis für freie Menschen.* Frankfurt: Suhrkamp; 1979.

[Feyerabend 1984] Feyerabend P: On the Limits of Research. *New Ideas in Psychology* 1984, 2: 3–7. doi: 10.1016/0732-118x(84)90027-8.

[Feyerabend 2005] Feyerabend P: Notes on Relativism. In *Truth: Engagements Across Philosophical Traditions.* Edited by Medina J, Wood D; 2005. doi: 10.1002/9780470776407. ch10.

[Feynman 1955] Feynman RP: Public address: The value of science. *Autumn Meeting of the National Academy of Sciences.* Caltech, Pasadena CA. November 2–4, 1955. Caltech. http://calteches.library.caltech.edu/40/2/Science.htm

[Feynman 1974] Feynman RP: Cargo Cult Science: Some remarks on science, pseudoscience, and learning how to not fool yourself. . *Engineering & science* 1974, 37:4.

[Feynman 1988] Feynman RP: "What Do You Care What Other People Think?": Further Adventures of a Curious Character. W. W. Norton; 1988.

[Feynman 2005] Feynman RP: *The Pleasure of Finding Things Out: The Best Short Works of Richard P. Feynman (Helix Books).* Basic Books; 2005.

[Firsov 2007] Firsov YA: Small polarons: Transport phenomena. In *Polarons in Advanced Materials. Volume 103.* Edited by Alexandrov AS. Dordrecht: Springer; 2007: 63–105.

[Fischer 1986] Fischer WE: On the neutronics of SINQ, a continuous spallation neutron source. *Nuclear Instruments & Methods in Physics Research Section a-Accelerators Spectrometers Detectors and Associated Equipment* 1986, 249:116–122. doi: 10.1016/0168-9002(86)90247-0.

[Fischer 1988] Fischer WE: Status-report of SINQ – a continuous spallation neutron source. In *10th Meeting of the International Colloquium on Advanced Neutron Sources (Icans 10); Oct 03–07;* Los Alamos Natl Lab, Los Alamos, NM. IOP Publishing Ltd; 1988: 67–78.

[Fischer 1989] Fischer WE: Status-report of SINQ – a continuous spallation neutron source. *Institute of Physics Conference Series* 1989:67–78.

[Fischer 1997] Fischer WE: SINQ – the spallation neutron source, a new research facility at PSI. *Physica B: Condensed Matter* 1997, 234–236:1202–1208. doi: 10.1016/s0921-4526(97)00260-3.

[Flachowsky 2017] Flachowsky S:"Die schwere Artillerie der Erzeugungsschlacht". Landwirtschaftliche Gas-Schlepper und die Mobilisierung alternativer Kraftstoffreserven im Vierteljahresplan. In *Ressourcenmobilisierung: Wissenschaftspolitik und Forschungspraxis im NS-Herrschaftssystem.* Edited by Flachowsky S, Hachtmann R, Schmaltz F. 1st edn. Göttingen: Verlag Wallstein; 2017: 691. Flachowsky S, Hachtmann R, Schmaltz F (Series Editor).

[Fleischer 1971] Fleischer A, Lander JJ: *Zinc-silver Oxide Batteries (Electrochemical Society) Hardcover.* John Wiley & Sons Inc; 1971.

[Fleischmann 1989a] Fleischmann M: CORRECTION. *Journal of Electroanalytical Chemistry* 1989a, 263:187–187.

[Fleischmann 1989b] Fleischmann M, Pons S: Electrochemically induced nuclear fusion of deuterium. *Journal of Electroanalytical Chemistry and Interfacial Electrochemistry* 1989b, 261:301–308. doi: 10.1016/0022-0728(89)80006-3.

[Fleischmann 1989c] Fleischmann M, Pons S, Hawkins M, Hoffman RJ: MEASUREMENT OF GAMMA-RAYS FROM COLD FUSION. *Nature* 1989c, 339:667–667. doi: 10.1038/339667a0.

[Fleischmann 1993] Fleischmann M, Pons S: Calorimetry of the Pd-D2o system – from simplicity via complications to simplicity. *Physics Letters A* 1993, 176:118–129. doi: 10.1016/0375-9601(93) 90327-V.

[Fleischmann 1994] Fleischmann M, Pons S, Preparata G: POSSIBLE THEORIES OF COLD-FUSION. *Nuovo Cimento Della Societa Italiana Di Fisica a-Nuclei Particles and Fields* 1994, 107:143–156. doi: 10.1007/bf02813078.

[Fleischmann 2006a] Fleischmann M: *Background to Cold Fusion: The Genesis of a Concept.* Singapore: World Scientific Publ Co Pte Ltd; 2006a. doi: 10.1142/9789812701510_0001.

[Fleischmann 2006b] Fleischmann M, Miles MH: The Instrument Function of Isoperibolic Calorimeters: Excess Enthalpy Generation Due to the Parasitic Reduction of Oxygen. Singapore: World Scientific Publ Co Pte Ltd; 2006b. doi: 10.1142/9789812701510_0022.

[Fletcher 1986] Fletcher JC: Actions to implement the recommendations of the presidential commission on the space shuttle challenger accident. Washington, DC: National Aeronautics and Space Administration (NASA); 1986.

[Flores 1997] Flores T, Junghans S, Wuttig M: Atomic mechanisms for the diffusion of Mn atoms incorporated in the Cu(100) surface: An STM study. *Surface Science* 1997, 371:1–13. doi: 10.1016/ s0039-6028(96)00978-8.

[Forecast 2002] Forecast W. DM2A3 Seehecht Torpedo – Archived 2/2003. 2002. Forecast International, 22 Commerce Road, Newtown, CT 06470. Newtown, CT 06470. November 2002

[Fortunati 2004] Fortunati GU: A brief history of risk assessment and management after the Seveso accident. In *Comparative Risk Assessment and Environmental Decision Making. Volume* 38. Edited by Linkov I, Ramadan AB. New York, Boston, Dordrecht, London, Moscow: Kluwer Academic Publishers; 2004: 423–429. *NATO Science Series IV Earth and Environmental Sciences.*

[Fox 2017] Fox M. Blog. 6 November 2017. Papa John's – better ingredients, better pizza, better supply chain management social media for business performance. 2018. https://smbp.uwaterloo.ca/ 2017/11/papa-johns-better-ingredients-better-pizza-better-supply-chain-management/

[Franke 2018] Franke A, Loades-Carter J. German 10 MW battery to be developed by UK-based RES Group. 10 January 2018. https://www.platts.com/latest-news/electric-power/london/ german-10-mw-battery-to-be-developed-by-uk-based-26868310

[Fraxedas 2014] Fraxedas J: *Water at Interfaces: A Molecular Approach.* Boca Raton: CRC Press; 2014. ISBN: 9781439861042

[Freitas 2007] Freitas M, Pegoretti VC, Pietre MK: Recycling manganese from spent Zn-MnO2 primary batteries. *Journal of Power Sources* 2007, 164:947–952. doi: 10.1016/j.jpowsour.2006.10.050.

[Friedman 2011] Friedman G: *The Geopolitics of the United States, Part 1: The Inevitable Empire.* Austin, TX: Stratfor Enterprises, LLC; 2011.

[Friedrich 2008] Friedrich S, Stan R, Lyudmila Z: Risk due to radiological terror attacks with natural radionuclides. In *Natural Radiation Environment. Volume* 1034. Edited by Paschoa AS, Steinhausler F. Melville: Amer Inst Physics; 2008: 3+. *AIP Conference Proceedings.*

[Fritzsch 2014] Fritzsch B: Electric organs: history and potential. *Science* 2014, 345:631–632. doi: 10.1126/science.345.6197.631-b.

[Frutschy 2013] Frutschy K, Chatwin T, Mao L, Smith CR, Bull R: Sodium nickel chloride battery design and testing. In *ASME 2012 International Mechanical Engineering Congress and Exposition; November 9–15, 2012*; Houston, TX, USA ASME; 2013: 429–438. doi: 10.1115/ IMECE2012-86379.

[Frutschy 2015] Frutschy K, Chatwin T, Bull R: Cell overcharge testing inside sodium metal halide battery. *Journal of Power Sources* 2015, 291:117–125. doi: 10.1016/j.jpowsour.2015.05.001.

[Fujii 2011] Fujii Y, Satake K, Sakai S, Shinohara M, Kanazawa T: Tsunami source of the 2011 off the Pacific coast of Tohoku Earthquake. *Earth Planets and Space* 2011, 63:815–820. doi: 10.5047/eps.2011.06.010.

[Fujishima 1972] Fujishima A, Honda K: Electrochemical photolysis of water at a semiconductor electrode. *Nature* 1972, 238:37+. doi: 10.1038/238037a0.

[Fung 2005] Energy density of hydrogen. http://hypertextbook.com/facts/2005/MichelleFung.shtml

[Furuya 1990] Furuya N, Yoshiba H: Electroreduction of nitrogen to ammonia on gas-diffusion electrodes loaded with inorganic catalyst. *Journal of Electroanalytical Chemistry* 1990, 291:269–272. doi: 10.1016/0022-0728(90)87195-P.

[Gaillard 2012] Gaillard N, Chang Y, Braun A, Deangelis A: Copper tungstate (CuWO 4)-based materials for photoelectrochemical hydrogen production. In *Materials Research Society Symposium Proceedings; 2012*. 2012: 19–24. doi: 10.1557/opl.2012.952.

[Gaillard 2013] Gaillard N, Chang Y, Deangelis A, Higgins S, Braun A: A nanocomposite photoelectrode made of 2.2 eV band gap copper tungstate (CuWO4) and multi-wall carbon nanotubes for solar-assisted water splitting. *International Journal of Hydrogen Energy* 2013, 38:3166–3176. doi: 10.1016/j.ijhydene.2012.12.104.

[Gajda-Schrantz 2013] Gajda-Schrantz K, Tymen S, Boudoire F, Toth R, Bora DK, Calvet W, Gratzel M, Constable EC, Braun A: Formation of an electron hole doped film in the alpha-Fe2O3 photoanode upon electrochemical oxidation. *Phys Chem Chem Phys* 2013, 15:1443–1451. doi: 10.1039/c2cp42597a.

[Galda 1992] Galda L, Macgregor P: THE GLASS ARK – THE STORY OF BIOSPHERE-2 – GENTRY,L, LIPTAK,K. *Reading Teacher* 1992, 46:236–245.

[Gallaway 2014] Gallaway JW, Gaikwad AM, Hertzberg B, Erdonmez CK, Chen-Wiegart YCK, Sviridov LA, Evans-Lutterodt K, Wang J, Banerjee S, Steingart DA: An in situ synchrotron study of zinc anode planarization by a bismuth additive. *Journal of The Electrochemical Society* 2014, 161:A275–A284. doi: 10.1149/2.037403jes.

[Galvani 1791] Galvani A: *De viribus electricitatis in motu musculari commentarius*. Bononiae: Ex Typographia Instituti Scientiarium; 1791:72.

[Galvani 1793] Galvani M, Volta A: Account of some discoveries made by Mr. Galvani, of Bologna; With experiments and observations on them. In two letters from Mr. Alexander Volta, F. R. S. Professor of Natural Philosophy in the University of Pavia, to Mr. Tiberius Cavallo, F. R. S. *Philosophical Transactions of the Royal Society of London* 1793, 83:10–44. doi: 10.1098/rstl.1793.0005.

[Gamow 1928] Gamow G: The quantum theory of nuclear disintegration. *Nature* 1928, 122:805–806. doi: 10.1038/122805b0.

[Gamow 1938] Gamow G: Nuclear Energy Sources and Stellar Evolution. *Physical Review* 1938, 53:595–604. doi: 10.1103/PhysRev.53.595.

[Gandomi 2016] Gandomi YA, Aaron DS, Zawodzinski TA, Mench MM: In situ potential distribution measurement and validated model for all-vanadium redox flow battery. *Journal of the Electrochemical Society* 2016, 163:A5188–A5201. doi: 10.1149/2.0211601jes.

[Gao 1996] Gao Y, Reimers JN, Dahn JR: Changes in the voltage profile of Li/Li1+xMn2−xO4cells as a function of x. *Physical Review B* 1996, 54:3878–3883. doi: 10.1103/PhysRevB.54.3878.

[Gardner 2000] Gardner FJ, Day MJ, Brandon NP, Pashley MN, Cassidy M: SOFC technology development at Rolls-Royce. *Journal of Power Sources* 2000, 86:122–129.doi: 10.1016/s0378-7753(99)00428-0.

[Gärtner 1959] Gärtner WW: Depletion-Layer Photoeffects in Semiconductors. *Physical Review* 1959, 116:84–87. doi: doi.org/10.1103/PhysRev.116.84.

[Gashev 1965] Gashev MA, Gustov GK Dyachenk, Kk, Komar EG, Malyshev IF, Monoszon NA, Popkovic Av, Ratnikov BK, .Rozhdest Bv, Rumyants.Nn, Saksagan.Gl, Spevakov.Rm, Stolov AM, Streltso.Ns, Yavno AK: Basic technical characteristics of experimental thermonuclear device Tokamak-3. *Journal of Nuclear Energy Part C-Plasma Physics Accelerators Thermonuclear Research* 1965, 7:491.

[Gast 2014] Gast M: Farbstoffsolarzellen. From DIY to SolarChic. *form Design Magazine* 2014: 104–109.

[Gast 2017] Gast M, Tedx U: Microbes – algae, bacteria and fungi utilized in an industrial context | Mareike Gast | TEDxUniHalle. In *TEDxUniHalle*. Halle, Germany: Youtube; 2017.

[Gaudernack 1998] Gaudernack B: Hydrogen production from fossil fuels. *Hydrogen Power: Theoretical and Engineering Solutions*. 1998: 75–89 doi: 10.1007/978-94-015-9054-9_10.

[Geitmann 2017a] Geitmann S. Blog. 15 May 2017. Elektrolyseur-Hersteller bringen sich in Stellung. Hydrogeit. 2018.

[Geitmann 2017b] Geitmann S. Blog. 13 December 2017. H2-Erzeugung per Elektrolyse förderfähig. H2Blog. 2018.

[Gelderman 2007] Gelderman K, Lee L, Donne SW: Flat-band potential of a semiconductor: Using the Mott–Schottky Equation. *Journal of Chemical Education* 2007, 84. doi: 10.1021/ed084p685.

[Georgiev 1945] Georgiev AM: *The Electrolytic Capacitor*. New York: Murray Hill Books Inc.; 1945.

[Gerischer 1968] Gerischer H, Tributsch H: Electrochemistry of ZnO monocrystal spectral sensitivity. *Berichte Der Bunsen-Gesellschaft Fur Physikalische Chemie* 1968, 72:437+. doi: 10.1002/bbpc.196800013.

[Gerovasili 2014] Gerovasili E, May JF, Sauer DU: Experimental evaluation of the performance of the sodium metal chloride battery below usual operating temperatures. *Journal of Power Sources* 2014, 251:137–144. doi: 10.1016/j.jpowsour.2013.11.046.

[Ghoneim 2012] Ghoneim AF: The political economy of food price policy in Egypt. In *Working Paper No 2012/96* Helsinki: UNU-WIDER: United Nations University - World Institute for Development Economics Research; 2012.

[Giacomini 2017] Giacomini: Hydrogen powered boiler developed by Giacomini. 2017. https://www.boilerguide.co.uk/hydrogen-powered-boiler-developed-by-giacomini;https://uk.giacomini.com/zero-impact-conditioning

[Giancoli 2008] Giancoli D: 25. Electric currents and resistance. In *Physics for Scientists and Engineers with Modern Physics* 4th ed. Upper Saddle River, NJ: Pearson; 2008: 1328.

[Girishkumar 2010] Girishkumar G, Mccloskey B, Luntz AC, Swanson S, Wilcke W: Lithium–air battery: Promise and challenges. *The Journal of Physical Chemistry Letters* 2010, 1:2193–2203. doi: 10.1021/jz1005384.

[Gittus 1989] Gittus J, Bockris J: EXPLANATIONS OF COLD FUSION. *Nature* 1989, 339:105–105. doi: 10.1038/339105b0.

[Godula-Jopek 2015] Godula-Jopek A: *Hydrogen Production*. Weinheim: Wiley VCH; 2015. doi: 10.1002/9783527676507.

[Goldman 1991] Goldman M, Gilad D, Ronen A, Melloul A: Mapping of Seawater Intrusion into the Coastal Aquifer of Israel by the Time Domain Eelectromagnetic Method. *Geoexploration* 1991, 28:153–174. doi: 10.1016/0016-7142(91)90046-f.

[Goldschmidt 1926] Goldschmidt VM: Die Gesetze der Krystallochemie. *Die Naturwissenschaften* 1926, 14:477–485. doi: 10.1007/BF01507527.

[Golnik 2003] Golnik, A: Energy Density of Gasoline [http://hypertextbook.com/facts/2003/ArthurGolnik.shtml]

[Gonsalves 2010] Gonsalves C: The Bhopal catastrophe: Politics, conspiracy and betrayal. *Economic and Political Weekly* 2010, 45:68–75.

[González 2010] González MS: Experimental Investigation of the Thermophysical Properties of New and Representative Materials from Room Temperature up to 1300°C (Berichte aus der Materialwissenschaft). 1st edn. Herzogenrath: Shaker Verlag GmbH; 2010.

[Goodenough 1955] Goodenough JB: Theory of the role of covalence in the perovskite-type manganites[La, M(II)]MnO3. *Physical Review* 1955, 100:564–573. doi: 10.1103/PhysRev.100.564.

[Goodenough 2008] Goodenough J: Goodenough-Kanamori rule. *Scholarpedia* 2008, 3:7382. doi: 10.4249/scholarpedia.7382.

[Gordon 1997] Gordon WE: HAARP facility in Alaska (Letter to Science). *Science* 1997, 275:1861–1861.

[Gorelik 2013] Gorelik G: Tamms Doktor and Andrej Sacharow. In *Andreĭ Sakharov A Life for Science and Freedom* 2013: 83–96. doi: 10.1007/978-3-0348-0474-5_6.

[Gosser 1993] Gosser DK: Cyclic Voltammetry; Simulation and Analysis of Reaction Mechanisms. New York: VCH; 1993.

[Goucher 1951] Goucher FS, Pearson GL, Sparks M, Teal GK, Shockley W: Theory and experiment for a Germaniump–njunction. *Physical Review* 1951, 81:637–638. doi: 10.1103/PhysRev.81.637.2.

[Grafov 1971] Grafov BM, Pekar EV: Onsager reciprocal relations in the electrode impedance theory. *Journal of Electroanalytical Chemistry and Interfacial Electrochemistry* 1971, 31:137–151. doi: 10.1016/s0022-0728(71)80052-9.

[Grass 2010] Grass ME, Karlsson PG, Aksoy F, Lundqvist M, Wannberg B, Mun BS, Hussain Z, Liu Z: New ambient pressure photoemission endstation at advanced light source beamline 9.3.2. *Review of Scientific Instruments* 2010, 81:053106. doi: 10.1063/1.3427218.

[Grätzel 2001] Grätzel M: Photoelectrochemical cells. *Nature* 2001, 414:338–344. doi: 10.1038/35104607.

[Grayer 1984] Grayer S, Halmann M: Electrochemical and photoelectrochemical reduction of molecular nitrogen to ammonia. *Journal of Electroanalytical Chemistry and Interfacial Electrochemistry* 1984, 170:363–368. doi: 10.1016/0022-0728(84)80059-5.

[Green 1953] Green RM, Foley MG: A Translation of Luigi Galvani's De viribus electricitatis in motu musculari commentarius. Commentary on the effect of electricity on muscular motion. *Journal of the American Medical Association* 1953, 153. doi: 10.1001/jama.1953.02940270095033.

[Greenough 1957] Greenough H: *Form and Function: Remarks on Art, Design, and Architecture*. Berkeley: University of California Press; 1957.

[Greenwood 1997] Greenwood NN, Earnshaw A: *Chemistry of the Elements.*. Amsterdam, Boston: Elsevier; 1997.

[Gregory 1972] Gregory DP, Ng DYC, Long GM: The hydrogen economy. In *Electrochemistry of Cleaner Environments*. Edited by Bockris JO. Boston, MA: Springer; 1972: 226–280. doi: 10.1007/978-1-4684-1950-4_8.

[Griffiths 1998] Griffiths DJ: 7. Electrodynamics. In *Introduction to Electrodynamics*. 3rd ed. Edited by Reeves A. Upper Saddle River, NJ: Prentice Hall; 1998: 286.

[Grove 2009] Grove WR: LXXII. On a gaseous voltaic battery. *The London, Edinburgh, and Dublin Philosophical Magazine and Journal of Science* 2009, 21:417–420. doi: 10.1080/14786444208621600.

[Grove 2012] Grove WR: On a gaseous voltaic battery. Philosophical Magazine 2012, 92:3753–3756. doi: 10.1080/14786435.2012.742293.

[Gueymard 2002] Gueymard CA, Myers D, Emery K: Proposed reference irradiance spectra for solar energy systems testing. *Solar Energy* 2002, 73:443–467. doi: 10.1016/s0038-092x(03)00005-7.

[Guiart 1956] Guiart J: CULTURE CONTACT AND THE JOHN FRUM MOVEMENT ON TANNA, NEW HEBRIDES. *Southwestern Journal of Anthropology* 1956, 12:105–116. doi: 10.1086/soutjanth.12.1.3628861.

[Gunnarsson 2002] Gunnarsson R: Some aspects of interfaces in perovskite manganites. *Licentiate Thesis*. Chalmers University of Technology, Göthenburg University, Microelectronics and Nanoscience;2002.

[Ha 2014] Ha S, Kim JK, Choi A, Kim Y, Lee KT: Sodium-metal halide and sodium-air batteries. *ChemPhysChem* 2014, 15:1971–1982. doi: 10.1002/cphc.201402215.

[Haas 1996] Haas O, Müller S, Wiesener K: Wiederaufladbare Zink/Luftsauerstoff-Batterien. *Chemie Ingenieur Technik* 1996, 68:524–542. doi: 10.1002/cite.330680505.

[Haga 2008] Haga K, Adachi S, Shiratori Y, Itoh K, Sasaki K: Poisoning of SOFC anodes by various fuel impurities. *Solid State Ionics* 2008, 179:1427–1431. doi: 10.1016/j.ssi.2008.02.062.

[Haga 2010] Haga K, Shiratori Y, Nojiri Y, Ito K, Sasaki K: Phosphorus poisoning of Ni-Cermet anodes in solid oxide fuel cells. *Journal of The Electrochemical Society* 2010, 157. doi: 10.1149/1.3489265.

[Hahn 2001] Hahn M, Bartsch M, Schnyder B, Kotz R, Carlen M, Ohler C, Evard D: A 24 V bipolar electrochemical double layer capacitor based on activated glassy carbon. *Power Sources for the New Millennium, Proceedings* 2001, 2000:220–228.

[Hall 2010] Hall G: *Hermann Minkowski and Special Relativity.* Dordrecht: Springer; 2010. doi: 10.1007/978-90-481-3475-5_3.

[Hallwachs 1888] Hallwachs W: Ueber den Einfluss des Lichtes auf electrostatisch geladene Körper. *Annalen der Physik und Chemie* 1888, 269:301–312. doi: 10.1002/andp.18882690206.

[Hamann 2005] Hamann CF, Vielstich W: *Elektrochemie.* 4th edn. Weinheim: Wiley-VCH Verlag GmbH & Co. KGaA; 2005.

[Hamelin 1996] Hamelin A: Cyclic voltammetry at gold single-crystal surfaces. Part 1. Behaviour at low-index faces. *Journal of Electroanalytical Chemistry* 1996, 407:1–11. doi: 10.1016/0022-0728(95)04499-x.

[Hamer 1965] Hamer WJ: *Standard Cells – Their Construction, Maintenance, and Characteristics* Washington, DC: United States Department of Commerce; 1965.

[Hammoudeh 1995] Hammoudeh S, Madan V: Expectations, target zones, and oil price dynamics. *Journal of Policy Modeling* 1995, 17:597–613. doi: 10.1016/0161-8938(95)00022-4.

[Han 2002] Han J, Lee SM, Chang H: Metal membrane-type 25-kW methanol fuel processor for fuel-cell hybrid vehicle. *Journal of Power Sources* 2002, 112:484–490. doi: 10.1016/s0378-7753(02)00440-8.

[Hankin 2017] Hankin A, Bedoya-Lora FE, Ong CK, Alexander JC, Petter F, Kelsall GH: From millimetres to metres: The critical role of current density distributions in photo-electrochemical reactor design. *Energy & Environmental Science* 2017. doi: 10.1039/c6ee03036j.

[Hardee 1975] Hardee KL, Bard AJ: Semiconductor Electrodes .1. Chemical vapor-deposition and application of polycrystalline N-type titanium-dioxide electrodes to photosensitized electrolysis of water. *Journal of the Electrochemical Society* 1975, 122:739–742. doi: 10.1149/1.2134312.

[Hardee 1976] Hardee KL, Bard AJ: Semiconductor electrodes. 5. application of chemically vapor-deposited iron-oxide films to photosensitized electrolysis. *Journal of the Electrochemical Society* 1976, 123:1024–1026. doi: 10.1149/1.2132984.

[Hardee 1977] Hardee KL, Bard AJ: Semiconductor electrodes. 10. photoelectrochemical behavior of several polycrystalline metal-oxide Electrodes in aqueous-solutions. *Journal of the Electrochemical Society* 1977, 124:215–224. doi: 10.1149/1.2133269.

[Hartmann 2012] Hartmann J, Neubert A: Designers' open. In *Das Festival für Design Leipzig.* Edited by Gmbh DO. Leipzig: Designers' Open GmbH; 2012.

[Hassmann 2001] Hassmann K: SOFC power plants, the Siemens-Westinghouse approach. *Fuel Cells* 2001, 1: 78–84.

[Hauschildt 2009] Allgemein – U-Boot-Technologie der Zukunft. https://www.globaldefence.net/technologie/allgemein-u-boot-technologie-der-zukunft/

[Hay 1976] Hay A: TOXIC CLOUD OVER SEVESO. *Nature* 1976, 262:636–638.

[Hebb 1952] Hebb MH: Electrical conductivity of silver sulfide. *The Journal of Chemical Physics* 1952, 20:185–190. doi: 10.1063/1.1700165.

[Hegedus 1986] Hegedus F, Green WV, Stiller P, Oliver BM, Green S, Herrnberger V, Victoria M, Stiefel U: The strength of the spallation neutron-flux of SINQ for radiation-damage fusion technology. *Journal of Nuclear Materials* 1986, 141:911–914. doi: 10.1016/0022-3115(86)90117-0.

[Heidegger 1976] Heidegger M: *Vom Wesen der Wahrheit*. Sechste Auflage edn. Frankfurt A.M.: Vittorio Klostermann; 1976.

[Heiland 1952] Heiland G: Zum Energieumsatz bei Licht- oder Elektronenbestrahlung dünner Zinkoxyd-Schichten. *Zeitschrift für Physik* 1952, 132.

[Heiland 1954] Heiland G: Zurn Einfluss von adsorbiertem Sauerstoff auf die elektrische Leitfahigkeit von Zinkoxydkristallen. *Zeitschrift für Physik* 1954, 138:459–464.

[Heiland 1955] Heiland G: Die elektrische Leitfahigkeit an der Oberfläche von Zinkoxydkristallen. *Zeitschrift für Physik* 1955, 142:415–432.

[Heiland 1957] Heiland G: Zur Theorie der Anreicherungsrandschicht an der Oberfläche von Halbleitern. *Zeitschrift für Physik* 1957, 148:28–33. doi: 10.1007/bf01327363.

[Heiland 1961] Heiland G: Photoconductivity of zinc oxide as a surface phenomenon. *Journal of Physics and Chemistry of Solids* 1961, 22:227–234.

[Heiland 1978] Heiland GK, D.: Interpretation of Surface Phenomena on ZnO by the Compensation Model. *phys stat sol* 1978, 49:27–37.

[Heinz 1996] Heinz K, Kottcke M, Löffler U, Döll R: Recent advances in LEED surface crystallography. *Surface Science* 1996, 357–358:1–9. doi: 10.1016/0039-6028(96)00048-9.

[Helberg 1947] Helberg C: The Vemork Action – A classic act of sabotage. Arlington, VA: Central Intelligence Agency, Center-for-the-study-of-Intelligence; 1947.

[Heller 1981] Heller A: Conversion of sunlight into electrical-power and photoassisted electrolysis of water in photoelectrochemical cells. *Accounts of Chemical Research* 1981, 14:154–162. doi: 10.1021/ar00065a004.

[Hellwig 2013] Hellwig CA: Modeling, simulation and experimental investigation of the thermal and electrochemical behavior of a $LiFePO_4$-based lithium-ion battery. *Dissertation*. Universität Stuttgart, Chemie; 2013.

[Helm 2001] Helm T. Friend of Fischer jailed for role in 1975 Opec killings. *The Telegraph*. 16 Feb 2001. http://www.telegraph.co.uk/news/worldnews/europe/1322890/Friend-of-Fischer-jailed-for-role-in-1975-Opec-killings.html

[Helmholtz 1850] Helmholtz H: Vorläufiger Bericht über die Fortpflanzungsgeschwindigkeit der Nervenreizung. Archiv für Anatomie, Physiologie und wissenschaftliche Medicin. In: Monatsbericht der Königlichen Akademie der Wissenschaften. *Archiv für Anatomie, Physiologie und wissenschaftliche Medicin In: Monatsbericht der Königlichen Akademie der Wissenschaften* 1850:71–73.

[Helmholtz 1853] Helmholtz H: Ueber einige Gesetze der Vertheilung elektrischer Ströme in körperlichen Leitern mit Anwendung auf die thierisch-elektrischen Versuche. *Annalen der Physik und Chemie* 1853, 165:211–233. doi: 10.1002/andp.18531650603.

[Helmholtz 1879] Helmholtz H: Studien über electrische Grenzschichten. *Annalen der Physik und Chemie* 1879, 243:337–382. doi: 10.1002/andp.18792430702.

[Hempelmann 1995] Hempelmann R, Karmonik C, Matzke T, Cappadonia M, Stimming U, Springer T, Adams MA: Quasi-elastic neutron-scattering study of proton diffusion in SrCe0.95Yb0.05HO.0202.985. *Solid State Ionics* 1995, 77:152–156. doi: 10.1016/0167-2738(94)00264-S.

[Hempelmann 2000] Hempelmann R: *Quasielastic Neutron Scattering and Solid State Diffusion*. Oxford Series on Neutron Scatt; 2000.

[Hendry 2009] Hendry AP, Huber SK, De Leon LF, Herrel A, Podos J: Disruptive selection in a bimodal population of Darwin's finches. *Proceedings of the Royal Society: Biological Sciences* 2009, 276:753–759. doi: 10.1098/rspb.2008.1321.

[Heo 2013] Heo J-Y, Yoo S-H: The public's value of hydrogen fuel cell buses: A contingent valuation study. *International Journal of Hydrogen Energy* 2013, 38: 4232–4240. doi: 10.1016/j.ijhydene.2013.01.166.

[Herdlitschka 2017] Herdlitschka M. Joint Press Release: New TOTAL hydrogen filling station in Karlsruhe unites mobility and renewable energy. 2017. Daimler Communications Future Powertrain & e-Mobility. http://media.daimler.com/marsMediaSite/en/instance/ko/Joint-Press-Release-New-TOTAL-hydrogen-filling-station-in-Karlsruhe-unites-mobility-and-renewable-energy.xhtml?oid=29153376

[Herle 2010] Herle JV, Wochele J, Wellinger M, Ludwig C, Nurk G, Holtappels P, Braun A, Diethelm S, Morandin M, Maréchal F, Amelio S, Biollaz S, Schildhauer T, Rhyner U. CCEM Final Report 2010 WoodGas SOFC – I. 2010. 15 Dec 2010.

[Hernández Redondo 2009] Hernández Redondo A: Copper(I) polypyridine complexes. the sensitizers of the future for dye-sensitized solar cells (DSSCs) Universität Basel, Chemie; 2009.

[Hertz 1887] Hertz H: Ueber einen Einfluss des ultravioletten Lichtes auf die electrische Entladung. *Annalen der Physik und Chemie* 1887, 267:983–1000. doi: 10.1002/andp.18872670827.

[Hervey 1994] Hervey JL: The 1973 oil crisis: One generation and counting. *Chicago Fed Letter* 1994, 86.

[Hila 2018] Lemon Battery. http://hilaroad.com/camp/projects/lemon/lemon_battery.html

[Hilbert 1910] Hilbert D: Hermann Minkowski. *Mathematische Annalen* 1910, 68:445–471. doi: 10.1007/bf01455870.

[Ho 2015] Ho CK, Sims CA, Christian JM: Evaluation of glare at the Ivanpah Solar Electric Generating System. In *International Conference on Concentrating Solar Power and Chemical Energy Systems, Solarpaces 2014. Volume 69.* Edited by Wang Z. Amsterdam: Elsevier Science Bv; 2015: 1296–1305. *Energy Procedia.* doi: 10.1016/j.egypro.2015.03.150.

[Ho 2016] Ho CK: Review of Avian mortality studies at concentrating solar power plants. In *Solarpaces 2015: International Conference on Concentrating Solar Power and Chemical Energy Systems. Volume 1734.* Edited by Rajpaul V, Richter C. Melville: Amer Inst Physics; 2016. *AIP Conference Proceedings.* doi: 10.1063/1.4949164.

[Hodgkin 1939] Hodgkin AL, Huxley AF: Action potentials recorded from inside a nerve fibre. *Nature* 1939, 144:710–711. doi: 10.1038/144710a0.

[Hodgkin 1945] Hodgkin AL, Huxley AF: Resting and action potentials in single nerve fibres. *Journal of Physiology* 1945, 104:176–195. doi: 10.1113/jphysiol.1945.sp004114.

[Hodgkin 1946] Hodgkin AL, Huxley AF: Potassium leakage from an active nerve fibre. *Nature* 1946, 158:376. doi: 10.1038/158376b0.

[Hodgkin 1949] Hodgkin AL, Huxley AF, Katz B: Ionic Currents Underlying Activity in the Giant Axon of the Squid. *Archives Des Sciences Physiologiques* 1949, 3:129–150.

[Hodgkin 1952a] Hodgkin AL, Huxley AF: Propagation of electrical signals along giant nerve fibres. *Proceedings of the Royal Society Series B-Biological Sciences* 1952a, 140:177–183. doi: 10.1098/rspb.1952.0054.

[Hodgkin 1952b] Hodgkin AL, Huxley AF, Katz B: Measurement of current-voltage relations in the membrane of the giant axon of Loligo. *Journal of Physiology* 1952b, 116:424–448. doi: 10.1113/jphysiol.1952.sp004716.

[Hoffman 2016] The Weston Standard Cell. http://conradhoffman.com/stdcell.htm

[Hoffmann 2015] Hoffmann J: Emissionsreduzierte Antriebe – alternative Energieerzeugung an Bord – Status Brennstoffzelle. 2015. Ag S. www.schiffsingenieure.de/images/pdf/20150910_SiemensvortragHH.pdf.

[Hofmann 2012] Hofmann M, Gilles R, Gao Y, Rijssenbeek JT, Muhlbauer MJ: Spatially resolved phase analysis in sodium metal halide batteries: Neutron diffraction and tomography. *Journal of the Electrochemical Society* 2012, 159:A1827–A1833. doi: 10.1149/2.058211jes.

[Hoikkala 1996] Hoikkala P: The Ohlone past and present – Native-Americans of the San-Francisco-Bay Region – Bean,Lj. *Journal of the West* 1996, 35:101–101.

[Holcomb 1969] Holcomb RW: Fusion power: optimism and a tokamak gap at dubna. *Science* 1969, 166:363–364. doi: 10.1126/science.166.3903.363.

[Holdsworth 2010] Holdsworth SB, Freddy; Braun, Artur; Chen, Qianli, Strässle, Thierry; Embs, Jan: *Measurement Cell for Neutron Studies Under High Pressure and High Temperature.* Dübendorf: Empa; 2010.

[Holmes 2015] Holmes CE: Marikana in translation: Print nationalism in South Africa's Multilingual Press. *African Affairs* 2015, 114:271–294. doi: 10.1093/afraf/adv001.

[Holzer 1998] Holzer F, Müller S, Haas O: Development and tests of cell components for a 12 V/20 Ah electrically rechargeable zinc/air battery. In *Proceedings of 38th Power Sources Conference*; Cherry Hill, NJ. 1998.

[Holzwarth 2012] Holzwarth A: AMPEA Workshop on „Artificial Photosynthesis", October 15–16, 2012, MPI-CEC, Mülheim a.d. Ruhr, Germany. In *AMPEA Workshop on „Artificial Photosynthesis"*; 15–16 October 2012; MPI-CEC, Mülheim a.d. Ruhr, Germany; 2012.

[Horne 2000] Horne CR, Bergmann U, Grush MM, Perera RCC, Ederer DL, Callcott TA, Cairns EJ, Cramer SP: Electronic structure of chemically-prepared LixMn2O4Determined by Mn X-ray absorption and emission spectroscopies. *The Journal of Physical Chemistry B* 2000, 104:9587–9596. doi: 10.1021/jp994475s.

[Hu 2016] Hu Y, Boudoire F, Hermann-Geppert I, Bogdanoff P, Tsekouras G, Mun BS, Fortunato G, Graetzel M, Braun A: Molecular origin and electrochemical influence of capacitive surface states on iron oxide photoanodes. *The Journal of Physical Chemistry C* 2016, 120:3250–3258. doi: 10.1021/acs.jpcc.5b08013.

[Hu 2016a] Hu Y: Defects on surface and interface for photoelectrochemical properties of hematite photoanodes. *Dissertation*. ÉCOLE POLYTECHNIQUE FÉDÉRALE DE LAUSANNE, À LA FACULTÉ DES SCIENCES DE BASE; 2016a.

[Hu 2016b] Hu Y, Boudoire F, Hermann-Geppert I, Bogdanoff P, Tsekouras G, Mun BS, Fortunato G, Graetzel M, Braun A: Molecular origin and electrochemical influence of capacitive surface states on iron oxide photoanodes. *The Journal of Physical Chemistry C* 2016b, 120:3250–3258. doi: 10.1021/acs.jpcc.5b08013.

[Huber 2014] Huber M. Smartphones: Länger telefonieren dank Akku-Ersatz? Aargauer Zeitung. 29 April 2014. https://www.aargauerzeitung.ch/leben/digital/smartphones-laenger-telefonieren-dank-akku-ersatz-127924616

[Hubert 1969] Hubert P: Thermonuclear future of the Tokamak magnetic confinement device. *Nuclear Fusion* 1969, 9:209–214. doi: 10.1088/0029-5515/9/3/003.

[Hueso 2013] Hueso KB, Armand M, Rojo T: High temperature sodium batteries: status, challenges and future trends. *Energy & Environmental Science* 2013, 6:734. doi: 10.1039/c3ee24086j.

[Hüfner 1995] Hüfner S: Photoelectron Spectroscopy. Principles and Applications. Berlin, New York: Springer; 1995.

[Hüfner 2003] Hüfner S: *Photoelectron Spectroscopy*. Berlin, Heidelberg: Springer Verlag; 2003. doi: 10.1007/978-3-662-09280-4.

[Huggins 1975] Huggins FE: Mössbauer Studies of Iron Minerals Under Pressures of up to 200 Kilobars. Oxford; 1970. Thesis, Massachusetts Institute of Technology, Cambridge MA, oai:dspace.mit.edu:1721.1/59592.

[Huggins 2008] Huggins FE, Shah N, Braun A, Huffman GP, Kelly K, Wagner D, Sarofim AF: Speciation of Iron, Carbon, and sulfur in diesel exhaust particulate from combustion of diesel fuel with and without addition of 0.1 Wt\% ferrocene. In *AIChE Annual Meeting, Conference Proceedings*; 2008.

[Hund 1951] Hund F: Anomale Mischkristalle im System ZrO2·Y2O3 Kristallbau der Nernst-Stifte. *Berichte der Bunsengesellschaft für physikalische Chemie* 1951, 55.doi: 10.1002/bbpc.19510550505.

[Hunt 1999] Hunt GW: Valve-regulated lead/acid battery systems. Power Engineering Journal 1999, 13:113–116. doi: 10.1049/pe:19990302.

[Huot 1997] Huot JY: *Advances in Zinc Batteries*. Montreal: Ecole Polytechnique Montreal; 1997.

[Hussain 2016] Hussain H, Torrelles X, Cabailh G, Rajput P, Lindsay R, Bikondoa O, Tillotson M, Grau-Crespo R, Zegenhagen J, Thornton G: Quantitative structure of an acetate Dye molecule analogue at the TiO2-acetic acid interface. *Journal of Physical Chemistry C Nanomater Interfaces* 2016, 120:7586–7590. doi: 10.1021/acs.jpcc.6b00186.

[Huxley 1950] Huxley AF, Stampfli R: *Direkte Bestimmung Des Membranpotentials Der Markhaltigen Nervenfaser in Ruhe Und Erregung. *Helvetica Physiologica Et Pharmacologica Acta* 1950, 8: 107–109.

[Huxley 1963] Huxley AF: Nobel lecture: The quantitative analysis of excitation and conduction in nerve. In *The Nobel Prize in Physiology or Medicine 1963*. Stockholm: Nobel Media AB; 1963.

[Hyde 2004] Hyde ME, Jacobs RMJ, Compton RG: An AFM study of the correlation of lead dioxide electrocatalytic activity with observed morphology. *Journal of Physical Chemistry B* 2004, 108:6381–6390. doi: 10.1021/jp031263t.

[Hyodo 1997] Hyodo T, Hayashi M, Mitsutake S, Miura N, Yamazoe N: Oxygen reduction activities of praseodymium manganites in alkaline solution. *Journal of the Ceramic Society of Japan* 1997, 105:412–417. doi: 10.2109/jcersj.105.412.

[Hyundai 2013] Hyundai: ix35 fuel cell – emergency response guide. 2013. Company HM. Hyundai Motor Company. https://www.nfpa.org/-/media/Files/Training/AFV/Emergency-Response-Guides/Hyundai/Hyundai-Tucson-Fuel-cell.ashx?la=en&hash=CDD885043A05DAEFA6DB21796B4CE443231F7420

[IBAarau 2016] IBAarau. Die erste Schweizer Wasserkraft- Elektrolyseanlage für Wasserstoff – Nachhaltig erzeugt am Wasserkraftwerk in Aarau. 2016. IBAarau AG. https://www.ibaarau.ch/upload/cms/user/broschuere_wasserstoff1.pdf

[Ibach 1988] Ibach H, Lüth H: *Festkörperphysik – Eine Einführung in die Grundlagen*. Springer; 1988.

[Ibach 1996] Ibach H, Giesen M, Flores T, Wuttig M, Treglia G: Vacancy generation at steps and the kinetics of surface alloy formation. *Surface Science* 1996, 364:453–466. doi: 10.1016/0039-6028(96)00635-8.

[Ibach 2005] Ibach H, Lüth H, Kohl C-D, Sander W: Nachruf auf Gerhard Heiland. *Physik Journal* 2005, 4:65.

[Ibl 1971] Ibl N, Adam E, Venczel J, Schalch E: Mass transfer in electrolysis with gas agitation. *Chemie Ingenieur Technik* 1971, 43:202+.

[Ibl 1982] Ibl N: Nomenclature for transport phenomena in electrolytic systems. *Electrochimica Acta* 1982, 27:629–642. doi: 10.1016/0013-4686(82)85051-2.

[Ihssen 2014] Ihssen J, Braun A, Faccio G, Gajda-Schrantz K, Thony-Meyer L: Light harvesting proteins for solar fuel generation in bioengineered photoelectrochemical cells. *Current Protein & Peptide Science* 2014, 15:374–384.

[Iht 2018] High pressure electrolyzers. http://www.iht.ch/technologie/electrolysis/industry/high-pressure-electrolysers.html

[Iino 2018] Iino K, Yoshioka R, Fuchigami M, Nakao M: Precautions at Fukushima that would have suppressed the accident severity. *Journal of Nuclear Engineering and Radiation Science* 2018, 4:14.doi: 10.1115/1.4039343.

[Imada 1998] Imada M, Fujimori A, Tokura Y: Metal-insulator transitions. *Reviews of Modern Physics* 1998, 70:1039–1263. doi: 10.1103/RevModPhys.70.1039.

[Inzelt 2013] Inzelt G, Lewenstam A, Scholz F: *Handbook of Reference Electrodes*. Heidelberg, New York, Dordrecht, London: Springer; 2013. doi: 10.1007/978-3-642-36188-3.

[Gitelson 2008] Gitelson IIL, Genry. M.: Creation of closed ecological life support systems: Results, critical problems and potentials. *Journal of Siberian Federal University, Biology 1*, 2008: 19–39.

[Islar 2016] Islar M, Busch H: "We are not in this to save the polar bears!" - the link between community renewable energy development and ecological citizenship. Innovation-the European Journal of Social Science Research 2016, 29:303–319. doi: 10.1080/13511610.2016.1188684.

[Itoh 2007] Itoh M, Saito M, Tajima N, Machida K: Ammonia synthesis using atomic hydrogen supplied from silver-palladium alloy membrane. *Materials Science Forum* 2007, 561–565:1597–1600. doi: 10.4028/www.scientific.net/MSF.561-565.1597.

[Jackson 1998] Jackson RH: The Costanoan/Ohlone Indians of the San Francisco and Monterey Bay area: A research guide. *Ethnohistory* 1998, 45:381–382.doi: 10.2307/483073.

[Jahan 2016] Jahan S, Jespersen E, Mukherjee S, Kovacevic M, Abdreyeva B, Bonini A, Calderon C, Cazabat C, Hsu Y-C, Lengfelder C, Luongo P, Mukhopadhyay T, Nayyar S, Tapia H: *Human Development Report 2016*. New York: Communications Development Incorporated, Washington, DC; 2016.

[James 2014] James WS. Jesse; Mathison, Steve: An Introduction to SAE hydrogen fueling standardization. In *Webinar: Introduction to SAE Hydrogen Fueling Standardization*. https://energy.gov/eere/fuelcells/webinar-introduction-sae-hydrogen-fueling-standardization: U.S. Department of Energy; 2014.

[Jan 2000] Jan H, Kim IS, Choi HS: Purifier-integrated methanol reformer for fuel cell vehicles. *Journal of Power Sources* 2000, 86:223–227.

[Janssen 1997] Janssen S, Mesot J, Holitzner L, Furrer A, Hempelmann R: FOCUS: A hybrid TOF-spectrometer at SINQ. *Physica B* 1997, 234:1174–1176. doi: 10.1016/s0921-4526(97)00209-3.

[Janssen 2000] Janssen S, Altorfer F, Holitzner L, Hempelmann R: Time-of-flight spectrometer FOCUS at SINQ: First results. *Physica B* 2000, 276:89–90. doi: 10.1016/s0921-4526(99)01253-3.

[Janz 1953] Janz GJ, Taniguchi H: The silver-silver halide electrodes. Preparation, stability, and standard potentials in aqueous and non-aqueous media. *Chemical Reviews* 1953, 53:397–437. doi: 10.1021/cr60166a002.

[Jenkins 1976] Jenkins GM, Kawamura K: *Polymeric Carbons – Carbon Fibre, Glass and Char*. 1st ed. Cambridge: Cambridge University Press; 1976.

[Jennings 2005] Jennings RC, Engelmann E, Garlaschi F, Casazza AP, Zucchelli G: Photosynthesis and negative entropy production. *Biochim Biophys Acta* 2005, 1709:251–255. doi: 10.1016/j.bbabio.2005.08.004.

[Jervis 2016] Jervis R, Brown LD, Neville TP, Millichamp J, Finegan DP, Heenan TMM, Brett DJL, Shearing PR: Design of a miniature flow cell for in situ x-ray imaging of redox flow batteries. *Journal of Physics D: Applied Physics* 2016, 49:434002. doi: 10.1088/0022-3727/49/43/434002.

[Jewulski 1990] Jewulski JR, Osif TL, Remick RJ. Solid-State Proton Conductors. 1990. Institute of Gas Technology IIT Center 3424 S. State Street Chicago, IL 60616. Morgantown WV. Energy USDO. DOE/MC/24218-2957. December 1990

[JFE Steel Co. 2017] JFE Steel Co. L: *Electrical Steel Sheets, JFE G-Core, JFE N-Core*. Corporation JS ed. Tokyo, Japan: JFE Steel Co., Ltd.; 2017.

[Jiang 1997] Jiang GM, Lin GH: Changes of photosynthetic capacity of some plant species under very high CO2 concentrations in Biosphere 2. *Chinese Science Bulletin* 1997, 42:859–864. doi: 10.1007/bf02882501.

[Jiang 1997] Jiang Y, Chai X, Yang W, Zhang D, Cao Y, Zhu Z, Li T, Lehn J-M: Frontier orbital interactions of electron pushing and drawing substituents with ferrocenyl group. *Science in China Series B: Chemistry* 1997, 40:236–244. doi: 10.1007/BF02877724.

[Jiang 2008] Jiang SP: Development of lanthanum strontium manganite perovskite cathode materials of solid oxide fuel cells: A review. *Journal of Materials Science* 2008, 43:6799–6833. doi: 10.1007/s10853-008-2966-6.

[Jiang 2010] Jiang P, Chen J-L, Borondics F, Glans P-A, West MW, Chang C-L, Salmeron M, Guo J: In situ soft X-ray absorption spectroscopy investigation of electrochemical corrosion of copper in aqueous NaHCO3 solution. *Electrochemistry Communications* 2010, 12:820–822. doi: 10.1016/j.elecom.2010.03.042.

[Jingang 2012] Jingang H, Charpentier JF, Tianhao T: State of the art of fuel cells for ship applications. In *2012 IEEE International Symposium on Industrial Electronics (ISIE); May 2012; Hangzhou, China*. 2012: 1456–1461. doi: 10.1109/isie.2012.6237306.

[Jodidio 2001] Jodidio P: Christo and Jeanne-Claude: Wrapped Reichstag, 1971–95. *Connaissance Des Arts* 2001:39–39.

[Joghee 2015] Joghee P, Malik JN, Pylypenko S, O'hayre R: A review on direct methanol fuel cells – In the perspective of energy and sustainability. *MRS Energy & Sustainability* 2015, 2. doi: 10.1557/mre.2015.4.

[John 2007] John W, Gaida B: *Technische Mathematik für die Galvanotechnik*. 9th edn. Bad Saulgau: Eugen G. Leuze Verlag; 2007.

[Johnson 1996] Johnson TR: The Ohlone past and present – native-Americans of the San-Francisco-Bay region – Bean,Lj. *American Indian Culture and Research Journal* 1996, 20:233–239.

[Johnson 1997] Johnson KL: The Ohlone, past and present: Native Americans of the San Francisco Bay Region. *Ethnohistory* 1997, 44:588–590.doi: 10.2307/483052.

[Jones 2011] Jones KS: *State of Solid-State Batteries Department of Materials Science and Engineering*. University of Florida Department of Materials Science and Engineering, University of Florida; 2011.

[Jordi 1995] Jordi H: A financing system for battery recycling in Switzerland. *Journal of Power Sources* 1995, 57:51–53. doi: 10.1016/0378-7753(95)02240-6.

[Jung 2000] Jung WH, Wakai H, Nakatsugawa H, Iguchi E: Small polarons in La[sub 2/3]TiO[sub 3−δ]. *Journal of Applied Physics* 2000, 88:2560. doi: 10.1063/1.1287755.

[Jung 2001a] Jung W-H: Transport mechanisms in La0.7Sr0.3FeO3: Evidence for small polaron formation. *Physica B: Condensed Matter* 2001a, 299:120–123. doi: 10.1016/s0921-4526(00)00590-1.

[Jung 2001b] Jung WH: Dielectric loss anomaly and polaron hopping conduction of Gd1Õ3Sr2Õ3FeO3. *Journal of Applied Physics* 2001b. doi: 10.1063/1.1388600※.

[Jung 2002] Jung WH: Non-adiabatic small polaron hopping conduction in Gd1/3Sr2/3FeO3. *Materials Letters* 2002.

[Jung 2006] Jung W-H: Polaron transport in n=3 Ruddlesden–Popper phase Sr4Mn1.5Fe1.5O9.72 ceramics. *Journal of Materials Science* 2006, 41:3143–3145. doi: 10.1007/s10853-006-6435-9.

[Jung 2007] Jung W-H: Variable range hopping conduction of polaron in n=2 Ruddlesden–Popper phase Sr3Fe1.5Co0.5O6.77. *Materials Letters* 2007, 61:2274–2276. doi: 10.1016/j.matlet.2006.08.066.

[Juranyi 2003] Juranyi F, Janssen S, Mesot J, Holitzner L, Kagi C, Tuth R, Burge R, Christensen M, Wilmer D, Hempelmann R: The new mica monochromator for the time-of-flight spectrometer FOCUS at SINQ. *Chemical Physics* 2003, 292:495–499. doi: 10.1016/s0301-0104(03)00175-7.

[Kaiser 2000] Kaiser A: Tetragonal tungsten bronze type phases (Sr1–xBax)0.6Ti0.2Nb0.8O3–δ: Material characterisation and performance as SOFC anodes. *Solid State Ionics* 2000, 135: 519–524. doi: 10.1016/S0167-2738(00)00432-X.

[Kaku 2016] Kaku H, Dong YR, Hanafusa K, Moriuchi K, Shigematsu T: Effect of Ti(IV) ion on Mn(III) stability in Ti/Mn electrolyte for redox flow battery. *ECS Transactions* 2016, 72:1–9. doi: 10.1149/07210.0001ecst.

[Kalfatovic 2002] Kalfatovic MR: Christo and Jeanne-Claude: A biography. *Library Journal* 2002, 127:96–97.

[Kalyanasundaram 2008] Kalyanasundaram K: Protogonists in chemistry. *Inorganica Chimica Acta* 2008, 361:561–571. doi: 10.1016/j.ica.2007.10.001.

[Kalyani 2013] Kalyani P, Anitha A: Biomass carbon & its prospects in electrochemical energy systems. *International Journal of Hydrogen Energy* 2013, 38:4034–4045. doi: 10.1016/j.ijhydene.2013.01.048.

[Kamra 1997] Kamra AK, Deshpande CG, Gopalakrishnan V: Effect of relative humidity on the electrical conductivity of marine air. *Quarterly Journal of the Royal Meteorological Society* 1997, 123:1295–1305. doi: 10.1002/qj.49712354108.

[Kamsali 2011] Kamsali N, Prasad BSN, Datta J: The electrical conductivity as an index of air pollution in the atmosphere. In *Advanced Air Pollution*. Edited by Nejadkoorki F. Rijeka, Croatia: Intech Open; 2011. doi: 10.5772/17163.

[Kanamori 1959] Kanamori J: Superexchange interaction and symmetry properties of electron orbitals. *Journal of Physics and Chemistry of Solids* 1959, 10:87–98. doi: 10.1016/0022-3697(59)90061-7.

[Kanellos 2014] Kanellos M. The MIT curse strikes again: Lilliputian systems runs out of gas. *Forbes*. 21 Aug. 2014. https://www.forbes.com/sites/michaelkanellos/2014/08/21/the-mit-curse-strikes-again-lilliputian-systems-runs-out-of-gas/#95c36c1e4c86

[Kangro 1949] Kangro W. Verfahren zur Speicherung von elektrischer Energie 1949. Patentamt D. DE1949P047135 19490628. https://worldwide.espacenet.com/publicationDetails/biblio?FT=D&date=19540628&DB=&locale=de_EP&CC=DE&NR=914264C&KC=C&ND=3

[Kangro 1954] Kangro W. Verfahren zur Speicherung von elektrischer Energie in Fluessigkeiten 1954. Patentamt D.DE1954K022841 19540714. https://worldwide.espacenet.com/publication Details/biblio?locale=de_EP&CC=DE&NR=1006479#

[Kangro 1962] Kangro W, Pieper H: Zur Frage der Speicherung von elektrischer Energie in Flüssigkeiten. *Electrochimica Acta* 1962, 7:435–448. doi: 10.1016/0013-4686(62)80032-2.

[Kannan 2009] Kannan AM, Renugopalakrishnan V, Filipek S, Li P, Audette GF, Munukutla L: Bio-batteries and bio-fuel cells: Leveraging on electronic charge transfer proteins. *Journal of Nanoscience and Nanotechnology* 2009, 9:1665–1678. doi: 10.1166/jnn.2009.SI03.

[Kanungo 2007] Kanungo SB: Preparation of battery-grade manganese dioxide through dispropor-tionation of partially oxidised hydrous manganese oxide in dilute sulphuric acid. *Journal of Chemical Technology & Biotechnology* 2007, 50:91–100. doi: 10.1002/jctb.280500110.

[Karle 1980] Karle J: Some developments in anomalous dispersion for the structural investigation of macromolecular systems in biology. *International Journal of Quantum Chemistry* 1980, 18:357–367.

[Karlin 2006] Karlin I, Braun A: 2006. CEMTEC – Computational engineering of multi-scale transport in small-scale surface based energy conversion. CCEM. 705. Zürich, Dübendorf, Villigen, Lausanne. 3 years. finished. https://www.google.ch/url?sa=t&rct=j&q=&esrc=s&source=web&cd=1&ved=0ahUKEwjF1uLZvPvZAhUEbqOKHRxyBdoQFggsMAA&url=http%3A%2F%2Fwww.ccem.ch%2FMediaBoard%2FCEMTEC_webupdate.pdf&usg=AOvVaw0gBz4eNoMm5jb-3NSEpEEH

[Karlsson 2010] Karlsson M, Engberg D, BjöRketun MRE, Matic A, WahnströM GR, Sundell PG, Berastegui P, Ahmed I, Falus P, Farago B, BöRjesson L, Eriksson S: Using neutron spin–echo to investigate proton dynamics in proton-conducting perovskites†. *Chemistry of Materials* 2010, 22:740–742. doi: 10.1021/cm901624v.

[Karmonik 1995] Karmonik C, Hempelmann R, Matzke T, Springer T: Proton diffusion in strontium cerate ceramics studied by quasi-elastic neutron-scattering and impedance spectroscopy. *Zeitschrift Fur Naturforschung Section a-a Journal of Physical Sciences* 1995, 50: 539–548.

[Karpinsky 2002] Karpinsky: The silver-zinc battery system: A 60 year retrospective, from Andre, to Sputnik, to Mars. 2002.

[Katsapov 2010] Katsapov GY, Braun A: Deuterium tracer experiments prove the thiophenic hydrogen involvement during the initial step of thiophene hydrodesulfurization. *Catalysis Letters* 2010, 138:224–230. doi: 10.1007/s10562-010-0400-6.

[Katsura 2017] Katsura T, Baba K, Yoshino T, Kogiso T: Electrical conductivity of the oceanic asthenosphere and its interpretation based on laboratory measurements. *Tectonophysics* 2017, 717:162–181. doi: 10.1016/j.tecto.2017.07.001.

[Katz 1970] Katz B: Nobel lecture: On the quantal mechanism of neural transmitter release. In *The Nobel Prize in Physiology or Medicine 1970*. Stockholm: Nobel Media AB 1970.

[Kaufmann 2014] Kaufmann RK, Banerjee S: A unified world oil market: Regions in physical, economic, geographic, and political space. *Energy Policy* 2014, 74:235–242. doi: 10.1016/j.enpol.2014.08.028.

[Kayelaby 2018] Kaye and Laby: 2.7.9 Physical properties of sea water. http://www.kayelaby.npl.co.uk/general_physics/2_7/2_7_9.html.

[Keiser 1976] Keiser H, Beccu KD, Gutjahr MA: Abschätzung der porenstruktur poröser elektroden aus impedanzmessungen. *Electrochimica Acta* 1976, 21:539–543. doi: 10.1016/0013-4686(76)85147-x.

[Kennedy 1977] Kennedy JH, Frese KW: Photooxidation of water at alpha-Fe2O3 electrodes. *Journal of the Electrochemical Society* 1977, 124:C130–C130.

[Kennedy 1978] Kennedy JH, Frese KW: Photooxidation of water at α-Fe[sub 2]O[sub 3] electrodes. *Journal of the Electrochemical Society* 1978, 125:709. doi: 10.1149/1.2131532.

[Kenning 2017] Kenning T. SDG&E and Sumitomo unveil largest vanadium redox flow battery in the US. *Energy Storage News*. 17 March 2017.

[Kerbalek 2010] Kerbalek I: The migration of the academic elites: Opportunities and threats. *Transylvanian Review* 2010, 19:146–158.

[Khan 2008] Khan N, Saleem Z, Abas N, IEEE: *Experimental study of earth batteries*. New York: IEEE; 2008.

[Kiehne 2003] Kiehne HA: *Battery Technology Handbook*. 2nd edn. Boca Raton, FL: Marcel Dekker Inc.; 2003.

[Kim 2010] Kim S-H, Ahn B-K, Lim T-W: Development of Hyundai-Kia's fuel cell stack. *MTZ worldwide* 2010, 71: 10–13.doi: 10.1007/BF03226996.

[Kim 2011] Kim SH, Choi W, Lee KB, Choi S: Advanced dynamic simulation of supercapacitors considering parameter variation and self-discharge. *Ieee Transactions on Power Electronics* 2011, 26:3377–3385. doi: 10.1109/tpel.2011.2136388.

[Kim 2016] Kim DK, Min HE, Kong IM, Lee MK, Lee CH, Kim MS, Song HH: Parametric study on interaction of blower and back pressure control valve for a 80-kW class PEM fuel cell vehicle. *International Journal of Hydrogen Energy* 2016, 41: 17595–17615. doi: 10.1016/j.ijhydene.2016.07.218.

[Kim 2016] Kim LH, Kim YJ, Hong H, Yang D, Han M, Yoo G, Song HW, Chae Y, Pyun J-C, Grossman AR, Ryu W: Patterned nanowire electrode array for direct extraction of photosynthetic electrons from multiple living algal cells. *Advanced Functional Materials* 2016, 26:7679–7689. doi: 10.1002/adfm.201602171.

[Kim 2016] Kim W, Edri E, Frei H: Hierarchical Inorganic Assemblies for Artificial Photosynthesis. *Accounts of Chemical Research* 2016, 49:1634–1645. doi: 10.1021/acs.accounts.6b00182.

[Kingatua 2017] Determining the Equivalent Series Resistance (ESR) of Capacitors. https://www.allaboutcircuits.com/technical-articles/determining-equivalent-series-resistance-esr-of-capacitors/

[Kinn 2009] Kinn J, Lee S, Hwang I, Lim T, *The Safety of Hydrogen Realease and Discharge from a Fuel Cell Vehicle*. Proceedings of the 7th International Conference on Fuel Cell Science, Engineering, and Technology, 2009, 783–787..

[Kinoshita 1988] Kinoshita K: *Carbon: Electrochemical and Physicochemical Properties*. New York: Wiley; 1988.

[Kinoshita 1996] Kinoshita K: Exploratory technology research program for electrochemical energy storage. Annual report for 1995. 1996. Lawrence Berkeley National Lab., CA (United States). Berkeley CA. LBNL–38842 ON: DE97001192. 1 Jun 1996

[Kirchhofer 2015]Kirchhofer ND, Rasmussen MA, Dahlquist FW, Minteer SD, Bazan GC: The photobioelectrochemical activity of thylakoid bioanodes is increased via photocurrent generation and improved contacts by membrane-intercalating conjugated oligoelectrolytes. *Energy & Environmental Science* 2015, 8:2698–2706. doi: 10.1039/c5ee01707f.

[Kitching 1983] Kitching B: Dont blame the Arabs – world wide inflation built in before the oil crisis. *Rivista Internazionale Di Scienze Economiche E Commerciali* 1983, 30:561–565.

[Klahr 2012] Klahr B, Gimenez S, Fabregat-Santiago F, Hamann T, Bisquert J: Water oxidation at hematite photoelectrodes: the role of surface states. *Journal of American Chemical Society* 2012, 134:4294–4302. doi: 10.1021/ja210755h.

[Klein 2004] Klein C, Nazeeruddin MK, Di Censo D, Liska P, Gratzel M: Amphiphilic ruthenium sensitizers and their applications in dye-sensitized solar cells. *Inorg Chem* 2004, 43: 4216–4226. doi: 10.1021/ic049906m.

[Klesius 2009] Klesius M: How things work: Flying fuel cells out of gas? Not a problem. *Air & Space Magazine* 2009..

[Kluiters 1999] Kluiters EC, Schmal D, Ter Veen WR, Posthumus K: Testing of a sodium nickel chloride (ZEBRA) battery for electric propulsion of ships and vehicles. *Journal of Power Sources* 1999, 80:261–264. doi: 10.1016/s0378-7753(99)00075-0.

[Kneen 2003] Kneen B. Cargill: Size is everything. *The Ecologist*. 33. 3. 1 April 2003. https://theecologist.org/2003/apr/01/cargill-size-everything

[Knight 2008] Knight C, Davidson J, Behrens S: Energy options for wireless sensor nodes. *Sensors (Basel)* 2008, 8:8037–8066. doi: 10.3390/s8128037.

[Kobayashi 2016] Kobayashi K, Sakka Y, Suzuki TS: Development of an electrochemical impedance analysis program based on the expanded measurement model. *Journal of the Ceramic Society of Japan* 2016, 124:943–949. doi: 10.2109/jcersj2.16120.

[Kodentsov 1998] Kodentsov AA, Rijnders MR, Van Loo FJJ: Periodic pattern formation in solid state reactions related to the Kirkendall effect. *Acta Materialia* 1998, 46:6521–6528. doi: 10.1016/s1359-6454(98)00309-7.

[Koepp 1960] Koepp H-M, Wendt H, Strehlow H: Der Vergleich der Spannungsreihen in verschiedenen Solventien. II*. *Berichte der Bunsengesellschaft für physikalische Chemie* 1960, 64. doi: 10.1002/bbpc.19600640406.

[Koetz 1985] Koetz ER, Neff H: Anodic iridium oxide film – an ups study of emersed electrodes. *Surface Science* 1985, 160:517–530. doi: 10.1016/0039-6028(85)90791-5.

[Koh 1973] Koh JCY, Fortini A: Prediction of thermal conductivity and electrical resistivity of porous metallic materials. *International Journal of Heat and Mass Transfer* 1973, 16:2013–2022. doi: 10.1016/0017-9310(73)90104-X.

[Kokam 2016] Kokam: Powered by Kokam's ultra high energy NMC batteries, solar impulse 2 completes first flight around the world by a zero-fuel solar airplane. 2016. PRNewswire. http://www.prnewswire.com/news-releases/powered-by-kokams-ultra-high-energy-nmc-batteries-solar-impulse-2-completes-first-flight-around-the-world-by-a-zero-fuel-solar-airplane-300313150.html

[Kolarevic 2015] Kolarevic B: Actualising (overlooked) material capacities. *Architectural Design* 2015, 85:128–133. doi: 10.1002/ad.1965.

[Kolesnikov 2009] Kolesnikov A, Kutcherov VG, Goncharov AF: Methane-derived hydrocarbons produced under upper-mantle conditions. *Nature Geoscience* 2009, 2:566–570. doi: 10.1038/ngeo591.

[Koller 2015] Koller M: Design of closed photobioreactors for algal cultivation. In *Algal Biorefineries*; Edited by Aleš Prokop, Rakesh K. Bajpai, Mark E. Zappi. Heidelberg, New York, Dordrecht, London: Springer, Cham; 2015: 133–186. doi: 10.1007/978-3-319-20200-6_4.

[Köngeter 2013] Köngeter J, Kohler B, Ebert M, Libisch C: *Talsperren in Deutschland*. 1st edn. Wiesbaden: Springer Vieweg; 2013. doi: 10.1007/978-3-8348-2107-2.

[Koopmans 1934] Koopmans T: Über die Zuordnung von Wellenfunktionen und Eigenwerten zu den Einzelnen Elektronen Eines Atoms. *Physica* 1934, 1:104–113. doi: 10.1016/S0031-8914(34)90011-2.

[Kopka 2013] Kopka R, Tarczynski W: Measurement system for determination of supercapacitor equivalent parameters. *Metrology and Measurement Systems* 2013, 20:581–590. doi: 10.2478/mms-2013-0049.

[Koretz 2001] Koretz B, Naimer N. 3300 mAh zinc-air batteries for portable consumer products. *Sixteenth Annual Battery Conference on Applications and Advances Proceedings of the Conference* (Cat No01TH8533) Long Beach, CA. IEEE.

[Kotowski 2015] Kotowski J: Personal Communication. Tom33. 2015. Electrical conductivity of pool water. https://chemistry.stackexchange.com/questions/28333/electrical-conductivity-of-pool-water

[Kötz 1984] Kötz R, Neff, H., Stucki, S. S: Anodic iridium oxide films. *Journal of The Electrochemical Society* 1984, 131. doi: 10.1149/1.2115548.

[Kötz 1996] Kötz R, Carlen M, Desilvestro H: 1996. Supercapacitor for Lok2000. ETH Rat. Villigen PSI. 4 years. finished.

[Kötz 1998] Kötz R, Alliata D, Barbero C, Bärtsch M, Braun A, Imhof R, Schnyder B, Sullivan M: Characterization of surface layers on glassy carbon formed by constant potential anodization. In Meeting Abstracts of the 193rd Meeting of Electrochemical Society, San Diego, USA, May 3–8, 1998; 1998: 85

[Kötz 2000] Kötz R, Carlen M: Principles and applications of electrochemical capacitors. *Electrochimica Acta* 2000, 45:2483–2498. doi: 10.1016/s0013-4686(00)00354-6.

[Kowalczyk 1976] Kowalczyk SP: Photoelectron spectroscopy and Auger electron spectroscopy of solids and surfaces. University of California Berkeley, Materials and Molecular Research Division, Lawrence Berkeley Laboratory and Department of Chemistry University of California, Berkeley, CA; 1976.

[Kowalski 2010] Kowalski L: Commentary letter in „Hot topics in cold fusion". *Physics Today* 2010, 63:1. doi: 10.1063/1.3455240.

[Krajnc 2015] Krajnc N: *Wood Fuels Handbook*. Rome Italy, and Pristina Albania: Food and Agriculture Organization of the United Nations (FAO); 2015.

[Kreuer 1996] Kreuer KD: Proton conductivity: Materials and applications. *Chemistry of Materials* 1996, 8:610–641. doi: 10.1021/cm950192a.

[Kreuer 2012] Kreuer KD, Wohlfarth A: Limits of proton conductivity. *Angew Chem Int Ed Engl* 2012, 51:10454–10456; author reply 10457–10458. doi: 10.1002/anie.201203887.

[Kubecka 2001] Kubecka P: A possible world record maximum natural ground surface temperature. *Weather* 2001, 56:218–221. doi: 10.1002/j.1477-8696.2001.tb06577.x.

[Kuboki 2005] Kuboki T, Okuyama T, Ohsaki T, Takami N: Lithium-air batteries using hydrophobic room temperature ionic liquid electrolyte. *Journal of Power Sources* 2005, 146:766–769. doi: 10.1016/j.jpowsour.2005.03.082.

[Kulka 2015] Kulka A, Braun A, Huang T-W, Wolska A, Klepka MT, Szewczyk A, Baster D, Zając W, Świerczek K, Molenda J: Evidence for Al doping in lithium sublattice of LiFePO4. *Solid State Ionics* 2015, 270:33–38. doi: 10.1016/j.ssi.2014.12.004.

[Kumar 2018] Kumar S, Nehra M, Kedia D, Dilbaghi N, Tankeshwar K, Kim KH: Carbon nanotubes: A potential material for energy conversion and storage. *Progress in Energy and Combustion Science* 2018, 64: 219–253. doi: 10.1016/j.pecs.2017.10.005.

[Kummer 1967] Kummer JT, Nweber N. Energy conversion device comprising a solid crystalline electrolyte and a solid reaction zone separator.1967. Patent US.US 3404036 A.

[Kundt 2014] Kundt W, Marggraf O: *Physikalische Mythen auf dem Prüfstand*. Berlin, Heidelberg: Springer Spektrum; 2014.

[Kusch 2013] Kusch R. Sonntagsfahrverbot: Autos mussten wegen Ölkrise zu Hause bleiben. Duetschlandfunk. http://www.deutschlandfunk.de/sonntagsfahrverbot-autos-mussten-wegen-oelkrise-zu-hause.871.de.html?dram:article_id=269708

[Kuyucak 1994] Kuyucak S, Chung S-H: Temperature dependence of conductivity in electrolyte solutions and ionic channels of biological membranes. *Biophysical Chemistry* 1994, 52:15–24. doi: 10.1016/0301-4622(94)00034-4.

[Kyriakou 2017] Kyriakou V, Garagounis I, Vasileiou E, Vourros A, Stoukides M: Progress in the electrochemical synthesis of ammonia. *Catalysis Today* 2017, 286:2–13. doi: 10.1016/j.cattod.2016.06.014.

[La Cour 1905] La Cour P: *Die Windkraft und ihre Anwendung zum Antrieb von Elektrizitäts-Werken*. Leipzig: Verlag von M.Heinsius Nachf; 1905.

[Lajnef 2013] Lajnef T, Abid S, Ammous A: Modeling, control, and simulation of a solar h1ydrogen/fuel cell hybrid energy system for grid-connected applications. *Advances in Power Electronics* 2013, 2013:1–9. doi: 10.1155/2013/352765.

[Lamar 2010] Lamar K, Greenhill KM: Weapons of mass migration: Forced displacement as an instrument of coercion; strategic insights, v. 9, issue 1 (Spring-Summer 2010). In *Calhoun: Institutional Archive of the Naval Postgraduate School*. Edited by Center on Contemporary Conflict NSaD, Naval Postgraduate School, Monterey, California. Monterey, CA: Faculty and Researcher Publications; 2010.

[Lander 1962] Lander JJ, Keralla JA. Development of sealed silver oxide-zinc secondary batteries. 1962. Flight Accessories Laboratory, Aeronautical Systems Division, Air Force Systems Command Wright-Patterson Air Force Base. Ohio. ASD-TDR-62–668. October 1962.

[Landsberger 2005] Landsberger S, Casey J, Dodoo-Amoo D: Determination of heavy metals and their leaching characteristics in DOE lead-lined gloves. *Journal of Radioanalytical and Nuclear Chemistry* 2005, 264:477–479. doi: 10.1007/s10967-005-0740-7.

[Láng 2012] Láng GG, Barbero CA: Introduction to probe beam deflection techniques. In *Laser Techniques for the Study of Electrode Processes*. Edited by Scholz, F. Heidelberg, Berlin: Springer; 2012: 159–165: *Monographs in Electrochemistry*]. doi: 10.1007/978-3-642-27651-4_9.

[Langlais 2010] Langlais B, Lesur V, Purucker ME, Connerney JEP, Mandea M: Crustal magnetic fields of terrestrial planets. *Space Science Reviews* 2010, 152:223–249. doi: 10.1007/s11214-009-9557-y.

[Larouche 1995] Larouche LH.Tarpley WG.Interview.1995.LaRouche Warns of Food & Metals Hoarding.22 August 1995. http://www.larouchepub.com/tv/tlc_programs_1995.html;

https://archive.org/details/LaRoucheWarnsOfTheOligarchyHoardingMetalsAndFood; https://archive.org/details/LaRoucheWarnsOfHoardingMetalsAndFood

[Leblebici 2019] Leblebici A, Mayor P, Rajman M, De Micheli G: *Nano-Tera.ch: Engineering the Future of Systems for Health, Environment and Energy.* Switzerland: Springer; 2019. doi: 10.1007/978-3-319-99109-2

[Leclanché 1866]Leclanché G.1866. Office FP. Patent Nr. 71 865.

[Leddy 2004] Leddy JB, Viola; Vanýsek, P.: *Historical Perspectives on the Evolution of Electrochemical Tools.* Pennington, NJ: The Electrochemical Society, Inc.; 2004.

[Lee 2014] Lee V, Berejnov V, West M, Kundu S, Susac D, Stumper J, Atanasoski RT, Debe M, Hitchcock AP: Scanning transmission X-ray microscopy of nanostructured thin film catalysts for proton-exchange-membrane fuel cells. *Journal of Power Sources* 2014, 263:163–174. doi: 10.1016/j.jpowsour.2014.04.020.

[Lee 2017] Lee W-J: A study on recycling the waste soot produced from marine diesel engines. *Journal of the Korean Society of Marine Engineering* 2017, 41:543–548. doi: 10.5916/jkosme.2017.41.6.543.

[Lehner 2014] Lehner M, Tichler, R, Steinmüller, H, Koppe, M: *Power-to-Gas: Technology and Business Models.* Heidelberg, New York, Dordrecht, London: Springer Cham; 2014. doi: 10.1007/978-3-319-03995-4.

[Lehto 2006] Lehto KA, Lehto HJ, Kanervo EA: Suitability of different photosynthetic organisms for an extraterrestrial biological life support system. *Research in Microbiology* 2006, 157: 69–76. doi: 10.1016/j.resmic.2005.07.011.

[Lejuste 2011] Lejuste M. Chevron leaving U.S. coal industry. MarketWatch. 29 Jan 2011. https://www.marketwatch.com/story/chevron-leaving-us-coal-industry-2011-01-29

[Lemofouet 2006] Lemofouet S, Rufer A: A hybrid energy storage system based on compressed air and supercapacitors with maximum efficiency point tracking (MEPT). *IEEE Transactions on Industrial Electronics* 2006, 53: 1105–1115.doi: 10.1109/TIE.2006.878323.

[Lemström 1904] Lemström S: *Electricity in Agriculture and Horticulture (Electroculture).* Talma Studios (6 Mar. 2017); 1904.

[Leszczynski 2013] Leszczynski PJ, Grochala W: Strong cationic oxidizers: Thermal decomposition, electronic structure and magnetism of their compounds. *Acta Chim Slov* 2013, 60:15.

[Levenspiel 1962] Levenspiel O: Chemical Reaction Engineering – An Introduction to the Design of Chemical Reactors. New York: Wiley; 1962.

[Lewis 1997] Lewis NS, Tributsch H, Nozik AJ: Biography: Heinz Gerischer. *The Journal of Physical Chemistry B* 1997, 101:2391–2391. doi: 10.1021/jp970544w.

[Li 1991] Li J, Downer NW, Smith GH: Evaluation of Surface-Bound Membranes with Electrochemical Impedance Spectroscopy. 1991.TSI Mason Research Institute Biochemistry Department. Worcester, MA 01460. Research OON. Technical Report No. 4.

[Li 1994] Li JG, Downer NW, Smith HG: Evaluation of surface-bound membranes with electrochemical impedance spectroscopy. *Biomembrane Electrochemistry* 1994, 235:491–510.

[Li 2014a] Li GS, Lu XC, Kim JY, Lemmon JP, Sprenkle VL: Improved cycling behavior of ZEBRA battery operated at intermediate temperature of 175 degrees C. *Journal of Power Sources* 2014a, 249:414–417. doi: 10.1016/j.jpowsour.2013.10.110.

[Li 2014b] Li Y, Dai H: Recent advances in zinc-air batteries. *Chem Soc Rev* 2014b, 43:5257–5275. doi: 10.1039/C4CS00015C.

[Li 2016] Li Q, Chen J, Fan L, Kong X, Lu Y: Progress in electrolytes for rechargeable Li-based batteries and beyond. *Green Energy & Environment* 2016, 1:18–42. doi: 10.1016/j.gee.2016.04.006.

[Lide 2006] Lide DR: *CRC Handbook of Chemistry and Physics.* 87th edn. Boca Raton: CRC Press; 2006. doi: 10.1021/ja069813z.

[Ligen 2018] Ligen Y: Design and operation of a grid to mobility demonstrator. In *European Battery, Hybrid and Fuel Cell Electric Vehicle Convention on Infrastructure*; 14th March 2018; *Geneva*. Geneva;14th March 2018.

[Lin 1990] Lin GH, Kainthla RC, Packham NJC, Bockris JOM: Electrochemical fusion: A mechanism speculation. *Journal of Electroanalytical Chemistry and Interfacial Electrochemistry* 1990, 280:207–211. doi: 10.1016/0022-0728(90)87098-5.

[Lin 2013] Lin CC, Wei CH, Chen CI, Shieh CJ, Liu YC: Characteristics of the photosynthesis microbial fuel cell with a Spirulina platensis biofilm. *Bioresour Technol* 2013, 135:640–643. doi: 10.1016/j.biortech.2012.09.138.

[Linden 2002] Linden D, Reddy TB: *Handbook of Batteries*. 3 edn. New York: McGraw-Hill; 2002.

[Lingenfelter 1986] Lingenfelter RE: *Death Valley & The Amargosa*. Berkeley and Los Angeles, CA: University of California Press; 1986.

[Lippert 2007] Lippert T, Montenegro MJ, Döbeli M, Weidenkaff A, Müller S, Willmott PR, Wokaun A: Perovskite thin films deposited by pulsed laser ablation as model systems for electrochemical applications. *Progress in Solid State Chemistry* 2007, 35:221–231. doi: 10.1016/j.progsolidstchem.2007.01.029.

[Liu 1997] Liu M, Finlayson TR, Smith TF: High-resolution dilatometry measurements of SrTiO3 along cubic and tetragonal axes. *Physical Review B* 1997, 55:3480–3484. doi: 10.1103/PhysRevB.55.3480.

[Liyu Li 2014] Liyu Li, Soowhan Kim, Zhenguo Yang, Wei Wang, Jianlu Zhang, Baowei Chen, Zimin Nie, Guanguang Xia P. Fe-V redox flow batteries. 2014. Patent Office USP.

[Lob 1914] Lob W: The question of electroculture. *Zeitschrift Fur Elektrochemie Und Angewandte Physikalische Chemie* 1914, 20:587–592.

[Löb 1914] Löb W: XXI. Hauptversammlung der Deutschen Bunsen-Gesellschaft für angewandte physikalische Chemie: vom 21. bis 24. Mai 1914 in Leipzig: ZUR FRAGE DER ELEKTROKULTUR. *Zeitschrift Fur Elektrochemie Und Angewandte Physikalische Chemie* 1914, 20:587–592. doi: 10.1002/bbpc.191400014.

[Long 2006] Long TC, Saleh N, Tilton RD, Lowry GV, Veronesi B: Titanium dioxide (P25) produces reactive oxygen species in immortalized brain microglia (BV2): Implications for nanoparticle neurotoxicity†. *Environmental Science & Technology* 2006, 40:4346–4352. doi: 10.1021/es060589n.

[Long 2016] Long L, Wang S, Xiao M, Meng Y: Polymer electrolytes for lithium polymer batteries. *Journal of Materials Chemistry A* 2016, 4:10038–10069. doi: 10.1039/c6ta02621d.

[Lovelock 2016] Lovelock K, Martin G: Eldercare work, migrant care workers, affective care and subjective proximity. *Ethn Health* 2016, 21: 379–396. doi: 10.1080/13557858.2015.1045407.

[Lu 2013a] Lu XC, Lemmon JP, Kim JY, Sprenkle VL, Yang ZG: High energy density Na-S/NiCl2 hybrid battery. *Journal of Power Sources* 2013a, 224:312–316. doi: 10.1016/j.jpowsour.2012.09.108.

[Lu 2013b] Lu XC, Li GS, Kim JY, Lemmon JP, Sprenkle VL, Yang ZG: A novel low-cost sodium-zinc chloride battery. *Energy & Environmental Science* 2013b, 6:1837–1843. doi: 10.1039/c3ee24244g.

[Lunan 2002] Lunan D: Lord Young of Dartington and the argo venture. *Space Policy* 2002, 18:163–165.

[Macclancy 2007] Macclancy J: Nakomaha: A counter-colonial life and its contexts. Anthropological approaches to biography. *Oceania* 2007, 77:191–214. doi: 10.1002/j.1834-4461.2007.tb00012.x.

[Macilwain 1994] Macilwain C: BIOSPHERE-2 WINS GROWING APPROVAL. *Nature* 1994, 370:495-495.

[Madden 1996] Madden TR, Mackie RL: What electrical measurements can say about changes in fault systems. *Proceedings of the National Academy of Sciences of the United States of America* 1996, 93:3776–3780. doi: 10.1073/pnas.93.9.3776.

[Magistri 2007] MagistriL, Bozzolo M, Tarnowski O, Agnew G, Massardo AF: Design and off-design analysis of a MW hybrid system based on Rolls-Royce integrated planar solid oxide fuel cells. Journal of Engineering for Gas Turbines and Power-Transactions of the Asme 2007, 129: 792–797. doi: 10.1115/1.1839917.

[Maguire 2000] Maguire E, Gharbage B, Marques FMB, Labrincha JA: Cathode materials for intermediate temperature SOFCs. *Solid State Ionics* 2000, 127:329–335. doi: 10.1016/S0167-2738(99)00286-6.

[Mahato 2015] Mahato N, Banerjee A, Gupta A, Omar S, Balani K: Progress in material selection for solid oxide fuel cell technology: A review. *Progress in Materials Science* 2015, 72:141–337. doi: 10.1016/j.pmatsci.2015.01.001.

[Mainar 2018] Mainar AR, Iruin E, Colmenares LC, Kvasha A, De Meatza I, Bengoechea M, Leonet O, Boyano I, Zhang Z, Blazquez JA: An overview of progress in electrolytes for secondary zinc-air batteries and other storage systems based on zinc. *Journal of Energy Storage* 2018, 15:304–328. doi: 10.1016/j.est.2017.12.004.

[Manage 2011] Manage MN, Hodgson D, Milligan N, Simons SJR, Brett DJL: A techno-economic appraisal of hydrogen generation and the case for solid oxide electrolyser cells. *International Journal of Hydrogen Energy* 2011, 36:5782–5796. doi: 10.1016/j.ijhydene.2011.01.075.

[Manke 2007] Manke I, Banhart J, Haibel A, Rack A, Zabler S, Kardjilov N, Hilger A, Melzer A, Riesemeier H: In situ investigation of the discharge of alkaline Zn–MnO2 batteries with synchrotron x-ray and neutron tomographies. *Applied Physics Letters* 2007, 90:214102. doi: 10.1063/1.2742283.

[Mansour 1990] Mansour AN, Dallek S: A new method for the quantitative-analysis of silver-oxide cathodes. *Journal of The Electrochemical Society* 1990, 137:1467–1471. doi: 10.1149/1.2086691.

[Marcus 1997] Marcus RA: Nobel lecture 1992: Electron transfer reactions in chemistry: Theory and experiment. In *Nobel Lectures, Chemistry 1991–1995*. Edited by Malmström BG. Singapore: World Scientific Publishing Co.; 1997: *Nobel Lectures, Chemistry 1991–1995*.

[Martin 1993] Martin BA: Telluric effects on a buried pipeline. *Corrosion* 1993, 49:343–350. doi: 10.5006/1.3316059.

[Martin 1994] Martin BA: CATHODIC PROTECTION OF A REMOTE RIVER PIPELINE. *Materials Performance* 1994, 33:12–15.

[Maruyama 2017] Maruyama J, Maruyama S, Fukuhara T, Hanafusa K: Efficient edge plane exposure on graphitic carbon fiber for enhanced flow-battery reactions. *Journal of Physical Chemistry C* 2017, 121:24425–24433. doi: 10.1021/acs.jpcc.7b07961.

[Marx 2013] Marx A, Adir N: Allophycocyanin and phycocyanin crystal structures reveal facets of phycobilisome assembly. *Biochim Biophys Acta* 2013, 1827:311–318. doi: 10.1016/j.bbabio.2012.11.006.

[Mashkina 2005] Mashkina E: Structures, ionic conductivity and atomic diffusion in A(Ti1-xFex)O3-δδδδ – derived perovskites (A=Ca, Sr, Ba) Friedrich-Alexander-Universität Erlangen-Nürnberg, 2005.

[Materlik 1984] Materlik G, Zegenhagen J: X-ray standing wave analysis with synchrotron radiation applied for surface and bulk systems. *Physics Letters A* 1984, 104:47–50. doi: 10.1016/0375-9601(84)90587-5.

[Materlik 1987] Materlik G, Schmah M, Zegenhagen J, Uelhoff W: Structure Determination of Adsorbates on Single-Crystal Electrodes with X-ray Standing Waves. *Berichte Der Bunsen-Gesellschaft-Physical Chemistry Chemical Physics* 1987, 91:292–296.

[Matteucci 1847] Matteucci C: Electro-physiological researches. Seventh and last series. Upon the relation between the intensity of the electric current, and that of the corresponding physiological effect. *Philosophical Transaction of the Royal Society London* 1847, 137:243–248.

[Matteucci 1850] Matteucci C: Electro-physiological researches. On induced contraction. Ninth Series. *Philosophical Transactions of the Royal Society London* 1850:645–649. doi: 10.1098/rstl.1850.0032.

[Matthes 2015] Matthes S, Matschke M, Cotta F, Grossmann J, Griehl C: Reliable production of microalgae biomass using a novel microalgae platform. *Journal of Applied Phycology* 2015, 27:1755–1762. doi: 10.1007/s10811-014-0502-4.

[Mattich 1998] Mattich C, Hasselwander K, Lommert H, Beyzavi AN: *Electrolytic zinc manufacture with Waelz treatment of neutral leach residues.* 1998.

[Matula 1979] Matula RA: Electrical resistivity of copper, gold, palladium, and silver. *Journal of Physical and Chemical Reference Data* 1979, 8:1147–1298. doi: 10.1063/1.555614.

[Matveev 1961] Matveev VV, Sokolov AD: HARD X-RAY RADIATION FROM TOKAMAK-2, A TOROIDAL SYSTEM. *Soviet Physics-Technical Physics* 1961, 5:1084–1088.

[Matweb 2018] AISI 1010 Steel, cold drawn. http://www.matweb.com/search/DataSheet.aspx?MatGUID=025d4a04c2c640c9b0eaaef28318d761

[Matzke 1996] Matzke T, Stimming U, Karmonik C, Soetratmo M, Hempelmann R, Guthoff F: Quasielastic thermal neutron scattering experiment on the proton conductor SrCeO. 95Yb0.05H0.02O2.985. *Solid State Ionics* 1996, 86-8:621–628. doi: 10.1016/0167-2738(96)00223-8.

[Mauritz 2004] Mauritz KA, Moore RB: State of understanding of nafion. *Chem Rev* 2004, 104: 4535–4585. doi: 10.1021/cr0207123.

[Maxwill 2012] Maxwill P. Elektroauto-Revolution 1912 – Summsumm statt Brummbrumm. SPIEGEL Online. 11 Juni 2012 . http://www.spiegel.de/einestages/elektroauto-revolution-vor-100-jahren-a-947600.html

[Mayer 2015] Mayer K, Wallenius M, Lutzenkirchen K, Horta J, Nicholl A, Rasmussen G, Van Belle P, Varga Z, Buda R, Erdmann N, Kratz JV, Trautmann N, Fifield LK, Tims SG, Frohlich MB, Steier P: Uranium from German Nuclear Power Projects of the 1940s–A Nuclear Forensic Investigation. *Angew Chem Int Ed Engl* 2015, 54:13452–13456. doi: 10.1002/anie.201504874.

[Mayneord 1979] Mayneord WV: John Alfred Valentine Butler. 14 February 1899–16 July 1977. *Biographical Memoirs of Fellows of the Royal Society* 1979, 25. doi: 10.1098/rsbm.1979.0004.

[Mcclendon 2015] Mcclendon E: "First Paris Fashions out of the Sky": The 1962 Telstar satellite's impact on the transatlantic fashion system. Fashion Theory 2015, 18:297–315. doi: 10.2752/175174114x13938552557844.

[McCrory 2013] McCrory CC, Jung S, Peters JC, Jaramillo TF: Benchmarking heterogeneous electrocatalysts for the oxygen evolution reaction. *Journal of American Chemical Society* 2013, 135:16977–16987. doi: 10.1021/ja407115p.

[McKeown 1999] McKeown DA, Hagans PL, Carette LPL, Russell AE, Swider KE, Rolison DR: Structure of hydrous ruthenium oxides: Implications for charge storage. *Journal of Physical Chemistry B* 1999, 103:4825–4832. doi: 10.1021/jp990096n.

[McNaught 2014] McNaught AD, Wilkinson A: IUPAC. Compendium of chemical terminology. In *IUPAC Compendium of Chemical Terminology – The Gold Book*. Edited by Nic M, Jirat J, Kosata B, Jenkins A. 2nd ed. (the "Gold Book"). Oxford: Blackwell Scientific Publications; 2014. doi: 10.1351 /goldbook.

[Mcnaught 1997-2014] Mcnaught AD, Wilkinson A: IUPAC. Compendium of chemical terminology. In *IUPAC Compendium of Chemical Terminology – The Gold Book*. Edited by Nic M, Jirat J, Kosata B, Jenkins A. 2nd ed. (the „Gold Book"). Oxford: Blackwell Scientific Publications; 1997–2014. doi: 10.1351/goldbook.

[Meador 1995] Meador WR: The Pecos project. *Journal of Power Sources* 1995, 57:37–40. doi: 10.1016/0378-7753(95)02236-8.

[Meier 1994] Meier CA: *Die Empirie des Unbewussten*. Verlag, Einsiedeln: Daimon; 1994.

[Melin 2011] Melin H, Stallard T, Miller S, Trafton LM, Encrenaz T, Geballe TR: SEASONAL VARIABILITY IN THE IONOSPHERE OF URANUS. *Astrophysical Journal* 2011, 729. doi: 10.1088/0004-637x/729/2/134.

[Mendelssohn 1964] Mendelssohn K: WALTHER NERNST – APPRECIATION. *Cryogenics* 1964, 4:129. doi: 10.1016/s0011-2275(64)80001-1.

[Menin 1997] Menin L, Schoepp B, Garcia D, Parot P, Vermeglio A: Characterization of the reaction center bound tetraheme cytochrome of Rhodocyclus tenuis. *Biochemistry* 1997, 36:12175–12182. doi: 10.1021/bi971162j.

[Merz 2016] Merz M, Schweiss P, Nagel P, Huang M-J, Eder R, Wolf T, Von Löhneysen H, Schuppler S: Of substitution and doping: Spatial and electronic structure in Fe Pnictides. *Journal of the Physical Society of Japan* 2016, 85. doi: 10.7566/jpsj.85.044707.

[Meshcheryakov 2017] Meshcheryakov V, Rossouw A, Grande L: Comberry Solid-state supercapacitors. In *IDTechEx Show Europe 2017 in Berlin*. Edited by Meshcheryakov V, Rossouw A, Grande L. Germany: Estrel Berlin Convention Center. http://armdevices.net/?s=comberry; 2017.

[Meszéna 1999] Meszéna G, Westerhoff HV: Non-equilibrium thermodynamics of light absorption. *Journal of Physics A: Mathematical and General* 1999, 32:301–311. doi: 10.1088/0305-4470/32/2/006.

[Metromagazine 2015] 41% of U.S. public transit buses use alt fuels, hybrid technology [http://www.metro-magazine.com/sustainability/news/293950/41-of-u-s-public-transit-buses-use-alt-fuels-hybrid-technology]

[Meuzelaar 2015] Meuzelaar N: *Flying Angel: Vanuatu, the Happiest Country You Never Heard of !*: Xlibris 2015.

[Michelson Livingston 1987] Michelson Livingston D: Michelson-Morley: The great failure. *The Scientist*. 13 July 1987. https://www.the-scientist.com/books-etc-/michelson-morley-the-great-failure-63642

[Miklos 1980] Patent.Miklos J, Mund K, Naschwitz W.Doppelschichtkondensator.1980. Office EP.DE 30 11 701. http://google.com/patents/EP0036602A2?cl=nl

[Miles 1990a] Miles MH, Park KH, Stilwell DE: Electrochemical calorimetric evidence for cold fusion in the palladium—deuterium system. Journal of Electroanalytical Chemistry and Interfacial Electrochemistry 1990a, 296:241–254. doi: 10.1016/0022-0728(90)87246-g.

[Miles 1990b] Miles MH, Park KH, Stilwell DE, Univ Utah NCFI: *ELECTROCHEMICAL CALORIMETRIC STUDIES OF THE COLD FUSION EFFECT.* Salt Lake City: Natl Cold Fusion Inst; 1990b.

[Miles 1994] Miles MH, Bush BF, Stilwell DE: Calorimetric principles and problems in measurements of excess power during Pd-D2O electrolysis. *The Journal of Physical Chemistry* 1994, 98: 1948–1952. doi: 10.1021/j100058a038.

[Miles 1998] Miles MH: Reply to "examination of claims of Miles et al. in Pons-Fleischmann-type cold fusion experiments". *Journal of Physical Chemistry B* 1998, 102:3642–3646. doi: 10.1021/jp961751j.

[Miles 2000] Miles MH: Calorimetric studies of Pd/D2O+LiOD electrolysis cells. *Journal of Electroanalytical Chemistry* 2000, 482:56–65. doi: 10.1016/s0022-0728(00)00018-8.

[Miles 2001] Miles MH, Imam MA, Fleischmann M: *Calorimetric Analysis of a Heavy Water Electrolysis Experiment Using a Pd-B Alloy Cathode.* Pennington: Electrochemical Society Inc; 2001.

[Milikh 2008] Milikh GM, Papadopoulos K, Shroff H, Chang CL, Wallace T, Mishin EV, Parrot M, Berthelier JJ: Formation of artificial ionospheric ducts. *Geophysical Research Letters* 2008, 35: 5.doi: 10.1029/2008gl034630.

[Milikh 2010] Milikh G, Vartanyan A: HAARP-induced ionospheric ducts. In *Modern Challenges in Nonlinear Plasma Physics: A Festschrift Honoring the Career of Dennis Papadopoulos.*

Volume 1320. Edited by Vassiliadis D, Fung SF, Shao X, Daglis IA, Huba JD. Melville: Amer Inst Physics; 2010: 185–191: *AIP Conference Proceedings*.

[Minakshi 2010] Minakshi M, Ionescu M: Anodic behavior of zinc in Zn-MnO2 battery using ERDA technique. International Journal of Hydrogen Energy 2010, 35:7618–7622. doi: 10.1016/j.ijhydene.2010.04.143.

[Minkel 2008] Minkel RJ: The 2003 Northeast blackout – five years later. *Scientific American* 2008.

[Mishra 1999] Mishra SK, Ceder G: Structural stability of lithium manganese oxides. *Physical Review B* 1999, 59:6120–6130. doi: 10.1103/PhysRevB.59.6120.

[Mitchell 2018] Mitchell M, Muftakhidinov B, Winchen T, Jędrzejewski-Szmek Z, Badger TG, Badshah400, Wilms A: markummitchell/engauge-digitizer: Version 10.6 Support for Flathub. 2018. doi: 10.5281/zenodo.1214854.

[Mitropoulos 2013] Mitropoulos LK, Prevedouros PD: Assessment of sustainability for transportation vehicles. *Transportation Research Record* 2013:88–97. doi: 10.3141/2344-10.

[Mitropoulos 2016] Mitropoulos LK, Prevedouros PD: Urban transportation vehicle sustainability assessment with a comparative study of weighted sum and fuzzy methods. *Journal of Urban Planning and Development* 2016, 142: 16. doi: 10.1061/(asce)up.1943-5444.0000336.

[Miyake 2001] Miyake S, Tokuda N. Vanadium redox-flow battery for a variety of applications. 2001 Power Engineering Society Summer Meeting, JUL 15–19. Vancouver, Canada. IEEE, New York. <Go to ISI>://WOS:000176406700095

[Mizrachi 2017] Mizrachi A. 7.62 mm NATO Ball. Ltd IS. http://www.imisystems.com/wp-content/uploads/2017/01/7.62-nato-ball.pdf

[Moment 1955] Moment GB: A translation of Luigi Galvani's "De Viribus Electricitatis in Motu Musculari Commentarius" -commentary on the effect of electricity on muscular motion. Robert Montraville GreenLuigi Galvani: Commentary on the effects of electricity on muscular motion. Margaret Glover Foley. *The Quarterly Review of Biology* 1955, 30:375–376. doi: 10.1086/401036.

[Mondragon 2009] Mondragon C: A boat to paradise: The cult of John Frum in Tanna (Vanuatu). *Contemporary Pacific* 2009, 21:389–392.

[Moore 2016] Moore S, Hackett EJ: The construction of technology and place: Concentrating solar power conflicts in the United States. *Energy Research & Social Science* 2016, 11:67–78. doi: 10.1016/j.erss.2015.08.003.

[Moore 2017] Moore G-J: Charge Carrier Dynamics in Semicondcutor-Electrolyte Interface of Hematite Based PEC Water-Splitting Cell. University of Pretoria, Physics, Faculty of Natural and Agricultural Sciences; 2017.

[Morcinek2013] Morcinek M. Kriegsspiele im Nordatlantik – U-boot soll US-Träger angreifen. N-TV. 7 Februar 2013. https://www.n-tv.de/panorama/U-Boot-soll-US-Traeger-angreifen-article10085286.html

[Moruzzi 1996] Moruzzi G: The electrophysiological work of Carlo Matteucci. *Brain Research Bulletin* 1996, 40:69–91. doi: 10.1016/0361-9230(96)00036-6.

[Moser 2013] Moser F, Fourgeot F, Rouget R, Crosnier O, Brousse T: In situ X-ray diffraction investigation of zinc based electrode in Ni-Zn secondary batteries. *Electrochimica Acta* 2013, 109:110–116. doi: 10.1016/j.electacta.2013.07.023.

[Moser 2014] Moser C, Baillat J, Braun A, Haussener S, Psaltis D: 2014. SHINE – developing an efficient and cost effective hydrogen production system using sunlight and water. Swiss National Science Foundation. Lausanne, Neuchatel, Dübendorf. 48 months. finished.

[Mosnier 1985] Mosnier J: A study of the physics of telluric current flow at very low frequencies in the Earth's crust. *Geophysical Journal International* 1985, 82:479–496. doi: 10.1111/j.1365-246X.1985.tb05147.x.

[Motavalli2010] Motavalli J. Hawaii's synergy: New hydrogen stations + GM fuel-cell cars. *CBS News*. 2 September 2010. http://www.cbsnews.com/news/hawaiis-synergy-new-hydrogen-stations-plus-gm-fuel-cell-cars/

[MSE5320 2010] Oxygen Ion Conductivity in Yttria Stabilized Zirconia. http://electronicstructure. wikidot.com/predicting-the-ionic-conductivity-of-ysz-from-ab-initio-calc. Ramamurthy "Rampi" Ramprasad, UConn, Storrs CT.

[Mukunda 2010] Mukunda HS, Dasappa S, Paul PJ, Rajan NKS, Yagnaraman M, Kumar DR, Deogaonkar M: Gasifier stoves – science, technology and field outreach. *Current Science* 2010, 98:627–638.

[Müller 1993] Müller S: 1993. Alkalische wiederaufladbare Zink-Luft Batterie. Schweizer Bundesamt für Energie. 2730. Villigen PSI, Schweiz. 604'360.45 CHF. 42 months. finished. https://www. aramis.admin.ch/Grunddaten/?ProjectID=3839

[Müller 1996] Müller S: 1996. Entwicklung eines eletrisch wiederaufladbaren Zink-Luft Batterie-Demonstrationsmoduls (12V/20 Ah) Schweizer Bundesamt für Energie. 17006. Villigen PSI, Schweiz. 235'565.05 CHF 30 Monate. finished. https://www.aramis.admin.ch/ Grunddaten/?ProjectID=4297

[Müller 1998a] Müller S, Haas O, Schlatter C, Comninellis C: Development of a 100 W rechargeable bipolar zinc/oxygen battery. *Journal of Applied Electrochemistry* 1998a, 28:305–310. doi: 10.1023/a:1003267700824.

[Müller 1998b] Müller S, Holzer F, Haas O: Optimized zinc electrode for the rechargeable zinc–air battery. *Journal of Applied Electrochemistry* 1998b, 28:895–898. doi: 10.1023/ A:1003464011815.

[Müller 2001] Müller R. AccuCell Datenblatt. Waiblingen: AccuCell GmbH; 2001.

[Musk 2014] Tesla Motors 2014. https://www.teslamotors.com/de_CH/blog/all-our-patent-are-belong-you

[Nabae 2005] Nabae Y, Yamanaka I, Takenaka S, Hatano M, Otsuka K: Direct oxidation of methane by Pd-Ni bimetallic catalyst over lanthanum chromite based anode for SOFC. *Chemistry Letters* 2005, 34:774–775. doi: 10.1246/cl.2005.774.

[Nabae 2008] Nabae Y, Yamanaka I, Hatano M, Otsuka K: Mechanism of suppression of carbon deposition on the Pd-Ni/Ce(Sm)O-2-La(Sr)CrO3 anode in dry CH4 fuel. *Journal of Physical Chemistry C* 2008, 112:10308–10315. doi: 10.1021/jp801496v.

[Nagaraj 2002] Nagaraj B, Wu T, Ogale SB, Venkatesan T, Ramesh R: Interface characterization of all-perovskite oxide field effect heterostructures. *Journal of Electroceramics* 2002, 8:233–241. doi: 10.1023/A:1020806402413.

[Naicker 2016] Naicker C, Bruchhausen S: Broadening conceptions of democracy and citizenship: The subaltern histories of rural resistance in Mpondoland and Marikana. *Journal of Contemporary African Studies* 2016, 34:388–403. doi: 10.1080/02589001.2016.1232882.

[Nakajima 1998] Nakajima M, Sawahata M, Yoshida S, Sato K, Kaneko H, Negishi A, Nozaki K: Vanadium redox flow battery with resources saving recycle ability I. Production of electrolytic solution for vanadium redox flow battery from boiler soot. *Denki Kagaku* 1998, 66: 600–608.

[Nakata 2015] Nakata A, Fukuda K, Murayama H, Tanida H, Yamane T, Arai H, Uchimoto Y, Sakurai K, Ogumi Z: Operando X-ray fluorescence imaging for zinc-based secondary batteries. *Electrochemistry* 2015, 83:849–851. doi: 10.5796/electrochemistry.83.849.

[Nandakumar 2017] Nandakumar T: In Germany, power from the people. *The Hindu*. 8 April. http:// www.thehindu.com/sci-tech/energy-and-environment/power-from-the-people/article17893853.ece

[Nasu 2015] Nasu H, Faunce T: Nanotechnology in Japan: A route to energy security after Fukushima? *Bulletin of the Atomic Scientists* 2015, 69:68–74. doi: 10.1177/0096340213501367.

[Nationalpost 2014] Nationalpost: Cultural studies: Why Kraftwerk's influence is pervasive in pop, from Bowie to new wave, and Kanye West to Arcade Fire. National Post. 30 March 2014. http://nationalpost.com/entertainment/music/cultural-studies-why-kraftwerks-influence-is-pervasive-in-pop-from-bowie-to-new-wave-and-kanye-west-to-arcade-fire/wcm/7422fad5-e1b5-4530-8e28-27b0f696118a

[Nations 1979] Nations U: Oils and gases from coal – a symposium of the United Nations Economic Commission for Europe. In *Oils and Gases from Coal – a Symposium of the United Nations Economic Commission for Europe; Katowice, Poland*. Pergamon Press; 1979. doi: 10.1016/B978-0-08-025678-8.50002-6.

[Ndlovu 2014] Ndlovu J. Overview of PGM processing standard bank conference. 11 November 2014. AngloAmerican Platinum. http://www.angloamericanplatinum.com/~/media/Files/A/Anglo-American-Platinum/investor-presentation/standardbankconference-anglo-american-platinum-processing-111114.pdf

[Nedbal 1992] Nedbal L, Samson G, Whitmarsh J: Redox state of a one-electron component controls the rate of photoinhibition of photosystem II. *Proceedings of the National Academy Science USA* 1992, 89:7929–7933. doi: 10.1073/pnas.89.17.7929.

[Neher 1992] Neher E, Sakmann B: The patch clamp technique. *Scientific American* 1992, 266:44–51. doi: 10.1038/scientificamerican0392-44.

[Nel 2017] Nel.Nel Hydrogen Electrolyser – The world's most efficient and reliable electrolyser. 2017. Electrolyser NH. http://nelhydrogen.com/product/electrolysers/

[Nelson 1991] Nelson M, Alvarezromo N, Maccallum T: The biosphere-2 project and its potential role in assisting space exploration. *Space Policy* 1991, 7: 157–160. doi: 10.1016/0265-9646(91)90027-f..

[Nelson 1992b] NelsonM, Leigh L, Alling A, Maccallum T, Allen J, Alvarezromo N: *Biosphere 2 Test Module – A Ground-Based Sunlight-Driven Prototype of a Closed Ecological Life-Support-System* 1992b. doi: 10.1016/0273-1177(92)90021-o.

[Nelson 2013] Nelson M, Dempster WF, Allen JP: Key ecological challenges for closed systems facilities. *Advances in Space Research* 2013, 52:86–96. doi: 10.1016/j.asr.2013.03.019.

[Nemsak 2014] Nemsak S, Shavorskiy A, Karslioglu O, Zegkinoglou I, Rattanachata A, Conlon CS, Keqi A, Greene PK, Burks EC, Salmassi F, Gullikson EM, Yang SH, Liu K, Bluhm H, Fadley CS: Concentration and chemical-state profiles at heterogeneous interfaces with sub-nm accuracy from standing-wave ambient-pressure photoemission. *Nat Commun* 2014, 5:5441. doi: 10.1038/ncomms6441.

[Nernst 1889] Nernst W: Die elektromotorische Wirksamkeit der Ionen. *Zeitschrift für Physikalische Chemie* 1889, 4:129–181.

[Nernst 1900] Nernst W, Dolezalek F: The polarisation of gas in the lead accumulator. *Zeitschrift Fur Elektrochemie* 1900, 6:549–550.

[Nernst 1901] Nernst W: Concerning the significance of electrical methods and theories for chemistry. *Physikalische Zeitschrift* 1901, 3:63–70.

[Nernst 1904] Nernst W, Barratt JOW: On the electric nerve stimulation by alternating current. *Zeitschrift Fur Elektrochemie Und Angewandte Physikalische Chemie* 1904, 10:664–679.

[Nernst 1908a] Nernst W: The theory of the electric stimulus. *Archiv Fur Die Gesamte Physiologie Des Menschen Und Der Tiere* 1908a, 122:275–314. doi: 10.1007/bf01677956.

[Nernst 1908b] Nernst W: The theory of electrical neural excitation. *Zeitschrift Fur Elektrochemie Und Angewandte Physikalische Chemie* 1908b, 14:545–549. doi: 10.1002/bbpc.19080143512.

[Nernst 1922] Nernst W: Zum Gültigkeitsbereich der Naturgesetze. *Naturwissenschaften* 1922, 21, 489–495.

[Newman 1962] Newman JS, Tobias CW: Theoretical analysis of current distribution in porous electrodes. *Journal of the Electrochemical Society* 1962, 109. doi: 10.1149/1.2425269.

[Newman 1966] Newman J: Schmidt number correction for the rotating disk. *The Journal of Physical Chemistry* 1966, 70:1327–1328. doi: 10.1021/j100876a509.

[Newman 1967] Newman J. Current distribution and mass transfer in electrochemical systems. 1967. Ernest O. Lawrence Radiation Laboratory. Berkeley. UCRL-.17294. 1 Jan 1967.

[Newman 1967] Newman J: *Current Distribution and Mass Transfer in Electrochemical Systems.* Berkeley CAL Lawrence Berkeley National Laboratory; 1967. Berkeley U. UCRL – 17294.

[Newman 2004] Newman J, Thomas-Alyea KE: *Electrochemical Systems.* 3rd edn. Wiley; 2004.

[Newman 2005] Newman MEJ: Power laws, Pareto distributions and Zipf's law. *Contemporary Physics* 2005, 46:323–351. doi: 10.1080/00107510500052444.

[Nica 2009] Nica E, Popescu GH, Mironescu A: Brain-drain effects over Romanian economy. *Metalurgia International* 2009, 14:94–96.

[Nieke 1969] Nieke H: Über die Halbleitereigenschaften des Kupferoxyduls. XVI. Kennlinie und Kapazität von Kupferoxydul-Gleichrichtern. *Annalen Der Physik* 1969, 478:251–270. doi: 10.1002/andp.19694780506.

[Nn0029386 2017] Nn0029386: Authorization to discharge under the national pollution discharge elimination system (NPDES) for Chevron mining Inc., Epa ed.: EPA; 2017.

[Norberg-Bohm] 2000 Norberg-Bohm V: Creating Incentives for environmentally enhancing technological change. Technological Forecasting and Social Change 2000, 65: 125–148. doi: 10.1016/s0040-1625(00)00076-7.

[Norskov 2016] Norskov JK, Chen J, Goldstein J, Miranda R, Fitzsimmons T, Stack R. Sustainable Ammonia Synthesis – Exploring the scientific challenges associated with discovering alternative, sustainable processes for ammonia production 2016. U.S. Department of Energy, Office of Science. Doe US. 1283146. February 18, 2016

[Norton 2017] Norton B, Balick M, Hobday R, Fournier C, Scartezzini JL, Solt J, Braun A: Sponsored collection | changing perspectives on daylight: Science, technology, and culture. *Science* 2017, 358:680.682–680. doi: 10.1126/science.358.6363.680-b.

[Notch 2015] Notch: *Carbon Black World Data Book 2015.* Amherst: Notch Consulting Inc; 2015.

[Nurk 2011] Nurk G, Holtappels P, Figi R, Wochele J, Wellinger M, Braun A, Graule T: A versatile salt evaporation reactor system for SOFC operando studies on anode contamination and degradation with impedance spectroscopy. *Journal of Power Sources* 2011, 196:3134–3140. doi: 10.1016/j.jpowsour.2010.11.023.

[Nurk 2013] Nurk G, Huthwelker T, Braun A, Ludwig C, Lust E, Struis RPWJ: Redox dynamics of sulphur with Ni/GDC anode during SOFC operation at mid- and low-range temperatures: An operando S K-edge XANES study. *Journal of Power Sources* 2013, 240:448–457. doi: 10.1016/j.jpowsour.2013.03.187.

[O'regan 1991] O'regan B, Grätzel M: A low-cost, high-efficiency solar cell based on dye-sensitized colloidal TiO2 films. *Nature* 1991, 353:737–740. doi: 10.1038/353737a0.

[O'Brien 2005] O'Brien TF, Bommaraju TV, Hine F: *Handbook of Chlor-Alkali Technology.* Boston, MA: Springer; 2005. doi: 10.1007/b113786.

[O'malley 1992] O'malley J: *Schaum's Outline of Theory and Problems of Basic Circuit Analysis*, ISBN 0-07-047824-4. 2nd edn. New York: McGraw-Hill; 1992: 19.

[Obrist 2009] Obrist BV, Chen DF, Ahrens A, Schunemann V, Scopelliti R, Hu XL: An iron carbonyl pyridonate complex related to the active site of the Fe -Hydrogenase (Hmd). *Inorganic Chemistry* 2009, 48:3514–3516. doi: 10.1021/ic900281g.

[Odum 1993] Odum EP: BIOSPHERE-2 – A NEW KIND OF SCIENCE. *Science* 1993, 260:878–879. doi: 10.1126/science.260.5110.878.

[Oestermeyer 1962] Patent. Oestermeyer CF, Havlick HT, Koenig HL, Raney RM: Deferred action battery.1962.Office USP.US 3036140 A. http://www.google.com/patents/US3036140

[Ohki 1984] Ohki S: Membrane-potential of squid axons – Comparison between the Goldman-Hodgkin-Katz Equation and the surface-diffusion potential equation. *Bioelectrochemistry and Bioenergetics* 1984, 13:439–451. doi: 10.1016/0302-4598(84)87045-2.

[Ohring 1995] Ohring M: *Engineering Materials Science*. San Diego, New York, Boston, Sydney: Academic Press; 1995.

[Olson 2016a] Olson C, Lenzmann F: The social and economic consequences of the fossil fuel supply chain. *MRS Energy & Sustainability* 2016a, 3. doi: 10.1557/mre.2016.7.

[Olson 2016b] Olson C, Lenzmann F: Bringing the social costs and benefits of electric energy from photovoltaics versus fossil fuels to light. *MRS Energy & Sustainability* 2016b, 3. doi: 10.1557/mre.2016.6.

[Onda 2004] Onda K, Kyakuno T, Hattori K, Ito K: Prediction of production power for high-pressure hydrogen by high-pressure water electrolysis. *Journal of Power Sources* 2004, 132:64–70. doi: 10.1016/j.jpowsour.2004.01.046.

[Ortega 2015] Ortega JD, Christian JM, Yellowhair JE, Ho CK: Coupled optical-thermal-fluid and structural analyses of novel light-trapping tubular panels for concentrating solar power receivers. In *High and Low Concentrator Systems for Solar Energy Applications X. Volume 9559.* Edited by Plesniak AP, Prescod AJ. Bellingham: Spie-Int Soc Optical Engineering; 2015. *Proceedings of SPIE*. doi: 10.1117/12.2188235.

[Oschman 2009] Oschman JL: Charge transfer in the living matrix. *Journal of Bodywork and Movement Therapies* 2009, 13:215–228. doi: 10.1016/j.jbmt.2008.06.005.

[Oswald 1933] Oswald W: *Beiträge zur Theorie der Elektrokultur*. Dissertation. Verlag von Gebrüder Borntrager, Berlin: ETH Zürich; 1933.

[Overton 2014] Overton TW: Ivanpah solar electric generating system earns POWER's highest honor. *Power* 2014, 158:26+.

[Ozer 1994a] Ozer N, He YX, Lampert CM: Ionic-conductivity of tantalum oxide-films prepared by sol-gel process for electrochromic devices. *Optical Materials Technology for Energy Efficiency and Solar Energy Conversion Xiii* 1994a, 2255:456–466. doi: 10.1117/12.185388.

[Ozer 1994b] Ozer N, He YX, Lampert CM: *Ionic Conductivity of Tantalum Oxide-Films Prepared by Sol-Gel Process for Electrochromic Devices*. Bellingham: SPIE – Int Soc Optical Engineering; 1994b. doi: 10.1117/12.185388.

[Ozer 1995] Ozer N, Lampert CM: Electrochemical lithium insertion in sol-gel deposited $LiNbO_3$ films. *Solar Energy Materials and Solar Cells* 1995, 39:367–375. doi: 10.1016/0927-0248(96)80002-x.

[Ozer 1997] Ozer N, Lampert CM: Structural and optical properties of sol-gel deposited proton conducting Ta_2O_5 films. *Journal of Sol-Gel Science and Technology* 1997, 8:703–709. doi: 10.1023/a:1018396900214.

[Ozgit 2014] Ozgit D, Hiralal P, Amaratunga GA: Improving performance and cyclability of zinc-silver oxide batteries by using graphene as a two dimensional conductive additive. *ACS Appl Mater Interfaces* 2014, 6:20752–20757. doi: 10.1021/am504932j.

[Pan 1995] Pan LS, Kania DR: *Diamond: Electronic Properties and Applications*, 1st ed. Boston: Springer; 1995. doi: 10.1007/978-1-4615-2257-7.

[Paladini 2018] Paladini C: Le marche du vehicule a hydrogen peine a demarrer. Le Nouvelliste, 1 Sep 2018. https://www.lenouvelliste.ch/articles/valais/martigny-region/martigny-le-marche-du-vehicule-a-hydrogene-peine-a-demarrer-780864

[Paoletti 2014] Paoletti P, Mahadevan L: Intermittent locomotion as an optimal control strategy. *Proceedings of the Royal Society a-Mathematical Physical and Engineering Sciences* 2014, 470. doi: 10.1098/rspa.2013.0535.

[Papadopoulos 2011] Papadopoulos K, Chang CL, Labenski J, Wallace T: First demonstration of HF-driven ionospheric currents. *Geophysical Research Letters* 2011, 38:4. doi: 10.1029/2011gl049263.

[Papatsiba 2006] Papatsiba V: Making higher education more European through student mobility? Revisiting EU initiatives in the context of the Bologna Process1. *Comparative Education* 2006, 42: 93–111. doi: 10.1080/03050060500515785.

[Park 1994] Park J: ELECTRON SPECTROSCOPIC STUDY OF 3d TRANSITION METAL OXIDES AND METAL-INSULATOR TRANSITIONS. University of Michigan; 1994.

[Parks 2014] Parks G,Boyd R, Cornish J, Remick R: Hydrogen station compression, storage, and dispensing technical status and costs: Systems integration. 2014. NREL; U.S. Department of Energy. Golden, CO. 2014–05–01

[Parsons 2011] Parsons R: Volcano curves in electrochemistry. In *Catalysis in Electrochemistry: From Fundamentals to Strategies for Fuel Cell Development*. Edited by Santos E, Schmickler W: 1st edn. Hoboken NJ: John Wiley & Sons, Inc.; 2011.

[Pashley 2005] Pashley RM, Rzechowicz M, Pashley LR, Francis MJ: De-gassed water is a better cleaning agent. *Journal of Physical Chemistry B* 2005, 109:1231–1238. doi: 10.1021/jp045975a.

[Paul 2008] Paul A, Lewis A: Safety goggles: Are they adequate to prevent eye injuries caused by rotating wire brushes? *Emergency Medicine Journal* 2008, 25:385. doi: 10.1136/emj.2007.054759.

[Pavlishchuk 2000] Pavlishchuk VV, Addison AW: Conversion constants for redox potentials measured versus different reference electrodes in acetonitrile solutions at 25°C. *Inorganica Chimica Acta* 2000, 298:97–102. doi: 10.1016/S0020-1693(99)00407-7.

[Pawar 2009] Pawar SD, Murugavel P, Lal DM: Effect of relative humidity and sea level pressure on electrical conductivity of air over Indian Ocean. *Journal of Geophysical Research* 2009, 114. doi: 10.1029/2007JD009716.

[Pedersen 2015] Pedersen T: HAARP, the most powerful ionosphere heater on Earth. *Physics Today* 2015, 68:72–73. doi: 10.1063/pt.3.3032.

[Peiris 2014] Peiris TA, Sagu JS, Wijayantha KG, Garcia-Canadas J: Electrochemical determination of the density of states of nanostructured NiO films. *ACS Appl Mater Interfaces* 2014, 6: 14988–14993. doi: 10.1021/am502827z.

[Pell 1996] Pell WG, Conway BE: Quantitative modeling of factors determining Ragone plots for batteries and electrochemical capacitors. *Journal of Power Sources* 1996, 63:255–266. doi: 10.1016/s0378-7753(96)02525-6.

[Pentland 2017] Pentland W. German utility plans to build world's biggest battery in a salt cavern. *Forbes*. 30 June 2017. https://www.forbes.com/sites/williampentland/2017/06/30/german-utility-plans-to-build-worlds-biggest-battery-in-a-salt-cavern/#5e175ead7feb

[Perez 2007] Perez MG, O'keefe MJ, O'keefe T, Ludlow D: Chemical and morphological analyses of zinc powders for alkaline batteries. *Journal of Applied Electrochemistry* 2007, 37:225–231. doi: 10.1007/s10800-006-9239-3.

[Perry 2002] Perry ML, Fuller TF: A historical perspective of fuel cell technology in the 20th century. *Journal of the Electrochemical Society* 2002, 149. doi: 10.1149/1.1488651.

[Perry 2016] Perry ML, Weber AZ: Advanced redox-flow batteries: A perspective. *Journal of the Electrochemical Society* 2016, 163:A5064–A5067. doi: 10.1149/2.0101601jes.

[Petchsingh 2016] Petchsingh C, Quill N, Joyce JT, Ní Eidhin D, Oboroceanu D, Lenihan C, Gao X, Lynch RP, Buckley DN: Spectroscopic measurement of state of charge in vanadium flow batteries with an analytical model of VIV–VV absorbance. *Journal of The Electrochemical Society* 2016, 163. doi: 10.1149/2.0091601jes.

[Peter 1990] Peter LM: Dynamic aspects of semiconductor photoelectrochemistry. *Chemical Reviews* 1990, 90:753–769. doi: 10.1021/cr00103a005.

[Peter 2016] Peter LM: Photoelectrochemistry: From basic principles to photocatalysis. In *Photocatalysis: Fundamentals and Perspectives*. Edited by Jenny Schneider DB, Jinhua Ye, Gianluca Li Puma, Dionysios D Dionysiou. RSC; 2016. doi: 10.1039/9781782622338-00001.

[Peterseim 2016] Peterseim JH, White S, Hellwig U: Novel solar tower structure to lower plant cost and construction risk. In *Solarpaces 2015: International Conference on Concentrating Solar Power and Chemical Energy Systems. Volume 1734*. Edited by Rajpaul V, Richter C. Melville: Amer Inst Physics; 2016. *AIP Conference Proceedings*. doi: 10.1063/1.4949172.

[Peterson 1907a] Peterson F: The galvanometer as a measurer of emotions. *British Medical Journal* 1907a, 1907:804–806.

[Peterson 1907b] Peterson F, Jung CG: Psycho-physical investigations with the galvanometer and pneumograph in normal and insane individuals. *Brain* 1907b, 30:153–218. doi: 10.1093/brain/30.2.153.

[Piccolino 1998] Piccolino M: Animal electricity and the birth of electrophysiology: The legacy of Luigi Galvani. *Brain Research Bulletin* 1998, 46:381–407. doi: 10.1016/s0361-9230(98)00026-4.

[Pickard 1976] Pickard WF: Generalizations of Goldman Hodgkin Katz Equation. *Mathematical Biosciences* 1976, 30:99–111. doi: 10.1016/0025-5564(76)90018-3.

[Pieper 1958] Pieper H: Zur Frage der Speicherung von elektrischer Energie in Flüssigkeiten. Braunschweig, Techn. Hochsch., 1958.

[Pieper 2008] Pieper J, Buchsteiner A, Dencher NA, Lechner RE, Hauss T: Transient protein softening during the working cycle of a molecular machine. *Physical Review Letters* 2008, 100:4. doi: 10.1103/PhysRevLett.100.228103.

[Pierson 1994] Pierson HO: Handbook of Carbon, Graphite, Diamonds and Fullerenes – Processing, Properties and Applications. 1st ed. Park Ridge, NJ: Noyes Publication; 1994.

[Pike 1929] Pike RD, West GH: Thermal characteristics and heat balance of a large oil-gas generator. *Industrial and Engineering Chemistry* 1929, 21:104–109. doi: 10.1021/ie50230a002.

[Pinaud 2013] Pinaud BA, Benck JD, Seitz LC, Forman AJ, Chen Z, Deutsch TG, James BD, Baum KN, Baum GN, Ardo S, Wang H, Miller E, Jaramillo TF: Technical and economic feasibility of centralized facilities for solar hydrogen production via photocatalysis and photoelectrochemistry. *Energy & Environmental Science* 2013, 6:1983. doi: 10.1039/c3ee40831k.

[Pinstrup-Andersen 2014] Pinstrup-Andersen P: *Food Price Policy in an Era of Market Instability*. Oxford Scholarship Online 2014. doi: 10.1093/acprof:oso/9780198718574.001.0001.

[Pizzini 1972] Pizzini S, Buzzanca G, Mari C, Rossi L, Torchio S: Preparation, structure and electrical properties of thick ruthenium dioxide films. *Materials Research Bulletin* 1972, 7:449–462. doi: 10.1016/0025-5408(72)90147-x.

[Planck 1901] Planck M: Ueber das Gesetz der Energieverteilung im Normalspectrum. *Annalen Der Physik* 1901, 309:553–563. doi: 10.1002/andp.19013090310.

[Plum 2010] Plum LM, Rink L, Haase H: The essential toxin: impact of zinc on human health. *International Journal of Environmental Research Public Health* 2010, 7:1342–1365. doi: 10.3390/ijerph7041342.

[Polo 2017] Polo J, Ballestrin J, Alonso-Montesinos J, Lopez-Rodriguez G, Barbero J, Carra E, Fernandez-Reche J, Bosch JL, Batlles FJ: Analysis of solar tower plant performance influenced by atmospheric attenuation at different temporal resolutions related to aerosol optical depth. *Solar Energy* 2017, 157:803–810. doi: 10.1016/j.solener.2017.09.003.

[Popper 2009] Popper KR: Reason or revolution? *European Journal of Sociology* 2009, 11. doi: 10.1017/s0003975600002071.

[Porter 1991] Porter FC: Zinc Handbook: Properties, Processing, and Use in Design (Mechanical Engineering). New York: Taylor & Francis; 1991.

[Potter 1911] Potter MC: Electrical effects accompanying the decomposition of organic compounds. *Proceedings of the Royal Society B: Biological Sciences* 1911, 84:260–276. doi: 10.1098/rspb.1911.0073.

[Pourbaix 1966] Pourbaix MJN: *Atlas of Electrochemical Equilibria in Aqueous Solutions*. Oxford, New York,: National Association of Corrosion Engineers International, Houston, Texas; Cebelcor, Brussels Oxford, New York, Pergamon Press; 1966.

[Pourbaix 1974] Pourbaix M: *Atlas of Electrochemical Equilibria in Aqueous Solutions*. 2nd edn. NACE – National Association of Corrosion Engineers; 1974.

[Priestley 1775] Priestley J: The history and present state of electricity: With original experiments; 1775. Reprinted in New York : Cambridge University Press, London : J. Dodsley, 2013.

[Qi 2016] Qi Z, Huang S, Younis A, Chu D, Li S: Nanostructured metal oxides-based electrode in supercapacitor applications. In *Supercapacitor Design and Applications*. London: Intech Open; 2016. doi: 10.5772/65155.

[Qiu 2012] Qiu G, Joshi AS, Dennison CR, Knehr KW, Kumbur EC, Sun Y: 3-D pore-scale resolved model for coupled species/charge/fluid transport in a vanadium redox flow battery. *Electrochimica Acta* 2012, 64:46–64. doi: 10.1016/j.electacta.2011.12.065.

[Quaino 2014] Quaino P, Juarez F, Santos E, Schmickler W: Volcano plots in hydrogen electrocatalysis – uses and abuses. *Beilstein Journal of Nanotechnology* 2014, 5:846–854. doi: 10.3762/bjnano.5.96.

[Queyreau 2010] Queyreau S, Monnet G, Devincre B: Orowan strengthening and forest hardening superposition examined by dislocation dynamics simulations. *Acta Materialia* 2010, 58: 5586–5595. doi: 10.1016/j.actamat.2010.06.028.

[Rabinowitch 1954] Rabinowitch E, Holt AS, Jacobs EE, Kromhout R: Die Spektroskopie des Chlorophylls, XIII. Internationaler Kongress fur Reine und Angewandte Chemie in Uppsala und Stockholm, 29 Julöy–5 August 1953. In *XIII Internationaler KongreB fur Reine und Angewandte Chemie; Uppsala und Stockholm*. Angewandte Chemie; 1954: 147–147. doi: 10.1002/ange.19540660510.

[Rabinowitch 1965] Rabinowitch EI, Govindjee: The role of chlorophyll in photosynthesis. *Scientific Anerican* 1965, 213:74–83.

[Raffaele 2006] Raffaele P: In John they trust (John Frum). *Smithsonian* 2006, 36:70+.

[Rakov 2007] Rakov VA, Uman MA: *Lightning: Physics and Effects*. Cambridge, UK: Cambridge University Press; 2007 ISBN: 978-0521035415.

[Ramana 2015] Ramana MV: Nuclear policy responses to Fukushima: Exit, voice, and loyalty. *Bulletin of the Atomic Scientists* 2015, 69:66–76. doi: 10.1177/0096340213477995.

[Rand 1841] Patent. Rand J. Improvement in the construction of vessels or apparatus for preserving paint. 1841. Office USP. US2252 A. http://www.google.com/patents/US2252#v=onepage&q&f=false

[Randles 1947] Randles JEB: Kinetics of rapid electrode reactions. *Discussions of the Faraday Society* 1947, 1:11–19.

[Randles 1952a] Randles JEB: Kinetics of rapid electrode reactions. Part 2.—Rate constants and activation energies of electrode reactions. *Transactions of the Faraday Society* 1952a, 48: 828–832. doi: 10.1039/tf9524800828.

[Randles 1952b] Randles JEB, Somerton KW: Kinetics of rapid electrode reactions. Part 3.—Electron exchange reactions. *Transactions of the Faraday Society* 1952b, 48:937–950. doi: 10.1039/tf9524800937.

[Randles 1952c] Randles JEB, Somerton KW: Kinetics of rapid electrode reactions. Part 4.—Metal ion exchange reaction at amalgam electrodes. *Transactions of the Faraday Society* 1952c, 48: 951–955. doi: 10.1039/tf9524800951.

[Randolph 2017] Oil Embargo, 1973–1974. https://history.state.gov/milestones/1969-1976/oil-embargo

[Rechenberg 1994] Rechenberg I: *Photobiologische Wasserstoffproduktion in Der Sahara (Werkstatt Bionik Und Evolutionstechnik)*. (German Edition). Friedrich Frommann Verlag Gunther 1994.

[Rechenberg 1998] Rechenberg I: Artificial bacterial algal symbiosis (Project ArBAS) – Sahara experiments. In *Biohydrogen*. Edited by Zaborsky OR. New York: Springer; 1998: 281–294. doi: 10.1007/b102384.

[Redaction 2012] Redaction L. Le règne de la lumière: lechoc lumineux de l'impressionnisme. Savoir. 7 Mai 2012. https://arts.savoir.fr/le-regne-de-la-lumiere-le-choc-lumineux-de-l-impressionnisme/

[Reed 1901] Reed CJ: Gas-polarization in lead accumulators. *Journal of Physical Chemistry* 1901, 5: 1–16. doi: 10.1021/j150028a001.

[Reed 1979] Reed TB, Jantzen D: *Generator Gas: The Swedish Experience from 1939–1945*. 3rd edn. Golden, CO: Solar Energy Research Institute, A Division of Midwest Research Institute; 1979.

[Reisner 2009] Reisner E, Powell DJ, Cavazza C, Fontecilla-Camps JC, Armstrong FA: Visible light-driven H(2) production by hydrogenases attached to dye-sensitized TiO(2) nanoparticles. *Jurnal of American Chemical Society* 2009, 131:18457–18466. doi: 10.1021/ja907923r.

[Renewables 2018] Dry Lake Wind Power Project. http://www.avangridrenewables.us/cs_drylake.html

[Renger 2008] Renger G, Renger T: Photosystem II: The machinery of photosynthetic water splitting. *Photosynthesis Research* 2008, 98:53–80. doi: 10.1007/s11120-008-9345-7.

[Rice 1928] Rice OK: Application of the fermi statistics to the distribution of electrons under fields in metals and the theory of electrocapillarity. *Physical Review* 1928, 31:1051–1059. doi: 10.1103/PhysRev.31.1051.

[Richter 2008a] Richter J: Mixed conducting ceramics for high temperature electrochemical devices. *Dissertation*. ETH Zürich, D-MAT; 2008a.

[Richter 2008b] Richter J, Braun A, Harvey AS, Holtappels P, Graule T, Gauckler LJ: Valence changes of manganese and praseodymium in Pr1–xSrxMn1–yInyO3–δ perovskites upon cation substitution as determined with XANES and ELNES. *Physica B: Condensed Matter* 2008b, 403:87–94. doi: 10.1016/j.physb.2007.08.011.

[Ricksher 1907] Ricksher C, Jung CG: Further investigations on the galvanic phenomenon and respiration in normal and insane individuals. *The Journal of Abnormal Psychology* 1907, 2:189–217. doi: 10.1037/h0073786.

[Riess 1991a] Riess I: Measurements of electronic and ionic partial conductivities in mixed conductors, without the use of blocking electrodes. *Solid State Ionics* 1991a, 44:207–214. doi: 10.1016/0167-2738(91)90009-z.

[Riess 1991b] Riess I: Measurement of ionic conductivity in semiconductors and metals. *Solid State Ionics* 1991b, 44:199–205. doi: 10.1016/0167-2738(91)90008-y.

[Rijssenbeek 2011] Rijssenbeek J, Wiegman H, Hall D, Chuah C, Balasubramanian G, Brady C, Ieee: *Sodium-Metal Halide Batteries in Diesel-Battery Hybrid Telecom Applications*. New York: Ieee; 2011.

[Risch 2017] Risch M, Stoerzinger KA, Han B, Regier TZ, Peak D, Sayed SY, Wei C, Xu ZJ, Shao-Horn Y: Redox Processes of manganese oxide in catalyzing oxygen evolution and reduction: An in situ soft x-ray absorption spectroscopy study. *The Journal of Physical Chemistry C* 2017. doi: 10.1021/acs.jpcc.7b05592.

[Rix 2003] Rix B: *Engineering Bulletin Capacitors*. San Diego: General Atomics Energy Products; 2003.

[Robertson 1975] Robertson PM, Schwager F, Ibl N: New cell for electrochemical processes. *Journal of Electroanalytical Chemistry* 1975, 65:883–900. doi: 10.1016/S0022-0728(75)80170-7.

[Rock 2016] Rock SE, Simpson DE, Turk MC, Rijssenbeek JT, Zappi GD, Roy D: Nucleation controlled mechanism of cathode discharge in a Ni/NiCl2 molten salt half-cell battery. *Journal of the Electrochemical Society* 2016, 163:A2282–A2292. doi: 10.1149/2.0641610jes.

[Roe 2016] Roe S, Menictas C, Skyllas-Kazacos M: A high energy density vanadium redox flow battery with 3 M vanadium electrolyte. *Journal of the Electrochemical Society* 2016, 163:A5023–A5028. doi: 10.1149/2.0041601jes.

[Rogers 1986] Rogers WPA, Neil A.; Acheson, David Campion; Covert, Eugene E.; Feynman, Richard P.; Hotz, Robert B.; Kutyna, Donald J.; Ride, Sally K.; Rummel, Robert W.; Sutter, Joseph F.; Walker, Arthur B. C.; Wheelon, Albert D.; Yeager, Charles E.; Keel, Jr., Alton G. Report of the Presidential Commission on the Space Shuttle Challenger Accident. 1986. By the Presidential Commission on the Space Shuttle Challenger Accident. Washington, DC.

[Rogers 2016] Rogers J. Battery damage grounds Solar Impulse 2 until 2016. FoxNews. 15 July 2015. http://www.foxnews.com/science/2015/07/15/battery-damage-grounds-solar-impulse-2-until-2016.html

[Rosenbaum 2010] Rosenbaum M, He Z, Angenent LT: Light energy to bioelectricity: Photosynthetic microbial fuel cells. *Current Opinion in Biotechnology* 2010, 21:259–264. doi: 10.1016/j.copbio.2010.03.010.

[Ross 2005] Ross D: The middle east predicament. *Foreign Affairs* 2005, 84:61+. doi: 10.2307/20034207.

[Rougier 1998] Rougier A, Striebel KA, Wen SJ, Cairns EJ: Cyclic voltammetry of pulsed laser deposited LixMn2 O 4 thin films. *Journal of the Electrochemical Society* 1998, 145. doi: 10.1149/1.1838750

[Rowe 1997] Rowe C , James D. Meindl, Hamilton W. Arnold, Robert W. Brodersen, Elton J. Cairns, Paul G. Cerjan, Walter L. Davis, Charles W. Gwyn, Deborah J. Jackson, Millard F. Rose, Alvin J. Salkind, Daniel P. Siewiorek, Nelson R. Sollenberger, William F. Weldon, Nancy K. Welker, Thornton CG, Robert J. Love, Duncan M. Brown, Cecelia L. Ray: *Energy-Efficient Technologies for the Dismounted Soldier*. National Academy Press. Washington, DC; 1997. doi: 10.17226/5905.

[Rozhkova 2012] Rozhkova EA, Braun A, Moore AL, Ariga K: Symposium D: From molecules to materials – pathways to artificial photosynthesis. In *Materials Research Society Spring Meeting 2012*. Warrendale, PA: Materials Research Society; 2012.

[Rubin 2015] Rubin S. Into the Sun: Environmentalists and shadowy labor group may sue over planned solar farm. *Monterey County Weekly*. 12 March. http://www.montereycountyweekly.com/news/local_news/environmentalists-and-shadowy-labor-group-may-sue-over-planned-solar/article_fc6c4b56-c82d-11e4-808d-b74aad84e585.html

[Ruetschi 1993] Ruetschi P: Energy-storage and the environment – the role of battery technology. *Journal of Power Sources* 1993, 42:1–7. doi: 10.1016/0378-7753(93)80132-9.

[Russo 1987] Russo RE, Mclarnon FR, Spear JD, Cairns EJ: Probe Beam Deflection for in situ Measurements of Concentration and Spectroscopic Behavior during Copper Oxidation and Reduction. *Journal of the Electrochemical Society* 1987, 134. doi: 10.1149/1.2100287.

[Ryu 2010] Ryu W, Bai SJ, Park JS, Huang Z, Moseley J, Fabian T, Fasching RJ, Grossman AR, Prinz FB: Direct extraction of photosynthetic electrons from single algal cells by nanoprobing system. *Nano Letters* 2010, 10:1137–1143. doi: 10.1021/nl903141j.

[(SAE) 2016] Society of Automotive Engineers (SAE): Fueling protocols for light duty and medium duty gaseous hydrogen surface vehicles. *SAE Journal* 2016, 2601. http://standards.sae.org/j2601_201612/

[Safran 1980] Safran SA: Phase diagrams for staged intercalation compounds. *Physical Review Letters* 1980, 44:937–940. doi: 10.1103/PhysRevLett.44.937.

[Saft 2013] Saft. Magnesium-silver chloride seawater activated battery. 2013.

[Saft 2013] Saft. Reserve silver-zinc combat battery. Levallois-Perret; 2013.

[Sakaebe 2014] Sakaebe H: ZEBRA Batteries. In *Encyclopedia of Applied Electrochemistry*. Edited by (Eds.) GKEA. New York: Springer; 2014: 2165–2169. doi: 10.1007/978-1-4419-6996-5_437.

[Sakharov 1975] Sakharov A: Nobel lecture – Peace, progress, human rights. Edited by Bonner Sakharova E. Stockholm: The Nobel Foundation; 1975.

[Salameh 2008] Salameh T. SOFC Auxiliary Power Units (APUs) for Vehicles. 2008. Dept. of Energy Sciences, Faculty of Engineering, Lund University, Box 118, 22100 Lund, Sweden 2008 TRRF05 Fuel Cell Technology. 4 December 2008

[Saltas 2013] Saltas V, Vallianatos F, Gidarakos E: Charge transport in diatomaceous earth studied by broadband dielectric spectroscopy. *Applied Clay Science* 2013, 80–81:226–235. doi: 10.1016/j.clay.2013.02.028.

[Salvador 1980] Salvador P: The behaviour of aluminum -doped n-TiO2 electrodes in the photoassisted oxidation of water. *Materials Research Bulletin* 1980, 15:1287–1294. doi: 10.1016/0025-5408(80)90033-1.

[Sambeth 1983] Sambeth J: What really happened at Seveso. *Chemical Engineering* 1983, 90:44.

[Sambeth 1983a] Sambeth J: The Seveso accident. *Chemosphere* 1983a, 12:681–686. doi: 10.1016/0045-6535(83)90227-8.

[Sambeth 1983b] Sambeth J: WHAT REALLY HAPPENED AT SEVESO. *Chemical Engineering* 1983b, 90:44.

[Sambeth 2004] Sambeth J: *Zwischenfall in Seveso: Ein Tatsachenroman*. Zurich: Unionsverlag; 2004.

[Samgin 2000] Samgin AL: Lattice-assisted proton motion in perovskite oxides. *Solid State Ionics* 2000, 136:291–295. doi: 10.1016/S0167-2738(00)00406-9.

[Samgin 2014] Samgin AL, Ezin AN: Room-temperature proton-hopping transport in rutile-type oxides in the field of resonant laser radiation. *Technical Physics Letters* 2014, 40:252–255. doi: 10.1134/S1063785014030262.

[Sammells 1988] Sammells AF, Semkow KW: The electrochemical generation of useful chemical species from lunar materials. *Journal of Power Sources* 1988, 22:285–291. doi: 10.1016/0378-7753(88)80023-5.

[Santoro 2017] Santoro C, Arbizzani C, Erable B, Ieropoulos I: Microbial fuel cells: From fundamentals to applications. A review. *Journal Power Sources* 2017, 356:225–244. doi: 10.1016/j.jpowsour.2017.03.109.

[Sapountzi 2017] Sapountzi FM, Gracia JM, Weststrate CJ, Fredriksson HOA, Niemantsverdriet JW: Electrocatalysts for the generation of hydrogen, oxygen and synthesis gas. *Progress in Energy and Combustion Science* 2017, 58:1–35. doi: 10.1016/j.pecs.2016.09.001.

[Sauerbrey 1959] Sauerbrey G: Verwendung von Schwingquarzen zur Wägung dünner Schichten und zur Mikrowägung. *Zeitschrift für Physik* 1959, 155:206–222. doi: 10.1007/bf01337937.

[Schiff 1876] Schiff M: Carlo Matteucci: His merits in physiological and medical physics. By Moritz Schiff, late of the florentine laboratory of physiology. *The Lancet* 1876, 108:117–118. doi: 10.1016/s0140-6736(02)49687-0.

[Schiller 2014] Schiller M: "Fun Fun Fun on the Autobahn": Kraftwerk Challenging Germanness. *Popular Music and Society* 2014, 37:618–637.doi: 10.1080/03007766.2014.908522.

[Schmickler 1984] Schmickler W, Henderson D: The interphase between jellium and a hard sphere electrolyte. A model for the electric double layer. *The Journal of Chemical Physics* 1984, 80: 3381–3386. doi: 10.1063/1.447092.

[Schmickler 2014] Schmickler W: *Electrochemical Theory: Double Layer*. Reference Module in Chemistry, Molecular Sciences and Chemical Engineering; 2014. doi: 10.1016/b978-0-12-409547-2.11149-7.

[Schmid 2018] Schmid M: *Programming in Igor Pro: A Comprehensive Introduction*. North Charleston: CreateSpace Independent Publishing Platform; 2018.

[Schmidgen 2004] Schmidgen H: Die Geschwindigkeit von Gefühlen und Gedanken. *NTM International Journal of History and Ethics of Natural Sciences, Technology and Medicine* 2004, 12:100–115. doi: 10.1007/s00048-004-0190-2.

[Scholl 1995] Scholl G, Bayer P, Melone A, Schmitz P: *Product Policy and the Environment: The Example of Batteries* Berlin, Heidelberg: Institut für ökologische Wirtschaftsforschung; 1995.

[Schönborn 1998] Schönborn H-B: *Lok 2000. Re 460/465 – modernste Elektrolok der Schweiz.* München: GeraMond Verlag; 1998.

[Schottky 1922] Schottky W: On the crisis regarding the notions of causality. *Naturwissenschaften* 1922, 10:982–982. doi: 10.1007/bf01565654.

[Schrantz 2017] Schrantz K, Wyss PP, Ihssen J, Toth R, Bora DK, Vitol EA, Rozhkova EA, Pieles U, Thony-Meyer L, Braun A: Hematite photoanode co-functionalized with self-assembling melanin and C-phycocyanin for solar water splitting at neutral pH. *Catalysis Today* 2017, 284:44–51. doi: 10.1016/j.cattod.2016.10.025.

[Schröder 2014] Schröder M, Kailasam K, Rudi S, Fündling K, Rieß J, Lublow M, Thomas A, Schomäcker R, Schwarze M: Applying thermo-destabilization of microemulsions as a new method for co-catalyst loading on mesoporous polymeric carbon nitride – towards large scale applications. *RSC Adv* 2014, 4:50017–50026. doi: 10.1039/C4RA10814K.

[Schuetze 1983] Schuetze SM: The discovery of the action-potential. *Trends in Neurosciences* 1983, 6:164–168. doi: 10.1016/0166-2236(83)90078-4.

[Schüler 2000] Patent.Schüler C, Kötz R: Mehrschichtige Elektrode für elektrochemische Anwendungen; 2000. (Dpma) DP-UM. https://www.google.ch/patents/DE19836651A1?cl=de

[Schutting 2011] Schutting S: Entwicklung von Zink-Elektroden für aufladbare Zink-Luft-Batterien Technische Universität Graz, Chemie; 2011.

[Schwalbe 2017] Bicycle tire inflation pressure. https://www.schwalbetires.com/tech_info/inflation_pressure

[Scott 2000] Scott JF, Bohn HG, Schenk W: Ionic Wiedemann–Franz law. *Applied Physics Letters* 2000, 77:2599–2600. doi: 10.1063/1.1318939.

[Seeberger 1992] Seeberger DA. Household batteries: Evaluation of collection methods 1992. USDOE, Washington, DC (United States). Hennepin County, Department of Environmental Management, Minneapolis, Minnesota; 31.12.1992.

[Semkow 1987] Semkow KW, Sammells AF: A lithium oxygen secondary battery. *Journal of The Electrochemical Society* 1987, 134:2084–2085. doi: 10.1149/1.2100826.

[Seong 2016] Seong J, Park J, Lee J, Ahn B, Yeom JH, Kim J, Hassinen T, Rhee S, Ko S, Lee D, Shin K-H: Practical design guidelines for the development of high-precision roll-to-roll slot-Die coating equipment and the process. *IEEE Transactions on Components, Packaging and Manufacturing Technology* 2016, 6:1677–1686. doi: 10.1109/tcpmt.2016.2613141.

[Sequeira 1984] Sequeira CaC, Hooper A: *Solid State Batteries*. 1984. doi: 10.1007/978-94-009-5167-9.

[Sequeira 2014] Sequeira C, Santos D: *Polymer Electrolytes – Fundamentals and Applications*. 1st edn. Woodhead Publishing; 2014.

[Sergi 2014] Sergi F, Andaloro L, Napoli G, Randazzo N, Antonucci V: Development and realization of a hydrogen range extender hybrid city bus. *Journal of Power Sources* 2014, 250: 286–295. doi: 10.1016/j.jpowsour.2013.11.006.

[Serway 1998] Serway RA: *Principles of Physics*. 2nd ed. Fort Worth, TX: London: Saunders College Pub.; 1998.

[Seyfarth 2006] Seyfarth EA: Julius Bernstein (1839-1917): pioneer neurobiologist and biophysicist. *Biol Cybern* 2006, 94:2–8. doi: 10.1007/s00422-005-0031-y.

[Shah 1976] Shah RP, Solomon HD. Energy conversion alternatives study (ECAS), general electric phase I final report. 1976. NASA Lewis Research Center. Cleveland, OH. NASA-CR 134948.

[Shannon 1976] Shannon RD: Revised effective ionic radii and systematic studies of interatomic distances in halides and chalcogenides. *Acta Crystallographica Section A* 1976, 32:751–767. doi: 10.1107/s0567739476001551.

[Sharma 1999] Sharma MM, Krishnan B, Zachariah S, Shah CU: Study to enhance the electrochemical activity of manganese dioxide by doping technique *Journal of Power Sources* 1999, 79. doi: 10.1016/S0378-7753(99)00045-2

[Shchukarev 2006a] Shchukarev A: XPS at solid–solution interface: Experimental approaches. *Surface and Interface Analysis* 2006a, 38:682–685. doi: 10.1002/sia.2162.

[Shchukarev 2006b] Shchukarev A: XPS at solid-aqueous solution interface. *Advances in Colloid and Interface Science* 2006b, 122:149–157. doi: 10.1016/j.cis.2006.06.015.

[Shchukarev 2008] Shchukarev A, Boily J-F: XPS study of the hematite–aqueous solution interface. *Surface and Interface Analysis* 2008, 40:349–353. doi: 10.1002/sia.2657.

[Shchukin 1995] Shchukin ED, Vidensky IV, Petrova IV: Luggin's capillary in studying the effect of electrochemical reaction on mechanical properties of solid surfaces. *Journal of Materials Science* 1995, 30:3111–3114. doi: 10.1007/BF01209224.

[Sheftel 1966] Sheftel ITP, Va. V.: Electrical conductivity and thermal electromotive force in the system of Mn, Co, Ni, and Cu oxides. *Izvestiya Akademii Nauk SSSR, Neorganicheskie Materialy* 1966, 2.

[Shen 1986] Shen W-M, Siripala W, Tomkiewicz M, Cahen D: Electrolyte electroreflectance study of surface optimization of n-CuInSe[sub 2] in photoelectrochemical solar cells. *Journal of the Electrochemical Society* 1986, 133. doi: 10.1149/1.2108502.

[Shibasaki 2017] Shibasaki R, Azuma T, Yoshida T, Teranishi H, Abe M: Global route choice and its modelling of dry bulk carriers based on vessel movement database: Focusing on the Suez Canal. *Research in Transportation Business and Management* 2017, 25:51–65. doi: 10.1016/j.rtbm.2017.08.003.

[Shibata 1998] Shibata T, Sasa S-I: Equilibrium chemical engines. *Journal of the Physical Society of Japan* 1998, 67:2666–2670. Doi: 10.1143/JPSJ.67.2666.

[Shimakawa 1997] Shimakawa Y, Numata T, Tabuchi J: Verwey-type transition and magnetic properties of the LiMn2O4Spinels. *Journal of Solid State Chemistry* 1997, 131:138–143. doi: 10.1006/jssc.1997.7366.

[Shimizu 1987] Shimizu M, Mori N, Kuno M, Mizunami K, Shigematsu T: Development of redox flow battery. *Journal of the Electrochemical Society* 1987, 134:C410–C410.

[Shimoda 2017] Shimoda N, Kobayashi Y, Kimura Y, Nakagawa G, Satokawa S: Electrochemical synthesis of ammonia using a proton conducting solid electrolyte and nickel cermet electrode. *Journal of the Ceramic Society of Japan* 2017, 125:252–256. doi: 10.2109/jcersj2.16286.

[Shipman 2017] Shipman MA, Symes MD: Recent progress towards the electrosynthesis of ammonia from sustainable resources. *Catalysis Today* 2017, 286:57–68 doi: 10.1016/j.cattod.2016.05.008.

[Shiraishi 1997] Shiraishi Y, Nakai I, Tsubata T, Himeda T, Nishikawa F: In situ transmission x-ray absorption fine structure analysis of the charge–discharge process in LiMn2O4, a rechargeable lithium battery material. *Journal of Solid State Chemistry* 1997, 133:587–590. doi: 10.1006/jssc.1997.7615.

[Siefert 2012] Siefert N, Ashok G, Asme: *Exergy & Economic Analysis of Two Different Fuel Cell Systems for generating electricity at Waste Water Treatment Plants.* New York: Amer Soc Mechanical Engineers; 2012.

[Siegel 2017] Siegel E. The Sun's energy doesn't come from fusing hydrogen into helium (Mostly). *Forbes* 2017.

[Siemens 2013] Ag S. Siemens AG Siemens: SINAVY PEM fuel cell for submarines. 2013. https://www.industry.siemens.com/verticals/global/de/marine/marineschiffe/energieverteilung/Documents/sinavy-pem-fuel-cell-en.pdf

[Sigrist 1979] Sigrist L, Dossenbach O, Ibl N: Mass-transport in electrolytic cells with gas sparging. *International Journal of Heat and Mass Transfer* 1979, 22:1393–1399. doi: 10.1016/0017-9310(79)90201-1.

[Silva 1998] Silva TM, Simões AMP, Ferreira MGS, Walls M, Da Cunha Belo M: Electronic structure of iridium oxide films formed in neutral phosphate buffer solution. *Journal of Electroanalytical Chemistry* 1998, 441:5–12. doi: 10.1016/s0022-0728(97)00300-8.

[Simon 2002] Simon B: *Undead Science: Science Studies and the Afterlife of Cold Fusion.* Rutgers University Press; 2002.

[Simon 2014] Simon P, Gogotsi Y, Dunn B: Materials science. Where do batteries end and supercapacitors begin? *Science* 2014, 343:1210–1211. doi: 10.1126/science.1249625.

[Singh 2012] Singh A: Habermas' wrapped Reichstag: Limits and exclusions in the discourse of post-secularism. *European Review* 2012, 20:131–147. doi: 10.1017/s1062798711000366.

[Singh 2017] Singh AR, Rohr BA, Schwalbe JA, Cargnello M, Chan K, Jaramillo TF, Chorkendorff I, Norskov JK: Electrochemical ammonia synthesis-the selectivity challenge. *ACS Catalysis* 2017, 7:706–709. doi: 10.1021/acscatal.6b03035.

[Sissell 1998] Sissell K: Mexico – 1,000 evacuated in Pemex ammonia release. *Chemical Week* 1998, 160:13–13.

[Sissell 2003] Sissell M: Explosion injures eight at Pemex vinyls complex. *Chemical Week* 2003, 165:14–14.

[Sloop 1978] Sloop JL: Liquid hydrogen as a propulsion fuel, 1945 – 1959. NASA – The Science and Technical Information Office. Washington, D.C. NASA SP–44041978.

[Sloop 2017] Sloop S, Xu T. Advances in direct recycling for lithium-ion batteries NDIA Event #7670 joint service power expo Virginia Beach, VA. May 1–4, 2017.

[Smith 1999] Smith DF, Gucinski JA: Synthetic silver oxide and mercury-free zinc electrodes for silver–zinc reserve batteries. *Journal of Power Sources* 1999, 80:66–71. doi: 10.1016/s0378-7753(98)00251-1.

[Smolinka 2012] Smolinka T, Garche J, Hebling C, Ehret O. Overview on water electrolysis for hydrogen production and storage – Results of the NOW study „Stand und Entwicklungspotenzial der Wasserelektrolyse zur Herstellung von H2 aus regenerativen Energien". Symposium – Water electrolysis and hydrogen as part of the future Renewable Energy System. Copenhagen/Denmark. 18 May 2012.

[Song 2006] Song Y, Xu H, Wei Y, Kunz HR, Bonville LJ, Fenton JM: Dependence of high-temperature PEM fuel cell performance on Nafion® content. *Journal of Power Sources* 2006, 154:138–144. doi: 10.1016/j.jpowsour.2005.04.001.

[Song 2011] Song Y, Penmasta V, Wang C: Recent development of miniatured enzymatic biofuel cells. In *Biofuel's Engineering Process Technology.* Edited by Bernardes MaDS. Rijeka (Croatia), Shanghai (China): Intechopen; 2011. doi: 10.5772/17445.

[Sourcewatch 2012] McKinley Mine. https://www.sourcewatch.org/index.php/McKinley_Mine

[Spahr 2010] Spahr EJ, Wen L, Stavola M, Boatner LA, Feldman LC, Tolk NH, Lupke G: Giant enhancement of hydrogen transport in rutile TiO2 at low temperatures. *Physical Review Letters* 2010, 104:205901. doi: 10.1103/PhysRevLett.104.205901.

[Specht 2001] Specht PI, Bider M: Entertainment P, With) VNFIA, With) FPIA, With) FTSIA, With) AMCaIA. Film or Broadcast. *Marilyn Monroe: The Final Days* 2001. http://www.imdb.com/title/tt0286809/

[Srinivasan 1996] Srinivasan S, Mosdale R: Fuel cells for the 21st Century progress, challenges and prognosis. *Bulletin of Electrochemistry* 1996, 12:170–180.

[Srivastava 1972] Srivastava GP, Huddar BB, Mani A: Electrical conductivity and potential gradient measurements in the free atmosphere over India. Pure and Applied Geophysics PAGEOPH 1972, 100:81–93. doi: 10.1007/bf00880229.

[Steel 2015] Steel W: Europe's largest battery energy storage project opens in Feldheim, Germany. Clean Technica. 21 Sep 2015. https://cleantechnica.com/2015/09/21/new-10-mw-storage-plant-opened-feldheim-germany-europes-largest/

[Stefan 2002a] Stefan IC, Mo YB, Antonio MR, Scherson DA: In situ Ru L-II, and Li-III edge X-ray absorption near edge structure of electrodeposited ruthenium dioxide films. *Journal of Physical Chemistry B* 2002, 106:12373–12375. doi: 10.1021/jp026300f.

[Steinberger-Wilckens 2007] Steinberger-Wilckens R, Tietz F, Smith MJ, Mougin J, Rietveld B, Bucheli O, Van Herle J, Rosenberg R, Zahid M, Holtappels P: Real-SOFC – a joint European effort in understanding SOFC degradation. *Solid Oxide Fuel Cells 10 (Sofc-X), Pts 1 and 2* 2007, 7:67–76. doi: 10.1149/1.2729075.

[Steinberger-Wilckens 2011] Steinberger-Wilckens R, Mai A, Bucheli O, Barfod R, Lefebvre-Joud F, Hjelm J, Braun A, Bronin D, Kiviaho J, Van Herle J, Atkinson A, Baranek P: 2011. Solid oxide fuel cells – integrating degradation effects into lifetime prediction models (SOFC-Life). European Commission FCH JU 256885 Europe. 2.400.000 € 36 months. finished. https://cordis.europa.eu/result/rcn/168374_en.html

[Steinfeld 1995] Steinfeld A: Solar thermal production of zinc and syngas via combined ZnO-reduction and CH4-reforming processes. *International Journal of Hydrogen Energy* 1995, 20:793–804. doi: 10.1016/0360-3199(95)00016-7.

[Stern 1924] Stern O: Zur Theorie der elektrolytischen Doppelschicht. *Zeitschrift Fur Elektrochemie* 1924, 30:508–516. doi: 10.1002/bbpc.192400182.

[Stiller 2014] Stiller C: Nutzung von konventionellem und grünem Wasserstoff in der chemischen Industrie. In *Wasserstoff und Brennstoffzelle*. 2014: 175–188. doi: 10.1007/978-3-642-37415-9_10.

[Stillings 2016] Stillings J: The evolution of Ivanpah solar. *Du* 2016:42–42.

[Stöhr 2016] Stöhr M: Aachener Batteriegroßspeicher: In seiner Art weltweit einzigartig. Aachener Nachrichten. 8 Juni 2016. http://www.aachener-nachrichten.de/lokales/aachen/aachener-batteriegrossspeicher-in-seiner-art-weltweit-einzigartig-1.1377569

[Storey 1976] Storey EL: Electroculture cuts food costs. *Abstracts of Papers of the American Chemical Society* 1976, 172:17–17.

[Stratton 1914] Stratton SW: *Copper Wire Tables*. 3rd ed. Washington, DC: Government Printing Office; 1914.

[Striebel 1999] Striebel KA, Rougier A, Horne CR, Reade RP, Cairns EJ: Electrochemical studies of substituted spinel thin films. *Journal of the Electrochemical Society* 1999, 146:4339–4347. doi: 10.1149/1.1392640.

[Strizki 2017] Hydrogen house project. http://hydrogenhouseproject.org/index.html

[Strong 1961] Strong FC: Faraday's laws in one equation. *Journal of Chemical Education* 1961, 38:98. doi: 10.1021/ed038p98.

[Sudworth 1994] Sudworth JL: ZEBRA batteries. *Journal of Power Sources* 1994, 51: 105–114. doi: 10.1016/0378-7753(94)01967-3.

[Suh 2006] Suh KW: Modeling, analysis and control of fuel cell hybrid power systems. The University of Michigan, Ann Arbor, Michigan, Department of Mechanical Engineering; 2006.

[Sullivan 1997] Sullivan MG, Bartsch M, Kotz R, Haas O: *An electrochemical capacitor using modified glassy carbon electrodes*. Proceedings of the Symposium on Electrochemical Capacitors II Vol.

96/25, 197-201, San Antonio TX 6-11 Oct 1996, Eds. Delnick, F. M., Ingersoll, D., Andrieu, X., Naoi, K. The Electrochemical Society Inc. 1997.

[Sullivan 2000] Sullivan MG, Schnyder B, Bartsch M, Alliata D, Barbero C, Imhof R, Kotz R: Electrochemically modified glassy carbon for capacitor electrodes characterization of thick anodic layers by cyclic voltammetry, differential electrochemical mass spectrometry, spectroscopic ellipsometry, X-ray photoelectron spectroscopy, FTIR, and AFM. *Journal of The Electrochemical Society* 2000, 147:2636–2643. doi: 10.1149/1.1393582.

[Sumi 2011] Sumi H, Lee YH, Muroyama H, Matsui T, Kamijo M, Mimuro S, Yamanaka M, Nakajima Y, Eguchi K: Effect of carbon deposition by carbon monoxide disproportionation on electrochemical characteristics at low temperature operation for solid oxide fuel cells. *Journal of Power Sources* 2011, 196:4451–4457. doi: 10.1016/j.jpowsour.2011.01.061.

[Sun 1989] Sun Z, Tomanek D: Cold fusion: How close can deuterium atoms come inside palladium? *Physical Review Letters* 1989, 63:59–61. doi: 10.1103/PhysRevLett.63.59.

[Sun 2015] Sun ZG, Lu HM, Hong QS, Fan L, Chen CB, Leng J: Evaluation of an alkaline electrolyte system for Al-air battery. *ECS Electrochemistry Letters* 2015, 4:A133–A136. doi: 10.1149/2.0031512eel.

[Suo 2015] Suo L, Borodin O, Gao T, Olguin M, Ho J, Fan X, Luo C, Wang C, Xu K: "Water-in-salt" electrolyte enables high-voltage aqueous lithium-ion chemistries. *Science* 2015, 350:938–943. doi: 10.1126/science.aab1595

[Symvouloi 2017] Symvouloi D: 2017. Demonstration of 4MW pressurized alkaline electrolyser for grid balancing services – demo4Grid. European Commission, FCH JU. 736351. Switzerland, Austria, Spain. 2,932,554.38 €. 5 years. active. http://www.fch.europa.eu/project/demonstration-4mw-pressurized-alkaline-electrolyser-grid-balancing-services; https://cordis.europa.eu/project/rcn/207243_en.html

[Szczesniak 1998] Szczesniak B, Cyrankowska M, Nowacki A: Corrosion kinetics of battery zinc alloys in electrolyte solutions. *Journal of Power Sources* 1998, 75:130–138. doi: 10.1016/s0378-7753(98)00108-6.

[Tabani 2010] Tabani M: The carnival of custom: Land dives, millenarian parades and other spectacular ritualizations in Vanuatu. *Oceania* 2010, 80:309–328. doi: 10.1002/j.1834-4461.2010.tb00088.x.

[Tamm 1932] Tamm I: On the possible bound states of electrons on a crystal surface. *Physical Z Soviet Union* 1932, 1:3.

[Tamm 1932a] Tamm I: Über eine mögliche Art der Elektronenbindung an Kristalloberflächen. *Zeitschrift für Physik* 1932a, 76:849–850. doi: 10.1007/bf01341581.

[Tamm 1932b] Tamm I: On the possible bound states of electrons on a crystal surface. *Physics Z Soviet Union* 1932b, 1:3.

[Taniguchi 1957] Taniguchi H, Janz GJ: Preparation and reproducibility of the thermal-electrolytic silver-silver chloride electrode. *Journal of the Electrochemical Society* 1957, 104:123. doi: 10.1149/1.2428506.

[Tarascon 2004] Tarascon JM, Delacourt C, Prakash AS, Morcrette M, Hegde MS, Wurm C, Masquelier C: Various strategies to tune the ionic/electronic properties of electrode materials. *Dalton Trans* 2004:2988–2994. doi: 10.1039/b408442j.

[Technology 2018] U212/U214 Submarines. https://www.naval-technology.com/projects/type_212/

[Thayer 1959] Thayer V: Heavy water at Aswan. *IAEA Bulletin* 1959, 39:20–21.

[Thiel 1995] Thiel G: Dynamics of chloride and potassium currents during the action-potential in chara studied with action-potential clamp. *European Biophysics Journal* 1995, 24:85–92.

[Tien 1968] Tien HT: Photoelectric effects in thin and bilayer lipid membranes in aqueous media. *The Journal of Physical Chemistry* 1968, 72:4512–4519. doi: 10.1021/j100859a024.

[Tien 1999] Tien HT, Ottova AL: From self-assembled bilayer lipid membranes (BLMs) to supported BLMs on metal and gel substrates to practical applications. *Colloids and Surfaces A: Physicochemical and Engineering Aspects* 1999, 149:217–233. doi: 10.1016/s0927-7757(98)00330-6.

[Tien 2000] Tien HT, Ottova-Leitmannova A: Chapter 9 Membrane photobiophysics and photobiology. In *Membrane Biophysics – Planar Lipid Bilayers and Spherical Liposomes. Volume 5.* Edited by Tien HT, Ottova-Leitmannova A; 2000: 493–576. *Membrane Science and Technology.* doi: 10.1016/s0927-5193(00)80032-7.

[Tikkanen 2014] Tikkanen M, Aro EM: Integrative regulatory network of plant thylakoid energy transduction. *Trends in Plant Science* 2014, 19:10–17. doi: 10.1016/j.tplants.2013.09.003.

[Timeanddate 2018] Calendar for Year 1872 (United Kingdom). https://www.timeanddate.com/calendar/index.html?country=9&year=1872

[Timofeeva 2013] Timofeeva EV, Katsoudas JP, Segre CU, Singh D. Rechargeable nanofluid electrodes for high energy density flow battery. TechConnect 2013, NSTI Nanotechnology Conference and Expo. Washington, DC. https://www.researchgate.net/publication/272563533_Rechargeable_Nanofluid_Electrodes_for_High_Energy_Density_Flow_Battery

[Tinguely 2009] Tinguely J-C: Low-temperature roll-to-roll processing of a mesoporous metal oxide layer for dye-sensitized solar cells. *Master of Science Thesis in Micro- and Nanotechnology.* Fachhochschule Vorarlberg, Master of Science Thesis in Micro- and Nanotechnology; 2009.

[Tinguely 2011] Tinguely J-C, Solarska R, Braun A, Graule T: Low-temperature roll-to-roll coating procedure of dye-sensitized solar cell photoelectrodes on flexible polymer-based substrates. *Semiconductor Science and Technology* 2011, 26:045007. doi: 10.1088/0268-1242/26/4/045007.

[Tinker 2000] Tinker LA, Striebel KA: Rechargeable zinc-air batteries for portable products. In *Energy Storage Systems in Electronics.* Edited by Osaka T, Datta M. Boca Raton: CRC Press; 2000: 604: *New Trends in Electrochemical Technology.*

[Tippee 2014] Tippee B: OPEC was destined not to cut output in Vienna meeting. *Oil & Gas Journal* 2014, 112:29–29.

[Titien 1970] Titien H, Verma SP: Electronic processes in bilayer lipid membranes. *Nature* 1970, 227:1232. doi: 10.1038/2271232a0.

[Tobler 1932] Tobler J: *Studien über Brennstoffketten.* ETH Zürich, 1932.

[Tokuda 1985] Tokuda K: Finite-element method approach to the problem of the IR-potential drop and overpotential measurements by means of a Luggin-Haber capillary. *Journal of the Electrochemical Society* 1985, 132. doi: 10.1149/1.2113584.

[Tollin 1960] Tollin G, Kearns DR, Calvin M: .Electrical properties of organic solids. I. Kinetics and mechanism of conductivity of metal-free phthalocyanine. *The Journal of Chemical Physics* 1960, 32:1013–1019. doi: 10.1063/1.1730843.

[Tomkiewicz 1980a] Tomkiewicz M: Surface states on chemically modified TiO2 electrodes. *Surface Science* 1980a, 101:286–294. doi: 10.1016/0039-6028(80)90622-6.

[Tomkiewicz 1980b] Tomkiewicz M: The nature of surface states on chemically modified TiO[sub 2] electrodes. *Journal of the Electrochemical Society* 1980b, 127:1518. doi: 10.1149/1.2129941.

[Tomkiewicz 1980c] Tomkiewicz M, Silberstein RP, Pollak FH: Determination of the potential Distribution dt the Cdse-electrolyte interface by electrolyte electroreflectance spectroscopy. *Bulletin of the American Physical Society* 1980c, 25:265–265.

[Torabi 2012] Torabi F, Aliakbar A: A single-domain formulation for modeling and simulation of zinc-silver oxide batteries. *Journal of The Electrochemical Society* 2012, 159:A1986–A1992. doi: 10.1149/2.038212jes.

[Toth 2016] Toth R, Walliser RM, Murray NS, Bora DK, Braun A, Fortunato G, Housecroft CE, Constable EC: A self-assembled, multicomponent water oxidation device. *Chemical Communications* 2016, 52:2940–2943. doi: 10.1039/c5cc09556e.

[Totlandsdal 2012] Totlandsdal AI, Herseth JI, Bolling AK, Kubatova A, Braun A, Cochran RE, Refsnes M, Ovrevik J, Lag M: Differential effects of the particle core and organic extract of diesel exhaust particles. *Toxicology Letter* 2012, 208: 262–268. doi: 10.1016/j.toxlet.2011.10.025.

[Toulemonde 1999] Toulemonde O, Millange F, Studer F, Raveau B, Park JH, Chen CT: Changes in the Jahn-Teller distortion at the metal-insulator transition in CMR manganites. *Journal of Physics: Condensed Matter* 1999, 11:109–120. doi: 10.1088/0953-8984/11/1/009.

[Trainham 1981] Trainham JA, Newman J: A comparison between flow-through and flow-by porous electrodes for redox energy storage. *Electrochimica Acta* 1981, 26:455–469. doi: 10.1016/0013-4686(81)87024-7.

[Transmissionline 2011] Electrical Properties of Wood Poles Used in Transmission Lines. http://www.transmission-line.net/2011/07/electrical-properties-of-wood-poles.html

[Trasatti 1971] Trasatti S, Buzzanca G: Ruthenium dioxide: A new interesting electrode material. Solid states structure and electrochemical behaviour. *Journal of Electroanalytical Chemistry* 1971, 29: A1–A5. doi: 10.1016/s0022-0728(71)80111-0.

[Trasatti 1981] Trasatti S: Effect of the nature of the metal on the dielectric-properties of polar liquids at the interface with electrodes – a phenomenological approach. *Journal of Electroanalytical Chemistry* 1981, 123:121–139. doi: 10.1016/S0022-0728(81)80047-2.

[Trasino 2009] Trasino F, Bozzolo M, Magistri L, Massardo AF, Asme: *Modelling and Performance Analysis of the Rolls-Royce Fuel Cell Systems Limited. 1 MW Plant.* New York: Amer Soc Mechanical Engineers; 2009.

[Trasino 2011] TrasinoF, Bozzolo M, Magistri L, Massardo AF: Modeling and performance analysis of the Rolls-Royce fuel cell systems limited: 1 MW plant. *Journal of Engineering for Gas Turbines and Power-Transactions of the Asme* 2011, 133:11. doi: 10.1115/1.4000600.

[Tributsch 2008] Tributsch H: Photovoltaic hydrogen generation. *International Journal of Hydrogen Energy* 2008, 33:5911–5930. doi: 10.1016/j.ijhydene.2008.08.017.

[Trueb 1998] Trueb LF, Rüetschi P: Batterien und Akkumulatoren – Mobile Energiequellen für heute und morgen. Berlin, Heidelberg: Springer; 1998.

[Tsekouras 2014] Tsekouras G, Braun A: Conductivity and oxygen reduction activity changes in lanthanum strontium manganite upon low-level chromium substitution. *Solid State Ionics* 2014, 266:19–24. doi: 10.1016/j.ssi.2014.08.006.

[Tsekouras 2015] Tsekouras G, Boudoire F, Pal B, Vondracek M, Prince KC, Sarma DD, Braun A: Electronic structure origin of conductivity and oxygen reduction activity changes in low-level Cr-substituted (La, Sr)MnO3. *Journal of Chemical Physics* 2015, 143:7. doi: 10.1063/1.4931033.

[Tsierkezos 2007] Tsierkezos NG: Cyclic voltammetric studies of ferrocene in nonaqueous solvents in the temperature range from 248.15 to 298.15 K. *Journal of Solution Chemistry* 2007, 36: 289–302. doi: 10.1007/s10953-006-9119-9.

[Tucker 2002] Tucker MC, Reimer JA, Cairns EJ: A Li-7 NMR study of capacity fade in metal-substituted lithium manganese oxide spinels. *Journal of the Electrochemical Society* 2002, 149: A574–A585. doi: 10.1149/1.1466856.

[Tummers 2006] Tummers B, Van Der Laan J, Huyser K: DataThief III., vol. III. pp. A data digitalization program; 2006: A data digitalization program.

[Turner 1982] Turner L: 4th annual opec seminar, Vienna, 24–26 November 1981. *Energy Policy* 1982, 10:161–161. doi: 10.1016/0301-4215(82)90031-3.

[Turner Jones 2011] Turner Jones MD : *Lonmin Plc – Process Division – Analyst Presentation.* Johannesburg: Lonmin Plc; 2011.

[Ugur 2006] Ugur U: Resistivity of steel, The Physics Factbook, retrieved and archived 16 June 2011; 2006.

[Ulleberg 1998] Ulleberg O: Stand-alone power systems for the future: Optimal design, operation and control of solar-hydrogen energy systems. Trondheim University of Science and Technology, Thermal Energy & Hydropower; 1998.

[Ullman 1980] Ullman DL: Discussion of "The Potential Distribution at the TiO[sub 2] Aqueous Electrolyte Interface" [Micha Tomkiewicz (pp. 1505–1510, Vol. 126, No. 9)]. *Journal of The Electrochemical Society* 1980, 127:1321. doi: 10.1149/1.2129886.

[UN383 2013] UN383: *Recommendations on the TRANSPORT OF DANGEROUS GOODS: Manual of Tests and Criteria.* 5th revised edn. Amendment 1 edn. United Nations Pubn; 2013.

[Valero-Bover 2014] Valero-Bover D, Olivella-Rosell P, Villafafila-Robles R, Cestau-Cubero S, *IEEE*: Performance Analysis of an Electric Vehicle Fleet for Commercial Purposes. *2014 IEEE International Electric Vehicle Conference (IEVC)* 2014.

[Vamanu 2014] Vamanu DV, Acasandrei VT, Vamanu BI: SAFETY RISKS IN SPENT NUCLEAR FUEL ROAD TRANSPORTATION: „BLACK SWANS" BY MALICIOUS INTENT. *Romanian Reports in Physics* 2014, 66:823–843.

[Van Der Kloot 2004] Van Der Kloot W: April 1915: Five future nobel prize-winners inaugurate weapons of mass destruction and the academic-industrial-military complex. *Notes and Records of the Royal Society of London* 2004, 58:149–160. doi: 10.1098/rsnr.2004.0053.

[Vanyan 2002] Vanyan LL, Egorov IV: Telluric currents near extended metal conductors. *Izvestiya-Physics of the Solid Earth* 2002, 38:905–912.

[Vanysek 2010] Vanysek P: Electrochemical series. In *CRC Handbook of Chemistry and Physics*, 91st edn. Edited by Haynes WM. Boca Raton: CRC Press; 2010.

[Vanýsek 2011] Vanýsek P: Electrochemical Series. In *Handbook of Chemistry and Physics.* 92nd edn. Edited by Haynes WM: Boca Raton: Chemical Rubber Company/CRC; 2011: 2656. [Haynes WM (Series Editor)].

[Varta 2018] Powersports Freshpack. https://www.varta-automotive.de/de-de/produkte/varta-powersports-freshpack.

[Vasquez 2006] Vasquez FA, Heaviside R, Tang ZJ, Taylor JS: Effect of free chlorine and chloramines on lead release in a distribution system. *Journal American Water Works Association* 2006, 98: 144–154.

[Vayssieres 2010] Vayssieres L: *On Solar Hydrogen & Nanotechnology.* Singapore: John Wiley & Sons (Asia) Pte Ltd; 2010. doi: 10.1002/9780470823996.

[VBG 2012] 1 Jahr Hybridbusse im ZVV: Erwartungen übertroffen. https://vbg.ch/2012/06/1-jahr-hybridbusse-im-zvv/

[Veneri 2017] Veneri O, Capasso C, Patalano S: Experimental study on the performance of a ZEBRA battery based propulsion system for urban commercial vehicles. *Applied Energy* 2017, 185: 2005–2018. doi: 10.1016/j.apenergy.2016.01.124.

[Venet 2013] Venet P: Electrolytic capacitors. In *Dielectric Materials for Electrical Engineering*, 1. Edited by Martinez-Vega J. Hoboken NJ: Wiley; 2013. doi: 10.1002/9781118557419.ch18.

[Veraguth 1907] Veraguth PDDO: Das psycho-galvanische Reflex-Phänomen. *European Neurology* 1907, 21:387–406. doi: 10.1159/000211619.

[Verdegaal 2015] Verdegaal WM, Becker S, Von Olshausen C: Power-to-liquids: Synthetic crude oil from CO2, water, and sunshine. *Chemie Ingenieur Technik* 2015, 87:340–346. doi: 10.1002/cite.201400098.

[Verhüllter Reichstag – Projekt für Berlin 1994] Protokoll der 211. Sitzung des Deutschen Bundestages. In *Deutscher Bundestag*, 211 edn. Berlin: Deutscher Bundestag; 1994: 18275–18288.

[Vogel 2015] Vogel B: An alternative to the Tesla Powerwall. *Swiss Federal Office of Energy* 2015:5.

[Von Helmolt 2007] Von Helmolt R, Eberle U: Fuel cell vehicles: Status 2007. *Journal of Power Sources* 2007, 165: 833–843. doi: 10.1016/j.jpowsour.2006.12.073.

[Von Kleist 1745] Von Kleist EGJ: *Acta Societatis Physicae Experimentalis*. Danzig; 1745, 204–212. http://www.pbc.gda.pl/dlibra/doccontent?id=20444&from=FBC.

[Von Sturm 1981] Von Sturm F: Secondary batteries – silver-zinc battery. *Comprehensive Treatise of Electrochemistry* 1981: 407–419. doi: 10.1007/978-1-4615-6687-8_14.

[Von Weizsäcker 1937] Von Weizsäcker CF: Über Elementumwandlungen im Innern der Sterne (On element conversions in the inside of stars. I.). *Physikalische Zeitschrift der Sowjetunion* 1937, 38.

[Von Weizsäcker 1938] Von Weizsäcker CF: Über Elementumwandlungen im Innern der Sterne (Element conversions in the interior of stars. II). *Physikalische Zeitschrift der Sowjetunion* 1938, 39.

[Wade 2011] Wade NJ: Carlo Matteucci, Giuseppe Moruzzi and stimulation of the senses: A visual appreciation. *Archives Italiennes de Biologie* 2011, 149 (Suppl.):18–28.

[Wadsack 2010] Wadsack KE, Acker T, Asme: POLICY SOLUTIONS FOR INCREASING ECONOMIC IMPACTS OF WIND DEVELOPMENT IN ARIZONA. *Es2010: Proceedings of Asme 4th International Conference on Energy Sustainability, Vol 1* 2010:103–111.

[Wagner 1931] Wagner C: Theory of Current Rectifiers. *Phys Z* 1931, 32:641–645.

[Wagner 1933] Wagner C: Über die Natur des elektrischen Leitvermögens von α-Silbersulfld. *Zeitschrift für Physikalische Chemie* 1933, 21B. doi: 10.1515/zpch-1933-2106.

[Wagner 1943] Wagner C: Über den Mechanismus der elektrischen Stromleitung im Nernststift. *Die Naturwissenschaften* 1943, 31:265–268. doi: 10.1007/bf01475685.

[Wagner 2012] Wagner H: Einfluss der Temperatur auf die elektrische Leitfähigkeit verdünnter, wässriger Lösungen. *VGB Powertech* 2012, 92.

[Wakeford 2015] Wakeford J, Swilling M: Implications of increasing world oil scarcity for national food security in South Africa. *Agrekon* 2015, 53:68–91. doi: 10.1080/03031853.2014.974626.

[Walde 1976] Walde CJ, Ruka RJ, Isenberg AO: Energy conversion alternatives study (ECAS), Westinghouse phase 1. Vol. X11: fuel cells. Final report. [Phosphoric acid, potassium hydroxide, molten carbonate, and stabilized zirconia electrolytes are compared]. 1976. Westinghouse Research Labs. Pittsburgh, PA (USA). N-76-23703; NASA-CR-134941-V-12; REPT-76-9E9-ECAS-RLV.12-V-12. 12 Dec 1976 .

[Walford 2002] Walford RL: Biosphere 2 as voyage of discovery: The serendipity from inside. *Bioscience* 2002, 52:259–263. doi: 10.1641/0006-3568(2002)052[0259:bavodt]2.0.co;2.

[Walsh 2010] Walsh T, Wilson G, O'Connor E: Local, European and global: An exploration of migration patterns of social workers into Ireland. *British Journal of Social Work* 2010, 40: 1978–1995. doi: 10.1093/bjsw/bcp141.

[Wang 1992] Wang JD, Jang CS, Hwang YH, Chen ZS: Lead contamination around a kindergarten near a battery recycling plant. *Bulletin of Environmental Contamination and Toxicology* 1992, 49:23–30.

[Wang 2014] Wang P, Dimitrijevic NM, Chang AY, Schaller RD, Liu Y, Rajh T, Rozhkova EA: Photoinduced electron transfer pathways in hydrogen-evolving reduced graphene oxide-boosted hybrid nano-bio catalyst. *ACS Nano* 2014, 8:7995–8002. doi: 10.1021/nn502011p.

[Wang 2016] Wang J-J, Hu Y, Toth R, Fortunato G, Braun A: A facile nonpolar organic solution process of a nanostructured hematite photoanode with high efficiency and stability for water splitting. *Journal of Materials Chemistry A* 2016, 4:2821–2825. doi: 10.1039/c5ta06439b.

[Wang 2017] Wang P, Chang AY, Novosad V, Chupin VV, Schaller RD, Rozhkova EA: Cell-free synthetic biology chassis for nanocatalytic photon-to-hydrogen conversion. *ACS Nano* 2017, 11: 6739–6745. doi: 10.1021/acsnano.7b01142.

[Warren-Rhodes 2013] Warren-Rhodes KA, Mckay CP, Boyle LN, Wing MR, Kiekebusch EM, Cowan DA, Stomeo F, Pointing SB, Kaseke KF, Eckardt F, Henschel JR, Anisfeld A, Seely M, Rhodes KL: Physical ecology of hypolithic communities in the central Namib Desert: The role of fog, rain, rock habitat, and light. *Journal of Geophysical Research: Biogeosciences* 2013, 118: 1451–1460. doi: 10.1002/jgrg.20117.

[Watanabe 1996] Watanabe Y, Matsumoto M, Takasu K: The market for utility-scale fuel cell plants. *Journal of Power Sources* 1996, 61:53–59. doi: 10.1016/s0378-7753(96)02337-3.

[Watson 1993] Watson T: CAN BASIC RESEARCH EVER FIND A GOOD HOME IN BIOSPHERE-2. *Science* 1993, 259:1688–1689. doi: 10.1126/science.259.5102.1688.

[Waysand 1984] Waysand G: Whether to make cruise-missiles, or walkmans – Electronics industry and military spending by governments *Quinzaine Litteraire* 1984:9–9.

[Weber 2015] Weber AZ, Van Nguyen T: Redox flow batteries–reversible fuel cells. *Journal of the Electrochemical Society* 2015, 163:Y1–Y1. doi: 10.1149/2.0331601jes.

[Webster 1997] Webster R: Regression and functional relations. *European Journal of Soil Science* 1997, 48: 557–566. doi: 10.1111/j.1365-2389.1997.tb00222.x.

[Wee 2012] Wee HM, Yang WH, Chou CW, Padilan MV: Renewable energy supply chains, performance, application barriers, and strategies for further development. *Renewable & Sustainable Energy Reviews* 2012, 16:5451–5465. doi: 10.1016/j.rser.2012.06.006.

[Weisman 1987] Weisman SRH, Sanjoy: Theory of Bhopal sabotage is offered. *The New York Times.* http://www.nytimes.com/1987/06/23/world/theory-of-bhopal-sabotage-is-offered.html?pagewanted=all

[Wellesley 2017] Wellesley L, Preston F, Lehne J, Bailey R: Chokepoints in global food trade: Assessing the risk. *Research in Transportation Business and Management* 2017, 25: 15–28. doi: 10.1016/j.rtbm.2017.07.007.

[West 1993] West AR: PHASE-DIAGRAMS OF INORGANIC MATERIALS – APPLICATIONS TO COMPLEX SOLID-SOLUTION SYSTEMS, SITE SUBSTITUTIONS AND STOICHIOMETRY-PROPERTY CORRELATIONS. *Journal of Materials Chemistry* 1993, 3:433–440. doi: 10.1039/jm9930300433.

[Weston 1893] Patent.Weston E: Voltaic Cell. 1893. Patent US. 494,827.

[Whaley 2010] Whaley JA, Mcdaniel AH, El Gabaly F, Farrow RL, Grass ME, Hussain Z, Liu Z, Linne MA, Bluhm H, Mccarty KF: Note: Fixture for characterizing electrochemical devices in-operando in traditional vacuum systems. *Review of Scientific Instruments* 2010, 81:086104. doi: 10.1063/1.3479384.

[Wiaux 1995] Wiaux JP, Waefler JP: Recycling zinc batteries: An economical challenge in consumer waste management. *Journal of Power Sources* 1995, 57:61–65. doi: 10.1016/0378-7753(95) 02242-2.

[Wiedemann 1853] Wiedemann GF, R.: I. Ueber die Wäremeitungsfähigkeit der Metalle. *Annalen der Physik und Chemie* 1853, 1853:497–531. doi: 10.1002/bbpc.19280340521.

[Wiemhöfer 2002] Wiemhöfer H-D, Bremes H-G, Nigge U, Zipprich W: Studies of ionic transport and oxygen exchange on oxide materials for electrochemical gas sensors. *Solid State Ionics* 2002, 150:63–77. doi: 10.1016/s0167-2738(02)00264-3.

[Wilken 2017] Wilken L, Dahlberg MG: Between international student mobility and work migration: experiences of students from EU's newer member states in Denmark. *Journal of Ethnic and Migration Studies* 2017, 43:1347–1361.doi: 10.1080/1369183x.2017.1300330.

[Williams 1972] Williams CW: One more argument for safety goggles – contact lens hazard. *Journal of Chemical Education* 1972, 49:515.

[Williford 2001] Williford RE, Stevenson JW, Chou SY, Pederson LR: Computer simulations of thermal expansion in lanthanum-based perovskites. *Journal of Solid State Chemistry* 2001, 156: 394–399. doi: 10.1006/jssc.2000.9011.

[Wilson 1900] Wilson CTR: Atmospheric electricity. *Nature* 1900, 62:149–151. doi: 10.1038/062149b0.

[Wilson 1901] Wilson CTR: "On the ionisation of atmospheric air." *Proceedings of the Royal Society of London* 1901, 68:151–161. doi: 10.1098/rspl.1901.0032.

[Wilson 1902a] Wilson CTR: "On the spontaneous ionisation of gases." *Proceedings of the Royal Society of London* 1902a, 69:277–282.

[Wilson 1902b] Wilson CTR: On the leakage of electricity through dust-free air. *Proceedings of the Cambridge Philosophical Society* 1902b, 11:32–32.

[Wilson 1903] Wilson CTR: Atmospheric electricity. *Nature* 1903, 68:102–104. doi: 10.1038/068102d0.

[Wilson 1904a] Wilson CTR: The condensation method of demonstrating the ionisation of air under normal conditions. *Philosophical Magazine* 1904a, 7:681–690. doi: 10.1080/14786440409463162.

[Wilson 1904b] Wilson CTR: On a sensitive gold leaf electrometer. *Proceedings of the Cambridge Philosophical Society* 1904b, 12:135–139.

[Wilson 1906a] Wilson CTR: On a portable gold-leaf electrometer for low or high potentials, and its application to measurements in atmospheric electricity. *Proceedings of the Cambridge Philosophical Society* 1906a, 13:184–188.

[Wilson 1906b] Wilson CTR: On the measurement of the earth-air current and on the origin of atmospheric electricity. *Proceedings of the Cambridge Philosophical Society* 1906b, 13: 363–382.

[Wilson 1908] Wilson CTR: On the measurement of the atmospheric electric potential gradient and the earth-air current. *Proceedings of the Royal Society of London Series a-Containing Papers of a Mathematical and Physical Character* 1908, 80:537–547. doi: 10.1098/rspa.1908.0048.

[Wilson 1909] Wilson CTR: On thunderstorm electricity. *Philosophical Magazine* 1909, 17:634–641.

[Wilson 1911] Wilson CTR: On a method of making visible the paths of ionising particles through a gas. *Proceedings of the Royal Society of London Series a-Containing Papers of a Mathematical and Physical Character* 1911, 85:285–288. doi: 10.1098/rspa.1911.0041.

[Wilson 1921] Wilson CTR: Investigations on lightning discharges and on the electric field of thunderstorms. *Philosophical Transactions of the Royal Society of London Series a-Containing Papers of a Mathematical or Physical Character* 1921, 221:73–115. doi: 10.1098/rsta.1921.0003.

[Wilson 1925] Wilson CTR: The electric field of a thundercloud and some of its effects. *Proceedings of the Physical Society of London* 1925, 37:32D–37D.

[Wing 2013] Wing J: Fuel cells and submarines – analyst view. *Fuel Cell Today*. 3 July 2013. http://www.fuelcelltoday.com/analysis/analyst-views/2013/13-07-03-fuel-cells-and-submarines

[Winkler 2008] Winkler W, Nehter P: Thermodynamics of fuel cells. In *Modeling Solid Oxide Fuel Cells*. Edited by Bove R., Ubertini S.Dordrecht: Springer; 2008: 13–50: *Fuel Cells and Hydrogen Energy*. doi: 10.1007/978-1-4020-6995-6_2.

[Winter 2018] WebElements Periodic Table of the Elements | Iron | compounds information. https://www.webelements.com/iron/

[Wolff 2015] Wolff T, Brinkmann T, Kerner M, Hindersin S: CO2 enrichment from flue gas for the cultivation of algae – a field test. *Greenhouse Gases-Science and Technology* 2015, 5: 505–512. doi: 10.1002/ghg.1510.

[Wood 1976] Wood JL: JOHN-FRUM-HE-COME – POLEMICAL WORK ABOUT A BLACK TRAGEDY – RICE,E. *Contemporary Sociology-a Journal of Reviews* 1976, 5:379–379. doi: 10.2307/2064157.

[Wood 1993] Wood I: THE INTERNATIONAL-MOTOR-SHOW MARCH 4–14,1993, PALEXPO GENEVA. *Design* 1993:42–44.

[Wood 2018] Wood WHJ, Macgregor-Chatwin C, Barnett SFH, Mayneord GE, Huang X, Hobbs JK, Hunter CN, Johnson MP: Dynamic thylakoid stacking regulates the balance between linear and cyclic photosynthetic electron transfer. *Nature Plants* 2018, 4:116–127. doi: 10.1038/s41477-017-0092-7.

[Wu 2010] Wu Y, Zhou C, Xiao J, Kurths J, Schellnhuber HJ: Evidence for a bimodal distribution in human communication. *Proceedings of the National Academy of Science USA* 2010, 107: 18803–18808. doi: 10.1073/pnas.1013140107.

[Wyczalek 1995] Wyczalek FA: Ultra Light Electric Vehicles (EV). *Journal of Circuits Systems and Computers* 1995, 5:81–91. doi: 10.1142/s0218126695000072.

[Xu 2016] Xu T, Yin CJ,Wodrich MD, Mazza S, Schultz KM, Scopelliti R, Hu X: A Functional model of [Fe]-hydrogenase. *Journal of American Chemical Society* 2016, 138:3270–3273. doi: 10.1021/jacs.5b12095.

[Xu2016] Xu FQ, Muneyoshi H: Product innovation versus business model innovation: The case of the walkman and the iPod. Proceedings of the 4th International Conference on Innovation and Entrepreneurship (ICIE 2016) 2016:281–285.

[Yagi 2005] Yagi M, Tomita E, Kuwabara T: Remarkably high activity of electrodeposited IrO2 film for electrocatalytic water oxidation. *Journal of Electroanalytical Chemistry* 2005, 579:83–88. doi: 10.1016/j.jelechem.2005.01.030.

[Yamahara 2005] Yamahara K, Sholklapper TZ, Jacobson CP, Visco SJ, De Jonghe LC: Ionic conductivity of stabilized zirconia networks in composite SOFC electrodes. *Solid State Ionics* 2005, 176:1359–1364. doi: 10.1016/j.ssi.2005.03.010.

[Yamanaka 2007] Yamanaka I, Ito T, Nabae Y, Hatano M: Effect of steam on direct oxidation of methane over Pd-Ni electrocatalyst supported on lanthanum chromite anode. In *Solid Oxide Fuel Cells 10. Volume Volume 7*. Edited by Eguchi K, Singhai SC, Yokokawa H, Mizusaki H. Pennington: Electrochemical Soc Inc; 2007: 1745+. *ECS Transactions*. doi: 10.1149/1.2729286.

[Yee 1993] Yee HS, Abruna HD: In situ x-ray absorption spectroscopy studies of copper underpotentially deposited in the absence and presence of chloride on platinum (111). *Langmuir* 1993, 9:2460–2469. doi: 10.1021/la00033a032.

[Yokokawa 2011] Yokokawa H, Sakai N, Horita T, Yamaji K, Brito ME: Electrolytes for solid-oxide fuel cells. *Mrs Bulletin* 2011, 30:591–595. doi: 10.1557/mrs2005.166.

[You 2009] You DJ, Zhang HM, Chen J: A simple model for the vanadium redox battery. *Electrochimica Acta* 2009, 54:6827–6836. doi: 10.1016/j.electacta.2009.06.086.

[Young 2000] Young JA: Getting students to wear safety goggles. *Journal of Chemical Education* 2000, 77:1214–1214.

[Zabaniotou 1999] Zabaniotou A, Kouskoumvekaki E, Sanopoulos D: Recycling of spent lead/acid batteries: The case of Greece. *Resources Conservation and Recycling* 1999, 25:301–317. doi: 10.1016/s0921-3449(98)00071-8.

[Zegenhagen 1991] Zegenhagen J: Surface structure analysis with X-ray standing waves. *Physica Scripta* 1991, T39:328–332. doi: 10.1088/0031-8949/1991/t39/051.

[Zegenhagen 2013] Zegenhagen J, Kazimirov A (Eds.): The X ray standing wave technique: Principles and applications. Singapore: World Scientific; 2013.

[Zener 1951] Zener C :Interaction between the d-Shells in the transition metals. II. ferromagnetic compounds of manganese with perovskite structure. *Physical Review* 1951, 82:403–405. doi: 10.1103/PhysRev.82.403.

[Zhang 2004] Zhang GQ, Zhang XG: MnO2/MCMB electrocatalyst for all solid-state alkaline zinc-air cells. *Electrochimica Acta* 2004, 49:873–877. doi: 10.1016/j.electacta.2003.09.039.

[Zhang 2010] Zhang C, Grass ME, Mcdaniel AH, Decaluwe SC, El Gabaly F, Liu Z, Mccarty KF, Farrow RL, Linne MA, Hussain Z, Jackson GS, Bluhm H, Eichhorn BW: Measuring fundamental properties in operating solid oxide electrochemical cells by using in situ X-ray photoelectron spectroscopy. *Nature Materials* 2010, 9:944–949. doi: 10.1038/nmat2851.

[Zhang 2012] Zhang LC, Sun X, Hu Z, Yuan CC, Chen CH: Rice paper as a separator membrane in lithium-ion batteries. *Journal of Power Sources* 2012, 204:149–154. doi: 10.1016/j.jpowsour.2011.12.028.

[Zhang 2013] Zhang LC, Hu Z, Wang L, Teng F, Yu Y, Chen CH: Rice paper-derived 3D-porous carbon films for lithium-ion batteries. *Electrochimica Acta* 2013, 89:310–316. doi: 10.1016/j.electacta.2012.11.042.

[Zheng 2016] Zheng F, Jiang J, Sun B, Zhang W, Pecht M: Temperature dependent power capability estimation of lithium-ion batteries for hybrid electric vehicles. *Energy* 2016, 113: 64–75. doi: 10.1016/j.energy.2016.06.010.

[Zhong 2015] Zhong C, Deng Y, Hu W, Qiao J, Zhang L, Zhang J: A review of electrolyte materials and compositions for electrochemical supercapacitors. *Chemical Society Review* 2015, 44: 7484–7539. doi: 10.1039/C5CS00303B.

[Zhou 2005] Zhou JS, Goodenough JB, Dabrowski B: Exchange interaction in the insulating phase of RNiO3. *Physical Review Letters* 2005, 95:127204. doi: 10.1103/PhysRevLett.95.127204.

[Zhu 2013] Zhu H, Bhavaraju S, Kee RJ: Computational model of a sodium–copper–iodide rechargeable battery. Electrochimica Acta 2013, 112:629–639. doi: 10.1016/j.electacta.2013.09.010.

[Zimmermann 2005] Zimmermann KF: European labour mobility: Challenges and potentials. *De Economist* 2005, 153: 425–450. doi: 10.1007/s10645-005-2660-x.ss

[Zinth 2015] Zinth V, Seidlmayer S, Zanon N, Crugnola G, Schulz M, Gilles R, Hofmann M: In situ spatially resolved neutron diffraction of a sodium metal halide battery. *Journal of the Electrochemical Society* 2015, 162:A384–A391. doi: 10.1149/2.0421503jes.

[Zinth 2016] Zinth V, Schulz M, Seidlmayer S, Zanon N, Gilles R, Hofmann M: Neutron tomography and radiography on a sodium metal halide cell under operating conditions. *Journal of the Electrochemical Society* 2016, 163:A838–A845. doi: 10.1149/2.0181606jes.

[Zintl 1939] Zintl E, Croatto U: Fluorite lattice with empty anionic spaces. *Zeitschrift für anorganische und allgemeine Chemie* 1939, 242:79–86. doi: 10.1002/zaac.19392420109.

[Zou 2009] Zou Y, Pisciotta J, Billmyre RB, Baskakov IV: Photosynthetic microbial fuel cells with positive light response. *Biotechnol Bioeng* 2009, 104:939–946. doi: 10.1002/bit.22466.

[Zou 2010] Zou Y, Pisciotta J, Baskakov IV: Nanostructured polypyrrole-coated anode for sun-powered microbial fuel cells. *Bioelectrochemistry* 2010, 79:50–56. doi: 10.1016/j.bioelechem.2009.11.001.

[Zuo 2006] Zuo CD, Zha SW, Liu ML, Hatano M, Uchiyama M: Ba(Zr0.1Ce0.7Y0.2)O3-delta as an electrolyte for low-temperature solid-oxide fuel cells. *Advanced Materials* 2006, 18:3318+. doi: 10.1002/adma.200601366.

[Zyryanov 2011] Zyryanov VN: Physical properties of seawater, including its three phases. In *UNESCO – Encyclopedia of Life Support Systems (EOLSS)*. Volume Water Sciences, Engineering and Technology Resources. Edited by Khublaryan MG. Paris (F), Abu Dhabi (UAE): UNESCO; 2011.

Index

https://doi.org/10.1515/9783110561838-016

www.ingramcontent.com/pod-product-compliance
Lightning Source LLC
Chambersburg PA
CBHW080345220326
41598CB00030B/4615

9 783110 561821